Logarithms

$y = \log_a x$ means $a^y = x$

$\log_a a^x = x$ $a^{\log_a x} = x$

$\log_a 1 = 0$ $\log_a a = 1$

$\log x = \log_{10} x$ $\ln x = \log_e x$

$\log_a xy = \log_a x + \log_a y$

$\log_a \left(\dfrac{x}{y} \right) = \log_a x - \log_a y$

$\log_a x^b = b \log_a x$ $\log_b x = \dfrac{\log_a x}{\log_a b}$

Geometric Formulas

Formulas for area A, circumference C, and volume V:

Triangle

$A = \frac{1}{2}bh$

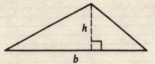

Circle

$A = \pi r^2$

$C = 2\pi r$

Sphere

$V = \frac{4}{3}\pi r^3$

$A = 4\pi r^2$

Cylinder

$V = \pi r^2 h$

Cone

$V = \frac{1}{3}\pi r^2 h$

Angle Measurement

π radians $= 180°$

$1° = \dfrac{\pi}{180}$ rad

$1 \text{ rad} = \dfrac{180°}{\pi}$

$s = r\theta$

(θ in radians)

Right Angle Trigonometry

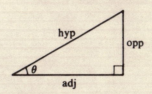

$\sin \theta = \dfrac{\text{opp}}{\text{hyp}}$ $\csc \theta = \dfrac{\text{hyp}}{\text{opp}}$

$\cos \theta = \dfrac{\text{adj}}{\text{hyp}}$ $\sec \theta = \dfrac{\text{hyp}}{\text{adj}}$

$\tan \theta = \dfrac{\text{opp}}{\text{adj}}$ $\cot \theta = \dfrac{\text{adj}}{\text{opp}}$

Trigonometric Functions

$\sin \theta = \dfrac{y}{r}$ $\csc \theta = \dfrac{r}{y}$

$\cos \theta = \dfrac{x}{r}$ $\sec \theta = \dfrac{r}{x}$

$\tan \theta = \dfrac{y}{x}$ $\cot \theta = \dfrac{x}{y}$

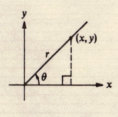

Graphs of the Trigonometric Functions

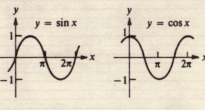

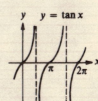

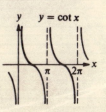

To The Student:

Many students experience difficulty with *precalculus* mathematics in calculus courses. Calculus requires that you understand and remember precalculus topics. For this reason, it may be helpful to retain this text as a reference in your calculus course. It has been written with this purpose in mind.

Some References To Calculus in This Text:

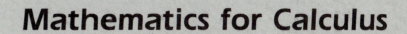

Mathematics for Calculus

Mathematics for Calculus

James Stewart
McMaster University

Lothar Redlin
The Pennsylvania State University

Saleem Watson
California State University, Long Beach

Brooks/Cole Publishing Company
Pacific Grove, California

Brooks/Cole Publishing Company

A Division of Wadsworth, Inc.

© 1989 by Wadsworth, Inc., Belmont, California 94002. All rights reserved. No part of this book may be reproduced, stored in a retrieval system, or transcribed, in any form or by any means—electronic, mechanical, photocopying, recording, or otherwise—without the prior written permission of the publisher, Brooks/Cole Publishing Company, Pacific Grove, California 93950, a division of Wadsworth, Inc.

Printed in the United States of America

10   9   8   7   6   5   4   3   2

Library of Congress Cataloging-in-Publication Data

Stewart, James, [date]
    Mathematics for calculus/James Stewart, Lothar Redlin, Saleem Watson.
      p. cm.
    Includes index.
    ISBN 0-534-10080-5
    1. Mathematics—1961– I. Redlin, Lothar. II. Watson, Saleem.
III. Title.
QA39.2.S75 1989
512.1—dc 19

88-32118
CIP

Sponsoring Editor: *Jeremy Hayhurst*
Editorial Assistant: *Virge Kelmser*
Production: *Cece Munson, The Cooper Company*
Production Coordinator: *Joan Marsh*
Manuscript Editor: *Carol Reitz*
Permissions Editor: *Carline Haga*
Interior Design: *Detta Penna and Vernon T. Boes*
Cover Design: *Vernon T. Boes*
Cover Photograph: *Rosenthal Fotostudio, M. Tessmann and A. F. Endress*
Cover Art Piece: *"Onda Construtta" by Marcello Morandini*
Cover Researchers: *Sue C. Howard and Heidi Wieland*
Interior Illustration: *Scientific Illustrators and Eric Bosch*
Typesetting: *Polyglot Pte. Ltd.*
Cover Printing: *The Lehigh Press, Inc.*
Printing and Binding: *Arcata Graphics/Fairfield*

Preface

What does a student really need to know to prepare for calculus? This question has motivated the writing of this text, and we hope that the text itself is in large part the answer to it.

Students often begin their study of calculus unprepared for the task that lies ahead. Often what they lack is not only technical skill but a correct concept of what mathematics is. A student who enters calculus with the idea that mathematics is a set of rules to be memorized will attempt, and fail at, the overwhelming task of learning mathematics by "memorizing all the rules." What the student needs to learn is how to *think* about mathematics.

We remain convinced that a central goal of any mathematics course should be the development of mathematical thinking. Accordingly, we present precalculus mathematics as a problem-solving endeavor. The first chapter introduces this reasoning, via problem-solving strategies, immediately after reviewing essential topics from intermediate algebra. The intention throughout the book is to encourage the student to organize his or her thoughts when tackling a mathematical problem. We have spent much effort determining how and where the problem-solving process should be applied for maximum benefit to the student. We have concluded that there is little benefit where there is little challenge. At the end of most chapters there is a set of challenging problems that we call Problems Plus, which are there to further exercise and develop problem-solving skills. In selecting these, as well as other exercises, we have kept the following advice from David Hilbert in mind: "A mathematical problem should be difficult in order to entice us, yet not completely inaccessible, lest it mock our efforts."

Exercises and Examples

Since mathematics is learned "by doing," a large selection of well-graded exercises is probably the most important feature any textbook can offer. The book contains more than 3200 exercises, and each set includes problems ranging from the essential routine ones to those that require a high level of problem-solving skill. We have further attempted to build the students' intuition through our explanations and examples. The book has been written as concisely as possible without sacrificing explanations important in developing an intuitive understanding of the subject matter. Geometric intuition is very important in good comprehension. We have made this book unusually rich in graphics and art.

Applications

The student should never get the impression that mathematics is a static subject isolated from the real world. In fact, some surprising discoveries in the sciences are the results of simple mathematical reasoning accessible at this level. We have made special effort to seek out and provide a variety of both familiar and little-known applications that illuminate the mathematics. For instance, measuring the distance from the earth to the sun is an almost-immediate consequence of the definition of the trigonometric functions (Section 5.2, exercise 68). Of course there are more modern applications, such as the use of elementary properties of ellipses in lithotripsy, the new medical technique of using shock waves to destroy kidney stones (Section 8.2).

Student and Instructor Aids

A *Student Solutions Manual* containing solutions to every odd-numbered exercise in the text has been prepared by Eric Bosch. An *Instructor's Manual* containing the remaining even-numbered solutions is also available. The even-numbered exercises are also available to the adopting instructor in electronic form using Brooks/Cole's unique test authoring system, EXP-Test. In addition, an excellent function plotting and graphics program called Plotpak, authored by John Mowbray, is available at no cost to adopting schools.

We thank David Dankort, McMaster University student, and the following California State University, Long Beach, students for their careful and consistent problem checking: Phyllis Panman, Denise Sobieralski, and David Crigger.

We wish to thank the manuscript reviewers who diligently read page after page of an unusually complete manuscript: Gail Kaplan, Sam Lesseig, Mary Jane McMaster, George D. Parker, Kenneth M. Shiskowski, Eugene Spiegel, and John Unbehaun.

Our production editor, Cece Munson; production coordinator, Joan Marsh; designer, Detta Penna; cover designer, Vernon Boes; editorial assistant, Virge Kelmser; and the rest of the crew at Brooks/Cole have all done an outstanding job. Our most heartfelt thanks are reserved for our editor, Jeremy Hayhurst, who guided us through every step of the project to its completion.

To the Student

This textbook has been written as a guide to mastering the mathematics you will need to know when you study calculus. Here are some suggestions to help you get the most out of your course.

First of all, you should read the appropriate section of text *before* you attempt your homework problems. Reading a mathematics text is quite different from reading a novel, a newspaper, or even another textbook. You may find that you have to re-read a passage several times before you understand it. Pay special attention to the examples, and work them out yourself with pencil and paper as you read. With this kind of preparation you will be able to do your homework much more quickly and with more understanding.

Do not make the mistake of trying to memorize every single rule or fact you may come across. Mathematics does not consist simply of memorization. Mathematics is a *problem-solving art*, not just a collection of facts. To master the subject you must solve problems—lots of problems. Do as many of the exercises as you can. Be sure to write your solutions in logical, step-by-step fashion.

Answers to the odd-numbered exercises, as well as all the answers to each chapter test, appear in the back of the book. If your answer differs from the one given, don't immediately assume that you are wrong. There may be an algebraic or trigonometric identity that connects the two answers and makes both correct. For example, if you get $1/(\sqrt{2} - 1)$ but the answer given is $1 + \sqrt{2}$, your answer is correct, because you can multiply both numerator and denominator of your answer by $\sqrt{2} + 1$ to change it to the given answer.

The symbol $\oslash$ is used to warn against committing an error. We have placed this symbol in the margin to point out situations where we have found that many of our students make the same mistake.

A thorough mastery of the topics in this book is absolutely essential for success in future mathematics courses, although don't be concerned if your class does not cover all of these topics. This is an unusually comprehensive precalculus text, which might be very useful as a reference when you take calculus. Chapter 7, in particular, will be helpful as a refresher should you go on to take linear algebra as well.

Calculators and Calculations

A calculator has become essential in most mathematics and science subjects. Precalculus courses are no exception. Although calculators are powerful tools, they need to be used with care. Here are a few guidelines to help you obtain meaningful answers from your calculator.

For this course you will need a *scientific* calculator—one that has, as a minimum, the following functions:

1. $+, -, \times, \div$
2. Powers and roots
3. Logarithms ($\log x$, $\ln x$)
4. Exponential function (e^x)
5. Trigonometric functions ($\sin, \cos, \tan$)
6. Inverse trigonometric functions (arc sin or $\sin^{-1}$ or inv sin, etc.)

In addition, a memory and at least some degree of programmability will be useful. Because so many different types of calculators are available, we give no specific instructions for operating one. You should read your owner's manual carefully, and work through the examples given in it to become familiar with the features of your calculator.

Most of the applied examples and exercises in this book involve approximate values. For example, one exercise states that the moon has a radius of 1074 miles. This does not mean that the moon's radius is *exactly* 1074 miles but simply that this is the radius rounded to the nearest mile.

One simple method for specifying the accuracy of a number is to state how many **significant digits** it has. The significant digits in a number are the ones from the first nonzero digit to the last nonzero digit (reading from left to right). Thus 1074 has four significant digits, 1070 has three, 1100 has two, and 1000 has one significant digit.

This rule may sometimes lead to ambiguities. For example, if a distance is 200 km to the nearest kilometer, then the number 200 really has three significant digits, not just one. This ambiguity is avoided if we use *scientific notation*—that is, if we express the number as a multiple of a power of 10:

$$2.00 \times 10^2$$

Number	Significant Digits
12,300	3
3000	1
3000.58	6
0.03416	4
4200	2
4.2×10^3	2
4.20×10^3	3

When working with approximate values, students often make the mistake of giving a final answer with **more** significant digits than the original data. This is incorrect because you cannot "create" precision by using a calculator. The final result can be no more accurate than the measurements given in the problem. For example, suppose we are told that the two shorter sides of a right triangle are measured to be 1.25 and 2.33 inches long. By the Pythagorean Theorem, we find, using a calculator, that the hypotenuse has length

$$\sqrt{1.23^2 + 2.33^2} \approx 2.644125564 \text{ in.}$$

But since the given lengths were expressed to three significant digits, the answer cannot be any more accurate. We can therefore say only that the hypotenuse is 2.64 in. long, rounding to the nearest hundredth.

In general, the final answer should be expressed with the same accuracy as the *least*-accurate measurement given in the statement of the problem. The following rules make this principle more precise.

Rules for Working with Approximate Data

1. When multiplying or dividing, round off the final result so that it has as many *significant digits* as the given value with the fewest number of significant digits.

2. When adding or subtracting, round off the final result so that it has its last significant digit in the *decimal place* in which the least-accurate given value has its last significant digit.

3. When taking powers or roots, round off the final result so that it has the same number of *significant digits* as the given value.

As an example, suppose that a rectangular table top is measured to be 122.64 in. by 37.3 in. We express its area and perimeter as follows:

$$\text{Area} = \text{length} \times \text{width} = 122.64 \times 37.3 \approx 4570 \text{ in.}^2 \quad \text{(\textit{three significant digits})}$$
$$\text{Perimeter} = 2(\text{length} + \text{width}) = 2(122.64 + 37.3) \approx 319.9 \text{ in.} \quad \text{(\textit{tenths digit})}$$

Note that in the formula for the perimeter, the value 2 is an exact value, not an approximate measurement. It therefore does not affect the accuracy of the final result. In general, if a problem involves only exact values, we may express the final answer with as many significant digits as we wish.

Note also that to make the final result as accurate as possible, you should wait until the last step to round off your answer. If necessary, use the memory feature of your calculator to retain the results of intermediate calculations.

Contents

1

2

Functions   66

3

Polynomials and Rational Functions   110

4

Exponential and Logarithmic Functions   167

5

6

7

Topics in Analytic Geometry   429

Sequences and Series   476

1

Fundamentals

One learns by doing the thing; for though you think you know it, you have
no certainty until you try.

Sophocles

A great discovery solves a great problem but there is a grain of discovery
in the solution of any problem. Your problem may be modest; but if it
challenges your curiosity and brings into play your inventive faculties, and if
you solve it by your own means, you may experience the tension and enjoy the
triumph of discovery.

George Polya

In this first chapter we review the basic ideas from algebra and coordinate geometry that will be needed throughout this book. In the last section we give some general guidelines for problem solving that you may find useful when attacking difficult problems.

Real Numbers

Let us recall the types of numbers that make up the real number system. We start with the **integers**:

$$\ldots, -3, -2, -1, 0, 1, 2, 3, 4, \ldots$$

Then we construct the **rational numbers**, which are ratios of integers. Thus any rational number r can be expressed as:

$$r = \frac{m}{n} \qquad \text{where } m \text{ and } n \text{ are integers and } n \neq 0$$

Examples are

$$\tfrac{1}{2} \qquad -\tfrac{3}{7} \qquad 46 = \tfrac{46}{1} \qquad 0.17 = \tfrac{17}{100}$$

(Recall that division by 0 is always ruled out, so expressions like $\frac{3}{0}$ and $\frac{0}{0}$ are undefined.) There are also real numbers, such as $\sqrt{2}$, that cannot be expressed as a ratio of integers and are therefore called **irrational numbers**. (See Example 5 in Section 1.9 for a proof that $\sqrt{2}$ is irrational.) It can be shown, with varying degrees of difficulty, that the following are also irrational numbers:

$$\sqrt{3} \qquad \sqrt{5} \qquad \sqrt[3]{2} \qquad \pi \qquad \frac{3}{\pi^2} \qquad \sin 1°$$

The set of all real numbers is usually denoted by the symbol R. When we use the word *number* without qualification, we will mean "real number."

Every number has a decimal representation. If the number is rational, the corresponding decimal is repeating. For example,

$$\tfrac{1}{2} = 0.5000\ldots = 0.5\overline{0} \qquad\qquad \tfrac{2}{3} = 0.66666\ldots = 0.\overline{6}$$
$$\tfrac{157}{495} = 0.3171717\ldots = 0.3\overline{17} \qquad\qquad \tfrac{9}{7} = 1.285714285714\ldots = 1.\overline{285714}$$

(The bar indicates that the sequence of digits repeats forever.) On the other hand, if the number is irrational, the decimal representation is nonrepeating:

$$\sqrt{2} = 1.414213562373095\ldots \qquad \pi = 3.141592653589793\ldots$$

If we stop the decimal expansion of any number at a certain place, we get an approximation to the number. For instance, we can write

$$\pi \approx 3.14159265$$

where the symbol $\approx$ is read "is approximately equal to." The more decimal places we retain, the better the approximation we get.

Figure 1
The real line

The real numbers can be represented by points on a line as in Figure 1. The positive direction (to the right) is indicated by an arrow. We choose an arbitrary reference point O, called the **origin**, which corresponds to the real number 0. Given any convenient unit of measurement, each positive number x is represented by the point on the line a distance of x units to the right of the origin, and each negative number $-x$ is represented by the point x units to the left of the origin. Thus every real number is represented by a point on the line, and every point P on the line corresponds to exactly one real number. The number associated with the point P is called the **coordinate** of P and the line is then called a **coordinate line**, or a **real number line**, or simply a **real line**. Often we identify the point with its coordinate and think of a number as being a point on the real line.

The real numbers are ordered. We say "a is less than b" and write $a < b$ if $b - a$ is a positive number. Geometrically this means that a lies to the left of b on the number line. (Equivalently, we say "b is greater than a" and write $b > a$.) The symbol $a \le b$ (or $b \ge a$) means that either $a < b$ or $a = b$ and is read "a is less than or equal to b." For instance, the following are true inequalities:

$$7 < 7.4 < 7.5 \qquad -3 > -\pi \qquad \sqrt{2} < 2 \qquad 2 \le 2$$

In what follows we need to use set notation. A **set** is a collection of objects, and these objects are called the **elements** of the set. If S is a set, the notation $a \in S$ means that a is an element of S, and $a \notin S$ means that a is not an element of S. For example, if Z represents the set of integers, then $-3 \in Z$ but $\pi \notin Z$.

Some sets can be described by listing their elements between braces. For instance, the set A that consists of all positive integers less than 7 can be written as

$$A = \{1, 2, 3, 4, 5, 6\}$$

We could also write A in set-builder notation as

$$A = \{x \mid x \text{ is an integer and } 0 < x < 7\}$$

which is read "A is the set of x such that x is an integer and $0 < x < 7$."

If S and T are sets, then their **union** $S \cup T$ is the set that consists of all elements that are in S *or* T (or in both S and T). The **intersection** of S and T is the set $S \cap T$ that consists of all elements that are in both S *and* T. In other words, $S \cap T$ is the common part of S and T. The **empty set**, denoted by $\varnothing$, is the set that contains no elements.

Example 1

If $S = \{1, 2, 3, 4, 5\}$, $T = \{4, 5, 6, 7\}$, and $V = \{6, 7, 8\}$, find the sets $S \cup T$, $S \cap T$, and $S \cap V$.

Solution

$$S \cup T = \{1, 2, 3, 4, 5, 6, 7\}$$
$$S \cap T = \{4, 5\}$$
$$S \cap V = \varnothing$$

Intervals

There are certain sets of real numbers, called **intervals**, that occur frequently in calculus and correspond geometrically to line segments. For example, if $a < b$, the **open interval** from a to b consists of all numbers between a and b and is denoted by the symbol (a, b). Using set-builder notation, we can write

$$(a, b) = \{x \mid a < x < b\}$$

Notice that the endpoints of the interval—namely, a and b—are excluded. This is indicated by the round brackets ( ) and the open circles in Figure 2.

The **closed interval** from a to b is the set

$$[a, b] = \{x \mid a \le x \le b\}$$

Here the endpoints of the interval are included. This is indicated by the square brackets [ ] and the solid circles in Figure 3. It is also possible to include only one endpoint in an interval, as shown in the table of intervals preceding Example 2.

We also need to consider infinite intervals such as

$$(a, \infty) = \{x \mid x > a\}$$

This does not mean that ∞ ("infinity") is a number. The notation (a, ∞) stands for the set of all numbers that are greater than a, so the symbol ∞ simply indicates that the interval extends indefinitely far in the positive direction.

The following table lists the nine possible types of intervals. When these intervals are discussed, it will always be assumed that $a < b$.

Figure 2
The open interval (a, b)

Figure 3
The closed interval $[a, b]$

Notation	Set Description	Picture
(a, b)	$\{x \mid a < x < b\}$	
$[a, b]$	$\{x \mid a \le x \le b\}$	
$[a, b)$	$\{x \mid a \le x < b\}$	
$(a, b]$	$\{x \mid a < x \le b\}$	
(a, ∞)	$\{x \mid x > a\}$	
$[a, \infty)$	$\{x \mid x \ge a\}$	
$(-\infty, b)$	$\{x \mid x < b\}$	
$(-\infty, b]$	$\{x \mid x \le b\}$	
$(-\infty, \infty)$	R (set of all real numbers)	

Example 2

Express the following intervals in terms of inequalities and graph the intervals:

(a) $[-1, 2)$ (b) $[1.5, 4]$ (c) $(-3, \infty)$

Solution

(a) $[-1, 2) = \{x \mid -1 \le x < 2\}$

(b) $[1.5, 4] = \{x \mid 1.5 \le x \le 4\}$

(c) $(-3, \infty) = \{x \mid x > -3\}$

Example 3

Graph the following sets:

(a) $(1, 3) \cap [2, 7]$ (b) $(-2, -1) \cup (1, 2)$

Solution

(a) The intersection of two intervals consists of the numbers that are in both intervals. Therefore

$$(1, 3) \cap [2, 7] = \{x \mid 1 < x < 3 \text{ and } 2 \le x \le 7\}$$
$$= \{x \mid 2 \le x < 3\}$$
$$= [2, 3)$$

and this interval is shown in Figure 4(a).

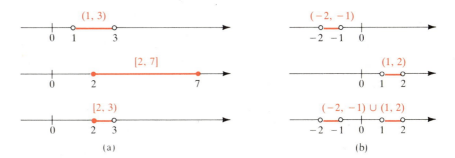

Figure 4   (a)   (b)

(b) The union of the intervals $(-2, -1)$ and $(1, 2)$ consists of the numbers that are in either $(-2, -1)$ or $(1, 2)$, so

$$(-2, -1) \cup (1, 2) = \{x \mid -2 < x < -1 \quad \text{or} \quad 1 < x < 2\}$$

This set is illustrated in Figure 4(b).   ●

Absolute Value

The **absolute value** of a number a, denoted by $|a|$, is the distance from a to 0 on the real number line. Distances are always positive or zero, so we have

$$|a| \ge 0 \qquad \text{for every number } a$$

Remembering that $-a$ is positive when a is negative, we have

$$
\begin{aligned}
|a| &= a && \text{if } a \ge 0 \\
|a| &= -a && \text{if } a < 0
\end{aligned}
$$

Example 4

(a) $|3| = 3$

(b) $|-3| = -(-3) = 3$

(c) $|0| = 0$

(d) $|\sqrt{2} - 1| = \sqrt{2} - 1$, since $\sqrt{2} > 1$

(e) $|3 - \pi| = -(3 - \pi) = \pi - 3$, since $\pi > 3$ ●

If a and b are any real numbers, then the distance between a and b is the absolute value of the difference—namely, $|a - b|$, which is also equal to $|b - a|$. See Figure 5.

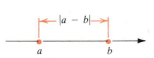

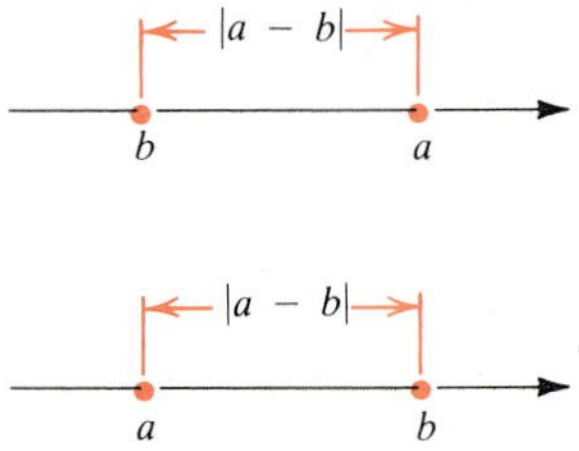

Figure 5

Length of a line segment $= |a - b|$

Example 5

The distance between the numbers -8 and 2 is

$$|-8 - 2| = |-10| = 10$$

This calculation can be checked by looking at Figure 6.   ●

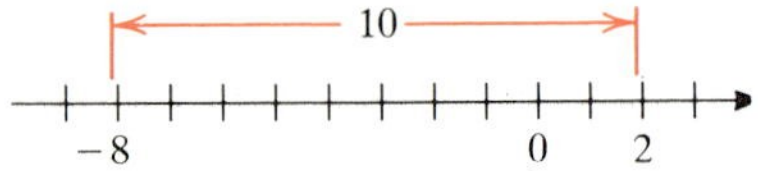

Figure 6

Exercises 1.1

In Exercises 1–10 state whether the given inequality is true or false.

1. $-6 < -10$

2. $-3 < -1$

3. $0.66 < \frac{2}{3}$

4. $\sqrt{2} > 1.41$

5. $\frac{10}{11} < \frac{12}{13}$

6. $-\pi > -3$

7. $8 \le 8$

8. $8 \le 9$

9. $\pi \ge 3$

10. $1.1 > 1.\bar{1}$

In Exercises 11–16 find the given set if $A = \{1, 2, 3, 4, 5, 6\}$, $B = \{2, 4, 6, 8\}$, and $C = \{7, 8, 9, 10\}$.

11. $A \cup B$

12. $A \cap B$

13. $B \cap C$

14. $B \cup C$

15. $A \cup C$

16. $A \cap C$

In Exercises 17–26 express the interval in terms of inequalities and graph the interval.

17. $(-3, 0)$

18. $(2, 8]$

19. $[2, 8)$

20. $[-6, -\frac{1}{2}]$

21. $[-1, 1]$

22. $(-4, \infty)$

23. $[2, \infty)$

24. $(-\infty, 1)$

25. $(-\infty, -2]$

26. $(0, \pi)$

In Exercises 27–34 express the inequality in interval notation and graph the corresponding interval.

27. $x \le 1$

28. $0 < x < 8$

29. $1 \le x \le 2$

30. $x < 3$

31. $-2 < x \le 1$

32. $x \ge -5$

33. $x > -1$

34. $-5 < x < 2$

In Exercises 35–42 graph the set.

35. $(-2, 0) \cup (-1, 1)$

36. $(-2, 0) \cap (-1, 1)$

37. $[-4, 6] \cap [0, 8)$

38. $[-4, 6) \cup [0, 8)$

39. $(-\infty, -4) \cup (4, \infty)$ **40.** $(-\infty, 6] \cap (2, 10)$

41. $(0, 7) \cap [2, \infty)$ **42.** $(-1, 0] \cup [1, 2)$

In Exercises 43–54 evaluate each expression.

43. $|100|$ **44.** $|-73|$

45. $|2 - 6|$ **46.** $|-8 - (-23)|$

47. $|-\pi|$ **48.** $|\pi - 10|$

49. $|\sqrt{3} - 3|$ **50.** $||-6| - |-4||$

51. $\dfrac{-1}{|-1|}$ **52.** $|-26 - 14|$

53. $|2 - |-12||$ **54.** $-1 - |1 - |-1||$

In Exercises 55–60 find the distance between the given numbers.

55. 2 and 17   **56.** -3 and 21

57. -14 and 12   **58.** 100 and -150

59. -38 and -57 **60.** -2.6 and -1.8

61. Show that the sum, difference, and product of rational numbers are rational numbers.

62. **(a)** Is the sum of two irrational numbers always an irrational number?
(b) Is the product of two irrational numbers always an irrational number?

Exponents and Radicals

Integer Exponents

If a is any real number and n is a positive integer, then the **nth power of a** is

$$a^n = \underbrace{a \cdot a \cdots \cdot a}_{n \text{ factors}}$$

The number a is called the **base** and n is called the **exponent**.

If the base is not zero and the exponent is zero or is negative, we use the following definition:

$$\text{If } a \neq 0 \text{ and } n > 0, \text{ then}$$
$$a^0 = 1 \qquad a^{-n} = \frac{1}{a^n}$$

Example 1

(a) $3^4 = 3 \cdot 3 \cdot 3 \cdot 3 = 81$

(b) $\left(\dfrac{4}{7}\right)^0 = 1$

(c) $(-2)^{-3} = \dfrac{1}{(-2)^3} = \dfrac{1}{(-2)(-2)(-2)} = \dfrac{1}{-8} = -\dfrac{1}{8}$

It is essential to be familiar with the following rules for working with exponents and bases.

Laws of Exponents

Let a and b be real numbers and m and n be integers. Then

1. $a^m a^n = a^{m+n}$

2. $\dfrac{a^m}{a^n} = a^{m-n}$ $(a \neq 0)$

3. $(a^m)^n = a^{mn}$

4. $(ab)^n = a^n b^n$

5. $\left(\dfrac{a}{b}\right)^n = \dfrac{a^n}{b^n}$ $(b \neq 0)$

In words, these five laws can be stated as follows:

1. To multiply two powers of the same number, we add the exponents.
2. To divide two powers of the same number, we subtract the exponents.
3. To raise a power to a new power, we multiply the exponents.
4. To raise a product to a power, we raise each factor to the power.
5. To raise a quotient to a power, we raise both numerator and denominator to the power.

All these laws of exponents follow from the definition of an nth power, but we will prove only the first law for the case where m and n are positive integers:

$$a^m a^n = \underbrace{(a \cdot a \cdot a \cdots \cdot a)}_{m \text{ factors}} \underbrace{(a \cdot a \cdot a \cdots \cdot a)}_{n \text{ factors}}$$

$$= \underbrace{a \cdot a \cdot a \cdots \cdot a}_{m + n \text{ factors}} = a^{m+n}$$

Example 2

(a) $x^4 x^7 = x^{4+7} = x^{11}$

(b) $y^4 y^{-7} = y^{4-7} = y^{-3} = \dfrac{1}{y^3}$

(c) $\dfrac{c^9}{c^5} = c^{9-5} = c^4$

(d) $\dfrac{d^2}{d^{10}} = d^{2-10} = d^{-8} = \dfrac{1}{d^8}$

(e) $(xy)^3 = x^3 y^3$

(f) $\left(\dfrac{x}{2}\right)^5 = \dfrac{x^5}{2^5} = \dfrac{x^5}{32}$

(g) $(b^4)^5 = b^{4 \cdot 5} = b^{20}$

(h) $(2^m)^3 = 2^{3m}$ ●

Example 3

Simplify:

(a) $(2a^3 b^2)(3ab^4)^3$

(b) $\left(\dfrac{x}{y}\right)^3 \left(\dfrac{y^2 x}{z}\right)^4$

(c) $\dfrac{6st^{-4}}{2s^{-2}t^2}$

Solution

(a) $(2a^3b^2)(3ab^4)^3 = (2a^3b^2)[3^3a^3(b^4)^3] = (2a^3b^2)(27a^3b^{12})$
$$= (2)(27)a^3a^3b^2b^{12} = 54a^6b^{14}$$

(b) $\left(\dfrac{x}{y}\right)^3\left(\dfrac{y^2x}{z}\right)^4 = \dfrac{x^3}{y^3} \cdot \dfrac{y^8x^4}{z^4} = (x^3x^4)\left(\dfrac{y^8}{y^3}\right)\dfrac{1}{z^4} = \dfrac{x^7y^5}{z^4}$

(c) $\dfrac{6st^{-4}}{2s^{-2}t^2} = \dfrac{6}{2} \cdot \dfrac{s}{s^{-2}} \cdot \dfrac{t^{-4}}{t^2} = 3s^{1-(-2)}t^{-4-2} = 3s^3t^{-6} = \dfrac{3s^3}{t^6}$ ●

Radicals

The symbol $\sqrt{}$ means "the positive square root of." Thus

$$\boxed{\sqrt{a} = b \quad \text{means} \quad b^2 = a \quad \text{and} \quad b \geq 0}$$

Since $a = b^2 \geq 0$, the symbol $\sqrt{a}$ makes sense only when $a \geq 0$.

Thus, for instance, $\sqrt{4} = 2$. It is true that the number 4 has two square roots, 2 and -2, but the notation $\sqrt{4}$ is reserved for the *positive* square root of 4 (sometimes called the *principal square root* of 4). We write the two solutions of the equation $x^2 = 5$ as $x = \sqrt{5}$ and $x = -\sqrt{5}$.

Notice that

$$\sqrt{4^2} = \sqrt{16} = 4 \quad \text{but} \quad \sqrt{(-4)^2} = \sqrt{16} = 4 = |-4|$$

 Thus the equation $\sqrt{a^2} = a$ is not always true; it is true only when $a \geq 0$. If $a < 0$, then $-a > 0$, so we have $\sqrt{a^2} = -a$. In terms of absolute values, however, we can always write

$$\boxed{\sqrt{a^2} = |a|}$$

Here are two rules for working with square roots:

$$\boxed{\begin{array}{c} \text{If } a \geq 0 \text{ and } b > 0, \text{ then} \\[2mm] \sqrt{ab} = \sqrt{a}\,\sqrt{b} \qquad \sqrt{\dfrac{a}{b}} = \dfrac{\sqrt{a}}{\sqrt{b}} \end{array}}$$

There is no similar rule for the square root of a sum. In fact, you should avoid making the following common error:

$$\sqrt{a + b} \neq \sqrt{a} + \sqrt{b}$$

For instance, we take $a = 9$ and $b = 16$ to see the error:

$$\sqrt{9 + 16} \stackrel{?}{=} \sqrt{9} + \sqrt{16}$$
$$\sqrt{25} \stackrel{?}{=} 3 + 4$$
$$5 \stackrel{?}{=} 7 \quad \textit{(wrong!)}$$

Example 4

(a) $\sqrt{72} = \sqrt{36 \cdot 2} = \sqrt{36} \cdot \sqrt{2} = 6\sqrt{2}$

(b) $\dfrac{\sqrt{18}}{\sqrt{2}} = \sqrt{\dfrac{18}{2}} = \sqrt{9} = 3$

(c) $\sqrt{x^2 y} = \sqrt{x^2}\,\sqrt{y} = |x|\sqrt{y}$ ●

In general, if n is any positive integer, then the principal nth roots are defined as follows:

$$\sqrt[n]{a} = b \qquad \text{means} \qquad b^n = a$$

where, if n is even, we must have $a \geq 0$ and $b \geq 0$.

Thus $\sqrt[3]{-8} = -2$, since $(-2)^3 = -8$, but $\sqrt{-8}$, $\sqrt[4]{-8}$, and $\sqrt[6]{-8}$ are not defined. Notice also that odd roots are unique but even roots are not.

The equation $x^5 = 31$ has only one real solution: $x = \sqrt[5]{31}$.

The equation $x^4 = 31$ has two real solutions: $x = \pm\sqrt[4]{31}$.

The following rules are used in working with nth roots. In each case we assume that all the roots exist.

$$\sqrt[n]{ab} = \sqrt[n]{a}\,\sqrt[n]{b} \qquad \sqrt[n]{\dfrac{a}{b}} = \dfrac{\sqrt[n]{a}}{\sqrt[n]{b}} \qquad \sqrt[m]{\sqrt[n]{a}} = \sqrt[mn]{a}$$

Example 5

(a) $\sqrt[3]{32} + \sqrt[3]{108} = \sqrt[3]{8 \cdot 4} + \sqrt[3]{27 \cdot 4} = 2\sqrt[3]{4} + 3\sqrt[3]{4} = 5\sqrt[3]{4}$

(b) $\sqrt[4]{81 x^8 y^4} = \sqrt[4]{81}\,\sqrt[4]{x^8}\,\sqrt[4]{y^4} = 3x^2 |y|$

(c) $\sqrt[3]{x^4} = \sqrt[3]{x^3 x} = \sqrt[3]{x^3}\,\sqrt[3]{x} = x\sqrt[3]{x}$ ●

When a denominator contains a radical, it is often useful to eliminate the radical by multiplying both numerator and denominator by an appropriate expression. This procedure, called **rationalizing the denominator**, is illustrated in the following example.

Example 6

Rationalize the denominator in the following expressions:

(a) $\dfrac{2}{\sqrt{3}}$ (b) $\sqrt[3]{\dfrac{x}{4}}$ (c) $\dfrac{1}{1 + \sqrt{2}}$

Solution

(a) $\dfrac{2}{\sqrt{3}} = \dfrac{2}{\sqrt{3}} \cdot \dfrac{\sqrt{3}}{\sqrt{3}} = \dfrac{2\sqrt{3}}{(\sqrt{3})^2} = \dfrac{2\sqrt{3}}{3}$

(b) $\sqrt[3]{\dfrac{x}{4}} = \sqrt[3]{\dfrac{x}{2^2} \cdot \dfrac{2}{2}} = \sqrt[3]{\dfrac{2x}{2^3}} = \dfrac{\sqrt[3]{2x}}{\sqrt[3]{2^3}} = \dfrac{\sqrt[3]{2x}}{2}$

(c)   Here it is appropriate to multiply both numerator and denominator by the *conjugate radical* $1 - \sqrt{2}$:

$$\frac{1}{1 + \sqrt{2}} = \frac{1}{1 + \sqrt{2}} \cdot \frac{1 - \sqrt{2}}{1 - \sqrt{2}} = \frac{1 - \sqrt{2}}{1 - \sqrt{2} + \sqrt{2} - 2} = \frac{1 - \sqrt{2}}{-1} = \sqrt{2} - 1$$

Rational Exponents

To define what is meant by fractional exponents we need to use radicals. The reason is that, in defining $a^{1/n}$ in a way that is consistent with the Laws of Exponents, we would have to have

$$(a^{1/n})^n = a^{(1/n)n} = a^1 = 1$$

Therefore we define

$$\boxed{a^{1/n} = \sqrt[n]{a}}$$

Notice that if n is even, we require that $a \geq 0$.

Finally we define powers for any rational exponent m/n in lowest terms, where m and n are integers and $n > 0$. If $\sqrt[n]{a}$ exists as a real number (and this will be the case for any number a if n is odd, and for $a \geq 0$ if n is even), we define

$$\boxed{a^{m/n} = (\sqrt[n]{a})^m \qquad \text{or equivalently} \qquad a^{m/n} = \sqrt[n]{a^m}}$$

With this definition it can be proved that the Laws of Exponents are still true for rational exponents.

Example 7

(a) $4^{3/2} = (\sqrt{4})^3 = 2^3 = 8$
    Alternate solution: $4^{3/2} = \sqrt{4^3} = \sqrt{64} = 8$

(b) $(125)^{-1/3} = \dfrac{1}{(125)^{1/3}} = \dfrac{1}{\sqrt[3]{125}} = \dfrac{1}{5}$

(c) $\dfrac{1}{\sqrt[3]{x^4}} = \dfrac{1}{x^{4/3}} = x^{-4/3}$

Example 8

(a) $(2\sqrt{x})(3\sqrt[3]{x}) = (2x^{1/2})(3x^{1/3}) = 6x^{1/2+1/3} = 6x^{5/6}$

(b) $(2a^3b^4)^{3/2} = 2^{3/2}(a^3)^{3/2}(b^4)^{3/2} = (\sqrt{2})^3 a^{3(3/2)}b^{4(3/2)} = 2\sqrt{2}\,a^{9/2}b^6$

(c) $\left(\dfrac{2x^{3/4}}{y^{1/3}}\right)^3\left(\dfrac{y^4}{x^{-1/2}}\right) = \dfrac{2^3(x^{3/4})^3}{(y^{1/3})^3}\cdot(y^4x^{1/2}) = \dfrac{8x^{9/4}}{y}\cdot y^4 x^{1/2} = 8x^{11/4}y^3$ ●

Example 9

$$\sqrt{x\sqrt{x\sqrt{x}}} = [x(xx^{1/2})^{1/2}]^{1/2} = [x(x^{3/2})^{1/2}]^{1/2}$$
$$= (x\cdot x^{3/4})^{1/2} = (x^{7/4})^{1/2} = x^{7/8}$$ ●

Exercises 1.2

Evaluate the numbers in Exercises 1–10.

1. $(-3)^5$

2. 4^{-3}

3. $2^{-3}5^4$

4. $3^{-2} - (1.7)^0$

5. $36^{1/2}$

6. $\sqrt[3]{-64}$

7. $125^{2/3}$

8. $9^{7/2}$

9. $\sqrt[5]{-32}$

10. $64^{-4/3}$

In Exercises 11–18 write the number as a power of 2

11. 128

12. $2^6\cdot 8^4$

13. $(2^9)^4$

14. $\dfrac{1}{4}$

15. $\dfrac{2^{3.1}}{2^{4.6}}$

16. $\sqrt{2}$

17. $4\sqrt{2}$

18. 1

In Exercises 19–50 eliminate negative exponents and simplify the expressions.

19. $t^7 t^{-2}$

20. $(4x^2)(6x^7)$

21. $(12x^2y^4)(\tfrac{1}{2}x^5 y)$

22. $(6y)^3$

23. $\dfrac{x^9(2x)^4}{x^3}$

24. $\dfrac{a^{-3}b^4}{a^{-5}b^5}$

25. $b^4(\tfrac{1}{3}b^2)(12b^{-8})$

26. $(2s^3t^{-1})(\tfrac{1}{4}s^6)(16t^4)$

27. $(rs)^3(2s)^{-2}(4r)^4$

28. $(2u^2v^3)^3(3u^3v)^{-2}$

29. $\dfrac{(6y^3)^4}{2y^5}$

30. $\dfrac{(2x^3)^2(3x^4)}{(x^3)^4}$

31. $\dfrac{(x^2y^3)^4(xy^4)^{-3}}{x^2y}$

32. $\left(\dfrac{c^4d^3}{cd^2}\right)\left(\dfrac{d^2}{c^3}\right)^3$

33. $\dfrac{(xy^2z^3)^4}{(x^3y^2z)^3}$

34. $(3ab^2c)\left(\dfrac{2a^2b}{c^3}\right)^{-2}$

35. $x^{2/3}x^{1/5}$

36. $(-2a^{3/4})(5a^{3/2})$

37. $(4b)^{1/2}(8b^{2/5})$

38. $(8x^6)^{-2/3}$

39. $(c^2d^3)^{-1/3}$

40. $(4x^6y^8)^{3/2}$

41. $(y^{3/4})^{2/3}$

42. $(a^{2/5})^{-3/4}$

43. $(2x^4y^{-4/5})^3(8y^2)^{2/3}$

44. $(x^{-5}y^3z^{10})^{-3/5}$

45. $\left(\dfrac{x^6y}{y^4}\right)^{5/2}$

46. $\left(\dfrac{-2x^{1/3}}{y^{1/2}z^{1/6}}\right)^4$

47. $\left(\dfrac{3a^{-2}}{4b^{-1/3}}\right)^{-1}$

48. $\dfrac{(y^{10}z^{-5})^{1/5}}{(y^{-2}z^3)^{1/3}}$

49. $\dfrac{(9st)^{3/2}}{(27s^3t^{-4})^{2/3}}$

50. $\left(\dfrac{a^2b^{-3}}{x^{-1}y^2}\right)^3\left(\dfrac{x^{-2}b^{-1}}{a^{3/2}y^{1/3}}\right)$

In Exercises 51–60 write the given expression as a power of x.

51. $x^a x^b x^c$

52. $((x^a)^b)^c$

53. $\sqrt[3]{x^5}$

54. $\dfrac{1}{\sqrt[7]{x^3}}$

55. $x^2\sqrt{x}$

56. $x\sqrt[3]{x}$

57. $\dfrac{(x^2)^n x^5}{x^n}$

58. $\dfrac{(x^2)^m(x^3)^n}{x^{m+n}x^{m-n}}$

59. $\sqrt{x\sqrt{x}}$

60. $\sqrt{x\sqrt{x\sqrt{x\sqrt{x}}}}$

Simplify the expressions in Exercises 61–78.

61. $\sqrt[3]{0.000001}$

62. $\sqrt[5]{\dfrac{-1}{243}}$

63. $\sqrt[3]{3}\sqrt[3]{9}$

64. $\dfrac{\sqrt{48}}{\sqrt{3}}$

65. $\sqrt{75}$

66. $\sqrt{8} + \sqrt{50}$

67. $\sqrt{245} - \sqrt{125}$

68. $\sqrt[3]{54} - \sqrt[3]{16}$

69. $\sqrt[4]{x^4}$

70. $\sqrt[3]{x^3 y^6}$

71. $\sqrt[3]{x^3 y}$

72. $\sqrt{x^4 y^3}$

73. $\sqrt[5]{a^6 b^7}$

74. $\sqrt[3]{a^2 b}\sqrt[3]{a^4 b}$

75. $\sqrt{x^2 y^6}$

76. $\sqrt[4]{x^4 y^2 z}$

77. $\sqrt[3]{\sqrt{64x}}$

78. $\sqrt[4]{r^{2n+1}} \times \sqrt[4]{r^{-1}}$

Rationalize the denominator in Exercises 79–90.

79. $\dfrac{1}{\sqrt{6}}$

80. $\sqrt{\dfrac{2}{3}}$

81. $\sqrt{\dfrac{3}{20}}$

82. $\sqrt[3]{\dfrac{4}{9}}$

83. $\sqrt{\dfrac{x^5}{2}}$

84. $\sqrt{\dfrac{x}{3y}}$

85. $\sqrt[3]{\dfrac{x^3}{y}}$

86. $\sqrt{\dfrac{1}{2x^3 y^5}}$

87. $\dfrac{1}{3 - \sqrt{5}}$

88. $\dfrac{3}{\sqrt{2} + \sqrt{5}}$

89. $\dfrac{2}{\sqrt{a} + 1}$

90. $\dfrac{1}{\sqrt{x} - \sqrt{y}}$

In Exercises 91–96 state whether or not the given equation is true for all values of x.

91. $\sqrt[6]{x^6} = x$

92. $\sqrt[3]{x^3} = x$

93. $x^3 x^{-1/3} = x^{8/3}$

94. $(x^3)^4 = x^7$

95. $\sqrt[4]{x^2} = \sqrt{x}$

96. $\sqrt{x^2 + 4} = |x| + 2$

97. Find a number x such that $3^x = 27^9$.

98. Find a number x such that $2^x = 4^{x-3}$.

99. Without using a calculator, determine which of the numbers $7^{1/4}$ and $4^{1/3}$ is larger.

100. Without using a calculator, determine which of the numbers $\sqrt[3]{5}$ and $\sqrt{3}$ is larger.

Section 1.3

Algebraic Expressions

Algebraic expressions such as

$$(1) \qquad 2x^2 - 3x + 4 \qquad ax + b \qquad \dfrac{y - 1}{y^2 + 2} \qquad \dfrac{cx^2 y + dy^2 z}{\sqrt{x^2 + y^2 + z^2}}$$

are obtained by starting with variables such as x, y, and z and constants such as 2, -3, a, b, c, and d, and combining them using addition, subtraction, multiplication, division, and roots. A **variable** is a letter that can represent any number in a given set of numbers, whereas a **constant** represents a fixed number. Usually we use letters near the end of the alphabet (like t, u, x, y, z) for variables and letters near the beginning of the alphabet (like a, b, c) for constants. The simplest types of algebraic expressions use only addition, subtraction, and multiplication and are called **polynomials**. (The first two expressions in (1) are polynomials.) They will be studied in greater detail in Chapter 3.

In working with algebraic expressions, we need to use the basic rules of algebra, which come from the following properties of real numbers:

$$a + b = b + a \qquad ab = ba \qquad \text{(Commutative Law)}$$
$$(a + b) + c = a + (b + c) \qquad (ab)c = a(bc) \qquad \text{(Associative Law)}$$
$$a(b + c) = ab + ac \qquad \text{(Distributive Law)}$$

In particular, putting $a = -1$ in the Distributive Law, we get

$$-(b + c) = -b - c$$

Example 1

(a) $4 - 3(x - 2) = 4 - 3x + 6 = 10 - 3x$

(b) $(x^3 - 2x^2 + 1) - (x^2 + 6x - 3) = x^3 - 2x^2 + 1 - x^2 - 6x + 3$
$$= x^3 - 3x^2 - 6x + 4$$

(c) $\sqrt{x}(x^2 + 2x + \sqrt{x}) = x^2\sqrt{x} + 2x\sqrt{x} + \sqrt{x}\sqrt{x} = x^{5/2} + 2x^{3/2} + x$ ●

If we use the Distributive Law three times, we get

$$(a + b)(c + d) = (a + b)c + (a + b)d = ac + bc + ad + bd$$

This says that we multiply two factors by multiplying each term in one factor by each term in the other factor and adding the products. Schematically, we have

$$(a + b)(c + d)$$

Example 2

(a) $(2x + 1)(3x - 5) = 6x^2 + 3x - 10x - 5 = 6x^2 - 7x - 5$

(b) $3(x - 1)(4x + 3) - 2(x + 6) = 3(4x^2 - x - 3) - 2x - 12$
$$= 12x^2 - 3x - 9 - 2x - 12$$
$$= 12x^2 - 5x - 21$$

(c) $(x^2 - 3)(x^3 + 2x + 1) = x^2(x^3 + 2x + 1) - 3(x^3 + 2x + 1)$
$$= x^5 + 2x^3 + x^2 - 3x^3 - 6x - 3$$
$$= x^5 - x^3 + x^2 - 6x - 3$$

(d) $(1 + \sqrt{x})(2 - 3\sqrt{x}) = 2 - 3\sqrt{x} + 2\sqrt{x} - 3(\sqrt{x})^2 = 2 - \sqrt{x} - 3x$ ●

Certain types of products occur so frequently that you should memorize them. You can verify the following formulas by performing the multiplications.

Special Product Formulas

1. $(a - b)(a + b) = a^2 - b^2$
2. $(a + b)^2 = a^2 + 2ab + b^2$
3. $(a - b)^2 = a^2 - 2ab + b^2$
4. $(a + b)^3 = a^3 + 3a^2b + 3ab^2 + b^3$
5. $(a - b)^3 = a^3 - 3a^2b + 3ab^2 - b^3$

Example 3

Use the special product formulas to find the following products:

(a) $(2x + 5)^2$ (b) $\left(2\sqrt{y} - \dfrac{1}{\sqrt{x}}\right)\left(2\sqrt{y} + \dfrac{1}{\sqrt{x}}\right)$ (c) $(x^2 - 2)^3$

Solution

(a)   Product Formula 2, with $a = 2x$ and $b = 5$, gives

$$(2x + 5)^2 = (2x)^2 + 2(2x)(5) + 5^2 = 4x^2 + 20x + 25$$

(b)   Using Product Formula 1 with $a = 2\sqrt{y}$ and $b = 1/\sqrt{x}$, we have

$$\left(2\sqrt{y} - \frac{1}{\sqrt{x}}\right)\left(2\sqrt{y} + \frac{1}{\sqrt{x}}\right) = (2\sqrt{y})^2 - \left(\frac{1}{\sqrt{x}}\right)^2 = 4y - \frac{1}{x}$$

(c)   Putting $a = x^2$ and $b = 2$ in Product Formula 5, we get

$$(x^2 - 2)^3 = (x^2)^3 - 3(x^2)^2(2) + 3x^2(2)^2 - 2^3 = x^6 - 6x^4 + 12x^2 - 8$$

●

Factoring

We have used the Distributive Law to expand algebraic expressions. We sometimes need to reverse this process (again using the Distributive Law) by factoring an expression as a product of simpler ones. The easiest case is when there is a common factor as follows:

$$\xrightarrow{\quad\text{expanding}\quad}$$
$$3x(x - 2) = 3x^2 - 6x$$
$$\xleftarrow{\quad\text{factoring}\quad}$$

To factor a quadratic of the form $x^2 + bx + c$, we note that

$$(x + r)(x + s) = x^2 + (r + s)x + rs$$

so we need to choose numbers r and s so that $r + s = b$ and $rs = c$.

Example 4

Factor $x^2 + 5x - 24$.

Solution

The two integers that add to give 5 and multiply to give -24 are -3 and 8. Therefore

$$x^2 + 5x - 24 = (x - 3)(x + 8)$$ ●

Example 5

Factor $2x^2 - 7x - 4$.

Solution

Even though the coefficient of x^2 is not 1, we can still look for factors of the form $2x + r$ and $x + s$, where $rs = -4$. After trial and error we find that

$$2x^2 - 7x - 4 = (2x + 1)(x - 4)$$ ●

Some special polynomials can be factored using the following formulas. The first one is just Product Formula 1 read backward. The other two can be verified by multiplying out the right sides.

Factoring Formulas	
$a^2 - b^2 = (a - b)(a + b)$	(difference of squares)
$a^3 - b^3 = (a - b)(a^2 + ab + b^2)$	(difference of cubes)
$a^3 + b^3 = (a + b)(a^2 - ab + b^2)$	(sum of cubes)

Example 6

Factor:

(a) $4x^2 - 25$ (b) $x^6 + 8$ (c) $2x^4 - 8x^2$

Solution

(a) Using the formula for a difference of squares, with $a = 2x$ and $b = 5$, we have

$$4x^2 - 25 = (2x)^2 - 5^2 = (2x - 5)(2x + 5)$$

(b) Using the formula for a sum of cubes, with $a = x^2$ and $b = 2$, we have

$$x^6 + 8 = (x^2)^3 + 2^3 = (x^2 + 2)(x^4 - 2x^2 + 4)$$

(c) $\qquad 2x^4 - 8x^2 = 2x^2(x^2 - 4)$ *(common factor)*

$$= 2x^2(x - 2)(x + 2) \quad \textit{(difference of squares)}$$ ●

When factoring expressions that involve fractional powers, we use the Laws of Exponents.

Example 7

Factor $3x^{3/2} - 9x^{1/2} + 6x^{-1/2}$.

Solution

We factor out the term with the smallest exponent:

$$3x^{3/2} - 9x^{1/2} + 6x^{-1/2} = 3x^{-1/2}(x^2 - 3x + 2)$$
$$= 3x^{-1/2}(x - 1)(x - 2) \quad \bullet$$

Example 8

$$x^3 + x^2 + 4x + 4 = (x^3 + x^2) + (4x + 4) \quad \textit{(grouping)}$$
$$= x^2(x + 1) + 4(x + 1) \quad \textit{(common factors)}$$
$$= (x^2 + 4)(x + 1) \quad \bullet$$

Fractional Expressions

In simplifying rational expressions (ratios of polynomials), we factor both numerator and denominator and cancel any common factors. In doing this we are using the following property of real numbers:

$$\frac{ac}{bc} = \frac{a}{b} \quad (b \neq 0, c \neq 0)$$

(This says that we can divide both numerator and denominator by the number c.) We assume that all fractions are defined; that is, we deal only with values of the variables such that the denominators are not zero.

Example 9

(a)
$$\frac{x^2 - 1}{x^2 + x - 2} = \frac{(x - 1)(x + 1)}{(x - 1)(x + 2)}$$
$$= \frac{x + 1}{x + 2}$$

(b)
$$\frac{2x^3 + 5x^2 - 3x}{6 - x - x^2} = \frac{x(2x^2 + 5x - 3)}{-(x^2 + x - 6)} = \frac{x(2x - 1)(x + 3)}{-(x - 2)(x + 3)}$$
$$= -\frac{x(2x - 1)}{x - 2} \quad \bullet$$

When multiplying and dividing fractional expressions, we use the following properties of real numbers. The second equation says that to divide two fractions, we invert the denominator and multiply.

$$\frac{a}{b} \cdot \frac{c}{d} = \frac{ac}{bd} \qquad \frac{\dfrac{a}{b}}{\dfrac{c}{d}} = \frac{a}{b} \cdot \frac{d}{c} = \frac{ad}{bc}$$

Example 10

Perform the indicated operations and simplify:

(a) $\dfrac{x^2 + 2x - 3}{x^2 + 8x + 16} \cdot \dfrac{3x + 12}{x - 1}$

(b) $\dfrac{x - 4}{x^2 - 4} \div \dfrac{x^2 - 3x - 4}{x^2 + 5x + 6}$

Solution

(a)   We first factor:

$$\frac{x^2 + 2x - 3}{x^2 + 8x + 16} \cdot \frac{3x + 12}{x - 1} = \frac{(x - 1)(x + 3)}{(x + 4)^2} \cdot \frac{3(x + 4)}{x - 1}$$

$$= \frac{3(x - 1)(x + 3)(x + 4)}{(x - 1)(x + 4)^2}$$

$$= \frac{3(x + 3)}{x + 4}$$

(b)
$$\frac{x - 4}{x^2 - 4} \div \frac{x^2 - 3x - 4}{x^2 + 5x + 6} = \frac{x - 4}{x^2 - 4} \cdot \frac{x^2 + 5x + 6}{x^2 - 3x - 4}$$

$$= \frac{(x - 4)(x + 2)(x + 3)}{(x - 2)(x + 2)(x - 4)(x + 1)}$$

$$= \frac{x + 3}{(x - 2)(x + 1)}$$ ●

In adding and subtracting rational expressions, we first find a common denominator and then use the following property of real numbers:

(2)
$$\boxed{\frac{a}{b} + \frac{c}{b} = \frac{a + c}{b}}$$

Although any common denominator will work, it is best to use the **least common denominator** (LCD). This is found by factoring each denominator and taking the product of the distinct factors, using the largest exponents that appear in any of the factors.

Example 11

Combine and simplify:

(a) $\dfrac{3}{x - 1} + \dfrac{x}{x + 2}$

(b) $\dfrac{1}{x^2 - 1} - \dfrac{2}{(x + 1)^2}$

(c) $\dfrac{1}{x^2 + 4x - 5} - \dfrac{1}{2x} + \dfrac{x + 1}{x^2 - x}$

Solution

(a)   Here the LCD is simply the product $(x - 1)(x + 2)$.

$$\frac{3}{x-1} + \frac{x}{x+2} = \frac{3(x+2)}{(x-1)(x+2)} + \frac{x(x-1)}{(x-1)(x+2)}$$

$$= \frac{3x+6+x^2-x}{(x-1)(x+2)} = \frac{x^2+2x+6}{(x-1)(x+2)}$$

(b) The LCD of $x^2 - 1 = (x-1)(x+1)$ and $(x+1)^2$ is $(x-1)(x+1)^2$, so we have

$$\frac{1}{x^2-1} - \frac{2}{(x+1)^2} = \frac{1}{(x-1)(x+1)} - \frac{2}{(x+1)^2} = \frac{(x+1) - 2(x-1)}{(x-1)(x+1)^2}$$

$$= \frac{x+1-2x+2}{(x-1)(x+1)^2} = \frac{3-x}{(x-1)(x+1)^2}$$

(c) $\dfrac{1}{x^2+4x-5} - \dfrac{1}{2x} + \dfrac{x+1}{x^2-x} = \dfrac{1}{(x-1)(x+5)} - \dfrac{1}{2x} + \dfrac{x+1}{x(x-1)}$

The LCD of the denominators is $2x(x-1)(x+5)$ and so

$$\frac{1}{(x-1)(x+5)} - \frac{1}{2x} + \frac{x+1}{x(x-1)} = \frac{2x - (x-1)(x+5) + (x+1)(2)(x+5)}{2x(x-1)(x+5)}$$

$$= \frac{2x - (x^2+4x-5) + (2x^2+12x+10)}{2x(x-1)(x+5)}$$

$$= \frac{x^2+10x+15}{2x(x-1)(x+5)} \qquad \bullet$$

Example 12

$$\frac{\dfrac{x}{y}+1}{1-\dfrac{y}{x}} = \frac{\dfrac{x+y}{y}}{\dfrac{x-y}{x}} = \frac{x+y}{y} \cdot \frac{x}{x-y} = \frac{x(x+y)}{y(x-y)} \qquad \bullet$$

Note: It is sometimes useful to use Equation 2 backward—that is, in the form

$$\frac{a+c}{b} = \frac{a}{b} + \frac{c}{b}$$

For example, in certain calculus problems it is advantageous to write

$$\frac{x+3}{x} = \frac{x}{x} + \frac{3}{x} = 1 + \frac{3}{x}$$

But remember to avoid the following common error:

$$\frac{a}{b+c} \neq \frac{a}{b} + \frac{a}{c}$$

(For instance, take $a = b = c = 1$ to see the error.)

The remaining examples show situations in calculus where facility with fractional expressions is required.

Example 13

Simplify $\dfrac{\dfrac{1}{(a+h)^2} - \dfrac{1}{a^2}}{h}$.

Solution

$$\dfrac{\dfrac{1}{(a+h)^2} - \dfrac{1}{a^2}}{h} = \dfrac{\dfrac{a^2 - (a+h)^2}{(a+h)^2 a^2}}{h} = \dfrac{a^2 - (a^2 + 2ah + h^2)}{(a+h)^2 a^2} \cdot \dfrac{1}{h}$$

$$= \dfrac{-2ah - h^2}{h(a+h)^2 a^2} = \dfrac{h(-2a - h)}{h(a+h)^2 a^2} = -\dfrac{2a+h}{(a+h)^2 a^2}$$

Example 14

Rationalize the numerator: $\dfrac{\sqrt{4+h} - 2}{h}$.

Solution

We multiply numerator and denominator by the conjugate radical $\sqrt{4+h} + 2$. The advantage of doing this is that we can use Product Formula 1 and the roots disappear.

$$\dfrac{\sqrt{4+h} - 2}{h} = \dfrac{\sqrt{4+h} - 2}{h} \cdot \dfrac{\sqrt{4+h} + 2}{\sqrt{4+h} + 2}$$

$$= \dfrac{(\sqrt{4+h})^2 - 2^2}{h(\sqrt{4+h} + 2)} = \dfrac{4 + h - 4}{h(\sqrt{4+h} + 2)}$$

$$= \dfrac{h}{h(\sqrt{4+h} + 2)} = \dfrac{1}{\sqrt{4+h} + 2}$$

Exercises 1.3

In Exercises 1–40 perform the indicated operations.

1. $3(2 - x) - 5(x + 2)$

2. $4(2x + 3) - 3(3x - 1)$

3. $(x^2 + 2x + 3) + (2x^2 - 3x + 4)$

4. $(x^2 + 2x + 3) - (2x^2 - 3x + 4)$

5. $(x^3 + 6x^2 - 4x + 7) - (3x^2 + 2x - 4)$

6. $4(x^2 - x + 2) - 5(x^2 - 2x + 1)$

7. $2(2 - 5t) + t^2(t - 1) - (t^4 - 1)$

8. $5(3t - 4) - (t^2 + 2) - 2t(t - 3)$

9. $\sqrt{x}(x - \sqrt{x})$

10. $x^{3/2}\left(\sqrt{x} - \dfrac{1}{\sqrt{x}}\right)$

11. $\sqrt[3]{y}(y^2 - 1)$

12. $\sqrt{x}(1 + \sqrt{x} - 2x)$

13. $(x + 6)(2x - 3)$

14. $(4x - 1)(3x + 7)$

15. $(3t - 2)(7t - 5)$

16. $(t + 6)(t + 5) - 3(t + 4)$

17. $(1 - 2y)^2$

18. $(3u - v)(3u + v)$

19. $(\sqrt{x} + \sqrt{y})(\sqrt{x} - \sqrt{y})$

20. $(3x + 4)^2$

21. $(2x - 5)(x^2 - x + 1)$

22. $(x^2 + 3)(5x - 6)$

23. $x(x - 1)(x + 2)$

24. $(1 + 2x)(x^2 - 3x + 1)$

25. $y^4(6 - y)(5 + y)$

26. $(t - 5)^2 - 2(t + 3)(8t - 1)$

27. $(2x^2 + 3y^2)^2$

28. $(\sqrt{r} - 2\sqrt{s})^2$

29. $(x^2 - a^2)(x^2 + a^2)$

30. $(3x - 4)^3$

31. $(1 + a^3)^3$

32. $(x - 1)(x^2 + x + 1)$

33. $\left(\sqrt{a} - \dfrac{1}{b}\right)\left(\sqrt{a} + \dfrac{1}{b}\right)$

34. $\left(c + \dfrac{1}{c}\right)^2$

35. $(x^2 + x - 2)(x^3 - x + 1)$

36. $(1 + x + x^2)(1 - x + x^2)$

37. $(1 - b)^2(1 + b)^2$

38. $(1 + x - x^2)^2$

39. $(3x^2y + 7xy^2) \div x$

40. $(16x^2y^3 - 12x^3y^4 + x^5y^5) \div xy^2$

41. Verify Product Formulas 1, 2, and 4.

42. Verify Product Formulas 3 and 5.

Factor the expressions in Exercises 43–74.

43. $2x + 12x^3$

44. $5ab - 8abc$

45. $x^2 + 7x + 6$

46. $x^2 - x - 6$

47. $x^2 - 2x - 8$

48. $2x^2 + 7x - 4$

49. $9x^2 - 36$

50. $8x^2 + 10x + 3$

51. $6x^2 - 5x - 6$

52. $x^2 + 10x + 25$

53. $t^3 + 1$

54. $4t^2 - 9s^2$

55. $4t^2 - 12t + 9$

56. $x^3 - 27$

57. $x^3 + 2x^2 + x$

58. $3x^3 - 27x$

59. $x^4 + 2x^3 - 3x^2$

60. $x^6 + 64$

61. $8x^3 - 125$

62. $x^4 + 2x^2 + 1$

63. $x^4 + x^2 - 2$

64. $x^3 + 3x^2 - 3x - 3$

65. $y^3 - 3y^2 - 4y + 12$

66. $y^3 - y^2 + y - 1$

67. $x^6 - y^6$

68. $x^8 - 1$

69. $x^{5/2} - x^{1/2}$

70. $3x^{-1/2} + 4x^{1/2} + x^{3/2}$

71. $x^{-3/2} + 2x^{-1/2} + x^{1/2}$

72. $(x - 1)^{7/2} - (x - 1)^{3/2}$

73. $(x^2 + 1)^{1/2} + 2(x^2 + 1)^{-1/2}$

74. $x^3(1 - x^3)^{1/2} - 3(1 - x^3)^{3/2}$

75. Verify the formula for a difference of cubes.

76. Verify the formula for a sum of cubes.

In Exercises 77–82 state whether or not the given equation is true for all values of the variables. (Disregard values that make denominators 0.)

77. $\dfrac{16 + a}{16} = 1 + \dfrac{a}{16}$

78. $\dfrac{b}{b - c} = 1 - \dfrac{b}{c}$

79. $\dfrac{2}{4 + x} = \dfrac{1}{2} + \dfrac{2}{x}$

80. $\dfrac{x + 1}{y + 1} = \dfrac{x}{y}$

81. $\dfrac{x}{x + y} = \dfrac{1}{1 + y}$

82. $\dfrac{1 + x + x^2}{x} = \dfrac{1}{x} + 1 + x$

In Exercises 83–126 simplify the expression.

83. $\dfrac{x^2 + 3x + 2}{x^2 + 5x + 6}$

84. $\dfrac{x^2 + x - 6}{x^2 - 4}$

85. $\dfrac{y - y^2}{y^2 - 1}$

86. $\dfrac{2y^2 - 9y - 18}{4y^2 + 16y + 15}$

87. $\dfrac{2x^3 - x^2 - 6x}{2x^2 - 7x + 6}$

88. $\dfrac{1 - x^2}{x^3 - 1}$

89. $\dfrac{t-3}{t^2+9} \cdot \dfrac{t+3}{t^2-9}$

90. $\dfrac{x^2-x-6}{x^2+2x} \cdot \dfrac{x^3+x^2}{x^2-2x-3}$

91. $\dfrac{2x^2+3x+1}{x^2+2x-15} \div \dfrac{x^2+6x+5}{2x^2-7x+3}$

92. $\dfrac{4y^2-9}{2y^2+9y-18} \div \dfrac{2y^2+y-3}{y^2+5y-6}$

93. $\dfrac{x/y}{z}$

94. $\dfrac{x}{y/z}$

95. $\dfrac{1}{x+5} + \dfrac{2}{x-3}$

96. $\dfrac{1}{x+1} + \dfrac{1}{x-1}$

97. $\dfrac{1}{x+1} - \dfrac{1}{x+2}$

98. $\dfrac{x}{x-4} - \dfrac{3}{x+6}$

99. $\dfrac{x}{(x+1)^2} + \dfrac{2}{x+1}$

100. $\dfrac{5}{2x-3} - \dfrac{3}{(2x-3)^2}$

101. $u+1+\dfrac{u}{u+1}$

102. $\dfrac{2}{a^2} - \dfrac{3}{ab} + \dfrac{4}{b^2}$

103. $\dfrac{1}{x^2} + \dfrac{1}{x^2+x}$

104. $\dfrac{1}{x} + \dfrac{1}{x^2} + \dfrac{1}{x^3}$

105. $\dfrac{2}{x+3} - \dfrac{1}{x^2+7x+12}$

106. $\dfrac{x}{x^2-4} + \dfrac{1}{x-2}$

107. $\dfrac{1}{x+3} + \dfrac{1}{x^2-9}$

108. $\dfrac{x}{x^2+x-2} - \dfrac{2}{x^2-5x+4}$

109. $\dfrac{2}{x} + \dfrac{3}{x-1} - \dfrac{4}{x^2-x}$

110. $\dfrac{x}{x^2-x-6} - \dfrac{1}{x+2} - \dfrac{2}{x-3}$

111. $\dfrac{1}{x^2+3x+2} - \dfrac{1}{x^2-2x-3}$

112. $\dfrac{1}{x+1} - \dfrac{2}{(x+1)^2} + \dfrac{3}{x^2-1}$

113. $\dfrac{\dfrac{x}{y} - \dfrac{y}{x}}{\dfrac{1}{x^2} - \dfrac{1}{y^2}}$

114. $x - \dfrac{y}{\dfrac{x}{y} + \dfrac{y}{x}}$

115. $\dfrac{1 + \dfrac{1}{c-1}}{1 - \dfrac{1}{c-1}}$

116. $1 + \dfrac{1}{1 + \dfrac{1}{1+x}}$

117. $\dfrac{\dfrac{5}{x-1} - \dfrac{2}{x+1}}{\dfrac{x}{x-1} + \dfrac{1}{x+1}}$

118. $\dfrac{\dfrac{a-b}{a} - \dfrac{a+b}{b}}{\dfrac{a-b}{b} + \dfrac{a+b}{a}}$

119. $\dfrac{x^{-2} - y^{-2}}{x^{-1} + y^{-1}}$

120. $\dfrac{x^{-1} + y^{-1}}{(x+y)^{-1}}$

121. $\dfrac{\dfrac{1}{a+h} - \dfrac{1}{a}}{h}$

122. $\dfrac{(x+h)^{-3} - x^{-3}}{h}$

123. $\dfrac{\dfrac{1-(x+h)}{2+(x+h)} - \dfrac{1-x}{2+x}}{h}$

124. $\dfrac{(x+h)^3 - 7(x+h) - (x^3-7x)}{h}$

125. $\sqrt{1 + \left(\dfrac{x}{\sqrt{1-x^2}}\right)^2}$ **126.** $\sqrt{1 + \left(x^3 - \dfrac{1}{4x^3}\right)^2}$

In Exercises 127–132 rationalize the numerator.

127. $\dfrac{\sqrt{r} + \sqrt{s}}{t}$

128. $\dfrac{\sqrt{3(x+h)+5} - \sqrt{3x+5}}{h}$

129. $\dfrac{\sqrt{x} - \sqrt{x+h}}{h\sqrt{x}\sqrt{x+h}}$ **130.** $\sqrt{x^2+1} - x$

131. $\sqrt{x^2+x+1} + x$ **132.** $\sqrt{x+1} - \sqrt{x}$

133. Factor $x^4 + 3x^2 + 4$. [*Hint:* Write the expression as $(x^4 + 4x^2 + 4) - x^2$ and note that this is a difference of squares.]

134. Factor $y^4 + 1$.

Equations

An **equation** is a statement that two mathematical expressions are equal. **A root** or **solution** of an equation is a value of the unknown (or variable) that makes the equation true. For instance, the equation $4x + 8 = 0$ has the solution $x = -2$, since $4(-2) + 8 = 0$. When we are asked to **solve** an equation, we are requested to find all the roots of the equation.

We solve an equation by transforming it into simpler equivalent equations with solutions that are obvious. We do this by adding or subtracting a quantity to both sides of the equation and by multiplying or dividing both sides by a nonzero quantity.

The simplest type of equation is a **linear equation**, or first-degree equation, which is an equation that is equivalent to one of the form $ax + b = 0$. The equation in the following example is linear.

Example 1

Solve the equation $\dfrac{x}{x + 1} = \dfrac{2x + 1}{2x - 3}$.

Solution

If $x \neq -1$ and $x \neq \frac{3}{2}$, we can multiply both sides of the equation by the LCD, which is $(x + 1)(2x - 3)$:

$$(x + 1)(2x - 3)\left(\frac{x}{x + 1}\right) = (x + 1)(2x - 3)\left(\frac{2x + 1}{2x - 3}\right)$$

$$(2x - 3)x = (x + 1)(2x + 1)$$

$$2x^2 - 3x = 2x^2 + 3x + 1$$

$$-3x = 3x + 1 \qquad \textit{(subtracting } 2x^2 \textit{ from both sides)}$$

$$-6x = 1 \qquad \textit{(subtracting } 3x \textit{ from both sides)}$$

$$x = -\frac{1}{6} \qquad \textit{(dividing both sides by } -6\textit{)}$$

These equations are equivalent and the solution is $-\frac{1}{6}$, as you can verify.   ●

Some equations can be solved using factoring and the following basic property of real numbers:

$$ab = 0 \qquad \text{if and only if} \qquad a = 0 \text{ or } b = 0$$

Example 2

Solve the equation $x^2 + 5x = 24$.

Solution

We first take the term 24 to the left side of the equation and then we factor the left side:

$$x^2 + 5x = 24$$
$$x^2 + 5x - 24 = 0$$
$$(x - 3)(x + 8) = 0$$
$$x - 3 = 0 \quad \text{or} \quad x + 8 = 0$$
$$x = 3 \qquad\qquad x = -8$$

The solutions are $x = 3$ and $x = -8$. ●

Example 3

Solve $x^5 - x = 0$.

Solution

$$x(x^4 - 1) = 0 \qquad \textit{(common factor)}$$
$$x(x^2 - 1)(x^2 + 1) = 0 \qquad \textit{(difference of squares)}$$
$$x(x - 1)(x + 1)(x^2 + 1) = 0 \qquad \textit{(difference of squares)}$$
$$x = 0 \quad \text{or} \quad x - 1 = 0 \quad \text{or} \quad x + 1 = 0 \quad \text{or} \quad x^2 + 1 = 0$$

The only real solutions are $x = 0$, $x = 1$, and $x = -1$. (The equation $x^2 + 1 = 0$ has no real solution because $x^2 \geq 0$ and so $x^2 + 1 \geq 1$.) ●

The equation in Example 2 is a quadratic equation. In general, a **quadratic equation** has the form $ax^2 + bx + c = 0$, where a, b, and c are constants and $a \neq 0$. If such an equation does not factor readily, we use the technique called **completing the square**. We note that any expression of the form $x^2 + bx$ can be completed to become a perfect square by adding the square of half the coefficient of x:

$$(x^2 + bx) + \left(\frac{b}{2}\right)^2 = x^2 + bx + \frac{b^2}{4} = \left(x + \frac{b}{2}\right)^2$$

Example 4

Solve the equation $x^2 - 8x + 13 = 0$ by completing the square.

Solution

We subtract 13 from both sides of the equation. Then we add the square of half the coefficient of x, namely $(-4)^2 = 16$, to both sides in order to complete the square.

$$x^2 - 8x + 13 = 0$$
$$x^2 - 8x = -13$$
$$x^2 - 8x + 16 = 16 - 13$$
$$(x - 4)^2 = 3$$
$$x - 4 = \sqrt{3} \qquad\qquad \text{or} \qquad x - 4 = -\sqrt{3}$$
$$x = 4 + \sqrt{3} \qquad\qquad\qquad x = 4 - \sqrt{3} \qquad ●$$

We can use the technique of completing the square to derive a formula for the roots of the general quadratic equation $ax^2 + bx + c = 0$. First, we divide both sides of the equation by a and move the constant to the right, giving

$$x^2 + \frac{b}{a}x = -\frac{c}{a}$$

We now complete the square by adding $(b/2a)^2$ to both sides of the equation:

$$x^2 + \frac{b}{a}x + \left(\frac{b}{2a}\right)^2 = -\frac{c}{a} + \left(\frac{b}{2a}\right)^2$$

and so

$$\left(x + \frac{b}{2a}\right)^2 = \frac{-4ac + b^2}{4a^2}$$

Taking square roots of both sides, we see that

$$x + \frac{b}{2a} = \pm\sqrt{\frac{-4ac + b^2}{4a^2}} = \pm\frac{\sqrt{b^2 - 4ac}}{2a}$$

Solving for x, we get

$$x = \frac{-b \pm \sqrt{b^2 - 4ac}}{2a}$$

The Quadratic Formula

The roots of the quadratic equation $ax^2 + bx + c = 0$ are

$$x = \frac{-b \pm \sqrt{b^2 - 4ac}}{2a}$$

The quadratic formula could be used to solve the equation in Example 2. You should carry out the details of this calculation.

Example 5

Find all solutions of each of the following equations:

(a) $3x^2 - 5x - 1 = 0$ (b) $x^2 + 2x + 2 = 0$

Solution

(a)   Using the quadratic formula with $a = 3$, $b = -5$, and $c = -1$, we get

$$x = \frac{-(-5) \pm \sqrt{(-5)^2 - 4(3)(-1)}}{2(3)} = \frac{5 \pm \sqrt{37}}{6}$$

If approximations are desired, we use a calculator and obtain

$$x = \frac{5 + \sqrt{37}}{6} \approx 1.8471 \qquad \text{or} \qquad x = \frac{5 - \sqrt{37}}{6} \approx -0.1805$$

(b) $$x = \frac{-2 \pm \sqrt{2^2 - 4 \cdot 2}}{2} = \frac{-2 \pm \sqrt{-4}}{2} = \frac{-2 \pm 2\sqrt{-1}}{2} = -1 \pm \sqrt{-1}$$

Since the square of any real number is nonnegative, $\sqrt{-1}$ is undefined in the real number system. The equation has no real solutions.   ●

In Section 3.4 we will study the complex number system, in which the square roots of negative numbers do exist. We will see that the complex solutions of the equation in Example 5(b) are $x = -1 + i$ and $x = -1 - i$, where $i = \sqrt{-1}$.

The quantity $\Delta = b^2 - 4ac$, which appears under the square root sign in the quadratic formula, is called the **discriminant** of the equation $ax^2 + bx + c = 0$. If $\Delta < 0$, then $\sqrt{b^2 - 4ac}$ is undefined, so the quadratic equation has no real solutions [as in Example 5(b)]. If $\Delta = 0$, then the quadratic formula becomes

$$x = \frac{-b \pm \sqrt{0}}{2a} = \frac{-b}{2a}$$

so there is only one solution. Finally, if $\Delta > 0$, then $\sqrt{b^2 - 4ac} > 0$, so

$$\frac{-b + \sqrt{b^2 - 4ac}}{2a} \quad \text{and} \quad \frac{-b - \sqrt{b^2 - 4ac}}{2a}$$

will have different values. This means that the quadratic equation will have two different solutions [as in Example 5(a)].

The following box contains a summary of what we have just proved about the discriminant.

The **discriminant** of the general quadratic equation $ax^2 + bx + c = 0$ is $\Delta = b^2 - 4ac$.

1. If $\Delta > 0$, then the equation has two different real solutions.
2. If $\Delta = 0$, then the equation has exactly one real solution.
3. If $\Delta < 0$, then the equation has no real roots.

Example 6

Use the discriminant to determine how many real solutions the following equations possess:

(a) $2x^2 + 13x - 38 = 0$ (b) $3x - 1 = 7x^2$

Solution

(a)   We have $a = 2$, $b = 13$, and $c = -38$, so the discriminant is

$$\Delta = b^2 - 4ac = (13)^2 - 4(2)(-38) = 473 > 0$$

Therefore there are two real solutions.

(b)   Rewriting the equation as $7x^2 - 3x + 1 = 0$, we see that $a = 7$, $b = -3$, and $c = 1$, so

$$\Delta = (-3)^2 - 4(7)(1) = -19 < 0$$

The equation has no real solutions.   ●

Some equations that at first glance may not appear to be quadratic can be changed into quadratic equations by performing simple algebraic operations on them, such as multiplying both sides by a common denominator or squaring both sides. Special care needs to be taken in solving such equations, as we see in the following example.

Example 7

Solve the following equations:

(a) $x + 3 = \dfrac{2x^2 - 7x + 3}{3 - x}$

(b) $x = 1 - \sqrt{2 - \dfrac{x}{2}}$

Solution

(a) Multiplying both sides of the equation by $x - 3$ to clear the denominator (which we can do as long as $x \neq 3$), we get $x^2 - 9 = -2x^2 + 7x - 3$, so

$$3x^2 - 7x - 6 = 0$$

and

$$x = \frac{7 \pm \sqrt{7^2 - 4 \cdot 3(-6)}}{2 \cdot 3} = \frac{7 \pm \sqrt{121}}{6} = \frac{7 \pm 11}{6} = 3 \quad \text{or} \quad -\frac{2}{3}$$

[Instead of using the quadratic formula, we could have factored the equation as $(x - 3)(3x + 2) = 0$.] But $x = 3$ does not satisfy the original equation (since division by 0 is impossible), so the only solution is $x = -\frac{2}{3}$.

(b) To eliminate the square root, we first write the equation as

$$x - 1 = -\sqrt{2 - \frac{x}{2}}$$

and then square both sides, giving

$$(x - 1)^2 = 2 - \frac{x}{2}$$

so

$$x^2 - 2x + 1 = 2 - \frac{x}{2}$$

and

$$2x^2 - 3x - 2 = 0$$
$$(x - 2)(2x + 1) = 0$$

Thus $x = 2$ and $x = -\frac{1}{2}$ are potential solutions. Substituting these into the original equation, we see that $x = -\frac{1}{2}$ is a solution (since this leads to $-\frac{1}{2} = 1 - \sqrt{2 + \frac{1}{4}}$, which is true) but that $x = 2$ is not a solution (since $2 \neq 1 - \sqrt{2 - 1}$). The only solution is $x = -\frac{1}{2}$. ●

 The reason that extraneous solutions are often introduced when we square both sides of an equation while solving it is that the operation of squaring can turn an inequality into an equality. For example, $-1 \neq 1$, but $(-1)^2 = 1^2$. This is exactly what we did unwittingly in Example 7(b) in the case when $x = 2$, and that is why the solutions we obtain initially in such cases are really only potential solutions that have to be checked by substituting them into the original equation.

In some cases fourth-degree (or higher) polynomial equations can be changed to quadratic equations by performing an algebraic substitution, as in the following example.

Example 8

Find the solutions of $x^4 - 2x^2 - 2 = 0$.

Solution

If we set $y = x^2$, then

$$x^4 - 2x^2 - 2 = (x^2)^2 - 2x^2 - 2$$
$$= y^2 - 2y - 2 = 0$$

and the solutions of the equation in y are

$$y = \frac{2 \pm \sqrt{2^2 + 4 \cdot 2}}{2} = 1 \pm \sqrt{3}$$

Since $x^2 = y$, $x = \pm\sqrt{y}$, and since $1 - \sqrt{3}$ is a negative number, we get

$$x = \pm\sqrt{1 + \sqrt{3}}$$

as the only solutions of the original equation.   ●

In the remaining examples we use the methods of this section to solve word problems.

Example 9

Bill left his house at 2:00 P.M. and rode his bicycle down Main Street at a speed of 12 mi/h. When his friend Mary arrived at his house at 2:10 P.M., Bill's mother told her the direction in which Bill had left and Mary cycled at a speed of 16 mi/h. When did Mary catch up with Bill?

Solution

Let t be the time, in hours, that it took Mary to catch up with Bill. Because Bill had a 10-min, or $\frac{1}{6}$-h, head start, he cycled for $(t + \frac{1}{6})$ h.

In problems involving distance, rate (speed), and time, it is helpful to organize the information in a table using the formula

$$\text{distance} = \text{rate} \times \text{time}$$

	Distance (mi)	Rate (mi/h)	Time (h)
Mary	$16t$	16	t
Bill	$12(t + \frac{1}{6})$	12	$t + \frac{1}{6}$

To find t we use the fact that at the instant Mary overt̶
cycled the same distance:

$$\text{distance traveled by Mary} = \text{distance traveled by Bill}$$
$$16t = 12(t + \tfrac{1}{6})$$
$$16t = 12t + 2$$
$$4t = 2$$
$$t = \tfrac{1}{2}$$

Mary caught up with Bill after half an hour—that is, at 2:40 P.M.   ●

Example 10

A rectangular supermarket, 270 ft by 180 ft, is to be built on a city block that has an area of 400,000 ft^2. There is to be a uniform strip around the supermarket for parking. How wide is the strip?

Solution

In a problem such as this that involves geometry, it is essential to draw a diagram,

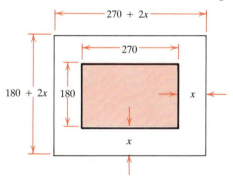

Figure 1

as in Figure 1. Let x be the width of the strip. The dimensions of the city block are $(180 + 2x)$ ft by $(270 + 2x)$ ft. Thus the area of the block is

$$(180 + 2x)(270 + 2x) = 400,000$$
$$4x^2 + 900x + 48,600 = 400,000$$
$$x^2 + 225x - 87,850 = 0$$

This is a quadratic equation, which we solve using the quadratic formula:

$$x = \frac{-225 \pm \sqrt{(225)^2 - 4(1)(-87,850)}}{2} = \frac{-225 \pm \sqrt{402,025}}{2}$$

We reject the negative root, since x must be positive. Thus

$$x = \frac{-225 + \sqrt{402,025}}{2} \approx 205$$

The width of the strip should be about 205 ft.   ●

More challenging word problems are considered in Section 1.9.

Exercises 1.4

4. $\frac{2}{3}y + \frac{1}{2}(y - 3) = -4$

5. $\frac{1}{x} = \frac{4}{3x} + 1$

6. $\frac{2x - 1}{x + 2} = \frac{4}{5}$

7. $\frac{2}{t + 6} = \frac{3}{t - 1}$

8. $\frac{1}{t - 1} + \frac{t}{3t - 2} = \frac{1}{3}$

9. $x^2 + 8x + 15 = 0$

10. $x^2 - 2x - 8 = 0$

11. $x^2 + 4x = 12$

12. $x^2 - 21 = 4x$

13. $2x^2 + 3x - 2 = 0$

14. $6x^2 + 5x + 1 = 0$

15. $x^3 + 2x^2 + x = 0$

16. $x^3 = x$

17. $x^4 - 25 = 0$

18. $x^9 - 256x = 0$

In Exercises 19–22 solve the quadratic equation by completing the square.

19. $x^2 + 2x - 2 = 0$

20. $x^2 - 4x + 2 = 0$

21. $x^2 + x - 1 = 0$

22. $2x^2 + 8x + 1 = 0$

In Exercises 23–50 find all real solutions of the equation.

23. $x^2 - 2x - 15 = 0$

24. $6x^2 + 7x - 3 = 0$

25. $x^2 + 36x - 1440 = 0$

26. $12x^2 + 140x + 375 = 0$

27. $3x^2 + 2x - 2 = 0$

28. $x^2 - 6x + 1 = 0$

29. $2y^2 - y - \frac{1}{2} = 0$

30. $\theta^2 - \frac{3}{2}\theta + \frac{9}{16} = 0$

31. $3x^2 + 6x + 4 = 0$

32. $4x = x^2 + 1$

33. $3 + 5z + z^2 = 0$

34. $w^2 = 3(w - 1)$

35. $x^2 - \sqrt{5}x + 1 = 0$

36. $\sqrt{6}x^2 + 2x - \sqrt{\frac{3}{2}} = 0$

37. $\frac{x^2}{x + 100} = 50$

38. $1 + \frac{2x}{(x + 3)(x + 4)} = \frac{2}{x + 3} + \frac{4}{x + 4}$

39. $\frac{x + 5}{x - 2} = \frac{5}{x + 2} + \frac{28}{x^2 - 4}$

40. $\frac{x}{2x + 7} - \frac{x + 1}{x + 3} = 1$

41. $\sqrt{2x + 1} + 1 = x$

42. $x - \sqrt{9 - 3x} = 0$

43. $\sqrt{5 - x} + 1 = \sqrt{x}$

44. $\sqrt{2x + 1} + \sqrt{x + 1} = 2$

45. $\sqrt{\sqrt{x - 5} + x} = 5$

46. $x^4 - 5x^2 + 4 = 0$

47. $2x^4 + 4x^2 + 1 = 0$

48. $x^6 - 2x^3 - 3 = 0$

49. $4(x + 1)^{1/2} - 5(x + 1)^{3/2} + (x + 1)^{5/2} = 0$

50. $x^{1/2} + 3x^{-1/2} = 10x^{-3/2}$

In Exercises 51–56 solve the equation for the indicated variable.

51. $P = 2L + 2W$, for W

52. $V = \frac{1}{3}\pi r^2 h$, for r

53. $s = \frac{a}{1 - r}$, for r

54. $A = \frac{h}{2}(a + b)$, for a

55. $a^2 + b^2 = c^2$, for b

56. $\frac{1}{R} = \frac{1}{R_1} + \frac{1}{R_2}$, for R_1

In Exercises 57 and 58 find the value(s) of k that will ensure that the indicated values of x are the solutions of the given quadratic equation.

57. $2x^2 + 5x - k = 0$; solutions $x = -3, \frac{1}{2}$

58. $kx^2 + x - 4 = 0$; solutions $x = -\frac{4}{3}, 1$

Using the discriminant, determine how many real solutions each of the equations in Exercises 59–64 will have, without solving the equations.

59. $5x^2 + 3x + 1 = 0$

60. $x^2 = 6x + 6$

61. $x^2 + 2.20x + 1.21 = 0$

62. $x^2 + 2.21x + 1.21 = 0$

63. $x^2 - rx + s = 0$ $(r > 0, \sqrt{r} > 2s)$

64. $x^2 + rx - s = 0$ $(s > 0)$

65. Find all values of k that will ensure that the given equation has the indicated number of real solutions.
  (a) $4x^2 + kx + 25 = 0$; one real solution
  (b) $5x^2 - 6x + k = 0$; no real solution
  (c) $kx^2 + 4x - \frac{1}{2} = 0$; two real solutions

66. A cash box contains $525 in $5 and $10 bills. The total number of bills is 74. How many bills of each denomination are there?

67. A merchant mixes tea that sells for $3.00 a pound with tea that sells for $2.75 a pound to get 80 lb of a mixture that sells for $2.90 a pound. How many pounds of each type of tea does she use?

68. What quantity of a 60% acid solution must be mixed with a 30% solution to produce 300 mL of a 50% solution?

69. Steve invested $6000, part at an interest rate of 9% per year and the rest at a rate of 8% per year. After a year the total interest on these investments was $525. How much did he invest at each rate?

70. Wendy took a trip from Davenport to Omaha, a distance of 300 mi. She traveled part of the way by bus and arrived at the train station just in time to complete her journey by train. The bus averaged 40 mi/h and the train 60 mi/h. The entire trip took $5\frac{1}{2}$ h. How long did she spend on the train?

71. Craig drove from Fresno to Bakersfield at 50 mi/h. Mike left at the same time but drove at 56 mi/h. When Mike arrived in Bakersfield, Craig had another 12 mi to drive. How far is it from Fresno to Bakersfield?

72. Find two numbers whose sum is 55 and whose product is 684.

73. The sum of the squares of two consecutive even integers is 1252. Find the integers.

74. A circle has a radius of 4 cm. By how much should the radius be increased so that the area is increased by 10 cm²?

75. A box with a square base and no top is to be made from a square piece of cardboard by cutting 4-in. squares from each corner and folding up the sides as in the figure. The box is to hold 100 in.³. How big a piece of cardboard is needed?

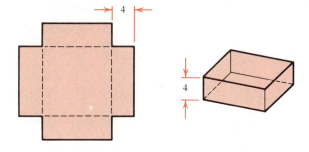

76. A pilot flew a jet from Montreal to Los Angeles, a distance of 2500 mi. On the return trip the speed was increased by 100 mi/h. The round trip took 9 h and 10 min. What was the speed from Montreal to Los Angeles?

Section 1.5

Inequalities

When working with inequalities, we use the following rules:

Rules for Inequalities

1. If $a < b$, then $a + c < b + c$.
2. If $a < b$ and $c < d$, then $a + c < b + d$.
3. If $a < b$ and $c > 0$, then $ac < bc$.
4. If $a < b$ and $c < 0$, then $ac > bc$.
5. If $0 < a < b$, then $1/a > 1/b$.

These rules apply more generally to any of the other order relations $>$, $\leq$, and $\geq$. Rule 1 says that we can add (or subtract) any number to (or from) both sides of an inequality, and Rule 2 says that two inequalities can be added. However, we have to be careful with multiplication. Rule 3 says that we can multiply (or divide) both sides of an inequality by a *positive* number, but Rule 4 says that if we multiply both sides of an inequality by a negative number, then we reverse the direction of the inequality. For example, if we take the inequality

$3 < 5$ and multiply by 2, we get $6 < 10$, but if we multiply by -2, we get $-6 > -10$. Finally, Rule 5 says that if we take reciprocals, then we reverse the direction of an inequality (provided the numbers are positive).

Rules 1–5 are useful in solving inequalities, such as $1 + x < 7x + 5$, in which a variable appears. This inequality is satisfied by some values of x but not by others. To **solve** an inequality means to find all values of the variable for which the inequality is true.

Example 1

Solve the inequality $1 + x < 7x + 5$ and sketch the solution set.

Solution

First we subtract 1 from each side of the inequality (using Rule 1 with $c = -1$):

$$x < 7x + 4$$

Then we subtract $7x$ from both sides (Rule 1 with $c = -7x$):

$$-6x < 4$$

Now we divide both sides by -6 (Rule 4 with $c = -\frac{1}{6}$):

$$x > -\frac{4}{6} = -\frac{2}{3}$$

These steps can be reversed, so the solution set consists of all numbers greater than $-\frac{2}{3}$. In other words, the solution of the inequality is the interval $(-\frac{2}{3}, \infty)$. It is graphed in Figure 1. ●

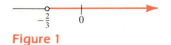

Figure 1

Example 2

Solve the inequalities $4 \le 3x - 2 < 13$.

Solution

The solution set consists of all values of x that satisfy both inequalities. Using Rules 1 and 3, we see that the following inequalities are equivalent:

$$4 \le 3x - 2 < 13$$
$$6 \le 3x < 15 \qquad \textit{(add 2)}$$
$$2 \le x < 5 \qquad \textit{(divide by 3)}$$

Therefore the solution set is $[2, 5)$. It is shown in Figure 2. ●

Figure 2

Example 3

Solve $2x + 1 \le 4x - 3 \le x + 7$.

Solution

This time we first solve the inequalities separately:

$$2x + 1 \le 4x - 3 \qquad 4x - 3 \le x + 7$$
$$4 \le 2x \qquad\qquad 3x \le 10$$
$$2 \le x \qquad\qquad\quad x \le \frac{10}{3}$$

Figure 3

Since x must satisfy both inequalities, we have

$$2 \leq x \leq \tfrac{10}{3}$$

Thus the solution set is the closed interval $[2, \tfrac{10}{3}]$ and is shown in Figure 3.

Example 4

Solve the inequality $x^2 - 5x + 6 \leq 0$.

Solution

First we factor the left side:

$$(x - 2)(x - 3) \leq 0$$

We know that the corresponding equation $(x - 2)(x - 3) = 0$ has the solutions 2 and 3. The numbers 2 and 3 divide the real line into three intervals:

$$(-\infty, 2) \qquad (2, 3) \qquad (3, \infty)$$

On each of these intervals we determine the signs of the factors. For instance,

$$x \in (-\infty, 2) \Rightarrow x < 2 \Rightarrow x - 2 < 0$$

Then we record these signs in the following chart:

Interval	$x - 2$	$x - 3$	$(x - 2)(x - 3)$
$x < 2$	$-$	$-$	$+$
$2 < x < 3$	$+$	$-$	$-$
$x > 3$	$+$	$+$	$+$

Another method for obtaining the information in the chart is to use **test values**. For instance, if we use the test value $x = 1$ for the interval $(-\infty, 2)$, then substitution in $x^2 - 5x + 6$ gives

$$1^2 - 5(1) + 6 = 2$$

The polynomial $x^2 - 5x + 6$ does not change sign on any of the three intervals, so we conclude that it is positive on $(-\infty, 2)$.

Then we read from the chart that $(x - 2)(x - 3)$ is negative when $2 < x < 3$. Thus the solution of the inequality $(x - 2)(x - 3) \leq 0$ is

$$\{x \mid 2 \leq x \leq 3\} = [2, 3]$$

Notice that we have included the endpoints 2 and 3 because we seek values of x such that the product is either negative or zero. The solution is illustrated in Figure 4.

Figure 4

Example 5

Solve $x^2 + 3x > 4$.

Solution 1

First we take all nonzero terms to one side of the inequality sign and factor the resulting expression:

$$x^2 + 3x - 4 > 0 \qquad \text{or} \qquad (x - 1)(x + 4) > 0$$

As in Example 4 we solve the corresponding equation $(x - 1)(x + 4) = 0$ and use the solutions $x = 1$ and $x = -4$ to divide the real line into three intervals $(-\infty, -4)$, $(-4, 1)$, and $(1, \infty)$. On each interval the product keeps a constant sign as shown in the following chart:

Interval	$x - 1$	$x + 4$	$(x - 1)(x + 4)$
$x < -4$	$-$	$-$	$+$
$-4 < x < 1$	$-$	$+$	$-$
$x > 1$	$+$	$+$	$+$

Then we read from the chart that the solution set is

$$\{x \mid x < -4 \quad \text{or} \quad x > 1\} = (-\infty, -4) \cup (1, \infty)$$

The solution is illustrated in Figure 5.

Figure 5

Solution 2

As in the first solution we write the inequality in the form

$$(x - 1)(x + 4) > 0$$

Then we use the fact that the product of two numbers is positive when both factors are positive or when both factors are negative.

CASE I: When $x - 1 > 0$ and $x + 4 > 0$, we have $x > 1$ and $x > -4$. But if $x > 1$, then $x > -4$ holds automatically. So the solution in this case is given by $x > 1$.

CASE II: When $x - 1 < 0$ and $x + 4 < 0$, we have $x < 1$ and $x < -4$. But if $x < -4$, then $x < 1$ holds automatically. So the solution in this case is given by $x < -4$.

Combining both cases, we see that the solution set is

$$\{x \mid x < -4 \quad \text{or} \quad x > 1\} = (-\infty, -4) \cup (1, \infty) \qquad \bullet$$

Example 6

Solve $\dfrac{1 + x}{1 - x} > 1$.

Solution 1

One method is to take all nonzero terms to the left side and use a common denominator:

$$\frac{1 + x}{1 - x} > 1$$

$$\frac{1 + x}{1 - x} - 1 > 0$$

$$\frac{1 + x - 1 + x}{1 - x} > 0$$

$$\frac{2x}{1 - x} > 0$$

The numerator is zero when $x = 0$ and the denominator is zero when $x = 1$. As before, we can set up a chart to determine the sign on each of the intervals $(-\infty, 0)$, $(0, 1)$, and $(1, \infty)$.

Interval	$2x$	$1 - x$	$2x/(1 - x)$
$x < 0$	$-$	$+$	$-$
$0 < x < 1$	$+$	$+$	$+$
$x > 1$	$+$	$-$	$-$

From the chart we see that the solution set is $\{x \mid 0 < x < 1\} = (0, 1)$.

Solution 2

Another method is to multiply both sides by $1 - x$, but in view of Rules 3 and 4 we must consider separately the cases where $1 - x$ is positive and negative.

CASE I: If $1 - x > 0$, that is, $x < 1$, then multiplying the given inequality by $1 - x$ gives

$$1 + x > 1 - x$$

which becomes $2x > 0$, that is, $x > 0$. So we have

$$0 < x < 1$$

CASE II: If $1 - x < 0$, that is, $x > 1$, then multiplying the given inequality by $1 - x$ gives

$$1 + x < 1 - x$$

which becomes $2x < 0$, that is, $x < 0$. But the conditions $x > 1$ and $x < 0$ are incompatible, so there is no solution in Case II.

Therefore the solution set is the open interval $(0, 1)$. ●

Exercises 1.5

In Exercises 1–30 solve the given inequalities in terms of intervals and illustrate the solution sets on the real number line.

1. $2x + 7 > 3$

2. $3x - 11 < 4$

3. $1 - x \leq 2$

4. $4 - 3x \geq 6$

5. $2x + 1 < 5x - 8$

6. $1 + 5x > 5 - 3x$

7. $-1 < 2x - 5 < 7$

8. $1 < 3x + 4 \leq 16$

9. $0 \leq 1 - x < 1$

10. $-5 \leq 3 - 2x \leq 9$

11. $4x < 2x + 1 \leq 3x + 2$

12. $2x + 3 < x + 4 < 3x - 2$

13. $1 - x \geq 3 - 2x \geq x - 6$

14. $x > 1 - x \geq 3 + 2x$

15. $(x - 1)(x - 2) > 0$

16. $(2x + 3)(x - 1) \geq 0$

17. $2x^2 + x \leq 1$

18. $x^2 < 2x + 8$

19. $x^2 > 3(x + 6)$

20. $3x^2 + 2x > 1$

21. $x^2 < 4$

22. $x^2 \geq 5$

23. $\dfrac{1}{x} < 4$

24. $-3 < \dfrac{1}{x} \leq 1$

25. $\dfrac{4}{x} < x$

26. $\dfrac{x}{x + 1} > 3$

27. $\dfrac{2x + 1}{x - 5} < 3$

28. $\dfrac{2 + x}{3 - x} \leq 1$

29. $\dfrac{x^2 - 1}{x^2 + 1} \geq 0$

30. $x^3 > x$

31. The relationship between the Celsius and Fahrenheit temperature scales is given by $C = \frac{5}{9}(F - 32)$, where C is the temperature in degrees Celsius and F is the temperature in degrees Fahrenheit. What interval on the Celsius scale corresponds to the temperature range $50 \leq F \leq 95$?

32. Use the relationship between C and F given in Exercise 31 to find the interval on the Fahrenheit scale corresponding to the temperature range $20 \leq C \leq 30$.

33. As dry air moves upward, it expands and in so doing cools at a rate of about 1°C for each 100-m rise, up to about 12 km.

(a) If the ground temperature is 20°C, write a formula for the temperature at height h.

(b) What range of temperature can be expected if a plane takes off and reaches a maximum height of 5 km?

34. Using calculus it can be proved that if a ball is thrown upward from the top of a building 128 ft high with an initial velocity of 16 ft/s, then the height h above the ground t seconds later will be

$$h = 128 + 16t - 16t^2$$

During what time interval will the ball be at least 32 ft above the ground?

35. Prove Rule 1. (*Hint:* Remember that $a < b$ means that $b - a > 0$.)

36. Prove Rule 3. (See the hint for Exercise 35 and remember that the product of two positive numbers is positive.)

37. Prove Rule 4. (Use the fact that the product of a positive number and a negative number is negative.)

38. Prove Rule 5.

In Exercises 39 and 40 solve the inequality for x assuming that a, b, and c are positive constants.

39. $a(bx - c) \geq bc$

40. $a \leq bx + c < 2a$

In Exercises 41 and 42 solve the inequality for x assuming that a, b, and c are negative constants.

41. $ax + b < c$

42. $\dfrac{ax + b}{c} \leq b$

43. Show that if $a < b$, then $a < \dfrac{a + b}{2} < b$.

44. Show that if $0 < a < b$, then $a^2 < b^2$.

45. Suppose that a, b, c, and d are positive numbers such that

$$\frac{a}{b} < \frac{c}{d}$$

Show that

$$\frac{a}{b} < \frac{a + c}{b + d} < \frac{a}{d}$$

Absolute Value

Recall from Section 1.1 that the absolute value of a number a is given by

$$|a| = \begin{cases} a & \text{if } a \geq 0 \\ -a & \text{if } a < 0 \end{cases}$$

and can be interpreted as the distance from a to the origin on the real number line.

Example 1

Express $|2x - 1|$ without using the absolute value symbol.

Solution

$$|2x - 1| = \begin{cases} 2x - 1 & \text{if } 2x - 1 \geq 0 \\ -(2x - 1) & \text{if } 2x - 1 < 0 \end{cases}$$

$$= \begin{cases} 2x - 1 & \text{if } x \geq \frac{1}{2} \\ 1 - 2x & \text{if } x < \frac{1}{2} \end{cases}$$

●

The following properties can be proved either from the definition of absolute value or by using the equation

$$|a| = \sqrt{a^2}$$

from Section 1.2.

Properties of Absolute Value

Suppose a and b are any real numbers and n is an integer. Then

1. $|ab| = |a||b|$

2. $\left|\dfrac{a}{b}\right| = \dfrac{|a|}{|b|}$ $(b \neq 0)$

3. $|a^n| = |a|^n$

Example 2

Simplify the expression $|2x - 6|$.

Solution

Using Property 1, we have

$$|2x - 6| = |2(x - 3)| = |2||x - 3| = 2|x - 3|$$

●

When solving equations or inequalities involving absolute values, it is often very helpful to use the following properties:

> Suppose $a > 0$. Then
>
> 4. $|x| = a$ if and only if $x = \pm a$
> 5. $|x| < a$ if and only if $-a < x < a$
> 6. $|x| > a$ if and only if $x > a$ or $x < -a$

For instance, the inequality $|x| < a$ says that the distance from x to the origin is less than a, and you can see from Figure 1 that this is true if and only if x lies between $-a$ and a.

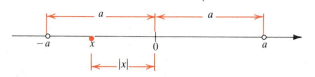

Figure 1

Example 3
Solve the equation $|2x - 5| = 3$.

Solution
By Property 4, $|2x - 5| = 3$ is equivalent to

$$2x - 5 = 3 \qquad \text{or} \qquad 2x - 5 = -3$$

So $2x = 8$ or $2x = 2$. Thus $x = 4$ or $x = 1$.   ●

Example 4
Solve the inequality $|x - 5| < 2$.

Solution 1
By Property 5, $|x - 5| < 2$ is equivalent to

$$-2 < x - 5 < 2$$

Therefore, adding 5 to each side, we have

$$3 < x < 7$$

and the solution set is the open interval $(3, 7)$.

Solution 2
Geometrically, the solution set consists of all numbers x whose distance from 5 is less than 2. From Figure 2 we see that this is the interval $(3, 7)$.   ●

Figure 2

Example 5
Solve the inequality $|3x + 2| \geq 4$.

Solution

By Properties 4 and 6, $|3x + 2| \geq 4$ is equivalent to

$$3x + 2 \geq 4 \qquad \text{or} \qquad 3x + 2 \leq -4$$

In the first case $3x \geq 2$, which gives $x \geq \frac{2}{3}$. In the second case $3x \leq -6$, which gives $x \leq -2$. So the solution set is

$$\{x \mid x \leq -2 \text{ or } x \geq \tfrac{2}{3}\} = (-\infty, -2] \cup [\tfrac{2}{3}, \infty) \qquad \bullet$$

Another important property of absolute value, called the Triangle In-equality, is used frequently not only in calculus but throughout mathematics in general.

The Triangle Inequality

If a and b are any real numbers, then

$$|a + b| \leq |a| + |b|$$

Observe that if the numbers a and b are both positive or both negative, then the two sides in the Triangle Inequality are actually equal. But if a and b have opposite signs, the left side involves a subtraction and the right side does not. This makes the Triangle Inequality seem reasonable, but we can prove it as follows.

Notice that

$$-|a| \leq a \leq |a|$$

is always true because a equals either $|a|$ or $-|a|$. If we write the corresponding statement for b, we have

$$-|b| \leq b \leq |b|$$

and adding these inequalities, we get

$$-(|a| + |b|) \leq a + b \leq |a| + |b|$$

If we now apply Properties 4 and 5 (with x replaced by $a + b$ and a by $|a| + |b|$), we obtain

$$|a + b| \leq |a| + |b|$$

which is what we wanted to show.

Example 6

If $|x - 4| < 0.1$ and $|y - 7| < 0.2$, use the Triangle Inequality to estimate $|(x + y) - 11|$.

Solution

In order to use the given information, we use the Triangle Inequality with

$$a = x - 4 \text{ and } b = y - 7:$$

$$|(x + y) - 11| = |(x - 4) + (y - 7)|$$
$$\leq |x - 4| + |y - 7|$$
$$< 0.1 + 0.2 = 0.3$$

Thus $\qquad |(x + y) - 11| < 0.3$ ●

Exercises 1.6

In Exercises 1–6 use the definition of absolute value to write the given expression without the absolute value symbol, as in Example 1.

1. $|x - 3|$ **2.** $|4x - 7|$ **3.** $|3x - 10|$

4. $|8 - 5x|$ **5.** $|x^2 + 1|$ **6.** $|x^2 - 1|$

In Exercises 7–12 simplify the expression, as in Example 2.

7. $|3x + 9|$ **8.** $|4x - 16|$ **9.** $|\tfrac{1}{2}x - \tfrac{5}{2}|$

10. $|-2x - 10|$ **11.** $|-x^2 - 9|$ **12.** $\left|\dfrac{x - 1}{1 - x}\right|$

In Exercises 13–18 solve the equation.

13. $|2x| = 3$ **14.** $|3x + 5| = 1$

15. $|x - 4| = 0.01$ **16.** $|x - 6| = -1$

17. $\left|\dfrac{2x - 1}{x + 1}\right| = 3$ **18.** $|x + 3| = |2x + 1|$

In Exercises 19–36 solve the given inequality.

19. $|x| < 3$ **20.** $|x| \geq 3$

21. $|x - 4| < 1$ **22.** $|x - 6| < 0.1$

23. $|x + 5| \geq 2$ **24.** $|x + 1| \geq 3$

25. $|2x - 3| \leq 0.4$ **26.** $|5x - 2| < 6$

27. $\left|\dfrac{x - 2}{3}\right| < 2$ **28.** $\left|\dfrac{x + 1}{2}\right| \geq 4$

29. $|x + 6| < 0.001$ **30.** $|x - a| < d$

31. $1 \leq |x| \leq 4$ **32.** $0 < |x - 5| < \tfrac{1}{2}$

33. $\left|\dfrac{x}{2 + x}\right| < 1$ **34.** $\left|\dfrac{2 - 3x}{1 + 2x}\right| \leq 4$

35. $|x| > |x - 1|$ **36.** $|2x - 5| \leq |x + 4|$

37. (a) Simplify the expression $|x| + |x - 2|$ in the following three cases:
     (i) $x < 0$ **(ii)** $0 < x < 2$ **(iii)** $x > 2$
   (b) Solve the inequality $|x| + |x - 2| < 3$.

38. (a) Simplify the expression $|x + 1| + |x - 2|$ in the following three cases:
     (i) $x < -1$ **(ii)** $-1 < x < 2$ **(iii)** $x > 2$
   (b) Solve the inequality $|x + 1| + |x - 2| < 7$.

39. Prove that $|ab| = |a||b|$.

40. Prove that $\left|\dfrac{a}{b}\right| = \dfrac{|a|}{|b|}$.

41. Suppose that $|x - 2| < 0.01$ and $|y - 3| < 0.04$. Show that $|(x + y) - 5| < 0.05$.

42. If $|a - 1| < 2$ and $|b - 1| < 3$, show that $|a + b - 2| < 5$.

43. Show that $|x - y| \leq |x| + |y|$ for all real numbers x and y.

44. Show that $|x - y| \geq |x| - |y|$ for all real numbers x and y. (*Hint:* Use the Triangle Inequality with $a = x - y$ and $b = y$.)

Section 1.7

Coordinate Geometry

Just as the points on a line can be identified with real numbers by assigning them coordinates, as described in Section 1.1, so the points in a plane can be identified with ordered pairs of real numbers. We start by drawing two perpendicular

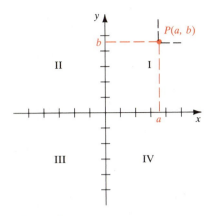

Figure 1

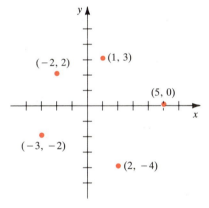

Figure 2

coordinate lines that intersect at the origin O on each line. Usually one line is horizontal with positive direction to the right and is called the **x-axis**; the other line is vertical with positive direction upward and is called the **y-axis**.

Any point P in the plane can be located by a unique ordered pair of numbers as follows. Draw lines through P perpendicular to the x- and y-axes. These lines will intersect the axes in points with coordinates a and b as shown in Figure 1. Then the point P is assigned the ordered pair (a, b). The first number a is called the **x-coordinate** (or **abscissa**) of P; the second number is called the **y-coordinate** (or **ordinate**) of P. We say that P is the point with coordinates (a, b) and we denote the point by the symbol $P(a, b)$. Several points are labeled with their coordinates in Figure 2.

By reversing the preceding process, we can start with an ordered pair (a, b) and arrive at the corresponding point P. Often we identify the point P with the ordered pair (a, b) and refer to "the point (a, b)." [Although the notation used for an open interval (a, b) is the same as the notation used for a point (a, b), you will be able to tell from the context which meaning is intended.]

This coordinate system is called the **rectangular coordinate system**, or the **Cartesian coordinate system** in honor of the French mathematician René Descartes (1596–1650), even though another Frenchman, Pierre Fermat (1601–1665), invented the principles of analytic geometry at about the same time as Descartes. The plane, supplied with this coordinate system, is called the **coordinate plane** or the **Cartesian plane** and is denoted by R^2.

The x- and y-axes are called the **coordinate axes** and divide the Cartesian plane into four quadrants, which are labeled I, II, III, and IV in Figure 1. Notice that the first quadrant consists of those points whose x- and y-coordinates are both positive.

Example 1

Describe and sketch the regions given by the following sets:

(a) $\{(x, y) \mid x \geq 0\}$ (b) $\{(x, y) \mid y = 1\}$ (c) $\{(x, y) \mid |y| < 1\}$

Solution

(a)   The points whose x-coordinates are 0 or positive lie on the x-axis or to the right of it [see Figure 3(a)].

Figure 3

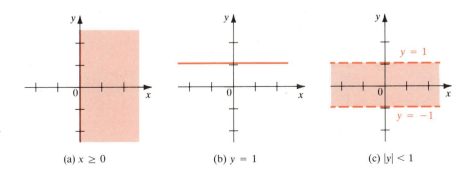

(a) $x \geq 0$ (b) $y = 1$ (c) $|y| < 1$

(b)   The set of all points with y-coordinate 1 is a horizontal line one unit above the x-axis [see Figure 3(b)].

(c)   Recall from Section 1.6 that

$$|y| < 1 \qquad \text{if and only if} \qquad -1 < y < 1$$

The given region consists of those points in the plane whose y-coordinates lie between -1 and 1. Thus the region consists of all points that lie between (but not on) the horizontal lines $y = 1$ and $y = -1$. [These lines are shown as broken lines in Figure 3(c) to indicate that the points on these lines do not lie in the set.]

Recall from Section 1.1 that the distance between points a and b on a number line is $|a - b| = |b - a|$. Thus the distance between points $P_1(x_1, y_1)$ and $P_3(x_2, y_1)$ on a horizontal line must be $|x_2 - x_1|$, and the distance between $P_2(x_2, y_2)$ and $P_3(x_2, y_1)$ on a vertical line must be $|y_2 - y_1|$ (see Figure 4).

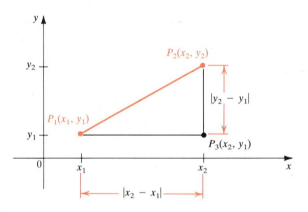

Figure 4

To find the distance $|P_1 P_2|$ between any two points $P_1(x_1, y_1)$ and $P_2(x_2, y_2)$, we note that triangle $P_1 P_2 P_3$ in Figure 4 is a right triangle, and so by the Pythagorean Theorem we have

$$|P_1 P_2| = \sqrt{|P_1 P_3|^2 + |P_2 P_3|^2} = \sqrt{|x_2 - x_1|^2 + |y_2 - y_1|^2}$$
$$= \sqrt{(x_2 - x_1)^2 + (y_2 - y_1)^2}$$

Distance Formula

The distance between the points $P_1(x_1, y_1)$ and $P_2(x_2, y_2)$ is

$$|P_1 P_2| = \sqrt{(x_2 - x_1)^2 + (y_2 - y_1)^2}$$

Example 2

Which of the points $P(1, -2)$ or $Q(8, 9)$ is closer to the point $A(5, 3)$?

Solution

The distance from P to A is

$$|PA| = \sqrt{(5-1)^2 + [3-(-2)]^2} = \sqrt{4^2 + 5^2} = \sqrt{41}$$

Similarly $|QA| = \sqrt{(5-8)^2 + (3-9)^2} = \sqrt{(-3)^2 + (-6)^2} = \sqrt{45}$

This shows that $|PA| < |QA|$, so P is closer to A.   ●

Let us now find the coordinates (x, y) of the midpoint M of the line segment that joins the point $P_1(x_1, y_1)$ to the point $P_2(x_2, y_2)$. Notice that triangles P_1AM and MBP_2 in Figure 5 are congruent because $|P_1M| = |MP_2|$ and corresponding angles are equal. It follows that $|P_1A| = |MB|$ and so

$$x - x_1 = x_2 - x$$

Solving this equation for x_1, we get

$$2x = x_1 + x_2 \qquad x = \frac{x_1 + x_2}{2}$$

Similarly, $y = (y_1 + y_2)/2$.

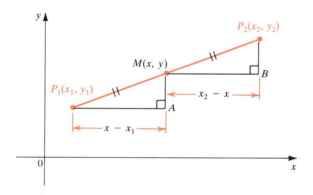

Figure 5

Midpoint Formula

The midpoint of the line segment from $P_1(x_1, y_1)$ to $P_2(x_2, y_2)$ is

$$\left(\frac{x_1 + x_2}{2}, \frac{y_1 + y_2}{2} \right)$$

Example 3

The midpoint of the line segment that joins $(-2, 5)$ and $(4, 9)$ is

$$\left(\frac{-2 + 4}{2}, \frac{5 + 9}{2} \right) = (1, 7)$$ ●

Equations and Graphs

Suppose we have an equation involving the variables x and y, such as

$$x^2 + y^2 = 1 \quad \text{or} \quad x = y^2 \quad \text{or} \quad y = \frac{2}{x}$$

The **graph** of such an equation in x and y is the set of all points (x, y) that satisfy the equation. The graph gives a visual representation of the equation. For example, we will see that the first equation represents a circle (later in this section), the second equation represents a parabola, and the third equation represents a hyperbola (see Chapter 8).

Example 4

Sketch the graph of the equation $x = y^2$.

Solution

There are infinitely many points on the graph and it is impossible to plot all of them. But we list some of these points in the following table and plot them in Figure 6. Then we connect the points by a smooth curve as shown.

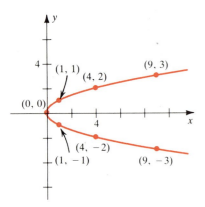

Figure 6

y	0	1	2	3	-1	-2	-3
$x = y^2$	0	1	4	9	1	4	9
(x, y)	$(0, 0)$	$(1, 1)$	$(4, 2)$	$(9, 3)$	$(1, -1)$	$(4, -2)$	$(9, -3)$

Example 5

Sketch the graph of the equation $2x - y = 3$.

Solution

If we solve the given equation for y, we get $y = 2x - 3$. This helps us calculate the y-coordinates in the following table:

x	-1	0	1	2	3	4
y	-5	-3	-1	1	3	5

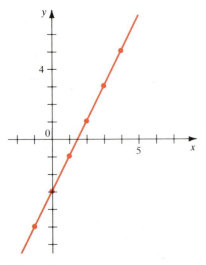

Figure 7

When we plot the points with these coordinates, it appears that they lie on a line, and we sketch the graph by joining the points in Figure 7. (In the next section we verify that the graph is indeed a line.)

The x-coordinates of the points where a graph intersects the x-axis are called the **x-intercepts** of the graph and are obtained by setting $y = 0$ in the equation of the graph. The y-coordinates of the points where a graph intersects the y-axis are called the **y-intercepts** of the graph and are obtained by setting $x = 0$ in the equation of the graph. In Example 5 the y-intercept is -3, and to find the x-intercept we set $y = 0$, which gives $2x = 3$ or $x = \frac{3}{2}$.

So far we have discussed how to find the graph of an equation in x and y. The converse problem is to find an *equation of a graph*—that is, an equation that represents a given curve in the xy-plane. Such an equation is satisfied by the coordinates of the points on the curve and by no other points. This is the other half of the basic principle of analytic geometry as formulated by Descartes and Fermat. The idea is that if a geometric curve can be represented by an algebraic equation, then the rules of algebra can be used to analyze the geometric problem.

As an example of this type of problem, let us find the equation of a circle with radius r and center (h, k). By definition, the circle is the set of all points $P(x, y)$ whose distance from the center $C(h, k)$ is r (see Figure 8). Thus P is on the circle if and only if $|PC| = r$. From the Distance Formula we have

$$\sqrt{(x - h)^2 + (y - k)^2} = r$$

or equivalently, squaring both sides,

$$(x - h)^2 + (y - k)^2 = r^2$$

This is the desired equation.

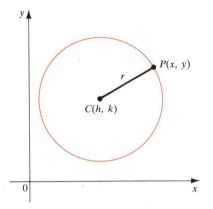

Figure 8

Equation of a Circle

The equation of a circle with center (h, k) and radius r is
$$(x - h)^2 + (y - k)^2 = r^2$$
In particular, if the center is the origin $(0, 0)$, the equation is
$$x^2 + y^2 = r^2$$

Example 6

Find an equation of the circle with radius 3 and center $(2, -5)$.

Solution

Using the equation of a circle with $r = 3$, $h = 2$, and $k = -5$, we obtain
$$(x - 2)^2 + (y + 5)^2 = 9$$ ●

Example 7

Find an equation of the circle that has the points $P(1, 8)$ and $Q(5, -6)$ as the endpoints of a diameter.

Solution

We first observe that the center is the midpoint of the diameter PQ, so by the Midpoint Formula the center is

$$\left(\frac{1+5}{2}, \frac{8-6}{2}\right) = (3, 1)$$

The radius r is the distance from P to the center, so

$$r^2 = (3-1)^2 + (1-8)^2 = 2^2 + (-7)^2 = 53$$

Therefore the equation of the circle is

$$(x-3)^2 + (y-1)^2 = 53 \qquad \bullet$$

Example 8

Sketch the graph of the equation $x^2 + y^2 + 2x - 6y + 7 = 0$ by first showing that it represents a circle and then finding its center and radius.

Solution

We first group the x-terms and y-terms as follows:

$$(x^2 + 2x) + (y^2 - 6y) = -7$$

Then we complete the square within each grouping, adding the square of half the coefficient of x and of half the coefficient of y to both sides of the equation:

$$(x^2 + 2x + 1) + (y^2 - 6y + 9) = -7 + 1 + 9$$

or

$$(x+1)^2 + (y-3)^2 = 3$$

Comparing this equation with the standard equation of a circle, we see that $h = -1, k = 3$, and $r = \sqrt{3}$, so the given equation represents a circle with center $(-1, 3)$ and radius $\sqrt{3}$. It is sketched in Figure 9. $\qquad \bullet$

$x^2 + y^2 + 2x - 6y + 7 = 0$

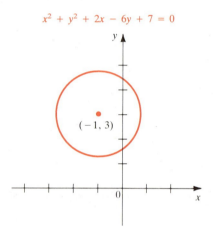

Figure 9

Symmetry

Notice in Figure 6 that the part of the curve $x = y^2$ in the fourth quadrant is a mirror image of the part of the curve in the first quadrant. This is because

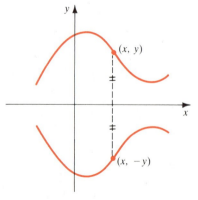

Symmetry about the x-axis

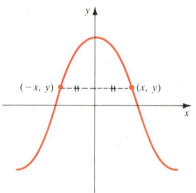

Symmetry about the y-axis

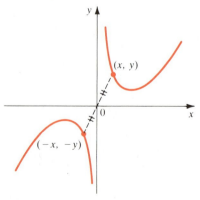

Symmetry about the origin

Figure 10

whenever (x, y) is on the curve, so is the point $(x, -y)$, and in this situation we say the curve is **symmetric with respect to the x-axis**. Similarly we say the curve is **symmetric with respect to the y-axis** if whenever the point (x, y) is on the curve, so is $(-x, y)$. A curve is called **symmetric with respect to the origin** if whenever (x, y) is on the curve, so is $(-x, -y)$. These three types of symmetry are illustrated in Figure 10.

Tests for Symmetry

1. A curve is symmetric with respect to the x-axis if its equation is unchanged when y is replaced by $-y$.

2. A curve is symmetric with respect to the y-axis if its equation is unchanged when x is replaced by $-x$.

3. A curve is symmetric with respect to the origin if its equation is unchanged when x is replaced by $-x$ and y is replaced by $-y$.

Example 9

Test the equation $y = x^2 - 4$ for symmetry and sketch the curve.

Solution

If x is replaced by $-x$ in the equation $y = x^2 - 4$, we get

$$y = (-x)^2 - 4 = x^2 - 4$$

and so the equation is unchanged. Therefore the curve is symmetric about the x-axis. We can sketch it by first plotting points just for $x > 0$ and then reflecting in the y-axis as in Figure 11.

x	y
0	-4
$\frac{1}{2}$	-3.75
1	-3
2	0
3	5

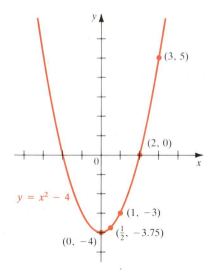

Figure 11

Example 10

The equation of the circle $x^2 + y^2 = 1$ remains unchanged when x is replaced by $-x$ and y is replaced by $-y$, so the circle exhibits all three types of symmetry. It is symmetric with respect to the x-axis, the y-axis, and the origin.   ●

Exercises 1.7

In Exercises 1–12 sketch the region given by the set.

1. $\{(x, y) \mid x < 0\}$ **2.** $\{(x, y) \mid y > 0\}$

3. $\{(x, y) \mid xy < 0\}$ **4.** $\{(x, y) \mid xy > 0\}$

5. $\{(x, y) \mid x = 3\}$ **6.** $\{(x, y) \mid y = -2\}$

7. $\{(x, y) \mid 1 < x < 2\}$ **8.** $\{(x, y) \mid 0 \le y \le 4\}$

9. $\{(x, y) \mid x \ge 1 \text{ and } y < 3\}$

10. $\{(x, y) \mid |y| > 1\}$ **11.** $\{(x, y) \mid |x| \le 2\}$

12. $\{(x, y) \mid |x| < 3 \text{ and } |y| > 2\}$

In Exercises 13–18, (a) find the distance between the given points and (b) find the midpoint of the line segment that joins the points.

13. $(1, 1), (4, 5)$ **14.** $(1, -3), (5, 7)$

15. $(6, -2), (-1, 3)$ **16.** $(1, -6), (-1, -3)$

17. $(2, 5), (4, -7)$ **18.** $(a, b), (b, a)$

19. Which of the points $A(6, 7)$ or $B(-5, 8)$ is closer to the origin?

20. Which of the points $C(-6, 3)$ or $D(3, 0)$ is closer to the point $E(-2, 1)$?

21. Show that the triangle with vertices $A(0, 2)$, $B(-3, -1)$, and $C(-4, 3)$ is isosceles.

22. Show that the triangle with vertices $A(6, -7)$, $B(11, -3)$, and $C(2, -2)$ is a right triangle using the converse of the Pythagorean Theorem. Find the area of the triangle.

23. Show that the points $A(-2, 9)$, $B(4, 6)$, $C(1, 0)$, and $D(-5, 3)$ are the vertices of a square.

24. Show that the points $A(-1, 3)$, $B(3, 11)$, and $C(5, 15)$ are collinear by showing that $|AB| + |BC| = |AC|$.

25. Find a point on the y-axis that is equidistant from the points $(5, -5)$ and $(1, 1)$.

26. Find the lengths of the medians of the triangle with vertices $A(1, 0)$, $B(3, 6)$, and $C(8, 2)$. (A median is a line segment from a vertex to the midpoint of the opposite side.)

In Exercises 27–46 make a table of values and sketch the graph of the given equation. Test each curve for symmetry. Find x- and y-intercepts for each curve.

27. $y = x - 1$ **28.** $y = 2x + 5$

29. $3x - y = 5$ **30.** $x + y = 3$

31. $y = 1 - x^2$ **32.** $y = x^2 + 2x$

33. $4y = x^2$ **34.** $8y = x^3$

35. $xy = 2$ **36.** $x + y^2 = 4$

37. $y = \sqrt{x}$ **38.** $x + \sqrt{y} = 4$

39. $x^2 + y^2 = 9$ **40.** $9x^2 + 9y^2 = 49$

41. $y = \sqrt{4 - x^2}$ **42.** $x = -\sqrt{25 - y^2}$

43. $y = |x|$ **44.** $x = |y|$

45. $y = 4 - |x|$ **46.** $y = |4 - x|$

In Exercises 47–52 find an equation of the circle that satisfies the given conditions.

47. Center $(3, -1)$; radius 5

48. Center $(-2, -8)$; radius 10

49. Center at the origin; passes through $(4, 7)$

50. Center $(-1, 5)$; passes through $(-4, -6)$

51. Endpoints of a diameter $P(-1, 3)$ and $Q(7, -5)$

52. Center $(7, -3)$; tangent to the x-axis

In Exercises 53–58 show that the given equation represents a circle and find the center and radius of the circle.

53. $x^2 + y^2 - 4x + 10y + 13 = 0$

54. $x^2 + y^2 + 6y + 2 = 0$

55. $x^2 + y^2 + x = 0$

56. $x^2 + y^2 + 2x + y + 1 = 0$

57. $2x^2 + 2y^2 - x + y = 1$

58. $16x^2 + 16y^2 + 8x + 32y + 1 = 0$

In Exercises 59 and 60 sketch the graph of the equation.

59. $x^2 + y^2 + 4x - 10y = 21$

60. $x^2 + y^2 - 16x + 12y + 200 = 0$

61. Under what conditions on the coefficients a, b, and c does the equation $x^2 + y^2 + ax + by + c = 0$ represent a circle? When that condition is satisfied, find the center and radius of the circle.

In Exercises 62 and 63 sketch the region given by the set.

62. $\{(x, y) \mid x^2 + y^2 \leq 1\}$ **63.** $\{(x, y) \mid x^2 + y^2 > 4\}$

64. Find the area of the region that lies outside the circle $x^2 + y^2 = 4$ but inside the circle $x^2 + y^2 - 4y - 12 = 0$.

Section 1.8

Lines

In this section we find equations of straight lines. We first need to measure the steepness of a line and so we define its slope.

> The **slope** of a nonvertical line that passes through the points $P_1(x_1, y_1)$ and $P_2(x_2, y_2)$ is
>
> $$m = \frac{y_2 - y_1}{x_2 - x_1}$$
>
> The slope of a vertical line is not defined.

Thus the slope of a line is the ratio of the change in y (the rise) to the change in x (the run) between two points on the line. From the similar triangles in Figure 1 we see that the slope is independent of which two points are chosen on the line:

$$\frac{y_2 - y_1}{x_2 - x_1} = \frac{y_2' - y_1'}{x_2' - x_1'}$$

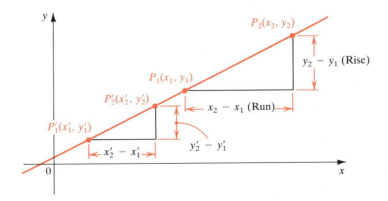

Figure 1

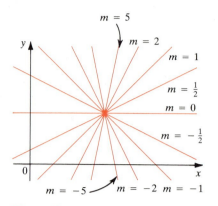

Figure 2

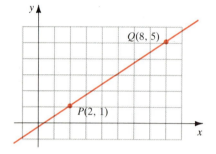

Figure 3

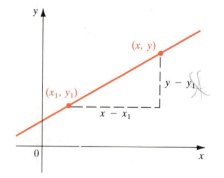

Figure 4

Figure 2 shows several lines labeled with their slopes. Notice that lines with positive slope slant upward to the right, whereas lines with negative slope slant downward to the right. Notice also that the steepest lines are the ones for which the absolute value of the slope is the largest and that a horizontal line has slope zero.

Example 1

Find the slope of the line that passes through the points $P(2, 1)$ and $Q(8, 5)$.

Solution

From the definition, the slope is

$$m = \frac{y_2 - y_1}{x_2 - x_1} = \frac{5 - 1}{8 - 2} = \frac{4}{6} = \frac{2}{3}$$

This says that for every three units we move to the right, the line rises two units. The line is drawn in Figure 3. ●

Now let us find the equation of the line that passes through a given point $P(x_1, y_1)$ and has slope m. A point $P(x, y)$ with $x \neq x_1$ lies on this line if and only if the slope of the line through P_1 and P is equal to m (see Figure 4); that is,

$$\frac{y - y_1}{x - x_1} = m$$

This equation can be rewritten in the form

$$y - y_1 = m(x - x_1)$$

and we observe that this equation is also satisfied when $x = x_1$ and $y = y_1$. Therefore it is an equation of the given line.

Point-Slope Form of the Equation of a Line

An equation of the line that passes through the point $P(x_1, y_1)$ and has slope m is

$$y - y_1 = m(x - x_1)$$

Example 2

(a)   Find an equation of the line through $(1, -3)$ with slope $-\frac{1}{2}$.

(b)   Sketch the line.

Solution

(a)   Using the point-slope form with $m = -\frac{1}{2}$, $x_1 = 1$, and $y_1 = -3$, we obtain an equation of the line as

$$y + 3 = -\tfrac{1}{2}(x - 1)$$

which we can rewrite as

$$2y + 6 = -x + 1 \quad \text{or} \quad x + 2y + 5 = 0$$

(b)   The fact that the slope is $-\frac{1}{2}$ tells us that when we move two units to the right, the line drops one unit. This enables us to sketch the line in Figure 5. ●

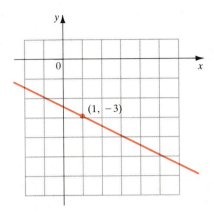

Figure 5

Example 3

Find an equation of the line through the points $(-1, 2)$ and $(3, -4)$.

Solution

The slope of the line is

$$m = \frac{-4 - 2}{3 - (-1)} = -\frac{3}{2}$$

Using the point-slope form with $x_1 = -1$ and $y_1 = 2$, we obtain

$$y - 2 = -\tfrac{3}{2}(x + 1)$$

which simplifies to

$$2y - 4 = -3x - 3 \quad \text{or} \quad 3x + 2y = 1 \qquad ●$$

Suppose a nonvertical line has slope m and y-intercept b (see Figure 6). This means it intersects the y-axis at the point $(0, b)$, so the point-slope form of the equation of the line, with $x = 0$ and $y = b$, becomes

$$y - b = m(x - 0)$$

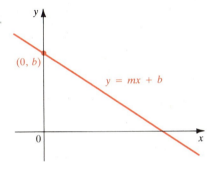

Figure 6

This simplifies to

$$\boxed{y = mx + b}$$

which is called the **slope-intercept form** of the equation of a line.

In particular, if a line is horizontal, its slope is $m = 0$, so its equation is $y = b$, where b is the y-intercept (see Figure 7). A vertical line does not have a slope, but we can write its equation as $x = a$, where a is the x-intercept, because the x-coordinate of every point on the line is a.

Observe that the equation of every line can be written in the form

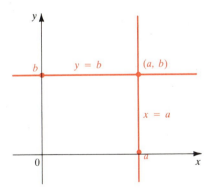

Figure 7
The equation of a vertical line through (a, b) is $x = a$. The equation of a horizontal line through (a, b) is $y = b$.

(1)
$$\boxed{Ax + By + C = 0}$$

because a vertical line has the equation $x = a$ or $x - a = 0$ ($A = 1$, $B = 0$, $C = -a$) and a nonvertical line has the equation $y = mx + b$ or $-mx + y - b = 0$ ($A = -m$, $B = 1$, $C = -b$). Conversely, if we start with a general first-degree equation, that is, an equation of the form (1) where A, B, and C are constants and A and B are not both 0, then we can show that it is the equation of a line. If $B = 0$, it becomes $Ax + C = 0$ or $x = -C/A$, which represents a vertical line with x-intercept $-C/A$. If $B \neq 0$, it can be rewritten by solving for y:

$$y = -\frac{A}{B}x - \frac{C}{B}$$

and we recognize this as being the slope-intercept form of the equation of a line ($m = -A/B$, $b = -C/B$). Therefore an equation of the form (1) is called a **linear equation** or the **general equation of a line**. For brevity, we often refer to "the line $Ax + By + C = 0$" instead of saying "the line whose equation is $Ax + By + C = 0$."

Example 4

Sketch the graph of the equation $3x - 5y = 15$.

Solution

Since the equation is linear, its equation is a line. To draw the graph it suffices to find two points on the line. It is easiest to find the intercepts. Substituting $y = 0$ (the equation of the x-axis) in the given equation, we get $3x = 15$, so $x = 5$ is the x-intercept. Substituting $x = 0$ in the equation, we find that the y-intercept is -3. This allows us to sketch the graph as in Figure 8.

Another method is to write the equation in slope-intercept form:

$$3x - 5y = 15$$
$$5y = 3x - 15$$
$$y = \tfrac{3}{5}x - 3$$

This is in the form $y = mx + b$, so the slope is $m = \tfrac{3}{5}$ and the y-intercept is $b = -3$. This information can be used to sketch the line.   ●

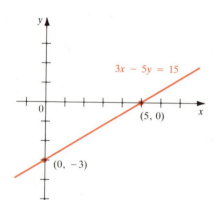

Figure 8

Parallel and Perpendicular Lines

Since slope measures the steepness of a line, it seems reasonable that parallel lines should have the same slope. In fact, we can prove this.

> Two nonvertical lines are parallel if and only if they have the same slope.

Proof

Let the lines l_1 and l_2 in Figure 9 have slopes m_1 and m_2. If the lines are parallel, then the right triangles ABC and DEF are similar, so

$$m_1 = \frac{|BC|}{|AC|} = \frac{|EF|}{|DF|} = m_2$$

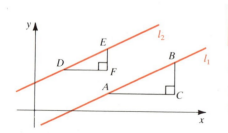

Figure 9

Conversely, if the slopes are equal, then the triangles will be similar, so $\angle BAC = \angle EDF$ and the lines are parallel. ●

Example 5

Find an equation of the line through the point $(5, 2)$ that is parallel to the line $4x + 6y + 5 = 0$.

Solution

The given line can be written in the form

$$y = -\tfrac{2}{3}x - \tfrac{5}{6}$$

which is in slope-intercept form with $m = -\tfrac{2}{3}$. Parallel lines have the same slope, so the required line has slope $-\tfrac{2}{3}$ and its equation in point-slope form is

$$y - 2 = -\tfrac{2}{3}(x - 5)$$

This simplifies to

$$3y - 6 = -2x + 10 \qquad \text{or} \qquad 2x + 3y = 16 \qquad ●$$

The condition for perpendicular lines is not so obvious.

> Two lines with slopes m_1 and m_2 are perpendicular if and only if $m_1 m_2 = -1$; that is, their slopes are negative reciprocals:
>
> $$m_2 = -\frac{1}{m_1}$$

Proof

In Figure 10 we show two lines intersecting at the origin. (If the lines intersect at some other point, we consider lines parallel to these that intersect at the origin. These lines have the same slopes as the original lines.)

If the lines l_1 and l_2 have slopes m_1 and m_2, then their equations are $y = m_1 x$ and $y = m_2 x$. Notice that $A(1, m_1)$ lies on l_1 and $B(1, m_2)$ lies on l_2. By the Pythagorean Theorem and its converse, $OA \perp OB$ if and only if

$$|OA|^2 + |OB|^2 = |AB|^2$$

By the Distance Formula, this becomes

$$(1^2 + m_1^2) + (1^2 + m_2^2) = (1 - 1)^2 + (m_2 - m_1)^2$$
$$2 + m_1^2 + m_2^2 = m_2^2 - 2m_1 m_2 + m_1^2$$
$$2 = -2m_1 m_2$$
$$m_1 m_2 = -1 \qquad ●$$

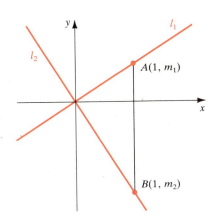

Figure 10

Example 6

Show that the points $P(3, 3)$, $Q(8, 17)$, and $R(11, 5)$ are the vertices of a right triangle.

Solution

The slopes of the lines PR and QR are

$$m_1 = \frac{5 - 3}{11 - 3} = \frac{1}{4} \quad \text{and} \quad m_2 = \frac{5 - 17}{11 - 8} = -4$$

Since $m_1 m_2 = -1$, these lines are perpendicular and so PQR is a right triangle.

●

Example 7

Find the equation of a line that is perpendicular to the line $4x + 6y + 5 = 0$ and passes through the origin.

Solution

In Example 5 we found that the slope of the line $4x + 6y + 5 = 0$ is $-\frac{2}{3}$. Thus the slope of a perpendicular line is the negative reciprocal—that is, $\frac{3}{2}$. Since the required line passes through $(0, 0)$, the point-slope form gives

$$y - 0 = \tfrac{3}{2}(x - 0) \quad \text{or} \quad y = \tfrac{3}{2}x$$

●

Example 8

(a) As dry air moves upward, it expands and cools. If the ground temperature is $20°C$ and the temperature at a height of 1 km is $10°C$, express the temperature T (in $°C$) in terms of the height h (in kilometers), assuming the expression is linear.

(b) Draw the graph of the linear equation. What does the slope represent?

(c) What is the temperature at a height of 2.5 km?

Solution

(a) Because we are assuming a linear relationship between h and T, the equation must be of the form

$$T = mh + b$$

where m and b are constants. When $h = 0$, we are given that $T = 20$, so

$$20 = m(0) + b \quad \text{or} \quad b = 20$$

Thus, we have

$$T = mh + 20$$

When $h = 1$, we have $T = 10$ and so

$$10 = m(1) + 20$$
$$m = 10 - 20 = -10$$

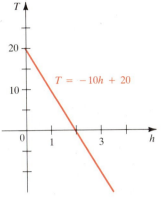

The required expression is

$$T = -10h + 20$$

(b) The graph is sketched in Figure 11.
The slope is $m = -10°/\text{km}$ and represents the rate of change of temperature with respect to distance.

(c) At a height of $h = 2.5$ km, the temperature is

$$T = -10(2.5) + 20 = -25 + 20 = -5°\text{C}$$

Figure 11

Exercises 1.8

In Exercises 1–4 find the slope of the line through P and Q.

1. $P(1, 5), Q(4, 11)$

2. $P(-1, 6), Q(4, -3)$

3. $P(-3, 3), Q(-1, -6)$

4. $P(-1, -4), Q(6, 0)$

In Exercises 5–20 find an equation of the line that satisfies the given conditions.

5. Through $(2, -3)$; slope 6

6. Through $(-1, 4)$; slope -3

7. Through $(1, 7)$; slope $\frac{2}{3}$

8. Through $(-3, -5)$; slope $-\frac{7}{2}$

9. Through $(2, 1)$ and $(1, 6)$

10. Through $(-1, -2)$ and $(4, 3)$

11. Slope 3; y-intercept -2

12. Slope $\frac{2}{5}$; y-intercept 4

13. x-intercept 1; y-intercept -3

14. x-intercept -8; y-intercept 6

15. Through $(4, 5)$; parallel to the x-axis

16. Through $(4, 5)$; parallel to the y-axis

17. Through $(1, -6)$; parallel to the line $x + 2y = 6$

18. y-intercept 6; parallel to the line $2x + 3y + 4 = 0$

19. Through $(-1, -2)$; perpendicular to the line $2x + 5y + 8 = 0$

20. Through $(\frac{1}{2}, -\frac{2}{3})$; perpendicular to the line $4x - 8y = 1$

21. (a) Sketch the line with slope $\frac{3}{2}$ that passes through the point $(-2, 1)$.
(b) Find the equation of this line.

22. (a) Sketch the line with slope -2 that passes through the point $(4, -1)$.
(b) Find the equation of this line.

In Exercises 23–30 find the slope and y-intercept of the line and draw its graph.

23. $x + 3y = 0$

24. $2x - 5y = 0$

25. $y = -2$

26. $2x - 3y + 6 = 0$

27. $3x - 4y = 12$

28. $y = 1.5$

29. $3x + 4y - 1 = 0$

30. $4x + 5y = 10$

31. Show that $A(1, 1), B(7, 4), C(5, 10)$, and $D(-1, 7)$ are vertices of a parallelogram.

32. Show that $A(-3, -1), B(3, 3)$, and $C(-9, 8)$ are vertices of a right triangle.

33. Show that $A(1, 1), B(11, 3), C(10, 8)$, and $D(0, 6)$ are vertices of a rectangle.

34. Use slopes to determine whether the given points are collinear (lie on a line).
(a) $(1, 1), (3, 9)$, and $(6, 21)$
(b) $(-1, 3), (1, 7)$, and $(4, 15)$

35. Find the equation of the perpendicular bisector of the line segment joining the points $A(1, 4)$ and $B(7, -2)$.

36. (a) Find equations for the sides of the triangle with vertices $P(1, 0)$, $Q(3, 4)$, and $R(-1, 6)$.
(b) Find equations for the medians of this triangle. Where do they intersect?

37. (a) Show that if the x- and y-intercepts of a line are nonzero numbers a and b, then the equation of the line can be put in the form

$$\frac{x}{a} + \frac{y}{b} = 1$$

This equation is called the **two-intercept form** of an equation of a line.
(b) Use part (a) to find an equation of the line whose x-intercept is 6 and whose y-intercept is -8.

38. The monthly cost of driving a car depends on the number of miles driven. Lynn found that in May it cost her \$380 to drive 480 mi and in June it cost her \$460 to drive 800 mi.

(a) Express the monthly cost C in terms of the distance driven d, assuming that a linear relationship gives a suitable model.
(b) Use part (a) to predict the cost of driving 1500 mi per month.
(c) Draw the graph of the linear equation. What does the slope of the line represent?
(d) What does the y-intercept of the graph represent?
(e) Why is a linear relationship a suitable model in this situation?

39. Jason and Debbie leave Detroit at 2:00 P.M. and drive at a constant speed west along I-90. They pass Ann Arbor, 40 mi from Detroit, at 2:50 P.M.
(a) Express the distance traveled in terms of the time elapsed.
(b) Draw the graph of the equation in part (a).
(c) What is the slope of this line? What does it represent?

Section 1.9

Problem Solving

There are no hard and fast rules that will ensure success in solving problems. However, it is possible to outline some general steps in the problem-solving process and to give some principles that may be useful in the solution of certain problems. These steps and principles are just common sense made explicit. They have been adapted from George Polya's book *How to Solve It*.

1. Understand the Problem

The first step is to read the problem and make sure that you understand it clearly. Ask yourself the following questions:

What is the unknown?

What are the given quantities?

What are the given conditions?

For many problems it is useful to

draw a diagram

and identify the given and required quantities on the diagram. Usually it is necessary to

introduce suitable notation.

In choosing symbols for the unknown quantities we often use letters such as a, b, c, m, n, x, and y, but in some cases it helps to use initials as suggestive symbols, for instance, V for volume or t for time.

2. Think of a Plan

Find a connection between the given information and the unknown that will enable you to calculate the unknown. If you do not see a connection immediately, the following ideas may be helpful in devising a plan.

a. *Try to recognize something familiar.* Relate the given situation to previous knowledge. Look at the unknown and try to recall a more familiar problem that has a similar unknown.

b. *Try to recognize patterns.* Some problems are solved by recognizing that some kind of pattern is occurring. The pattern could be geometric, or numerical, or algebraic. If you can see regularity or repetition in a problem, you might be able to guess what the continuing pattern is and then prove it.

c. *Use analogy.* Try to think of an analogous problem—that is, a similar problem, a related problem—but one that is easier than the original problem. If you can solve the similar, simpler problem, then it might give you the clues you need to solve the original, more difficult problem. For instance, if a problem involves very large numbers, you could first try a similar problem with smaller numbers. Or if the problem is in three-dimensional geometry, you could look for a similar problem in two-dimensional geometry. Or if the problem you start with is a general one, you could first try a special case.

d. *Introduce something extra.* It may sometimes be necessary to introduce something new, an auxiliary aid, to help make the connection between the given and the unknown. For instance, in geometry the auxiliary aid could be a new line drawn in a diagram. In algebra it could be a new unknown that is related to the original unknown.

e. *Take cases.* We may sometimes have to split a problem into several cases and give a different argument for each of the cases. We used this strategy in dealing with absolute value in Section 1.6 (see Examples 1, 3, and 5 and Exercises 37 and 38). In particular, in Exercise 37 we had to consider the three cases $x > 0, 0 < x < 2$, and $x > 2$ separately. We also considered different cases in solving certain inequalities (see Examples 5 and 6 in Section 1.5).

f. *Work backward.* Sometimes it is useful to imagine that your problem is solved and work backward, step by step, until you arrive at the given data. Then you may be able to reverse your steps and thereby construct a solution to the original problem. This procedure is commonly used in solving equations. For instance, in solving the equation $3x - 5 = 7$, we suppose that x is a number that satisfies $3x - 5 = 7$ and work backward. We add 5 to each side of the equation and then divide each side by 3 to get $x = 4$. Since each of these steps can be reversed, we have solved the problem.

g. *Indirect reasoning.* Sometimes it is appropriate to attack a problem indirectly. For instance, in a counting argument it might be best to count the total number of objects and subtract the number of objects that do *not* have the required property. Another example of indirect reasoning is **proof by contradiction**, in which we assume that the desired conclusion is false and eventually arrive at a contradiction.

h. *Mathematical induction.* In proving statements that involve a positive integer n, it is frequently helpful to use the Principle of Mathematical Induction, which is discussed in Section 9.6.

3. Carry Out the Plan   In Step 2 a plan was devised. In carrying out that plan, we have to check each stage of the plan and write the details that prove that each stage is correct.

4. Look Back   Having completed our solution, it is wise to look back over it, partly to see if there are errors in the solution and partly to see if there is an easier way to solve the problem. Another reason for looking back is that it will familiarize us with the method of solution and this may be useful for solving a future problem. Descartes said, "Every problem that I solved became a rule which served afterwards to solve other problems."

These principles of problem solving are illustrated in the following examples. To become a proficient problem solver you have to gain experience by solving a large number of problems. In addition to the nonstandard exercises in this section, you will see a collection of challenging problems, called Problems Plus, at the end of the remaining chapters. These problems relate to the material of the chapter but require a higher level of problem-solving skill.

Example 1

Find the final digit in the number 3^{459}.

Solution

(Analogy)   First notice that 3^{459} is a very large number—far too large for a calculator. Therefore we attack this problem by first looking at analogous problems. A similar, but simpler, problem would be to find the final digit in 3^9 or 3^{59}. In fact, let us start with the exponents $1, 2, 3, \ldots$ and see what happens.

Number	Final Digit
3^1	3
3^2	9
3^3	7
3^4	1
3^5	3
3^6	9
3^7	7
3^8	1

(Pattern)   By now you can see a pattern. The final digits occur in a cycle with length 4: $3, 9, 7, 1, 3, 9, 7, 1, 3, 9, 7, 1, \ldots$. Which number occurs in the 459*th* position? If we divide 459 by 4, the remainder is 3. So the final digit is the third number in the cycle—namely, 7.

The final digit in the number 3^{459} is 7.

Example 2

It takes John three hours to mow the lawn, but Mary can mow it in two hours. How long does it take them to mow the lawn together?

Solution

Although this problem resembles the word problems in Section 1.4, it requires more thought. We begin by letting t be the unknown quantity, the amount of time, in hours, that it would take John and Mary to mow the lawn together.

   To use the given information, the key idea is to ask what fraction of the job can be done in one hour. If together it takes t hours, then in one hour John and Mary could accomplish $1/t$ of the task. Likewise, in one hour John could mow $\frac{1}{3}$ of the lawn and Mary could mow $\frac{1}{2}$ of the lawn. So together they could mow

$$\frac{1}{3} + \frac{1}{2}$$

of the lawn in one hour. Thus we have

$$\frac{1}{3} + \frac{1}{2} = \frac{1}{t}$$
$$2t + 3t = 6$$
$$5t = 6$$
$$t = \frac{6}{5}$$

The time required is $\frac{6}{5}$ h, that is, 1 h and 12 min.   ●

Example 3

Solve the equation $|2x - |3x + 1|| = 1$.

Solution

(Work backward)   We assume that x is a number such that $|2x - |3x + 1|| = 1$. It follows from Property 4 of absolute values that there are two cases:

(Take cases)

$$2x - |3x + 1| = 1 \qquad \text{or} \qquad 2x - |3x + 1| = -1$$

$$|3x + 1| = 2x - 1 \qquad\qquad\qquad |3x + 1| = 2x + 1$$

(Take cases again)

$3x + 1 = 2x - 1$	or	$3x + 1 = 1 - 2x$		$3x + 1 = 2x + 1$	or	$3x + 1 = -2x - 1$
$x = -2$		$5x = 0$		$x = 0$		$5x = -2$
		$x = 0$				$x = -\frac{2}{5}$

Thus working backward gives three potential solutions: $x = 0$, $x = -2$, and $x = -\frac{2}{5}$. But is it possible to reverse the steps in each case? Let us check by trying to verify that these are indeed solutions:

Try $x = 0$: $|2(0) - |3(0) + 1|| = |0 - 1| = 1$ 0 is a solution

Try $x = -2$: $|2(-2) - |3(-2) + 1|| = |-4 - 5| = 9$ -2 is not a solution

Try $x = -\frac{2}{5}$: $|2(-\frac{2}{5}) - |3(-\frac{2}{5}) + 1|| = |-\frac{4}{5} - \frac{1}{5}| = 1$ $-\frac{2}{5}$ is a solution

Therefore the only solutions are 0 and $-\frac{2}{5}$.   ●

Example 4

Express the hypotenuse h of a right triangle in terms of its area A and its perimeter P.

Solution

(Understand the problem)   Let us first sort out the information by identifying the unknown quantity and the data:

$$\text{Unknown:} \quad h$$

$$\text{Given quantities:} \quad A, P$$

(Draw a diagram)   It helps to draw a diagram and we do so in Figure 1.

(Connect the given with the unknown)   In order to connect the given quantities to the unknown, we introduce two extra variables a and b, which are the lengths of the other two sides of the triangle. This enables us to express the given condition, which is that the triangle is right-angled, by the Pythagorean Theorem:

$$h^2 = a^2 + b^2$$

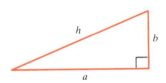

Figure 1

The other connections among the variables come by writing expressions for the area and perimeter:

$$A = \tfrac{1}{2}ab \qquad P = a + b + h$$

Since A and P are given, notice that we now have three equations in the three unknowns a, b, and h:

(1) $$h^2 = a^2 + b^2$$

(2) $$A = \tfrac{1}{2}ab$$

(3) $$P = a + b + h$$

Although we have the correct number of equations, they are not easy to solve in a straightforward fashion. But if we use the problem-solving strategy of trying *(Relate to the familiar)* to recognize something familiar, then we can solve these equations by an easier method. Look at the right sides of Equations 1, 2, and 3. Do these expressions remind you of anything familiar? Notice that they contain the ingredients of one of the special product formulas from Section 1.3:

$$(a + b)^2 = a^2 + 2ab + b^2$$

Using this idea, we express $(a + b)^2$ in two ways. From Equations 1 and 2 we have

$$(a + b)^2 = (a^2 + b^2) + 2ab = h^2 + 4A$$

From Equation 3 we have

$$(a + b)^2 = (P - h)^2 = P^2 - 2Ph + h^2$$

Thus
$$h^2 + 4A = P^2 - 2Ph + h^2$$

$$2Ph = P^2 - 4A$$

$$h = \frac{P^2 - 4A}{2P}$$

This is the required expression.

Example 5

Show that $\sqrt{2}$ is an irrational number.

Solution

(Indirect reasoning)   Here is a situation where it is necessary to use proof by contradiction. Suppose that $\sqrt{2}$ is a rational number. Recall from Section 1.1 that this means that

$$(4) \qquad\qquad \sqrt{2} = \frac{m}{n}$$

where m and n are integers and $n \neq 0$. In fact we can assume that m and n have no common factors; that is, any common factors have already been divided out.

Squaring both sides of Equation 4, we have

$$(5) \qquad\qquad 2 = \frac{m^2}{n^2} \qquad \text{or} \qquad m^2 = 2n^2$$

This equation shows that m^2 is an even number. It follows that m is an even number. [If m were odd, then $m = 2k + 1$ for some integer k, so $m^2 = (2k + 1)^2 = 4k^2 + 4k + 1$, which is 1 more than an even number and therefore odd.] But, since m is even, it must be of the form $m = 2k$, where k is an integer. Putting this expression into Equation 5, we have

$$(2k)^2 = 2n^2 \qquad 4k^2 = 2n^2 \qquad n^2 = 2k^2$$

This shows that n^2 is even and so n is even.

We have shown that both m and n are even. But this contradicts our assumption that m and n have no common factor. We have arrived at a contradiction, so we conclude that our hypothesis that $\sqrt{2}$ is rational is false. Thus $\sqrt{2}$ is irrational.   ●

You may find it useful to refer to this section from time to time as you solve the exercises in the remaining chapters of this book.

Exercises 1.9

1. Find the final digit in the number 947^{362}.

2. How many digits does the number $8^{15} \cdot 5^{37}$ have?

In Exercises 3–6 solve the equation.

3. $|5 - |x - 1|| = 3$

4. $||2x + 1| + 5| = 10$

5. $|4x - |x + 1|| = 3$

6. $||3x + 1| - x| = 2$

7. Use your calculator to find the value of the expression
$$\sqrt{3 + 2\sqrt{2}} - \sqrt{3 - 2\sqrt{2}}$$
The answer looks very simple. Show that the calculated value is correct.

8. Use your calculator to evaluate
$$\frac{\sqrt{2} + \sqrt{6}}{\sqrt{2} + \sqrt{3}}$$
Show that the calculated value is correct.

9. The sum of two numbers is 4 and their product is 1. Find the sum of their cubes.

10. Draw the graph of the equation
$$x^2 y - y^3 - 5x^2 + 5y^2 = 0$$
without making a table of values.

11. Bob and Jim, next-door neighbors, use hoses from both houses to fill Bob's swimming pool. They know it takes eighteen hours using both hoses. They also know that Bob's hose, used alone, can fill the pool in six hours less than Jim's hose alone. How much time is required by each hose alone?

12. Alice, Barb, and Cheryl, when working together, do a job in six hours less than Alice alone, in one hour less than Barb alone, and in half the time needed by Cheryl when working alone. How long would it take Alice and Barb to do the job together?

13. A man drives from home to work at a speed of 50 mi/h. The return trip from work to home is covered at the more leisurely pace of 30 mi/h. What is the average speed for the round trip?

14. A spoonful of cream is taken from a cup of cream and put into a cup of coffee and stirred. Then a spoonful of this mixture is taken and put into the cup of cream. Is there now more cream in the coffee cup or more coffee in the cup of cream?

15. In a right triangle, the hypotenuse has length 5 cm and another side has length 3 cm. What is the length of the altitude that is perpendicular to the base?

16. The perimeter of a right triangle is 60 cm and the altitude perpendicular to the hypotenuse is 12 cm. Find the lengths of the three sides.

17. If the lengths of the sides of a right triangle, in increasing order, are a, b, and c, show that $c^3 > a^3 + b^3$.

18. A point P is located in the interior of a rectangle so that the distance from P to one corner is 5 cm, from P to the opposite corner is 14 cm, and from P to a third corner is 10 cm. What is the distance from P to the fourth corner?

19. Draw the graph of the equation $|x| + |y| = 1 + |xy|$.

20. Sketch the region in the plane consisting of all points (x, y) such that

$$|x| + |y| \leq 1$$

21. Sketch the region in the plane consisting of all points (x, y) such that

$$|x - y| + |x| - |y| \leq 2$$

22. Find every positive integer that gives a perfect square if 132 is added to it and another perfect square if 200 is added to it.

23. How many integers are there from one to 1 million (inclusive) that are either perfect squares or perfect cubes (or both)?

24. How many integers less than 1000 are divisible neither by 5 nor by 7?

25. Player A has a higher batting average than Player B for the first half of the baseball season. Player A also has a higher batting average than Player B for the second half of the season. Prove, or disprove, that Player A has a higher batting average than Player B for the entire season.

26. Prove that at any party there are two people who know the same number of people. Assume that if A knows B, then B knows A. Assume also that everyone knows himself or herself. (*Hint:* Use indirect reasoning.)

27. **(a)** Prove that $\sqrt{6}$ is an irrational number.
    (b) Prove that $\sqrt{2} + \sqrt{3}$ is an irrational number.

28. Three tangent circles of radius 10 cm are drawn. All centers lie on the line AB. The tangent AC to the right-hand circle is drawn, intersecting the middle circle at D and E. Find the length $|DE|$.

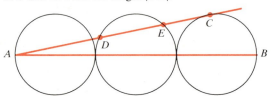

29. A car with tires that have a radius of 15 in. was driven on a trip and the odometer indicated that the distance traveled was 400 mi. Two weeks later, with snow tires installed, the odometer indicated that the distance for the return trip, over the same route, was 390 mi. Find the radius of the snow tires.

30. The positive integers are written in order starting with 1:

$$1234567891011121314151617181920 21 \ldots$$

What digit is in the 300,000*th* position?

CHAPTER 1 REVIEW

Define, state, or discuss each of the following.

1. Integers

2. Rational and irrational numbers

3. Real number line

4. Open interval; closed interval

5. Absolute value of a number

6. Base and exponent

7. Laws of exponents

8. Principal nth root

9. Rationalizing the denominator

10. Rational exponents

11. Variable and constant

12. Special product formulas for $(a + b)^2, (a - b)^2,$ $(a + b)^3, (a - b)^3$

13. Difference of squares formula

14. Difference of cubes formula

15. Sum of cubes formula

16. Rules for inequalities

17. Properties of absolute value

18. Triangle Inequality

19. Rectangular coordinate system

20. Distance Formula

21. Midpoint Formula

22. Graph of an equation

23. Equation of a circle

24. Tests for symmetry

Review Exercises

In Exercises 1 and 2 express the interval in terms of inequalities and graph the interval.

1. $(-1, 3]$

2. $(-\infty, 4]$

In Exercises 3 and 4 express the inequality in interval notation and graph the corresponding interval.

3. $x > 2$

4. $1 \le x \le 6$

Evaluate the numbers in Exercises 5–12.

5. $|3 - |-9||$

6. $1 - |1 - |-1||$

7. $2^{-3} - 3^{-2}$

8. $\sqrt[3]{-125}$

9. $216^{-1/3}$

10. $64^{2/3}$

11. $\dfrac{\sqrt{242}}{\sqrt{2}}$

12. $\sqrt[4]{4} \cdot \sqrt[4]{324}$

Simplify the expressions in Exercises 13–20.

13. $(2x^3 y)^2 (3x^{-1} y^2)$

14. $(a^2)^{-3} (a^3 b)^2 (b^3)^4$

15. $\dfrac{x^4 (3x)^2}{x^3}$

16. $\left(\dfrac{r^2 s^{4/3}}{r^{1/3} s}\right)^6$

17. $\sqrt[3]{(x^3 y)^2 y^4}$

18. $\sqrt{x^2 y^4}$

19. $\sqrt{\dfrac{x}{3}}$

20. $\dfrac{\sqrt{x} + 1}{\sqrt{x} - 1}$

Factor the expressions in Exercises 21–28.

21. $x^2 + 3x - 10$

22. $6x^2 + x - 12$

23. $25 - 16t^2$

24. $2y^6 - 32y^2$

25. $x^6 - 1$

26. $y^3 - 2y^2 - y + 2$

27. $x^{-1/2} - 2x^{1/2} + x^{3/2}$

28. $a^4 b^2 + ab^5$

In Exercises 29–48 perform the indicated operations.

29. $(2x + 1)(3x - 2) - 5(4x - 1)$

30. $(2y - 7)(2y + 7)$

31. $(2a^2 - b)^2$

32. $(1 + x)(2 - x) - (3 - x)(3 + x)$

33. $(x - 1)(x - 2)(x - 3)$

34. $(2x + 1)^3$

35. $\sqrt{x}(\sqrt{x} + 1)(2\sqrt{x} - 1)$

36. $\dfrac{x^3 + 2x^2 + 3x}{x}$

37. $\dfrac{x^2 - 2x - 3}{2x^2 + 5x + 3}$

38. $\dfrac{t^3 - 1}{t^2 - 1}$

39. $\dfrac{x^2 - 2x - 15}{x^2 - 6x + 5} \div \dfrac{x^2 - x - 12}{x^2 - 1}$

40. $x - \dfrac{1}{x + 1}$

41. $\dfrac{1}{x - 1} - \dfrac{x}{x^2 + 1}$

42. $\dfrac{2}{x} + \dfrac{1}{x - 2} + \dfrac{3}{(x - 2)^2}$

43. $\dfrac{1}{x - 1} - \dfrac{2}{x^2 - 1}$

44. $\dfrac{1}{x + 2} + \dfrac{1}{x^2 - 4} - \dfrac{2}{x^2 - x - 2}$

45. $\dfrac{\dfrac{1}{x} - \dfrac{1}{2}}{x - 2}$

46. $\dfrac{\dfrac{1}{x} - \dfrac{1}{x + 1}}{\dfrac{1}{x} + \dfrac{1}{x + 1}}$

47. $\dfrac{3(x + h)^2 - 5(x + h) - (3x^2 - 5)x}{h}$

48. $\dfrac{\sqrt{x + h} - \sqrt{x}}{h}$

In Exercises 49–56 solve the equation.

49. $\frac{1}{2}x + 4(1 - 3x) = 5x + 3$

50. $\dfrac{x + 1}{x - 1} = \dfrac{2x - 1}{2x + 1}$

51. $2x^2 + x = 1$

52. $4x^3 - 25x = 0$

53. $3x^2 + 4x - 1 = 0$

54. $\dfrac{1}{x} + \dfrac{2}{x - 1} = 3$

55. $x^{-1/2} - 2x^{1/2} + x^{3/2} = 0$

56. $|2x - 5| = 9$

57. The owner of a store sells raisins for \$3.20 a pound and nuts for \$2.40 a pound. He decides to mix the raisins and nuts and sell 50 lb of the mixture for \$2.72 a pound. What quantities of raisins and nuts should he use?

58. The hypotenuse of a right triangle has length 20 cm. The sum of the lengths of the other two sides is 28 cm. Find the lengths of the other two sides of the triangle.

In Exercises 59–67 solve the inequalities in terms of intervals and illustrate the solution sets on the real number line.

59. $-1 < 2x + 5 \le 3$ **60.** $3 - x \le 2x - 7$

61. $x^2 + 4x - 12 > 0$ **62.** $x^2 \le 1$

63. $\dfrac{2x + 5}{x + 1} \le 1$ **64.** $|x - 5| \le 3$

65. $|x - 4| < 0.02$ **66.** $|2x + 1| \ge 1$

67. $|x - 1| < |x - 3|$ (Try to solve this one by interpreting the quantities as distances.)

68. The volume of a sphere is given by $V = \frac{4}{3}\pi r^3$, where r is the radius. Find the interval of values of the radius if the volume is supposed to be between 8 ft³ and 12 ft³, inclusively.

69. (a) Find the distance between the points $P(5, -3)$ and $Q(-1, 5)$.

(b) Find the midpoint of the line segment PQ.

70. Sketch the region consisting of all points (x, y) such that $|x| \ge 1$ and $|y| \ge 2$.

71. Which of the points $A(4, 4)$ or $B(5, 3)$ is closer to the point $C(-1, -3)$?

72. Show that the points $A(1, 2)$, $B(2, 6)$, and $C(9, 0)$ are the vertices of a right triangle. Find the area of the triangle.

73. Find an equation of the circle that has center $(2, -5)$ and radius $\sqrt{2}$.

74. Find an equation of the circle that has center $(-5, -1)$ and passes through the origin.

75. Find an equation of the line that passes through the points $(-1, -6)$ and $(2, -4)$.

76. Find an equation of the line that passes through the point $(1, 7)$ and is perpendicular to the line $x - 3y + 16 = 0$.

In Exercises 77–84 sketch the graph of the given equation and test for symmetry.

77. $y = 2 - 3x$ **78.** $2x - y + 1 = 0$

79. $y = 16 - x^2$ **80.** $8x + y^2 = 0$

81. $x = \sqrt{y}$ **82.** $y = -\sqrt{1 - x^2}$

83. $2x^2 + 2y^2 = 5$

84. $x^2 + y^2 + 2x - 4y - 31 = 0$

In Exercises 85–91 state whether or not the given equation is true for all values of the variables. (Disregard values that make denominators 0.)

85. $(x + y)^3 = x^3 + y^3$

86. $\dfrac{1 + \sqrt{a}}{1 - a} = \dfrac{1}{1 - \sqrt{a}}$

87. $\dfrac{12 + y}{y} = \dfrac{12}{y} + 1$

88. $\sqrt[3]{a + b} = \sqrt[3]{a} + \sqrt[3]{b}$

89. $\sqrt{a^2} = a$

90. $\dfrac{1}{x + 4} = \dfrac{1}{x} + \dfrac{1}{4}$

91. $x^3 + y^3 = (x + y)(x^2 + xy + y^2)$

92. What is wrong with the following argument?

$$a = b$$
$$a^2 = ab$$
$$a^2 - b^2 = ab - b^2$$
$$(a + b)(a - b) = b(a - b)$$
$$a + b = b$$

Now put $a = b = 1$: $2 = 1$

93. Factor $x^4 + x^2y^2 + y^4$. (*Hint:* Add and subtract x^2y^2.)

94. Show that $(28^{1/2} + 27^{1/2})^{2/3} + (28^{1/2} - 27^{1/2})^{2/3} = 5$.
(*Hint:* Try the product of the two terms.)

Chapter 1 Test

1. Graph the intervals $[-2, 1]$ and $(6, \infty)$ on a real number line.

2. Evaluate $16^{-3/4}$.

3. Express $\dfrac{(x^2)^a(\sqrt{x})^b}{x^{a+b}x^{a-b}}$ as a power of x.

4. Simplify:

(a) $\sqrt{200} - \sqrt{8}$ **(b)** $\dfrac{x^2 + 3x + 2}{x^2 - x - 2}$

(c) $\dfrac{x}{x^2 - 4} + \dfrac{1}{x + 2}$

5. Factor $x^4 + 27x$.

6. Expand $(2 - x^2)^3$.

7. Solve the inequality $|x - 6| < 3$.

8. Solve the following equations:

(a) $\dfrac{1}{x - 2} = \dfrac{2}{x - 3}$ **(b)** $x^2 + x - 1 = 0$

(c) $x^{1/2} - 3x^{3/2} + 2x^{5/2} = 0$

9. (a) Find the distance between the points $A(4, -5)$ and $B(-1, 7)$.
   (b) Find the midpoint of the line segment AB.

10. Find an equation of the circle with center $(1, -2)$ that passes through the point $(6, 2)$.

11. Find an equation of the line that passes through the point $(-2, 5)$ and is parallel to the line $6x - 3y + 1 = 0$.

12. Sketch the graph of the equation $x = y^3$.

13. Adult tickets at a movie theater cost \$6.50 and student tickets cost \$4.00. If the total receipts for one evening were \$1275.00 and twice as many adult tickets were sold as student tickets, how many student tickets were sold?

2

Functions

That flower of modern mathematical thought—the notion of a function.
Thomas J. McCormack

When we cannot use the compass of mathematics or the torch of experience...it is certain we cannot take a single step forward.
Voltaire

The two fundamental ideas in all of mathematics are those of number and function. In this chapter we study functions and their graphs as well as ways of obtaining new functions from old ones.

Functions

The area A of a circle depends on the radius r of the circle. The rule that connects r and A is given by the equation $A = \pi r^2$. With each positive number r there is associated one value of A, and we say that A is a function of r.

The number N of bacteria in a culture depends on the time t. If the culture starts with 5000 bacteria and the population doubles every hour, then after t hours the number of bacteria will be $N = (5000)2^t$. This is the rule that connects t and N. For each value of t there is a corresponding value of N, and we say that N is a function of t.

The cost C of mailing a first-class letter depends on the weight w of the letter. Although there is no single neat formula that connects w and C, the post office has a rule for determining C when w is known.

In each of these examples there is a rule whereby, given a number (r, t, or w), another number (A, N, or C) is assigned. In each case we say that the second number is a function of the first number.

A **function f** is a rule that assigns to each element x in a set A exactly one element, called $f(x)$, in a set B.

We usually consider functions for which the sets A and B are sets of real numbers. The set A is called the **domain** of the function. The symbol $f(x)$ is read "f of x" and is called the **value of f at x**, or the **image of x under f**. The **range** of f is the set of all possible values of $f(x)$ as x varies throughout the domain—that is, $\{f(x) \mid x \in A\}$.

It is helpful to think of a function as a **machine** (see Figure 1). If x is in the domain of the function f, then when x enters the machine, it is accepted as an input and the machine produces an output $f(x)$ according to the rule of the function. Thus we can think of the domain as the set of all possible inputs and the range as the set of all possible outputs.

The preprogrammed functions in a calculator are good examples of a function as a machine. For example, the $\sqrt{x}$ key on your calculator is such a function. First you input x into the display. Then you press the key labeled $\sqrt{x}$. If $x < 0$, then x is not in the domain of this function; that is, x is not an acceptable input, and the calculator will indicate an error. If $x \geq 0$, then an approximation to $\sqrt{x}$ will appear in the display, correct to a certain number of decimal places. [Thus the $\sqrt{x}$ key on your calculator is not quite the same as the exact mathematical function f defined by $f(x) = \sqrt{x}$.]

Another way to picture a function is by an **arrow diagram** as in Figure 2. Each

Figure 1

Machine diagram for a function f

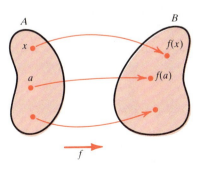

Figure 2

Arrow diagram for f

arrow connects an element of A to an element of B. The arrow indicates that $f(x)$ is associated with x, $f(a)$ is associated with a, and so on.

Example 1

The squaring function assigns to each real number x its square x^2. It is defined by the equation

$$f(x) = x^2$$

(a)   Evaluate $f(3)$, $f(-2)$, and $f(\sqrt{5})$.
(b)   Find the domain and range of f.
(c)   Draw a machine diagram and an arrow diagram for f.

Solution

(a)   The values of f are found by substituting for x in the equation $f(x) = x^2$:

$$f(3) = 3^2 = 9 \qquad f(-2) = (-2)^2 = 4 \qquad f(\sqrt{5}) = (\sqrt{5})^2 = 5$$

(b)   The domain of f is the set R of all real numbers. The range of f consists of all values of $f(x)$—that is, all numbers of the form x^2. But $x^2 \geq 0$ for all numbers x, and any nonnegative number c is a square, since $c = (\sqrt{c})^2 = f(\sqrt{c})$. Therefore the range of f is $\{y \mid y \geq 0\} = [0, \infty)$.

(c)   Machine and arrow diagrams for this function are shown in Figure 3.

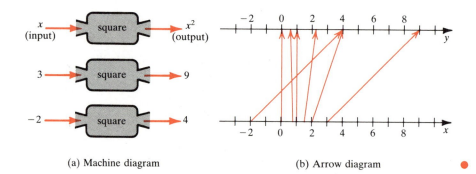

Figure 3

(a) Machine diagram   (b) Arrow diagram

Example 2

If we define a function g by

$$g(x) = x^2, \qquad 0 \leq x \leq 3$$

then the domain of g is given as the closed interval $[0, 3]$. This is different from the function f in Example 1 because in considering g we are restricting our attention to those values of x between 0 and 3. The range of g is

$$\{x^2 \mid 0 \leq x \leq 3\} = \{y \mid 0 \leq y \leq 9\} = [0, 9]$$

Example 3

If $f(x) = 2x^2 + 3x - 1$, evaluate:

(a) $f(a)$ (b) $f(-a)$

(c) $f(a + h)$ (d) $\dfrac{f(a + h) - f(a)}{h}, \quad h \neq 0$

[Expressions like the one in part (d) occur frequently in calculus where they are called *difference quotients*.]

Solution

(a) $f(a) = 2a^2 + 3a - 1$

(b) $f(-a) = 2(-a)^2 + 3(-a) - 1 = 2a^2 - 3a - 1$

(c) $f(a + h) = 2(a + h)^2 + 3(a + h) - 1$

$\qquad\qquad = 2(a^2 + 2ah + h^2) + 3(a + h) - 1$

$\qquad\qquad = 2a^2 + 4ah + 2h^2 + 3a + 3h - 1$

(d) $\dfrac{f(a + h) - f(a)}{h} = \dfrac{(2a^2 + 4ah + 2h^2 + 3a + 3h - 1) - (2a^2 + 3a - 1)}{h}$

$\qquad\qquad = \dfrac{4ah + 2h^2 + 3h}{h} = 4a + 2h + 3$ ●

In Examples 1 and 2 the domain of the function was given explicitly. But if a function is given by a formula and the domain is not stated explicitly, *the convention is that the domain is the set of all real numbers for which the formula makes sense and defines a real number.*

We should distinguish between a function f and the number $f(x)$, which is the value of f at x. Nonetheless, it is common to abbreviate an expression such as

the function f defined by $f(x) = x^2 + x$

to

the function $f(x) = x^2 + x$

Example 4

Find the domain of the function $f(x) = \dfrac{1}{x^2 - x}$.

Solution

Since

$$f(x) = \frac{1}{x^2 - x} = \frac{1}{x(x - 1)}$$

and division by 0 is not allowed, we see that $f(x)$ is not defined when $x = 0$ or $x = 1$. Thus the domain of f is

$$\{x \mid x \neq 0, x \neq 1\}$$

which could also be written in interval notation as

$$(-\infty, 0) \cup (0, 1) \cup (1, \infty)$$ ●

Example 5

Find the domain of the function $g(t) = \dfrac{t}{\sqrt{t+1}}$.

Solution

The square root of a negative number is not defined (as a real number), so we require that $t + 1 \geq 0$. Also the denominator cannot be 0; that is, $\sqrt{t+1} \neq 0$. Thus $g(t)$ exists when $t + 1 > 0$; that is, $t > -1$. So the domain of g is

$$\{t \mid t > -1\} = (-1, \infty)$$

Example 6

Find the domain of $h(x) = \sqrt{2 - x - x^2}$.

Solution

Since the square root of a negative number is not defined (as a real number), the domain of h consists of all values of x such that

$$2 - x - x^2 \geq 0$$

We solve this inequality using the methods of Section 1.5. Since

$$2 - x - x^2 = (2 + x)(1 - x)$$

the product will change sign when $x = -2$ or 1, as indicated in the following chart:

Interval	$2 + x$	$1 - x$	$(2 + x)(1 - x)$
$x \leq -2$	$-$	$+$	$-$
$-2 \leq x \leq 1$	$+$	$+$	$+$
$x \geq 1$	$+$	$-$	$-$

Therefore the domain of h is

$$\{x \mid -2 \leq x \leq 1\} = [-2, 1]$$

The symbol that represents an arbitrary number in the *domain* of a function f is called an **independent variable**. The symbol that represents a number in the *range* of f is called a **dependent variable**. For example, the squaring function of Example 1 could be defined by saying that each number x is assigned the number y by the rule $y = x^2$. Then x is the independent variable and y is the dependent variable. In the example on bacteria at the beginning of this section, t is the independent variable, N is the dependent variable, and they are connected by the equation $N = (5000)2^t$. In general, a function describes how one quantity or variable depends on another. For instance, we say that population is a function of time, and pressure is a function of temperature.

Example 7

Express the perimeter of a square as a function of its area.

Solution

Let P, A, and s represent the perimeter, area, and side length of the square, respectively. We start with the known formulas

$$P = 4s \qquad \text{and} \qquad A = s^2$$

Substituting $s = \sqrt{A}$ from the second equation into the first, we get

$$P = 4\sqrt{A}$$

This equation expresses P as a function of A. Since A must be positive, the domain is given by $A > 0$ or $(0, \infty)$.

In setting up expressions for functions in applied situations, it is useful to recall some of the problem-solving principles from Section 1.9 and adapt them to the present situation.

Steps in Setting Up Applied Functions

1. *Understand the problem.* The first step is to read the problem carefully until it is clearly understood. Ask yourself: What is the unknown? What are the given quantities? What are the given conditions?

2. *Draw a diagram.* In most problems it is useful to draw a diagram and identify the given and required quantities on the diagram.

3. *Introduce notation.* If the problem asks you for an expression for a certain quantity, assign a symbol to that quantity (let us call it Q for now). Also select symbols $(a, b, c, \ldots, x, y)$ for other unknown quantities and label the diagram with these symbols. It may help to use initials as suggestive symbols—for example, A for area, h for height, t for time.

4. Express Q in terms of some of the symbols from Step 3.

5. If Q has been expressed as a function of more than one variable in Step 4, use the given information to find relationships (in the form of equations) among these variables. Then use these equations to eliminate all but one of the variables in the expression for Q. Thus Q will be expressed as a function of one variable.

Example 8

A can holds 1 L (liter) of oil. Express the surface area of the can as a function of its radius.

Solution

(Notation)   Let r be the radius and h the height of the can, in centimeters (see Figure 4). Then the area of the top is πr^2, the area of the bottom is also πr^2, and the area of the

sides is the circumference times the height—that is, $2\pi rh$. So the total surface area is

$$A = 2\pi r^2 + 2\pi rh$$

To express A as a function of r, we need to eliminate h and we do this by using the fact that the volume is given as 1 L, which we take to be 1000 cm³. Thus

$$\pi r^2 h = 1000$$

Substituting $h = 1000/\pi r^2$ into the expression for A, we have

$$A = 2\pi r^2 + 2\pi r\left(\frac{1000}{\pi r^2}\right) = 2\pi r^2 + \frac{2000}{r}$$

Therefore the equation

$$A = 2\pi r^2 + \frac{2000}{r}, \qquad r > 0$$

expresses A as a function of r.

(Draw a diagram)

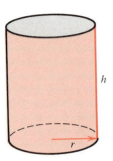

Figure 4

Exercises 2.1

1. If $f(x) = x^2 - 3x + 2$, find $f(1)$, $f(-2)$, $f(\tfrac{1}{2})$, $f(\sqrt{5})$, $f(a)$, $f(-a)$, and $f(a + b)$.

2. If $f(x) = x^3 + 2x^2 - 3$, find $f(0)$, $f(3)$, $f(-3)$, $f(-x)$, and $f(1/a)$.

3. If $g(x) = \dfrac{1 - x}{1 + x}$, find $g(2)$, $g(-2)$, $g(\pi)$, $g(a)$, $g(a - 1)$, and $g(-a)$.

4. If $h(t) = t + \dfrac{1}{t}$, find $h(1)$, $h(\pi)$, $h(t + 1)$, $h(t) + h(1)$, and $h(x)$.

5. If $f(x) = 2x^2 + 3x - 4$, find $f(0)$, $f(2)$, $f(\sqrt{2})$, $f(1 + \sqrt{2})$, $f(-x)$, $f(x + 1)$, $2f(x)$, and $f(2x)$.

6. If $f(x) = 2 - 3x$, find $f(1)$, $f(-1)$, $f\left(\dfrac{1}{3}\right)$, $f\left(\dfrac{x}{3}\right)$, $f(3x)$, $f(x^2)$, and $[f(x)]^2$.

In Exercises 7–10 find $f(a)$, $f(a) + f(h)$, $f(a + h)$, and $\dfrac{f(a + h) - f(a)}{h}$, where a and h are real numbers and $h \neq 0$.

7. $f(x) = 1 + 2x$

8. $f(x) = x^2 - 3x + 4$

9. $f(x) = 3 - 5x + 4x^2$

10. $f(x) = x^3 + x + 1$

In Exercises 11–14 find $f(2 + h)$, $f(x + h)$, and $\dfrac{f(x + h) - f(x)}{h}$, where $h \neq 0$.

11. $f(x) = 8x - 1$

12. $f(x) = x - x^2$

13. $f(x) = \dfrac{1}{x}$

14. $f(x) = \dfrac{x}{x + 1}$

In Exercises 15 and 16 draw a machine diagram and an arrow diagram for the given function.

15. $f(x) = \sqrt{x}, \quad 0 \le x \le 4$

16. $f(x) = \dfrac{2}{x}, \quad 1 \le x \le 4$

In Exercises 17–26 find the domain and range of the given function.

17. $f(x) = 2x, \; -1 \le x \le 5$

18. $f(x) = 2x + 7, \; -1 \le x \le 6$

19. $f(x) = 6 - 4x, \; -2 \le x \le 3$

20. $f(x) = x^2 + 1, \; -1 \le x \le 5$

21. $g(x) = 2 - x^2$

22. $g(x) = \sqrt{7 - 3x}$

23. $h(x) = \sqrt{2x - 5}$

24. $h(x) = 1 - \sqrt{x}$

25. $F(x) = 3 + \sqrt{1 - x^2}$

26. $G(x) = \sqrt{x^2 - 9}$

In Exercises 27–48 find the domain of the given function.

27. $f(x) = \dfrac{1}{x+4}$

28. $f(x) = \dfrac{2}{3x-5}$

29. $f(x) = \dfrac{x+2}{x^2-1}$

30. $f(x) = \dfrac{x^4}{x^2+x-6}$

31. $f(x) = \dfrac{x+2}{x^2+1}$

32. $f(x) = \dfrac{x^3+1}{x^3-x}$

33. $f(x) = \sqrt{5-2x}$

34. $f(x) = \sqrt[4]{x-9}$

35. $f(t) = \sqrt[3]{t-1}$

36. $f(t) = \sqrt{t^2+1}$

37. $g(x) = \dfrac{\sqrt{2+x}}{3-x}$

38. $g(x) = \dfrac{\sqrt{x}}{2x^2+x-1}$

39. $F(x) = \dfrac{x}{\sqrt{x-10}}$

40. $F(x) = x^2 - \dfrac{x}{\sqrt{9-2x}}$

41. $G(x) = \sqrt{x} + \sqrt{1-x}$

42. $G(x) = \sqrt{x+2} - 2\sqrt{x-3}$

43. $f(x) = \sqrt{4x^2-1}$

44. $f(x) = \sqrt{2-3x^2}$

45. $g(x) = \sqrt[4]{x^2-6x}$

46. $g(x) = \sqrt{x^2-2x-8}$

47. $\phi(x) = \sqrt{\dfrac{x}{\pi-x}}$

48. $\phi(x) = \sqrt{\dfrac{x^2-2x}{x-1}}$

In Exercises 49–62 find a formula for the described function and state its domain.

49. A rectangle has a perimeter of 20 ft. Express the area A of the rectangle as a function of the length x of one of its sides.

50. A rectangle has an area of 16 m². Express the perimeter P of the rectangle as a function of the length x of one of its sides.

51. Express the area A of an equilateral triangle as a function of the length x of a side.

52. Express the surface area A of a cube as a function of its volume V.

53. Express the radius r of a circle as a function of its area A.

54. Express the area A of a circle as a function of the circumference C.

55. An open rectangular box with a volume of 12 ft³ has a square base. Express the surface area A of the box as a function of the length x of a side of the base.

56. A woman 5 ft tall is standing near a street lamp that is 12 ft tall. Express the length L of her shadow as a function of her distance d from the base of the lamp.

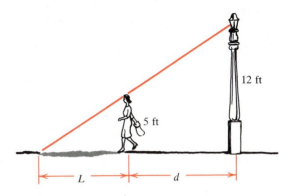

57. A Norman window has the shape of a rectangle surmounted by a semicircle. If the perimeter of the window is 30 ft, express the area A of the window as a function of the width x of the window.

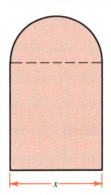

58. A box with an open top is to be constructed from a rectangular piece of cardboard with dimensions 12 in. by 20 in. by cutting out equal squares of side x at each corner and then folding up the sides as in the figure. Express the volume V of the box as a function of x.

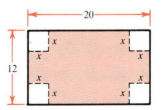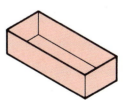

59. A farmer has 2400 ft of fencing and wants to fence off a rectangular field that borders a straight river. He needs no fence along the river. Express the area A of the field in terms of the width x of the field (see the figure on page 74).

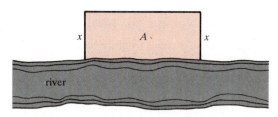

river

60. A rectangle is inscribed in a semicircle of radius r as in the figure. Express the area A of the rectangle as a function of the height h of the rectangle.

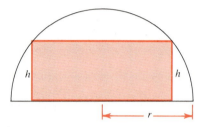

61. Two ships leave a port at the same time. One sails south at 15 mi/h and the other sails east at 20 mi/h. Express the distance d between the ships as a function of t, the time (in hours) after their departure.

62. A man is at a point A on a bank of a straight river, 2 mi wide, and wants to reach point B, 7 mi downstream on the opposite bank, by first rowing his boat to a point P on the opposite bank and then walking the remaining distance x to B. He can row at 4 mi/h and walk at 3 mi/h. Express the total time T that he takes to go from A to B as a function of x.

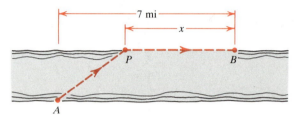

Section 2.2

Graphs of Functions

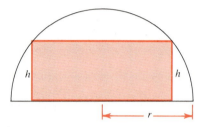

In the preceding section we saw how to picture functions using machine diagrams and arrow diagrams. A third method for visualizing a function is its graph. If f is a function with domain A, its **graph** is the set of ordered pairs

$$\{(x, f(x)) \mid x \in A\}$$

In other words, the graph of f consists of all points (x, y) in the coordinate plane such that $y = f(x)$ and x is in the domain of f. Thus the equation of the graph of f is $y = f(x)$.

Figure 1
Graph of f

The graph of a function f gives us a useful picture of the behavior or "life history" of a function. Since the y-coordinate of any point (x, y) on the graph is $y = f(x)$, we can read the value of $f(x)$ from the graph as being the height of the graph above the point x (see Figure 1). The graph of f also allows us to picture the domain and range of f on the x-axis and y-axis as in Figure 2.

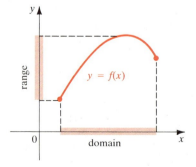

Figure 2

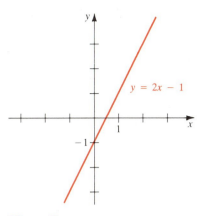

Figure 3

Graph of $f(x) = 2x - 1$

Example 1

Sketch the graph of the function $f(x) = 2x - 1$.

Solution

The equation of the graph is $y = 2x - 1$, and we recognize this as being the equation of a line with slope 2 and y-intercept -1. This enables us to sketch the graph of f in Figure 3. ●

   In general, a function f defined by an equation of the form $f(x) = mx + b$ is called a **linear function** because the equation of its graph is $y = mx + b$, which represents a line with slope m and y-intercept b. A special case of the linear function occurs when the slope is $m = 0$. The function $f(x) = b$, where b is a given number, is called a **constant function** because all its values are the same number—namely, b. Its graph is the horizontal line $y = b$.

Example 2

Sketch the graph of $f(x) = x^2$.

Solution

The equation of the graph is $y = x^2$. We draw it in Figure 4 by setting up a table of values and plotting points as in Chapter 1. Recall from Example 1 of Section 2.1 that the domain is R and the range is $[0, \infty)$.

x	$y = x^2$
0	0
$\pm\frac{1}{2}$	$\frac{1}{4}$
± 1	1
± 2	4
± 3	9

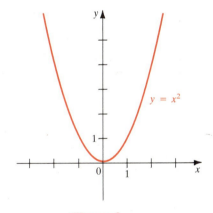

Figure 4

Graph of $f(x) = x^2$ ●

Example 3

Sketch the graph of $f(x) = x^3$.

Solution

We list some functional values and the corresponding points on the graph in the

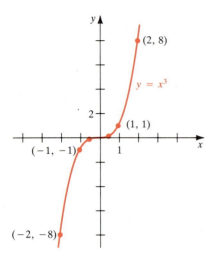

Figure 5
Graph of $f(x) = x^3$

following table:

x	0	$\frac{1}{2}$	1	2	$-\frac{1}{2}$	-1	-2
$f(x) = x^3$	0	$\frac{1}{8}$	1	8	$-\frac{1}{8}$	-1	-8
(x, x^3)	$(0,0)$	$(\frac{1}{2}, \frac{1}{8})$	$(1,1)$	$(2,8)$	$(-\frac{1}{2}, -\frac{1}{8})$	$(-1,-1)$	$(-2,-8)$

Then we plot points to obtain the graph shown in Figure 5.   ●

Example 4

If $f(x) = \sqrt{4 - x^2}$, sketch the graph of f and find its domain and range.

Solution

The equation of the graph is $y = \sqrt{4 - x^2}$, from which it follows that $y \geq 0$. Squaring this equation, we get

$$y^2 = 4 - x^2 \quad \text{or} \quad x^2 + y^2 = 4$$

which we recognize as the equation of a circle with center the origin and radius 2. But, since $y \geq 0$, the graph of f consists of just the upper half of this circle. From Figure 6 we see that the domain is the closed interval $[-2, 2]$ and the range is $[0, 2]$. We could also have found the domain as follows:

$$\begin{aligned} \text{domain} &= \{x \mid 4 - x^2 \geq 0\} = \{x \mid x^2 \leq 4\} \\ &= \{x \mid |x| \leq 2\} = \{x \mid -2 \leq x \leq 2\} \\ &= [-2, 2] \end{aligned}$$

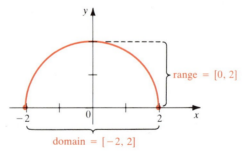

Figure 6
Graph of $f(x) = \sqrt{4 - x^2}$ ●

The graph of a function is a curve in the xy-plane. But the question arises: Which curves in the xy-plane are graphs of functions? This is answered by the following test.

The Vertical Line Test

A curve in the plane is the graph of a function if and only if no vertical line intersects the curve more than once.

We can see from Figure 7 why the Vertical Line Test is true. If each vertical line $x = a$ intersects a curve only once at (a, b), then exactly one functional value

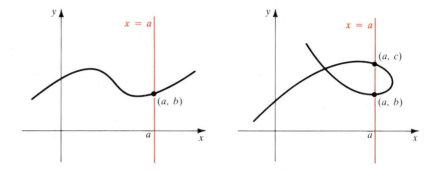

Figure 7

is defined by $f(a) = b$. But if a line $x = a$ intersects the curve twice at (a, b) and (a, c), then the curve cannot represent a function because a function cannot assign two different values to a.

Example 5

Using the Vertical Line Test, we see that the curves in Figures 8(b) and (c) represent functions, whereas those in parts (a) and (d) do not.

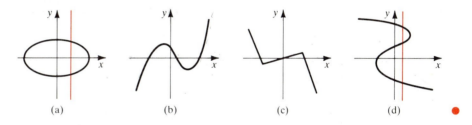

(a)   (b)   (c)   (d)   ●

Figure 8

Example 6

Sketch the graph of $f(x) = \sqrt{x}$.

Solution

First we note that the domain is $\{x \mid x \geq 0\} = [0, \infty)$. Then we plot the points given by the following table and use them to draw the sketch in Figure 9.

x	$y = \sqrt{x}$
0	0
1	1
2	$\sqrt{2}$
3	$\sqrt{3}$
4	2
5	$\sqrt{5}$

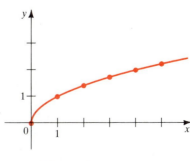

Figure 9
Graph of $f(x) = \sqrt{x}$

●

Note: The equation of the graph of the square root function in Example 6 is $y = \sqrt{x}$. If we square this equation, we get $y^2 = x$, but we have to remember that $y \geq 0$. We saw in Section 1.7 that the equation $x = y^2$ represents the parabola shown in Figure 10(a). By the Vertical Line Test this parabola does not represent a function of x. But we can regard $y^2 = x$ as representing *two* functions of x; the upper and lower halves of this parabola are the graphs of the functions $f(x) = \sqrt{x}$ and $g(x) = -\sqrt{x}$. [See Figures 10(b) and (c).]

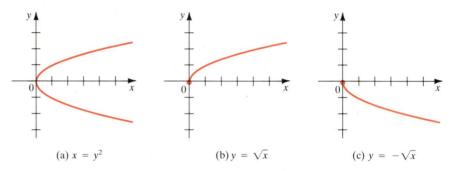

Figure 10

(a) $x = y^2$ (b) $y = \sqrt{x}$ (c) $y = -\sqrt{x}$

Increasing and Decreasing Functions

It is very useful to know where the graph of a function rises and where it falls. The graph in Figure 11 rises from A to B, falls from B to C, and rises again from C to D. The function f is said to be increasing on the interval $[a, b]$, decreasing on $[b, c]$, and increasing again on $[c, d]$. Notice that if x_1 and x_2 are any two numbers between a and b with $x_1 < x_2$, then $f(x_1) < f(x_2)$. We use this as the defining property of an increasing function.

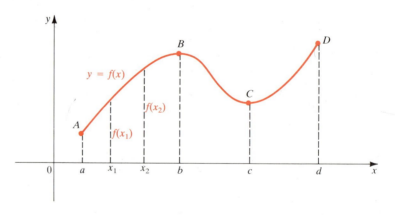

Figure 11

A function f is called **increasing** on an interval I if

$$f(x_1) < f(x_2) \quad \text{whenever} \quad x_1 < x_2 \text{ in } I$$

It is called **decreasing** on I if

$$f(x_1) > f(x_2) \quad \text{whenever} \quad x_1 < x_2 \text{ in } I$$

For instance, the functions in Examples 1, 3, and 6 are all increasing on their domains. In Example 2, f is decreasing on $(-\infty, 0]$ and increasing on $[0, \infty)$. In Example 4, f is increasing on $[-2, 0]$ and decreasing on $[0, 2]$.

Example 7

State the intervals on which the function whose graph is shown in Figure 12 is increasing or decreasing.

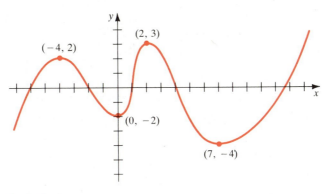

Figure 12

Solution

The function is increasing on $(-\infty, -4]$, $[0, 2]$, and $[7, \infty)$. It is decreasing on $[-4, 0]$ and $[2, 7]$. ●

Functions Defined by More Than One Equation

The functions we have looked at so far have been defined by means of simple formulas. But there are many functions that are not given by such formulas. Here are some examples: the cost of mailing a first-class letter as a function of its weight, the population of New York City as a function of time, and the cost of a taxi ride as a function of distance. The following examples give more illustrations.

Example 8

A function f is defined by

$$f(x) = \begin{cases} 1 - x & \text{if } x \leq 1 \\ x^2 & \text{if } x > 1 \end{cases}$$

Evaluate $f(0)$, $f(1)$, and $f(2)$ and sketch the graph.

Solution

Remember that a function is a rule. For this particular function the rule is the following: First look at the value of the input x. If it happens that $x \leq 1$, then the value of $f(x)$ is $1 - x$. On the other hand, if $x > 1$, then the value of $f(x)$ is x^2.

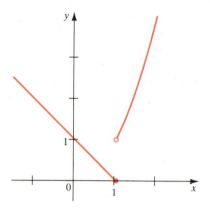

Figure 13

Since $0 \leq 1$, we have $f(0) = 1 - 0 = 1$.
Since $1 \leq 1$, we have $f(1) = 1 - 1 = 0$.
Since $2 > 1$, we have $f(2) = 2^2 = 4$.

How do we draw the graph of f? We observe that if $x \leq 1$, then $f(x) = 1 - x$, so the part of the graph of f that lies to the left of the vertical line $x = 1$ must coincide with the line $y = 1 - x$, which has slope -1 and y-intercept 1. If $x > 1$, then $f(x) = x^2$, so the part of the graph of f that lies to the right of the line $x = 1$ must coincide with the graph of $y = x^2$, which we sketched in Example 2. This enables us to sketch the graph in Figure 13. The solid dot indicates that the point is included on the graph; the open dot indicates that the point is excluded from the graph.

Example 9

Sketch the graph of the absolute value function $y = |x|$.

Solution

Recall from Section 1.1 that

$$|x| = \begin{cases} x & \text{if } x \geq 0 \\ -x & \text{if } x < 0 \end{cases}$$

Using the same method as in Example 8, we see that the graph of f coincides with the line $y = x$ to the right of the y-axis and coincides with the line $y = -x$ to the left of the y-axis (see Figure 14).

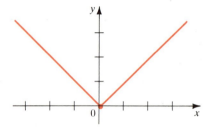

Figure 14
Graph of $f(x) = |x|$

Example 10

The cost of a long-distance daytime phone call from Toronto to New York City is 69 cents for the first minute and 58 cents for each additional minute (or part of a minute). Draw the graph of the cost C (in dollars) of the phone call as a function of time t (in minutes).

Solution

Let $C(t)$ be the cost for t minutes. Since $t > 0$, the domain of the function is $(0, \infty)$. From the given information we have

$$\begin{array}{ll} C(t) = 0.69 & \text{if } 0 < t \leq 1 \\ C(t) = 0.69 + 0.58 = 1.27 & \text{if } 1 < t \leq 2 \\ C(t) = 0.69 + 2(0.58) = 1.85 & \text{if } 2 < t \leq 3 \\ C(t) = 0.69 + 3(0.58) = 2.43 & \text{if } 3 < t \leq 4 \end{array}$$

and so on. The graph is shown in Figure 15.

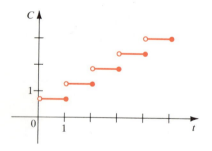

Figure 15

Symmetry

If a function f satisfies $f(-x) = f(x)$ for every number x in its domain, then f is called an **even function**. For instance, the function $f(x) = x^2$ of Example 2 is even because

$$f(-x) = (-x)^2 = x^2 = f(x)$$

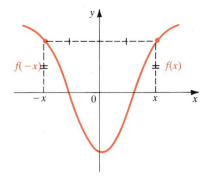

Figure 16
An even function

The geometric significance of an even function is that its graph is symmetric with respect to the y-axis (see Figure 16). This means that if we have plotted the graph of f for $x \geq 0$, we obtain the entire graph simply by reflecting in the y-axis.

If f satisfies $f(-x) = -f(x)$ for every number x in its domain, then f is called an **odd function**. For example, the function $f(x) = x^3$ of Example 3 is odd because

$$f(-x) = (-x)^3 = -x^3 = -f(x)$$

The graph of an odd function is symmetric about the origin (see Figure 17). If we already have the graph of f for $x \geq 0$, we can obtain the entire graph by rotating through 180° about the origin. For instance, in Example 3 we need only have plotted the graph of $y = x^3$ for $x \geq 0$ and then rotated that part about the origin.

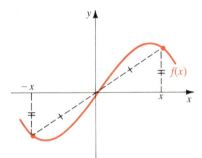

Figure 17
An odd function

Exercises 2.2

1. The graph of a function is given.
   (a) State the values of $f(-2)$, $f(0)$, $f(2)$, and $f(3)$.
   (b) State the domain of f.
   (c) State the range of f.
   (d) On what intervals is f decreasing? On what interval is it increasing?

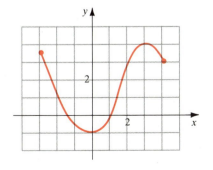

2. The graph of a function g is given.
   (a) State the values of $g(-3)$, $g(1)$, $g(2)$, and $g(3)$.
   (b) State the domain of g.
   (c) State the range of g.

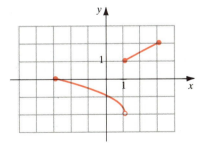

In Exercises 3 and 4 determine which of the curves are graphs of functions of x.

3. (a) 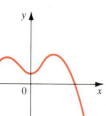 **(b)** 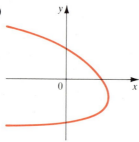 **(c)** 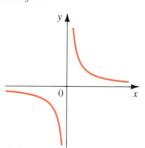 **(d)**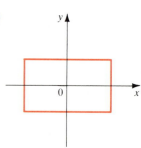

4. (a) 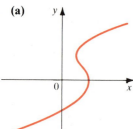 **(b)** 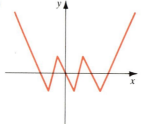 **(c)** 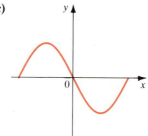 **(d)**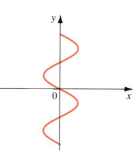

In Exercises 5–8 state whether the given curve is the graph of a function of x. If it is, state the domain and range of the function.

5. 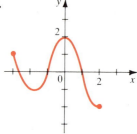 **6.** 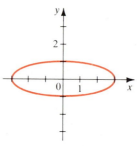 **7.** 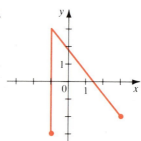 **8.**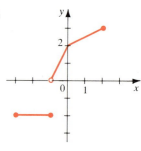

In Exercises 9–16, (a) sketch the graph of f, (b) find the domain of f, and (c) state the intervals on which f is increasing or decreasing.

9. $f(x) = 1 - x$

10. $f(x) = \frac{1}{2}(x + 1)$

11. $f(x) = x^2 - 4x$

12. $f(x) = x^2 - 4x + 1$

13. $f(x) = 1 + \sqrt{4 - x}$

14. $f(x) = \sqrt{x^2 - 4}$

15. $f(x) = -\sqrt{25 - x^2}$

16. $f(x) = x^3 - 3x + 1$

17. Draw the graphs of the functions $y = x$, $y = x^3$, and $y = x^5$ using the same axes.

18. Draw the graphs of the functions $y = x^2$, $y = x^4$, and $y = x^6$ using the same axes.

In Exercises 19–51 sketch the graph of the function.

19. $f(x) = 2$

20. $f(x) = -3$

21. $f(x) = 3 - 2x$

22. $f(x) = \dfrac{x + 3}{2}$, $-2 \le x \le 2$

23. $f(x) = -x^2$

24. $f(x) = x^2 - 4$

25. $f(x) = x^2 + 2x - 1$

26. $f(x) = -x^2 + 6x - 7$

27. $g(x) = 1 - x^3$

28. $g(x) = 2x^2 - x^4$

29. $g(x) = \sqrt{-x}$

30. $g(x) = \sqrt{6 - 2x}$

31. $F(x) = \dfrac{1}{x}$

32. $F(x) = \dfrac{2}{x+4}$

33. $G(x) = |x| + x$

34. $G(x) = |x| - x$

35. $H(x) = |2x|$

36. $H(x) = |x+1|$

37. $f(x) = |2x - 3|$

38. $f(x) = \dfrac{x}{|x|}$

39. $f(x) = \dfrac{x^2 - 1}{x - 1}$

40. $f(x) = \dfrac{x^2 + 5x + 6}{x + 2}$

41. $f(x) = \begin{cases} x+1 & \text{if } x \neq 1 \\ 1 & \text{if } x = 1 \end{cases}$

42. $f(x) = \begin{cases} x+3 & \text{if } x \neq -2 \\ 4 & \text{if } x = -2 \end{cases}$

43. $f(x) = \begin{cases} 0 & \text{if } x < 2 \\ 1 & \text{if } x \geq 2 \end{cases}$

44. $f(x) = \begin{cases} -1 & \text{if } x < -1 \\ 1 & \text{if } -1 \leq x \leq 1 \\ -1 & \text{if } x > 1 \end{cases}$

45. $f(x) = \begin{cases} x & \text{if } x \leq 0 \\ x+1 & \text{if } x > 0 \end{cases}$

46. $f(x) = \begin{cases} 2x+3 & \text{if } x < -1 \\ 3-x & \text{if } x \geq -1 \end{cases}$

47. $f(x) = \begin{cases} -1 & \text{if } x < -1 \\ x & \text{if } -1 \leq x \leq 1 \\ 1 & \text{if } x > 1 \end{cases}$

48. $f(x) = \begin{cases} |x| & \text{if } |x| \leq 1 \\ 1 & \text{if } |x| > 1 \end{cases}$

49. $f(x) = \begin{cases} x+2 & \text{if } x \leq -1 \\ x^2 & \text{if } x > -1 \end{cases}$

50. $f(x) = \begin{cases} 1 - x^2 & \text{if } x \leq 2 \\ 2x - 7 & \text{if } x > 2 \end{cases}$

51. $f(x) = \begin{cases} -1 & \text{if } x \leq -1 \\ 3x+2 & \text{if } |x| < 1 \\ 7 - 2x & \text{if } x \geq 1 \end{cases}$

52. A taxi company charges $2.00 for the first mile (or part of a mile) and 20 cents for each succeeding tenth of a mile (or part). Express the cost C (in dollars) of a ride as a function of the distance x traveled (in miles) for $0 < x < 2$ and sketch the graph of this function.

In Exercises 53–56 find a function whose graph is the given curve.

53. The line segment joining the points $(-2, 1)$ and $(4, -6)$

54. The line segment joining the points $(-3, -2)$ and $(6, 3)$

55. The bottom half of the parabola $x + (y - 1)^2 = 0$

56. The top half of the circle $(x - 1)^2 + y^2 = 1$

In Exercises 57–62 determine whether f is even, odd, or neither. If f is even or odd, use symmetry to sketch its graph.

57. $f(x) = x^{-2}$

58. $f(x) = x^{-3}$

59. $f(x) = x^2 + x$

60. $f(x) = x^4 - 4x^2$

61. $f(x) = x^3 - x$

62. $f(x) = 3x^3 + 2x^2 + 1$

63. **(a)** Sketch the graph of the function $f(x) = x^2 - 1$.
  (b) Sketch the graph of the function $g(x) = |x^2 - 1|$.

64. **(a)** Sketch the graph of the function $f(x) = |x| - 1$.
  (b) Sketch the graph of the function $g(x) = ||x| - 1|$.

65. The **greatest integer function** is defined by $[\![x]\!] = $ the largest integer that is less than or equal to x. (For instance, $[\![4]\!] = 4$, $[\![4.8]\!] = 4$, $[\![\pi]\!] = 3$, $[\![-\frac{1}{2}]\!] = -1$.) Sketch the graph of the greatest integer function.

66. Sketch the graph of the function $f(x) = x - [\![x]\!]$, where $[\![x]\!]$ is the greatest integer function defined in Exercise 65.

Transformations of Functions

In this section we see how to reduce the amount of work required to graph certain functions by using the following transformations: shifting, reflecting, and stretching.

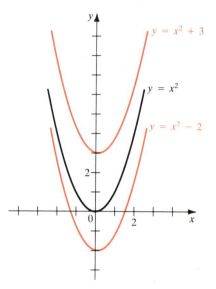

Figure 1

Example 1

Sketch the graphs of the functions (a) $f(x) = x^2 + 3$ and (b) $g(x) = x^2 - 2$.

Solution

(a) We start with the graph of the function $y = x^2$ from Example 2 of the preceding section. Then the equation $y = x^2 + 3$ indicates that the y-coordinate of a point on the graph of f is 3 more than the y-coordinate of the corresponding point on the curve $y = x^2$. This means that we obtain the graph of $y = x^2 + 3$ simply by shifting the graph of $y = x^2$ up by 3 units as in Figure 1.

(b) Similarly we get the graph of $g(x) = x^2 - 2$ by moving the parabola $y = x^2$ down by 2 units.   ●

In general we have the following rule for the translation (or shifting) of functions.

Vertical Shifts of Graphs

Let $c > 0$. To obtain the graph of

$y = f(x) + c$, move the graph of $y = f(x)$ c units up
$y = f(x) - c$, move the graph of $y = f(x)$ c units down

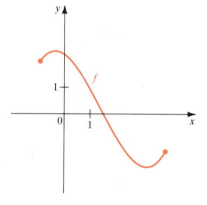

Figure 2

Example 2

Given the graph of $y = f(x)$ in Figure 2, sketch the graphs of the functions (a) $g(x) = f(x - 5)$ and (b) $h(x) = f(x + 8)$.

Solution

(a) The equation $g(x) = f(x - 5)$ says that the value of g at x is the same as the value of f at $x - 5$. Thus the value of g at a number is the same as the value of f, 5 units to the left of the number (see Figure 3). Therefore the graph of g is just the graph of f shifted 5 units to the right.

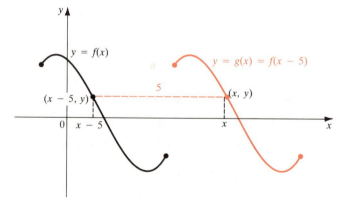

Figure 3

(b)   Similar reasoning shows that the graph of $h(x) = f(x + 8)$ is the graph of $y = f(x)$ shifted 8 units to the left (see Figure 4).

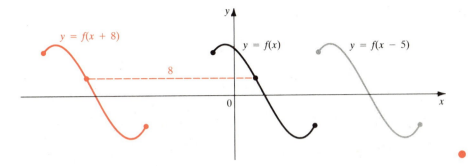

Figure 4

Horizontal Shifts of Graphs

Let $c > 0$. To obtain the graph of

$y = f(x - c)$, move the graph of $y = f(x)$ c units to the right
$y = f(x + c)$, move the graph of $y = f(x)$ c units to the left

Example 3

Sketch the graph of the function $f(x) = (x + 4)^2$.

Solution

We start with the graph of $y = x^2$ and move it 4 units to the left as in Figure 5.

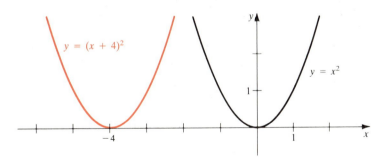

Figure 5

Example 4

Sketch the graph of the function $y = \sqrt{x - 3} + 4$.

Solution

We take the graph of the square root function $y = \sqrt{x}$ from Example 6 in Section 2.2 and move it 3 units to the right to get the graph of $y = \sqrt{x - 3}$.

Then we move this graph 4 units up to obtain the graph of $y = \sqrt{x - 3} + 4$ shown in Figure 6.

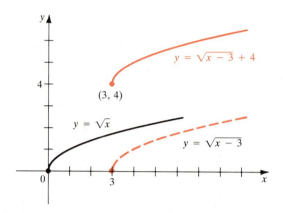

Figure 6

Vertical and horizontal shifts are illustrated in Figure 7.

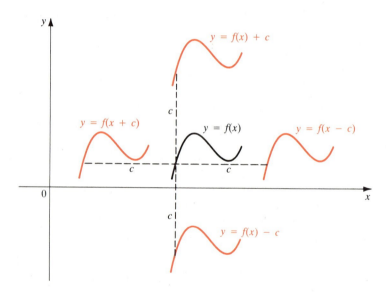

Figure 7

Example 5

Sketch the graph of (a) $y = 3x^2$ and (b) $y = \frac{1}{3}x^2$.

Solution

(a)   If we start with the parabola $y = x^2$ and multiply the y-coordinate of each point by 3, we get the curve $y = 3x^2$ shown in Figure 8(a). It is a narrower parabola than $y = x^2$ and is obtained by stretching vertically by a factor of 3.

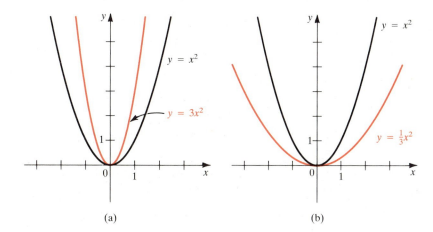

Figure 8

(a)   (b)

(b)   Here we multiply the y-coordinate of each point on $y = x^2$ by $\frac{1}{3}$ to get the parabola $y = \frac{1}{3}x^2$ shown in Figure 8(b). This wider parabola is obtained by shrinking the original parabola in the vertical direction.   ●

Example 6

Sketch the graph of $y = -x^2$.

Solution

We start with the parabola $y = x^2$ and multiply the y-coordinate of every point by -1. The point (x, y) is replaced by the point $(x, -y)$ and so the original graph is reflected in the x-axis (see Figure 9).   ●

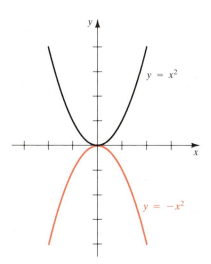

Figure 9

We summarize the effect of multiplying a function f by a constant as follows.

Vertical Stretching, Shrinking, and Reflecting

Let $a > 1$. To obtain the graph of

$y = af(x)$, stretch the graph of $y = f(x)$ vertically by a factor of a

$y = \dfrac{1}{a}f(x)$, shrink the graph of $y = f(x)$ vertically by a factor of a

$y = -f(x)$, reflect the graph of f in the x-axis

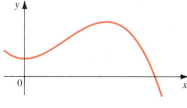

Figure 10

Example 7

Given the graph of f in Figure 10, draw the graphs of the following functions:

(a) $y = 2f(x)$ (b) $y = \frac{1}{2}f(x)$ (c) $y = -f(x)$

(d) $y = -2f(x)$ (e) $y = -\frac{1}{2}f(x)$

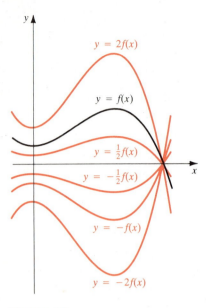

Figure 11

Solution

The graphs are shown in Figure 11. Notice, for instance, that the graph of $y = -2f(x)$ is obtained by stretching by a factor of 2 and reflecting in the x-axis.　●

We illustrate the effect of combining shifts, reflections, and stretching in the following example.

Example 8

Sketch the graph of the function $f(x) = 1 - 2(x - 3)^2$.

Solution

Starting with the graph of $y = x^2$, we first shift 3 units to the right to get the graph of $y = (x - 3)^2$. Then we reflect in the x-axis and stretch by a factor of 2 to get the graph of $y = -2(x - 3)^2$. Finally we shift up 1 unit to get the graph of $y = 1 - 2(x - 3)^2$ shown in Figure 12.

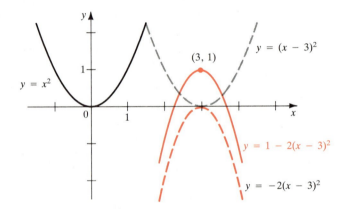

Figure 12

Exercises 2.3

In Exercises 1–11 suppose that the graph of f is given. Describe how the graphs of the following functions can be obtained from the graph of f.

1. $y = f(x) - 10$

2. $y = f(x - 10)$

3. $y = f(x + 1)$

4. $y = f(x) + 1$

5. $y = 8f(x)$

6. $y = -f(x)$

7. $y = -6f(x)$

8. $y = -\frac{1}{5}f(x)$

9. $y = f(x - 2) - 3$

10. $y = -2f(x - 3)$

11. $y = \frac{1}{2}f(x) + 9$

12. The graph of f is given. Draw the graphs of the following functions:

(a) $y = f(x + 4)$　　**(b)** $y = f(x) + 4$
(c) $y = 2f(x)$　　**(d)** $y = -\frac{1}{2}f(x) + 3$

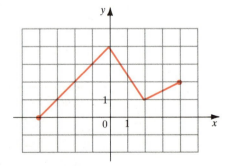

13. (a) Draw the graph of $f(x) = 1/x$ by plotting points.
   (b) Use the graph of f in part (a) to draw the graphs of the following functions:

   (i) $y = -\dfrac{1}{x}$ **(ii)** $y = \dfrac{1}{x-1}$

   (iii) $y = \dfrac{2}{x+2}$ **(iv)** $y = 1 + \dfrac{1}{x-3}$

14. (a) Draw the graph of $g(x) = \sqrt[3]{x}$ by plotting points.
   (b) Use the graph of f in part (a) to draw the graphs of the following functions:

   (i) $y = \sqrt[3]{x} - 2$ **(ii)** $y = \sqrt[3]{x+2} + 2$

   (iii) $y = 1 - \sqrt[3]{x}$

Sketch the graphs of the functions in Exercises 15–30, not by plotting points, but by starting with the graphs of standard functions and applying transformations.

15. $f(x) = (x-2)^2$ **16.** $f(x) = (x+7)^2$

17. $f(x) = -(x+1)^2$ **18.** $f(x) = 1 - x^2$

19. $f(x) = x^3 + 2$ **20.** $f(x) = -x^3$

21. $y = 1 + \sqrt{x}$ **22.** $y = 2 - \sqrt{x+1}$

23. $y = \frac{1}{2}\sqrt{x+4} - 3$ **24.** $y = 3 - 2(x-1)^2$

25. $y = 5 + (x+3)^2$ **26.** $y = \frac{1}{3}x^3 - 1$

27. $y = |x| - 1$ **28.** $y = |x - 1|$

29. $y = |x+2| + 2$ **30.** $y = 2 - |x|$

31. (a) The graph of f is given. Use it to graph the following functions:

   (i) $y = f(2x)$ **(ii)** $y = f(\frac{1}{2}x)$
   (iii) $y = f(-x)$ **(iv)** $y = -f(-x)$

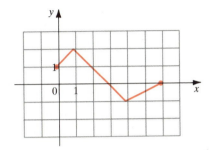

   (b) If $a > 1$, how are the graphs of the following functions obtained from the graph of f?
   (i) $y = f(ax)$ **(ii)** $y = f(x/a)$
   (iii) $y = f(-x)$

32. Use the graph of the greatest integer function (see Exercise 65 in Section 2.2) to draw the graphs of the following functions:

   (a) $y = [\![x + 2]\!]$ **(b)** $y = [\![x - 3.5]\!]$
   (c) $y = 2[\![x]\!]$ **(d)** $y = [\![2x]\!]$

Section 2.4

Quadratic Functions and Their Extreme Values

A **quadratic function** is a function f of the form

$$f(x) = ax^2 + bx + c$$

where a, b, and c are real numbers and $a \neq 0$.

In particular, if we take $a = 1$ and $b = c = 0$, we get the simple quadratic function $f(x) = x^2$ whose graph is the standard parabola that we drew in Example 2 of Section 2.2. In fact, the graphs of all quadratic functions have a similar shape and we will see that they can all be obtained from the graph of $y = x^2$ by the transformations given in Section 2.3.

Example 1

Sketch the graph of the quadratic function $f(x) = -2x^2 + 3$.

Solution

As in Section 2.3, we start with the graph of $y = x^2$. We stretch vertically by a factor of 2 to get the narrower parabola $y = 2x^2$. Then we reflect in the x-axis to get the graph of $y = -2x^2$. Finally we shift this graph up 3 units to obtain the graph of $f(x) = -2x^2 + 3$ shown in Figure 1.

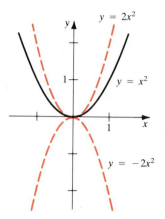 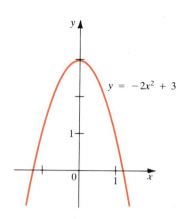

Figure 1

Steps in graphing $y = -2x^2 + 3$

If $f(x) = ax^2 + bx + c$, where $b \neq 0$, we first have to complete the square to put the function in a more convenient form for graphing.

Example 2

Sketch the graph of $y = x^2 + 4x$.

Solution

We complete the square by adding and subtracting the square of half the coefficient of x:

$$y = x^2 + 4x$$
$$= (x^2 + 4x + 4) - 4$$
$$= (x + 2)^2 - 4$$

In this form we see that the given graph is obtained by moving the graph of $y = x^2$ two units to the left and four units down. The point $(-2, -4)$ is called the *vertex* of the parabola. For greater accuracy in graphing we could find the x-intercepts by solving the equation $x^2 + 4x = 0$. This gives $x(x + 4) = 0$, so the x-intercepts are 0 and -4. The graph is sketched in Figure 2.

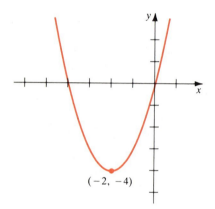

Figure 2

$y = x^2 + 4x$

Example 3

Graph the function $f(x) = -2x^2 + 4x - 5$.

Solution

Since the coefficient of x^2 is not 1, we must factor it from the terms involving x before we complete the square:

$$f(x) = -2x^2 + 4x - 5$$
$$= -2(x^2 - 2x) - 5$$
$$= -2(x^2 - 2x + 1) - 5 + 2$$
$$= -2(x - 1)^2 - 3$$

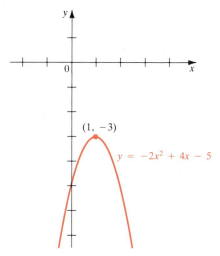

(1, −3)

$y = -2x^2 + 4x - 5$

Figure 3

(Notice that we had to add 2 because the 1 in the parentheses was multiplied by −2.) This form of the function tells us that we get the graph of f by taking the parabola $y = x^2$, shifting it 1 unit to the right, stretching it by a factor of 2, reflecting in the x-axis, and moving it 3 units down. Notice that the vertex is at $(1, -3)$ and the parabola opens downward. We sketch the graph in Figure 3 after noting that the y-intercept is $f(0) = -5$.

From the graph we see that there is no x-intercept. Another way to see this is to compute the discriminant:

$$\Delta = b^2 - 4ac = 4^2 - 4(-2)(-5) = -24$$

From Section 1.4 we know that a negative discriminant means that there is no real solution of the equation $-2x^2 + 4x - 5 = 0$, so f has no x-intercept. ●

If we start with a general quadratic function $f(x) = ax^2 + bx + c$ and complete the square as in Examples 2 and 3, we will arrive at an expression in the standard form:

$$y = a(x - h)^2 + k$$

We know from Section 2.3 that the graph of this function is obtained from the graph of $y = x^2$ by a horizontal shift, a stretch, and a vertical shift. The **vertex** of the resulting parabola is the point (h, k). If $a > 0$, the parabola opens upward and is as shown in Figure 4. But if $a < 0$, then a reflection in the x-axis is also involved, so the parabola opens downward as in Figure 5.

Observe from Figure 4 that if $a > 0$, then the lowest point on the parabola is the vertex (h, k), so the minimum value of the function occurs when $x = h$ and this **minimum value** is $f(h) = k$. Even without the picture we could note that $(x - h)^2 \geq 0$ for all x, so $a(x - h)^2 \geq 0$ (since $a > 0$) and therefore

$$f(x) = a(x - h)^2 + k \geq k \qquad \text{for all } x$$

and $f(h) = k$. Similarly, if $a < 0$, then the highest point on the parabola is (h, k), so the **maximum value** of f is $f(h) = k$.

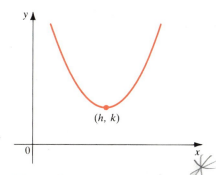

(h, k)

Figure 4
$y = a(x - h)^2 + k, a > 0, h > 0,$
$k > 0$

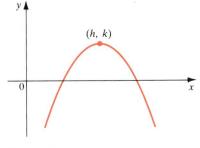

(h, k)

Figure 5
$y = a(x - h)^2 + k, a < 0, h > 0,$
$k > 0$

Example 4

Sketch the graph of the function $f(x) = -x^2 + x + 1$ and find its maximum value and intercepts.

Solution

First we complete the square:

$$\begin{aligned}
y &= -x^2 + x + 1 \\
&= -(x^2 - x) + 1 \\
&= -(x^2 - x + \tfrac{1}{4}) + 1 + \tfrac{1}{4} \\
&= -(x - \tfrac{1}{2})^2 + \tfrac{5}{4}
\end{aligned}$$

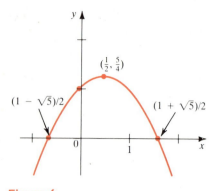

Figure 6
$f(x) = -x^2 + x + 1$

From this standard form we see that the graph is a parabola that opens downward and has vertex $(\frac{1}{2}, \frac{5}{4})$. The maximum value occurs at the vertex and is

$$f(\tfrac{1}{2}) = \tfrac{5}{4}$$

The y-intercept is $f(0) = 1$. To find the x-intercepts we use the quadratic formula to solve the equation $-x^2 + x + 1 = 0$:

$$x = \frac{-1 \pm \sqrt{1^2 - 4(-1)(1)}}{-2} = \frac{1 \pm \sqrt{5}}{2}$$

The graph of f is sketched in Figure 6.   ●

Example 5

Find the minimum value of the function $f(t) = 5t^2 - 30t + 49$.

Solution

Completing the square, we have

$$
\begin{aligned}
f(t) &= 5t^2 - 30t + 49 \\
&= 5(t^2 - 6t) + 49 \\
&= 5(t^2 - 6t + 9) + 49 - 45 \\
&= 5(t - 3)^2 + 4
\end{aligned}
$$

The minimum value of f occurs at the vertex of the graph and is

$$f(3) = 4$$ ●

In solving applied maximum and minimum problems, our initial task is to find an expression for an appropriate quadratic function as in Section 2.1.

Example 6

Among all pairs of numbers whose sum is 100, find a pair whose product is as large as possible.

Solution

Let the numbers be x and y. We know that $x + y = 100$, so $y = 100 - x$. Thus their product is

$$P = xy = x(100 - x)$$

and so we must find the maximum value of the quadratic function

$$P(x) = 100x - x^2$$

We do this by completing the square:

$$
\begin{aligned}
P(x) &= -x^2 + 100x \\
&= -(x^2 - 100x) \\
&= -(x^2 - 100x + 2500) + 2500 \\
&= -(x - 50)^2 + 2500
\end{aligned}
$$

From this standard form we see that the maximum value is 2500 and it occurs when $x = 50$. Then $y = 100 - 50 = 50$, so the two numbers are 50 and 50.

●

Example 7

A farmer wants to enclose a rectangular field by a fence and divide it into two smaller rectangular fields by a fence parallel to one side of the field. He has 3000 yd of fencing. Find the dimensions of the field so that the total enclosed area is a maximum.

Solution

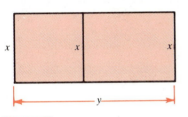

Figure 7

We draw a diagram as in Figure 7 and let x and y be the dimensions of the field (in yards). Then the area of the field is $A = xy$, but we need to eliminate y by using the fact that the total length of the fencing is 3000. Therefore

$$x + x + x + y + y = 3000$$
$$3x + 2y = 3000$$
$$2y = 3000 - 3x$$
$$y = 1500 - \tfrac{3}{2}x$$

This enables us to express A in terms of x alone:

$$A = x(1500 - \tfrac{3}{2}x) = 1500x - \tfrac{3}{2}x^2$$

If we now complete the square, we get

$$A = -\tfrac{3}{2}x^2 + 1500x$$
$$= -\tfrac{3}{2}(x^2 - 1000x)$$
$$= -\tfrac{3}{2}(x^2 - 1000x + 250,000) + 375,000$$
$$= -\tfrac{3}{2}(x - 500)^2 + 375,000$$

This expression shows that the maximum occurs when $x = 500$. Then $y = 1500 - \tfrac{3}{2}(500) = 750$. Thus the field should be 500 yd by 750 yd.   ●

Example 8

A hockey team plays in an arena with a seating capacity of 15,000 spectators. With ticket prices set at \$12, average attendance at a game has been 11,000. A market survey indicates that for each dollar that ticket prices are lowered, the average attendance will increase by 1000. How should the owners of the team set ticket prices so as to maximize their revenue from ticket sales?

Solution

Let x be the selling price of the ticket. Then $12 - x$ is the amount the ticket price has been lowered, so the number of tickets sold is

$$11,000 + 1000(12 - x) = 23,000 - 1000x$$

The revenue is

$$R(x) = x(23{,}000 - 1000x)$$
$$= -1000x^2 + 23{,}000x$$
$$= -1000(x^2 - 23x)$$
$$= -1000(x^2 - 23x + \tfrac{529}{4}) + 250 \cdot 529$$
$$= -1000(x - 11.5)^2 + 132{,}250$$

This shows that the revenue is maximized when $x = 11.5$. The owners should set the ticket price at $11.50.

Exercises 2.4

In Exercises 1–10 sketch the graph of the given parabola and state the coordinates of its vertex and its intercepts.

1. $y = \frac{1}{2}x^2 - 1$

2. $y = 4 - x^2$

3. $y = -3x^2 - 2$

4. $y = x^2 + 6x$

5. $y = x^2 - 5x$

6. $y = x^2 - 2x + 2$

7. $y = -x^2 - 6x - 8$

8. $y = x^2 + 4x - 4$

9. $y = 2x^2 - 20x + 57$

10. $y = -3x^2 + 6x - 2$

In Exercises 11–20 sketch the graph of the given quadratic function and find its maximum or minimum value.

11. $f(x) = 2x - x^2$

12. $f(x) = x + x^2$

13. $f(x) = x^2 + 2x - 1$

14. $f(x) = x^2 - 8x + 8$

15. $f(x) = -x^2 - 3x + 3$

16. $f(x) = 1 - 6x - x^2$

17. $g(x) = 3x^2 - 12x + 13$

18. $g(x) = 2x^2 + 8x + 11$

19. $h(x) = 1 - x - x^2$

20. $h(x) = 3 - 4x - 4x^2$

In Exercises 21–24 find the maximum or minimum value of the given function.

21. $f(x) = x^2 + x + 1$

22. $f(x) = 1 + 3x - x^2$

23. $f(t) = 100 - 50t - 7t^2$

24. $f(t) = 10t^2 + 40t + 113$

25. Find a function whose graph is a parabola with vertex $(1, -2)$ and that passes through the point $(4, 16)$.

26. Find a function whose graph is a parabola with vertex $(3, 4)$ and that passes through the point $(1, -8)$.

27. Find a function whose graph is a parabola and that passes through the points $(1, -1)$, $(-1, -3)$, and $(3, 9)$.

28. Find the domain and range of the function $f(x) = x^2 - 2x - 3$.

29. If a ball is thrown directly upward into the air with a velocity of 40 ft/s, its height in feet after t seconds is given by $y = 40t - 16t^2$. What is the maximum height attained by the ball?

30. The effectiveness of a television commercial depends on how many times a viewer watches it. After some experiments, an advertising agency found that if the effectiveness E is measured on a scale of 0 to 10, then

$$E(n) = \tfrac{2}{3}n - \tfrac{1}{90}n^2$$

where n is the number of times a viewer watches a given commercial. For a commercial to have maximum effectiveness, how many times should a viewer watch it?

31. Find two numbers whose difference is 100 and whose product is as small as possible.

32. Find two positive numbers whose sum is 100 and the sum of whose squares is a minimum.

33. Among all rectangles that have a perimeter of 20 ft, find the dimensions of the one with the largest area.

34. A farmer wants to enclose a rectangular pen so that it has an area of 100 m². Find the dimensions of the pen that will require the minimum amount of fencing to enclose.

35. A farmer has 2400 ft of fencing and wants to fence off a rectangular field that borders a straight river. He needs no fence along the river. (See the figure for Exercise 59 in Section 2.1.) What are the dimensions of the field that has the largest area?

36. Find the area of the largest rectangle that can be inscribed in a right triangle with legs of lengths 3 cm and

4 cm if two sides of the rectangle lie along the legs as in the figure.

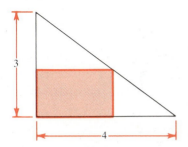

37. A farmer with 750 ft of fencing wants to enclose a rectangular area and then divide it into four pens with fencing parallel to one side of the rectangle. What is the largest possible total area of the four pens?

38. A student makes and sells necklaces on the beach during the summer months. The material for each necklace costs her $6 and she has been selling about 20 per day at $10 each. She has been wondering whether or not to raise her prices, so she takes a survey and finds that for every dollar increase she would lose two sales a day. What should her selling price be in order to maximize profits?

39. Show, by completing the square, that the vertex of the parabola $y = ax^2 + bx + c$ is

$$\left(\frac{-b}{2a}, \frac{4ac - b^2}{4a} \right)$$

40. Find the maximum value of the function $f(x) = 3 + 4x^2 - x^4$. (*Hint:* Let $t = x^2$.)

Section 2.5

Combining Functions

Two functions f and g can be combined to form new functions $f + g, f - g, fg$, and f/g in a manner similar to the way we add, subtract, multiply, and divide real numbers.

If we define the sum $f + g$ by the equation

(1) $(f + g)(x) = f(x) + g(x)$

then the right side of Equation 1 makes sense if both $f(x)$ and $g(x)$ are defined— that is, if x belongs to the domain of f and also to the domain of g. If the domain of f is A and the domain of g is B, then the domain of $f + g$ is the intersection of these domains—that is, $A \cap B$.

Notice that the $+$ sign on the left side of Equation 1 stands for the operation of addition of *functions*, but the $+$ sign on the right side of the equation stands for addition of the *numbers* $f(x)$ and $g(x)$.

Similarly, we can define the difference $f - g$ and the product fg, and their domains will also be $A \cap B$. But in defining the quotient f/g, we must remember not to divide by 0.

Algebra of Functions

Let f and g be functions with domains A and B. Then the functions $f + g, f - g, fg$, and f/g are defined as follows:

$$(f + g)(x) = f(x) + g(x) \qquad \text{domain} = A \cap B$$
$$(f - g)(x) = f(x) - g(x) \qquad \text{domain} = A \cap B$$
$$(fg)(x) = f(x)g(x) \qquad \text{domain} = A \cap B$$
$$\left(\frac{f}{g} \right)(x) = \frac{f(x)}{g(x)} \qquad \text{domain} = \{x \in A \cap B \,|\, g(x) \neq 0\}$$

Example 1

If $f(x) = \sqrt{x}$ and $g(x) = \sqrt{4 - x^2}$, find the functions $f + g, f - g, fg$, and f/g.

Solution

The domain of $f(x) = \sqrt{x}$ is $[0, \infty)$. The domain of $g(x) = \sqrt{4 - x^2}$ consists of all numbers x such that $4 - x^2 \geq 0$; that is, $x^2 \leq 4$. Taking square roots of both sides, we get $|x| \leq 2$, or $-2 \leq x \leq 2$, so the domain of g is the interval $[-2, 2]$. The intersection of the domains of f and g is

$$[0, \infty) \cap [-2, 2] = [0, 2]$$

Thus we have

$$(f + g)(x) = \sqrt{x} + \sqrt{4 - x^2} \qquad\qquad 0 \leq x \leq 2$$

$$(f - g)(x) = \sqrt{x} - \sqrt{4 - x^2} \qquad\qquad 0 \leq x \leq 2$$

$$(fg)(x) = \sqrt{x}\sqrt{4 - x^2} = \sqrt{4x - x^3} \qquad 0 \leq x \leq 2$$

$$\left(\frac{f}{g}\right)(x) = \frac{\sqrt{x}}{\sqrt{4 - x^2}} = \sqrt{\frac{x}{4 - x^2}} \qquad 0 \leq x < 2$$

Notice that the domain of f/g is the interval $[0, 2)$ because we must exclude the points where $g(x) = 0$; that is, $x = \pm 2$. ●

The graph of the function $f + g$ is obtained from the graphs of f and g by **graphical addition**. This means that we add corresponding y-coordinates as in Figure 1. Figure 2 shows the result of using this procedure to graph the function $f + g$ from Example 1.

Composition of Functions

There is another way of combining two functions to get a new function. For example, suppose that $y = f(u) = \sqrt{u}$ and $u = g(x) = x^2 + 1$. Since y is a function of u and u is, in turn, a function of x, it follows that y is ultimately a function of x. We compute this by substitution:

$$y = f(u) = f(g(x)) = f(x^2 + 1) = \sqrt{x^2 + 1}$$

The procedure is called *composition* because the new function is composed of the two given functions f and g.

In general, given any two functions f and g, we start with a number x in the domain of g and find its image $g(x)$. If this number $g(x)$ is in the domain of f, then we can calculate the value of $f(g(x))$. The result is a new function $h(x) = f(g(x))$ obtained by substituting g into f. It is called the *composition* (or *composite*) of f and g and is denoted by $f \circ g$ ("f circle g").

Given two functions f and g, the **composite function** $f \circ g$ (also called the **composition** of f and g) is defined by

$$(f \circ g)(x) = f(g(x))$$

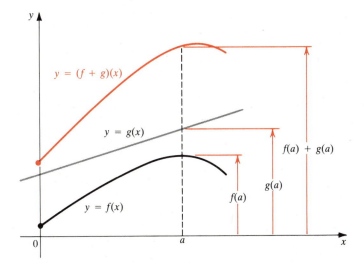

Figure 1

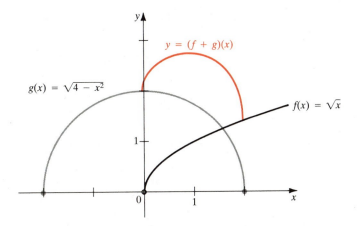

Figure 2

The domain of $f \circ g$ is the set of all x in the domain of g such that $g(x)$ is in the domain of f. In other words, $(f \circ g)(x)$ is defined whenever both $g(x)$ and $f(g(x))$ are defined. The best way to picture $f \circ g$ is by a machine diagram (Figure 3) or an arrow diagram (Figure 4).

Figure 3
The $f \circ g$ machine is composed of the g machine (first) and then the f machine

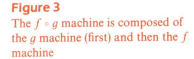

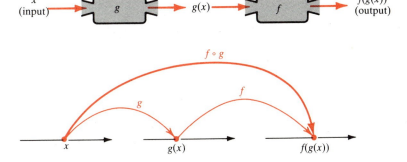

Figure 4
Arrow diagram for $f \circ g$

Example 2

If $f(x) = x^2$ and $g(x) = x - 3$, find the composite functions $f \circ g$ and $g \circ f$ and their domains.

Solution

We have

$$(f \circ g)(x) = f(g(x)) \qquad \text{(definition of } f \circ g)$$
$$= f(x - 3) \qquad \text{(definition of } g)$$
$$= (x - 3)^2 \qquad \text{(definition of } f)$$

and

$$(g \circ f)(x) = g(f(x)) \qquad \text{(definition of } g \circ f)$$
$$= g(x^2) \qquad \text{(definition of } f)$$
$$= x^2 - 3 \qquad \text{(definition of } g)$$

The domains of both $f \circ g$ and $g \circ f$ are R (the set of all real numbers). ●

   Note: You can see from Example 2 that, in general, $f \circ g \neq g \circ f$. Remember that the notation $f \circ g$ means that the function g is applied first and then f is applied second. In Example 2, $f \circ g$ is the function that *first* subtracts 3 and *then* squares; $g \circ f$ is the function that *first* squares and *then* subtracts 3.

Example 3

If $f(x) = \sqrt{x}$ and $g(x) = \sqrt{2 - x}$, find the following functions and their domains: (a) $f \circ g$, (b) $g \circ f$, (c) $f \circ f$, and (d) $g \circ g$.

Solution

(a)
$$(f \circ g)(x) = f(g(x))$$
$$= f(\sqrt{2 - x})$$
$$= \sqrt{\sqrt{2 - x}}$$
$$= \sqrt[4]{2 - x}$$

The domain of $f \circ g$ is

$$\{x \mid 2 - x \geq 0\} = \{x \mid x \leq 2\} = (-\infty, 2]$$

(b)
$$(g \circ f)(x) = g(f(x))$$
$$= g(\sqrt{x})$$
$$= \sqrt{2 - \sqrt{x}}$$

For $\sqrt{x}$ to be defined we must have $x \geq 0$. For $\sqrt{2 - \sqrt{x}}$ to be defined we must have $2 - \sqrt{x} \geq 0$; that is, $\sqrt{x} \leq 2$ or $x \leq 4$. Thus we have $0 \leq x \leq 4$, so the domain of $g \circ f$ is the closed interval $[0, 4]$.

(c)
$$(f \circ f)(x) = f(f(x))$$
$$= f(\sqrt{x})$$
$$= \sqrt{\sqrt{x}}$$
$$= \sqrt[4]{x}$$

The domain of $f \circ f$ is $[0, \infty)$.

(d)
$$(g \circ g)(x) = g(g(x))$$
$$= g(\sqrt{2 - x})$$
$$= \sqrt{2 - \sqrt{2 - x}}$$

This expression is defined when $2 - x \geq 0$, that is, $x \leq 2$, and $2 - \sqrt{2 - x} \geq 0$. This latter inequality is equivalent to $\sqrt{2 - x} \leq 2$, or $2 - x \leq 4$; that is, $x \geq -2$. Thus $-2 \leq x \leq 2$, so the domain of $g \circ g$ is $[-2, 2]$. ●

It is possible to take the composition of three or more functions. For instance, the composite function $f \circ g \circ h$ is found by first applying h, then g, and then f, as follows:

$$(f \circ g \circ h)(x) = f(g(h(x)))$$

Example 4

Find $f \circ g \circ h$ if $f(x) = x/(x + 1)$, $g(x) = x^{10}$, and $h(x) = x + 3$.

Solution

$$(f \circ g \circ h)(x) = f(g(h(x)))$$
$$= f(g(x + 3))$$
$$= f((x + 3)^{10})$$
$$= \frac{(x + 3)^{10}}{(x + 3)^{10} + 1}$$ ●

So far we have used composition to build up complicated functions from simpler ones. But in calculus it is useful to be able to decompose a complicated function into simpler ones, as in the following example.

Example 5

Given $F(x) = \sqrt[4]{x + 9}$, find functions f and g such that $F = f \circ g$.

Solution

Since the formula for F says to first add 9 and then take the fourth root, we let

$$g(x) = x + 9 \quad \text{and} \quad f(x) = \sqrt[4]{x}$$

Then

$$(f \circ g)(x) = f(g(x))$$
$$= f(x + 9)$$
$$= \sqrt[4]{x + 9}$$
$$= F(x) \qquad \bullet$$

Exercises 2.5

In Exercises 1–6 find $f + g$, $f - g$, fg, and f/g and their domains.

1. $f(x) = x^2 - x$, $g(x) = x + 5$

2. $f(x) = x^3 + 2x^2$, $g(x) = 3x^2 - 1$

3. $f(x) = \sqrt{1 + x}$, $g(x) = \sqrt{1 - x}$

4. $f(x) = \sqrt{9 - x^2}$, $g(x) = \sqrt{x^2 - 1}$

5. $f(x) = \sqrt{x}$, $g(x) = \sqrt[3]{x}$

6. $f(x) = \sqrt[4]{x + 1}$, $g(x) = \sqrt{x + 2}$

In Exercises 7 and 8 find the domain of the given function.

7. $F(x) = \dfrac{\sqrt{4 - x} + \sqrt{3 + x}}{x^2 - 2}$

8. $F(x) = \sqrt{1 - x} + \sqrt{x - 2}$

In Exercises 9–12 use the graphs of f and g and the method of graphical addition to sketch the graph of $f + g$.

9. $f(x) = x^3$, $g(x) = x$

10. $f(x) = x^2$, $g(x) = 1 - x$

11. $f(x) = x$, $g(x) = \dfrac{1}{x}$

12. $f(x) = x^3$, $g(x) = -x^2$

In Exercises 13–24 use $f(x) = 3x - 5$ and $g(x) = 2 - x^2$ to evaluate the expression.

13. $f(g(0))$

14. $g(f(1))$

15. $f(f(4))$

16. $g(g(3))$

17. $(f \circ g)(-2)$

18. $(g \circ f)(-2)$

19. $(f \circ f)(-1)$

20. $(g \circ g)(2)$

21. $(f \circ g)(x)$

22. $(g \circ f)(x)$

23. $(f \circ f)(x)$

24. $(g \circ g)(x)$

In Exercises 25–30 use the given graphs of f and g to evaluate the expression.

25. $f(g(2))$

26. $g(f(0))$

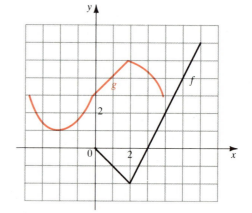

27. $(g \circ f)(4)$

28. $(f \circ g)(0)$

29. $(g \circ g)(-2)$

30. $(f \circ f)(4)$

In Exercises 31–40 find the functions $f \circ g$, $g \circ f$, $f \circ f$, and $g \circ g$, and their domains.

31. $f(x) = 2x + 3$, $g(x) = 4x - 1$

32. $f(x) = 6x - 5$, $g(x) = \dfrac{x}{2}$

33. $f(x) = 2x^2 - x$, $g(x) = 3x + 2$

34. $f(x) = \sqrt{x - 1}$, $g(x) = x^2$

35. $f(x) = \dfrac{1}{x}$, $g(x) = x^3 + 2x$

36. $f(x) = \dfrac{1}{x - 1}$, $g(x) = \dfrac{x - 1}{x + 1}$

37. $f(x) = \sqrt[3]{x}$, $g(x) = 1 - \sqrt{x}$

38. $f(x) = \sqrt{x^2 - 1}$, $g(x) = \sqrt{1 - x}$

39. $f(x) = \dfrac{x + 2}{2x + 1}$, $g(x) = \dfrac{x}{x - 2}$

40. $f(x) = \dfrac{1}{\sqrt{x}}$, $g(x) = x^2 - 4x$

In Exercises 41–44 find $f \circ g \circ h$.

41. $f(x) = x - 1$, $g(x) = \sqrt{x}$, $h(x) = x - 1$

42. $f(x) = \dfrac{1}{x}$, $g(x) = x^3$, $h(x) = x^2 + 2$

43. $f(x) = x^4 + 1$, $g(x) = x - 5$, $h(x) = \sqrt{x}$

44. $f(x) = \sqrt{x}$, $g(x) = \dfrac{x}{x - 1}$, $h(x) = \sqrt[3]{x}$

In Exercises 45–50 express the given function in the form $f \circ g$.

45. $F(x) = (x - 9)^5$

46. $F(x) = \sqrt{x} + 1$

47. $G(x) = \dfrac{x^2}{x^2 + 4}$

48. $G(x) = \dfrac{1}{x + 3}$

49. $H(x) = |1 - x^3|$

50. $H(x) = \sqrt{1 + \sqrt{x}}$

In Exercises 51–54 express the given function in the form $f \circ g \circ h$.

51. $F(x) = \dfrac{1}{x^2 + 1}$

52. $F(x) = \sqrt[3]{\sqrt{x} - 1}$

53. $G(x) = (4 + \sqrt[3]{x})^9$

54. $G(x) = \dfrac{2}{(3 + \sqrt{x})^2}$

55. A stone is dropped into a lake, creating a circular ripple that travels outward at a speed of 60 cm/s. Express the area of this circle as a function of time t (in seconds).

56. A spherical balloon is being inflated. If the radius of the balloon is increasing at a rate of 1 cm/s, express the volume of the balloon as a function of time t (in seconds).

57. If $f(x) = 3x + 5$ and $h(x) = 3x^2 + 3x + 2$, find a function g such that $f \circ g = h$.

58. If $f(x) = x + 4$ and $h(x) = 4x - 1$, find a function g such that $g \circ f = h$.

One-to-One Functions and Their Inverses

Let us compare the functions f and g whose arrow diagrams are shown in Figure 1. Note that f never takes on the same value twice (any two numbers in A have different images), whereas g does take on the same value twice (both 2 and 3 have the same image, 4). In symbols,

$$g(2) = g(3)$$

but $$f(x_1) \neq f(x_2) \qquad \text{whenever } x_1 \neq x_2$$

Functions that have this latter property are called *one-to-one*.

> A function with domain A is called a **one-to-one function** if no two elements of A have the same image; that is,
>
> $$f(x_1) \neq f(x_2) \qquad \text{whenever } x_1 \neq x_2$$

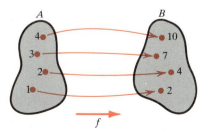
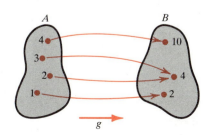

Figure 1

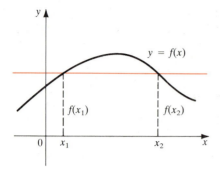

Figure 2
This function is not one-to-one
because $f(x_1) = f(x_2)$

If a horizontal line intersects the graph of f in more than one point, then we see from Figure 2 that there are numbers $x_1 \neq x_2$ such that $f(x_1) = f(x_2)$. This means that f is not one-to-one. Therefore we have the following geometric method for determining whether or not a function is one-to-one.

Horizontal Line Test

A function is one-to-one if and only if no horizontal line intersects its graph more than once.

Example 1

Is the function $f(x) = x^3$ one-to-one?

Solution 1

If $x_1 \neq x_2$, then $x_1^3 \neq x_2^3$ (two different numbers cannot have the same cube). Therefore $f(x) = x^3$ is one-to-one.

Solution 2

From Figure 3 we see that no horizontal line intersects the graph of $f(x) = x^3$ more than once. Therefore, by the Horizontal Line Test, f is one-to-one. ●

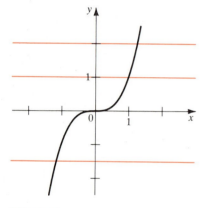

Figure 3
$f(x) = x^3$ is one-to-one

Example 2

Is the function $g(x) = x^2$ one-to-one?

Solution 1

This function is not one-to-one because, for instance,

$$g(1) = 1 = g(-1)$$

and so 1 and -1 have the same image.

Solution 2

From Figure 4 we see that there are horizontal lines that intersect the graph of g more than once. Therefore, by the Horizontal Line Test, g is not one-to-one.

●

Notice that the function f of Example 1 is increasing and is also one-to-one. In fact, it can be proved that every increasing function and every decreasing function are one-to-one (see Exercise 35).

One-to-one functions are important because they are precisely the functions that possess inverse functions according to the following definition.

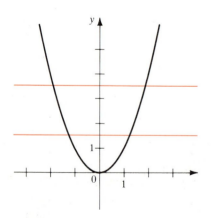

Figure 4
$g(x) = x^2$ is not one-to-one

Let f be a one-to-one function with domain A and range B. Then its **inverse function** f^{-1} has domain B and range A and is defined by

(1) $\qquad\qquad f^{-1}(y) = x \iff f(x) = y$

for any y in B.

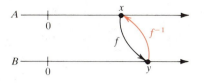

Figure 5

This definition says that if f takes x into y, then f^{-1} takes y back into x. (If f was not one-to-one, then f^{-1} would not be uniquely defined.) The arrow diagram in Figure 5 indicates that f^{-1} reverses the effect of f. Note the following:

$$\text{domain of } f^{-1} = \text{range of } f$$
$$\text{range of } f^{-1} = \text{domain of } f$$

For example, the inverse function of $f(x) = x^3$ is $f^{-1}(x) = x^{1/3}$ because if $y = x^3$, then

$$f^{-1}(y) = f^{-1}(x^3) = (x^3)^{1/3} = x$$

Caution: Do not mistake the -1 in f^{-1} for an exponent. Thus

$$f^{-1}(x) \quad \text{does not mean} \quad \frac{1}{f(x)}$$

The reciprocal $1/f(x)$ could, however, be written as $[f(x)]^{-1}$.

The letter x is traditionally used as the independent variable, so when we concentrate on f^{-1} rather than on f, we usually reverse the roles of x and y in (1) and write

(2)
$$f^{-1}(x) = y \quad \Leftrightarrow \quad f(y) = x$$

By substituting for y in (1) and substituting for x in (2), we get the following equations:

(3)
$$f^{-1}(f(x)) = x \text{ for every } x \text{ in } A$$
$$f(f^{-1}(x)) = x \text{ for every } x \text{ in } B$$

The first equation says that if we start with x, apply f, and then apply f^{-1}, we arrive back at x, where we started. Thus f^{-1} undoes what f does. The second equation says that f undoes what f^{-1} does.

For example, if $f(x) = x^3$, then $f^{-1}(x) = x^{1/3}$ and the equations in (3) become

$$f^{-1}(f(x)) = (x^3)^{1/3} = x$$
$$f(f^{-1}(x)) = (x^{1/3})^3 = x$$

These equations simply say that the cube function and the cube root function cancel each other out.

Let us now see how to compute inverse functions. If we have a function $y = f(x)$ and are able to solve this equation for x in terms of y, then according to (1) we must have $x = f^{-1}(y)$. If we then interchange x and y, we have $y = f^{-1}(x)$, which is the desired equation.

How to Find the Inverse Function of a One-to-One Function f

1. Write $y = f(x)$.
2. Solve this equation for x in terms of y (if possible).
3. Interchange x and y. The resulting equation is $y = f^{-1}(x)$.

Note that Steps 2 and 3 could be reversed. In other words, it is possible to interchange x and y first and then solve for y in terms of x.

Example 3

Find the inverse function of $f(x) = x^3 + 2$.

Solution

We first write

$$y = x^3 + 2$$

Then we solve this equation for x:

$$x^3 = y - 2$$
$$x = \sqrt[3]{y - 2}$$

Finally we interchange x and y:

$$y = \sqrt[3]{x - 2}$$

Therefore the inverse function is $f^{-1}(x) = \sqrt[3]{x - 2}$.

This formula for f^{-1} seems reasonable if we state the rules for f and f^{-1} in words. The instructions for $f(x) = x^3 + 2$ are "Cube, then add 2," whereas the instructions for $f^{-1}(x) = \sqrt[3]{x - 2}$ are "Subtract 2, then take the cube root." ●

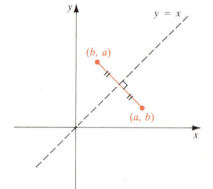

Figure 6

The principle of interchanging x and y to find the inverse function also gives us the method for obtaining the graph of f^{-1} from the graph of f. If $f(a) = b$, then $f^{-1}(b) = a$. Thus the point (a, b) is on the graph of f if and only if the point (b, a) is on the graph of f^{-1}. But we get the point (b, a) from the point (a, b) by reflecting in the line $y = x$ (see Figure 6). Therefore, as illustrated in Figure 7:

The graph of f^{-1} is obtained by reflecting the graph of f in the line $y = x$.

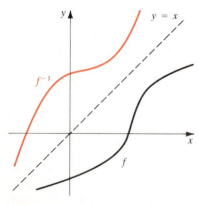

Figure 7

Example 4

(a) Sketch the graph of $f(x) = \sqrt{x - 2}$.

(b) Use the graph of f to sketch the graph of f^{-1}.

(c) Find an equation for f^{-1}.

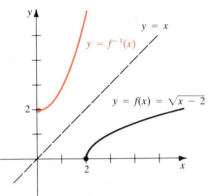

Figure 8

Solution

(a) As in Section 2.3 we sketch the graph of $y = \sqrt{x - 2}$ by taking the graph of the square root function (Example 6 in Section 2.2) and moving it 2 units to the right.

(b) The graph of f^{-1} is obtained by taking the graph of f from part (a) and reflecting it in the line $y = x$ as in Figure 8.

(c) Solve $y = \sqrt{x - 2}$ for x:

$$\sqrt{x - 2} = y \qquad \text{(note that } y \geq 0\text{)}$$
$$x - 2 = y^2$$
$$x = y^2 + 2, \qquad y \geq 0$$

Interchange x and y:

$$y = x^2 + 2, \qquad x \geq 0$$

Thus

$$f^{-1}(x) = x^2 + 2, \qquad x \geq 0$$

This expression shows that the graph of f^{-1} is the right half of the parabola $y = x^2 + 2$ and, checking with Figure 8, that seems reasonable. ●

Exercises 2.6

In Exercises 1–6 the graph of a function f is given. Determine whether or not f is one-to-one.

1.

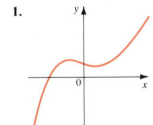

2.

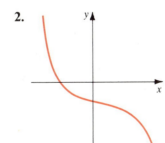

3.

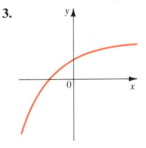

4.

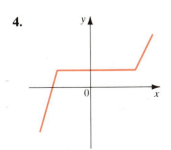

5.

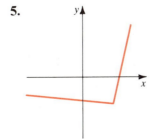

6.

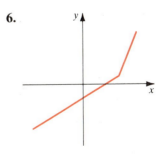

In Exercises 7–12 determine whether or not the given function is one-to-one.

7. $f(x) = 7x - 3$ **8.** $f(x) = x^2 - 2x + 5$

9. $g(x) = \sqrt{x}$ **10.** $g(x) = |x|$

11. $h(x) = x^4 + 5$

12. $h(x) = x^4 + 5, \quad 0 \le x \le 2$

13. If f is one-to-one and $f(2) = 7$, evaluate $f^{-1}(7)$.

14. Find $f^{-1}(4)$ if f is a one-to-one function and $f(1) = 4$.

15. If $f(x) = 5 - 2x$, find $f^{-1}(3)$.

16. If $g(x) = x^2 + 4x, x \ge -2$, find $g^{-1}(5)$.

In Exercises 17–22 show that f is one-to-one and find its inverse function.

17. $f(x) = 4x + 7$ **18.** $f(x) = \dfrac{x - 2}{x + 2}$

19. $f(x) = \dfrac{1 + 3x}{5 - 2x}$ **20.** $f(x) = 5 - 4x^3$

21. $f(x) = \sqrt{2 + 5x}$

22. $f(x) = x^2 + x, \quad x \ge -\frac{1}{2}$

In Exercises 23–28 find the inverse function of f and then verify that f^{-1} and f satisfy the equations in (3).

23. $f(x) = 3 - 5x$

24. $f(x) = \dfrac{1}{x + 2}, \quad x > -2$

25. $f(x) = 4 - x^2, \quad x \ge 0$

26. $f(x) = \sqrt{2x - 1}$

27. $f(x) = 4 + \sqrt[3]{x}$

28. $f(x) = (2 - x^3)^5$

In Exercises 29–34, (a) sketch the graph of f, (b) use the graph of f to sketch the graph of f^{-1}, and (c) find an equation for f^{-1}.

29. $f(x) = 2x + 1$ **30.** $f(x) = 6 - x$

31. $f(x) = 1 + \sqrt{1 + x}$

32. $f(x) = 9 - x^2, \quad 0 \le x \le 3$

33. $f(x) = x^4, \quad x \ge 0$

34. $f(x) = 1 - x^3$

35. Prove that if f is an increasing function on its entire domain, then f is one-to-one.

36. For what values of the number m is the linear function $f(x) = mx + b$ one-to-one? For those values of m, find f^{-1}.

CHAPTER 2 REVIEW

Define, state, or discuss the following.

1. Function
2. Domain of a function
3. Range of a function
4. Independent and dependent variables
5. Arrow diagram
6. Graph of a function
7. Linear function
8. Constant function
9. Vertical Line Test
10. Increasing function
11. Decreasing function
12. Even function
13. Odd function
14. Vertical shifts of graphs
15. Horizontal shifts of graphs
16. Vertical stretching, shrinking, and reflecting
17. Quadratic functions
18. Sum, difference, product, and quotient of functions
19. Composition of functions
20. One-to-one functions
21. Horizontal Line Test
22. Inverse function
23. Procedure for finding an inverse function
24. Graph of an inverse function

Review Exercises

1. If $f(x) = 1 + \sqrt{x - 1}$, find $f(5)$, $f(9)$, $f(a + 1)$, $f(-x)$, $f(x^2)$, and $[f(x)]^2$.

2. The graph of a function is given.
   (a) State the values of $f(-2)$ and $f(2)$.
   (b) State the domain of f.
   (c) State the range of f.
   (d) On what intervals is f increasing? On what intervals is it decreasing?
   (e) Is f one-to-one?

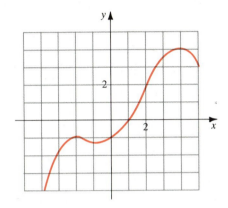

3. Which of the following figures are graphs of functions? Which of the functions are one-to-one?

(a)

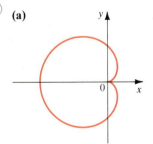

(b)

(c)

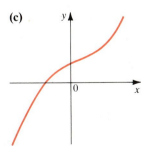

(d)

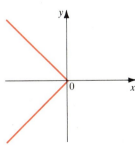

4. Find the domain and range of the function $f(x) = 2 + \sqrt{x + 3}$.

5. Find the domain and range of the function $F(t) = t^2 + 2t + 5$.

In Exercises 6–10 find the domain of the function.

6. $f(x) = \dfrac{2x + 1}{2x - 1}$

7. $f(x) = 3x - \dfrac{2}{\sqrt{x + 1}}$

8. $f(x) = \dfrac{\sqrt[3]{2x + 1}}{\sqrt[3]{2x + 2}}$

9. $g(x) = \dfrac{2x^2 + 5x + 3}{2x^2 - 5x - 3}$

10. $h(x) = \sqrt{4 - x} + \sqrt{x^2 - 1}$

In Exercises 11–28 sketch the graph of the function.

11. $f(x) = 1 - 2x$

12. $f(x) = \frac{1}{3}(x - 5)$, $2 \le x \le 8$

13. $g(x) = \dfrac{1}{x^2}$

14. $G(x) = \dfrac{1}{(x - 3)^2}$

15. $h(x) = \sqrt[3]{x}$

16. $H(x) = x^3 - 3x + 1$

17. $f(t) = 1 - \frac{1}{2}t^2$

18. $g(t) = t^2 - 2t$

19. $f(x) = x^2 - 6x + 6$

20. $f(x) = 3 - 8x - 2x^2$

21. $y = 1 - \sqrt{x}$

22. $y = -|x - 5|$

23. $y = \frac{1}{2}(x + 1)^3$

24. $y = 2 + \sqrt{x + 3}$

25. $f(x) = \begin{cases} 1 - x & \text{if } x < 0 \\ 1 & \text{if } x \ge 0 \end{cases}$

26. $f(x) = \begin{cases} 1 - 2x & \text{if } x \le 0 \\ 2x - 1 & \text{if } x > 0 \end{cases}$

27. $f(x) = \begin{cases} x + 6 & \text{if } x < -2 \\ x^2 & \text{if } x \ge -2 \end{cases}$

28. $f(x) = \begin{cases} -x & \text{if } x < 0 \\ 2x - x^2 & \text{if } 0 \le x < 2 \\ 1 & \text{if } x \ge 2 \end{cases}$

29. Suppose that the graph of f is given. Describe how the graphs of the following functions can be obtained from the graph of f.
   (a) $y = f(x) + 8$
   (b) $y = f(x + 8)$

(c) $y = 1 + 2f(x)$ **(d)** $y = f(x - 2) - 2$
(e) $y = -f(x)$ **(f)** $y = f^{-1}(x)$

30. The graph of f is given. Draw the graphs of the following functions:
   (a) $y = f(x - 8)$ **(b)** $y = -f(x)$
   (c) $y = 2 - f(x)$ **(d)** $y = \frac{1}{2}f(x) - 1$
   (e) $y = f^{-1}(x)$ **(f)** $y = f^{-1}(x + 3)$

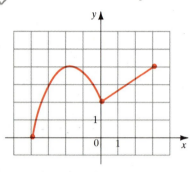

31. Find the maximum value of the function
$f(x) = 1 - x - x^2$.

32. Find the minimum value of the function
$g(x) = 2x^2 + 3x - 5$.

33. Determine whether f is even, odd, or neither even nor odd.
   (a) $f(x) = 2x^5 - 3x^2 + 2$
   (b) $f(x) = x^3 - x^7$
   (c) $f(x) = \dfrac{1 - x^2}{1 + x^2}$ **(d)** $f(x) = \dfrac{1}{x + 2}$

34. If $f(x) = x^2 - 3x + 2$ and $g(x) = 4 - 3x$, find:
   (a) $f + g$ **(b)** $f - g$ **(c)** fg
   (d) $\dfrac{f}{g}$ **(e)** $f \circ g$ **(f)** $g \circ f$

In Exercises 35 and 36 find the functions $f \circ g$, $g \circ f$, $f \circ f$, and $g \circ g$, and their domains.

35. $f(x) = 3x - 1,\ g(x) = 2x - x^2$

36. $f(x) = \sqrt{x},\ g(x) = \dfrac{2}{x - 4}$

37. Find $f \circ g \circ h$, where $f(x) = \sqrt{1 - x}$, $g(x) = 1 - x^2$, and $h(x) = 1 + \sqrt{x}$.

38. If $T(x) = \dfrac{1}{\sqrt{1 + \sqrt{x}}}$, find functions f, g, and h such that $f \circ g \circ h = T$.

In Exercises 39–42 determine whether or not the function is one-to-one.

39. $f(x) = 3 + x^3$ **40.** $g(x) = 2 - 2x + x^2$

41. $h(x) = \dfrac{1}{x^4}$

42. $r(x) = 2 + \sqrt{x + 3}$

43. Show that $f(x) = 3x - 2$ is a one-to-one function and find its inverse function.

44. If $f(x) = 1 + \sqrt[5]{x - 2}$, find f^{-1}.

45. **(a)** Sketch the graph of the function
$f(x) = x^2 - 4,\ x \geq 0$.
   (b) Use part (a) to sketch the graph of f^{-1}.
   (c) Find an equation for f^{-1}.

46. **(a)** Show that the function $f(x) = 1 + \sqrt[4]{x}$ is one-to-one.
   (b) Sketch the graph of f.
   (c) Use part (b) to sketch the graph of f^{-1}.
   (d) Find an equation for f^{-1}.

47. An isosceles triangle has a perimeter of 8 cm. Express the area A of the triangle as a function of the length b of the base of the triangle.

48. A rectangle is inscribed in an equilateral triangle with a perimeter of 30 cm as in the figure.
   (a) Express the area A of the rectangle as a function of the width of the rectangle.
   (b) Find the dimensions of the rectangle with the largest area.

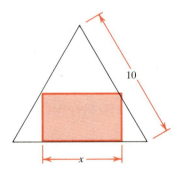

49. A piece of wire 10 m long is cut into two pieces. One piece, of length x, is bent into the shape of a square. The other piece is bent into the shape of an equilateral triangle.
   (a) Express the total area enclosed as a function of x.
   (b) For what value of x is this total area a minimum?

50. A baseball team plays in a stadium that holds 55,000 spectators. With ticket prices at $10, the average attendance has been 27,000. A market survey indicates that for every dollar that ticket prices are lowered, attendance will increase by 3000. How should ticket prices be set to maximize revenue?

Chapter 2 Test

1. State which of the following curves are graphs of functions. State which of the functions is one-to-one.

(a)

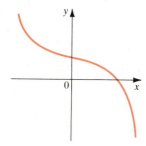

(b)

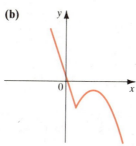

(c)

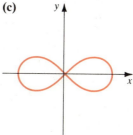

2. Find the domain of the function $f(x) = \dfrac{x^2 - 1}{\sqrt{x - 2}}$.

3. **(a)** Sketch the graph of the function $f(x) = x^3$.
   (b) Use part (a) to graph the function $g(x) = (x + 2)^3 - 3$.

4. **(a)** Sketch the graph of the function $f(x) = x^2 - 8x + 28$.
   (b) What is the minimum value of f?

5. Let $f(x) = \begin{cases} 1 - x^2 & \text{if } x \leq 0 \\ 2x + 1 & \text{if } x > 0 \end{cases}$.
   (a) Evaluate $f(-2)$ and $f(1)$.
   (b) Sketch the graph of f.

6. How is the graph of $y = 2 - f(x - 3)$ obtained from the graph of f?

7. If $f(x) = x^2 + 2x - 1$ and $g(x) = 2x - 3$, find $f \circ g$ and $g \circ f$.

8. **(a)** If $f(x) = \sqrt{3 - x}$, find the inverse function f^{-1}.
   (b) Sketch the graphs of f and f^{-1} on the same coordinate axes.

9. **(a)** If 800 ft of fencing are available to build three adjacent pens as in the diagram, express the total area as a function of x.

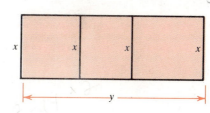

   (b) What value of x will maximize the total area?

Problems Plus

1. Find the domain of the function
$$f(x) = \sqrt{1 - \sqrt{2 - \sqrt{3 - x}}}$$

2. Sketch the graph of the function $f(x) = |x^2 - 4|x| + 3|$.

3. Sketch the graph of the function $g(x) = |x^2 - 1| - |x^2 - 4|$.

4. If $f_0(x) = x^2$ and $f_{n+1}(x) = f_0(f_n(x))$ for $n = 0, 1, 2, \ldots$, find a formula for $f_n(x)$.

5. If $f_0(x) = \dfrac{1}{1 - x}$ and $f_{n+1} = f_0 \circ f_n$ for $n = 0, 1, 2, \ldots$, find $f_{1000}(1000)$.

6. If $f_0(x) = \dfrac{1}{2 - x}$ and $f_{n+1} = f_0 \circ f_n$ for $n = 0, 1, 2, \ldots$, find $f_{100}(3)$.

7. Sketch the region in the plane defined by the equation
$$[\![x]\!]^2 + [\![y]\!]^2 = 1$$
where $[\![x]\!]$ denotes the greatest integer function (see Exercise 65 in Section 2.2).

Polynomials and Rational Functions

Each problem that I solved became a rule which served afterwards to solve other problems.

René Descartes

110

We have previously studied constant, linear, and quadratic functions, which have the equations $f(x) = b$, $f(x) = mx + b$, and $f(x) = ax^2 + bx + c$. All these functions are special cases of an important class of functions called polynomials. A **polynomial P of degree n** is a function of the form

$$P(x) = a_n x^n + a_{n-1} x^{n-1} + \cdots + a_1 x + a_0$$

where $a_n \neq 0$. Polynomials are constructed using the operations of addition, subtraction, and multiplication. If we introduce division as well, we obtain the set of rational functions. A **rational function r** is a function of the form

$$r(x) = \frac{P(x)}{Q(x)}$$

where P and Q are polynomials.

Virtually all the functions used in mathematics and the sciences are evaluated numerically with the use of polynomials, as you will discover when you study infinite series in calculus. In this chapter we study this important class of functions by first learning how to factor polynomials and solve polynomial equations. We then use this knowledge to help us graph polynomial functions. Finally, we study the graphs of rational functions and learn to solve inequalities involving rational functions.

Section 3.1

Dividing Polynomials

Long division for polynomials is very much like the familiar process of long division for numbers. For example, to divide $6x^2 - 26x + 12$ (the **dividend**) by $x - 4$ (the **divisor**), we arrange our work as follows:

$$
\begin{array}{r}
6x \ - \ 2 \\
x - 4 \overline{\smash{\big)}\ 6x^2 - 26x + 12} \\
\underline{6x^2 - 24x} \\
-2x + 12 \\
\underline{-2x + \ 8} \\
4
\end{array}
$$

The **quotient** at the top is obtained by first dividing the initial term of the dividend by the initial term of the divisor $[6x^2/x = 6x]$, which gives the first term of the quotient. This is then multiplied by the divisor $[6x(x - 4) = 6x^2 - 24x]$, and the result is written below the corresponding terms of the dividend and subtracted from those terms. After the next term of the dividend [the $+12$] is brought down, the whole process is repeated, with the new line $[-2x + 12]$ now being treated as the dividend. The process stops if, after the subtraction has been performed, the result has degree *smaller* than the degree of the divisor and there is nothing in the original dividend left to bring down. The last line of the process will contain the **remainder**, and the result of the division can be interpreted in either of the following two ways:

$$\frac{6x^2 - 26x + 12}{x - 4} = 6x - 2 + \frac{4}{x - 4}$$

or

$$6x^2 - 26x + 12 = (x - 4)(6x - 2) + 4$$

We summarize what happens in this or any such long division problem with the following theorem.

The Division Algorithm

If $P(x)$ and $D(x)$ are polynomials, with $D(x) \neq 0$, then there exist unique polynomials $Q(x)$ and $R(x)$ such that

$$P(x) = D(x) \cdot Q(x) + R(x)$$

where $R(x)$ is either 0 or of degree less than the degree of $D(x)$. $P(x)$ and $D(x)$ are called the **dividend** and **divisor**, respectively, $Q(x)$ is the **quotient**, and $R(x)$ is the **remainder**.

Example 1

Let $P(x) = 8x^4 + 6x^2 - 3x + 1$ and $D(x) = 2x^2 - x + 2$. Find polynomials $Q(x)$ and $R(x)$ such that $P(x) = D(x) \cdot Q(x) + R(x)$.

Solution

We use long division after first inserting the term $0x^3$ into the dividend to ensure that the columns will line up correctly in the long division process.

$$
\begin{array}{r}
4x^2 + 2x \\
2x^2 - x + 2 \overline{\smash{)}\,8x^4 + 0x^3 + 6x^2 - 3x + 1} \\
\underline{8x^4 - 4x^3 + 8x^2} \\
4x^3 - 2x^2 - 3x \\
\underline{4x^3 - 2x^2 + 4x} \\
-7x + 1
\end{array}
$$

So $8x^4 + 6x^2 - 3x + 1 = (2x^2 - x + 2)(4x^2 + 2x) + (-7x + 1)$ ●

The Remainder Theorem and the Factor Theorem

If the divisor in the Division Algorithm is of the form $x - c$ for some real number c, then the remainder must be a constant (since the degree of the remainder is less than the degree of the divisor). If we call this constant r, then

$$P(x) = (x - c) \cdot Q(x) + r$$

Setting $x = c$ in this equation, we get $P(c) = (c - c) \cdot Q(x) + r = 0 + r = r$. This proves the next theorem.

Remainder Theorem

If the polynomial $P(x)$ is divided by $x - c$, then the remainder is the value $P(c)$.

Example 2

Let $P(x) = x^3 - 4x^2 + 3x + 5$. If we divide $P(x)$ by $x - 2$ using long division, we obtain a quotient of $x^2 - 2x - 1$ and a remainder of 3. Thus by the Remainder Theorem, the value of $P(2)$ should be 3. To verify this, we calculate

$$P(2) = 2^3 - 4 \cdot 2^2 + 3 \cdot 2 + 5 = 8 - 16 + 6 + 5 = 3 \qquad \bullet$$

If $R(x) = 0$ in the Division Algorithm, then $P(x) = D(x) \cdot Q(x)$ and we say that $D(x)$ and $Q(x)$ are **factors** of $P(x)$. If $D(x) = x - c$ is a factor of $P(x)$, then obviously $P(c) = (c - c) \cdot Q(c) = 0$. On the other hand, if $P(c) = 0$, then by the Remainder Theorem,

$$P(x) = (x - c) \cdot Q(x) + 0 = (x - c) \cdot Q(x)$$

so that $x - c$ is a factor of $P(x)$. Thus we have proved the following theorem.

Factor Theorem

$P(c) = 0$ if and only if $x - c$ is a factor of $P(x)$.

Example 3

Let $P(x) = x^3 - 7x + 6$. Show that $P(1) = 0$, and use this fact to factor $P(x)$ completely.

Solution

Substituting, we see that $P(1) = 1^3 - 7 \cdot 1 + 6 = 0$. By the Factor Theorem, this means that $x - 1$ is a factor of $P(x)$. Using long division, we see that

$$P(x) = (x - 1)(x^2 + x - 6)$$

Since $x^2 + x - 6 = (x - 2)(x + 3)$, we have the complete factorization:

$$P(x) = (x - 1)(x - 2)(x + 3) \qquad \bullet$$

If $P(c) = 0$, then we say that c is a **zero** of the polynomial $P(x)$, or a **root** of the polynomial equation $P(x) = 0$. Thus the zeros of the polynomial in Example 3 are 1, 2, and -3.

Example 4

Find a polynomial $F(x)$ of degree 4 that has zeros $-3, 0, 1$, and 5.

Solution

By the Factor Theorem, $x - (-3), x - 0, x - 1$, and $x - 5$ must all be factors of the desired polynomial, so let

$$F(x) = (x + 3)(x - 0)(x - 1)(x - 5) = x^4 - 3x^3 - 13x^2 + 15x$$

Since $F(x)$ is to have degree 4, any other solution of the problem must be a constant multiple of the polynomial we have chosen (because multiplication by any polynomial other than a constant will increase the degree). ●

In Section 1.3 we developed some techniques and formulas that allowed us to factor certain special kinds of polynomials. We have seen in Example 3 that the Factor Theorem can help us to factor polynomials for which those techniques and formulas do not work. We will exploit this use of the Factor Theorem more fully in the next section.

Synthetic Division

The Remainder Theorem does not at first glance seem to provide a particularly useful way to evaluate polynomials. After all, to find the remainder when $x - c$ is divided into a polynomial requires long division, and the long division process appears much more complicated than simply substituting c in place of x and evaluating the polynomial. However, it turns out that if the divisor is of the form $x - c$, the long division process can be substantially simplified. For example, let us divide $2x^3 - 7x^2 + 5$ by $x - 3$:

$$
\begin{array}{r}
2x^2 - x - 3 \\
x - 3 \overline{\smash{\big)}\ 2x^3 - 7x^2 + 0x + 5} \\
\underline{2x^3 \boxed{-6}x^2 } \\
-x^2 + 0x \\
\underline{-x^2 + \boxed{3}x } \\
-3x + 5 \\
\underline{-3x + \boxed{9}} \\
\boxed{-4}
\end{array}
$$

Notice that the powers of x act simply as place-holders in this format, and we can certainly omit writing them down. Moreover, all the coefficients in the lines underneath the dividend except those in boxes are repeated either in the dividend or in the quotient. (For example, the 0 and the 5 were brought down from the dividend, and the 2, -1, and -3 are repeated in the quotient line in the same column.) This means we can omit these as well. If we now write the quotient line, together with the remainder, on the bottom instead of the top, the long division format simplifies to

$$
\begin{array}{r|rrrr}
-3 & 2 & -7 & 0 & 5 \\
   &   & -6 & 3 & 9 \\
\hline
   & 2 & -1 & -3 & -4
\end{array}
$$

Compare this carefully with the long division. The top line contains the coefficients of the divisor and dividend, while the bottom line contains the coefficients of the quotient and remainder. The middle line contains the remaining numbers in boxes in the long division format. Note that each entry in the bottom line is the difference of the two numbers immediately above it.

We can simplify the format even further by changing the sign of the entry that represents the divisor and of each entry in the second row. This makes the bottom row the sum rather than the difference of the two rows above it, and addition is easier to keep track of than subtraction. The division process performed above can thus be done as follows.

Step 1. Begin with a table that contains the coefficients of the divisor and dividend. (*Note:* Remember to supply zeros for missing coefficients in the dividend and to change the sign of the constant in the divisor.) Bring down the first coefficient of the dividend.

$$
\begin{array}{r|rrrr}
3 & 2 & -7 & 0 & 5 \\
\hline
 & 2 & & &
\end{array}
$$

Step 2. Multiply the 2 in the bottom row by the 3 in the upper left corner, and write the result in the second space of the second row. Then add this result to the number above it and write the sum in the bottom row.

$$
\begin{array}{r|rrrr}
3 & 2 & -7 & 0 & 5 \\
 & & 6 & & \\
\hline
 & 2 & -1 & &
\end{array}
$$

Step 3. Multiply the new element in the bottom row (the -1) by the 3 in the upper left corner, write the result in the next space of the second row, add it to the number above it, and write the sum in the bottom row.

$$
\begin{array}{r|rrrr}
3 & 2 & -7 & 0 & 5 \\
 & & 6 & -3 & \\
\hline
 & 2 & -1 & -3 &
\end{array}
$$

Step 4. Continue the process of multiplying the last element in the bottom row by the number in the upper left corner, putting the result in the second row, and then adding, until the table is complete.

$$
\begin{array}{r|rrrr}
3 & 2 & -7 & 0 & 5 \\
 & & 6 & -3 & -9 \\
\hline
 & 2 & -1 & -3 & -4
\end{array}
$$

The last entry in the bottom row is the remainder and the remaining entries are the coefficients of the quotient, so

$$2x^3 - 7x^2 + 5 = (x - 3)(2x^2 - x - 3) - 4.$$

The remainder is also the value of the dividend when $x = 3$, by the Remainder Theorem, so if $P(x) = 2x^3 - 7x^2 + 5$, then $P(3) = -4$.

This process, though somewhat complicated to derive and describe, is really very simple to carry out in practice. The procedure is summarized in the following box.

Synthetic Division of
$a_n x^n + a_{n-1} x^{n-1} + \cdots + a_1 x + a_0$ by $x - c$

$$
\begin{array}{c|cccccccc}
c & a_n & a_{n-1} & a_{n-2} & a_{n-3} & \cdots & a_2 & a_1 & a_0 \\
  &   & cb_{n-1} & cb_{n-2} & cb_{n-3} & \cdots & cb_2 & cb_1 & cb_0 \\
\hline
  & b_{n-1} & b_{n-2} & b_{n-3} & b_{n-4} & \cdots & b_1 & b_0 & r
\end{array}
$$

Each number in the bottom row is obtained by adding the numbers above it. In particular, $b_{n-1} = a_n$. The quotient is
$b_{n-1} x^{n-1} + b_{n-2} x^{n-2} + \cdots + b_1 x + b_0$, and the remainder is r.

Example 5

Find the quotient and remainder when $3x^5 + 5x^4 - 4x^3 + 7x + 3$ is divided by $x + 2$.

Solution

Since $x + 2 = x - (-2)$, the synthetic division table for this problem takes the following form:

$$
\begin{array}{c|cccccc}
-2 & 3 & 5 & -4 & 0 & 7 & 3 \\
   &   & -6 & 2 & 4 & -8 & 2 \\
\hline
   & 3 & -1 & -2 & 4 & -1 & 5
\end{array}
$$

The quotient is $3x^4 - x^3 - 2x^2 + 4x - 1$, and the remainder is 5. Thus

$$3x^5 + 5x^4 - 4x^3 + 7x + 3 = (x + 2)(3x^4 - x^3 - 2x^2 + 4x - 1) + 5$$

or $\dfrac{3x^5 + 5x^4 - 4x^3 + 7x + 3}{x + 2} = 3x^4 - x^3 - 2x^2 + 4x - 1 + \dfrac{5}{x + 2}$ ●

Example 6

Let $P(x) = x^5 - 5x^4 + 20x^2 - 10x + 10$. Find $P(4)$.

Solution

Since $P(4)$ is the remainder when $P(x)$ is divided by $x - 4$ (by the Remainder Theorem), we can use synthetic division.

$$
\begin{array}{c|cccccc}
4 & 1 & -5 & 0 & 20 & -10 & 10 \\
  &   & 4 & -4 & -16 & 16 & 24 \\
\hline
  & 1 & -1 & -4 & 4 & 6 & 34
\end{array}
$$

The remainder is 34, so $P(4) = 34$.   ●

To convince yourself that synthetic division is an easier and more practical way of evaluating polynomials than substitution, you should evaluate $(4)^5 - 5(4)^4 + 20(4)^2 - 10(4) + 10$ directly and compare with the solution of Example 6.

Exercises 3.1

In Exercises 1–25 find the quotient and remainder.

1. $\dfrac{4x^2 - 3x + 2}{x - 2}$

2. $\dfrac{x^2 - 12x - 22}{x + 2}$

3. $\dfrac{x^3 - 3x^2 + 5x - 2}{x - 3}$

4. $\dfrac{2x^3 + 6x^2 - 40x + 1}{x - 7}$

5. $\dfrac{x^5 + x^3 + x}{x + 1}$

6. $\dfrac{x^6 - x^4 + x^2 - 1}{x - 1}$

7. $\dfrac{x^6 + 6x^5 + 15x^4 + 20x^3 + 15x^2 + 6x + 1}{x + 1}$

8. $\dfrac{x^3 - 9x^2 + 27x - 27}{x - 3}$

9. $\dfrac{x^3 + 6x + 3}{x^2 - 2x + 2}$

10. $\dfrac{3x^4 - 5x^3 - 20x - 5}{x^2 + x + 3}$

11. $\dfrac{6x^3 + 2x^2 + 22x}{2x^2 + 5}$

12. $\dfrac{9x^2 - x + 5}{3x^2 - 7x}$

13. $\dfrac{x^5 + x^4 + x^3 + x^2 + x + 1}{x^2 + x + 1}$

14. $\dfrac{x^6 + x^4 + x^2 + 1}{x^2 + x + 1}$

15. $\dfrac{4x^5 + 2x^4 - 8x^3 + x^2 + 5x + 1}{2x^3 + x^2 - 3x - 1}$

16. $\dfrac{x^6 - 27}{x^2 - 3}$

17. $\dfrac{6x - 10}{x^3 - 4x + 7}$

18. $\dfrac{2x^5 - 7x^4 - 13}{4x^2 - 6x + 8}$

19. $\dfrac{2x^3 + 3x^2 - 2x + 1}{x - \frac{1}{2}}$

20. $\dfrac{6x^4 + 10x^3 + 5x^2 + x + 1}{x + \frac{2}{3}}$

21. $\dfrac{x^{101} - 1}{x - 1}$

22. $\dfrac{x^{100} - 1}{x + 1}$

(*Hint:* In Exercises 21 and 22 you are obviously not being asked to carry out more than 100 steps of a division process. Think! Use the Remainder Theorem to help you, and look for a pattern when finding the quotient.)

23. $\dfrac{x^2 - x - 3}{2x - 4}$ *Hint:*   Note that $\dfrac{x^2 - x - 3}{2x - 4} = \dfrac{1}{2}\left(\dfrac{x^2 - x - 3}{x - 2}\right).$

24. $\dfrac{x^3 + 3x^2 + 4x + 3}{3x + 6}$

25. $\dfrac{x^4 - 2x^3 + x + 2}{2x - 1}$

In Exercises 26–39 use synthetic division and the Remainder Theorem to evaluate $P(c)$.

26. $P(x) = x^2 - 2x - 7, c = 3$

27. $P(x) = 3x^2 + 9x + 5, c = -4$

28. $P(x) = 2x^2 + 9x + 1, c = \frac{1}{2}$

29. $P(x) = 2x^2 + 9x + 1, c = 0.1$

30. $P(x) = x^3 + 3x^2 - 7x + 6, c = 2$

31. $P(x) = 2x^3 - 21x^2 + 9x - 200, c = 11$

32. $P(x) = 5x^4 + 30x^3 - 40x^2 + 36x + 14, c = -7$

33. $P(x) = 6x^5 + 10x^3 + x + 1, c = -2$

34. $P(x) = x^7 - 3x^2 - 1, c = 3$

35. $P(x) = -2x^6 + 7x^5 + 40x^4 - 7x^2 + 10x + 112, c = -3$

36. $P(x) = 3x^3 + 4x^2 - 2x + 1, c = \frac{2}{3}$

37. $P(x) = x^3 - x + 1, c = \frac{1}{4}$

38. $P(x) = x^3 + 2x^2 - 3x - 8, c = \sqrt{3}$

39. $P(x) = -2x^3 + 3x^2 + 4x + 6, c = 1 + \sqrt{2}$

40. Let $P(x) = 6x^7 - 40x^6 + 16x^5 - 200x^4 - 60x^3 - 69x^2 + 13x - 139$. Calculate $P(7)$ by using synthetic division, and by substituting $x = 7$ into the polynomial and evaluating directly. Which method do you find preferable?

In Exercises 41–44 use the Factor Theorem to show that $x - c$ is a factor of $P(x)$ for the given values of c.

41. $P(x) = x^3 - 3x^2 + 3x - 1, c = 1$

42. $P(x) = x^3 + 2x^2 - 3x - 10, c = 2$

43. $P(x) = 2x^3 + 7x^2 + 6x - 5, c = \frac{1}{2}$

44. $P(x) = x^4 + 3x^3 - 16x^2 - 27x + 63, c = 3, -3$

In Exercises 45 and 46 show that the given value(s) of c are zeros of $P(x)$, and find all other zeros of $P(x)$.

45. $P(x) = x^3 - x^2 - 11x + 15, c = 3$

46. $P(x) = 3x^4 - x^3 - 21x^2 - 11x + 6, c = \frac{1}{3}, -2$

47. Find a polynomial of degree 3 that has zeros 1, -2, and 3, and in which the coefficient of x^2 is 3.

48. Find a polynomial of degree 4 with integer coefficients that has zeros 1, -1, 2, and $\frac{1}{2}$.

49. Find a polynomial of degree 4 that has integer coefficients, constant coefficient 24, and zeros 3, 4, and -2.

50. Find the remainder when the polynomial $6x^{1000} - 17x^{562} + 12x + 26$ is divided by $x + 1$.

51. Is $x - 1$ a factor of $x^{567} - 3x^{400} + x^9 + 2$?

52. If we divide the polynomial $P(x) = x^4 + kx^2 - kx + 2$ by $x + 2$, the remainder is 36. What must the value of k be?

53. For what values of k will $x - 3$ be a factor of $P(x) = k^2x^2 + 2kx - 12$?

54. The quadratic equation $x^2 - 2kx + 12 = 0$ has two positive roots, one of which is three times the other. What is the value of k, and what are the two roots of the equation?

Section 3.2

Rational Roots

In the preceding section we saw that if c is a root of a polynomial equation (or equivalently, a zero of the polynomial), then $x - c$ is a factor of the polynomial. This means that factoring a polynomial into linear factors is really equivalent to finding its zeros. But we need some sort of criterion to tell us which numbers we should check when we are looking for roots, since it is unlikely that we would find all the roots by trying numbers chosen at random. In this section we develop a systematic technique for finding all the *rational* zeros of a polynomial.

Let $P(x) = a_nx^n + a_{n-1}x^{n-1} + \cdots + a_1x + a_0$ be a polynomial with integer coefficients (with $a_n \neq 0$ and $a_0 \neq 0$), and suppose that p/q is a rational number in lowest terms for which

$$a_n\left(\frac{p}{q}\right)^n + a_{n-1}\left(\frac{p}{q}\right)^{n-1} + \cdots + a_1\left(\frac{p}{q}\right) + a_0 = 0$$

so that p/q is a **rational zero** of $P(x)$. We are going to show that p divides a_0 evenly and that q divides a_n evenly.

First we multiply both sides of the equation by q^n, giving

(1) $a_np^n + a_{n-1}p^{n-1}q + a_{n-2}p^{n-2}q^2 + \cdots + a_1pq^{n-1} + a_0q^n = 0$

Subtracting a_0q^n from both sides of the equation and factoring p from the left, we get

$$p(a_np^{n-1} + a_{n-1}p^{n-2}q + a_{n-2}p^{n-3}q^2 + \cdots + a_1q^{n-1}) = -a_0q^n$$

Now we can see that p divides the left side of the equation evenly, so it must divide the right side as well. But since p/q is in lowest terms, p and q have no integer factors in common, and so neither do p and q^n. Thus all the integer factors of p must go evenly into a_0, which means that p divides a_0.

If we now take Equation 1, subtract a_np^n from both sides, and then factor q from the left side, we get

$$q(a_{n-1}p^{n-1} + a_{n-2}p^{n-2}q + \cdots + a_1pq^{n-2} + a_0q^{n-1}) = -a_np^n$$

Using the same reasoning as before, we can now show that q must divide a_n. This proves the following theorem.

Rational Roots Theorem

If p/q is a rational root in lowest terms of the polynomial equation with integer coefficients

$$a_n x^n + a_{n-1} x^{n-1} + \cdots + a_1 x + a_0 = 0$$

(where $a_n \neq 0$ and $a_0 \neq 0$), then p divides a_0 and q divides a_n evenly.

If $a_n = 1$ in the Rational Roots Theorem, then q must be 1 or -1, so p/q is an integer that divides a_0. Thus we have the following corollary of the Rational Roots Theorem.

Corollary

Any rational root of the polynomial equation with integer coefficients and leading coefficient 1

$$x^n + a_{n-1} x^{n-1} + \cdots + a_1 x + a_0 = 0$$

must be an integer that divides a_0.

In this section, we will be using synthetic division (described in Section 3.1) both for evaluating and for dividing polynomials. For those who prefer not to use synthetic division, simple substitution can be used to evaluate polynomials and long division can be used to divide them.

Example 1

Find all rational roots of the following equation, and then solve it completely:

$$x^3 - 17x + 4 = 0$$

Solution

Since the leading coefficient is 1, all the rational roots are integer divisors of 4, so the possible candidates for rational solutions are ± 1, ± 2, and ± 4. Working through the list, we see that 1 and 2 are not roots but that 4 is because $P(4) = 0$.

$$
\begin{array}{r|rrrr}
1 & 1 & 0 & -17 & 4 \\
  &   & 1 & 1 & -16 \\
\hline
  & 1 & 1 & -16 & -12
\end{array}
\qquad
\begin{array}{r|rrrr}
2 & 1 & 0 & -17 & 4 \\
  &   & 2 & 4 & -26 \\
\hline
  & 1 & 2 & -13 & -22
\end{array}
\qquad
\begin{array}{r|rrrr}
4 & 1 & 0 & -17 & 4 \\
  &   & 4 & 16 & -4 \\
\hline
  & 1 & 4 & -1 & 0
\end{array}
$$

Thus $x^3 - 17x + 4 = (x - 4)(x^2 + 4x - 1)$. We now use the quadratic formula to find the zeros of the quotient:

$$x = \frac{-4 \pm \sqrt{4^2 + 4 \cdot 1 \cdot 1}}{2} = -2 \pm \sqrt{5}$$

So the only rational root is 4, and the remaining roots are $-2 + \sqrt{5}$ and $-2 - \sqrt{5}$, which are irrational.

Example 2

Find all solutions of the equation

$$2x^4 + x^3 - 17x^2 - 16x + 12 = 0$$

Solution

By the Rational Roots Theorem, if p/q is a rational root of the equation, p must divide 12 and q must divide 2. The divisors of 12 are ± 1, ± 2, ± 3, ± 4, ± 6, and ± 12. The divisors of 2 are ± 1 and ± 2. Thus the only possible rational roots are ± 1, ± 2, ± 3, ± 4, ± 6, ± 12, $\pm \frac{1}{2}$, and $\pm \frac{3}{2}$. Checking through the list of possibilities, we find that 1 and 2 are not roots but that 3 is. We divide by $x - 3$ to get

$$2x^4 + x^3 - 17x^2 - 16x + 12 = (x - 3)(2x^3 + 7x^2 + 4x - 4)$$

Rather than checking further through the list of possibilities, we will try instead to factor the quotient $2x^3 + 7x^2 + 4x - 4$. The possible rational zeros of this are ± 1, ± 2, ± 4, and $\pm \frac{1}{2}$. We have already found that 1 and 2 are not zeros of the original polynomial so they certainly cannot be zeros of the quotient either, and there is no need to check them again. Working through the list, we see that $\frac{1}{2}$ is a root.

$$2x^3 + 7x^2 + 4x - 4 = (x - \tfrac{1}{2})(2x^2 + 8x + 8)$$

The quotient here factors easily:

$$2x^2 + 8x + 8 = 2(x^2 + 4x + 4) = 2(x + 2)^2$$

So the original polynomial factors into

$$2(x + 2)^2(x - \tfrac{1}{2})(x - 3) = (x + 2)^2(2x - 1)(x - 3)$$

The roots are -2, $\frac{1}{2}$, and 3. There are no irrational roots.   ●

Descartes' Rule of Signs

In some cases the following rule, discovered by the French philosopher René Descartes around 1637, is helpful in eliminating candidates from lengthy lists of possible rational roots. Before we state the rule (which we do without proof), we must first explain what is meant by a *variation in sign*. If $P(x)$ is a polynomial with real coefficients, written with descending powers of x (and omitting powers with coefficient 0), then a **variation in sign** occurs whenever adjacent coefficients have opposite signs. For example,

$$P(x) = 5x^7 - 3x^5 - x^4 + 2x^2 + x - 3$$

has three variations in sign.

Descartes' Rule of Signs

If $P(x)$ is a polynomial with real coefficients, then:

1. The number of positive real zeros of $P(x)$ either is equal to the number of variations in sign in $P(x)$ or is less than that by an even whole number.
2. The number of negative real zeros of $P(x)$ either is equal to the number of variations in sign in $P(-x)$ or is less than that by an even whole number.

Example 3

Use Descartes' Rule of Signs to determine the possible number of positive and negative real roots of the equation

$$3x^6 + 4x^5 + 3x^3 - x - 3 = 0$$

Solution

The polynomial has one variation in sign and so there is one positive root. If $P(x)$ is the polynomial in the equation, then

$$P(-x) = 3(-x)^6 + 4(-x)^5 + 3(-x)^3 - (-x) - 3$$
$$= 3x^6 - 4x^5 - 3x^3 + x - 3$$

which has three variations in sign. There are either three or one negative roots, making a total of either four or two real roots. The remaining roots will be *complex numbers*, which we will study in Section 3.5.   ●

In the next example we see how Descartes' Rule of Signs can be used to shorten the list of possible rational zeros of a polynomial provided by the Rational Roots Theorem. From now on, whenever an example involves using synthetic division several times with the same dividend, we will use an abbreviated version of the synthetic division table in which we omit the middle line and write the top line only once, with the result of each division written below this. For example, the three divisions we performed in Example 1 would be combined as follows:

$$
\begin{array}{r|rrrr}
 & 1 & 0 & -17 & 4 \\
\hline
1 & 1 & 1 & -16 & -12 \\
2 & 1 & 2 & -13 & -22 \\
4 & 1 & 4 & -1 & 0 \\
\end{array}
$$

Example 4

Find all rational zeros of the polynomial

$$P(x) = 3x^4 + 22x^3 + 55x^2 + 52x + 12$$

Solution

The possible rational zeros of $P(x)$ are ± 1, ± 2, ± 3, ± 4, ± 6, ± 12, $\pm \frac{1}{3}$, $\pm \frac{2}{3}$, and $\pm \frac{4}{3}$. But $P(x)$ has no variations in sign, and hence no positive real zeros, so we can eliminate all the positive numbers in our list. Now

$$P(-x) = 3x^4 - 22x^3 + 55x^2 - 52x + 12$$

has four variations in sign, and so the equation has four, two, or zero negative real roots.

We now use synthetic division to check the candidates for rational roots until we find one that gives a remainder of 0.

$$
\begin{array}{r|rrrrr}
 & 3 & 22 & 55 & 52 & 12 \\
\hline
-1 & 3 & 19 & 36 & 16 & -4 \\
-2 & 3 & 16 & 23 & 6 & 0 \\
\end{array}
$$

We see that -2 is a zero, so we continue by factoring the quotient. Because -1 has been eliminated, the list of possible rational zeros of the quotient at this point is -2, -3, -6, $-\frac{1}{3}$, and $-\frac{2}{3}$.

$$
\begin{array}{r|rrrr}
 & 3 & 16 & 23 & 6 \\
\hline
-2 & 3 & 10 & 3 & 0 \\
\end{array}
$$

So $P(x) = (x + 2)(x + 2)(3x^2 + 10x + 3) = (x + 2)^2(x + 3)(3x + 1)$, and the zeros are -2, -3, and $-\frac{1}{3}$. ●

Example 5

Find all rational zeros of the polynomial

$$Q(x) = x^4 - 3x^3 - 3x^2 + 11x - 6$$

Solution

The possible rational zeros are all integers: ± 1, ± 2, ± 3, and ± 6. $Q(x)$ has three variations in sign, so there are either three or one positive real zeros.

$$Q(-x) = x^4 + 3x^3 - 3x^2 - 11x - 6$$

has one variation in sign, so there is definitely one negative real zero. We will therefore check the negative candidates first. We find that $Q(-1) = -16$, so -1 is not a zero. But $Q(-2) = 0$, and

$$Q(x) = (x + 2)(x^3 - 5x^2 + 7x - 3)$$

So -2 is the only negative real zero and we need to check no further among the negative candidates. We continue by factoring the quotient, using 1 and 3 as the only possible rational zeros. We find that $x - 1$ is a factor, with

$$x^3 - 5x^2 + 7x - 3 = (x - 1)(x^2 - 4x + 3)$$

So
$$Q(x) = (x + 2)(x - 1)(x^2 - 4x + 3)$$
$$= (x + 2)(x - 1)(x - 1)(x - 3)$$
$$= (x + 2)(x - 1)^2(x - 3)$$

and the zeros are -2, 1, and 3. ●

Upper and Lower Bounds for Roots

We say that a is a **lower bound** and b is an **upper bound** for the roots of a polynomial equation if every real root c of the equation satisfies $a \leq c \leq b$. The next theorem helps us find such bounds for any polynomial equation.

The Upper and Lower Bounds Theorem

Let $P(x)$ be a polynomial with real coefficients.

1. If we divide $P(x)$ by $x - b$ (where $b > 0$) using synthetic division, and if the row that contains the quotient and remainder has no negative entries, then b is an upper bound for the real roots of $P(x) = 0$.

2. If we divide $P(x)$ by $x - a$ (where $a < 0$) using synthetic division, and if the row that contains the quotient and remainder has entries that are alternately nonpositive and nonnegative, then a is a lower bound for the real roots of $P(x) = 0$.

A proof of this theorem is suggested in Exercises 39 and 40.

Example 6

Show that all the real roots of the equation $x^4 - 3x^2 + 2x - 5 = 0$ lie between -3 and 2.

Solution

We divide the polynomial by $x - 2$ and $x + 3$ using synthetic division:

$$
\begin{array}{r|rrrrr}
 & 1 & 0 & -3 & 2 & -5 \\
2 & 1 & 2 & 1 & 4 & 3 \\
-3 & 1 & -3 & 6 & -16 & 43
\end{array}
$$

$\leftarrow$ all positive
$\leftarrow$ alternately positive and negative

By the Upper and Lower Bounds Theorem, -3 is a lower bound and 2 is an upper bound for the roots. Since neither -3 nor 2 is a root, all the real roots lie between them. ●

Example 7

Factor completely the polynomial

$$P(x) = 2x^5 + 5x^4 - 8x^3 - 14x^2 + 6x + 9$$

Solution

The possible rational zeros of $P(x)$ are $\pm\frac{1}{2}$, ± 1, $\pm\frac{3}{2}$, ± 3, $\pm\frac{9}{2}$, and ± 9. We check the positive candidates first, beginning with the smallest.

$$
\begin{array}{r|rrrrrr}
 & 2 & 5 & -8 & -14 & 6 & 9 \\
\hline
\frac{1}{2} & 2 & 6 & -5 & -\frac{33}{2} & -\frac{9}{4} & \frac{63}{8} \\
1 & 2 & 7 & -1 & -15 & -9 & 0
\end{array}
$$

So 1 is a zero of $P(x)$. We continue with the quotient:

$$
\begin{array}{r|rrrrr}
 & 2 & 7 & -1 & -15 & -9 \\
\hline
1 & 2 & 9 & 8 & -7 & -16 \\
\frac{3}{2} & 2 & 10 & 14 & 6 & 0 \qquad \leftarrow \text{all nonnegative}
\end{array}
$$

We see that $\frac{3}{2}$ is both a zero and an upper bound for the zeros of $P(x)$, so we do not need to check any further for positive zeros.

$$
\begin{aligned}
P(x) &= (x-1)(x-\tfrac{3}{2})(2x^3 + 10x^2 + 14x + 6) \\
&= (x-1)(2x-3)(x^3 + 5x^2 + 7x + 3)
\end{aligned}
$$

By Descartes' Rule of Signs, $x^3 + 5x^2 + 7x + 3$ has no positive zeros, so the only possible rational zeros are -1 and -3.

$$
\begin{array}{r|rrrr}
 & 1 & 5 & 7 & 3 \\
\hline
-1 & 1 & 4 & 3 & 0
\end{array}
$$

Therefore

$$
P(x) = (x-1)(2x-3)(x+1)(x^2 + 4x + 3) = (x-1)(2x-3)(x+1)^2(x+3)
$$

●

Example 8

Find all rational solutions of the equation

$$
\tfrac{1}{4}x^5 + \tfrac{1}{4}x^4 - 3x^3 - \tfrac{3}{2}x^2 - 5x + 4 = 0
$$

Solution

First we clear the denominators by multiplying both sides of the equation by 4 so that we can use the Rational Roots Theorem:

$$
x^5 + x^4 - 12x^3 - 6x^2 - 20x + 16 = 0
$$

The possible rational roots are ± 1, ± 2, ± 4, ± 8, and ± 16, and we check the positive ones first:

$$
\begin{array}{r|rrrrrr}
 & 1 & 1 & -12 & -6 & -20 & 16 \\
\hline
1 & 1 & 2 & -10 & -16 & -36 & -20 \\
2 & 1 & 3 & -6 & -18 & -56 & -96 \\
4 & 1 & 5 & 8 & 26 & 84 & 352 \qquad \leftarrow \text{all positive}
\end{array}
$$

We have found an upper bound for the real roots, so there is no need to check 8 and 16. There are no positive rational roots. Now we check the negative candidates:

$$
\begin{array}{r|rrrrrr}
 & 1 & 1 & -12 & -6 & -20 & 16 \\
\hline
-1 & 1 & 0 & -12 & 6 & -26 & 42 \\
-2 & 1 & -1 & -10 & 14 & -48 & 112 \\
-4 & 1 & -3 & 0 & -6 & 4 & 0 \\
\end{array}
$$

$\leftarrow$ alternately nonpositive and nonnegative

We see that -4 is a root and also a lower bound for the real roots, so we do not need to check -8 and -16. The only rational root is -4. ●

It turns out that the equation in Example 8 does have two other real roots, but they are irrational. We have found bounds for the real roots, so we know these must lie between -4 and 4. In the next section we will learn how to find decimal approximations for irrational roots.

To summarize, the following procedure gives a systematic, efficient method for finding all the rational roots of a polynomial equation.

Step 1.   List all possible rational roots (in order) using the Rational Roots Theorem.

Step 2.   Use Descartes' Rule of Signs to see how many positive and negative real roots the equation may have. (In some cases the positive or negative rational candidates may be completely eliminated by this step.)

Step 3.   Check (in order) the positive and the negative candidates for rational roots provided by Step 1. Stop when you find a root, reach an upper or lower bound, or have found the maximum number of positive or negative roots predicted by Descartes' Rule of Signs.

Step 4.   If you find a root, repeat the process with the quotient. Remember that you do not need to check possible roots that have not worked at a previous stage. Once you reach a quotient that is quadratic or in some other way easily factorable, use the quadratic formula or other techniques you have learned to find the remaining roots.

Exercises 3.2

In Exercises 1–4 list all possible rational roots given by the Rational Roots Theorem (but do not check to see which are actually roots).

1. $x^4 - 3x^2 + 100x - 7 = 0$

2. $x^{30} - 4x^{25} + 6x^2 + 60 = 0$

3. $4x^5 - 3x^3 + x^2 - x + 15 = 0$

4. $6x^4 - x^3 + x^2 - 24 = 0$

In Exercises 5–11 use Descartes' Rule of Signs to determine how many positive and how many negative real zeros the given

polynomial can have, and then determine the possible total number of real zeros.

5. $x^4 + 5x^3 - x^2 + 4x - 3$

6. $3x^5 - 4x^4 + 8x^3 - 5$

7. $2x^6 + 5x^4 - x^3 - 5x - 1$

8. $x^4 + x^3 + x^2 + x + 12$

9. $x^5 + 4x^3 - x^2 + 6x$

10. $125x^{32} + 76x^{25} - 12x^{12} - 1$

11. $4x^7 - 3x^5 + 5x^4 + x^3 - 3x^2 + 2x - 5$

In Exercises 12 and 13 show that the given values for a and b are lower and upper bounds, respectively, for the real roots of the equation.

12. $8x^3 + 10x^2 - 39x + 9 = 0; a = -3, b = 2$

13. $3x^4 - 17x^3 + 24x^2 - 9x + 1 = 0; a = 0, b = 6$

Find integers that are upper and lower bounds for the real roots of the equations in Exercises 14 and 15.

14. $6x^3 - 13x^2 + x + 2 = 0$

15. $4x^4 + 20x^3 + 35x^2 + 32x + 15 = 0$

In Exercises 16–34 find all rational roots of each equation, and then find the irrational roots, if any. Whenever appropriate, use the Rational Roots Theorem, the Upper and Lower Bounds Theorem, Descartes' Rule of Signs, the quadratic formula, or other factoring techniques.

16. $x^3 - 2x^2 - x + 2 = 0$

17. $x^3 - x^2 - 8x + 12 = 0$

18. $x^4 + 6x^3 + 7x^2 - 6x - 8 = 0$

19. $x^4 - x^3 - 23x^2 - 3x + 90 = 0$

20. $x^5 - 7x^4 + 9x^3 + 23x^2 - 50x + 24 = 0$

21. $4x^3 - 7x + 3 = 0$

22. $20x^3 + 44x^2 + 23x + 3 = 0$

23. $4x^4 - 25x^2 + 36 = 0$

24. $6x^4 - 7x^3 - 8x^2 + 5x = 0$

25. $2x^4 + 15x^3 + 31x^2 + 20x + 4 = 0$

26. $3x^4 + 10x^3 - 14x^2 - 20x + 16 = 0$

27. $5x^4 - 27x^3 + 40x^2 - 22x + 4 = 0$

28. $\frac{1}{3}x^4 + 4x^3 + \frac{50}{3}x^2 + 28x + 15 = 0$

29. $x^4 - \frac{11}{4}x^3 - \frac{11}{2}x^2 + \frac{5}{4}x + 3 = 0$

30. $\frac{1}{12}x^4 - \frac{7}{6}x^3 + \frac{71}{12}x^2 - \frac{77}{6}x + 10 = 0$

31. $8x^5 - 14x^4 - 22x^3 + 57x^2 - 35x + 6 = 0$

32. $x^5 - \frac{1}{5}x^4 - 5x^3 + x^2 + 4x - \frac{4}{5} = 0$

33. $\frac{2}{3}x^6 - x^5 - \frac{13}{3}x^4 + \frac{29}{3}x^3 - 9x^2 + \frac{32}{3}x - 4 = 0$

34. $x^6 - 1 = 0$

Show that the polynomials in Exercises 35–38 do not have any rational zeros.

35. $x^3 - x - 2$ **36.** $2x^4 - x^3 + x + 2$

37. $3x^3 - x^2 - 6x + 12$ **38.** $x^{50} - 5x^{25} + x^2 - 1$

39. Let $P(x)$ be a polynomial with real coefficients, and let $b > 0$. Use the Division Algorithm to write $P(x) = (x - b) \cdot Q(x) + r$. Suppose that $r \geq 0$ and that all the coefficients in $Q(x)$ are nonnegative. Let $z > b$.
   (a) Show that $P(z) > 0$.
   (b) Prove the first part of the Upper and Lower Bounds Theorem.

40. Prove the second part of the Upper and Lower Bounds Theorem, following the pattern of the proof in Exercise 39.

41. Show that the equation

$$x^5 - x^4 - x^3 - 5x^2 - 12x - 6 = 0$$

   has exactly one rational root, and then prove that it must have either two or four irrational roots.

42. Consider the equation $x^3 + kx + 4 = 0$, where k is a real number.
   (a) Find all values of k for which this equation has at least one rational root.
   (b) For which values of k does this equation have one rational root and two irrational roots?
   (c) What is the largest number of rational roots this equation can have?

43. For what real values of k does the equation $x^3 - kx^2 + k^2x - 3 = 0$ have a rational root?

Section 3.3

Irrational Roots

For linear and quadratic equations we have formulas for the exact roots, including the irrational ones. For equations of higher degree, we described a method for finding all the rational roots in the preceding section. Although there are formulas analogous to the quadratic formula for the roots of the general third- and fourth-degree equations, they are complicated, difficult to remember, and cumbersome to use. For example, the formula for one of the three roots of the general cubic equation $ax^3 + bx^2 + cx + d = 0$ is

$$x = \frac{-b}{3a} + \frac{1}{a}\sqrt[3]{\frac{-b^3}{27} + \frac{abc}{6} - \frac{ad}{2} + a\sqrt{\frac{a^2d^2}{4} - \frac{b^2c^2}{108} + \frac{b^3d}{27} - \frac{abcd}{6}}}$$
$$+ \frac{1}{a}\sqrt[3]{\frac{-b^3}{27} + \frac{abc}{6} - \frac{ad}{2} - a\sqrt{\frac{a^2d^2}{4} - \frac{b^2c^2}{108} + \frac{b^3d}{27} - \frac{abcd}{6}}}$$

For equations of degree 5 and higher, no such formulas exist. In fact, the French mathematician Evariste Galois proved in 1832 (shortly before his death at the age of 21) that it is *impossible* to construct formulas involving radicals and the usual algebraic operations for the roots of the general nth-degree polynomial equation for $n \geq 5$.

In some special cases the exact values of the irrational zeros of a higher-degree polynomial can be found using the quadratic formula, as we saw in Example 1 and Exercises 25, 26, 27, and 31 in Section 3.2. In this section we will demonstrate an effective method for finding decimal approximations to the irrational roots of any polynomial equation to any desired accuracy. The technique depends on the following special case of the Intermediate Value Theorem, which you will study more fully in a calculus course.

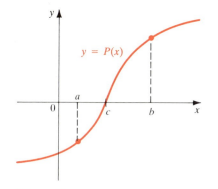

Figure 1

Intermediate Value Theorem for Polynomials

If P is a polynomial with real coefficients and if $P(a)$ and $P(b)$ have opposite signs, then there is at least one value c between a and b for which $P(c) = 0$.

Although we will not prove this theorem, Figure 1 shows why it is at least intuitively plausible. A part of the graph of $y = P(x)$ is shown, with the case $P(a) < 0$ and $P(b) > 0$ illustrated.

Example 1

Show that the following equation has exactly one positive irrational root, and find the decimal value of this root correct to the nearest hundredth:

$$P(x) = x^3 + x^2 - 2x - 3 = 0$$

Solution

By Descartes' Rule of Signs, the equation has exactly one positive real root. The only possible positive rational roots are 1 and 3, but since $P(1) = -3$ and $P(3) = 27$, neither actually turns out to be a root. So the positive real root must be irrational, and moreover it must lie between 1 and 3, since $P(1)$ and $P(3)$ are opposite in sign. We calculate $P(x)$ for enough values of x between 1 and 3 to locate this root between successive tenths. Comparing the values of $P(x)$ at $x = 1$ and $x = 3$ leads us to guess that the zero we are looking for is much closer to 1 than to 3, so we begin our search at $x = 1.3$.

x	$P(x)$
1.3	-1.713
1.4	-1.096
1.5	-0.375 ⎫ opposite in sign
1.6	0.456 ⎭

This means that the root lies somewhere between 1.5 and 1.6. We now attempt to locate it between successive hundredths.

x	$P(x)$
1.53	-0.138
1.54	-0.056 ⎫ opposite in sign
1.55	0.026 ⎭

So the root lies between 1.54 and 1.55. To see which of these is closer to the actual value, we calculate the value of P at 1.545, halfway between the two possibilities. We find that

$$P(1.545) \approx -0.015$$

Since $P(1.545)$ and $P(1.55)$ are opposite in sign, the root lies between 1.545 and 1.55 and thus is closer to 1.55 than 1.54. So the positive irrational root is 1.55, rounded to the nearest hundredth. ●

The process described in Example 1 can of course be continued to obtain the value of the root to any number of decimal places. The calculations involved are long and tedious, however. In your study of calculus you will learn more efficient methods for approximating the roots of polynomial equations.

Example 2

A right triangle has an area of 8 m², and its hypotenuse is 1 m longer than one of the legs. Find the lengths of all sides of the triangle correct to two decimal places.

Solution

Let the length of one leg be x. Then the hypotenuse has length $x + 1$, and the other leg (by the Pythagorean Theorem) has length $\sqrt{(x + 1)^2 - x^2} = \sqrt{2x + 1}$ (see Figure 2). So the area is $\frac{1}{2}x\sqrt{2x + 1} = 8$, and

$$x\sqrt{2x + 1} = 16$$
$$x^2(2x + 1) = 256$$
$$2x^3 + x^2 - 256 = 0$$

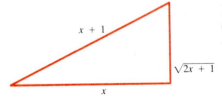

Figure 2

We are interested only in positive roots, and the equation has exactly one positive real root. Working through the rational possibilities, we get the following table:

x	$2x^3 + x^2 - 256$	
$\frac{1}{2}$	$-\frac{511}{2}$	
1	-253	
2	-236	
4	-112	opposite in sign
8	832	

Because of the sign change, the positive real root must lie between 4 and 8, and it must be irrational. We locate the root between successive integers, then tenths, and then hundredths as in Example 1.

x	$2x^3 + x^2 - 256$	
4	-112	sign change
5	19	

The root lies between 4 and 5. Because 19 is closer to 0 than -112, we guess that the root is closer to 5 than to 4.

x	$2x^3 + x^2 - 256$	
4.7	-26.264	
4.8	-11.776	sign change
4.9	3.308	

x	$2x^3 + x^2 - 256$	
4.87	-1.280	sign change
4.88	0.243	

Since the value of the polynomial in our equation is -0.520 when $x = 4.875$, the root is closer to 4.88 than to 4.87. The legs of the triangle are 4.88 m and 3.28 m long, and the hypotenuse is 5.88 m long, all correct to two decimal places.   ●

Exercises 3.3

In Exercises 1–6 find the indicated irrational root correct to two decimal places.

1. The positive root of $x^3 + 3x^2 + 3x - 1 = 0$

2. The real root of $x^3 + 4x^2 + 2 = 0$

3. The largest positive root of $x^4 - 8x^3 + 18x^2 - 8x - 2 = 0$

4. The negative root of $x^3 - 2x^2 - x + 3 = 0$

5. The positive root of $x^4 + 2x^3 + x^2 - 1 = 0$

6. The negative root of $x^4 + 2x^3 + x^2 - 1 = 0$

In Exercises 7–16 find all rational and irrational roots. Find the exact values of irrational roots using the quadratic formula whenever possible (as in Example 1 in Section 3.2). Otherwise find approximate values correct to two decimal places.

7. $x^3 + 2x^2 - 6x - 4 = 0$

8. $3x^3 + 10x^2 + 6x + 1 = 0$

9. $2x^5 + 3x^4 + x^2 + x - 1 = 0$

10. $x^4 - 4x^3 + 2x^2 - 5x + 10 = 0$

11. $x^3 - 3x^2 + 3 = 0$

12. $x^3 + 2x^2 - 4x - 7 = 0$

13. $x^5 + x^4 - 4x^3 - x^2 + 5x - 2 = 0$

14. $x^5 + 11x^4 + 43x^3 + 73x^2 + 52x + 12 = 0$

15. $2x^4 - 7x^3 + 9x^2 + 5x - 4 = 0$

16. $5x^4 + 13x^3 + 9x^2 - 16x + 4 = 0$

17. A rectangle with an area of 10 ft² has a diagonal that is 2 ft longer that one of its sides. What are the dimensions of the rectangle (correct to three decimal places)?

18. An open box with a volume of 1500 cm³ is to be constructed by taking a 20 cm by 40 cm piece of cardboard, cutting squares of side length x from each corner, and folding up the sides. Show that this can be done in two different ways, and find the exact dimensions of the box in each case.

19. A rocket consists of a right circular cylinder of height 20 m surmounted by a cone whose height and diameter are equal and whose radius is the same as that of the cylindrical section. What should this radius be (correct to two decimal places) if the total volume is to be $500\pi/3$ m³?

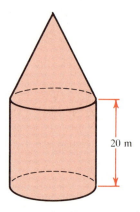

20 m

20. A rectangular box with a volume of $2\sqrt{2}$ ft³ has a square base. The diagonal of the box (between a pair of opposite corners) is 1 ft longer than each side of the base.

(a) If the base has sides of length x feet, show that

$$x^6 - 2x^5 - x^4 + 8 = 0$$

(b) Show that there are two different boxes that satisfy the given conditions. Find the dimensions in each case, to the nearest hundredth of a foot.

Complex Numbers

Throughout this chapter we have encountered polynomial equations that have no solutions in the real number system. A simple example of this is the equation $x^2 + 1 = 0$. This has no real solution because the square of any real number is nonnegative, so $x^2 + 1 \geq 1$ if x is real.

The **complex number system** was created to provide an environment in which every polynomial equation will have a solution. **Complex numbers** are expressions of the form $a + bi$, where a and b are real and the symbol i represents a number with the property $i^2 = -1$. The solutions of the equation $x^2 + 1 = 0$ are $\pm i$ in this number system.

In the complex number $z = a + bi$, a is called the **real part** and b is called the **imaginary part** of z. Note that the real and imaginary parts are both real numbers. Two complex numbers are **equal** if their real and imaginary parts are the same. Any real number can be thought of as a complex number with imaginary part 0; that is, $a = a + 0i$. Complex numbers with a nonzero imaginary part are called

imaginary numbers; complex numbers of the form bi (with real part 0) are called **pure imaginary numbers**.

Although we use the term *imaginary* in this context, imaginary numbers should not be thought of as any less "real" (in the ordinary rather than the mathematical sense of that word) than negative or irrational numbers. All numbers (except possibly the positive integers) are creations of the human mind—the numbers -1 and $\sqrt{2}$ no less than the number i. We study complex numbers because they complete, in a useful and elegant fashion, the study of the solution of polynomial equations.

Arithmetic Operations on Complex Numbers

Complex numbers are added, subtracted, multiplied, and divided just as we would any number of the form $a + b\sqrt{c}$. The only difference we must keep in mind is that $i^2 = -1$. Thus, in particular, the following calculation should be valid:

$$(a + bi) \cdot (c + di) = ac + (ad + bc)i + bdi^2$$
$$= ac + (ad + bc)i + bd(-1)$$
$$= (ac - bd) + (ad + bc)i$$

We therefore define the sum, difference, and product of complex numbers as follows:

$$(a + bi) + (c + di) = (a + c) + (b + d)i$$
$$(a + bi) - (c + di) = (a - c) + (b - d)i$$
$$(a + bi) \cdot (c + di) = (ac - bd) + (ad + bc)i$$

Example 1

Express the following in the form $a + bi$:

(a) $(3 + 5i) + (4 - 2i)$ (b) $(3 + 5i) - (4 - 2i)$

(c) $(3 + 5i) \cdot (4 - 2i)$

Solution

(a) $(3 + 5i) + (4 - 2i) = (3 + 4) + (5 - 2)i = 7 + 3i$

(b) $(3 + 5i) - (4 - 2i) = (3 - 4) + (5 - (-2))i = -1 + 7i$

(c) $(3 + 5i) \cdot (4 - 2i) = (3 \cdot 4 - 5(-2)) + (3(-2) + 5 \cdot 4)i = 22 + 14i$ ●

Division of complex numbers is much like rationalizing the denominator of a radical expression, which we considered in Section 1.2. For the complex number $z = a + bi$, we define its **complex conjugate** to be $\bar{z} = a - bi$. Note that $z \cdot \bar{z} = (a + bi) \cdot (a - bi) = a^2 + b^2$. So the product of a complex number and its conjugate is always a nonnegative real number. We use this property to divide complex numbers, as in the following example.

Example 2

Express the following in the form $a + bi$:

(a) $\dfrac{3 + 5i}{1 - 2i}$

(b) $\dfrac{7 + 3i}{4i}$

Solution

(a) We multiply both numerator and denominator by the complex conjugate of the denominator to make the new denominator a real number. The complex conjugate of $1 - 2i$ is $\overline{1 - 2i} = 1 + 2i$.

$$\frac{3 + 5i}{1 - 2i} = \left(\frac{3 + 5i}{1 - 2i}\right) \cdot \left(\frac{1 + 2i}{1 + 2i}\right) = \frac{-7 + 11i}{5} = -\frac{7}{5} + \frac{11}{5}i$$

(b)

$$\frac{7 + 3i}{4i} = \left(\frac{7 + 3i}{4i}\right) \cdot \left(\frac{-4i}{-4i}\right) = \frac{12 - 28i}{16} = \frac{3}{4} - \frac{7}{4}i \qquad \bullet$$

In the next section, we will need the following facts about the complex conjugate operation:

1. $\overline{z + w} = \bar{z} + \bar{w}$
2. $\overline{z \cdot w} = \bar{z} \cdot \bar{w}$
3. $\overline{z^n} = (\bar{z})^n$
4. $\bar{z} = z$ if and only if z is real

We now prove the first of these facts; the proofs of the rest are similar (see Exercises 39–43). If $z = a + bi$ and $w = c + di$ (where a, b, c, and d are real), then

$$\overline{z + w} = \overline{(a + bi) + (c + di)}$$
$$= \overline{(a + c) + (b + d)i}$$
$$= (a + c) - (b + d)i$$
$$= (a - bi) + (c - di)$$
$$= \bar{z} + \bar{w}$$

Just as every positive real number r has two square roots ($\sqrt{r}$ and $-\sqrt{r}$), every negative number has two square roots. Both are pure imaginary numbers, for if $r > 0$ is real, then

$$(i\sqrt{r})^2 = i^2 r = -r$$

and

$$(-i\sqrt{r})^2 = (-1)^2 i^2 r = -r$$

We call $i\sqrt{r}$ the **principal square root** of $-r$, and we use the symbol $\sqrt{-r}$ to denote the principal square root. The other square root is then $-\sqrt{-r} = -i\sqrt{r}$. Note that the two square roots of a negative real number are complex conjugates of each other.

Example 3

Evaluate (a) $\sqrt{-1}$, (b) $\sqrt{-16}$, and (c) $\sqrt{-3}$.

Solution

(a) $\quad \sqrt{-1} = i\sqrt{1} = i$

(b) $\quad \sqrt{-16} = i\sqrt{16} = 4i$

(c) $\quad \sqrt{-3} = i\sqrt{3}$ ●

Note that we will usually write $i\sqrt{b}$ instead of $\sqrt{b}i$ to avoid confusion with $\sqrt{bi}$.

Special care must be taken when performing calculations that involve the square roots of negative numbers. Although $\sqrt{a} \cdot \sqrt{b} = \sqrt{ab}$ when a and b are positive, this is not true when both are negative. For example, $\sqrt{-2} \cdot \sqrt{-3} = i\sqrt{2} \cdot i\sqrt{3} = i^2\sqrt{6} = -\sqrt{6}$, but $\sqrt{(-2) \cdot (-3)} = \sqrt{6}$, so

$$\sqrt{-2} \cdot \sqrt{-3} \neq \sqrt{(-2) \cdot (-3)}$$

When multiplying radicals of negative numbers, express them first in the form $i\sqrt{r}$ (where $r > 0$) to avoid possible error.

Example 4

Evaluate $(\sqrt{12} - \sqrt{-3})(3 + \sqrt{-4})$ and express in the standard form for complex numbers.

Solution

$$\begin{aligned}
(\sqrt{12} - \sqrt{-3})(3 + \sqrt{-4}) &= (\sqrt{12} - i\sqrt{3})(3 + i\sqrt{4}) \\
&= (2\sqrt{3} - i\sqrt{3})(3 + 2i) \\
&= (6\sqrt{3} + 2\sqrt{3}) + i(2 \cdot 2\sqrt{3} - 3\sqrt{3}) \\
&= 8\sqrt{3} + i\sqrt{3}
\end{aligned}$$
●

Graphing Complex Numbers

To graph real numbers or sets of real numbers, we have been using the number line, which has just one dimension. Complex numbers, however, have two components: the real part and the imaginary part. This suggests the following method for graphing complex numbers. We need two axes: one for the real part and one for the imaginary part. We call these the **real axis** and the **imaginary axis**, respectively. The plane determined by these two axes is called the **complex plane**. To graph the complex number $a + bi$, we plot the ordered pair of numbers (a, b) in this plane, as indicated in the following example.

Example 5

Graph the complex numbers $z_1 = 2 + 3i$, $z_2 = 5 - 2i$, and $z_1 + z_2$.

Solution

We have $z_1 + z_2 = (2 + 3i) + (5 - 2i) = 7 + i$. The graph is shown in Figure 1.

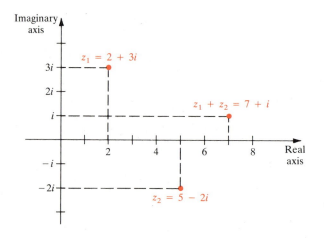

Figure 1

Example 6

Graph the following sets of complex numbers:

(a) $S = \{a + bi \mid a \geq 0\}$

(b) $T = \{a + bi \mid a < 1, b \geq 0\}$

Solution

(a) S is the set of complex numbers whose real part is nonnegative. Its graph is shown in Figure 2(a).

(b) T is the set of complex numbers for which the real part is less than 1 and the imaginary part is nonnegative. See Figure 2(b) for the graph.

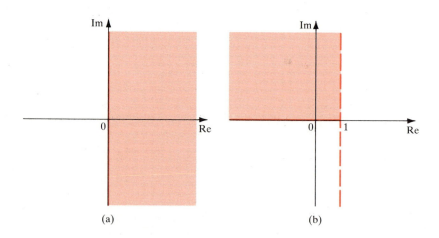

Figure 2   (a)   (b)

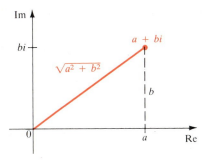

Figure 3

Recall from Section 1.1 that the absolute value of a real number can be thought of as its distance from the origin on the real number line. We define absolute value for complex numbers in a similar fashion. From Figure 3 we can see, using the Pythagorean Theorem, that the distance between $a + bi$ and the origin in the complex plane is $\sqrt{a^2 + b^2}$. This leads to the following definition:

The **modulus** (or **absolute value**) of the complex number $z = a + bi$ is

$$|z| = \sqrt{a^2 + b^2}$$

Example 7

Find the moduli of the complex numbers $3 + 4i$ and $8 - 5i$.

Solution

$$|3 + 4i| = \sqrt{3^2 + 4^2} = \sqrt{25} = 5$$
$$|8 - 5i| = \sqrt{8^2 + (-5)^2} = \sqrt{89}$$

●

Example 8

Graph the following sets of complex numbers:

(a) $C = \{z \mid |z| = 1\}$

(b) $D = \{z \mid |z| \leq 1\}$

Solution

(a) C is the set of complex numbers whose distance from the origin is 1. Thus C is a circle of radius 1 with center at the origin.

(b) D is the set of complex numbers whose distance from the origin is less than or equal to 1. Thus D is the disk that consists of all complex numbers on and inside the circle C of part (a).

The graphs of C and D are shown in Figure 4.

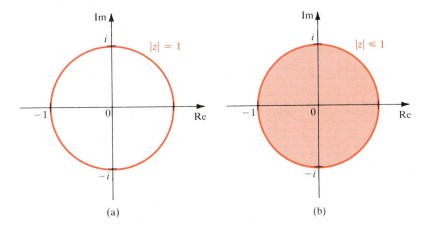

Figure 4   (a)   (b)   ●

Exercises 3.4

Find the real and imaginary parts of the complex numbers in Exercises 1–6.

1. $5 - 6i$

2. $i\sqrt{7}$

3. $\dfrac{4 + i\sqrt{3}}{2}$

4. $\dfrac{3 - \sqrt{-6}}{\sqrt{3}}$

5. $\sqrt{2} + \sqrt{-3} - \sqrt{4} - \sqrt{-5}$

6. $\dfrac{2 + 4i}{\sqrt{-16}}$

Evaluate the expressions in Exercises 7–38 and write the results in the form $a + bi$.

7. $(3 + 2i) + (7 - 3i)$

8. $(6 - 4i) + (-3 + 7i)$

9. $(4 - \tfrac{1}{2}i) + (\tfrac{3}{2} + 5i)$

10. $(1 + i) - (2 - 3i)$

11. $(-12 + 8i) - (7 + 4i)$

12. $3i - (4 - i)$

13. $4 \cdot (2 + 6i)$

14. $2i(\tfrac{1}{2} - i)$

15. $(3 - i)(4 + i)$

16. $(4 - 7i)(1 + 3i)$

17. $(3 - 4i)(5 - 12i)$

18. $(\tfrac{2}{3} + 12i)(\tfrac{1}{6} + 24i)$

19. $\dfrac{1}{i}$

20. $\dfrac{1}{1 + i}$

21. $\dfrac{2 + 3i}{1 - 5i}$

22. $\dfrac{5 - i}{3 + 4i}$

23. $\dfrac{26 + 39i}{2 - 3i}$

24. $\dfrac{3}{4 - 3i}$

25. $\dfrac{4i}{3 - 2i}$

26. $(2 - 6i)^{-1}$

27. i^3

28. $(2i)^4$

29. i^{100}

30. i^{1002}

31. $\sqrt{-25}$

32. $\sqrt{\dfrac{-9}{4}}$

33. $\sqrt{-3}\sqrt{-12}$

34. $\sqrt{\tfrac{1}{3}}\sqrt{-27}$

35. $(3 - \sqrt{-5})(1 + \sqrt{-15})$

36. $\dfrac{1 - \sqrt{-1}}{1 + \sqrt{-1}}$

37. $\dfrac{2 + \sqrt{-7}}{1 + \sqrt{-14}}$

38. $(\sqrt{3} - \sqrt{-4})(\sqrt{6} - \sqrt{-5})$

Prove the statements in Exercises 39–44 for the complex numbers z and w. (Hint: Express z as $a + bi$ and w as $c + di$.)

39. $\overline{z - w} = \bar{z} - \bar{w}$

40. $\overline{zw} = \bar{z} \cdot \bar{w}$

41. $\overline{z^2} = (\bar{z})^2$

42. $\overline{z^3} = (\bar{z})^3$

43. $\bar{z} = z$ if and only if z is real

44. $|z|^2 = z\bar{z}$

In Exercises 45–52 graph the complex number and find its modulus.

45. $3i$

46. -3

47. $5 + 2i$

48. $7 - 3i$

49. $\sqrt{3} + i$

50. $-1 - \dfrac{\sqrt{3}}{3}i$

51. $\dfrac{3 + 4i}{5}$

52. $\dfrac{-\sqrt{2} + i\sqrt{2}}{2}$

In Exercises 53 and 54 sketch the complex number z, and also sketch $2z$, $-z$, and $\tfrac{1}{2}z$ on the same complex plane.

53. $z = 1 + i$

54. $z = 2 - 3i$

In Exercises 55 and 56 sketch z_1, z_2, and $z_1 + z_2$.

55. $z_1 = 2 - i, z_2 = 2 + i$

56. $z_1 = -1 + i, z_2 = 2 - 3i$

In Exercises 57–64 sketch the given set of complex numbers.

57. $\{z = a + bi \,|\, a \le 0, b \ge 0\}$

58. $\{z = a + bi \,|\, a > 1, b > 1\}$

59. $\{z = a + bi \,|\, a + b < 2\}$

60. $\{z = a + bi \,|\, a \ge b\}$

61. $\{z \,|\, |z| = 3\}$

62. $\{z \,|\, |z| \ge 1\}$

63. $\{z \,|\, |z| < 2\}$

64. $\{z \,|\, 2 \le |z| \le 5\}$

In Exercises 65 and 66 prove the given statements, assuming $z = x + yi$ and $|z| \ne 0$.

65. $\dfrac{z}{\bar{z}} + \dfrac{\bar{z}}{z} = \dfrac{2(x^2 - y^2)}{x^2 + y^2}$

66. $\dfrac{z}{\bar{z}} - \dfrac{\bar{z}}{z} = \dfrac{4xyi}{x^2 + y^2}$

Complex Roots

We have already seen that the solutions of the quadratic equation $ax^2 + bx + c = 0$, where $a \neq 0$, are

$$x = \frac{-b \pm \sqrt{b^2 - 4ac}}{2a}$$

If $b^2 - 4ac < 0$, then the equation has no real solutions. But in the complex number system, this equation will always have solutions because negative numbers have square roots in this expanded setting.

Example 1

Solve the following equations:

(a) $x^2 + 9 = 0$ (b) $x^2 + 4x + 5 = 0$

Solution

(a) $x^2 + 9 = 0$ means $x^2 = -9$, so $x = \pm\sqrt{-9} = \pm i\sqrt{9} = \pm 3i$. The solutions are $3i$ and $-3i$.

(b)   By the quadratic formula,

$$x = \frac{-4 \pm \sqrt{4^2 - 4 \cdot 5}}{2}$$

$$= \frac{-4 \pm \sqrt{-4}}{2}$$

$$= \frac{-4 \pm 2i}{2}$$

so the solutions are $-2 + i$ and $-2 - i$.   ●

Example 2

Find all roots of the equation $x^6 - 1 = 0$.

Solution

$$x^6 - 1 = (x^3)^2 - 1$$
$$= (x^3 - 1)(x^3 + 1)$$
$$= (x - 1)(x^2 + x + 1)(x + 1)(x^2 - x + 1)$$

Setting each factor equal to zero, we see that 1 and -1 are real solutions of the equation. Using the quadratic formula on the remaining factors, we get

$$x = \frac{-1 \pm \sqrt{1 - 4}}{2} = \frac{-1 \pm \sqrt{-3}}{2}$$

and
$$x = \frac{1 \pm \sqrt{1-4}}{2} = \frac{1 \pm \sqrt{-3}}{2}$$

so the roots of the equation are 1, -1, $-\frac{1}{2} + i\frac{\sqrt{3}}{2}$, $-\frac{1}{2} - i\frac{\sqrt{3}}{2}$, $\frac{1}{2} + i\frac{\sqrt{3}}{2}$, and $\frac{1}{2} - i\frac{\sqrt{3}}{2}$. ●

The Fundamental Theorem of Algebra

It is a remarkable fact that adding just $\sqrt{-1}$ and its real multiples to the real number system, which is what we did when we created the complex numbers, is sufficient to provide a number system in which every polynomial equation has a root. Although we will not prove this fact (a proof requires mathematical expertise well beyond the scope of this book), it nevertheless forms the basis for much of our work in solving polynomial equations. This theorem was proved by the German mathematician C. F. Gauss in 1799.

Fundamental Theorem of Algebra

Every polynomial
$$P(x) = a_n x^n + a_{n-1} x^{n-1} + \cdots + a_1 x + a_0 \quad (n \geq 1, a_n \neq 0)$$
with complex coefficients has at least one complex zero.

Note that since any real number is also a complex number, the theorem applies to polynomials with real coefficients as well.

Since every zero c of a polynomial corresponds to a factor of the form $x - c$ (the Factor Theorem), the Fundamental Theorem of Algebra ensures that we can factor any polynomial $P(x)$ of degree n as follows:

$$P(x) = (x - c_1) \cdot Q_1(x)$$

[where $Q_1(x)$ is of degree $n - 1$ and c_1 is a zero of $P(x)$]. But now applying the Fundamental Theorem to the quotient $Q_1(x)$ gives us the factorization

$$P(x) = (x - c_1) \cdot (x - c_2) \cdot Q_2(x)$$

where $Q_2(x)$ is of degree $n - 2$ and c_2 is a zero of $Q_1(x)$. Continuing this process for n steps, we will get a final quotient $Q_n(x)$ of degree 0, which is therefore a nonzero constant, which we will call a. This proves the following corollary of the Fundamental Theorem of Algebra:

Complete Factorization Theorem

If $P(x)$ is a polynomial of degree $n > 0$, then there exist complex numbers $a, c_1, c_2, \ldots, c_n$ (with $a \neq 0$) such that
$$P(x) = a(x - c_1)(x - c_2) \cdots (x - c_n)$$

The number a is clearly the coefficient of x^n in $P(x)$. The numbers $c_1, c_2, \ldots, c_n$ are the zeros of $P(x)$ (by the Factor Theorem). These need not all be different. If the factor $x - c$ appears k times in the complete factorization of $P(x)$, we say that c is a zero of **multiplicity k**.

$P(x)$ can have no zeros other than $c_1, c_2, \ldots, c_n$ because if

$$P(c) = a(c - c_1)(c - c_2) \cdots (c - c_n) = 0$$

then at least one of the factors $c - c_i$ must be 0, so that $c = c_i$ for some $i \in \{1, 2, \ldots, n\}$. We have thus shown the following:

Zeros Theorem

Every polynomial of degree $n \geq 1$ has exactly n zeros, provided a zero with multiplicity k is counted k times.

Example 3

Find a polynomial $P(x)$ that satisfies the given description.

(a) A polynomial of degree 3, with zeros 1, 2, and -4, and constant coefficient 16

(b) A polynomial of degree 4, with zeros i, $-i$, 2, and -2, and value 25 when $x = 3$

Solution

(a) From the description, we see that $P(x)$ has the complete factorization $a(x - 1)(x - 2)(x - (-4))$ for some a. Thus

$$
\begin{aligned}
P(x) &= a(x^2 - 3x + 2)(x + 4) \\
&= a(x^3 + x^2 - 10x + 8) \\
&= ax^3 + ax^2 - 10ax + 8a
\end{aligned}
$$

Since the constant coefficient is to be 16, we see that $a = 2$, so

$$P(x) = 2x^3 + 2x^2 - 20x + 16$$

(b)
$$
\begin{aligned}
P(x) &= a(x - i)(x - (-i))(x - 2)(x - (-2)) \\
&= a(x^2 + 1)(x^2 - 4) \\
&= a(x^4 - 3x^2 - 4)
\end{aligned}
$$

$P(3) = a(3^4 - 3 \cdot 3^2 - 4) = 50a = 25$, so $a = \frac{1}{2}$ and $P(x) = \frac{1}{2}x^4 - \frac{3}{2}x^2 - 2$.

Example 4

Find all four zeros of $P(x) = x^4 + 2x^2 + 9$.

Solution

$$P(x) = x^4 + 2x^2 + 9$$
$$= x^4 + 6x^2 + 9 - 4x^2$$
$$= (x^2 + 3)^2 - (2x)^2$$
$$= (x^2 - 2x + 3)(x^2 + 2x + 3) \quad \textit{(difference of squares formula)}$$

Using the quadratic formula on each factor, we get

$$x = \frac{2 \pm \sqrt{4 - 12}}{2} = 1 \pm i\sqrt{2}$$

and

$$x = \frac{-2 \pm \sqrt{4 - 12}}{2} = -1 \pm i\sqrt{2}$$

so the zeros of $P(x)$ are $1 + i\sqrt{2}$, $1 - i\sqrt{2}$, $-1 + i\sqrt{2}$, and $-1 - i\sqrt{2}$. ●

Example 5

Find all five zeros of $Q(x) = 3x^5 - 2x^4 - x^3 - 12x^2 - 4x$.

Solution

By Descartes' Rule of Signs, $Q(x)$ has one positive real zero and either three or one negative real zeros. Since $x = 0$ is obviously a zero (but is neither positive nor negative), this proves that there are either five real and no imaginary zeros, or three real and two imaginary zeros. Checking through the list of possible rational zeros of $Q(x)/x$, we see that $Q(2) = 0$ and $Q(-\frac{1}{3}) = 0$, so by the Factor Theorem, $x - 2$ and $x + \frac{1}{3}$ are factors. Thus

$$Q(x) = x(x - 2)(x + \tfrac{1}{3})(3x^2 + 3x + 6) = x(x - 2)(3x + 1)(x^2 + x + 2)$$

The roots of the quadratic factor are

$$x = \frac{-1 \pm \sqrt{1 - 8}}{2} = -\frac{1}{2} \pm i\frac{\sqrt{7}}{2}$$

so the zeros of $Q(x)$ are 0, 2, $-\frac{1}{3}$, $-\frac{1}{2} + i\frac{\sqrt{7}}{2}$, and $-\frac{1}{2} - i\frac{\sqrt{7}}{2}$. ●

As you may have noticed from the examples so far, the imaginary roots of polynomial equations with real coefficients come in pairs. Whenever $a + bi$ is a root, so is its complex conjugate $a - bi$. This is always the case, as we show in the following theorem:

Conjugate Roots Theorem

If the polynomial $P(x)$ of degree $n > 0$ has real coefficients, and if the complex number z is a root of the equation $P(x) = 0$, then so is its complex conjugate $\bar{z}$.

Proof

Let
$$P(x) = a_n x^n + a_{n-1} x^{n-1} + \cdots + a_1 x + a_0$$

where each coefficient is real. If $P(z) = 0$, then

$$
\begin{aligned}
P(\bar{z}) &= a_n(\bar{z})^n + a_{n-1}(\bar{z})^{n-1} + \cdots + a_1 \bar{z} + a_0 \\
&= \overline{a_n} \overline{z^n} + \overline{a_{n-1}} \overline{z^{n-1}} + \cdots + \overline{a_1} \bar{z} + \overline{a_0} \quad \textcolor{red}{\textit{(since the coefficients are real)}} \\
&= \overline{a_n} \overline{z^n} + \overline{a_{n-1}} \overline{z^{n-1}} + \cdots + \overline{a_1 z} + \overline{a_0} \\
&= \overline{a_n z^n + a_{n-1} z^{n-1} + \cdots + a_1 z + a_0} \\
&= \overline{P(z)} = \bar{0} = 0
\end{aligned}
$$

Note that we have used the properties of the complex conjugate that we studied in the preceding section. This derivation shows that the conjugate $\bar{z}$ is also a root of $P(x) = 0$, and we have proved the theorem. ●

Example 6

Find a polynomial $P(x)$ of degree 5 that has integer coefficients and zeros $\frac{1}{2}, 3 - i$, and $2i$.

Solution

Since $3 - i$ and $2i$ are zeros, so are $3 + i$ and $-2i$ by the Conjugate Roots Theorem. This means that $P(x)$ has all the factors in the following product:

$$
\begin{aligned}
(x - \tfrac{1}{2})(x - (3 - i))(x - (3 + i))(x - 2i)(x + 2i) \\
= (x - \tfrac{1}{2})(x^2 - 6x + 10)(x^2 + 4) \\
= x^5 - \tfrac{13}{2}x^4 + 17x^3 - 31x^2 + 52x - 20
\end{aligned}
$$

Multiplying by 2 to make all coefficients integers, we get

$$P(x) = 2x^5 - 13x^4 + 34x^3 - 62x^2 + 104x - 40$$

Any other solution would have to be an integral multiple of this one. ●

Example 7

Find all roots of the equation $x^4 - 12x^3 + 56x^2 - 120x + 96 = 0$, given that one root is $3 + i\sqrt{3}$.

Solution

The complex conjugate of $3 + i\sqrt{3}$ must also be a root, so that both $x - (3 + i\sqrt{3})$ and $x - (3 - i\sqrt{3})$, and hence their product, must divide the polynomial in the equation.

$$(x - (3 + i\sqrt{3})) \cdot (x - (3 - i\sqrt{3})) = x^2 - 6x + 12$$

We divide $x^2 - 6x + 12$ into the given polynomial:

$$
\begin{array}{r}
x^2 - 6x + 8 \\
x^2 - 6x + 12\overline{)x^4 - 12x^3 + 56x^2 - 120x + 96} \\
\underline{x^4 - 6x^3 + 12x^2} \\
-6x^3 + 44x^2 - 120x \\
\underline{-6x^3 + 36x^2 - 72x} \\
8x^2 - 48x + 96 \\
\underline{8x^2 - 48x + 96} \\
0
\end{array}
$$

The quotient factors into $(x - 2)(x - 4)$, so the roots of the equation are $2, 4, 3 + i\sqrt{3}$, and $3 - i\sqrt{3}$.   ●

Exercises 3.5

Find all solutions of the equations in Exercises 1–10.

1. $x^2 + 4 = 0$

2. $25x^2 + 9 = 0$

3. $x^2 - 8x + 17 = 0$

4. $x^2 + \frac{1}{2}x + \frac{1}{4} = 0$

5. $3x^2 - 5x + 4 = 0$

6. $10x^2 + 5x + 2 = 0$

7. $t + 3 + \dfrac{3}{t} = 0$

8. $\theta^3 + \theta^2 + \theta = 0$

9. $z^2 - iz = 0$

10. $2z^3 + iz^2 = 0$

In Exercises 11–16 find polynomials with real coefficients that satisfy the given conditions.

11. $P(x)$ has degree 3, integer coefficients, and zeros $\frac{2}{3}$ and $3i$.

12. $Q(x)$ has degree 5 and zeros $1 + i\sqrt{3}$ and 4, -1 is a zero of multiplicity 2, and $Q(0) = 64$.

13. $R(x)$ has degree 6 and integer coefficients, and -1 is a zero of multiplicity 6.

14. $S(x)$ has degree 4 and zeros $1 + i$ and $3 - 4i$, and the coefficient of x^2 is 7.

15. $T(x)$ has constant coefficient 8 and is of the smallest possible degree consistent with having $1 + i$ as a zero of multiplicity 2.

16. $U(x)$ has degree 3, with $U(2i) = 0$, $U(2) = 0$, and $U(1) = -10$.

In Exercises 17–20 show that the indicated value(s) of z are solutions of the given equation, and then find all solutions.

17. $3x^4 + 2x^3 + 13x^2 + 22x - 6 = 0, z = 1 + i$

18. $x^4 - 13x^2 + 10x + 35 = 0, z = \frac{5}{2} - \frac{\sqrt{3}}{2}i$

19. $x^5 + x^4 + 7x^3 - x^2 + 12x - 20 = 0, z = 2i,$
$z = -1 - 2i$

20. $2x^6 - x^5 + 30x^4 + 25x^3 + 88x^2 + 306x - 180 = 0,$
$z = 3i$

Find all solutions of the equations in Exercises 21–32.

21. $16x^4 - 81 = 0$

22. $x^3 - 64 = 0$

23. $x^6 - 729 = 0$

24. $x^4 + 4 = 0$

25. $x^4 + x^2 + 25 = 0$

26. $x^6 + 7x^3 - 8 = 0$

27. $x^3 - 2x^2 + 2x - 1 = 0$

28. $x^3 + 7x^2 + 18x + 18 = 0$

29. $3x^4 + 10x^3 + 13x^2 - 2x - 8 = 0$

30. $8x^4 + 36x^3 + 42x^2 + 9x + 10 = 0$

31. $3x^5 - 23x^4 + 57x^3 - 59x^2 + 30x - 8 = 0$

32. $8x^6 - 24x^5 - 8x^4 + 23x^3 + 3x^2 + x - 3 = 0$

Find the complete factorizations of the polynomials in Exercises 33–38.

33. $2x^3 + 7x^2 + 12x + 9$

34. $x^4 - x^3 + 7x^2 - 9x - 18$

35. $x^3 + 27$ **36.** $4x^4 - 625$ **37.** $x^6 - 64$

38. $x^5 - x^4 + 3x^3 - 3x^2 + 2x - 2$

39. Show that every polynomial with real coefficients and odd degree has at least one real root. (*Hint:* Use the Conjugate Roots Theorem.)

In Exercises 40 and 41 find the value of the polynomial at the given complex number.

40. $2x^3 + ix^2 - 3ix - (50 + 5i)$, 3

41. $3x^3 - x^2 + x - 4$, $2i$

42. (a) Show that $2i$ and $1 - i$ are both solutions of the equation

$$x^2 - (1 + i)x + (2 + 2i) = 0$$

but that $-2i$ and $1 + i$ are not.

(b) Explain why this does not violate the Conjugate Roots Theorem.

43. (a) Find the polynomial with *real* coefficients of the smallest possible degree that has i and $1 + i$ as zeros, and in which the coefficient of the highest power is 1.

(b) Find the polynomial with *complex* coefficients of the smallest possible degree that has i and $1 + i$ as zeros, and in which the coefficient of the highest power is 1.

Section 3.6

Graphing Polynomials

The graphs of polynomials of degrees 0 and 1 are lines, and the graphs of polynomials of degree 2 are parabolas. Accurate graphing techniques for these polynomials were considered in Chapters 1 and 2. The higher the degree of a polynomial, the more complicated its graph, and to draw accurately the graph of a polynomial function of degree 3 or higher requires the techniques of calculus. Nevertheless we will be able to use the knowledge we have gained about the zeros of such functions to give us a general idea of how their graphs must look. We begin by graphing the simplest polynomial functions, those of the form $y = x^n$.

Example 1

Graph the functions (a) $y = x$, (b) $y = x^2$, (c) $y = x^3$, (d) $y = x^4$, and (e) $y = x^5$.

Solution

We are already familiar from our previous work with some of these graphs, but we include them for completeness.

x	x^2	x^3	x^4	x^5
0.1	0.01	0.001	0.0001	0.00001
0.2	0.04	0.008	0.0016	0.00032
0.5	0.25	0.125	0.0625	0.03125
0.7	0.49	0.343	0.2401	0.16807
1.0	1.0	1.0	1.0	1.0
1.2	1.44	1.728	2.0736	2.48832
1.5	2.25	3.375	5.0625	7.59375
2.0	4.0	8.0	16.0	32.0

If n is odd, $(-x)^n = -x^n$, so x^n is an odd function, and if n is even, $(-x)^n = x^n$, so x^n is an even function (see Section 2.2). We use these facts to determine the values for negative x from the table. Plotting the points leads to the graphs in Figure 1.

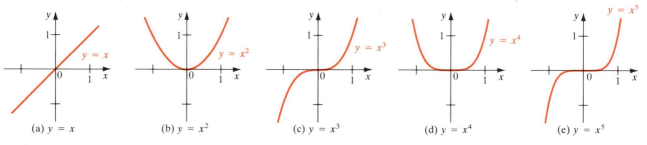

(a) $y = x$ (b) $y = x^2$ (c) $y = x^3$ (d) $y = x^4$ (e) $y = x^5$

Figure 1

As Example 1 suggests, when n is odd, the graph of $y = x^n$ has the same general shape as $y = x^3$, and when n is even, the graph of $y = x^n$ has more or less the same U-shape as $y = x^2$. However, note that as the degree n becomes larger, the graphs become flatter around the origin and steeper elsewhere.

Example 2

Sketch the graphs of the functions (a) $y = -x^3$, (b) $y = (x - 2)^4$, and (c) $y = -2x^5 + 4$.

Solution

We use the graphs in Example 1 and transform them using the techniques of Chapter 2.

(a)   The function $y = -x^3$ is the negative of $y = x^3$, so we simply reflect the graph in Figure 1(c) about the x-axis to obtain Figure 2.

(b)   The graph of the function $y = (x - 2)^4$ has the same shape as $y = x^4$.

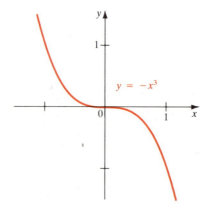

Figure 2

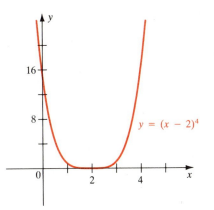

Figure 3

Replacing the x by $x - 2$ has the effect of shifting the graph of $y = x^4$ in Figure 1(d) to the right by 2 units (see Figure 3).

(c)   We begin with the graph of $y = x^5$ in Figure 1(e). Multiplying the function by 2 stretches the graph vertically. The negative sign reflects the graph about the x-axis, so at this stage we have the graph in Figure 4. Finally, adding 4 to the function shifts the graph up 4 units along the y-axis (see Figure 5). Since $-2x^5 + 4 = 0$ when $x^5 = 2$, the graph crosses the x-axis where $x = \sqrt[5]{2}$.

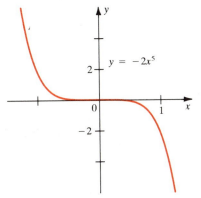

Figure 4

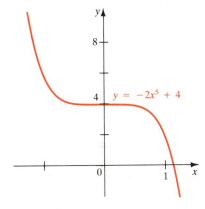

Figure 5

In the next example we use the zeros of the polynomial to help us sketch its graph. Note that if $P(c) = 0$, then the graph of $y = P(x)$ crosses the x-axis where $x = c$; that is, c is an x-intercept of the graph.

Example 3

Sketch the graph of the function $y = P(x) = x^3 - x^2 - 4x + 4$.

Solution

$$
\begin{aligned}
P(x) &= x^3 - x^2 - 4x + 4 \\
&= x^2(x - 1) - 4(x - 1) \\
&= (x^2 - 4)(x - 1) \\
&= (x + 2)(x - 2)(x - 1)
\end{aligned}
$$

The zeros of $P(x)$ are $-2, 2,$ and 1, so the graph crosses the x-axis where x takes on these values. In the table we calculate $P(x)$ for values of x in the vicinity of the zeros.

x	-3	-2	-1	0	1	$\frac{3}{2}$	2	3
$P(x)$	-20	0	6	4	0	$-\frac{7}{8}$	0	10

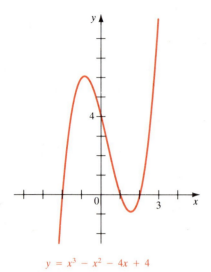

$$y = x^3 - x^2 - 4x + 4$$

Figure 6

Plotting the points given in the table and completing the sketch, we get the graph in Figure 6.   ●

The graph in Example 3 has two **turning points**, which are points at which a particle moving along the graph would change from moving upward to moving downward, or vice versa. The turning points are thus "peaks" and "valleys" on the graph. Without using calculus we cannot find the exact locations of such points. We cannot even prove that we have in fact found all the turning points that a graph may have, but one can show (again, using calculus) that a polynomial function of degree n can have at most $n - 1$ turning points. This means that in this case we have indeed found all of them, and by plotting sufficiently many points one can locate them approximately, as we did in Example 3.

Example 4

Sketch the graph of the function $P(x) = -2x^4 - x^3 + 6x^2 - x - 2$.

Solution

Using the techniques of Section 3.2, we factor

$$P(x) = -(x + 2)(2x + 1)(x - 1)^2$$

so the zeros of $P(x)$ are -2, $-\frac{1}{2}$, and 1. Using synthetic division (or a programmable calculator), we can calculate the values of $P(x)$ at various points in the vicinity of these zeros.

x	-2.5	-2	-1	-0.5	0	1	1.5	2
$P(x)$	-24.5	0	4	0	-2	0	-3.5	-20

Plotting these points and completing the graph, we obtain the graph shown in Figure 7. The graph cannot have any more turning points than the three we have found, since the degree of the polynomial is 4.   ●

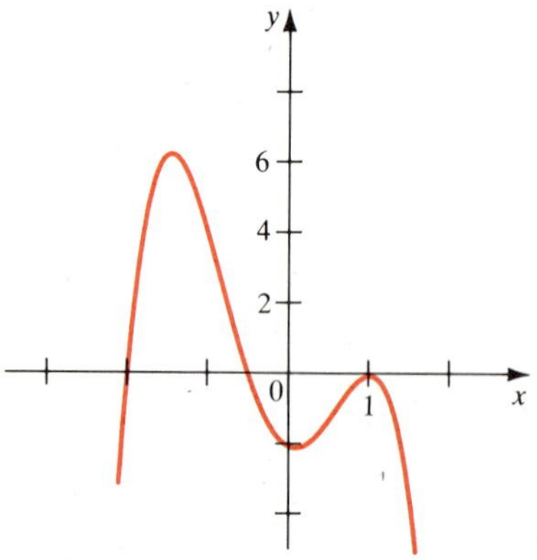

As a further aid in graphing polynomials, note that for large $|x|$, the values of $a_n x^n + a_{n-1} x^{n-1} + \cdots + a_1 x + a_0$ are close to $a_n x^n$ because

$$a_n x^n + a_{n-1} x^{n-1} + \cdots + a_1 x + a_0 = a_n x^n \left[1 + \frac{a_{n-1}}{a_n x} + \frac{a_{n-2}}{a_n x^2} + \cdots + \frac{a_0}{a_n x^n} \right]$$

and if $|x|$ is large, the quantity inside the brackets is close to 1 in value (since the reciprocals of large numbers are close to 0). So, for example, the graph of $y = P(x) = 3x^3 - 2x^2 - 2x + 4$ will look very much like the graph of $y = 3x^3$ for large $|x|$, and this in turn is just the graph of $y = x^3$ stretched vertically by a factor of 3. Of course, when $|x|$ is small (say, $-5 < x < 5$), the two graphs are quite different, since $y = P(x)$ crosses the x-axis three times, at $x = -\sqrt{2}$, $\sqrt{2}$, and 2, whereas $y = 3x^3$ crosses the x-axis only at the origin.

$$y = -2x^4 - x^3 + 6x^2 - x - 2$$

Figure 7

Exercises 3.6

Sketch the graphs of the functions in Exercises 1–8 by transforming the graph of the appropriate function of the form $y = x^n$. Indicate all x- and y-intercepts on your graph.

1. $y = -3x^4$

2. $y = (x + 3)^3$

3. $y = x^3 + 5$

4. $y = -x^6 + 8$

5. $y = -(x - 1)^4 + 1$

6. $y = 3x^5 - 9$

7. $y = 4(x - 2)^5 - 4$

8. $y = 3x^4 - 27$

Sketch the graphs of the functions in Exercises 9–22 by first plotting all x-intercepts, the y-intercept, and sufficiently many other points to detect the shape of the curve, and then filling in the rest of the graph.

9. $y = (x - 2)(x + 3)(x - 4)$

10. $y = (2x - 3)(x - 5)(x + 1)$

11. $y = (x + 2)^2(x - 3)^2$

12. $y = \frac{1}{4}(x + 1)^3(x - 3)$

13. $y = x^3 + 3x^2 - 4x - 12$

14. $y = x^3 - 4x^2 - 7x + 10$

15. $y = 2x^3 + 5x^2 + 4x + 1$

16. $y = 3x^3 - 14x^2 + 17x - 6$

17. $y = x^4 + 2x^3 - 15x^2 + 4x + 20$

18. $y = 3x^4 + 4x^3 + x^2$

19. $y = 5x^3 + 3x^2 + 8x - 4$

20. $y = x^4 - 2x^3 - 6x^2 + 2x + 5$

21. $y = x^5 - 9x^3$

22. $y = x^4 - 3x^2 - 4$

23. **(a)** On the same pair of axes, draw graphs (as accurately as possible) of the functions $y = x^3 - 2x^2 - x + 2$ and $y = -x^2 + 5x + 2$.
   (b) Based on your graph, at how many points do the two graphs appear to intersect?
   (c) Find the coordinates of all the intersection points.

24. Portions of the graphs of $y = x^2$, $y = x^3$, $y = x^4$, $y = x^5$, and $y = x^6$ are plotted in the accompanying figures. Determine which function belongs to which graph.

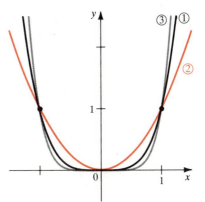

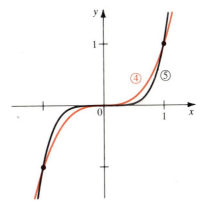

25. The steps in this problem will provide a proof of the fact that it is impossible for the graphs of two different polynomials, each of degree less than or equal to n, to intersect at more than n points.
   (a) Let $P(x) = a_n x^n + a_{n-1} x^{n-1} + \cdots + a_1 x + a_0$ and $Q(x) = b_n x^n + b_{n-1} x^{n-1} + \cdots + b_1 x + b_0$. Suppose that the graphs of $y = P(x)$ and $y = Q(x)$ intersect in the $n + 1$ points (x_1, y_1), $(x_2, y_2), \ldots, (x_n, y_n)$, and (x_{n+1}, y_{n+1}). Let $F(x) = P(x) - Q(x)$. Show that $F(x)$ has at least $n + 1$ zeros.
   (b) Show that $F(x) = 0$ for all x.
   (c) Conclude that $a_i = b_i$ for $i = 0, 1, 2, \ldots, n$ [so that $P(x)$ and $Q(x)$ are the same polynomial].

26. Recall that a function f is **odd** if $f(-x) = -f(x)$ and is **even** if $f(-x) = f(x)$, for all real x.
   (a) Show that if f and g are both odd, then so is the function $f + g$.
   (b) Show that if f and g are both even, then so is the function $f + g$.

(c) Show that if f is odd and g is even, and neither has constant value zero, then the function $f + g$ is neither even nor odd.
(d) Show that a polynomial $P(x)$ that contains only odd powers of x is an odd function.
(e) Show that a polynomial $P(x)$ that contains only even powers of x is an even function.

(f) Show that if a polynomial $P(x)$ contains both odd and even powers of x, then it is neither an odd nor an even function.
(g) Express the function

$$P(x) = x^5 + 6x^3 - x^2 - 2x + 5$$

as the sum of an odd function and an even function.

Rational Functions

A **rational function** is a function of the form

$$y = r(x) = \frac{P(x)}{Q(x)}$$

where P and Q are polynomials. We assume that $P(x)$ and $Q(x)$ have no factors in common (except possibly constants). Note that although polynomial functions are defined for all real values of x, rational functions are not defined for those values of x for which the denominator $Q(x)$ is zero. The x-intercepts (if any) of r are the zeros of the numerator $P(x)$, since a fraction is zero only when its numerator is zero.

Example 1

Find the domain, the x-intercepts, and the y-intercept of the function

$$r(x) = \frac{x^2 - 2x - 3}{2x^2 - x}$$

Solution

We factor the numerator and denominator to write

$$r(x) = \frac{(x + 1)(x - 3)}{x(2x - 1)}$$

The function is defined for all x except those for which the denominator is 0, so the domain of $r(x)$ consists of all real numbers except 0 and $\frac{1}{2}$.

The x-intercepts are the zeros of the numerator, so the graph of $y = r(x)$ crosses the x-axis where $x = -1$ and $x = 3$. The y-intercept is the value of the function when $x = 0$. Since the given function $r(x)$ is not defined for $x = 0$, the graph has no y-intercept.   ●

The most important feature that distinguishes the graphs of rational functions is **asymptotes**. Before we formulate a precise definition of this word, we consider an example to illustrate the concept.

Example 2

Draw a graph of the function $y = \dfrac{x - 3}{x - 2}$.

Solution

The function is not defined for $x = 2$, so we first examine the nature of the function for values of x near 2.

x	y
1	2
1.5	3
1.9	11
1.95	21
1.99	101
1.999	1001

x	y
3	0
2.5	-1
2.1	-9
2.05	-19
2.01	-99
2.001	-999

We see from the first table that as x approaches 2 from the left (that is, gets progressively closer to 2 while remaining less than 2), the values of y increase without bound. In fact, by taking x sufficiently close to 2 on the left, we can make y larger than any given number. We describe this situation by saying "y approaches infinity as x approaches 2 from the left," and we write this phrase using the notation

$$y \to \infty \text{ as } x \to 2^-$$

Similarly, the other table shows that as x approaches 2 from the right, the values of y decrease without bound, and by taking x sufficiently close to 2 on the right, we can make y smaller than any given negative number. In this situation we say that y approaches negative infinity as x approaches 2 from the right, and we write

$$y \to -\infty \text{ as } x \to 2^+$$

The graph of $y = r(x)$ therefore has the shape around $x = 2$ shown in Figure 1. Now we examine the behavior of the function as x becomes progressively larger in absolute value (for both negative and positive x).

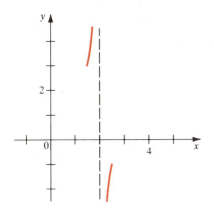

Figure 1

x	y
10	0.8750
100	0.9898
1000	0.9990
10,000	0.9999

x	y
-10	1.0833
-100	1.0098
-1000	1.0010
$-10,000$	1.0001

As $|x|$ becomes larger and larger, the values of y get progressively closer to 1. This means that the graph of $y = r(x)$ will approach the horizontal line $y = 1$ as x increases or decreases without bound. We express this by saying "y approaches 1 as x approaches infinity or negative infinity," and we write

$$y \to 1 \text{ as } x \to \infty \qquad \text{and} \qquad y \to 1 \text{ as } x \to -\infty$$

This means we can complete the graph of $y = r(x)$ as shown in Figure 2. •

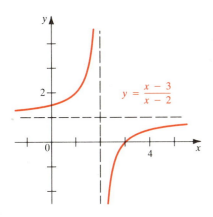

$$y = \frac{x - 3}{x - 2}$$

Figure 2

The line $x = 2$ is called a **vertical asymptote** of the graph in Example 2, and the line $y = 1$ is called a **horizontal asymptote**. An asymptote of a function is, informally speaking, a line that the graph of the function gets closer and closer to as one travels out along that line in either direction. More formally, we make the following definitions.

The line $x = a$ is a **vertical asymptote** of the function $y = f(x)$ if $y \to \infty$ or $y \to -\infty$ as x approaches a from the right or the left.

If the function $y = f(x)$ is a rational function, then its vertical asymptotes are the lines $x = a$, where a is a zero of the denominator of the function, because only when the denominator is 0 does the function fail to have a real, finite value.

The line $y = b$ is a **horizontal asymptote** of the function $y = f(x)$ if $y \to b$ as $x \to \infty$ or as $x \to -\infty$.

In the next example, we show the most efficient method for finding the horizontal asymptote of a rational function.

Example 3

Find all vertical and horizontal asymptotes and the x- and y-intercepts of the following function. Use this information to graph the function.

$$y = r(x) = \frac{x^2 - x - 6}{2x^2 + 5x - 3}$$

Solution

Factoring the numerator and denominator, we see that

$$r(x) = \frac{(x - 3)(x + 2)}{(2x - 1)(x + 3)}$$

The x-intercepts of the graph are the zeros of the numerator: $x = 3$ and $x = -2$. The y-intercept is $r(0) = (-6)/(-3) = 2$.

The horizontal asymptote (if it exists) will be the value that y approaches as $x \to \pm\infty$. To help us find this value, let us begin by dividing both the numerator and the denominator of $r(x)$ by x^2 (the highest power of x that appears in the function):

$$y = r(x) = \frac{1 - \dfrac{1}{x} - \dfrac{6}{x^2}}{2 + \dfrac{5}{x} - \dfrac{3}{x^2}}$$

Any function of the form c/x^n approaches 0 as $x \to \pm\infty$ (if n is a positive integer). For example, in the table we examine the values of $6/x^2$ (which appears in the above quotient) as x increases.

x	$6/x^2$	
10	0.06	
100	0.0006	
1000	0.000006	
10,000	0.00000006	← approaching 0

This means that as $x \to \pm\infty$,

$$y \to \frac{1 - 0 - 0}{2 + 0 - 0} = \frac{1}{2}$$

so that $y = \frac{1}{2}$ is the horizontal asymptote.

The vertical asymptotes occur where the function is undefined (or in other words, where the denominator is 0). This means that the vertical asymptotes here are $x = \frac{1}{2}$ and $x = -3$. To be able to graph the function, we need to know whether $y \to \infty$ or $y \to -\infty$ on each side of these vertical lines. Since there are only two choices, we need only determine the sign of y for values of x near the vertical asymptotes. As $x \to \frac{1}{2}^+$, the values of x are slightly larger than $\frac{1}{2}$, so $x - 3 < 0$, $x + 2 > 0$, $2x - 1 > 2 \cdot \frac{1}{2} - 1 = 0$, and $x + 3 > 0$. This means that as $x \to \frac{1}{2}^+$, y is the quotient of one negative and three positive factors and therefore must be negative. So $y \to -\infty$ as $x \to \frac{1}{2}^+$. We can represent what happens here and at other sides of the vertical asymptotes schematically as in the following table:

as $x \to$	$\frac{1}{2}^+$	$\frac{1}{2}^-$	-3^+	-3^-
sign of $y = \dfrac{(x-3)(x+2)}{(2x-1)(x+3)}$	$\dfrac{(-)(+)}{(+)(+)}$	$\dfrac{(-)(+)}{(-)(+)}$	$\dfrac{(-)(-)}{(-)(+)}$	$\dfrac{(-)(-)}{(-)(-)}$
$y \to$	$-\infty$	∞	$-\infty$	∞

Putting together this information about intercepts, asymptotes, and the behavior of the function near asymptotes, we obtain the partial graph in Figure 3, where the asymptotes have been plotted as broken lines.

All we need to do now is to plot a few more points and fill in the rest of the graph as in Figure 4.

x	-4	-1	1	2	4	6
y	1.56	0.67	-1.5	-0.27	0.12	0.24

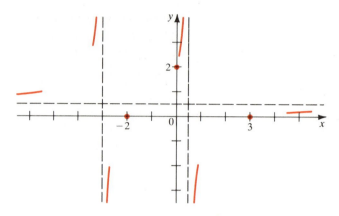

Figure 3

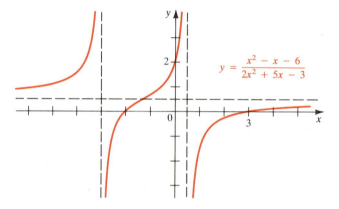

$$y = \frac{x^2 - x - 6}{2x^2 + 5x - 3}$$

Figure 4

We summarize the procedure to be followed in graphing rational functions by the following sequence of steps.

Sketching Graphs of Rational Functions

1. Factor the numerator and denominator.

2. Find the x-intercepts by determining the zeros of the numerator, and the y-intercept from the value of the function at $x = 0$.

3. Find the horizontal asymptote (if any) by dividing both numerator and denominator by the highest power of x that appears in the denominator and then letting $x \to \pm\infty$.

4. Find the x-intercepts by determining the zeros of the denominator, and then see if $y \to \infty$ or $y \to -\infty$ on each side of each vertical asymptote.

5. Sketch a partial graph using the information provided by the first four steps of this procedure. Then plot as many additional points as required to fill in the rest of the graph of the function.

Example 4

Graph the function $y = r(x) = \dfrac{x^2}{x^3 - 2x^2 - x + 2}$.

Solution

The graph passes through the origin. There are no other x- or y-intercepts. Dividing numerator and denominator by x^3 gives

$$r(x) = \frac{\dfrac{1}{x}}{1 - \dfrac{2}{x} - \dfrac{1}{x^2} + \dfrac{2}{x^3}}$$

so as $x \to \pm\infty$, $y \to 0/(1 - 0 - 0 + 0) = 0$. The horizontal asymptote is $y = 0$ (the x-axis).

Factoring the denominator, we get

$$r(x) = \frac{x^2}{(x - 2)(x - 1)(x + 1)}$$

so the vertical asymptotes are $x = 2$, $x = 1$, and $x = -1$. The table shows the behavior of the function around the asymptotes.

as $x \to$	2^+	2^-	1^+	1^-	-1^+	-1^-
sign of $y = \dfrac{x^2}{(x - 2)(x - 1)(x + 1)}$	$\dfrac{(+)}{(+)(+)(+)}$	$\dfrac{(+)}{(-)(+)(+)}$	$\dfrac{(+)}{(-)(+)(+)}$	$\dfrac{(+)}{(-)(-)(+)}$	$\dfrac{(+)}{(-)(-)(+)}$	$\dfrac{(+)}{(-)(-)(-)}$
$y \to$	∞	$-\infty$	$-\infty$	∞	∞	$-\infty$

Calculating the coordinates of a few more points now provides us with enough information to sketch the graph of the function in Figure 5.

x	-2	$-\frac{1}{2}$	$\frac{1}{2}$	$\frac{3}{2}$	3
y	$-\frac{1}{3}$	$\frac{2}{15}$	$\frac{2}{9}$	$-3\frac{3}{5}$	$\frac{9}{8}$

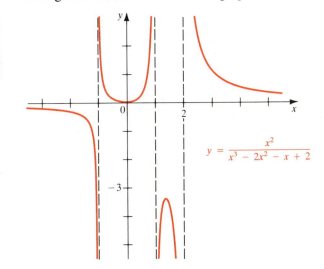

$$y = \frac{x^2}{x^3 - 2x^2 - x + 2}$$

Figure 5

Let

$$r(x) = \frac{P(x)}{Q(x)} = \frac{a_n x^n + a_{n-1} x^{n-1} + \cdots + a_1 x + a_0}{b_m x^m + b_{m-1} x^{m-1} + \cdots + b_1 x + b_0}$$

be a rational function. If the degrees of P and Q are the same (so that $n = m$), then we can see by dividing both numerator and denominator by x^n that $y = a_n/b_m$ is the horizontal asymptote of $y = r(x)$. This was the case in Example 3. If the degree of P is less than the degree of Q (that is, $n < m$), then dividing numerator and denominator by x^m shows that $y = 0$ is the horizontal asymptote, as in Example 4. Finally, if $n > m$, the same procedure shows that the function has no horizontal asymptotes, as the next example illustrates.

Example 5

Graph the function $y = r(x) = \dfrac{x^3 + 1}{x - 2}$.

Solution

Factoring the numerator, we see that

$$r(x) = \frac{(x + 1)(x^2 - x + 1)}{x - 2}$$

so the x-intercept is -1, the y-intercept is $-\frac{1}{2}$, and the only vertical asymptote is $x = 2$. By performing the same kind of analysis as in the previous two examples, we can show that $y \to \infty$ as $x \to 2^+$ and $y \to -\infty$ as $x \to 2^-$.

Dividing numerator and denominator by x gives

$$y = \frac{x^2 + \dfrac{1}{x}}{1 - \dfrac{2}{x}}$$

As $x \to \infty$, the terms $1/x$ and $2/x$ approach 0, so since $x^2 \to \infty$ and the denominator approaches 1, $y \to \infty$. Similarly, as $x \to -\infty$, $y \to \infty$, so the function has no horizontal asymptote.

Plotting the points listed in the table allows us to complete the graph of the function in Figure 6.

x	-3	-2	1	3	4
y	5.2	1.75	-2	28	32.5

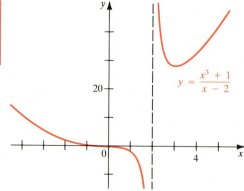

Figure 6

As with the graphs of polynomial functions, we cannot tell exactly where the turning point of the right branch of the function in Example 5 lies without the use of calculus. But from the asymptotic behavior of the function, we know that such a turning point must exist, and we could locate it more precisely by plotting more points in the vicinity of $x = 3$.

If $r(x) = P(x)/Q(x)$ is a rational function in which the degree of the numerator is one more than the degree of the denominator, we can use the Division Algorithm to express the function in the form

$$r(x) = ax + b + \frac{R(x)}{Q(x)}$$

where the degree of R is less than the degree of Q and $a \neq 0$. This means that as $x \to \pm\infty$, $R(x)/Q(x) \to 0$, so for large values of $|x|$, the graph of $y = r(x)$ approaches the graph of the line $y = ax + b$. In this situation we say that $y = ax + b$ is a **slant asymptote** or an **oblique asymptote**.

Example 6

Graph the function

$$r(x) = \frac{x^2 - 4x - 5}{x - 3}$$

Solution

Since the degree of the numerator is one more than the degree of the denominator, the function will have a slant asymptote. After dividing, we find

$$r(x) = x - 1 - \frac{8}{x - 3}$$

so the slant asymptote is the line $y = x - 1$. The line $x = 3$ is a vertical asymptote, and it is easy to see that $r(x) \to -\infty$ as $x \to 3^+$ and $r(x) \to \infty$ as $x \to 3^-$.

Factoring the numerator in the original formula for r gives

$$r(x) = \frac{(x + 1)(x - 5)}{x - 3}$$

so the x-intercepts are -1 and 5, and the y-intercept is $\frac{5}{3}$. Plotting the asymptotes, intercepts, and the additional points listed in the table allows us to complete the graph of the function in Figure 7.

x	-2	1	2	4	6
y	-1.4	4	9	-5	2.33

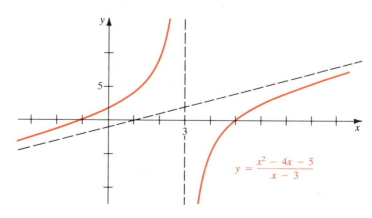

$$y = \frac{x^2 - 4x - 5}{x - 3}$$

Figure 7

When graphing a rational function in which the degree of the numerator is one plus the degree of the denominator, we must modify Step 3 of the guidelines given on page 152. Instead of finding the horizontal asymptote (which does not exist in this case), we find the slant asymptote using the method described in Example 6.

Exercises 3.7

In Exercises 1–6 find the x- and y-intercepts of the functions.

1. $y = \dfrac{x - 6}{x^2 - 2x + 1}$

2. $y = \dfrac{2}{x^3 - 2x + 8}$

3. $y = \dfrac{x^2 - 2x - 15}{x^4 + 16}$

4. $y = \dfrac{x^4 + 16}{x^2 - 2x - 15}$

5. $y = \dfrac{x^3 - 27}{x^3 - x}$

6. $y = \dfrac{x^2 + 10}{2x}$

In Exercises 7–16 find all asymptotes (including vertical, horizontal, and slant).

7. $y = \dfrac{5}{2x + 3}$

8. $y = \dfrac{3x + 1}{x - 3}$

9. $y = \dfrac{x^2}{x^2 - x - 6}$

10. $y = \dfrac{2x - 5}{x^2 + x + 1}$

11. $y = \dfrac{6}{x^4 + 2x^2 + 2}$

12. $y = \dfrac{(x - 1)(x - 2)(x - 3)}{(x - 4)(x - 5)}$

13. $y = \dfrac{3x^4 + 5x^3 + 2}{x^3 - 64}$

14. $y = \dfrac{x^3 + 3x^2}{x^2 + x + 16}$

15. $y = \dfrac{2x^3 - x^2 - 8x + 4}{x + 3}$

16. $y = \dfrac{6x^4 + x^2 - 1}{x^2 + 64}$

In Exercises 17–42 find intercepts and asymptotes, and then graph the given rational functions.

17. $y = \dfrac{4}{x + 2}$

18. $y = \dfrac{x + 9}{x - 3}$

19. $y = \dfrac{4x - 15}{2x + 3}$

20. $y = \dfrac{3x - 7}{x^2 - 6x - 7}$

21. $y = \dfrac{18}{(x - 3)^2}$

22. $y = \dfrac{x - 2}{(x + 2)^2}$

23. $y = \dfrac{(x - 1)(x + 2)}{(2x + 3)(x - 3)}$

24. $y = \dfrac{x(x + 4)}{(x^2 - 1)(x - 2)}$

25. $y = \dfrac{6x^3}{x^3 - 1}$

26. $y = \dfrac{x^3 - 8}{x^3 + 8}$

27. $y = \dfrac{x^2 - 4x + 3}{x^2 - 6x + 8}$

28. $y = \dfrac{6x^2 + 5x - 4}{3x^2 + 3x - 6}$

29. $y = \dfrac{x^2 + 4}{2x^2 + x - 1}$

30. $y = \dfrac{x^2 + x + 2}{2x - 1}$

31. $y = \dfrac{x^2 + 5x + 4}{x - 3}$

32. $y = \dfrac{x^3 + 2}{x + 2}$

33. $y = \dfrac{x^4 - 4}{x^2 - 4}$

34. $y = \dfrac{x^3 - 8}{x^2 - 8}$

35. $y = \dfrac{x^4 + 16}{2x^2 - x - 6}$

36. $y = \dfrac{16}{1 - x^4}$

37. $y = \dfrac{16}{1 + x^4}$

38. $y = \dfrac{x^3 - 3x^2 - x + 3}{3x^3 - 12x}$

39. $y = \dfrac{4x^3 + 8x^2 - x - 2}{x^2 + 2x - 3}$

40. $y = \dfrac{x^3 - x^2 - x + 1}{x^2 + 4x + 4}$

41. $y = \dfrac{4x^2 - 12x + 8}{2x^3 - 7x^2 - 5x + 4}$

42. $y = \dfrac{6x^3 - 24}{2x^3 + 11x^2 + 20x + 12}$

In this chapter we adopted the convention that in rational functions, the numerator and denominator do not share common factors. In Exercises 43–47 we study the graphs of rational functions that do not satisfy this rule.

43. Show that the graph of

$$r(x) = \frac{3x^2 - 3x - 6}{x - 2}$$

is the line $y = 3x + 3$ with the point $(2, 9)$ removed. (*Hint:* Divide. What is the domain of r?)

In Exercises 44–47 graph the given rational functions.

44. $y = \dfrac{x^2 + x - 20}{x + 5}$

45. $y = \dfrac{2x^2 - 5x - 3}{x^2 - 2x - 3}$

46. $y = \dfrac{x^2 - 3x + 2}{x^2 - 4x + 4}$

47. $y = \dfrac{2x - 5}{2x^3 - 5x^2 - 8x + 20}$

In Exercises 48–51 construct a rational function $y = P(x)/Q(x)$ that has the indicated properties, and in which the degrees of P and Q are as small as possible.

48. The function has vertical asymptote $x = 3$, horizontal asymptote $y = 0$, and y-intercept $-\frac{1}{3}$, and never crosses the x-axis.

49. The function has vertical asymptotes $x = 1$ and $x = -4$, horizontal asymptote $y = 1$, and x-intercepts 2 and 3.

50. The function has horizontal asymptote $y = 2$ but no vertical asymptotes. The origin is the only x-intercept, and i is a zero of $Q(x)$.

51. The function has slant asymptote $y = 3x - 6$ and vertical asymptote $x = \frac{1}{2}$, and its graph passes through the origin.

52. Show that the function

$$y = \frac{x^6 + 10}{x^4 + 8x^2 + 15}$$

has no horizontal, vertical, or slant asymptotes, and no x-intercepts.

Section 3.8

Polynomial and Rational Inequalities

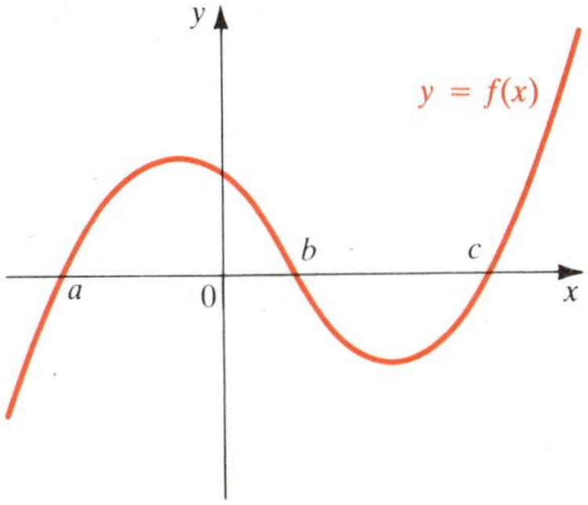

Figure 1

Functions with graphs that consist of a single unbroken curve are called **continuous**. We have seen from our work in Section 3.6 that polynomials are continuous functions. One important property of such functions, which we will use in this section to help us solve inequalities, is that if a and b are successive x-intercepts of the continuous function f, then either $f(x) > 0$ or $f(x) < 0$ for all values of x between a and b. For example, in Figure 1 $f(x) > 0$ if $a < x < b$ and $f(x) < 0$ if $b < x < c$. Moreover, assuming that the graph shows all the x-intercepts of the function, $f(x) < 0$ for $x < a$ and $f(x) > 0$ for $x > c$.

The x-intercepts of a polynomial are its zeros. We will use our knowledge of finding zeros of polynomials, together with the property of continuous functions we have just described, to solve polynomial inequalities. We have already encountered linear and quadratic inequalities in Section 1.5, but we are now able to attack more general types of inequalities.

Example 1

Find all values of x for which $x^2 + 3x - 10 > 0$.

Solution 1

Using the techniques of Section 2.4, we see that the graph of

$$y = x^2 + 3x - 10 = (x + 5)(x - 2)$$
$$= (x + \tfrac{3}{2})^2 - \tfrac{49}{4}$$

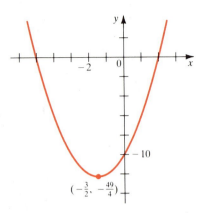

Figure 2

is a parabola with vertex $(-\frac{3}{2}, -\frac{49}{4})$, x-intercepts -5 and 2, and y-intercept -10. Its graph is shown in Figure 2. We are looking for values of x for which $y = x^2 + 3x - 10 > 0$. From the figure, we see that this is the case where

$$x < -5 \quad \text{or} \quad x > 2$$

Solution 2

Factoring the inequality, we get

$$(x + 5)(x - 2) > 0$$

The polynomial has zeros at -5 and 2, so it will be positive on one or more of the intervals $(-\infty, -5)$, $(-5, 2)$, and $(2, \infty)$ by what we have just learned about continuous functions.

Interval	$x + 5$	$x - 2$	$(x + 5)(x - 2)$
$x < -5$	$-$	$-$	$+$
$-5 < x < 2$	$+$	$-$	$-$
$2 < x$	$+$	$+$	$+$

To determine on which interval(s) the function is positive, it is enough to check the sign of the factors at a single, conveniently chosen point in each interval. To construct the preceding table, for example, we picked the test values $x = -6, 0$, and 3. The inequality is satisfied if x lies in the first or third of these intervals. The solution is

$$(-\infty, -5) \cup (2, \infty)$$ ●

Example 2

Solve the inequality $x^2(3x - 4) \le 5x - 2$.

Solution

We first express the inequality in the form $P(x) \le 0$ so that we can use the factoring technique of Example 1 to solve the inequality:

$$3x^3 - 4x^2 - 5x + 2 \le 0$$

By using the methods of Section 3.2, we can show that the zeros of the polynomial are $-1, \frac{1}{3}$, and 2, so the inequality factors as follows:

$$(x + 1)(3x - 1)(x - 2) \le 0$$

The polynomial will be negative on one or more of the intervals $(-\infty, -1)$,

$(-1, \frac{1}{3})$, $(\frac{1}{3}, 2)$, and $(2, \infty)$, so we check the sign of the polynomial at a test point in each of these intervals.

Interval	$x + 1$	$3x - 1$	$x - 2$	$(x + 1)(3x - 1)(x - 2)$
$x < -1$	$-$	$-$	$-$	$-$
$-1 < x < \frac{1}{3}$	$+$	$-$	$-$	$+$
$\frac{1}{3} < x < 2$	$+$	$+$	$-$	$-$
$2 < x$	$+$	$+$	$+$	$+$

Note that when we substitute test values, we do not need to actually calculate the products in the last column of this table. We are interested only in the *sign* of the product, which can be determined by counting negative factors. The table shows us that $(-\infty, -1)$ and $(\frac{1}{3}, 2)$ are in the solution set. Finally, we note that the zeros of the polynomial also belong to the solution because the inequality is of the form $\leq$. The complete solution is therefore

$$(-\infty, -1] \cup [\tfrac{1}{3}, 2] \qquad \bullet$$

From the examples in Section 3.7 we can see that, in general, rational functions are not continuous. The vertical asymptotes of such functions break up their graphs into separate branches. Thus when we are solving inequalities that involve rational functions, we must take their vertical asymptotes (as well as their x-intercepts) into account when setting up the intervals on which the function does not change sign. This is the reason for the following definition:

> If $r(x) = P(x)/Q(x)$ is a rational function, then the **cut points** for r are the values of x at which either $P(x)$ or $Q(x)$ is 0.

If a and b are two successive cut points for the rational function r, then either $r(x) > 0$ or $r(x) < 0$ for all x between a and b. We use this fact to solve rational inequalities in the following examples.

Example 3

Solve the inequality $\dfrac{2x + 5}{x - 5} < 0$.

Solution

The cut points of the given rational function are $-\frac{5}{2}$ and 5, so we must check the intervals $(-\infty, -\frac{5}{2})$, $(-\frac{5}{2}, 5)$, and $(5, \infty)$. We will determine on which of these intervals $(2x + 5)/(x - 5) < 0$ by checking the sign of the function at test values chosen from each interval.

Interval	$2x + 5$	$x - 5$	$(2x + 5)/(x - 5)$
$x < -\frac{5}{2}$	$-$	$-$	$+$
$-\frac{5}{2} < x < 5$	$+$	$-$	$-$
$5 < x$	$+$	$+$	$+$

Only numbers from the middle interval satisfy the inequality. Since neither $-\frac{5}{2}$ nor 5 is itself a solution, the complete solution is $(-\frac{5}{2}, 5)$. ●

Example 4

Find the solution set of

$$\frac{2x + 1}{x - 3} - \frac{x}{x + 1} \geq 1$$

Solution

If this was an equation rather than an inequality, we could multiply both sides by $(x - 3)(x + 1)$ to clear the denominators, and then solve. But multiplying an inequality by a negative number reverses its direction (see Section 1.2). Since we do not know whether $(x - 3)(x + 1)$ will turn out to be positive or negative, we cannot simplify in this way. Instead we subtract 1 from both sides (to make the right side 0) and then combine the fractions.

$$\frac{2x + 1}{x - 3} - \frac{x}{x + 1} - 1 \geq 0$$

$$\frac{(2x + 1)(x + 1) - x(x - 3) - (x - 3)(x + 1)}{(x - 3)(x + 1)} \geq 0$$

$$\frac{(2x^2 + 3x + 1) - (x^2 - 3x) - (x^2 - 2x - 3)}{(x - 3)(x + 1)} \geq 0$$

$$\frac{8x + 4}{(x - 3)(x + 1)} \geq 0$$

$$\frac{4(2x + 1)}{(x - 3)(x + 1)} \geq 0$$

The cut points are -1, $-\frac{1}{2}$, and 3, so the intervals to be considered are $(-\infty, -1), (-1, -\frac{1}{2}), (-\frac{1}{2}, 3)$, and $(3, \infty)$. This leads to the following table:

Interval	$x + 1$	$2x + 1$	$x - 3$	$4(2x + 1)/[(x - 3)(x + 1)]$
$x < -1$	$-$	$-$	$-$	$-$
$-1 < x < -\frac{1}{2}$	$+$	$-$	$-$	$+$
$-\frac{1}{2} < x < 3$	$+$	$+$	$-$	$-$
$3 < x$	$+$	$+$	$+$	$+$

The numbers in the intervals $(-1, -\frac{1}{2})$ and $(3, \infty)$ belong to the solution. Checking the cut points, we see that $-\frac{1}{2}$ is a solution but that -1 and 3 are not, since the function is undefined there. The complete solution is

$$(-1, -\tfrac{1}{2}] \cup (3, \infty) \qquad \bullet$$

To summarize, we solve inequalities involving polynomials or rational functions by performing the following sequence of steps:

Solving Inequalities

1. If necessary, rewrite the inequality so that one side is 0. (Take care not to multiply by expressions that have an undetermined sign.)

2. Factor the polynomial (or the numerator and denominator of the rational function) into linear factors, and use these to find the cut points.

3. List the intervals determined by the cut points.

4. Check the sign of the polynomial or rational function on each interval by calculating the sign at some convenient test number chosen from the interval.

5. Check whether the inequality is satisfied by some or all of the cut points themselves. (This may happen if the inequality involves $\geq$ or $\leq$.)

6. Combine the information obtained in Steps 4 and 5 to get the complete solution set.

In the next example we introduce a short-hand form for the table we have been using in our work so far to do Step 4.

Example 5

Solve
$$\frac{x^2 + 2x - 5}{(x - 5)^2} \geq 0.$$

Solution

Using the quadratic formula, we see that the zeros of the numerator are $1 + \sqrt{6}$ and $1 - \sqrt{6}$ (or about 3.45 and -1.45). Thus the cut points are $1 - \sqrt{6}, 1 + \sqrt{6}$, and 5, and the rational function factors into

$$\frac{(x - (1 - \sqrt{6}))(x - (1 + \sqrt{6}))}{(x - 5)^2}$$

Instead of making a table as before, we place the cut points on a number line and determine the sign of the function on each interval created by them (see Figure 3).

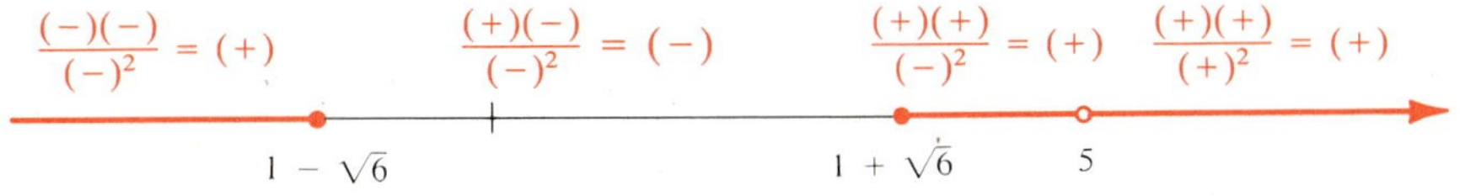

Figure 3

The numbers $1 - \sqrt{6}$ and $1 + \sqrt{6}$ do satisfy the inequality, but 5 does not. We have indicated this using filled-in and open circles in Figure 3. The solution set is thus

$$(-\infty, 1 - \sqrt{6}] \cup [1 + \sqrt{6}, 5) \cup (5, \infty)$$

Exercises 3.8

Solve the inequalities in Exercises 1–54.

1. $x^2 - x - 20 > 0$

2. $x^2 + 5x \leq 0$

3. $x^2 + 3x - 9 \geq 0$

4. $x^2 + 2x + 3 < 0$

5. $x^2 - 16 \leq 0$

6. $x^2 > 25$

7. $3x^2 + 11x - 4 \leq 0$

8. $6x^2 + 5x + 1 > 0$

9. $x(x - 2) < 8$

10. $3 + x^2 \leq 4x$

11. $x + 14 \geq 3x^2$

12. $2 > x(11 - 5x)$

13. $x^2 + 2 > x$

14. $2x^2 + 5x + 4 \leq 0$

15. $x^2 + 2x - 7 \leq 0$

16. $3x^2 - x + 1 > 0$

17. $9x < 2x^2 + 7$

18. $(x - 3)(x + 5)(2x + 11) < 0$

19. $(x - 1)(x + 2)(x - 3)(x + 4) \geq 0$

20. $(x + 5)^2(x + 3)(x - 1) > 0$

21. $(2x - 7)^4(x - 1)^3(x + 1) \leq 0$

22. $x^3 + 4x^2 - 4x - 16 \geq 0$

23. $2x^3 - x^2 - 18x + 9 < 0$

24. $2x^3 - x^2 + 18x - 9 < 0$

25. $x^4 + 3x^3 - x - 3 \geq 0$

26. $x(1 - x^2)^3 > 7(1 - x^2)^3$

27. $x^4 - 11x^2 - 18 < 0$

28. $4x^4 - 25x^2 + 36 \leq 0$

29. $x^3 + 6x > 5x^2$

30. $x^3 + x^2 - 17x + 15 \geq 0$

31. $x^2(7 - 6x) \leq 1$

32. $x^4 + 3x^3 - 3x^2 + 3x - 4 < 0$

33. $\dfrac{x - 1}{x - 10} < 0$

34. $\dfrac{3x - 7}{x + 2} \leq 0$

35. $\dfrac{2x + 5}{x^2 + 2x - 35} \geq 0$

36. $\dfrac{4x^2 - 25}{x^2 - 9} > 0$

37. $\dfrac{x}{x^2 + 2x - 2} \leq 0$

38. $\dfrac{x + 1}{2x^2 - 4x + 1} > 0$

39. $\dfrac{x^2 + 2x - 3}{3x^2 - 7x - 6} > 0$

40. $\dfrac{x - 1}{x^3 + 1} \geq 0$

41. $\dfrac{x^3 + 3x^2 - 9x - 27}{x + 4} \leq 0$

42. $\dfrac{x^2 - 16}{x^4 - 16} < 0$

43. $\dfrac{x - 3}{2x + 5} \geq 1$

44. $\dfrac{1}{x} + \dfrac{1}{x + 1} < \dfrac{2}{x + 2}$

45. $2 + \dfrac{1}{1 - x} \leq \dfrac{3}{x}$

46. $\dfrac{1}{x - 3} + \dfrac{1}{x + 2} \geq \dfrac{2x}{x^2 + x - 2}$

47. $\dfrac{(x - 1)^2}{(x + 1)(x + 2)} > 0$

48. $\dfrac{x^2 - 2x + 1}{x^3 + 3x^2 + 3x + 1} \leq 0$

49. $\dfrac{6}{x - 1} - \dfrac{6}{x} \geq 1$

50. $\dfrac{x}{2} \geq \dfrac{5}{x + 1} + 4$

51. $\dfrac{x + 2}{x + 3} < \dfrac{x - 1}{x - 2}$

52. $\dfrac{1}{x + 1} + \dfrac{1}{x + 2} \leq \dfrac{1}{x + 3}$

53. $\dfrac{(1 - x)^2}{\sqrt{x}} \geq 4\sqrt{x}(x - 1)$

54. $\frac{2}{3}x^{-1/3}(x + 2)^{1/2} + \frac{1}{2}x^{2/3}(x + 2)^{-1/2} < 0$

In Exercises 55–58 find all values of x for which the graph of f_1 lies above the graph of f_2.

55. $f_1(x) = x^2$, $f_2(x) = 3x + 10$

56. $f_1(x) = \dfrac{1}{x}$, $f_2(x) = \dfrac{1}{x - 1}$

57. $f_1(x) = 4x$, $f_2(x) = \dfrac{1}{x}$

58. $f_1(x) = x^3 + x^2$, $f_2(x) = \dfrac{1}{x}$

Find the domains of the functions in Exercises 59–62.

59. $f(x) = \sqrt{6 + x - x^2}$ **60.** $g(x) = \sqrt{\dfrac{5 + x}{5 - x}}$

61. $h(x) = \sqrt[4]{x^4 - 1}$ **62.** $k(x) = \dfrac{1}{\sqrt{x^4 - 5x^2 + 4}}$

63. Solve

$$\frac{x^2 + (a - b)x - ab}{x + c} \le 0$$

where $0 < a < b < c$.

CHAPTER 3 REVIEW

Define, state, or discuss the following.

1. The quadratic formula
2. The discriminant
3. Polynomial of degree n
4. Rational function
5. Dividend, divisor, quotient, and remainder
6. Division Algorithm
7. Remainder Theorem
8. Synthetic division
9. Factor Theorem
10. Zero of a polynomial
11. Root of a polynomial equation
12. Rational Roots Theorem
13. Descartes' Rule of Signs
14. Upper and lower bounds for roots
15. Upper and Lower Bounds Theorem
16. Intermediate Value Theorem for Polynomials
17. Finding approximate values for irrational zeros of polynomials
18. Complex number
19. Real part and imaginary part
20. Imaginary numbers, pure imaginary numbers
21. Complex conjugate
22. Fundamental Theorem of Algebra
23. Complete Factorization Theorem
24. Multiplicity of a zero
25. Zeros Theorem
26. Conjugate Roots Theorem
27. The graph of $y = x^n$, n a positive integer
28. Vertical, horizontal, and slant asymptotes
29. Cut points for a rational function

Review Exercises

In Exercises 1–8 find the quotient and remainder, using synthetic division whenever possible.

1. $\dfrac{x^3 - x^2 + x - 11}{x - 3}$ **2.** $\dfrac{x^4 + 30x + 12}{2x + 6}$

3. $\dfrac{x^3 - x^2 - 11x + 6}{x^2 + 2x - 5}$

4. $\dfrac{x^5 - 3x^4 + 3x^3 + 20x - 6}{x^2 + 2x - 6}$

5. $\dfrac{x^4 - 25x^2 + 4x + 15}{x + 5}$ **6.** $\dfrac{2x^3 - x^2 - 5}{x - \frac{3}{2}}$

7. $\dfrac{x^4 + x^3 - 2x^2 - 3x - 1}{x - \sqrt{3}}$

8. $\dfrac{15x - 7}{5x + 12}$

In Exercises 9 and 10 find the indicated value of the given polynomial using the Remainder Theorem.

9. $P(x) = 2x^3 - 9x^2 - 7x + 13$; find $P(5)$.

10. $Q(x) = x^4 + 4x^3 + 7x^2 + 10x + 15$; find $Q(-3)$.

11. Show that $\frac{1}{2}$ is a zero of the polynomial $2x^4 + x^3 - 5x^2 + 10x - 4$.

12. Show using the Factor Theorem that $x + 4$ is a factor of the polynomial $x^5 + 4x^4 - 7x^3 - 23x^2 + 23x + 12$.

13. What is the remainder when the polynomial $x^{500} + 6x^{201} - x^2 - 2x + 4$ is divided by $x - 1$?

In Exercises 14 and 15 list all possible rational roots (without testing to see if they actually are roots) and then determine the possible number of positive and negative real roots using Descartes' Rule of Signs.

14. $x^5 - 6x^3 - x^2 + 2x + 18 = 0$

15. $6x^4 + 3x^3 + x^2 + 3x + 4 = 0$

Evaluate the complex numbers in Exercises 16–21, writing the result in the form $a + bi$. Then find the modulus (absolute value) of the result.

16. $(3 - 5i) - (6 + 4i)$ **17.** $(2 + 7i)(6 - i)$

18. i^{45} **19.** $\dfrac{2 + i}{4 - 3i}$

20. $(1 - \sqrt{-3})(2 + \sqrt{-4})$ **21.** $\sqrt{-5} \cdot \sqrt{-20}$

22. Find a polynomial of degree 3 with constant coefficient 12 and with zeros $-\frac{1}{2}$, 2, and 3.

23. Find a polynomial of degree 4 that has integer coefficients and zeros $3i$ and 4, with 4 a double root.

In Exercises 24–33 find all rational, irrational, and imaginary roots (and state their multiplicities). Use Descartes' Rule of Signs, the Upper and Lower Bounds Theorem, the quadratic formula, or other factoring techniques to help you whenever possible.

24. $x^3 - 3x^2 - 13x + 15 = 0$

25. $2x^3 + 5x^2 - 6x - 9 = 0$

26. $x^4 + 6x^3 + 17x^2 + 28x + 20 = 0$

27. $x^4 + 7x^3 + 9x^2 - 17x - 20 = 0$

28. $x^5 - 3x^4 - x^3 + 11x^2 - 12x + 4 = 0$

29. $x^4 = 81$

30. $x^6 = 64$

31. $18x^3 + 3x^2 - 4x - 1 = 0$

32. $6x^4 - 18x^3 + 6x^2 - 30x + 36 = 0$

33. $x^4 + 15x^2 + 54 = 0$

34. Show that $2 - i\sqrt{2}$ is a root of $x^4 - 5x^3 + 8x^2 + 2x - 12 = 0$, and find all other roots.

35. Show that i and $-1 + i\sqrt{6}$ are roots of $x^6 + x^5 + 7x^4 - 4x^3 + 13x^2 - 5x + 7 = 0$, and find all other roots.

36. The polynomial $x^4 - x^2 - x - 2 = 0$ has one positive real root. Show that it is irrational, and find a decimal approximation for it, correct to two decimal places.

37. **(a)** Show that the polynomial $x^4 + 2x^2 - x - 1$ has exactly two real zeros, and that neither is rational.
   (b) Find decimal approximations of each of the two real roots, correct to the nearest hundredth.

Graph the polynomials and rational functions in Exercises 38–46. Show clearly all x- and y-intercepts and asymptotes.

38. $y = x^3 - 5x^2 - 4x + 20$

39. $y = x^3 - 3x^2$

40. $y = 2x^3 + 7x^2 + 2x - 3$

41. $y = x^4 - x^3 - 5x^2 + 4x + 4$

42. $y = \dfrac{3x - 12}{x + 1}$ **43.** $y = \dfrac{x - 2}{x^2 - 2x - 8}$

44. $y = \dfrac{2x^2 - 6x - 7}{x - 4}$ **45.** $y = \dfrac{x^2 - 9}{2x^2 + 1}$

46. $y = \dfrac{x^3 + 27}{x + 4}$

Solve the inequalities in Exercises 47–52.

47. $2x^2 \geq x + 3$ **48.** $\dfrac{x - 4}{x^2 - 4} \leq 0$

49. $\dfrac{5}{x^3 - x^2 - 4x + 4} < 0$ **50.** $\sqrt{x} > x^{3/2} - \dfrac{7}{\sqrt{x}}$

51. $\dfrac{3x + 1}{x + 2} \leq \dfrac{2}{3}$ **52.** $\dfrac{1}{x - 1} + \dfrac{2}{x + 1} \geq \dfrac{3}{x}$

53. Find the domains of the functions:
   (a) $f(x) = \sqrt{24 - x - 3x^2}$
   (b) $g(x) = \dfrac{1}{\sqrt[4]{x - x^4}}$

54. Find a polynomial of degree 3 with integer coefficients that has $2 - \sqrt[3]{2}$ as one of its zeros. How many other real zeros does the polynomial have?

55. (a) Show that -1 is a root of the equation
$2x^4 + 5x^3 + x + 4 = 0$.
(b) Use the information from part (a) to show that
$2x^3 + 3x^2 - 3x + 4 = 0$ has no positive real
roots. [*Hint:* Compare the coefficients of the
latter polynomial to the synthetic division table
from part (a).]

56. Find the coordinates of all points of intersection of the
graphs of

$$y = x^4 + x^2 + 24x$$

and $$y = 6x^3 + 20$$

Chapter 3 Test

1. Let $P(x) = 2x^5 - 9x^4 - 20x^3 + 11x^2 + 7x + 13$.
(a) Use synthetic division to find $P(6)$.
(b) Find the remainder when $P(x)$ is divided by $x + 2$.

2. Let $P(x) = 2x^4 - 17x^3 + 53x^2 - 72x + 36$.
(a) List all possible rational zeros of P given by the
Rational Roots Theorem.
(b) Use Descartes' Rule of Signs to determine how
many positive and how many negative real roots
the equation $P(x) = 0$ may have.
(c) Show that 9 is an upper bound for the real roots of
$P(x) = 0$ but is not itself a root.
(d) Use the information from your answers to parts (a),
(b), and (c) to construct a new, shorter list of possible rational zeros of P.
(e) Determine which of the possible rational zeros actually are zeros of P.
(f) Factor P in the form given in the Complete Factorization Theorem.

3. (a) State the Factor Theorem.
(b) Use the Remainder Theorem to prove the Factor
Theorem.

4. Find a fifth-degree polynomial with integer coefficients
that has zeros $1 + 2i$ and -1, with -1 being a zero of
multiplicity 3.

5. Let $P(x) = x^{23} - 5x^{12} + 8x - 1$,
$Q(x) = 3x^4 + x^2 - x - 15$,
and $R(x) = 4x^6 + x^4 + 2x^2 + 16$.
(a) Explain why an even integer could not possibly be
a zero of any of these three polynomials.
(b) Does R have any real zeros? Why or why not?
(c) How many real zeros does Q have? Why?
(d) Show that P has no rational zeros.

6. Find a decimal approximation for the positive root of
the equation $x^4 + 2x^2 - x - 4 = 0$ correct to the
nearest tenth.

7. Find all roots of the equation
$2x^4 - 17x^3 + 42x^2 - 25x + 4 = 0$.

8. Find all roots of the equation
$x^4 + x^3 + 5x^2 + 4x + 4 = 0$, given that
$-2i$ is one of its roots.

9. Graph the function $f(x) = x^3 - x^2 - 9x + 9$, showing
clearly all x- and y-intercepts.

10. Let $r(x) = \dfrac{2x - 1}{x^2 - x - 2}$, $s(x) = \dfrac{x^3 + 27}{x^2 + 4}$,
$t(x) = \dfrac{x^3 - 9x}{x + 2}$, and $u(x) = \dfrac{x^2 + x - 6}{x^2 - 25}$.
(a) Which of these four rational functions has a horizontal asymptote?
(b) Which has a slant asymptote?
(c) Which has no vertical asymptote?
(d) Graph $y = u(x)$, showing clearly any asymptotes
and x- and y-intercepts the function may have.

11. Evaluate and write your answer in the form $a + bi$:
(a) $\dfrac{6 - 2i}{2 + 3i}$ **(b)** $(2 - i)^3$ **(c)** i^{4003}

12. Solve $x \le \dfrac{6 - x}{2x - 5}$.

13. Find the domain of f:

$$f(x) = \frac{1}{\sqrt{4 - 2x - x^2}}$$

Problems Plus

1. Suppose that the equation
$$x^2 + px + 8 = 0$$
has two distinct real roots, r_1 and r_2. Show that
$$|r_1 + r_2| > 4\sqrt{2}$$

2. If the equation
$$x^4 + ax^2 + bx + c = 0$$
has roots 1, 2, and 3, find c.

3. If we know that the equation
$$x^5 + ax^3 + bx^2 + cx + 42 = 0$$
has roots 1, 2, and 3, what are the other two roots?

4. Prove that the value of the polynomial
$$P(x) = x^8 - x^5 + x^2 - x + 1$$
is positive for any real number x.

5. Graph the following sets of complex numbers:
   (a) $\{z \,|\, |z - i| \le 1\}$
   (b) $\{z \,|\, |z + 3| - |z + 3i| < 0\}$

6. The equation $x^3 = 1$ has only one real solution, so it must have two imaginary solutions. Let ω be one of the imaginary solutions. Show that the expression
$$(1 - \omega + \omega^2)(1 + \omega - \omega^2)$$
is a real number, and find its value.

4

Exponential and Logarithmic Functions

I keep the subject constantly before me and wait till the first dawnings open little by little into the full light.

Isaac Newton

So far we have studied relatively simple functions such as polynomials and rational functions. We now turn our attention to two of the most important functions in mathematics: the exponential function and its inverse function, the logarithmic function. We use these functions to describe exponential growth in biology and economics and radioactive decay in physics and chemistry.

Section 4.1

Exponential Functions

In Section 1.2 we defined a^x if $a > 0$ and x is a rational number, but we have not yet defined irrational powers. For instance, what is meant by $2^{\sqrt{3}}$ or 5^{π}? To help us answer this question we first look at the graph of the function $y = 2^x$, where x is rational. A representation of this graph is shown in Figure 1.

We want to enlarge the domain of $y = 2^x$ to include both rational and irrational numbers. There are holes in the graph in Figure 1. We want to fill in the holes by defining $f(x) = 2^x$, where $x \in R$, so that f is an increasing function.

In particular, since

$$1.7 < \sqrt{3} < 1.8$$

we must have

$$2^{1.7} < 2^{\sqrt{3}} < 2^{1.8}$$

Similarly, using better approximations for $\sqrt{3}$, we obtain better approximations for $2^{\sqrt{3}}$:

$$1.73 < \sqrt{3} < 1.74 \quad \Rightarrow \quad 2^{1.73} < 2^{\sqrt{3}} < 2^{1.74}$$
$$1.732 < \sqrt{3} < 1.733 \quad \Rightarrow \quad 2^{1.732} < 2^{\sqrt{3}} < 2^{1.733}$$
$$1.7320 < \sqrt{3} < 1.7321 \quad \Rightarrow \quad 2^{1.7320} < 2^{\sqrt{3}} < 2^{1.7321}$$
$$1.73205 < \sqrt{3} < 1.73206 \quad \Rightarrow \quad 2^{1.73205} < 2^{\sqrt{3}} < 2^{1.73206}$$

Using advanced mathematics, it can be shown that there is exactly one number that is greater than all of the numbers

$$2^{1.7}, 2^{1.73}, 2^{1.732}, 2^{1.7320}, 2^{1.73205}, \ldots$$

and less than all of the numbers

$$2^{1.8}, 2^{1.74}, 2^{1.733}, 2^{1.7321}, 2^{1.73206}, \ldots$$

We define $2^{\sqrt{3}}$ to be this number. Using the above approximation process, we can compute it correct to six decimal places:

$$2^{\sqrt{3}} \approx 3.321997$$

Similarly, we can define 2^x (or a^x, if $a > 0$), where x is any irrational number. The graph of $f(x) = 2^x$, where $x \in R$, is shown in Figure 2. You can see that it is an increasing function, but to demonstrate just how quickly it increases, let us perform the following thought experiment. Suppose we start with a piece of paper a thousandth of an inch thick and we fold it in half 50 times. Each time we fold the paper in half, the thickness of the paper doubles, so the thickness of the resulting paper would be $2^{50}/1000$ inches. How thick do you think that is? It works out to be more than 17 million miles!

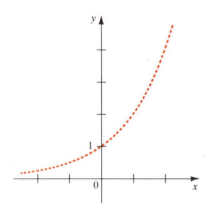

Figure 1
Representation of $y = 2^x$, x rational

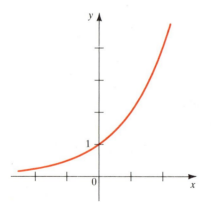

Figure 2
$y = 2^x$, x real

> If $a > 0$, the **exponential function with base a** is defined by
>
> $$f(x) = a^x$$
>
> for every real number x.

It can be proved that the Laws of Exponents are still true when the exponents are real numbers.

Example 1

Draw the graphs of the following functions: (a) $f(x) = 3^x$ and (b) $g(x) = (\frac{1}{3})^x$.

Solution

We calculate values of $f(x)$ and $g(x)$ and plot points to sketch the graphs in Figure 3.

x	$f(x) = 3^x$	$g(x) = (\frac{1}{3})^x$
-3	$\frac{1}{27}$	27
-2	$\frac{1}{9}$	9
-1	$\frac{1}{3}$	3
0	1	1
1	3	$\frac{1}{3}$
2	9	$\frac{1}{9}$
3	27	$\frac{1}{27}$

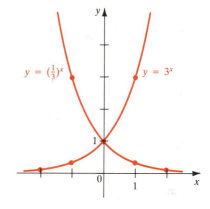

Figure 3

Notice that

$$g(x) = \left(\frac{1}{3}\right)^x = \frac{1}{3^x} = 3^{-x} = f(-x)$$

and so the graph of g could have been obtained from the graph of f by reflecting in the y-axis. ●

Figure 4 shows the graphs of the exponential function $f(x) = a^x$ for various values of the base a. Notice that all of these graphs pass through the same point $(0, 1)$ because $a^0 = 1$ for $a \neq 0$.

You can see from Figure 4 that there are basically three kinds of exponential functions $y = a^x$. If $0 < a < 1$, the exponential function decreases rapidly. If $a = 1$, it is constant. If $a > 1$, the function increases rapidly, and the larger the base the more rapid the increase. These three cases are illustrated in Figure 5.

Notice that, if $a > 1$, the graph of $y = a^x$ approaches zero as x decreases through negative values and so the x-axis is a horizontal asymptote. If

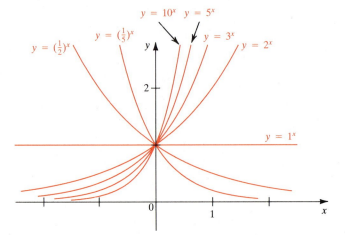

Figure 4

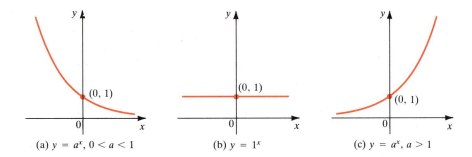

Figure 5

(a) $y = a^x$, $0 < a < 1$ (b) $y = 1^x$ (c) $y = a^x$, $a > 1$

$0 < a < 1$, the graph approaches zero as x increases indefinitely and again the x-axis is a horizontal asymptote. In both cases the graph never touches the x-axis, since $a^x > 0$ for all x. Thus, for $a \neq 1$, the exponential function $f(x) = a^x$ has domain R and range $(0, \infty)$.

In the next two examples we show how to graph certain functions, not by plotting points but by taking the basic graphs of the exponential functions in Figures 4 and 5 and applying the shifting and reflecting transformations of Section 2.3.

Example 2

Use the graph of $y = 2^x$ to sketch the graphs of the following functions:
(a) $y = 3 + 2^x$ and (b) $y = -2^x$.

Solution

(a)   The graph of $y = 3 + 2^x$ is obtained by starting with the graph of $y = 2^x$ in Figure 6(a) and shifting it three units upward. Notice from Figure 6(b) that the line $y = 3$ is a horizontal asymptote.

(b)   Again we start with the graph of $y = 2^x$, but here we reflect in the x-axis to get the graph of $y = -2^x$ shown in Figure 6(c).

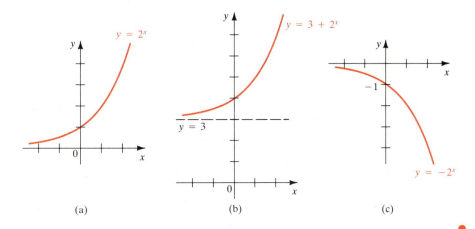

Figure 6

(a)   (b)   (c)

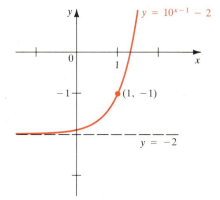

Figure 7

Example 3

(a) Use the graph of $y = 10^x$ to sketch the graph of $y = 10^{x-1} - 2$.

(b) State the asymptote, the domain, and the range of this function.

Solution

(a) Recall from Section 2.3 that we get the graph of $y = f(x - a)$ from the graph of $y = f(x)$ by shifting a units to the right. Thus we get the graph of $y = 10^{x-1} - 2$ by shifting the graph of $y = 10^x$ one unit to the right and two units down as in Figure 7.

(b) The horizontal asymptote is $y = -2$, the domain is R, and the range is

$$\{y \mid y > -2\} = (-2, \infty)$$

The Exponential Function with Base e

Any positive number can be used as the base for an exponential function, but some bases are used more frequently than others. We will see in the remaining sections of this chapter that the bases 2 and 10 are convenient for certain applications. But the most important base from the point of view of calculus is the number denoted by the letter e. This number is an irrational number, so we cannot write its exact value, but we can approximate it by using the following procedure.

The table in the margin shows the values, correct to five decimal places, of the function $f(n) = (1 + 1/n)^n$ for increasingly large values of n.

The number e is the number that the values of $(1 + 1/n)^n$ approach as n becomes large. (In calculus this idea is made more precise through the concept of a limit.) From the table it appears that, correct to five decimal places, we have

$$e \approx 2.71828$$

In fact it can be shown that the approximate value to 20 decimal places is

$$e \approx 2.71828182845904523536$$

n	$(1 + \frac{1}{n})^n$
1	2.00000
5	2.48832
10	2.59374
100	2.70481
1000	2.71692
10,000	2.71815
100,000	2.71827
1,000,000	2.71828
10,000,000	2.71828

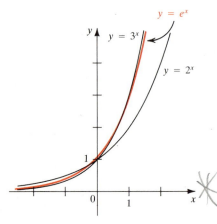

Figure 8

The notation e for this number was chosen by the Swiss mathematician Leonhard Euler in 1727 probably because it is the first letter of the word *exponential*.

Since $2 < e < 3$, the graph of the exponential function $y = e^x$ lies between the graphs of $y = 2^x$ and $y = 3^x$ as in Figure 8. Since e is the most important base for exponential functions, scientific calculators have a special key for the function $y = e^x$.

Example 4

Sketch the graph of the function $f(x) = e^{-x}$.

Solution

We start with the graph of $y = e^x$ and reflect in the y-axis to obtain $y = e^{-x}$ as in Figure 9.

In the next example we use our knowledge of exponential functions to solve a type of equation that occurs frequently in calculus.

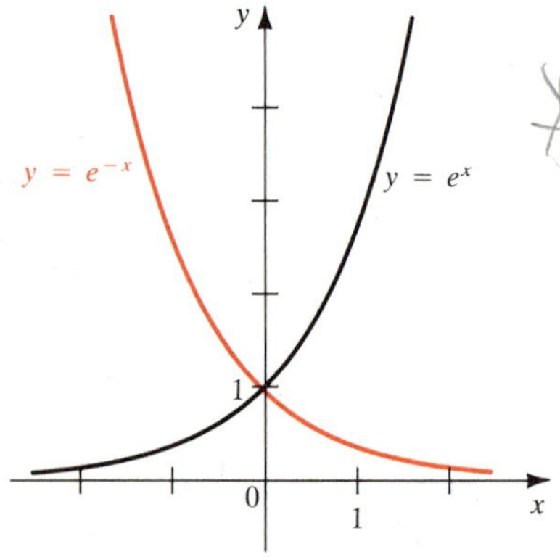

Example 5

Solve the equation $3x^2 e^x + x^3 e^x = 0$.

Solution

First we factor the left side of the equation:

$$3x^2 e^x + x^3 e^x = 0$$
$$(3x^2 + x^3)e^x = 0$$
$$x^2(3 + x)e^x = 0$$

Remember that the graph of every exponential function lies above the x-axis and so $e^x > 0$ for all x. Therefore we can divide both sides of the equation by e^x:

$$x^2(3 + x) = 0$$
$$x = 0 \quad \text{or} \quad x = -3$$

Figure 9

Exercises 4.1

In Exercises 1–4 sketch the graph of the function by making a table of values. Use a calculator if necessary.

1. $f(x) = 6^x$

2. $f(x) = (\frac{3}{2})^x$

3. $g(x) = (\frac{1}{4})^x$

4. $h(x) = (1.1)^x$

5. On the same axes graph the functions $y = 4^x$ and $y = 7^x$.

6. On the same axes graph the functions $y = (\frac{2}{3})^x$ and $y = (\frac{4}{3})^x$.

In Exercises 7–24 graph the given function, not by plotting points but by starting from the graphs in Figures 4, 5, and 8. State the domain, range, and asymptote of each function.

7. $f(x) = -5^x$

8. $f(x) = 10^{-x}$

9. $g(x) = 2^x - 3$

10. $g(x) = 2^{x-3}$

11. $y = 4 + (\frac{1}{2})^x$

12. $y = 6 - 3^x$

13. $y = 10^{x+3}$

14. $y = -(\frac{1}{5})^x$

15. $y = -e^x$ **16.** $y = 1 - e^x$

17. $y = e^{-x} - 1$ **18.** $y = -e^{-x}$

19. $y = 5^{-2x}$ **20.** $y = 1 + 2^{x+1}$

21. $y = 5 - 2^{x-1}$ **22.** $y = 2 + 5(1 - 10^{-x})$

23. $y = 2^{|x|}$ **24.** $y = 2^{-|x|}$

25. Use a calculator to help graph the function $f(x) = e^{-x^2}$ for $x \geq 0$. Then use the fact that f is an even function to draw the rest of the graph.

26. Compare the functions $f(x) = x^2$ and $g(x) = 2^x$ by evaluating each of them for $x = 0, 1, 2, 3, 4, 5, 6, 7, 8, 9, 10, 15,$ and 20. Then draw the graphs of f and g on the same set of axes for $-4 \leq x \leq 6$.

Solve the equations in Exercises 27–30.

27. $x^2 2^x - 2^x = 0$ **28.** $x^2 10^x - x10^x = 2(10^x)$

29. $4x^3 e^{-3x} - 3x^4 e^{-3x} = 0$

30. $x^2 e^x + xe^x - e^x = 0$

31. Solve the inequality: $x^2 e^x - 2e^x < 0$.

32. If $f(x) = 10^x$, show that
$$\frac{f(x + h) - f(x)}{h} = 10^x \left(\frac{10^h - 1}{h} \right)$$

33. The hyperbolic cosine function is defined by
$$\cosh(x) = \frac{e^x + e^{-x}}{2}$$

Sketch the graphs of the functions $y = \frac{1}{2}e^x$ and $y = \frac{1}{2}e^{-x}$ on the same axes and use graphical addition (see Section 2.5) to draw the graph of $y = \cosh(x)$.

34. The hyperbolic sine function is defined by
$$\sinh(x) = \frac{e^x - e^{-x}}{2}$$

Graph this function using graphical addition as in Exercise 33.

Use the definitions in Exercises 33 and 34 to prove the identities in Exercises 35–38.

35. $\cosh(-x) = \cosh(x)$ **36.** $\sinh(-x) = -\sinh(x)$

37. $[\cosh(x)]^2 - [\sinh(x)]^2 = 1$

38. $\sinh(x + y) = \sinh(x)\cosh(y) + \cosh(x)\sinh(y)$

Section 4.2

Application: Exponential Growth and Decay

The exponential function occurs frequently in mathematical models of nature and society. In particular, we will see in this section that it occurs in the description of population growth, radioactive decay, and compound interest.

Exponential Growth

Example 1

Under ideal conditions a certain bacteria population is known to double every three hours. Suppose that there are initially 1000 bacteria. (a) What is the size of the bacteria population after 15 h? (b) What is the size after t hours?

Solution

Let $n = n(t)$ be the number of bacteria after t hours. Then

$$n(0) = 1000$$
$$n(3) = 2 \cdot 1000$$
$$n(6) = 2(2 \cdot 1000) = 2^2 \cdot 1000$$
$$n(9) = 2(2^2 \cdot 1000) = 2^3 \cdot 1000$$
$$n(12) = 2(2^3 \cdot 1000) = 2^4 \cdot 1000$$
$$n(15) = 2(2^4 \cdot 1000) = 2^5 \cdot 1000$$

(a)   After 15 h the number of bacteria is $n(15) = 2^5 \cdot 1000 = 32{,}000$.

(b)   From the pattern given here, it appears that the number of bacteria after t hours is

$$n(t) = 2^{t/3} \cdot 1000$$ ●

In general, suppose that the initial size of a population is n_0 and the **doubling period** is d; that is, the time required for the population to double is d. Then the size of the population at time t is

(1)
$$\boxed{n(t) = n_0 2^{t/d}}$$

Example 2

A biologist makes a sample count of bacteria in a culture and finds that the population doubles every 20 minutes. The estimated bacteria count after two hours is 32,000. (a) What was the initial size of the culture at time $t = 0$? (b) What is the estimated count after 10 h? (c) What is the count after 2.5 h? (d) Sketch the graph of the bacteria population function.

Solution

(a)   The doubling period is $d = 20$ min $= \frac{1}{3}$ h, so the population at time t is

$$n(t) = n_0 2^{t/(1/3)} = n_0 2^{3t}$$

where n_0 is the initial size. We are given that $n(2) = 32{,}000$, so

$$n_0 2^{3(2)} = 32{,}000$$

$$n_0 = \frac{32{,}000}{2^6} = \frac{32{,}000}{64} = 500$$

The initial count was 500 bacteria.

(b)   From part (a) we have

$$n(t) = (500)2^{3t}$$

The population after 10 h is

$$n(10) = (500)2^{30} \approx 5.4 \times 10^{11}$$

(c)   The population after 2.5 h is

$$n(2.5) = (500)2^{7.5} \approx 9.1 \times 10^4$$

(d)   After noting that the domain is $t \geq 0$ and

$$n(t) = (500)2^{3t} = (500)(2^3)^t = (500)8^t$$

we draw the graph of $n(t)$ in Figure 1.   ●

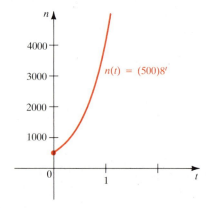

$n(t) = (500)8^t$

Figure 1

Radioactive Decay

Radioactive substances decompose by spontaneously emitting radiation. The **half-life** of a radioactive material is the period of time during which any given amount decays until half of it remains.

Example 3

An isotope of strontium, ^{90}Sr, has a half-life of 25 years. (a) Find the mass of ^{90}Sr that remains from a sample of 24 mg after 125 years. (b) Find the mass that remains after t years. (c) Sketch the graph of the mass as a function of time.

Solution

(a)   Let $m(t)$ be the mass, in milligrams, that remains after t years. Then

$$m(0) = 24$$

$$m(25) = \frac{1}{2}(24)$$

$$m(50) = \frac{1}{2}\left(\frac{1}{2} \cdot 24\right) = \frac{1}{2^2}(24)$$

$$m(75) = \frac{1}{2}\left(\frac{1}{2^2} \cdot 24\right) = \frac{1}{2^3}(24)$$

$$m(100) = \frac{1}{2}\left(\frac{1}{2^3} \cdot 24\right) = \frac{1}{2^4}(24)$$

$$m(125) = \frac{1}{2}\left(\frac{1}{2^4} \cdot 24\right) = \frac{1}{2^5}(24)$$

The mass of ^{90}Sr that remains after 125 years is

$$m(125) = \frac{1}{2^5}(24) = \frac{24}{32} = \frac{3}{4}\text{ mg}$$

(b)   From the pattern in part (a) it appears that the mass that remains after t years is

$$m(t) = \frac{1}{2^{t/25}}(24) = (24)2^{-t/25}$$

(c)   The graph of $m(t)$ is sketched in Figure 2.   ●

In general we have

(2)
$$\boxed{m(t) = m_0 2^{-t/h}}$$

where m is the mass of a radioactive substance that remains after time t, m_0 is the initial mass, and h is the half-life (measured in the same unit of time as t). Some radioactive materials decay very slowly, having half-lives of thousands of years. Others decay very rapidly, with half-lives of less than a second.

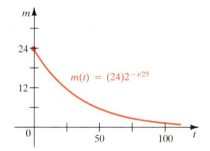

Figure 2

Example 4

Polonium-210 has a half-life of 140 days. If a sample has a mass of 300 mg, find the mass after 50 days.

Solution

Since $m_0 = 300$ and $h = 140$, the mass that remains after t days is

$$m(t) = (300)2^{-t/140} \qquad \textit{(from Equation 2)}$$

so

$$m(50) = (300)2^{-50/140} \approx 234$$

After 50 days the mass that remains is about 234 mg. ●

Example 5

After 30 h a sample of Plutonium-243 (^{243}Pu) has decayed to $\frac{1}{64}$ of its original mass. Find the half-life of ^{243}Pu.

Solution

If h is the half-life and m_0 is the original amount, then

$$m(t) = m_0 2^{-t/h}$$

We are given that $m(30) = \frac{1}{64}m_0$, so

$$m_0 2^{-30/h} = \frac{1}{64}m_0$$

$$2^{-30/h} = \frac{1}{64} = 2^{-6}$$

$$-\frac{30}{h} = -6$$

$$h = \frac{30}{6} = 5$$

The half-life of ^{243}Pu is 5 h. ●

Compound Interest

If an amount of money P, called the **principal**, is invested at a simple interest rate r, then the interest after one time period is Pr and the amount of money is

$$A = P + Pr = P(1 + r)$$

For instance, if $P = \$1000$ and the interest rate is 12% per year, then $r = 0.12$ and the amount after 1 year is $\$1000(1.12) = \$1120$.

If the interest is reinvested, then the new principal is $P(1 + r)$ and the amount after another time period is

$$A = P(1 + r)(1 + r) = P(1 + r)^2$$

Similarly, after a third time period the amount is $P(1 + r)^3$, and in general after k periods it is

$$A = P(1 + r)^k$$

Notice that this is an exponential function with base $1 + r$.

For example, if the interest rate is 12% per year compounded semiannually, then the time period is 6 months and the interest rate per time period is 6%, or 0.06.

If interest is compounded n times per year, then in each time period the interest rate is r/n and there are nt time periods in t years, so the amount after t years is

(3)

$$A = P\left(1 + \frac{r}{n}\right)^{nt}$$

Example 6

A sum of $1000 is invested at an interest rate of 12% per year. Find the amount in the account after 3 years if interest is compounded (a) annually, (b) semiannually, (c) quarterly, (d) monthly, and (e) daily.

Solution

We use Formula 3 with $P = \$1000$, $r = 0.12$, and $t = 3$.

(a)   With annual compounding, $n = 1$:

$$A = 1000(1.12)^3 = \$1404.93$$

(b)   With semiannual compounding, $n = 2$:

$$A = 1000\left(1 + \frac{0.12}{2}\right)^{2(3)} = 1000(1.06)^6 = \$1418.52$$

(c)   With quarterly compounding, $n = 4$:

$$A = 1000\left(1 + \frac{0.12}{4}\right)^{4(3)} = 1000(1.03)^{12} = \$1425.76$$

(d)   With monthly compounding, $n = 12$:

$$A = 1000\left(1 + \frac{0.12}{12}\right)^{12(3)} = 1000(1.01)^{36} = \$1430.77$$

(e)   With daily compounding, $n = 365$:

$$A = 1000\left(1 + \frac{0.12}{365}\right)^{365(3)} = \$1433.24$$ ●

We see from Example 6 that the interest paid increases as the number of compounding periods (n) increases. In general, let us see what happens as n

increases indefinitely. If we let $m = n/r$, then

$$A = P\left(1 + \frac{r}{n}\right)^{nt} = P\left[\left(1 + \frac{r}{n}\right)^{n/r}\right]^{rt} = P\left[\left(1 + \frac{1}{m}\right)^{m}\right]^{rt}$$

Recall that as m becomes large, the quantity $(1 + 1/m)^m$ approaches the number e. Thus the amount approaches

(4)

$$A = Pe^{rt}$$

When interest is paid according to Formula 4, we refer to **continuous compounding of interest**.

Example 7

Find the amount after 3 years if $1000 is invested at an interest rate of 12% per year compounded continuously.

Solution

Using Equation 4 with $P = \$1000$, $r = 0.12$, and $t = 3$, we have

$$A = 1000e^{(0.12)3} = 1000e^{0.36} = \$1433.33$$

Exercises 4.2

1. A bacteria culture contains 1500 bacteria initially and doubles every half-hour. Find the size of the bacteria population **(a)** after t hours, **(b)** after 20 min, and **(c)** after 24 h.

2. The doubling period of a bacteria population is 15 min. At time $t = 1.25$ h, an estimate of 80,000 was taken. **(a)** What was the initial population? **(b)** What is the size after t hours? **(c)** What is the size after 3 h? **(d)** What is the size after 40 min? **(e)** Sketch the graph of the population function.

3. A bacteria culture starts with 10,000 bacteria and the population triples every hour. **(a)** Find the number of bacteria after t hours. **(b)** Find the number of bacteria after 2.5 h.

4. A bacteria culture starts with 5000 bacteria. After 3 h the estimated count is 80,000. **(a)** Find the doubling period. **(b)** What is the population after 3.5 h?

5. The population of the world is doubling about every 35 years. In 1987 the total population was 5 billion. If the doubling period remains at 35 years, find the projected world population **(a)** for the year 2001 and **(b)** for the year 2100.

6. The population of a certain city grows at a rate of 5% per year. The population in 1988 was 421,000. If the growth rate remains constant, find the projected population of the city for the years 2000 and 2030.

7. The half-life of Radium-226 is 1590 years. If a sample has a mass of 150 mg, find a formula for the mass that remains after t years. Then find the mass that remains after 1000 years.

8. An isotope of sodium, ^{24}Na, has a half-life of 15 h. Find the amount that remains from a 2-g sample after **(a)** 5 h, **(b)** 20 h, and **(c)** 4 days.

9. The half-life of Palladium-100 (^{100}Pd) is 4 days. After 20 days a sample of ^{100}Pd has been reduced to a mass of 0.375 g. **(a)** What was the initial mass of the sample? **(b)** What is the mass after 3 days? **(c)** What is the mass after 3 weeks?

10. A certain amount of Vanadium-48 (^{48}V) decays to one-eighth of its original mass in 48 days. What is the half-life of ^{48}V?

11. If $10,000 is invested at an interest rate of 10% per year, compounded semiannually, find the value of the investment after **(a)** 5 years, **(b)** 10 years, and **(c)** 15 years.

12. If $4000 is borrowed at a rate of 16% interest per year, compounded quarterly, find the amount due at the end of **(a)** 4 years, **(b)** 6 years, and **(c)** 8 years.

13. If $3000 is invested at a rate of 9% per year, find the amount of the investment at the end of 5 years if the interest is compounded **(a)** annually, **(b)** semiannually, **(c)** monthly, **(d)** weekly, **(e)** daily, **(f)** hourly, and **(g)** continuously.

14. Which of the following would be the better investment: **(a)** an account paying $9\frac{1}{4}$% per year compounded semiannually or **(b)** an account paying 9% per year compounded continuously?

15. The **present value** of a sum of money is the amount that must be invested now, at a given rate of interest, to produce the desired sum at a later date. Find the present value of $10,000 if interest is paid at a rate of 9% compounded semiannually for 3 years.

Section 4.3

Logarithmic Functions

If $a > 0$, where $a \neq 1$, then the exponential function $f(x) = a^x$ is a one-to-one function by the Horizontal Line Test (see Figure 1) and therefore has an inverse function.

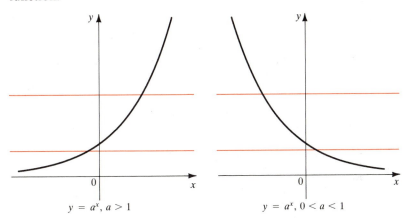

Figure 1

$f(x) = a^x$ is one-to-one

$y = a^x, a > 1$

$y = a^x, 0 < a < 1$

The inverse function f^{-1} of the exponential function $f(x) = a^x$ is called the **logarithmic function with base a** and is denoted by $\log_a$. Recall from Section 2.6 that the inverse function f^{-1} is defined by

$$f^{-1}(x) = y \quad \Leftrightarrow \quad f(y) = x$$

Thus the definition of the logarithmic function as the inverse of the exponential function means the following:

(1)
$$\log_a x = y \quad \Leftrightarrow \quad a^y = x$$

In words, this says that

$\log_a x$ is the exponent to which the base a must be raised to give x

In using Equation 1 to switch back and forth between the logarithmic form $\log_a x = y$ and the exponential form $a^y = x$, it is helpful to notice that in both cases the base is the same:

$$\underset{\underset{\text{base}}{\uparrow}}{\log_a x} = \overset{\overset{\text{exponent}}{\downarrow}}{y} \quad \Leftrightarrow \quad \overset{\overset{\text{exponent}}{\downarrow}}{a^y} = x$$
$$\underset{\text{base}}{}$$

Example 1

Evaluate (a) $\log_3 81$, (b) $\log_{16} 4$, and (c) $\log_{10} 0.0001$.

Solution

(a) $\log_3 81 = 4$ because $3^4 = 81$

(b) $\log_{16} 4 = \frac{1}{2}$ because $16^{1/2} = 4$

(c) $\log_{10} 0.0001 = -4$ because $10^{-4} = 0.0001$ ●

Example 2

Express in exponential form:

(a) $\log_2(\frac{1}{2}) = -1$ (b) $\log_{10} 100,000 = 5$ (c) $\log_3 z = t$

Solution

(a) $\log_2(\frac{1}{2}) = -1 \;\Rightarrow\; 2^{-1} = \frac{1}{2}$

(b) $\log_{10} 100,000 = 5 \;\Rightarrow\; 10^5 = 100,000$

(c) $\log_3 z = t \;\Rightarrow\; 3^t = z$ ●

Example 3

Express in logarithmic form:

(a) $1000 = 10^3$ (b) $2^{-3} = \frac{1}{8}$ (c) $s = 5^r$

Solution

(a) $10^3 = 1000 \;\Rightarrow\; \log_{10} 1000 = 3$

(b) $2^{-3} = \frac{1}{8} \;\Rightarrow\; \log_2(\frac{1}{8}) = -3$

(c) $s = 5^r \;\Rightarrow\; \log_5 s = r$ ●

Example 4

Solve for x: $\log_2(25 - x) = 3$.

Solution

The first step is to rewrite the equation in exponential form:

$$\log_2(25 - x) = 3$$
$$2^3 = 25 - x$$
$$8 = 25 - x$$
$$x = 25 - 8 = 17$$ ●

Example 5

Solve for x: $3^{x+2} = 7$.

Solution

First we rewrite the equation in logarithmic form:

$$\log_3 7 = x + 2$$
$$x = \log_3 7 - 2 \qquad \bullet$$

Graphs of Logarithmic Functions

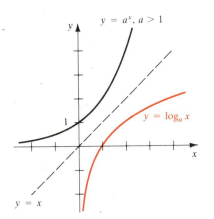

Figure 2

Recall that if a one-to-one function f has domain A and range B, then its inverse function f^{-1} has domain B and range A. Since the exponential function $f(x) = a^x$, where $a \neq 1$, has domain R and range $(0, \infty)$, we conclude that its inverse function, $f^{-1}(x) = \log_a x$ has domain $(0, \infty)$ and range R.

The graph of $f^{-1}(x) = \log_a x$ is obtained by reflecting the graph of $f(x) = a^x$ in the line $y = x$. Figure 2 shows the case where $a > 1$. (The most important logarithmic functions have base $a > 1$.) The fact that $y = a^x$ ($a > 1$) is a very rapidly increasing function is reflected in the fact that $y = \log_a x$ is a very slowly increasing function (see Exercise 71).

Notice that since $a^0 = 1$, we have

$$\boxed{\log_a 1 = 0}$$

and so the x-intercept of the function $y = \log_a x$ is 1. Notice also that since $y = a^x$ has the x-axis as a horizontal asymptote, the curve $y = \log_a x$ has the y-axis as a vertical asymptote. In fact, using the notation of Section 3.7, we can write

$$\log_a x \rightarrow -\infty \qquad \text{as} \qquad x \rightarrow 0^+$$

for the case $a > 1$.

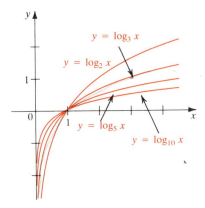

Figure 3

Figure 3 shows the relationship among the graphs of the logarithmic functions with bases 2, 3, 5, and 10. These graphs were drawn by reflecting the graphs of $y = 2^x$, $y = 3^x$, $y = 5^x$, and $y = 10^x$ (see Figure 4 in Section 4.1) in the line $y = x$.

In the next two examples we graph logarithmic functions by starting with the basic graphs in Figure 3 and using the transformations of Section 2.3.

Example 6

Sketch the graphs of the functions:

(a) $y = -\log_2 x$ \qquad\qquad (b) $y = \log_2(-x)$

Solution

(a) We start with the graph of $y = \log_2 x$ in Figure 4(a) and reflect in the x-axis to get the graph of $y = -\log_2 x$ in Figure 4(b).

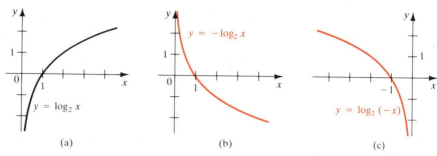

Figure 4

(a)   (b)   (c)

(b) To obtain the graph of $y = \log_2(-x)$ we reflect the graph of $y = \log_2 x$ in the y-axis. See Figure 4(c). ●

Example 7

Find the domain of the function $f(x) = \log_{10}(x - 3)$ and sketch its graph.

Solution

The domain of $y = \log_a x$ is the interval $(0, \infty)$, so $\log_{10} x$ is defined only when $x > 0$. Therefore the domain of $f(x) = \log_{10}(x - 3)$ is

$$\{x \mid x - 3 > 0\} = \{x \mid x > 3\} = (3, \infty)$$

The graph of f is obtained from the graph of $y = \log_{10} x$ by shifting three units to the right. Notice from Figure 5 that the line $x = 3$ is a vertical asymptote.

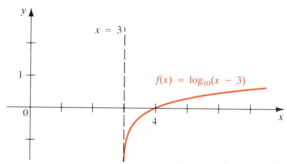

Figure 5

●

In Section 2.6 we saw that a function f and its inverse function f^{-1} satisfy the equations

$$f^{-1}(f(x)) = x \qquad [x \in \text{dom}(f)]$$
$$f(f^{-1}(x)) = x \qquad [x \in \text{dom}(f^{-1})]$$

When applied to $f(x) = a^x$ and $f^{-1}(x) = \log_a x$, these equations become

(2)

$$\boxed{\begin{aligned} \log_a(a^x) &= x \qquad (x \in R) \\ a^{\log_a x} &= x \qquad (x > 0) \end{aligned}}$$

For instance, we have

$$\log_{10}(10^x) = x \qquad \text{and} \qquad 2^{\log_2 x} = x$$

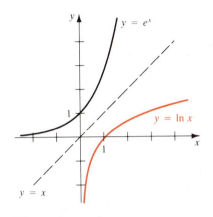

Figure 6

Natural Logarithms

Of all possible bases a for logarithms, it turns out that the most convenient for the purposes of calculus is the number e, which was defined in Section 4.1. The logarithm with base e is called the **natural logarithm** and is given a special name and notation:

$$\ln x = \log_e x$$

(The abbreviation ln is short for *logarithmus naturalis*.)

Thus the natural logarithmic function $y = \ln x$ is the inverse function of the exponential function $y = e^x$; they are both graphed in Figure 6.

If we put $a = e$ and write ln for $\log_e$ in (1) and (2), then the defining properties of the natural logarithm become

(3)
$$\ln x = y \quad \Leftrightarrow \quad e^y = x$$

(4)
$$\ln(e^x) = x \qquad (x \in R)$$
$$e^{\ln x} = x \qquad (x > 0)$$

In particular, it is worth noting that

$$\ln e = 1 \qquad \text{and} \qquad \ln 1 = 0$$

Example 8

Solve for x: $\ln x = 8$.

Solution 1

From (3) we see that

$$\ln x = 8 \qquad \text{means} \qquad e^8 = x$$

Therefore $x = e^8$.

Solution 2

Start with the equation

$$\ln x = 8$$

and apply the natural exponential function to both sides of the equation:

$$e^{\ln x} = e^8$$

The second equation in (4) says that $e^{\ln x} = x$. Therefore $x = e^8$.   ●

Example 9

Solve the equation $e^{3-2x} = 4$.

Solution

We take natural logarithms of both sides of the equation and use (4):

$$\ln(e^{3-2x}) = \ln 4$$
$$3 - 2x = \ln 4$$
$$2x = 3 - \ln 4$$
$$x = \tfrac{1}{2}(3 - \ln 4)$$

Example 10

Find the domain of the function $f(x) = \ln(4 - x^2)$.

Solution

As with any logarithmic function, $\ln x$ is defined when $x > 0$. Thus the domain of f is

$$\{x \mid 4 - x^2 > 0\} = \{x \mid x^2 < 4\} = \{x \mid |x| < 2\}$$
$$= \{x \mid -2 < x < 2\} = (-2, 2)$$

Exercises 4.3

In Exercises 1–8 express the equation in exponential form.

1. $\log_2 64 = 6$

2. $\log_5 1 = 0$

3. $\log_{10} 0.01 = -2$

4. $\log_8 4 = \frac{2}{3}$

5. $\log_8 512 = 3$

6. $\log_2(\frac{1}{16}) = -4$

7. $\log_a b = c$

8. $\log_r v = w$

In Exercises 9–16 express the equation in logarithmic form.

9. $2^3 = 8$

10. $10^5 = 100{,}000$

11. $10^{-4} = 0.0001$

12. $81^{1/2} = 9$

13. $4^{-3/2} = 0.125$

14. $6^{-1} = \frac{1}{6}$

15. $r^s = t$

16. $10^m = n$

Evaluate the expressions in Exercises 17–36.

17. $\log_6 6^4$

18. $\log_2 32$

19. $\log_4 64$

20. $\log_8 8^{17}$

21. $\log_9 9$

22. $\log_6 1$

23. $\log_3(\frac{1}{27})$

24. $\log_5 125$

25. $\log_{10} \sqrt{10}$

26. $3^{\log_3 8}$

27. $\log_5 0.2$

28. $\log_4 8$

29. $\log_8 0.25$

30. $\log_9 \sqrt{3}$

31. $2^{\log_2 37}$

32. $\log_4 0.5$

33. $\ln e^4$

34. $e^{\ln \pi}$

35. $e^{\ln \sqrt{5}}$

36. $\ln \frac{1}{e}$

In Exercises 37–52 solve the given equation for x.

37. $\log_2 x = 10$

38. $\log_5 x = 4$

39. $\log_{10}(3x + 5) = 2$

40. $\log_3(2 - x) = 3$

41. $\log_x 16 = 4$

42. $\log_x 6 = \frac{1}{2}$

43. $2^{1-x} = 3$

44. $3^{2x-1} = 5$

45. $\ln x = 10$

46. $\ln(2 + x) = 1$

47. $e^{12x} = 17$

48. $e^{1-4x} = 2$

49. $\log_2(x^2 - x - 2) = 2$

50. $2^{2/\log_5 x} = \frac{1}{16}$

51. $\log_2(\log_3 x) = 4$

52. $10^{5x} = 3$

53. Draw the graph of $y = 4^x$. Then use it to draw the graph of $y = \log_4 x$.

54. Most scientific calculators have keys for both LN and LOG $(= \log_{10})$. Use such a calculator to draw the graphs of $y = \ln x$ and $y = \log_{10} x$ on the same set of axes.

In Exercises 55–64 graph the given function, not by plotting points but by starting from the graphs in Figures 2, 3, and 6. State the domain, range, and asymptote of each function.

55. $f(x) = \log_2(x - 4)$

56. $f(x) = -\log_{10} x$

57. $g(x) = \log_5(-x)$

58. $g(x) = \ln(x + 2)$

59. $y = 2 + \log_3 x$

60. $y = \log_3(x - 1) - 2$

61. $y = 1 - \log_{10} x$

62. $y = 1 + \ln(-x)$

63. $y = |\ln x|$

64. $y = \ln|x|$

In Exercises 65–70 find the domain of the function.

65. $f(x) = \log_{10}(2 + 5x)$

66. $f(x) = \log_2(10 - 3x)$

67. $g(x) = \log_3(x^2 - 1)$

68. $g(x) = \ln(x - x^2)$

69. $h(x) = \ln x + \ln(2 - x)$

70. $h(x) = \sqrt{x - 2} - \log_5(10 - x)$

71. Suppose that the graph of $y = 2^x$ is drawn on a coordinate grid where the unit of measurement is an inch. Show that at a distance 2 ft to the right of the origin the height of the graph is about 265 mi. If the graph of $y = \log_2 x$ is drawn on the same set of axes, how far to the right of the origin do we have to go before the height of the curve reaches 2 ft?

72. Solve the inequality: $3 \le \log_2 x \le 4$.

73. Solve the inequality: $2 < 10^x < 5$.

74. Which is larger, $\log_4 17$ or $\log_6 24$?

75. **(a)** Find the domain of the function $f(x) = \log_2(\log_{10} x)$.
   (b) Find the inverse function of f.

76. **(a)** Find the domain of the function $f(x) = \ln(\ln(\ln x))$.
   (b) Find the inverse function of f.

77. Solve the equation $4^x - 2^{x+1} = 3$. (*Hint:*   First write the equation as a quadratic equation in 2^x.)

78. **(a)** Find the inverse of the function $f(x) = \dfrac{2^x}{1 + 2^x}$.
   (b) What is the domain of the inverse function?

Section 4.4

Laws of Logarithms

Since logarithms are exponents, the Laws of Exponents give rise to the Laws of Logarithms. These properties give the logarithmic functions a wide range of application, as we will see in the next section.

Laws of Logarithms

Suppose that $x > 0$, $y > 0$, and r is any real number. Then

1. $\log_a(xy) = \log_a x + \log_a y$

2. $\log_a\left(\dfrac{x}{y}\right) = \log_a x - \log_a y$

3. $\log_a(x^r) = r \log_a x$

Proof

We make use of the equation $\log_a(a^x) = x$ that was given as Equation 2 in Section 4.3.

1.   Let $$\log_a x = b \quad \text{and} \quad \log_a y = c$$

When written in exponential form, these equations become

$$a^b = x \quad \text{and} \quad a^c = y$$

Thus
$$\begin{aligned}
\log_a(xy) &= \log_a(a^b a^c) \\
&= \log_a(a^{b+c}) \\
&= b + c \\
&= \log_a x + \log_a y
\end{aligned}$$

2.   Using Law 1, we have

$$\log_a x = \log_a\left[\left(\frac{x}{y}\right)y\right] = \log_a\left(\frac{x}{y}\right) + \log_a y$$

so
$$\log_a\left(\frac{x}{y}\right) = \log_a x - \log_a y$$

3.   Let $\log_a x = b$. Then $a^b = x$, so

$$\log_a(x^r) = \log_a(a^b)^r = \log_a(a^{rb}) = rb = r\log_a x \qquad \bullet$$

We state the Laws of Logarithms in words as follows:

1.   The logarithm of a product is the sum of the logarithms of the factors.
2.   The logarithm of a quotient is the difference of the logarithms of the factors.
3.   The logarithm of a power of a number is the exponent times the logarithm of the number.

As the following examples illustrate, these laws are used in both directions. Since the domain of any logarithmic function is the interval $(0, \infty)$, we assume that all quantities whose logarithms occur are positive.

Example 1

Use the Laws of Logarithms to rewrite the following:

(a) $\log_2(6x)$ (b) $\log_{10}\sqrt{5}$ (c) $\log_5 x^3 y^6$ (d) $\ln\dfrac{ab}{\sqrt[3]{c}}$

Solution

(a) $\log_2(6x) = \log_2 6 + \log_2 x$

(b) $\log_{10}\sqrt{5} = \log_{10} 5^{1/2} = \frac{1}{2}\log_{10} 5$

(c) $\log_5(x^3 y^6) = \log_5 x^3 + \log_5 y^6 = 3\log_5 x + 6\log_5 y$

(d) $\ln\dfrac{ab}{\sqrt[3]{c}} = \ln ab - \ln\sqrt[3]{c} = \ln a + \ln b - \ln c^{1/3} = \ln a + \ln b - \frac{1}{3}\ln c$ $\bullet$

Example 2

Simplify the following:

(a) $\log_4 2 + \log_4 32$ (b) $\log_2 80 - \log_2 5$ (c) $-\frac{1}{3}\log_{10} 8$

Solution

(a) $\log_4 2 + \log_4 32 = \log_4(2 \cdot 32) = \log_4 64 = 3$

(b) $\log_2 80 - \log_2 5 = \log_2(\frac{80}{5}) = \log_2 16 = 4$

(c) $-\frac{1}{3}\log_{10} 8 = \log_{10} 8^{-1/3} = \log_{10} \frac{1}{2}$

●

Example 3

Express $3 \ln s + \frac{1}{2}\ln t - 4 \ln(t^2 + 1)$ as a single logarithm.

Solution

$$
\begin{aligned}
3 \ln s + \tfrac{1}{2}\ln t - 4 \ln(t^2 + 1) &= \ln s^3 + \ln t^{1/2} - \ln(t^2 + 1)^4 \\
&= \ln(s^3 t^{1/2}) - \ln(t^2 + 1)^4 \\
&= \ln\left(\frac{s^3 \sqrt{t}}{(t^2 + 1)^4}\right)
\end{aligned}
$$

●

Although the Laws of Logarithms tell us how to compute the logarithm of a product or a quotient, there is no corresponding rule for the logarithm of a sum or a difference. For instance,

$$\log_a(x + y) \neq \log_a x + \log_a y$$

In fact we know that the right side is equal to $\log_a(xy)$.

Similarly we *cannot* write

$$\frac{\log_{10} 6}{\log_{10} 2} \qquad \text{as} \qquad \log_{10}\left(\frac{6}{2}\right)$$

Nor can we write

$$(\log_2 x)^3 \qquad \text{as} \qquad 3 \log_2 x$$

Example 4

Solve the equation $\log_{10}(x + 2) + \log_{10}(x - 1) = 1$.

Solution

Using Law 1, we rewrite the equation as

(1) $\log_{10}(x + 2)(x - 1) = 1$

or $\log_{10}(x + 2)(x - 1) = \log_{10} 10$

Since the function $\log_{10}$ is one-to-one, we have

$$
\begin{aligned}
(x + 2)(x - 1) &= 10 \qquad \textit{(or raise 10 to both sides of Equation 1)}\\
x^2 + x - 12 &= 0 \\
(x + 4)(x - 3) &= 0 \\
x = -4 \quad &\text{or} \quad x = 3
\end{aligned}
$$

Let us check to see if these values satisfy the original equation. If $x = -4$, we have

$$\log_{10}(x + 2) = \log_{10}(-4 + 2) = \log_{10}(-2)$$

which is undefined, so $x = -4$ is an extraneous root. You can check that $x = 3$ satisfies the equation. Thus the only solution is $x = 3$.   ●

Common Logarithms

Logarithms with base 10 are called **common logarithms** and are often denoted by omitting the base:

$$\boxed{\log x = \log_{10} x}$$

It is standard practice to use common logarithms to solve exponential equations. For this reason they are found on scientific calculators.

Example 5

Find the solution of the equation $3^{x+2} = 7$ correct to six decimal places.

Solution

In Example 5 in Section 4.3 we obtained the solution in the form $x = \log_3 7 - 2$, but we cannot find the exact value of $\log_3 7$ and calculators do not have a $\log_3$ key. So instead we solve the given equation by taking common logarithms of both sides and using Law 3:

$$3^{x+2} = 7$$
$$\log(3^{x+2}) = \log 7$$
$$(x + 2)\log 3 = \log 7$$
$$x + 2 = \frac{\log 7}{\log 3}$$
$$x = \frac{\log 7}{\log 3} - 2 \approx -0.228756$$ ●

Notice that in solving Example 5 we could have used natural logarithms instead of common logarithms. In fact, using the same steps, we get

$$x = \frac{\ln 7}{\ln 3} - 2 \approx -0.228756$$

Change of Base

For some purposes it is useful to be able to change from logarithms in one base to logarithms in another base. Suppose that we are given $\log_a x$ and want to find $\log_b x$. Let

$$y = \log_b x$$

We write this in exponential form and take logarithms, with base a, of both sides:

$$b^y = x$$
$$\log_a(b^y) = \log_a x$$
$$y \log_a b = \log_a x$$
$$y = \frac{\log_a x}{\log_a b}$$

Thus we have proved the following formula:

Change of Base Formula

$$\log_b x = \frac{\log_a x}{\log_a b}$$

In particular if we put $x = a$, then $\log_a a = 1$ and the change of base formula becomes

$$\log_b a = \frac{1}{\log_a b}$$

Example 6

Evaluate $\log_8 5$ correct to six decimal places.

Solution

There is no $\log_8$ key on a calculator, but we use the change of base formula with $b = 8$ and $a = 10$ to convert to common logarithms:

$$\log_8 5 = \frac{\log_{10} 5}{\log_{10} 8} = \frac{\log 5}{\log 8} \approx 0.773976$$

Exercises 4.4

In Exercises 1–24 use the Laws of Logarithms to rewrite each expression in a form with no logarithms of products, quotients, or powers.

1. $\log_2 x(x - 1)$

2. $\log_5\left(\dfrac{x}{2}\right)$

3. $\log 7^{23}$

4. $\ln(\pi x)$

5. $\log_2(AB^2)$

6. $\log_6 \sqrt[4]{17}$

7. $\log_3(x\sqrt{y})$

8. $\log_2(xy)^{10}$

9. $\log_5 \sqrt[3]{x^2 + 1}$

10. $\log_a \dfrac{x^2}{yz^3}$

11. $\ln \sqrt{ab}$

12. $\ln \sqrt[3]{3r^2 s}$

13. $\log \dfrac{x^3 y^4}{z^6}$

14. $\log \dfrac{a^2}{b^4 \sqrt{c}}$

15. $\log_2 \dfrac{x(x^2 + 1)}{\sqrt{x^2 - 1}}$

16. $\log_5 \sqrt{\dfrac{x - 1}{x + 1}}$

17. $\ln\left(x\sqrt{\dfrac{y}{z}}\right)$

18. $\ln \dfrac{3x^2}{(x + 1)^{10}}$

19. $\log \sqrt[4]{x^2 + y^2}$

20. $\log \dfrac{x}{\sqrt[3]{1 - x}}$

21. $\log \sqrt[3]{\dfrac{x^2 + 4}{(x^2 + 1)(x^3 - 7)^2}}$

22. $\log \sqrt{x\sqrt{y\sqrt{z}}}$

23. $\ln \dfrac{\sqrt{x}\, z^4}{\sqrt[3]{y^2 + 6y + 17}}$

24. $\log \dfrac{10^x}{x(x^2 + 1)(x^4 + 2)}$

In Exercises 25–34 evaluate the expression.

25. $\log_5 \sqrt{125}$

26. $\log_2 112 - \log_2 7$

27. $\log 2 + \log 5$

28. $\log \sqrt{0.1}$

29. $\log_4 192 - \log_4 3$

30. $\log_{12} 9 + \log_{12} 16$

31. $\ln 6 - \ln 15 + \ln 20$

32. $e^{3 \ln 5}$

33. $10^{2 \log 4}$

34. $\log_2 8^{33}$

In Exercises 35–44 rewrite the expression as a single logarithm.

35. $\log_3 5 + 5 \log_3 2$

36. $\log 12 + \frac{1}{2} \log 7 - \log 2$

37. $\log_2 A + \log_2 B - 2 \log_2 C$

38. $\log_5(x^2 - 1) - \log_5(x - 1)$

39. $4 \log x - \frac{1}{3} \log(x^2 + 1) + 2 \log(x - 1)$

40. $\ln(a + b) + \ln(a - b) - 2 \ln c$

41. $\ln 5 + 2 \ln x + 3 \ln(x^2 + 5)$

42. $\frac{1}{2}[\log_5 x + 2 \log_5 y - 3 \log_5 z]$

43. $\frac{1}{3} \log(2x + 1) + \frac{1}{2}[\log(x - 4) - \log(x^4 - x^2 - 1)]$

44. $\log_a b + c \log_a d - r \log_a s$

In Exercises 45–54 state whether or not the given equation is an identity.

45. $\log_2(x - y) = \log_2 x - \log_2 y$

46. $\log_5 \dfrac{a}{b^2} = \log_5 a - 2 \log_5 b$

47. $\log 2^z = z \log 2$

48. $(\log P)(\log Q) = \log P + \log Q$

49. $\dfrac{\log a}{\log b} = \log a - \log b$

50. $(\log_2 7)^x = x \log_2 7$

51. $\log_a a^a = a$

52. $\log(x - y) = \dfrac{\log x}{\log y}$

53. $-\ln \dfrac{1}{A} = \ln A$

54. $r \ln s = \ln(s^r)$

55. For what value of x is it true that $\log(x + 3) = \log x + \log 3$?

56. For what value of x is it true that $(\log x)^3 = 3 \log x$?

In Exercises 57–64 solve the equation for x.

57. $\log_2 3 + \log_2 x = \log_2 5 + \log_2(x - 2)$

58. $2 \log x = \log 2 + \log(3x - 4)$

59. $\log x + \log(x - 1) = \log 4x$

60. $\log_5 x + \log_5(x + 1) = \log_5 20$

61. $\log_5(x + 1) - \log_5(x - 1) = 2$

62. $\log x + \log(x - 3) = 1$

63. $\log_9(x - 5) + \log_9(x + 3) = 1$

64. $\ln(x - 1) + \ln(x + 2) = 1$

In Exercises 65–72 find the solution of the equation correct to six decimal places.

65. $2^{x-5} = 3$

66. $8^{1-x} = 5$

67. $10^{3-2x} = 13$

68. $3^{x/14} = 0.1$

69. $5^{-x/100} = 2$

70. $e^{3-5x} = 16$

71. $2^{3x+1} = 3^{x-2}$

72. $7^{x/2} = 5^{1-x}$

In Exercises 73–76 use the change of base formula and a calculator to evaluate the logarithm correct to six decimal places.

73. $\log_2 7$

74. $\log_5 2$

75. $\log_3 11$

76. $\log_6 92$

77. Use the change of base formula to show that
$$\log e = \dfrac{1}{\ln 10}.$$

78. Simplify $(\log_2 5)(\log_5 7)$.

79. Show that $-\ln(x - \sqrt{x^2 - 1}) = \ln(x + \sqrt{x^2 - 1})$.

80. Solve the equation $(x - 1)^{\log(x - 1)} = 100(x - 1)$.

81. Solve the inequality $\log(x - 2) + \log(9 - x) < 1$.

82. Solve the equation $\log_2 x + \log_4 x + \log_8 x = 11$.

83. Find the error:

$$\log 0.1 < 2 \log 0.1$$
$$= \log(0.1)^2$$
$$= \log 0.01$$
$$\log 0.1 < \log 0.01$$
$$0.1 < 0.01$$

Section 4.5

Applications of Logarithms

Logarithms were invented by John Napier (1550–1617) to eliminate the tedious calculations involved in multiplying, dividing, and taking powers and roots of the large numbers that occur in astronomy and other sciences. With the advent of computers and calculators, logarithms are no longer important for such calculations. However, logarithms are useful for other reasons. They arise in problems of exponential growth and decay because logarithmic functions are inverses of exponential functions. Because of the Laws of Logarithms, they also turn out to be useful in the measurement of the loudness of sounds and the intensity of earthquakes, as well as many other phenomena.

Exponential Growth

We saw in Section 4.2 that if the initial size of an animal or bacteria population is n_0 and the doubling period is d, then the size of the population at time t is

(1)
$$n(t) = n_0 2^{t/d}$$

Now that we are equipped with logarithms, we are in a position to answer questions concerning the time at which the population will reach a certain level.

Example 1

A bacteria culture starts with 10,000 bacteria and the doubling period is 40 min. After how many minutes will there be 50,000 bacteria?

Solution

Using Equation 1 with $n_0 = 10,000$ and $d = 40$, we have

$$n(t) = 10,000 \cdot 2^{t/40}$$

We want to find the value of t such that $n(t) = 50,000$. Thus we have to solve the exponential equation

$$10,000 \cdot 2^{t/40} = 50,000$$

or
$$2^{t/40} = 5$$

Taking common logarithms of both sides, we have

$$\log(2^{t/40}) = \log 5$$

$$\frac{t}{40}\log 2 = \log 5$$

$$t = 40\frac{\log 5}{\log 2} \approx 93$$

The population will reach 50,000 in about 93 min.   ●

Although we used common logarithms in Example 1, natural logarithms would have worked just as well.

Radioactive Decay

In Section 4.2 we worked with the equation

(2) $$m(t) = m_0 2^{-t/h}$$

where $m(t)$ is the mass of a radioactive substance after time t, m_0 is the initial mass, and h is the half-life. Now we can determine the time required for decay to a certain amount.

Example 2

The half-life of Polonium-210 is 140 days. How long will it take a 300-mg sample to decay to a mass of 200 mg?

Solution

Using (2) with $m_0 = 300$ and $h = 140$, we see that the mass that remains after t days is

$$m(t) = (300)2^{-t/140}$$

The mass will be 200 mg when

$$(300)2^{-t/140} = 200$$

$$2^{-t/140} = \frac{200}{300} = \frac{2}{3}$$

$$-\frac{t}{140}\log 2 = \log\left(\frac{2}{3}\right)$$

$$t = -140\frac{\log(2/3)}{\log 2} \approx 82$$

The time required is about 82 days.   ●

Newton's Law of Cooling

Newton's Law of Cooling states that the rate of cooling of an object is proportional to the temperature difference between the object and its surroundings, provided that the temperature difference is not too large. Using calculus, it can be deduced from this law that the temperature of the object at time t is

(3)
$$T(t) = T_s + D_0 e^{-kt}$$

where T_s is the temperature of the surroundings, D_0 is the initial temperature difference, and k is a positive constant that is associated with the cooling object.

Example 3

A cup of coffee has a temperature of $200°F$ and is in a room that has a temperature of $70°F$. After 10 min the temperature of the coffee is $150°F$. (a) What is the temperature of the coffee after 15 min? (b) When will the coffee have cooled to $100°F$?

Solution

(a)   The temperature of the room is $T_s = 70°F$ and the initial temperature difference is

$$D_0 = 200 - 70 = 130°F$$

so, by Equation 3, the temperature after t minutes is

$$T(t) = 70 + 130e^{-kt}$$

When $t = 10$, the temperature is $T(10) = 150$, so we have

$$70 + 130e^{-10k} = 150$$
$$130e^{-10k} = 80$$
$$e^{-10k} = \frac{80}{130} = \frac{8}{13}$$
$$-10k = \ln\left(\frac{8}{13}\right)$$
$$k = -\frac{1}{10}\ln\left(\frac{8}{13}\right)$$

Putting this value of k into the expression for $T(t)$, we get

$$T(t) = 70 + 130e^{\ln(8/13)(t/10)}$$

So the temperature of the coffee after 15 min is

$$T(15) = 70 + 130e^{\ln(8/13)(15/10)} \approx 133°F$$

(b)   The temperature will be $100°F$ when

$$70 + 130e^{\ln(8/13)(t/10)} = 100$$
$$130e^{\ln(8/13)(t/10)} = 30$$
$$e^{\ln(8/13)(t/10)} = \frac{30}{130} = \frac{3}{13}$$

Take the natural logarithms of both sides:

$$\ln\left(\frac{8}{13}\right) \cdot \left(\frac{t}{10}\right) = \ln\left(\frac{3}{13}\right)$$

$$t = 10\frac{\ln(3/13)}{\ln(8/13)} \approx 30$$

The coffee will have cooled to 100°F after about half an hour.   ●

Compound Interest

In Section 4.2 we computed the amount of an investment or a loan using the formula

(4)
$$A = P\left(1 + \frac{r}{n}\right)^{nt}$$

where P is the principal, r is the interest rate, and interest is compounded n times per year for t years. Now we can use logarithms to determine the time it would take for the principal to increase to a given amount.

_____

Example 4

A sum of $5000 is invested at an interest rate of 9% per year compounded semiannually. How long will it take for the money to double?

Solution

In Formula 4 we have $P = \$5000$, $A = \$10,000$, $r = 0.09$, and $n = 2$, and we solve for t:

$$5000(1 + 0.045)^{2t} = 10,000$$
$$(1.045)^{2t} = 2$$
$$2t \log 1.045 = \log 2$$
$$t = \frac{\log 2}{2 \log 1.045} \approx 7.9$$

Thus the money will double during the eighth year.   ●

Logarithmic Scales

When physical quantities can vary over large ranges it is often convenient to take their logarithms in order to have a more manageable set of numbers. We discuss three such situations: the pH scale in chemistry; the Richter scale, which measures the intensity of earthquakes; and the decibel scale, which measures the loudness of sounds. Other quantities that are measured on logarithmic scales include light intensity, information capacity, and radiation.

The pH Scale. Chemists measured the acidity of a solution by giving its hydrogen ion concentration until Sorensen, in 1909, proposed a more convenient measure. He defined

(5)
$$pH = -\log_{10}[H^+] = -\log[H^+]$$

where $[H^+]$ is the concentration of hydrogen ions measured in moles per liter. He did this to avoid very small numbers and negative exponents. For instance,

$$\text{if} \quad [H^+] = 10^{-4} \text{ M}, \quad \text{then} \quad pH = -\log_{10}(10^{-4}) = -(-4) = 4$$

Solutions with a pH of 7 are called *neutral*; those with pH < 7 are called *acidic*; and those with pH > 7 are called *basic*. Notice that when the pH increases by one unit, $[H^+]$ decreases by a factor of 10.

Example 5

(a)   The hydrogen ion concentration of a sample of human blood was measured to be $[H^+] = 3.16 \times 10^{-8}$ M. Find the pH and classify the sample as acidic or basic.

(b)   The most acidic rainfall ever measured was in Scotland in 1974 and the pH was 2.4. Find the hydrogen ion concentration.

Solution

(a)   A calculator gives

$$pH = -\log[H^+] = -\log(3.16 \times 10^{-8}) \approx 7.5$$

Since this is greater than 7, the blood is basic.

(b)   Writing Equation 5 in exponential form, we have

$$\log[H^+] = -pH \quad \Rightarrow \quad [H^+] = 10^{-pH}$$

and so
$$[H^+] = 10^{-2.4} \approx 4.0 \times 10^{-3} \text{ M}$$ ●

The Richter Scale.   In 1935 the American geologist Charles Richter (1900–1984) defined the magnitude of an earthquake to be

(6)
$$M = \log\frac{I}{S}$$

where I is the intensity of the earthquake (measured by the amplitude of a seismograph 100 km from the epicenter of the earthquake) and S is the intensity of a "standard" earthquake (where the amplitude is only 1 micron $= 10^{-4}$ cm). Notice that the magnitude of the standard earthquake is

$$M = \log\frac{S}{S} = \log 1 = 0$$

Richter studied many earthquakes that had occurred between 1900 and 1950. The largest had magnitude 8.9 on the Richter scale; the smallest had magnitude 0. This corresponds to a ratio of intensities of 800 million, so the Richter scale provides more manageable numbers to work with. For instance, an earthquake of magnitude 6 is ten times stronger than an earthquake of magnitude 5.

Example 6

The 1906 earthquake in San Francisco had a magnitude of 8.3 on the Richter scale. In the same year the strongest earthquake ever recorded occurred on the Colombia–Ecuador border and was four times as intense. What was the magnitude of the Colombia–Ecuador earthquake on the Richter scale?

Solution

If I is the intensity of the San Francisco earthquake, then from Equation 6 we have

$$\log \frac{I}{S} = 8.3$$

The intensity of the Colombia–Ecuador earthquake was $4I$, so its magnitude was

$$\log \frac{4I}{S} = \log 4 + \log \frac{I}{S} = \log 4 + 8.3 \approx 8.9$$

Example 7

The 1979 San Francisco earthquake had a magnitude of only 5.9 on the Richter scale. How many times more intense was the 1906 earthquake (see Example 6) than the 1979 earthquake?

Solution

If I_1 and I_2 are the intensities of the 1906 and 1979 earthquakes, respectively, then we are required to find I_1/I_2. To relate this to the definition of magnitude in Equation 6 we divide numerator and denominator by S and we first find the common logarithm of I_1/I_2:

$$\log\left(\frac{I_1}{I_2}\right) = \log \frac{I_1/S}{I_2/S}$$

$$= \log \frac{I_1}{S} - \log \frac{I_2}{S}$$

$$= 8.3 - 5.9 = 2.4$$

Therefore
$$\frac{I_1}{I_2} = 10^{\log(I_1/I_2)} = 10^{2.4} \approx 251$$

The 1906 earthquake was about 250 times as intense as the 1979 earthquake.

The Decibel Scale. The ear is sensitive to an extremely large range of sound intensities. We take as a reference intensity $I_0 = 10^{-12}$ watt/m^2 at a frequency of 1000 hertz (Hz), which measures a sound that is just barely audible (the threshold of hearing). The psychological sensation of loudness varies with the logarithm of the intensity (the Weber–Fechner Law) and so the **intensity level** β, measured in decibels, is defined as

(7)
$$\beta = 10 \log \frac{I}{I_0}$$

Notice that the intensity level of the barely audible reference sound is

$$\beta = 10 \log \frac{I_0}{I_0} = 10 \log 1 = 0 \text{ dB}$$

Example 8

Find the intensity level of a jet plane taking off if the intensity was measured at 100 W/m^2 at a distance of 40 m from the jet.

Solution

From Equation 7 we see that the intensity level is

$$\beta = 10 \log \frac{I}{I_0} = 10 \log \frac{10^2}{10^{-12}}$$

$$= 10 \log 10^{14} = 140 \text{ dB}$$

The following table lists decibel intensity levels for some common sounds ranging from the threshold of hearing to the jet takeoff of Example 8. The threshold of pain is about 120 dB.

Source of Sound	β(dB)
Jet takeoff (40 m away)	140
Jackhammer	130
Rock concert (2 m from speakers)	120
Subway	100
Heavy traffic	80
Ordinary traffic	70
Normal conversation	50
Whisper	30
Rustling leaves	10–20
Threshold of hearing	0

Exercises 4.5

1. The hydrogen ion concentrations of samples of three substances are given. Calculate the pH.
   (a) lemon juice: $[H^+] = 5.0 \times 10^{-3}$ M
   (b) tomato juice: $[H^+] = 3.2 \times 10^{-4}$ M
   (c) seawater: $[H^+] = 5.0 \times 10^{-9}$ M

2. An unknown substance has a hydrogen ion concentration of $[H^+] = 3.1 \times 10^{-8}$ M. Find the pH and classify the substance as acidic or basic.

3. The pH readings of samples of the following substances are given. Calculate the hydrogen ion concentrations.
   (a) vinegar: pH = 3.0   (b) milk: pH = 6.5

4. The pH readings of glasses of beer and water are given. Find the hydrogen ion concentrations.
   (a) beer: pH = 4.6   (b) water: pH = 7.3

5. The hydrogen ion concentrations in cheese range from

4.0×10^{-7} M to 1.6×10^{-5} M. Find the corresponding range of pH readings.

6. The pH readings for wine vary from 2.8 to 3.8. Find the corresponding range of hydrogen ion concentrations.

7. A bacteria culture contains 1500 bacteria initially and doubles every half hour. After how many minutes will there be 4000 bacteria?

8. If a bacteria culture starts with 8000 bacteria and doubles every 20 min, when will the population reach a level of 30,000?

9. A bacteria culture starts with 20,000 bacteria. After 1 h the count is 52,000. What is the doubling period?

10. The count in a bacteria culture was 400 after 2 h and 25,600 after 6 h.
(a) What was the initial size of the culture?
(b) Find the doubling period.
(c) Find the size after 4.5 h.
(d) When will the population be 50,000?

11. The population of the world is doubling about every 35 years. In 1987 the total population was 5 billion. If the doubling period remains at 35 years, in what year will the world population reach 50 billion?

12. The population of California was 10,586,223 in 1950 and 23,668,562 in 1980.
(a) Assuming exponential growth during that period, find the doubling period.
(b) Use these data to predict the population of California in the year 2000.

13. The half-life of Strontium-90 is 25 years. How long will it take a 50-mg sample to decay to a mass of 32 mg?

14. Radium-221 has a half-life of 30 s. How long will it take for 95% of a sample to decompose?

15. If 250 mg of a radioactive element decays to 200 mg in 48 h, find the half-life of the element.

16. After 3 days a sample of Radon-222 decayed to 58% of its original amount.
(a) What is the half-life of Radon-222?
(b) How long will it take the sample to decay to 20% of its original amount?

17. A roast turkey is taken from an oven when its temperature has reached 185°F and is placed on a table in a room where the temperature is 75°F.
(a) If the temperature of the turkey is 150°F after half an hour, what is the temperature after 45 min?
(b) When will the turkey have cooled to 100°F?

18. A kettle full of water is brought to a boil in a room with temperature 20°C. After 15 min the temperature of the

water has decreased from 100°C to 75°C. Find the temperature after another 10 min.

19. Find the length of time required for an investment of $5000 to amount to $8000 at an interest rate of 9.5% per year compounded quarterly.

20. Nancy wants to invest $4000 in savings certificates that bear an interest rate of 9.75% compounded semiannually. How long a time period should she choose in order to save an amount of $5000?

21. How long would it take for an investment to double in value if the interest rate is 8.5% per year compounded continuously?

22. A sum of $1000 was invested for four years and the interest was compounded semiannually. If this sum amounted to $1435.77 after the four years, what was the interest rate?

23. If one earthquake is 20 times as intense as another, how much larger is its magnitude on the Richter scale?

24. The 1906 earthquake in San Francisco had a magnitude of 8.3 on the Richter scale. At the same time in Japan there was an earthquake with magnitude 4.9 that caused only minor damage. How many times more intense was the San Francisco earthquake than the Japanese earthquake?

25. The Alaska earthquake of 1964 had a magnitude of 8.6 on the Richter scale. How many times more intense was this than the 1906 San Francisco earthquake?

26. The intensity of the sound of rush-hour traffic at a busy intersection was measured at 2.0×10^{-5} W/m². Find the intensity level in decibels.

27. The intensity level of the sound of a subway train was measured at 98 dB. Find the intensity in watts per square meter.

28. The noise from a power mower was measured at 106 dB. The noise level at a rock concert was measured at 120 dB. Find the ratio of the intensity of the rock music to that of the power mower.

29. The figure on page 199 shows an electric circuit that contains a battery producing a voltage of 60 volts, a resistor with a resistance of 13 ohms, and an inductor with an inductance of 5 henries. Using calculus it can be shown that the current $I = I(t)$ (in amperes) t seconds after the switch is closed is

$$I = \frac{60}{13}(1 - e^{-13t/5})$$

Use this equation to express the time t as a function of the current I.

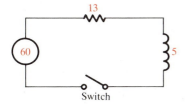

Switch

30. A **learning curve** is a graph of a function $P(t)$ that measures the performance of someone learning a skill as a function of the training time t. At first, the rate of learning is rapid. Then, as performance increases and approaches a maximal value M, the rate of learning decreases. It has been found that the function

$$P(t) = M - Ce^{-kt}$$

where k and C are positive constants and $C < M$, is a reasonable model for learning. **(a)** Sketch the graph of P. **(b)** Express the learning time t as a function of the performance level P.

31. It is a law of physics that the intensity of sound is inversely proportional to the square of the distance d from the source:

$$I = \frac{k}{d^2}$$

(a) Use this and Equation 7 to show that the decibel levels β_1 and β_2 at distances d_1 and d_2 from a sound

source are related by the equation

$$\beta_2 = \beta_1 + 20 \log \frac{d_1}{d_2}$$

(b) The intensity level at a rock concert is 120 dB at a distance of 2 m from the speakers. Find the intensity level at a distance of 10 m.

32. The table shows the mean distances d of the planets from the sun (taking the unit of measurement to be the distance from the earth to the sun) and their periods T (time of revolution in years). Try to discover a relationship between T and d. (*Hint:* Consider their logarithms.)

Planet	d	T
Mercury	0.387	0.241
Venus	0.723	0.615
Earth	1.000	1.000
Mars	1.523	1.881
Jupiter	5.203	11.861
Saturn	9.541	29.457
Uranus	19.190	84.008
Neptune	30.086	164.784
Pluto	39.507	248.35

CHAPTER 4 REVIEW

Define, state, or discuss the following.

1. The exponential function with base a

2. The number e

3. The exponential function with base e

4. Graphs of exponential functions

5. Doubling period

6. Population function

7. Half-life

8. Mass of a radioactive material

9. Compound interest

10. Continuous compounding of interest

11. The logarithmic function with base a

12. Natural logarithms

13. Common logarithms

14. Laws of Logarithms

15. Change of base formula

16. Newton's Law of Cooling

17. pH

18. The Richter scale

19. The decibel scale

Review Exercises

In Exercises 1–12 sketch the graph of the function. State the domain, range, and asymptote.

1. $f(x) = \dfrac{1}{2^x}$

2. $g(x) = 3^{x-2}$

3. $y = 5 - 10^x$

4. $y = 1 + 5^{-x}$

5. $f(x) = \log_3(x - 1)$

6. $g(x) = \log(-x)$

7. $y = 2 - \log_2 x$

8. $y = 3 + \log_5(x + 4)$

9. $F(x) = e^x - 1$

10. $G(x) = \frac{1}{2}e^{x-1}$

11. $y = 2 \ln x$

12. $y = \ln(x^2)$

In Exercises 13 and 14 find the domain of the function.

13. $f(x) = 10^{x^2} + \log(1 - 2x)$

14. $g(x) = \ln(2 + x - x^2)$

In Exercises 15–18 write the equation in exponential form.

15. $\log_2 1024 = 10$

16. $\log_6 37 = x$

17. $\log x = y$

18. $\ln c = 17$

In Exercises 19–22 write the equation in logarithmic form.

19. $2^6 = 64$

20. $49^{-1/2} = \frac{1}{7}$

21. $10^x = 74$

22. $e^k = m$

In Exercises 23–38 evaluate the expression without using a calculator.

23. $\log_2 128$

24. $\log_8 1$

25. $10^{\log 45}$

26. $\log 0.000001$

27. $\ln(e^6)$

28. $\log_4 8$

29. $\log_3(\frac{1}{27})$

30. $2^{\log_2 13}$

31. $\log_5 \sqrt{5}$

32. $e^{2\ln 7}$

33. $\log 25 + \log 4$

34. $\log_3 \sqrt{243}$

35. $\log_2 16^{23}$

36. $\log_5 250 - \log_5 2$

37. $\log_8 6 - \log_8 3 + \log_8 2$

38. $\log \log 10^{100}$

In Exercises 39–44 rewrite the expression in a form with no logarithms of products, quotients, or powers.

39. $\log AB^2C^3$

40. $\log_2 x\sqrt{x^2 + 1}$

41. $\ln \sqrt{\dfrac{x^2 - 1}{x^2 + 1}}$

42. $\log \dfrac{4x^3}{y^2(x - 1)^5}$

43. $\log_5 \dfrac{x^2(1 - 5x)^{3/2}}{\sqrt{x^3 - x}}$

44. $\ln \dfrac{\sqrt[3]{x^4 + 12}}{(x + 16)\sqrt{x - 3}}$

In Exercises 45–50 rewrite the expression as a single logarithm.

45. $\log 6 + 4 \log 2$

46. $\log x + \log x^2 y + 3 \log y$

47. $\frac{3}{2}\log_2(x - y) - 2 \log_2(x^2 + y^2)$

48. $\log_5 2 + \log_5(x + 1) - \frac{1}{3}\log_5(3x + 7)$

49. $\log(x - 2) + \log(x + 2) - \frac{1}{2}\log(x^2 + 4)$

50. $\frac{1}{2}[\ln(x - 4) + 5 \ln(x^2 + 4x)]$

In Exercises 51–60 solve the equation for x without using a calculator.

51. $\log_2(1 - x) = 4$

52. $2^{3x-5} = 7$

53. $5^{5-3x} = 26$

54. $\ln(2x - 3) = 14$

55. $e^{3x/4} = 10$

56. $2^{1-x} = 3^{2x+5}$

57. $\log x + \log(x + 1) = \log 12$

58. $\log_8(x + 5) - \log_8(x - 2) = 1$

59. $x^2 e^{2x} + 2xe^{2x} = 8e^{2x}$

60. $2^{3x} = 5$

In Exercises 61–64 use a calculator to find the solution of the equation correct to six decimal places.

61. $5^{-2x/3} = 0.63$

62. $2^{3x-5} = 7$

63. $5^{2x+1} = 3^{4x-1}$

64. $e^{-15k} = 10,000$

65. Evaluate $\log_4 15$ correct to six decimal places.

66. Solve the inequality $0.2 \le \log x < 2$.

67. Which is larger, $\log_4 258$ or $\log_5 620$?

68. Find the inverse function of the function $f(x) = 2^{3^x}$ and state its domain and range.

69. The hydrogen ion concentration of fresh white eggs was measured as $[H^+] = 1.3 \times 10^{-8}$ M. Find the pH and classify as acidic or basic.

70. The pH of lime juice is 1.9. Find the hydrogen ion concentration.

71. A bacteria culture starts with 1000 bacteria and doubles every 25 min.
   (a) Find the population after 1 h.
   (b) When will the population reach 5000?

72. A bacteria culture contains 10,000 bacteria initially. After 1 h the bacteria count is 25,000.
   (a) Find the doubling period.
   (b) Find the population after 3 h.

73. Uranium-234 has a half-life of 2.7×10^5 years.
   (a) Find the amount that remains from a 10-mg sample after 1000 years.
   (b) How long would it take this sample to decompose until its mass is 7 mg?

74. A sample of Bismuth-210 decayed to 33% of its original mass after 8 days.

(a) Find the half-life of this element.
(b) Find the mass that remains after 12 days.

75. If $12,000 is invested at an interest rate of 10% per year, find the amount of the investment at the end of 3 years if the interest is compounded (a) semiannually, (b) monthly, (c) daily, and (d) continuously.

76. A sum of $5000 is invested at an interest rate of $8\frac{1}{2}\%$ per year compounded semiannually.
(a) Find the amount of the investment after a year and a half.

(b) After what period of time will the investment amount to $7000?

77. If an earthquake has magnitude 6.5 on the Richter scale and another is 35 times as intense, what is its magnitude on the Richter scale?

78. The noise from a jackhammer was measured at 132 dB. The sound of whispering was measured at 28 dB. Find the ratio of the intensity of the jackhammer to that of the whispering.

Chapter 4 Test

1. Graph the functions $y = 2^x$ and $y = \log_2 x$ on the same axes.

2. Sketch the graph of the function $f(x) = \log(x - 3)$ and state the domain, range, and asymptote of f.

3. Evaluate:
(a) $\log_8 4$
(b) $\log_6 4 + \log_6 9$

4. Use the Laws of Logarithms to rewrite the following expression with no logarithms of products, quotients, powers, or roots:

$$\log \sqrt{\frac{x^2 - 1}{x^3(y^2 + 1)^5}}$$

5. Write the following expression as a single logarithm:

$$\ln x - 2 \ln(x^2 + 1) + \tfrac{1}{2}\ln(3 - x^4)$$

6. Solve for x without using a calculator:
(a) $2^x = 10$
(b) $\ln(3 - x) = 4$

7. The initial size of a bacteria culture is 1000. After 1 h the count is 8000.
(a) Find the doubling period.
(b) Find the population after t hours.
(c) Find the population after 1.5 h.
(d) When will the population reach 15,000?
(e) Sketch the graph of the population function.

8. Find the domain of the function
$f(x) = \log(x + 4) + \log(8 - 5x)$.

Problems Plus

1. Prove that the number $\log_2 5$ is an irrational number.

2. Evaluate $(\log_2 3)(\log_3 4)(\log_4 5) \cdots (\log_{31} 32)$.

3. Show that if $x > 0$ and $x \neq 1$, then

$$\frac{1}{\log_2 x} + \frac{1}{\log_3 x} + \frac{1}{\log_5 x} = \frac{1}{\log_{30} x}$$

4. Solve for x: $(\log_a x)(\log_5 x) = \log_a 5$.

5. Solve the inequality: $\log(x^2 - 2x - 2) \leq 0$.

6. Solve the inequality: $\log_{1/2}(1 + x) + \log_2\left(1 + \frac{1}{x}\right) \geq 1$.

5

Trigonometric Functions

Trigonometry contains the science of continually undulating magnitude...

Augustus de Morgan

Mathematics compares the most diverse phenomena and discovers the secret analogies which unite them.

Joseph Fourier

In this chapter we define certain ratios, called trigonometric ratios, that relate the angles and the sides of a triangle. Although these ratios will first be defined for acute angles, they have a natural extension to all angles. The trigonometric ratios are powerful tools for solving problems that involve the measurement of distance. Among other applications, they are used in calculating distances that are otherwise difficult or impossible to measure directly.

In the second part of this chapter we study the trigonometric ratios from a different point of view. Since the measure of any angle (in radians) is a real number, we are able to define the trigonometric *functions* of a real number. Their graphs immediately suggest that they have certain repetitive (or periodic) properties. For this reason the trigonometric functions have some striking applications to harmonic motion.

Section 5.1

Angles

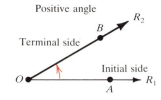

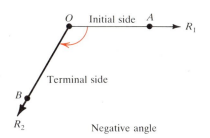

Figure 1

The fundamental concepts needed for the study of trigonometry are those of angle and angle measure. We review these concepts here.

Angle AOB consists of two rays R_1 and R_2 with a common vertex O. The **measure** of this angle is the amount of rotation about the vertex required to move R_1 onto R_2. Intuitively, this is how much the angle "opens." In this case R_1 is called the **initial side** and R_2 is called the **terminal side** of the angle. If the rotation is counterclockwise, then the measure of the angle is considered positive, and if the rotation is clockwise, the measure is considered negative (see Figure 1).

One unit of measurement for angles is the **degree**. An angle of measure $1°$ is formed by rotating the initial side $\frac{1}{360}$ of a complete revolution. This choice for the size of a degree is arbitrary and is made for historical reasons. In calculus and other branches of mathematics another method of measuring angles is used. The amount an angle opens is measured along the arc of a circle with center at the vertex of the angle. More precisely, if a circle of radius 1 is drawn with the vertex O of the angle as its center, then we define the measure of the angle in **radians** to be the length of the arc of this circle that is subtended by the angle.

The number π is the ratio of the circumference of *any* circle to its diameter. It follows that the circumference C of a circle of radius r is $C = 2\pi r$. The circumference of a circle of radius 1 is 2π and so a complete revolution has measure 2π radians, a straight angle has measure π radians, and a right angle has measure $\pi/2$ radians (see Figure 2). The relationship between degrees and radians is given by

$$\pi \text{ rad} = 180°$$

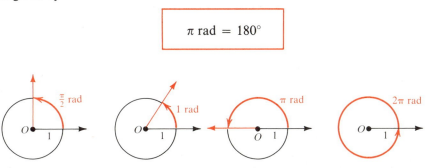

Figure 2

So $1° = \pi/180$ rad and 1 rad $= (180/\pi)°$. To get some idea of the size of a radian notice that

$$1 \text{ rad} \approx 57.296°$$

and $$1° \approx 0.01745 \text{ rad}$$

Example 1

(a)   Find the radian measure of $60°$.

(b)   Express $\pi/6$ rad in degrees.

Solution

(a)   To convert degrees to radians we multiply by $\pi/180$:

$$60° = 60\left(\frac{\pi}{180}\right) \text{rad} = \frac{\pi}{3} \text{rad}$$

(b)   To convert radians to degrees we multiply by $180/\pi$:

$$\frac{\pi}{6} \text{rad} = \left(\frac{\pi}{6}\right)\left(\frac{180}{\pi}\right) = 30° \qquad \bullet$$

An angle whose radian measure is θ subtends an arc that is the fraction $\theta/2\pi$ of the circumference of a circle. Thus in a circle of radius r, the length s of an arc subtended by the angle θ (see Figure 3) is

$$s = \frac{\theta}{2\pi} \times (\text{circumference of circle})$$

$$= \frac{\theta}{2\pi}(2\pi r) = \theta r$$

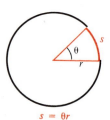

$s = \theta r$

Figure 3

(1)

> If θ is measured in radians, then
>
> $$s = \theta r \qquad \text{or} \qquad \theta = \frac{s}{r}$$

Equation 1 gives us a way of defining radian measure using any circle. This equation says that the radian measure of an angle θ that subtends an arc of length s in a circle of radius r is $\theta = s/r$.

Example 2

(a)   Find the length of an arc of a circle with radius 10 m that subtends a central angle of $30°$.

(b)   A central angle θ in a circle of radius 4 m subtends an arc of length 6 m. Find the measure of θ in radians.

Solution

(a)   From Example 1(b) we see that $30° = \pi/6$ rad. So the length of the arc is

$$s = \theta r = \frac{\pi}{6}(10) = \frac{5\pi}{3}$$

(b)   From (1) we have

$$\theta = \frac{s}{r} = \frac{6}{4} = \frac{3}{2} \text{ rad} \qquad \bullet$$

The area of a circle of radius r is $A = \pi r^2$. A sector of this circle with central angle θ has an area that is the fraction $\theta/2\pi$ of the area of the entire circle. So the area of this sector is

$$A = \frac{\theta}{2\pi} \times (\text{area of circle})$$

$$= \frac{\theta}{2\pi}(\pi r^2) = \tfrac{1}{2}r^2\theta$$

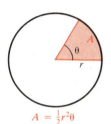

$A = \tfrac{1}{2}r^2\theta$

Figure 4

> The area of a sector with central angle whose radian measure is θ in a circle of radius r is
>
> (2) $\qquad\qquad\qquad\qquad A = \tfrac{1}{2}r^2\theta$

Example 3

(a)   Find the area of a sector of a circle with central angle $60°$ if the radius of the circle is 3 m.

(b)   A sector of a circle of radius 100 m has an area of 10,000 m^2. Find the central angle of the sector.

Solution

(a)   We can use Equation 2, but first we must find the central angle of the sector in radians: $60° = 60(\pi/180)$ rad $= \pi/3$ rad. The area of the sector is

$$A = \frac{1}{2}(3)^2\left(\frac{\pi}{3}\right) = \frac{3\pi}{2} \text{ m}^2$$

(b)   We need to find θ. Solving for θ in Equation 2, we get

$$\theta = \frac{2A}{r^2}$$

In our case we have

$$\theta = \frac{2(10,000)}{100^2} = 2 \text{ rad}$$

Thus the central angle of the sector has measure 2 rad. $\qquad \bullet$

We say that an angle *AOB* is in **standard position** if it is drawn in the *xy*-plane with its vertex at the origin and its initial side on the positive *x*-axis. Figure 5 gives some examples of angles in standard position.

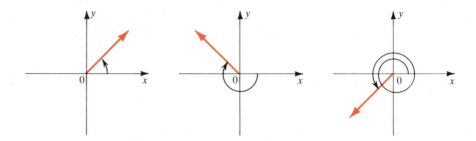

Figure 5
Angles in standard position

Two angles in standard position are called **coterminal** if their sides coincide. Thus, the angles 30° and 390°, the angles $\pi/4$ and $9\pi/4$, and the angles $-50°$ and 310° are all examples of coterminal angles (see Figure 6). In general, the measures of angles that are coterminal differ by an integer multiple of 2π rad (or 360°).

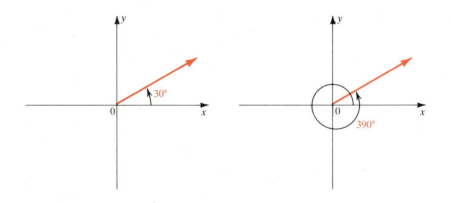

Figure 6
Coterminal angles

Example 4

Find an angle with measure between 0° and 360° that is coterminal with the angle of measure 1290° in standard position.

Solution

We can subtract 360° as many times as we wish from 1290° and the resulting angle will be coterminal with 1290°. Thus $1290° - 360° = 930°$ is coterminal with 1290° and so is the angle $1290° - 2(360)° = 1290° - 720° = 570°$. To find the angle that we want between 0° and 360° we need to subtract 360° from 1290° as many times as possible. An efficient way to do this is to find out how many times 360° goes into 1290° (that is, to divide 1290 by 360) and the remainder will be the angle we are looking for. We see that 360 goes into 1290 three times with a remainder of 210. Thus 210° is the desired angle (see Figure 7).

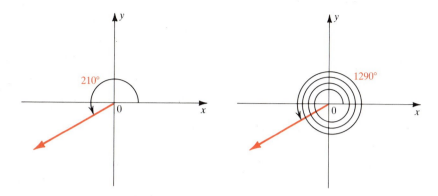

Figure 7

Exercises 5.1

In Exercises 1–8 find the radian measure of an angle with the given degree measure.

1. 40°   **2.** 330°   **3.** 72°

4. −30°   **5.** 45°   **6.** −80°

7. 765°   **8.** −150°

In Exercises 9–16 find the degree measure of an angle with the given radian measure.

9. $\dfrac{3\pi}{4}$ **10.** $-\dfrac{7\pi}{2}$ **11.** $\dfrac{5\pi}{6}$

12. 2   **13.** 1.5   **14.** $\dfrac{2\pi}{9}$

15. $-\dfrac{\pi}{12}$ **16.** $\dfrac{\pi}{5}$

In Exercises 17–26 determine the quadrant in which the given angle lies if placed in standard position. Sketch the angle.

17. 300°   **18.** 135°   **19.** $\dfrac{3\pi}{4}$

20. $\dfrac{11\pi}{6}$ **21.** 250°   **22.** −550°

23. 3.5   **24.** −3.5   **25.** 970°

26. 1888°

In Exercises 27–36 determine whether or not the pairs of angles with the given measures are coterminal.

27. 50°, 340°   **28.** −30°, 330°   **29.** 70°, 430°

30. $\dfrac{32\pi}{3}, \dfrac{11\pi}{3}$ **31.** 22π, 2π   **32.** $\dfrac{5\pi}{6}, \dfrac{10\pi}{3}$

33. 155°, 875°   **34.** 260°, −460°   **35.** 57°, 777°

36. 1770°, −30°

In Exercises 37–44 find an angle between 0° and 360° that is coterminal with the given angle.

37. 733°   **38.** 361°   **39.** 2223°

40. −100°   **41.** −800°   **42.** 1270°

43. −1°   **44.** 450°

In Exercises 45–52 find an angle between 0 and 2π that is coterminal with the given angle.

45. $\dfrac{12\pi}{5}$ **46.** $-\dfrac{7\pi}{3}$ **47.** 87π

48. 10   **49.** $\dfrac{17\pi}{4}$ **50.** $\dfrac{51\pi}{2}$

51. −23   **52.** $\dfrac{17\pi}{6}$

53. Find the length of the arc s in the figure.

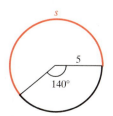

54. Find the angle θ in the figure.

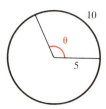

55. Find the radius r of the circle in the figure.

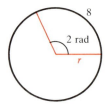

56. Find the length of an arc of a circle with radius 10 m that subtends a central angle of 45°.

57. What is the length of an arc of a circle of radius 2 mi, if the arc subtends a central angle of 2 rad?

58. A central angle θ in a circle of radius 5 m is subtended by an arc of length 6 m. Find the measure of θ in degrees and in radians.

59. An arc of length 100 m subtends a central angle θ in a circle of radius 50 m. Find the measure of θ in degrees and in radians.

60. A circular arc of length 3 ft subtends a central angle of 25°. Find the radius of the circle.

61. Find the radius of the circle if an arc of length 6 m on the circle subtends a central angle of $\pi/6$ rad.

62. How many revolutions will a car wheel of diameter 30 in. make as the car travels a distance of 1 mi?

63. How many revolutions will a car wheel of diameter 28 in. make over a period of half an hour if the car is traveling at 60 mi/h?

64. Pittsburgh, Pennsylvania, and Miami, Florida, are approximately on the same meridian. Pittsburgh has a latitude of 40.5°N and Miami 25.5°N. Find the distance between these two cities. (The radius of the earth is 3960 mi.)

65. Memphis, Tennessee, and New Orleans, Louisiana, lie approximately on the same meridian. Memphis has lat-

itude 30°N and New Orleans 35°N. Find the distance between these two cities. (The radius of the earth is 3960 mi.)

66. Find the distance that the earth travels in one day in its path around the sun. Assume that a year has 365 days and that the path of the earth around the sun is a circle of radius 93 million miles. (The path of the earth around the sun is actually an *ellipse* with the sun at one focus. This ellipse, however, has very small eccentricity so that it is very nearly circular.)

67. The Greek mathematician Eratosthenes measured the circumference of the earth from the following data. Eratosthenes noticed that on a certain day the sun shone directly down a deep well in Syene, a city in ancient Egypt. At the same time in Alexandria, 500 mi away from Syene (on the same meridian), the rays of the sun shone at an angle of 7.2° to the zenith. Use this information and the figure to find the radius and the circumference of the earth. (The data used in this problem are more accurate than those available to Eratosthenes.)

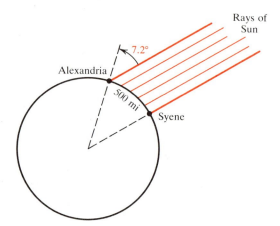

In Exercises 68–70 assume that the earth is a sphere with radius 3960 mi.

68. Find the distance along an arc on the surface of the earth that subtends a central angle of 1 minute (a *minute* is $\frac{1}{60}$ of a degree). This distance is called a **nautical mile**.

69. New York and Los Angeles are 2450 mi apart. Find the angle that the arc between these two cities subtends at the center of the earth.

70. A rope is wrapped around the earth suspended everywhere 1 ft above the equator. If the rope is now wrapped snugly around the equator, how much excess rope is there? How much excess rope would there be if this same experiment is done around a beach ball of radius 1 ft instead of around the earth?

71. Find the area of the sector shown in the figure.

(a)   **(b)**

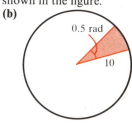

72. Find the area of a sector with central angle 1 rad in a circle of radius 10 m.

73. A sector of a circle has a central angle of 60°. Find the area of the sector if the radius of the circle is 3 mi.

74. The area of a sector of a circle with a central angle of 2 rad is 16 m². Find the radius of the circle.

75. Find the radius of a circle if the area of a quarter of the circle is 9π ft².

76. Find the central angle of a sector of a circle of radius 8 m if the area of the sector is 5 m².

77. A sector of a circle of radius 24 mi has an area of 288 mi². Find the central angle of the sector.

78. The area of a circle is 72 cm². Find the area of a sector of this circle that subtends a central angle of $\pi/6$ rad.

*If an object is moving in a circular path of radius r at a constant rate, then its **speed** (or **linear speed**) v is the distance traveled along the circumference of the circle per unit of time. So speed is*

$$v = \frac{s}{t}$$

*where s is the distance traveled along the arc in time t. The **angular speed** ω of this object is the number of radians that the line segment that joins the object with the center of the circle turns through per unit of time. So*

$$\omega = \frac{\theta}{t}$$

where θ is the number of radians turned through in time t. Since $s = \theta r$, we have the following relationship between linear speed and angular speed: $\omega = \theta/t = s/rt = v/r$, or

$$\omega = \frac{v}{r} \qquad or \qquad v = r\omega$$

79. A wheel turns at the rate of 600 rpm. What is the angular speed of the wheel?

80. A wheel turns at the rate of 1000 rpm. What is the angular speed of the wheel? What is the linear speed of a point on the wheel that is 10 in. from the center of the wheel?

81. A toy train is traveling around a circular track that is 2 m in diameter. The linear speed of the train is $\frac{1}{2}$ m/s. **(a)** What is its angular speed? **(b)** How many times per minute does the train go around the track?

82. Find the linear speed of the earth in its orbit around the sun. (See Exercise 66 for information needed to do this problem.)

83. A belt drives the alternator in a car at 3000 rpm. If the radius of the pulley on the alternator is 3 in., what is the linear speed of the belt?

84. What is your linear speed (as you sit at your desk on the surface of the earth at the equator) as the earth turns on its axis? Assume that a day is 24 h long and that the radius of the earth is 3960 mi.

85. A satellite travels in a circular orbit 600 mi above the surface of the earth. If the satellite completes one revolution every 105 min, what is the linear speed of the satellite in miles per hour? What is its angular speed? (The radius of the earth is 3960 mi.)

86. A car with 28-in. tires is traveling at 60 mi/h.
   (a) How many revolutions per minute do the tires make?
   (b) What is the angular speed of a point on the tread of the tire?

87. Two wheels of radii 1 m and 2 m are attached by a belt that turns both wheels.
   (a) If the linear speed of a point on the belt is 3 m/s, what is the linear speed of points on the edges of the two wheels?
   (b) If the angular speed of the larger wheel is 10 rad/min, what is the angular speed of the smaller wheel?

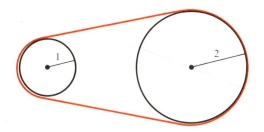

88. The following data are calculated from the information given in the manual of a new car. The ratio tells how many times the engine turns for every turn of the drive wheels of the car. (This ratio is related to but is not the same as what is commonly called the gear ratio.)

Gear	Ratio
1st	12.68 to 1
2nd	7.28 to 1
3rd	6.20 to 1
4th	4.00 to 1
5th	2.84 to 1

The wheels of this car have a diameter of 24 in.

(a) If the engine turns at 3000 rpm and the car is in second gear, at what rate are the wheels turning?
(b) How fast is the car in part (a) moving?
(c) The driver of this car decides to shift from second gear to third gear while traveling at 20 mi/h. How fast is the engine turning before and after he shifts?
(d) If the car is traveling at 60 mi/h in fifth gear, how fast is the engine turning?

The Pythagorean Theorem is fundamental to the study of trigonometry. Exercises 89–93 deal with this theorem.

89. Find the missing side in the given right triangle.

(a)   **(b)**

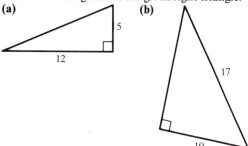

90. Find x, y, and z in the figure.

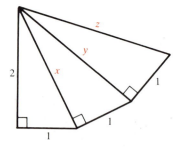

91. Find the area of the isosceles triangle in the figure.

(a)   **(b)**

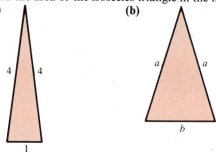

92. Find a formula for the area of an equilateral triangle of side a.

93. Three circles with radii 1, 2, and 3 ft are externally tangent to one another, as shown in the figure. Find the area of the sector of the circle of radius 1 that is cut off by the line segments that join the center of that circle to the centers of the other two circles.

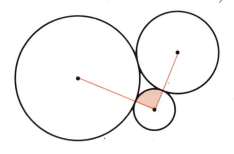

Trigonometry of Right Triangles

Let *ABC* be any right triangle with θ as one of its acute angles. We define the following **trigonometric ratios** (see Figure 1):

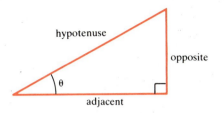

Figure 1

$$\sin\theta = \frac{\text{opposite}}{\text{hypotenuse}} \qquad \csc\theta = \frac{\text{hypotenuse}}{\text{opposite}}$$

$$\cos\theta = \frac{\text{adjacent}}{\text{hypotenuse}} \qquad \sec\theta = \frac{\text{hypotenuse}}{\text{adjacent}}$$

$$\tan\theta = \frac{\text{opposite}}{\text{adjacent}} \qquad \cot\theta = \frac{\text{adjacent}}{\text{opposite}}$$

The symbols we use for these ratios are abbreviations for their full names: **sine, cosine, tangent, cosecant, secant,** and **cotangent.**

Since any two right triangles with angle θ are similar, these ratios are the same regardless of the size of the triangle. They depend only on the angle θ and not on the size of the triangle. For example, the two triangles in Figure 2 are similar. If we compare the values of $\sin\theta$, $\cos\theta$, and $\tan\theta$, we see that they are the same for the small triangle and the large triangle.

Small Triangle

$$\sin\theta = \frac{7}{25}$$

$$\cos\theta = \frac{24}{25}$$

$$\tan\theta = \frac{7}{24}$$

Large Triangle

$$\sin\theta = \frac{7000}{25,000} = \frac{7}{25}$$

$$\cos\theta = \frac{24,000}{25,000} = \frac{24}{25}$$

$$\tan\theta = \frac{7000}{24,000} = \frac{7}{24}$$

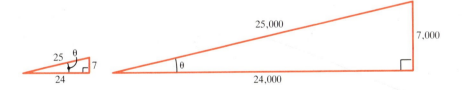

Figure 2

One of the important uses of trigonometry is a consequence of this property. As we will see later in this section, one can measure the distances to certain stars by reasoning that an enormously large triangle with the star at one vertex has the same trigonometric ratios as a much smaller triangle where these ratios can actually be measured (see Exercise 72). First, here are some examples of how the trigonometric ratios can be calculated.

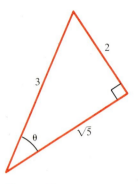

Figure 3

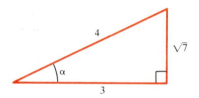

Figure 4

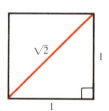

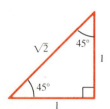

Figure 5

Figure 6

Example 1

Find the six trigonometric ratios of the angle θ in Figure 3.

Solution

$$\sin \theta = \frac{2}{3} \qquad \csc \theta = \frac{3}{2}$$

$$\cos \theta = \frac{\sqrt{5}}{3} \qquad \sec \theta = \frac{3}{\sqrt{5}}$$

$$\tan \theta = \frac{2}{\sqrt{5}} \qquad \cot \theta = \frac{\sqrt{5}}{2}$$

●

Example 2

If $\cos \alpha = \frac{3}{4}$, find the other five trigonometric ratios of α.

Solution

Since $\cos \alpha$ is defined as the ratio of the adjacent side to the hypotenuse, we draw a triangle with a hypotenuse of length 4 and a side with length 3 adjacent to α. If the opposite side is x, then by the Pythagorean Theorem, $3^2 + x^2 = 4^2$ or $x^2 = 7$, $x = \sqrt{7}$. We now use the triangle in Figure 4 to write

$$\sin \alpha = \frac{\sqrt{7}}{4} \qquad \csc \alpha = \frac{4}{\sqrt{7}}$$

$$\sec \alpha = \frac{4}{3}$$

$$\tan \alpha = \frac{\sqrt{7}}{3} \qquad \cot \alpha = \frac{3}{\sqrt{7}}$$

●

For certain angles the exact values of the trigonometric ratios are easily found. For these special angles, 30°, 45°, and 60°, the values of the trigonometric ratios can be calculated from two right triangles that we now construct. The first triangle is obtained by drawing a diagonal in a square of side 1 (see Figure 5). By the Pythagorean Theorem this diagonal has length $\sqrt{2}$. The resulting triangle has angles 45°, 45°, and 90° (or $\pi/4$, $\pi/4$, and $\pi/2$).

To get the second triangle we start with an equilateral triangle ABC of side 2 and draw the perpendicular bisector BD of the base. By the Pythagorean Theorem the length of BD is $\sqrt{3}$. Since BD also bisects angle ABC, we obtain a triangle with angles 30°, 60°, and 90° (or $\pi/6$, $\pi/3$, and $\pi/2$). See Figure 6.

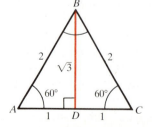

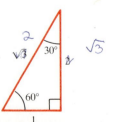

From the triangles in Figures 5 and 6 the values of the trigonometric ratios for the angles $\pi/6$, $\pi/4$, and $\pi/3$ can now be found:

(1)

$$\sin\frac{\pi}{6} = \frac{1}{2} \qquad \sin\frac{\pi}{4} = \frac{1}{\sqrt{2}} = \frac{\sqrt{2}}{2} \qquad \sin\frac{\pi}{3} = \frac{\sqrt{3}}{2}$$

$$\cos\frac{\pi}{6} = \frac{\sqrt{3}}{2} \qquad \cos\frac{\pi}{4} = \frac{1}{\sqrt{2}} = \frac{\sqrt{2}}{2} \qquad \cos\frac{\pi}{3} = \frac{1}{2}$$

$$\tan\frac{\pi}{6} = \frac{1}{\sqrt{3}} = \frac{\sqrt{3}}{3} \qquad \tan\frac{\pi}{4} = \frac{1}{1} = 1 \qquad \tan\frac{\pi}{3} = \sqrt{3}$$

It is useful to remember these special trigonometric ratios because they occur often. Of course, these ratios can be remembered easily if we remember the triangles from which they are obtained.

Although we have listed the values of sine, cosine, and tangent of these special angles, the values of the other ratios are readily obtained from these. We must notice first that the following **reciprocal relations** follow immediately from the definitions of these ratios:

(2)

$$\csc\theta = \frac{1}{\sin\theta} \qquad \sec\theta = \frac{1}{\cos\theta} \qquad \cot\theta = \frac{1}{\tan\theta}$$

$$\tan\theta = \frac{\sin\theta}{\cos\theta} \qquad \cot\theta = \frac{\cos\theta}{\sin\theta}$$

From these relations and from (1) we see, for example, that $\csc(\pi/3) = 2/\sqrt{3}$ and $\cot(\pi/3) = 1/\sqrt{3}$.

How can the trigonometric ratios of other angles be found? One method that the ancients used is to carefully draw a right triangle with the desired angle and measure the trigonometric ratios for that angle directly from the triangle (see Exercise 46). This method is not satisfactory because it is not very accurate. However, there are mathematical methods (called *numerical methods*) that are used to find the trigonometric ratios to any desired accuracy. These methods involve calculus and you will study some of them when you take a calculus course. Some of these techniques are based on the relations that exist among the trigonometric ratios. (In Chapter 6 we will see how these relations can be used to find the trigonometric ratios of many angles.) Using such methods, mathematicians have constructed extensive tables of the values of the trigonometric ratios.

The mathematical methods used to find the trigonometric ratios are programmed directly into scientific calculators. For instance, when the SIN key is pressed, the calculator computes the value of the sine of the given angle. In this book we will always use a calculator to find the values of the trigonometric ratios for angles other than the special angles mentioned earlier. Notice that most calculators give the values of sine, cosine, and tangent. The other ratios can be easily calculated from these using the reciprocal relations (2).

We follow the convention that when we write sin t, where t is a real number, we mean the sine of the angle whose radian measure is t. For instance, sin 1 means the sine of the angle whose radian measure is 1. When using a calculator to find an approximate value for this number, we set our calculator to radian mode and find that

$$\sin 1 \approx 0.841471$$

If we want to find the sine of the angle whose measure is 1°, we would write sin 1° and with our calculator in degree mode, we get

$$\sin 1° \approx 0.0174524$$

Solving Right Triangles

A triangle has six parts: three angles and three sides. To **solve a triangle** means to determine all of its parts from the information known about the triangle—that is, to determine the lengths of the three sides and the measures of the three angles.

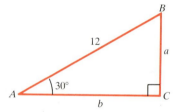

Figure 7

Example 3

Solve triangle ABC shown in Figure 7.

Solution

It is clear that $\angle B = 60°$. To find a, we look for an equation that relates a to the lengths and angles that we already know. In this case we have

$$\sin 30° = \frac{a}{12}$$

so

$$a = 12 \sin 30° = 12\left(\frac{1}{2}\right) = 6$$

Similarly,

$$\cos 30° = \frac{b}{12}$$

so

$$b = 12 \cos 30° = 12\left(\frac{\sqrt{3}}{2}\right) = 6\sqrt{3} \qquad \bullet$$

It is useful to know that, given the information in Figure 8, the lengths of the legs of the triangle are

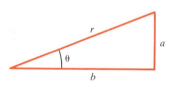

Figure 8

(3) $\qquad\qquad a = r \sin \theta$

(4) $\qquad\qquad b = r \cos \theta$

Example 4

Triangle ABC is a right triangle with hypotenuse 42 cm and one leg 21 cm long. Solve the triangle.

Solution

We sketch the triangle as in Figure 9. Now, to find the measure of angle A, notice that

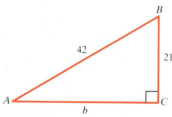

Figure 9

$$\sin A = \frac{21}{42} = \frac{1}{2}$$

We know that the acute angle whose sine is $\frac{1}{2}$ is 30°, so $\angle A = 30°$. It follows that $\angle B = 60°$ because the sum of the angles of any triangle is 180°. To find the length of side b we use Equation 4:

$$b = 42\cos 30° = 42\left(\frac{\sqrt{3}}{2}\right)$$

$$= 21\sqrt{3} \qquad \bullet$$

   The ability to solve right triangles using the trigonometric ratios is fundamental to many problems in navigation, surveying, astronomy, and measuring distances. The applications we consider in this section always involve right triangles, but as we will see in Section 5.4, trigonometry is also useful in solving triangles that are not right triangles.

   To state the next examples we need some terminology. If an observer is looking at an object, then the line from the observer's eye to the object is called the **line of sight** (see Figure 10). If the object being observed is above the horizontal, then the angle that the line of sight makes with the horizontal is called the **angle of elevation**. If the object is below the horizontal, then the angle between the line of sight and the horizontal is called the **angle of depression**.

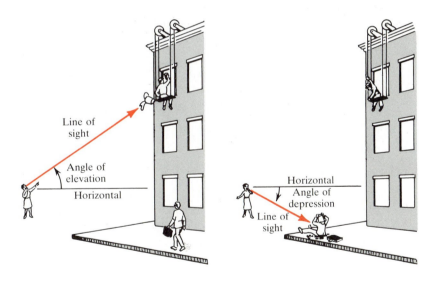

Figure 10

   The next example gives a remarkable application of trigonometry to the problem of measurement. Although the example is simple, it is typical of the method of applying the trigonometric ratios to such problems. In the example, the height of a tall tree is measured without the need to climb the tree!

Example 5

A giant redwood tree casts a shadow that is 532 ft long. Find the height of the tree if the angle of elevation of the sun is 25.7°.

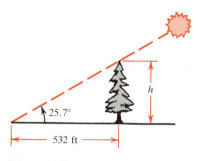

Figure 11

Solution

Let the height of the tree be h. From Figure 11 we see that

$$\tan 25.7° = \frac{h}{532}$$

so

$$h = 532 \tan 25.7°$$

$$\approx 532(0.4813) \approx 256$$

Therefore the height of the tree is about 256 ft.   •

The very same method used in this example can be used to find the distance to the sun! Finding this distance was a triumph of mathematical reasoning. The Greek mathematician Aristarchus of Samos (*circa* 310–230 B.C.) first calculated the distance to the sun by this method. You will be asked to rediscover how he did this in Exercise 68.

Example 6

From a point on the ground 500 ft from the base of a building, it is observed that the angle of elevation to the top of the building is 24° and the angle of elevation to the top of a flagpole on the building is 27°. Find the height of the building and the length of the flagpole.

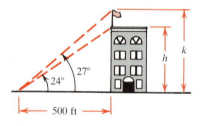

Figure 12

Solution

Figure 12 illustrates the situation. The height of the building is found in the same way that we found the height of the tree in Example 5.

$$\frac{h}{500} = \tan 24°$$

so

$$h = 500 \tan 24°$$

$$\approx 500(0.4452) \approx 223$$

The height of the building is approximately 223 ft. To find the length of the flagpole, let us first find the height from the ground to the top of the flagpole:

$$\frac{k}{500} = \tan 27°$$

Thus

$$k = 500 \tan 27°$$

$$\approx 500(0.5095) \approx 255$$

To find the length of the flagpole we subtract h from k. So the length of the flagpole is approximately $255 - 223 = 32$ ft.   •

Example 7

A pilot is flying over Washington, D.C. The pilot does not know his elevation but he sights the Washington Monument, which he knows to be 555 ft high. He measures the angle of depression to the top of the monument to be 17° and the

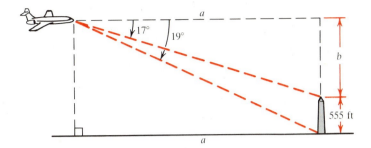

Figure 13

angle of depression to the bottom of the monument to be 19°. Find the elevation of the pilot and the distance between the plane and the monument along the ground.

Solution

From Figure 13 we see that

(5)
$$\tan 17° = \frac{b}{a}$$

(6)
$$\tan 19° = \frac{b + 555}{a}$$

Notice that we have two unknowns in this problem and neither of the two right triangles in Figure 13 can be solved independently. What we can do is solve Equations 5 and 6 together. First, we rewrite these two equations as follows:

(7)
$$b = a \tan 17°$$

(8)
$$555 + b = a \tan 19°$$

Now eliminate b by subtracting Equation 7 from Equation 8 to get

$$555 = a \tan 19° - a \tan 17°$$
$$= a(\tan 19° - \tan 17°)$$

Solving for a, we get
$$a = \frac{555}{\tan 19° - \tan 17°}$$

$$\approx \frac{555}{0.3443 - 0.3057} = \frac{555}{0.0386} \approx 14,400$$

To find b we substitute the value we found for a in Equation 7, say, to get

$$b = a \tan 17°$$
$$\approx 14,379(0.3057) \approx 4396$$

Thus the distance of the pilot from the monument along the ground is 14,400 ft or 2.73 mi. The elevation of the pilot is $b + 555 \approx 4950$ ft. ●

The trigonometric ratios can also be used to find angles, as the following example shows.

Example 8

A 40-ft ladder leans against a building. If the base of the ladder is 6 ft from the base of the building, what is the angle formed by the ladder and the building?

Solution

If θ is the angle between the ladder and the building, as shown in Figure 14, then

$$\sin \theta = \frac{6}{40} = 0.15$$

So θ is the angle whose sine is 0.15. To find the angle θ we use a calculator and the INV SIN key. Making sure our calculator is set to degrees, we get

$$\theta \approx 8.6°$$

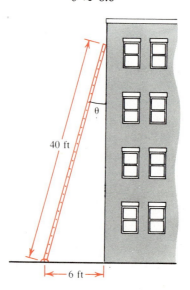

Figure 14

On some calculators the INV SIN key is called ARC SIN or SIN^{-1}. When this key is pressed, the calculator works "backward" and computes the angle that corresponds to the given value of sin. The same is true for the INV COS and INV TAN keys. These inverse trigonometric functions will be studied in more detail in Section 6.6.

Exercises 5.2

In Exercises 1–6 find the values of the six trigonometric ratios of the angle θ in the triangle shown.

1.

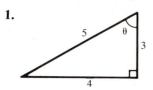

2.

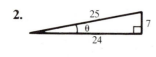

3.

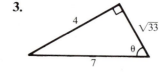

4.

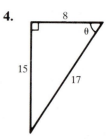

5.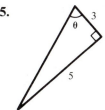

6.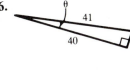

26.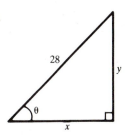

27.

In Exercises 7–10 find (a) sin α and cos β, (b) tan α and cot β, and (c) sec α and csc β.

7.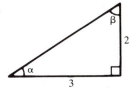

8.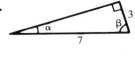

28.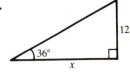

29.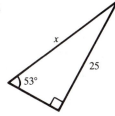

9.

10.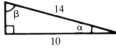

In Exercises 30–33 use Equations 3 and 4 of this section to find x and y in terms of θ.

In Exercises 11–19 find the other five trigonometric ratios of θ.

30.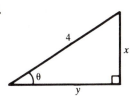

31.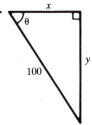

11. $\sin \theta = \dfrac{3}{5}$ **12.** $\cos \theta = \dfrac{2}{7}$ **13.** $\cot \theta = 1$

14. $\tan \theta = \sqrt{3}$ **15.** $\sec \theta = 7$ **16.** $\csc \theta = \dfrac{13}{12}$

17. $\tan \theta = \dfrac{a}{b}$ **18.** $\sin \theta = c$ **19.** $\cos \theta = d$

32.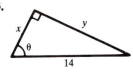

33.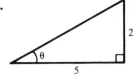

In Exercises 20–23 find the six trigonometric functions of θ from the information given. The letter a represents a real number.

20.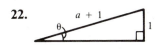

21.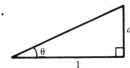

In Exercises 34–37 find the angle θ from the information given. State your answer correct to five decimal places.

22.

23.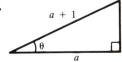

34.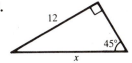

35.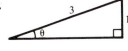

In Exercises 24–29 find the side labeled x. In Exercises 28 and 29 state your answer correct to five decimal places.

24.

25.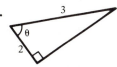

36.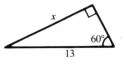

37.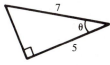

In Exercises 38–45 evaluate the given expression.

38. $\sin\dfrac{\pi}{6} + \cos\dfrac{\pi}{6}$ **39.** $\sin 30° \csc 30°$

40. $\sin 30° \cos 60° + \sin 60° \cos 30°$

41. $(\sin 60°)^2 + (\cos 60°)^2$ **42.** $\log_3\left(\tan\dfrac{\pi}{3}\right)$

43. $\log_2\left(\sin\dfrac{\pi}{4}\right)$ **44.** $(\cos 30°)^2 - (\sin 30°)^2$

45. $\left(\sin\dfrac{\pi}{3}\cos\dfrac{\pi}{4} - \sin\dfrac{\pi}{4}\cos\dfrac{\pi}{3}\right)^2$

46. From the figure shown estimate:
    (a) $\sin 15°$, $\cos 15°$, and $\tan 15°$
    (b) $\sin 40°$, $\cos 40°$, and $\tan 40°$

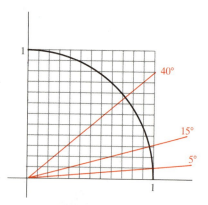

In Exercises 47–52 solve the given right triangle.

47.

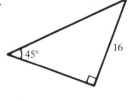

48.

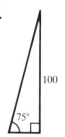

49.

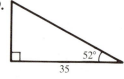

50.

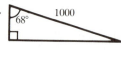

51.

52.

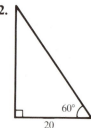

53. The angle of elevation to the top of the Empire State Building in New York is found to be 11° from the ground at a distance of 1 mi from the bottom of the building. Using this information, find the height of the Empire State Building.

54. A plane is flying within sight of the Gateway Arch in St. Louis, Missouri, at an elevation of 35,000 ft. The pilot would like to estimate her distance from the Gateway Arch. She finds that the angle of depression to a point on the ground below the Gateway Arch is 22°. What is the distance between the plane and the Gateway Arch? What is the distance between a point on the ground directly below the plane and the Gateway Arch?

55. A laser beam is to be directed toward the center of the moon, but the beam is 0.5° from its intended path. How far has the beam diverged from its assigned target when it reaches the moon? (The distance from the earth to the moon is 240,000 mi.) The radius of the moon is about 1000 mi. Will the beam strike the moon?

56. From the top of a 200-ft lighthouse, the angle of depression to a ship in the ocean is 23°. How far is the ship from the base of the lighthouse?

57. A 20-ft ladder leans against a building so that the angle between the ground and the ladder is 72°. How high does the ladder reach on the building?

58. A 20-ft ladder is leaning against a building. If the base of the ladder is 6 ft from the base of the building, what is the angle of elevation of the ladder? How high does the ladder reach on the building?

59. A 96-ft tree casts a shadow that is 120 ft long. What is the angle of elevation of the sun?

60. A 600-ft guy wire is attached to the top of a communication tower. If the wire makes an angle of 65° with the ground, how high is the communication tower?

61. A girl flying a kite lets out 450 ft of string. She estimates the angle of elevation of the kite to be 50°. How high is the kite above the ground?

62. The height of a steep cliff is to be measured from a point on the opposite side of the river. Find the height of the cliff from the information given in the figure.

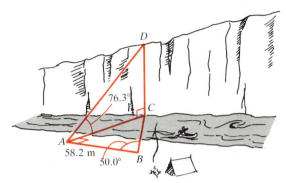

63. A high tower is 325 ft from a building. From a window in the building it is observed that the angle of elevation to the top of the tower is 39° and the angle of depression to the bottom of the tower is 25°. How high is the tower? How high is the window?

64. An airplane is flying at an elevation of 5150 ft directly above a straight highway. Two cars are on the highway on opposite sides of the plane, and the angle of depression to one car is 35° and to the other is 52°. How far apart are the two cars?

65. What is the answer to Exercise 64 if the two cars are on the highway on the same side of the plane and the angle of depression to one car is 38° and to the other car is 52°?

66. A hot-air balloon is above a straight road. To estimate their height above the ground, the balloonists measure the angle of depression to two consecutive mile posts on the road on the same side of the balloon. The angles of depression are found to be 20° and 22°. How high is the balloon?

67. To estimate the height of a mountain above a level plain, the angle of elevation to the top of the mountain is measured to be 32°. One thousand feet closer to the mountain along the plain, it is found that the angle of elevation is 35°. Estimate the height of the mountain.

68. When the moon is exactly half full, the angle formed by the earth, moon, and sun is a right angle (see the figure). At that time the angle formed by the sun, earth, and moon is measured to be 89.85°. If the distance from the earth to the moon is 240,000 mi, estimate the distance from the earth to the sun.

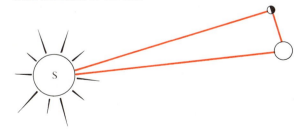

69. To find the distance to the sun as in Exercise 68, we needed to know the distance to the moon. Here is a way of estimating that distance: When the moon is seen at the zenith at a point A on the earth, it is observed to be at the horizon from point B (see the figure). Points A and B are 6155 mi apart and the radius of the earth is 3960 mi. **(a)** Find the angle θ in degrees. **(b)** Estimate the distance from point A to the moon.

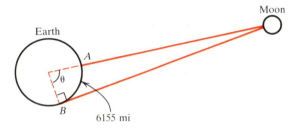

70. In Exercise 67 of Section 5.1 a method was given for finding the radius of the earth. Here is a more modern method of doing this. From a satellite 600 mi above the earth it is observed that the angle formed by the vertical and the line of sight to the horizon is 60.276°. Use this information to find the radius of the earth.

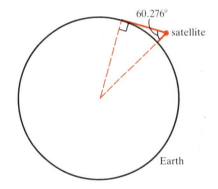

71. As viewed from the earth, the angle subtended by the full moon is 0.518°. Use this information and the fact that the distance AB from the earth to the moon is 236,900 mi to find the radius of the moon.

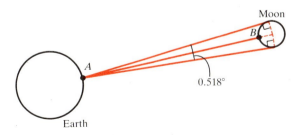

72. To find the distance to nearby stars, the method of parallax is used. The idea is to find a triangle with the star at one vertex and with a base as large as possible. To do this, the star is observed at two different times exactly 6

months apart and its apparent change in position is recorded. From these two observations $\angle E_1 SE_2$ can be calculated. (The times are chosen so that $\angle E_1 SE_2$ is as large as possible, which guarantees that $\angle E_1 OS$ is $90°$.) The angle $E_1 SO$ is called the *parallax* of the star. The star Alpha Centauri, the nearest one to us, has a parallax of $0.000211°$. Estimate the distance to this star. (Take the distance from the earth to the sun to be 9.3×10^7 mi.)

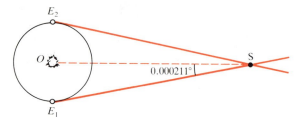

In Exercises 73–76 find x correct to one decimal place.

73.

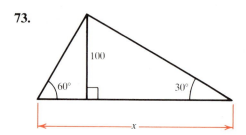

74.

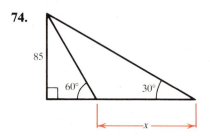

75.   **76.**

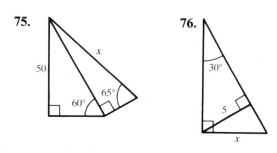

In Exercises 77 and 78 find x in terms of θ.

77.   **78.**

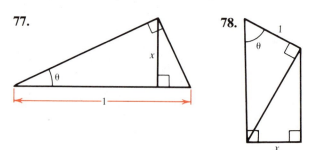

79. Express the lengths a, b, c, and d in the figure in terms of the trigonometric ratios of θ.

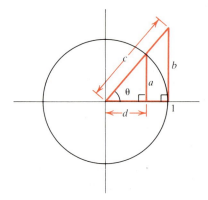

80. Express the lengths a and b in the figure in terms of the trigonometric ratios of θ.

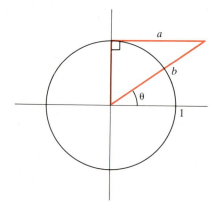

Two facts from geometry are needed for Exercises 81–83.
(1) An angle inscribed in a semicircle is a right angle.
(2) The central angle subtended by a chord of a circle is twice the angle subtended by that chord on the circle.

81. Express the length *d* in the diagram in terms of θ.

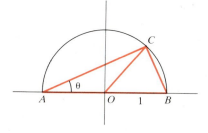

82. From the figure show that $\sin 2\theta = 2 \sin \theta \cos \theta$.
(*Hint:* Find the area of triangle *ABC* in two ways.)

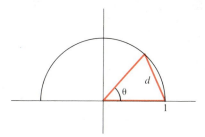

83. **(a)** Show from the figure that

$$\tan \theta = \frac{\sin 2\theta}{1 + \cos 2\theta}$$

(b) Use part (a) to find the exact values of tan 22.5° and tan 15°.

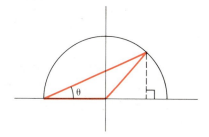

Trigonometric Ratios of Any Angle

In the preceding section we defined the trigonometric ratios for acute angles in terms of right triangles. In this section we show how the trigonometric ratios can be defined for any angle.

Let θ be an acute angle in standard position and let $P(x, y)$ be a point on the terminal side of this angle (see Figure 1). In triangle *POQ* the opposite side has length *y* and the adjacent side has length *x*. By the Pythagorean Theorem the hypotenuse has length $r = \sqrt{x^2 + y^2}$. So $\sin \theta = y/r$, $\cos \theta = x/r$, $\tan \theta = y/x$, and the others can be found in the same way by using the *x*- and *y*-coordinates of the point $P(x, y)$. Notice again that the values of the trigonometric ratios do

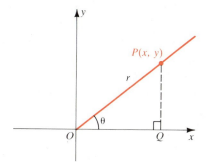

Figure 1

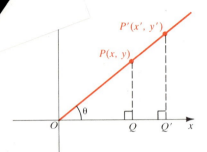

Figure 2

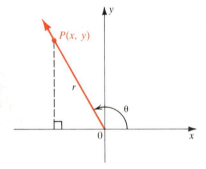

Figure 3

not depend on the choice of the point $P(x, y)$. This is so because if $P'(x', y')$ is another point on the terminal side, as in Figure 2, then triangles POQ and $P'OQ'$ are similar.

If θ is *any* angle in standard position, let $P(x, y)$ be a point on the terminal side and let $r = \sqrt{x^2 + y^2}$ be the distance from the origin to the point $P(x, y)$ (see Figure 3). By extending the definition for an acute angle, we define the six trigonometric ratios for θ as follows:

$$\sin \theta = \frac{y}{r} \qquad\qquad \csc \theta = \frac{r}{y} \quad (y \neq 0)$$

$$\cos \theta = \frac{x}{r} \qquad\qquad \sec \theta = \frac{r}{x} \quad (x \neq 0)$$

$$\tan \theta = \frac{y}{x} \quad (x \neq 0) \qquad \cot \theta = \frac{x}{y} \quad (y \neq 0)$$

Since division by zero is an undefined operation, certain trigonometric ratios are not defined for certain angles. For example, $\tan(90°) = y/x$ is undefined because $x = 0$. The angles for which the trigonometric ratios may be undefined are the angles for which either the x- or y-coordinate of a point on the terminal side of the angle is zero. These are of course angles that are coterminal with the coordinate axes. Here is a table of the trigonometric ratios for these angles:

θ	$\sin \theta$	$\cos \theta$	$\tan \theta$	$\csc \theta$	$\sec \theta$	$\cot \theta$
$0°$	0	1	0	undefined	1	undefined
$90°$	1	0	undefined	1	undefined	0
$180°$	0	-1	0	undefined	-1	undefined
$270°$	-1	0	undefined	-1	undefined	0

To use this table, or any other table of trigonometric values, it is important to notice the following obvious fact: *The values of a trigonometric ratio are the same for angles that are coterminal.* Thus $\cos(450°)$ has the same value as $\cos(90°)$, and $\tan(135°)$ has the same value as $\tan(-225°)$.

The reciprocal relations we mentioned in the preceding section still hold for the trigonometric ratios of any angle whenever these ratios are defined. This fact again follows from the definitions of these ratios.

From the definition we see that the values of the trigonometric ratios are all positive if the angle θ has its terminal side in the first quadrant. This is so because x and y are positive in that quadrant. [Of course, r is always positive, since it is simply the distance from the origin to the point $P(x, y)$.] If the terminal side of θ is in the second quadrant, however, then x is negative and y is positive. So in that quadrant $\sin\theta$ and $\csc\theta$ are positive and all the other trigonometric ratios have negative values. The following table tells which trigonometric values are positive in which quadrants:

Quadrant containing θ	Positive trigonometric ratios
I	all
II	$\sin\theta$, $\csc\theta$
III	$\tan\theta$, $\cot\theta$
IV	$\cos\theta$, $\sec\theta$

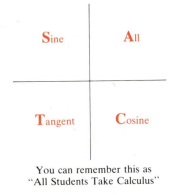

You can remember this as
"All Students Take Calculus"

Figure 4

The mnemonic device shown in Figure 4 makes it easy to remember the signs of the trigonometric ratios. For example, the letter C in quadrant IV reminds us that the values of $\cos\theta$ are positive in that quadrant (of course, $\sec\theta$ is also positive in quadrant IV because $\sec\theta = 1/\cos\theta$) and the rest are negative.

We now turn our attention to finding the values of the trigonometric ratios for any angle.

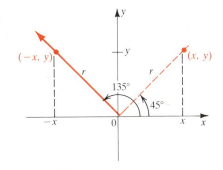

Figure 5

Example 1

Find (a) $\cos 135°$ and (b) $\tan 390°$.

Solution

(a) From Figure 5 we see that $\cos 135° = -x/r$. But $\cos 45° = x/r$, and since $\cos 45° = \sqrt{2}/2$, we have

$$\cos 135° = -\frac{\sqrt{2}}{2}$$

(b) Notice that the angles $390°$ and $30°$ are coterminal. From Figure 6 it is clear that

$$\tan 390° = \tan 30°$$

and since $\tan 30° = \sqrt{3}/3$, we have

$$\tan 390° = \frac{\sqrt{3}}{3}$$

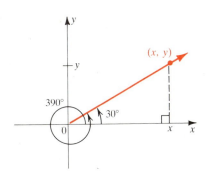

Figure 6

From these examples we see that it is not difficult to find trigonometric ratios of angles that are not acute. The reason for this is that the trigonometric ratios for angles that are not acute are the same, except possibly for the sign, as the corresponding trigonometric ratios of an acute angle. That acute angle is called

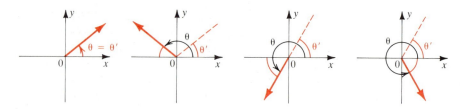

Figure 7
The reference angle θ' for the angle θ

the **reference angle**. In fact, *the reference angle is the acute angle formed by the terminal side of the angle under consideration and the x-axis.* Figure 7 shows that to find the reference angle it is useful to know which quadrant the terminal side of an angle is in.

Example 2

Find the reference angles for (a) $\theta = 5\pi/3$ and (b) $\theta = 870°$.

Solution

(a)   The reference angle is the acute angle formed by the terminal side of the angle $5\pi/3$ and the x-axis in Figure 8. Since the terminal side of this angle is in quadrant IV, the reference angle is

$$2\pi - \frac{5\pi}{3} = \frac{\pi}{3}$$

(b)   The angles $870°$ and $150°$ are coterminal [because $870 - 2(360) = 150$]. Thus the terminal side of this angle is in the second quadrant (see Figure 9). So the reference angle is $180° - 150° = 30°$.

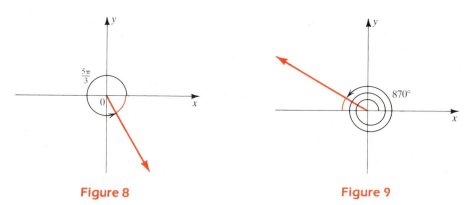

Figure 8   **Figure 9**

The method used in Example 1 to find the trigonometric ratios for any angle θ is summarized in the following three steps:

Step 1   Find the reference angle associated with the angle θ.

Step 2   Determine the sign of the trigonometric ratio of θ.

Step 3   Find the value of the trigonometric ratio of θ using Steps 1 and 2.

Example 3

Find (a) sin 240° and (b) cot 495°.

Solution

$$\begin{array}{c|c} S & A \\ \hline T & C \end{array}$$ sin 240° *is negative.*

(a)   This angle has its terminal side in quadrant III as shown in Figure 10. The reference angle is thus $240° - 180° = 60°$ and the value of $\sin 240°$ is negative. Thus

$$\sin 240° = -\sin 60° = -\frac{\sqrt{3}}{2}$$

$$\underset{\text{sign}}{\uparrow} \quad \underset{\text{reference angle}}{\uparrow}$$

$$\begin{array}{c|c} S & A \\ \hline T & C \end{array}$$ tan 495° *is negative,*
so cot 495° *is negative.*

(b)   The angle 495° is coterminal with the angle 135° and the terminal side of this angle is in quadrant II as shown in Figure 11. So the reference angle is $180° - 135° = 45°$ and the value of cot 495° is negative. We have

$$\cot 495° = \cot 135° = -\cot 45° = -1$$

$$\underset{\text{coterminal angles}}{\uparrow} \qquad \underset{\text{sign}}{\uparrow} \quad \underset{\text{reference angle}}{\uparrow}$$

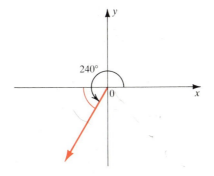

Figure 10

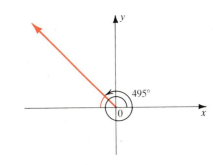

Figure 11

Example 4

Find (a) $\sin\left(\dfrac{16\pi}{3}\right)$ and (b) $\sec\left(-\dfrac{\pi}{4}\right)$.

Solution

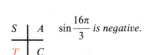

$\sin\dfrac{16\pi}{3}$ *is negative.*

(a)   The angle $16\pi/3$ is coterminal with $4\pi/3$ and these angles are in quadrant III (see Figure 12). Thus the reference angle is $4\pi/3 - \pi = \pi/3$. Since the value of sine is negative in that quadrant, we have

$$\sin\frac{16\pi}{3} = \sin\frac{4\pi}{3} = -\sin\frac{\pi}{3} = -\frac{\sqrt{3}}{2}$$

$$\underset{\text{coterminal angles}}{\uparrow} \qquad \underset{\text{sign}}{\uparrow} \quad \underset{\text{reference angle}}{\uparrow}$$

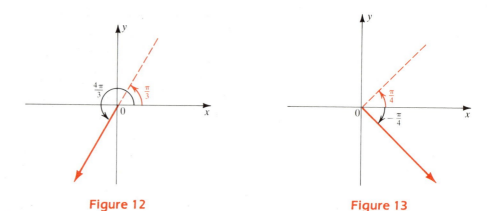

Figure 12   **Figure 13**

$$\begin{array}{c|c} S & A \\ \hline T & C \end{array}$$ $\cos\left(-\dfrac{\pi}{4}\right)$ *is positive,*

so $\sec\left(-\dfrac{\pi}{4}\right)$ *is positive.*

(b) The angle $-\pi/4$ is in quadrant IV and its reference angle is $\pi/4$ (see Figure 13). Since secant is positive in this quadrant, we get

$$\sec\left(-\frac{\pi}{4}\right) = +\sec\frac{\pi}{4} = \sqrt{2}$$

$$\uparrow \qquad \uparrow$$

sign   reference angle ●

Trigonometric Identities

The trigonometric ratios are related to one another through several important equations called **trigonometric identities**. We have already encountered the reciprocal identities:

$$\csc\theta = \frac{1}{\sin\theta} \qquad \sec\theta = \frac{1}{\cos\theta} \qquad \cot\theta = \frac{1}{\tan\theta}$$

$$\tan\theta = \frac{\sin\theta}{\cos\theta} \qquad \cot\theta = \frac{\cos\theta}{\sin\theta}$$

These identities continue to hold for any angle θ, provided both sides of the equations are defined.

One of the most important trigonometric identities is obtained by applying the Pythagorean Theorem to the triangle in Figure 14.*

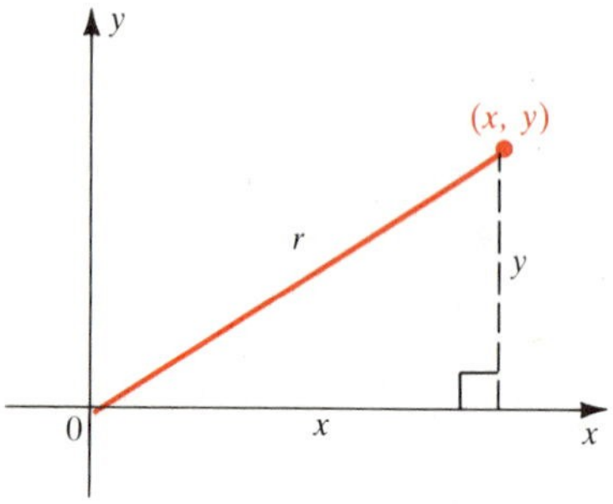

$$\sin^2\theta + \cos^2\theta = \left(\frac{y}{r}\right)^2 + \left(\frac{x}{r}\right)^2 = \frac{x^2 + y^2}{r^2} = \frac{r^2}{r^2} = 1$$

(1)
$$\sin^2\theta + \cos^2\theta = 1$$

Figure 14

---

* We follow the usual convention of writing $\sin^2\theta$ for $(\sin\theta)^2$. In general, we write $\sin^n\theta$ for $(\sin\theta)^n$ for all integers n except $n = -1$. The exponent $n = -1$ will be assigned another meaning in Section 6.7. Of course the same convention applies to the other five trigonometric ratios.

Although the proof of this identity has been indicated for acute angles, you should check that a similar argument shows that it holds for all angles θ.

Dividing both sides of Equation 1 by $\cos^2\theta$ (provided $\cos\theta \neq 0$), we get

$$\frac{\sin^2\theta}{\cos^2\theta} + \frac{\cos^2\theta}{\cos^2\theta} = \frac{1}{\cos^2\theta}$$

or

$$\left(\frac{\sin\theta}{\cos\theta}\right)^2 + 1 = \left(\frac{1}{\cos\theta}\right)^2$$

Since $\sin\theta/\cos\theta = \tan\theta$ and $1/\cos\theta = \sec\theta$, this becomes

$$\tan^2\theta + 1 = \sec^2\theta$$

Similarly, dividing both sides of Equation 1 by $\sin^2\theta$ (provided $\sin\theta \neq 0$) gives us

$$1 + \cot^2\theta = \csc^2\theta$$

We summarize these three important identities, sometimes called the **Pythagorean identities**:

(1)
(2)
(3)

$$\sin^2\theta + \cos^2\theta = 1$$
$$\tan^2\theta + 1 = \sec^2\theta$$
$$1 + \cot^2\theta = \csc^2\theta$$

These identities can be used to write any trigonometric ratio in terms of any other. For example, from Equation 1 we get

$$\sin\theta = \pm\sqrt{1 - \cos^2\theta}$$

where the sign depends on the quadrant. For instance, if θ is in quadrant II, then $\sin\theta$ is positive and hence

$$\sin\theta = \sqrt{1 - \cos^2\theta}$$

whereas if θ is in quadrant III, $\sin\theta$ is negative and so

$$\sin\theta = -\sqrt{1 - \cos^2\theta}$$

Example 5

Write $\tan\theta$ in terms of $\sin\theta$ where θ is in quadrant II.

Solution

Since $\tan\theta = \sin\theta/\cos\theta$, we need to write $\cos\theta$ in terms of $\sin\theta$. By Equation 1,

$$\cos\theta = \pm\sqrt{1 - \sin^2\theta}$$

and since $\cos\theta$ is negative in quadrant II, the negative sign applies here. Thus

$$\tan\theta = \frac{\sin\theta}{\cos\theta} = \frac{\sin\theta}{-\sqrt{1 - \sin^2\theta}}$$

●

Example 6

If $\tan \theta = \frac{2}{3}$ and θ is in quadrant III, find $\cos \theta$.

Solution 1

We need to write $\cos \theta$ in terms of $\tan \theta$. From Equation 2 we have $\sec \theta = \pm\sqrt{\tan^2\theta + 1}$ and in quadrant III $\sec \theta$ is negative, so

$$\sec \theta = -\sqrt{\tan^2\theta + 1}$$

Thus,
$$\cos \theta = \frac{1}{\sec \theta} = \frac{1}{-\sqrt{\tan^2\theta + 1}}$$

$$= \frac{1}{-\sqrt{(\frac{2}{3})^2 + 1}} = \frac{1}{-\sqrt{\frac{13}{9}}} = -\frac{3}{\sqrt{13}}$$

Solution 2

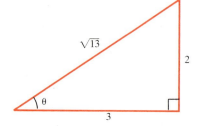

This problem can be solved more easily using the method of Example 2 of Section 5.2. Recall that except for the sign, the value of the trigonometric ratios of any angle are the same as those of an acute angle (the reference angle). So, ignoring the sign for the moment, let us draw a right triangle with an acute angle θ that satisfies $\tan \theta = \frac{2}{3}$ (see Figure 15). By the Pythagorean Theorem the hypotenuse of this triangle has length $\sqrt{13}$. From the triangle we immediately see that $\cos \theta = 3/\sqrt{13}$. Since θ is in quadrant III, $\cos \theta$ is negative and so $\cos \theta = -3/\sqrt{13}$. ●

Figure 15

Example 7

If $\sec \theta = 2$ and θ is in quadrant IV, find the other five trigonometric ratios of θ.

Solution

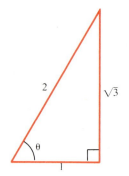

We draw a triangle as in Figure 16 so that $\sec \theta = 2$. Taking into account that θ is in quadrant IV, we get

$$\sin \theta = -\frac{\sqrt{3}}{2} \qquad \csc \theta = -\frac{2}{\sqrt{3}}$$

$$\cos \theta = \frac{1}{2}$$

$$\tan \theta = -\sqrt{3} \qquad \cot \theta = -\frac{1}{\sqrt{3}} \qquad ●$$

Figure 16

Areas of Triangles

The area of a triangle is $A = \frac{1}{2}(\text{base}) \times (\text{height})$. If we know two sides and the included angle of a triangle, then we can find the height using the trigonometric ratios, and from this we can find the area.

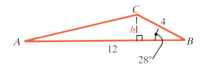

Figure 17

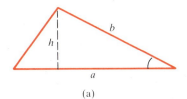

(a)

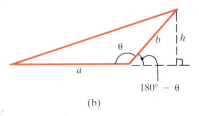

(b)

Figure 18

Example 8

Find the area of the triangle with sides of length 4 and 12 and included angle 28°.

Solution

We first sketch the triangle as in Figure 17. If we take AB to be the base of the triangle and h to be the height, then

$$\frac{h}{4} = \sin 28° \qquad \text{or} \qquad h = 4\sin 28°$$

Thus the area of the triangle is

$$A = \tfrac{1}{2}bh = \tfrac{1}{2}(12)(4\sin 28°) \approx 11.3$$
●

In general, if θ is an acute angle, then the area of the triangle in Figure 18(a) is given by

$$A = \tfrac{1}{2}ab\sin\theta$$

because the height of the triangle is $h = b\sin\theta$. If the angle θ is not acute, then from Figure 18(b) we see that the height of the triangle is

$$h = b\sin(180° - \theta) = b\sin\theta$$

This is so because the reference angle of θ is the angle $180° - \theta$. Thus in this case also the area of the triangle is

$$A = \tfrac{1}{2}(\text{base}) \times (\text{height}) = \tfrac{1}{2}ab\sin\theta$$

The area of a triangle with sides of lengths a and b and with included angle θ is

(4) $$A = \tfrac{1}{2}ab\sin\theta$$

Example 9

Find the area of triangle ABC if $|AB| = 10$, $|BC| = 3$, and the angle included by these two sides is 120°.

Solution

By Equation 4 the area of the triangle is

$$A = \tfrac{1}{2}ab\sin\theta$$
$$= \tfrac{1}{2}(10)(3)\sin(120°)$$
$$= 15\sin(60°) \qquad \textit{(reference angle)}$$
$$= 15\frac{\sqrt{3}}{2}$$

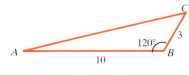

Figure 19

●

Exercises 5.3

In Exercises 1–24 find the reference angle for the given angle.

1. 225°

2. −35°

3. 181°

4. 290°

5. 750°

6. 570°

7. −230°

8. −310°

9. −1234°

10. 335°

11. −95°

12. 165°

13. $\dfrac{3\pi}{5}$

14. $\dfrac{7\pi}{6}$

15. $-\dfrac{2\pi}{3}$

16. $\dfrac{17\pi}{3}$

17. $-\dfrac{\pi}{4}$

18. 3

19. 1.7

20. −7

21. −0.7

22. $\dfrac{23\pi}{11}$

23. $\dfrac{23}{11}$

24. $\dfrac{17\pi}{7}$

In Exercises 25–28 determine whether the given expression is positive or negative if θ is in the given quadrant.

25. $\sin\theta\cos\theta$, θ in quadrant II

26. $\tan^2\theta\sec\theta$, θ in quadrant IV

27. $\dfrac{\sin\theta\sec\theta}{\cot\theta}$, θ in quadrant III

28. $\sin\theta\csc\theta$, θ in any quadrant

29. If $\sec\theta > 0$ and $\tan\theta < 0$, in which quadrant is θ?

30. If $\csc\theta > 0$ and $\cos\theta < 0$, in which quadrant is θ?

In Exercises 31–60 find the exact values of the trigonometric ratios.

31. $\sin 150°$

32. $\cos 225°$

33. $\sin 135°$

34. $\tan 330°$

35. $\sin(-60°)$

36. $\sec(-60°)$

37. $\csc(-630)°$

38. $\cot 210°$

39. $\cos 570°$

40. $\sec 120°$

41. $\tan 750°$

42. $\cos 660°$

43. $\sin 450°$

44. $\tan 450°$

45. $\csc 330°$

46. $\sin\left(\dfrac{2\pi}{3}\right)$

47. $\sin\left(\dfrac{5\pi}{3}\right)$

48. $\sin\left(\dfrac{3\pi}{2}\right)$

49. $\cos\left(\dfrac{7\pi}{3}\right)$

50. $\cos\left(-\dfrac{7\pi}{3}\right)$

51. $\tan\left(\dfrac{5\pi}{6}\right)$

52. $\sec\left(\dfrac{17\pi}{3}\right)$

53. $\csc\left(\dfrac{5\pi}{4}\right)$

54. $\cot\left(-\dfrac{\pi}{4}\right)$

55. $\cos\left(\dfrac{7\pi}{4}\right)$

56. $\tan\left(\dfrac{5\pi}{2}\right)$

57. $\csc(17\pi)$

58. $\tan\left(\dfrac{7\pi}{2}\right)$

59. $\sin\left(\dfrac{11\pi}{6}\right)$

60. $\cos\left(\dfrac{7\pi}{6}\right)$

In Exercises 61–68 write the first trigonometric ratio in terms of the second for θ in the given quadrant.

61. $\tan\theta$ in terms of $\sin\theta$, θ in quadrant III

62. $\cot\theta$ in terms of $\sin\theta$, θ in quadrant II

63. $\cos\theta$ in terms of $\sin\theta$, θ in quadrant IV

64. $\sec\theta$ in terms of $\tan\theta$, θ in quadrant II

65. $\cos\theta$ in terms of $\tan\theta$, θ in quadrant II

66. $\csc\theta$ in terms of $\cot\theta$, θ in quadrant III

67. $\sin\theta$ in terms of $\sec\theta$, θ in quadrant IV

68. $\tan\theta$ in terms of $\sec\theta$, θ in quadrant IV

In Exercises 69–76 find the remaining five trigonometric ratios of θ from the information given.

69. $\sin\theta = \frac{3}{4}$, θ in quadrant II

70. $\cos\theta = -\frac{7}{12}$, θ in quadrant III

71. $\tan\theta = -\frac{3}{4}$, θ in quadrant IV

72. $\sec\theta = 5$, θ in quadrant IV

73. $\csc\theta = 2$, θ in quadrant I

74. $\cot\theta = \frac{1}{4}$, θ in quadrant III

75. $\cos\theta = -\frac{2}{7}$, θ in quadrant II

76. $\tan\theta = -4$, θ in quadrant II

77. If $\theta = \pi/3$, find **(a)** $\sin 2\theta$ and $2\sin\theta$, **(b)** $\sin\dfrac{\theta}{2}$ and $\dfrac{\sin\theta}{2}$, and **(c)** $\sin^2\theta$ and $\sin\theta^2$.

In Exercises 78–81 find the area of the triangle from the information given. State your answer correct to one decimal place.

78. The triangle has sides of lengths 7 and 9 and included angle 72°.

79. The triangle has sides of lengths 10 and 22 and included angle 10°.

80. The triangle has sides of lengths 1 and 2 and included angle 150°.

81. The triangle is an equilateral triangle with side of length 10.

82. A triangle has an area of 16 in.² and two of the sides have lengths 5 in. and 7 in. Find the angle included by these two sides.

83. Find the area of the shaded region in the figure.

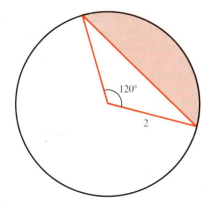

84. Deduce the following identities from the figure.

(a) $\sin^2\theta = \dfrac{1 - \cos 2\theta}{2}$ (*Hint:* Find the length of *DB* in two different ways.)

(b) $\cos^2\theta = \dfrac{1 + \cos 2\theta}{2}$ (*Hint:* Find the length of *AD* in two different ways.)

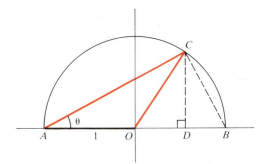

85. Use the identities in Exercise 84 to find the exact values of **(a)** sin 15° and cos 15°, **(b)** sin 22.5° and cos 22.5°, and **(c)** sin 7.5° and cos 7.5°.

Applications: The Law of Sines and the Law of Cosines

In the preceding sections the trigonometric ratios were used to solve right triangles. Here we see how trigonometry is used to solve oblique triangles—that is, triangles with no right angles. This is important because in many applications the triangles involved are not right triangles. Here is an example.

Example 1

The path of a satellite orbiting the earth causes it to pass directly overhead in both Phoenix and Los Angeles. At an instant when the satellite is between these two cities, its angles of elevation are simultaneously observed to be 60° at Phoenix and 75° at Los Angeles. How far is the satellite from Los Angeles if the distance between the observation stations in Phoenix and Los Angeles is 340 mi? In other words, we need to find the distance *AC* in Figure 1.   ●

To solve problems like this we need to know a relationship between the sides and the angles of any triangle. Such relationships exist; two of these are the Law of Sines and the Law of Cosines. To state these laws (or formulas) more easily we follow the convention of labeling the angles of a triangle by *A*, *B*, *C*, and the lengths of the corresponding opposite sides by *a*, *b*, *c*, as in Figure 2.

Figure 1

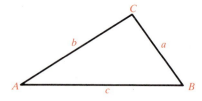

Figure 2

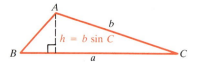

Figure 3

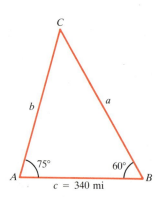

Figure 4

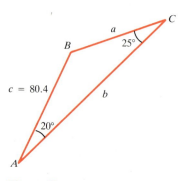

Figure 5

The Law of Sines

The **Law of Sines** says that in any triangle the lengths of the sides are proportional to the sines of the corresponding opposite angles.

The Law of Sines

$$\frac{\sin A}{a} = \frac{\sin B}{b} = \frac{\sin C}{c}$$

To see why this is true refer to Figure 3. The area of triangle ABC is $\frac{1}{2}ab \sin C$ by Equation 4 of Section 5.3. But the area of this triangle is also $\frac{1}{2}ac \sin B$ and $\frac{1}{2}bc \sin A$. Thus

$$\tfrac{1}{2}bc \sin A = \tfrac{1}{2}ac \sin B = \tfrac{1}{2}ab \sin C$$

Multiplying by $2/abc$, we get

$$\frac{\sin A}{a} = \frac{\sin B}{b} = \frac{\sin C}{c}$$

We can now solve the problem in Example 1.

Solution to Example 1

Whenever two angles in a triangle are known, the third angle can be determined immediately from the fact that the sum of the angles of a triangle is $180°$. In this case $\angle C = 180° - (75° + 60°) = 45°$ (see Figure 4). We have

$$\frac{\sin B}{b} = \frac{\sin C}{c}$$

or

$$\frac{\sin 60°}{b} = \frac{\sin 45°}{340}$$

Solving for b, we get $\qquad b = \dfrac{340 \sin 60°}{\sin 45°} \approx 416$

Thus the distance of the satellite from Los Angeles is approximately 416 mi.

●

Example 2

Solve the triangle in Figure 5.

Solution

First, $\angle B = 180° - (20° + 25°) = 135°$. Since side c is known, we use the relation

$$\frac{\sin A}{a} = \frac{\sin C}{c}$$

to find side a. Thus

$$a = \frac{c \sin A}{\sin C} = \frac{80.4 \sin 20°}{\sin 25°} \approx 65.1$$

Similarly, to find b use

$$\frac{\sin B}{b} = \frac{\sin C}{c}$$

so

$$b = \frac{c \sin B}{\sin C} = \frac{80.4 \sin 135°}{\sin 25°} \approx 134.5$$ ●

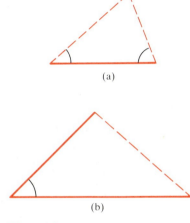

(a)

(b)

Figure 6

To solve a triangle it is necessary to have enough information about the sides and angles to completely determine the triangle. To decide whether this is the case it is often helpful to sketch the triangle using the information provided. For instance, if we are given two angles and the included side of a triangle, then it is clear that one and only one triangle is formed [see Figure 6(a)]. Similarly, if two sides and the included angle are known, then a unique triangle is determined in this way [Figure 6(b)]. But if, for example, we know all three angles of a triangle, we cannot solve such a triangle, since there are many triangles with the same three angles. (All these triangles would be similar, of course.) So we will not consider this case.

In general, a triangle is determined by three of its parts (angles and sides) as long as at least one of these parts is a side. So the possibilities are as follows:

Case 1   SAA   A side and two angles
Case 2   SSA   Two sides and the angle opposite one of those sides
Case 3   SAS   Two sides and the included angle
Case 4   SSS   Three sides

In each of these cases, except Case 2, a unique triangle is determined by the information given. Case 2 presents a situation where there may be two triangles, one triangle, or no triangles with the given properties. For this reason Case 2 is sometimes called the **ambiguous case**. To see how this can be, we sketch in Figure 7 the possibilities if angle A and sides a and b are given. In part (a) no solution is possible, since side a is too short to complete the triangle. In part (b) the solution is a right triangle. In part (c) there are two possible solutions, and in part (d) there is a unique triangle with the given properties. We illustrate some of these possibilities in the following examples.

Figure 7
The ambiguous case

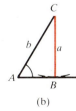
(a)

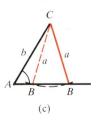
(b)

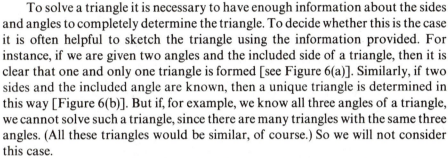
(c)

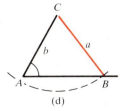
(d)

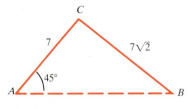

Figure 8

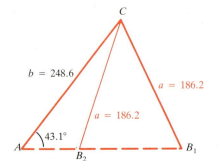

Figure 9

Example 3

In triangle ABC, $\angle A = 45°$, $a = 7\sqrt{2}$, and $b = 7$. Solve the triangle.

Solution

We first sketch the triangle with the information we have. Our sketch in Figure 8 is necessarily tentative, since we do not yet know what the other angles of the triangle are. Nevertheless sketching the triangle will help us see what the possibilities can be.

We first find $\angle B$. By the Law of Sines,

$$\frac{\sin A}{a} = \frac{\sin B}{b}$$

so

$$\sin B = \frac{b\sin A}{a} = \frac{7}{7\sqrt{2}}\sin 45° = \frac{1}{\sqrt{2}}\frac{\sqrt{2}}{2} = \frac{1}{2}$$

Which angles B have $\sin B = \frac{1}{2}$? From the preceding section we know that there are two such angles that are smaller than 180°. They are 30° and 150°. (We consider only angles that are smaller than 180°, since no triangle can contain an angle of 180° or larger.) Which of these possibilities is compatible with what we know about triangle ABC? Observe that since $\angle A = 45°$, we cannot have $\angle B = 150°$ because $45° + 150° > 180°$. So $\angle B = 30°$ and the remaining angle is $\angle C = 180° - (30° + 45°) = 105°$.

It remains to find side c. Again using the Law of Sines, we have

$$\frac{\sin B}{b} = \frac{\sin C}{c}$$

so

$$c = \frac{b\sin C}{\sin B} = \frac{7\sin 105°}{\sin 30°} = \frac{7\sin 105°}{\frac{1}{2}} = 14\sin 105°$$

An approximation for this is $c \approx 13.5$.   ●

In this example there were two possibilities for angle B and it turned out that one of these possibilities was not compatible with the rest of the information we knew about the triangle. Notice that if $\sin A < 1$ we must check the angle and its supplement as possibilities because any angle smaller than 180° can be in the triangle. To decide whether either possibility works, we see whether the resulting sum of the angles of the triangle exceeds 180°. It can happen, as in part (c) of Figure 7, that both possibilities are compatible with the given information. In that case there are two different triangles that are solutions to the problem.

Example 4

Solve the triangle with $\angle A = 43.1°$, $a = 186.2$, and $b = 248.6$.

Solution

First sketch the triangle from the given information as in Figure 9. We see that side a may be drawn in two possible positions to complete the triangle. From the

Law of Sines, we get

$$\sin B = \frac{b \sin A}{a} = \frac{248.6 \sin 43.1°}{186.2} \approx 0.91225$$

There are two possible angles B between $0°$ and $180°$ such that $\sin B = 0.91225$. Using a calculator, we find that one of these angles is approximately $65.8°$. The other is approximately $180° - 65.8° = 114.2°$. We denote these two angles by B_1 and B_2 so that

$$\angle B_1 \approx 65.8° \quad \text{and} \quad \angle B_2 \approx 114.2°$$

Thus there are two possible triangles that satisfy the given conditions: triangle $A_1 B_1 C_1$ and triangle $A_2 B_2 C_2$. We first solve triangle $A_1 B_1 C_1$. Clearly

$$\angle C_1 \approx 180° - (43.1° + 65.8°) = 71.1°$$

By the Law of Sines, we get

$$c_1 = \frac{a_1 \sin C_1}{\sin A_1} \approx \frac{186.2 \sin 71.1°}{\sin 43.1°} \approx 257.8$$

To solve triangle $A_2 B_2 C_2$, notice that

$$\angle C_2 \approx 180° - (43.1° + 114.2°) = 22.7°$$

By the Law of Sines,

$$c_2 = \frac{a_2 \sin C_2}{\sin A_2} \approx \frac{186.2 \sin 22.7°}{\sin 43.1°} \approx 105.2$$

Triangles $A_1 B_1 C_1$ and $A_2 B_2 C_2$ are sketched in Figure 10.   ●

The next example presents a situation where there is no triangle compatible with the given data.

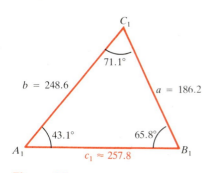

INV SIN
or
ARC SIN
or
SIN⁻¹

Figure 10

Example 5

In triangle ABC, $\angle A = 42°$, $a = 70$, and $b = 122$. Solve the triangle.

Solution

Let us try to find $\angle B$. By the Law of Sines, we have

$$\frac{\sin A}{a} = \frac{\sin B}{b}$$

so

$$\sin B = \frac{b \sin A}{a} = \frac{122 \sin 42°}{70} \approx 1.17$$

Since the sine of an angle is never greater than 1, we conclude that there is no triangle that satisfies the conditions in this problem.   ●

The Law of Cosines

In general, the Law of Sines cannot be used directly to solve triangles in Cases 3 and 4. However, triangles in these two cases can be solved by the **Law of Cosines**.

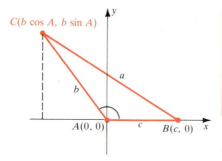

$C(b \cos A,\, b \sin A)$

$A(0, 0)$ c $B(c, 0)$ x

Figure 11

The Law of Cosines

$$a^2 = b^2 + c^2 - 2bc \cos A$$
$$b^2 = a^2 + c^2 - 2ac \cos B$$
$$c^2 = a^2 + b^2 - 2ab \cos C$$

To prove the Law of Cosines, place triangle ABC so that angle A is at the origin as shown in Figure 11. The coordinates of the vertices B and C of the triangle are $(c, 0)$ and $(b \cos A, b \sin A)$, respectively. (You should check that the coordinates of these points would be the same if we had drawn angle A to be an acute angle.) Using the Distance Formula, we get

$$\begin{aligned} a^2 &= (b \cos A - c)^2 + (b \sin A - 0)^2 \\ &= b^2 \cos^2 A - 2bc \cos A + c^2 + b^2 \sin^2 A \\ &= b^2 (\cos^2 A + \sin^2 A) - 2bc \cos A + c^2 \\ &= b^2 + c^2 - 2bc \cos A \end{aligned}$$

where in the last step we have used the fact that $\sin^2 \theta + \cos^2 \theta = 1$.

The other two formulas are obtained in the same way by placing each of the other vertices of the triangle at the origin and repeating the above argument.

In words, the Law of Cosines says that the square of any side of a triangle is equal to the sum of the squares of the other two sides minus twice the product of those two sides times the cosine of the included angle.

Notice that if one of the angles of the triangle, say $\angle C$, is a right angle, then $\cos C = 0$ and the Law of Cosines reduces to the Pythagorean Theorem: $c^2 = a^2 + b^2$. Thus the Pythagorean Theorem is a special case of the Law of Cosines.

Example 6

In triangle ABC it is known that $\angle A = 46.53°$, $b = 10.5$, and $c = 18.0$. Solve the triangle.

Solution

We can find a using the Law of Cosines:

$$\begin{aligned} a^2 &= b^2 + c^2 - 2bc \cos A \\ &= (10.5)^2 + (18.0)^2 - 2(10.5)(18.0)(\cos 46.53°) \approx 174.20 \end{aligned}$$

Thus $a \approx 13.19832$. The two remaining angles can now be found using the Law of Sines. We have

$$\sin B = \frac{b \sin A}{a} \approx \frac{10.5 \sin 46.53°}{13.19832} \approx 0.57736$$

So B is the angle whose sine is 0.57736. For this we use a calculator to get $\angle B \approx 35.3°$. Since angle B can have measure between $0°$ and $180°$, another possibility for angle B is $\angle B \approx 180° - 35.3° = 144.7°$. It is a simple matter to choose between these two possibilities, since the largest angle in a triangle must be

INV SIN
or
ARC SIN
or
SIN^{-1}

opposite the longest side. So the correct choice is $\angle B \approx 35.3°$. In this case $\angle C \approx 180° - (46.5° + 35.3°) = 98.2°$ and indeed the largest angle $\angle C$ is opposite the longest side $c = 18$.

To summarize: $\angle B \approx 35.3°$, $\angle C \approx 98.2°$, and $a \approx 13.2$.   ●

Example 7

The sides of a triangle are $a = 5$, $b = 8$, and $c = 12$ (see Figure 12). Find the angles of the triangle.

Solution

We find $\angle A$. From the Law of Cosines, $a^2 = b^2 + c^2 - 2bc \cos A$. Solving for $\cos A$, we get

$$\cos A = \frac{b^2 + c^2 - a^2}{2bc} = \frac{8^2 + 12^2 - 5^2}{2(8)(12)} = \frac{183}{192} = 0.953125$$

Using a calculator, we find that $\angle A \approx 18°$. In the same way the equations

$$\cos B = \frac{a^2 + c^2 - b^2}{2ac} = \frac{5^2 + 12^2 - 8^2}{2(5)(12)} = 0.875$$

$$\cos C = \frac{a^2 + b^2 - c^2}{2ab} = \frac{5^2 + 8^2 - 12^2}{2(5)(8)} = -0.6875$$

give $\angle B \approx 29°$ and $\angle C \approx 133°$. Of course, once two angles are calculated, the third can be found more easily from the fact that the sum of the angles of a triangle is $180°$. However, it is a good idea to calculate all three angles using the Law of Cosines and then add the three angles as a check on the computations.   ●

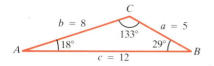

Figure 12

INV COS
or
ARC COS
or
COS⁻¹

Exercises 5.4

In Exercises 1–6 sketch each triangle approximately to scale and determine the number of triangles that satisfy the given conditions.

1. $\angle A = 40°, a = 30, b = 10$

2. $\angle B = 25°, b = 25, c = 30$

3. $\angle A = 30°, a = 50, b = 100$

4. $a = 10, b = 50, c = 20$

5. $\angle C = 38°, b = 45, c = 22$

6. $\angle A = 65°, \angle B = 24°, b = 10$

In Exercises 7–12 use the Law of Sines to find the indicated side or angle.

7.

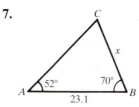

8.

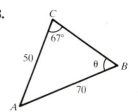

9.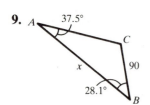

10.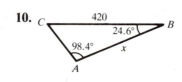

29.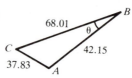

30.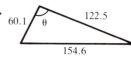

11.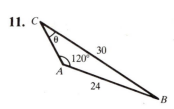

12.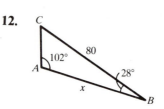

31.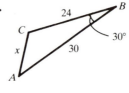

32.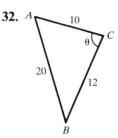

In Exercises 13–18 sketch each triangle and then solve the triangle using the Law of Sines.

13. $\angle A = 50°, \angle B = 68°, c = 230$

14. $\angle A = 23°, \angle B = 110°, c = 50$

15. $\angle A = 30°, \angle C = 65°, b = 10$

16. $\angle A = 22°, \angle B = 95°, a = 420$

17. $\angle B = 29°, \angle C = 51°, b = 44$

18. $\angle B = 10°, \angle C = 100°, c = 115$

In Exercises 19–26 solve for all possible triangles that satisfy the given conditions using the Law of Sines.

19. $a = 28, b = 15, \angle A = 110°$

20. $a = 30, c = 40, \angle A = 37°$

21. $a = 20, c = 45, \angle A = 125°$

22. $b = 45, c = 42, \angle C = 38°$

23. $b = 25, c = 30, \angle B = 25°$

24. $a = 75, b = 100, \angle A = 30°$

25. $a = 50, b = 100, \angle A = 50°$

26. $a = 100, b = 80, \angle A = 135°$

In Exercises 27–32 use the Law of Cosines to determine the indicated side or angle.

27.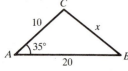

28.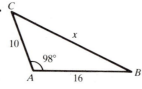

In Exercises 33–40 find the indicated side or angle.

33.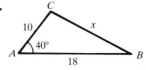

34.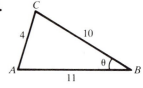

35.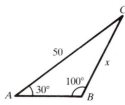

36.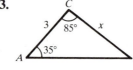

37.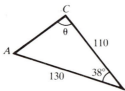

38.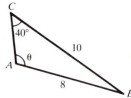

39.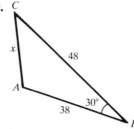

40.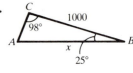

In Exercises 41–52 solve the given triangles.

41.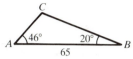

42.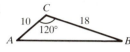

43.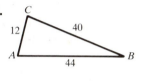

44.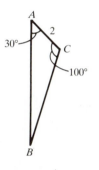

45. $a = 65, c = 50, \angle C = 52°$

46. $a = 50, b = 65, \angle A = 55°$

47. $b = 125, c = 162, \angle B = 40°$

48. $a = 73.5, \angle B = 61°, \angle C = 83°$

49. $a = 3.0, b = 4.0, \angle C = 53°$

50. $b = 60, c = 30, \angle A = 70°$

51. $a = 20, b = 25, c = 22$

52. $a = 10, b = 12, c = 16$

53. In order to find the distance across a small lake, the measurements in the figure were made by a surveyor. Find the distance across the lake using this information.

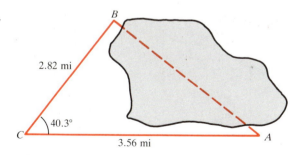

54. The path of a satellite orbiting the earth causes it to pass directly over two tracking stations A and B that are 50 mi apart. When the satellite is on one side of the two stations, the angles of elevation at A and B are measured to be 87.0° and 84.2°, respectively. **(a)** How far is the

satellite from station A? **(b)** How high is the satellite above the ground?

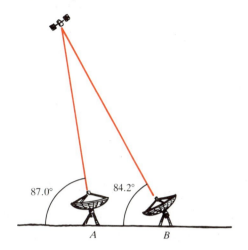

55. A tree on a hillside casts a shadow 215 ft down the hill. If the angle of inclination of the hillside is 22° to the horizontal and if the angle of elevation of the sun is 52°, find the height of the tree.

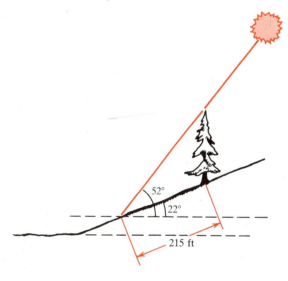

56. Points A and B are separated by a lake. To find the distance between them, a surveyor locates a point C on land such that $\angle CAB = 48.6°$. She also measures CA as 312 ft and CB as 527 ft. Find the distance between A and B.

57. Two boats leave the same port at the same time. One travels at a speed of 30 mi/h in the direction N50°E and the other travels at a speed of 26 mi/h in a direction

S70°E (see the figure). How far apart are the two boats after 1 h?

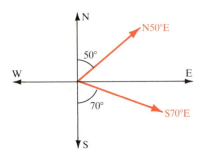

58. In the figure find **(a)** $\angle BCD$ and **(b)** $\angle DCA$.

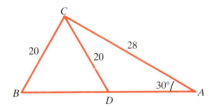

59. In this problem we develop criteria for determining whether there are two solutions, one solution, or no solutions to a triangle in the ambiguous case. Suppose we are given angle A and sides a and b. Sketch triangles like the ones in Figure 7 to show the following:
   (a) If $a \geq b$, then there is exactly one solution.
   (b) If $a < b$, then there are three possibilities:
      If $a < b \sin A$, then there are no solutions.
      If $a = b \sin A$, then there is exactly one solution.
      If $a > b \sin A$, then there are two solutions.

60. Refer to Exercise 59. If $A = 30°$ and $b = 100$, find the range of values of a for which **(a)** there are two solutions, **(b)** there is one solution, and **(c)** there are no solutions.

61. Given $\angle A = 40°$, $a = 15$, and $b = 20$.
   (a) Show that there are two triangles ABC and $A'B'C'$ that satisfy these conditions.
   (b) Show that the areas of these triangles are proportional to the sines of the angles C and C'; that is,

$$\frac{\text{area of } \triangle ABC}{\text{area of } \triangle A'B'C'} = \frac{\sin C}{\sin C'}$$

62. Show that, given the three angles A, B, C of a triangle and one side, say a, the area of the triangle is

$$\text{area} = \frac{a^2 \sin B \sin C}{2 \sin A}$$

63. Let r be the radius of the circle in which triangle ABC is inscribed (see the figure). Prove that

$$2r = \frac{a}{\sin A} = \frac{b}{\sin B} = \frac{c}{\sin C}$$

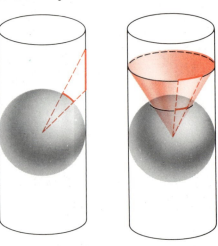

(*Hint:* $\angle AOD = \angle ACB$.)

64. Let S be the area of triangle ABC. Use Exercise 63 to show that $S = abc/4r$.

65. In order to draw a map of the spherical earth on a flat paper, several ingenious methods are known. One of these is the Mercator projection. In this method each point on the spherical earth is projected onto a circumscribed cylinder tangent to the sphere at every point on the equator by a line through the center of the earth. (See the figure.) By what factor are the following distances distorted using this method of projection?
   (a) The distance between 20° and 21° North latitude along a meridian
   (b) The distance between 40° and 41° North latitude along a meridian
   (c) The distance between 80° and 81° North latitude along a meridian
   (d) The distance between two points that are 1° apart on the 20th parallel
   (e) The distance between two points that are 1° apart on the 40th parallel
   (f) The distance between two points that are 1° apart on the 80th parallel

66. A surveyor wishes to find the distance between two points A and B on the opposite side of a river. She chooses two points C and D on her side of the river that are 20 m apart and measures the angles shown in the figure. Use this information to find the distance between the points A and B.

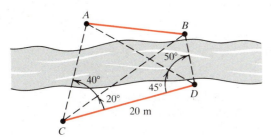

67. When two bubbles cling together in midair, their common surface is part of a sphere whose center D lies on the line that passes through the centers of the bubbles. Also, the angles BCA and ACD both have measure 60°.

(a) Show that the radius r of the common face is given by

$$r = \frac{ab}{a - b}$$

(b) If the radii of the two bubbles are 4 cm and 3 cm, respectively, find the radius of the common face.

(c) What shape does the common face have if the two bubbles have radii of the same length?

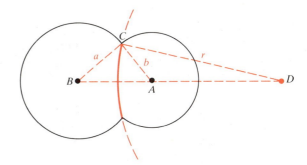

Section 5.5

Trigonometric Functions

A function on the real line is a rule that assigns to each real number another real number. So far the rules we have encountered to define functions have been in terms of algebraic expressions. This was the case when we defined the polynomial functions, the rational functions, and the exponential functions. In this section we define some new functions using the trigonometric ratios. The idea is this: the measure of any angle in radians is a real number and the trigonometric ratios act on that real number to give us another real number. This allows us to define the following *function*:

> The function **sin** is the rule that assigns to each real number t the real number sin t; namely, the sine of the angle whose radian measure is t.

Notice that the name of this function is sin. The value of this function at the real number t, as for any other function, is denoted by sin(t). But, by convention, we usually omit the brackets and denote the value of the function sin at the real number t by sin t. The other five trigonometric *functions* are defined in an analogous fashion.

There is a subtle difference between the trigonometric ratios we studied in the preceding sections and the trigonometric functions. The difference is one of point of view. In calculus we will be interested in the trigonometric functions as functions of a real number without the necessity of considering angles. But the difference between the two points of view is most striking when we consider the applications of the trigonometric ratios and the trigonometric functions. The trigonometric ratios were used in the preceding section to solve problems involving the measurement of distances. Historically this was the motivation for inventing the trigonometric ratios. However, when viewed as functions, the trigonometric functions behave in a way that makes them ideal for describing certain kinds of motion called harmonic motion. This application of the trigonometric functions will be studied in Section 5.7.

To see how the trigonometric functions can also be defined without the need to mention angles, let us consider points (x, y) in the plane that have distance 1 from the origin. (We use the number 1 for convenience only; any number r will do.) The set of such points is a circle of radius 1 called the **unit circle**. Given any positive real number t, let us mark off a distance of t units along the unit circle starting at the point $(1, 0)$ and moving in a counterclockwise direction if t is positive and in a clockwise direction if t is negative (see Figure 1). In this way we arrive at a point (x, y), and, since this point depends on t, let us call it $P(t)$. So $P(t) = (x, y)$. We can now define the six trigonometric functions using the x- and y-coordinates of the point $P(t)$ as follows:

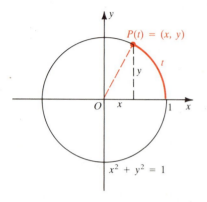

Figure 1

(1)

$$\sin t = y \qquad\qquad \csc t = \frac{1}{y} \quad (y \neq 0)$$

$$\cos t = x \qquad\qquad \sec t = \frac{1}{x} \quad (x \neq 0)$$

$$\tan t = \frac{y}{x} \quad (x \neq 0) \qquad \cot t = \frac{x}{y} \quad (y \neq 0)$$

Notice that we did not need to mention angles to define the trigonometric functions in this way. But it is important to notice that $\sin t$ still has the same value as the sine of the angle whose radian measure is t. To check this, observe that an arc of length t along the unit circle subtends an angle of radian measure t. So $\sin t = y/r = y/1 = y$. You should check that the definitions we have given for the other five trigonometric functions are in the same way consistent with the trigonometric ratios in terms of radian measure. Because the trigonometric functions are defined in terms of the unit circle they are sometimes called the **circular functions**.

Notice that since $\sin t$ and $\cos t$ are the x- and y-coordinates of a point on the unit circle, their values are always between -1 and 1:

$$-1 \leq \sin t \leq 1 \qquad \text{and} \qquad -1 \leq \cos t \leq 1$$

Example 1

Let $P(t) = (\sqrt{3}/3, \sqrt{2}/\sqrt{3})$. Show that this point is on the unit circle and find $\sin t$ and $\cos t$.

Solution

To show that this point is on the unit circle we need to show that it satisfies the equation of the unit circle $x^2 + y^2 = 1$. But

$$\left(\frac{\sqrt{3}}{3}\right)^2 + \left(\frac{\sqrt{2}}{\sqrt{3}}\right)^2 = \frac{3}{9} + \frac{2}{3} = \frac{1}{3} + \frac{2}{3} = 1$$

so this point is on the unit circle. From Equations 1 it follows immediately that

$$\sin t = \frac{\sqrt{2}}{\sqrt{3}} \quad \text{and} \quad \cos t = \frac{\sqrt{3}}{3} \qquad \bullet$$

Example 2

The x-coordinate of a point $P(t) = (x, y)$ on the unit circle is $\sqrt{3}/2$. If this point is in quadrant IV, find the values of the six trigonometric functions of t.

Solution

Since the point $P(t) = (x, y)$ is on the unit circle, we have

$$\left(\frac{\sqrt{3}}{2}\right)^2 + y^2 = 1$$

Solving for y, we get

$$y^2 = 1 - \frac{3}{4} = \frac{1}{4}$$

$$y = \pm\frac{1}{2}$$

Since this point is in quadrant IV, its y-coordinate must be negative, so $y = -\frac{1}{2}$. Now that we know both the x- and y-coordinates of the point $P(t)$, we can find the values of the trigonometric functions at t from Equations 1:

$$\sin t = -\frac{1}{2} \qquad\qquad \csc t = -2$$

$$\cos t = \frac{\sqrt{3}}{2} \qquad\qquad \sec t = \frac{2}{\sqrt{3}}$$

$$\tan t = \frac{-\frac{1}{2}}{\frac{\sqrt{3}}{2}} = -\frac{1}{\sqrt{3}} = -\frac{\sqrt{3}}{3} \qquad \cot t = \frac{\frac{\sqrt{3}}{2}}{-\frac{1}{2}} = -\sqrt{3} \qquad \bullet$$

More Identities

The reciprocal identities and the Pythagorean identities, which we have already encountered for the trigonometric ratios, are still valid for the trigonometric functions. Here we will discover some additional identities.

Let us consider the relationship between the trigonometric functions of t and

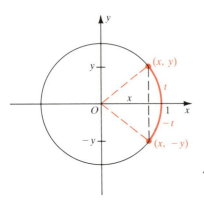

Figure 2

those of $-t$. From Figure 2 we see that

$$\sin(-t) = -y = -\sin t$$
$$\cos(-t) = x = \cos t$$
$$\tan(-t) = \frac{-y}{x} = -\frac{y}{x} = -\tan t$$

These equations show that the functions $\sin t$ and $\tan t$ are odd, whereas the function $\cos t$ is even. We summarize these three equations in the following box:

Odd-Even Identities

(2)
$$\sin(-t) = -\sin t$$
$$\cos(-t) = \cos t$$
$$\tan(-t) = -\tan t$$

From Figure 3 we see that

$$\sin(t + \pi) = -\sin t$$
$$\cos(t + \pi) = -\cos t$$

so
$$\tan(t + \pi) = \frac{\sin(t + \pi)}{\cos(t + \pi)} = \frac{-\sin t}{-\cos t} = \tan t$$

We summarize these three properties:

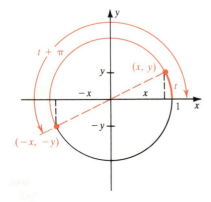

Figure 3

Reduction Formulas

(3)
$$\sin(t + \pi) = -\sin t$$
$$\cos(t + \pi) = -\cos t$$
$$\tan(t + \pi) = \tan t$$

Equations 3 give a relation between the trigonometric functions of t and those of $t + \pi$. Equations of this type are called **reduction formulas**.

Let us derive reduction formulas for $t + \pi/2$. Figure 4 shows a length t along the arc of the unit circle and a length $t + \pi/2$ along the same circle. Both lengths are marked off along the circle starting at $(1, 0)$ and moving in a counterclockwise direction. It is easy to check that triangle AOB is congruent to triangle CDO. So if $P(t) = (x, y)$, then $P(t + \pi/2) = (-y, x)$. From this we get

$$\sin\left(t + \frac{\pi}{2}\right) = x = \cos t$$

$$\cos\left(t + \frac{\pi}{2}\right) = -y = -\sin t$$

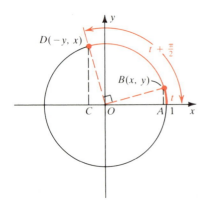

Figure 4

so $\qquad \tan\left(t + \dfrac{\pi}{2}\right) = \dfrac{\sin\left(t + \dfrac{\pi}{2}\right)}{\cos\left(t + \dfrac{\pi}{2}\right)} = \dfrac{\cos t}{-\sin t} = -\cot t$

We summarize these three reduction formulas:

Reduction Formulas

(4)
$$\sin\left(t + \frac{\pi}{2}\right) = \cos t$$
$$\cos\left(t + \frac{\pi}{2}\right) = -\sin t$$
$$\tan\left(t + \frac{\pi}{2}\right) = -\cot t$$

There are many other reduction formulas. Some of them are included in the exercises at the end of this section.

Periodic Properties

The trigonometric functions repeat their values in a regular fashion. To see exactly how this happens we first notice that the circumference of the unit circle is 2π. So if t is any real number, then the points $P(t)$ and $P(t + 2\pi)$ are the same point. Similarly $P(t) = P(t + 2(2\pi)) = P(t + 3(2\pi))$, and in general $P(t) = P(t + 2k\pi)$ for $k = \pm 1, \pm 2, \pm 3, \ldots$. It follows that the values of the trigonometric functions are unchanged by adding a multiple of 2π to t.

(5)
$$\begin{aligned} \sin(t + 2k\pi) &= \sin t &&\text{for} &&k = \pm 1, \pm 2, \pm 3, \ldots \\ \cos(t + 2k\pi) &= \cos t &&\text{for} &&k = \pm 1, \pm 2, \pm 3, \ldots \end{aligned}$$

There is a name for functions that repeat their values in this fashion; they are called periodic.

> A function f is called **periodic** if there is a positive real number a such that $f(t + a) = f(t)$ for every t. The least such positive number (if it exists) is called the **period** of f.

The period of the sine and cosine functions is 2π. The last equation in (3) shows that the tangent function is also periodic. The period of the tangent function is π.

Graphs of Sine and Cosine Functions

The properties of the sine and cosine functions that we derived in this section will help us sketch their graphs. The fact that sine is an odd function means that its graph is symmetric with respect to the origin. The cosine function is even, so its graph is symmetric with respect to the y-axis. Since these functions are periodic of period 2π, their values repeat in successive intervals of length 2π, so we need only draw their graphs on any interval of length 2π.

Since the sine and cosine functions are defined for all real numbers, the domain of each of these functions is the set of real numbers. To sketch the graph of $y = \sin t$ we could try to make a table of values and use those points to draw the graph. But of course no such table can be complete. Instead, let us look more closely at the definition of the sine function. Recall that $\sin t$ is the y-coordinate of the point $P(t) = (x, y)$ on the unit circle that is a distance of t units along the circumference of the unit circle starting at $(1, 0)$ (see Figure 5). How does the y-coordinate of this point vary as t increases? It is easy to see that the y-coordinate of $P(t)$ increases to 1 and then decreases to -1 repeatedly as the point $P(t)$ winds around the unit circle. In fact, as t increases from 0 to $\pi/2$, $y = \sin t$ increases from 0 to 1. As t increases further from $\pi/2$ to π, the value of $y = \sin t$ decreases from 1 to 0. The following table shows the variation of the function sin:

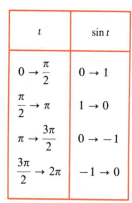

t	$\sin t$
$0 \to \dfrac{\pi}{2}$	$0 \to 1$
$\dfrac{\pi}{2} \to \pi$	$1 \to 0$
$\pi \to \dfrac{3\pi}{2}$	$0 \to -1$
$\dfrac{3\pi}{2} \to 2\pi$	$-1 \to 0$

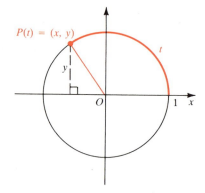

Figure 5

To draw the graph of this function more accurately, we find a few other values of $\sin t$.

t	0	$\dfrac{\pi}{6}$	$\dfrac{\pi}{4}$	$\dfrac{\pi}{3}$	$\dfrac{\pi}{2}$	$\dfrac{2\pi}{3}$	$\dfrac{3\pi}{4}$	$\dfrac{5\pi}{6}$	π	$\dfrac{7\pi}{6}$	$\dfrac{5\pi}{4}$	$\dfrac{4\pi}{3}$	$\dfrac{3\pi}{2}$	$\dfrac{5\pi}{3}$	$\dfrac{7\pi}{4}$	$\dfrac{11\pi}{6}$	2π
$\sin t$	0	$\dfrac{1}{2}$	$\dfrac{\sqrt{2}}{2}$	$\dfrac{\sqrt{3}}{2}$	1	$\dfrac{\sqrt{3}}{2}$	$\dfrac{\sqrt{2}}{2}$	$\dfrac{1}{2}$	0	$-\dfrac{1}{2}$	$-\dfrac{\sqrt{2}}{2}$	$-\dfrac{\sqrt{3}}{2}$	-1	$-\dfrac{\sqrt{3}}{2}$	$-\dfrac{\sqrt{2}}{2}$	$-\dfrac{1}{2}$	0

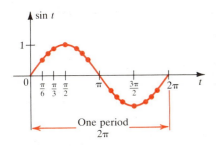

Figure 6

Graph of $\sin t$ for $0 \le t \le 2\pi$

Using this information we sketch in Figure 6 the graph of the function $\sin t$ for t between 0 and 2π. This is a graph of one period of the sine function. Using the fact that this function is periodic with period 2π, we get the complete graph in Figure 7 by continuing this same pattern to the left and to the right in every successive interval of length 2π.

The graph of the function $\cos t$ can be drawn by finding how the cosine function varies as t increases in the same way that we drew the graph of $\sin t$. But we follow a different method that exploits a simple relationship between the sine and cosine functions and a graphing technique from Chapter 2. Recall the reduction formula in Equations 4

$$\cos t = \sin\left(t + \frac{\pi}{2}\right)$$

This immediately shows that the graph of $\cos t$ is the same as the graph of $\sin t$ shifted $\pi/2$ units to the left. Figure 8(a) shows the graph of one period of cosine for $0 \le t \le 2\pi$. The complete graph of cosine is in Figure 8(b).

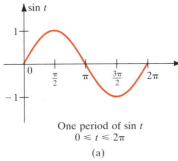

One period of $\sin t$

$0 \le t \le 2\pi$

(a)

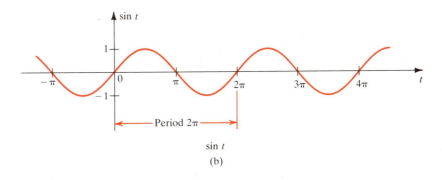

Period 2π

$\sin t$

(b)

Figure 7

Graph of $\sin t$

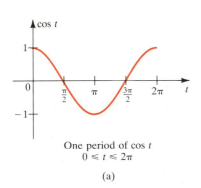

One period of $\cos t$

$0 \le t \le 2\pi$

(a)

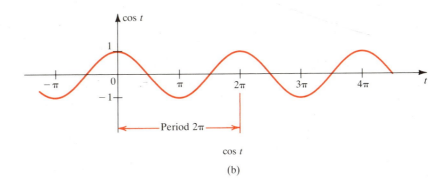

Period 2π

$\cos t$

(b)

Figure 8

Graph of $\cos t$

Exercises 5.5

In Exercises 1–10 show that the point $P(t)$ is on the unit circle and find the values of $\sin t$, $\cos t$, and $\tan t$.

1. $P(t) = \left(\dfrac{3}{5}, \dfrac{4}{5}\right)$

2. $P(t) = \left(-\dfrac{3}{5}, \dfrac{4}{5}\right)$

3. $P(t) = \left(\dfrac{6}{7}, -\dfrac{\sqrt{13}}{7}\right)$

4. $P(t) = \left(-\dfrac{1}{3}, -\dfrac{2\sqrt{2}}{3}\right)$

5. $P(t) = \left(\dfrac{40}{41}, \dfrac{9}{41}\right)$

6. $P(t) = \left(-\dfrac{3}{5}, -\dfrac{4}{5}\right)$

7. $P(t) = \left(\dfrac{1}{3}, \dfrac{2\sqrt{2}}{3}\right)$

8. $P(t) = \left(-\dfrac{6}{7}, \dfrac{\sqrt{13}}{7}\right)$

9. $P(t) = \left(-\dfrac{5}{13}, -\dfrac{12}{13}\right)$

10. $P(t) = \left(\dfrac{\sqrt{5}}{5}, \dfrac{2\sqrt{5}}{5}\right)$

In Exercises 11–16 find $\sin t$ and $\cos t$ if $P(t)$ is a point on the unit circle that satisfies the given properties.

11. The x-coordinate of $P(t)$ is 3/5 and $P(t)$ is in quadrant I.

12. The y-coordinate of $P(t)$ is $-1/3$ and $P(t)$ is in quadrant III.

13. The x-coordinate of $P(t)$ is 2/3 and the y-coordinate is negative.

14. The y-coordinate of $P(t)$ is $-\sqrt{5}/5$ and the x-coordinate is positive.

15. The x-coordinate of $P(t)$ is $\sqrt{2}/3$ and t is between $3\pi/2$ and 2π.

16. The x-coordinate of $P(t)$ is $-2/5$ and t is between $\pi/2$ and π.

In Exercises 17–22 find the point $P(t)$ on the unit circle that corresponds to the given value of t.

17. $t = \dfrac{\pi}{4}$

18. $t = \dfrac{\pi}{2}$

19. $t = -\dfrac{\pi}{3}$

20. $t = \dfrac{7\pi}{6}$

21. $t = \pi$

22. $t = -\dfrac{5\pi}{6}$

In Exercises 23–26 use methods similar to those in the text to show that the following reduction formulas hold.

23. **(a)** $\sin\left(t - \dfrac{\pi}{2}\right) = -\cos t$

    (b) $\cos\left(t - \dfrac{\pi}{2}\right) = \sin t$

    (c) $\tan\left(t - \dfrac{\pi}{2}\right) = -\cot t$

24. **(a)** $\sin\left(t + \dfrac{3\pi}{2}\right) = -\cos t$

    (b) $\cos\left(t + \dfrac{3\pi}{2}\right) = \sin t$

    (c) $\tan\left(t + \dfrac{3\pi}{2}\right) = -\cot t$

25. **(a)** $\sin\left(\dfrac{\pi}{2} - t\right) = \cos t$

    (b) $\cos\left(\dfrac{\pi}{2} - t\right) = \sin t$

    (c) $\tan\left(\dfrac{\pi}{2} - t\right) = \cot t$

26. **(a)** $\sin(\pi - t) = \sin t$
    (b) $\cos(\pi - t) = -\cos t$
    (c) $\tan(\pi - t) = -\tan t$

In Exercises 27–32 find $f(g(t))$ for the given functions f and g and simplify the result using the trigonometric identities.

27. $f(x) = \sqrt{1 - x^2}$, $g(t) = \sin t$ $(0 < t < \pi/2)$

28. $f(x) = \sqrt{9 - x^2}$, $g(t) = 3\cos t$ $(0 < t < \pi)$

29. $f(x) = \sqrt{x^2 - 16}$, $g(t) = 4\sec t$ $(0 < t < \pi/2)$

30. $f(u) = \dfrac{1}{(7 + u^2)^{3/2}}$, $g(t) = \sqrt{7}\tan t$ $(0 < t < \pi/2)$

31. $f(u) = \dfrac{u}{(u^2 - 1)^{3/2}}$, $g(t) = \sec t$ $(0 < t < \pi/2)$

32. $f(x) = \dfrac{x^4}{(1 + x^2)^2}$, $g(t) = \tan t$ $(0 < t < \pi/2)$

In Exercises 33–36 determine whether the function whose graph is shown is periodic. If so, determine its period from the graph.

33.

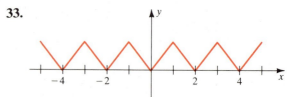

34.

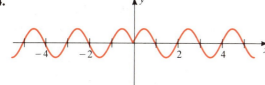

35.

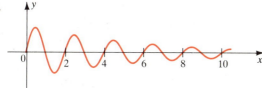

36.

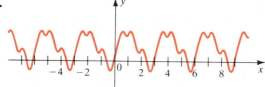

37. Use the periodicity of the functions sin, cos, and tan to show the following:
  (a) The function csc is periodic with period 2π.
  (b) The function sec is periodic with period 2π.
  (c) The function cot is periodic with period π.

38. If f is a periodic function with period p, show that the function $1/f$ is also periodic with period p.

In Exercises 39–46 determine the period of the function.

39. $f(x) = 1 + \sin x$

40. $f(x) = 2\sin x + 5\cos x$

41. $f(x) = \sin^2 x$

42. $f(x) = \cos(3x)$

43. $f(x) = e^{\sin x}$

44. $f(x) = 2^{\sin 2x}$

45. $f(x) = \tan x + \cot x$

46. $f(x) = \sin(\cos x)$

In Exercises 47–54 determine whether the given function is even, odd, or neither.

47. $f(x) = x^2 \sin x$

48. $f(x) = x^2 \cos(2x)$

49. $f(x) = \sin x \cos x$

50. $f(x) = e^x \sin x$

51. $f(x) = |x| \cos x$

52. $f(x) = x \sin^3 x$

53. $f(x) = x^3 + \cos x$

54. $f(x) = \cos(\sin x)$

55. **(a)** Make a table showing the variation of the cosine function for $0 \le t \le 2\pi$ similar to the one done for the sine function in the text.
  (b) Make a table of values for the function $\cos t$ using the same values for t as in the table done for $\sin t$ in the text.
  (c) Use the information found in parts (a) and (b) to draw the graph of the function $\cos t$ as accurately as possible.

In Exercises 56–65 use the graphing techniques discussed in Section 2.3 to obtain the graphs of the functions from the graphs of sin *and* cos.

56. $f(t) = 2 + \sin t$

57. $g(t) = -\sin t$

58. $h(t) = \sin(t + \pi)$

59. $f(t) = 2 + \sin(t + \pi)$

60. $g(t) = -1 + \cos t$

61. $h(t) = 4 - \cos t$

62. $g(t) = |\sin t|$

63. $h(t) = |\cos t|$

64. $g(t) = \dfrac{\sin t}{|\sin t|}$

65. $h(t) = \sqrt{\sin t - 2}$

Trigonometric Graphs

In this section we use the graphing techniques that we studied in Chapter 2 to sketch the graphs of functions that involve the trigonometric functions. These graphs are important for understanding the applications to physical situations that we discuss in the next section. They are also beautiful graphs that are interesting in their own right.

It is traditional to use the letter x instead of the letter t to denote the variable in the domain of a function. So from here on we use the letter x and write $y = \sin x$, $y = \cos x$, $y = \tan x$, and so on to denote these functions. This should cause no confusion, since any symbol whatsoever can be used to denote a variable.

Graphs of the Tangent and Cotangent Functions

Unlike the sine and cosine functions, the domain of the tangent function is not the set of all real numbers. Since

$$\tan x = \frac{\sin x}{\cos x}$$

it follows that $\tan x$ is not defined whenever $\cos x = 0$—that is, for $x = \pm \pi/2$, $\pm 3\pi/2, \pm 5\pi/2, \ldots$. Thus the domain of the function tan is the set of all real numbers except

$$x = \frac{2k + 1}{2}\pi, \qquad k = 0, \pm 1, \pm 2, \pm 3, \ldots$$

Since we know that the function tan is periodic with period π, we need to sketch the graph only on an interval of length π and the graph will repeat the same pattern to the left and to the right. We will sketch the graph on the interval between $-\pi/2$ and $\pi/2$. Since $\pi/2$ is not in the domain of this function, we need to be careful when we sketch the graph at points near $\pi/2$. As x gets near $\pi/2$, the value of $\tan x$ gets large. To see this notice that as x gets close to $\pi/2$, $\cos x$ gets small and $\sin x$ gets close to 1. So $\tan x = \sin x/\cos x$ is large. The following table gives values of $\tan x$ for x close to $\pi/2$ (≈ 1.570796):

t	0	$\dfrac{\pi}{6}$	$\dfrac{\pi}{4}$	$\dfrac{\pi}{3}$	1.4	1.5	1.55	1.57	1.5707
$\tan t$	0	0.577	1	1.732	5.798	14.101	48.078	1255.77	10,381.3

By choosing x close enough to $\pi/2$ we can make the value of $\tan x$ as large as we please. In fact, $\tan x \to \infty$ as $x \to \frac{\pi}{2}^-$. Thus $x = \frac{\pi}{2}$ is a vertical asymptote (see Section 3.7).

With the information we have so far we draw the graph of $\tan x$ for $0 \le x < \pi/2$ in Figure 1. To sketch the rest of the graph in the interval $-\pi/2 < t < \pi/2$ we recall that tan is an odd function. Thus $\tan(-x) = -\tan x$ and the graph is symmetric with respect to the origin.

The complete graph of tan in Figure 2 is now obtained from the fact that this function is periodic with period π.

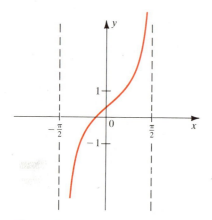

Figure 1

Graph of $y = \tan x$ for

$$-\frac{\pi}{2} \le x \le \frac{\pi}{2}$$

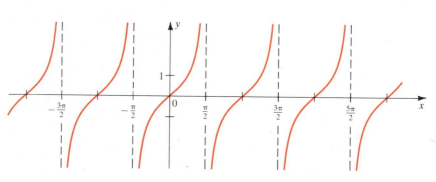

Figure 2

Graph of $y = \tan x$

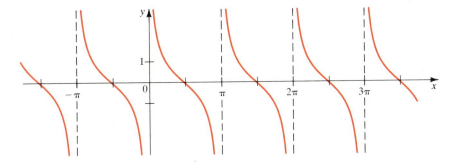

Figure 3
Graph of $y = \cot x$

The graph of cot in Figure 3 is obtained from the graph of tan by using the reduction formula

$$\cot x = -\tan\left(x + \frac{\pi}{2}\right)$$

from Section 5.5. Thus the graph of $\cot x$ is the same as the graph of $-\tan x$ shifted $\pi/2$ units to the left.

Graphs of the Secant and Cosecant Functions

To draw the graph of the cosecant function we use the fact that

(1) $$\csc x = \frac{1}{\sin x}$$

Thus we can find the values of $\csc x$ by taking the reciprocals of the corresponding values of $\sin x$. This is possible except at the points where $\sin x = 0$. Thus the domain of the function csc is the set of all real numbers x except

$$x = n\pi, \qquad n = 0, \pm 1, \pm 2, \pm 3, \ldots$$

Let us first examine the graph on the interval $0 < x < \pi$. Again we need to examine carefully the values of the function near 0 and π, since these are not in the domain of the function. As $x \to 0^+$, $\sin x \to 0$, and so it follows from Equation 1 that $\csc x \to \infty$. Similarly, as $x \to \pi^-$, $\sin x \to 0$, and so again from Equation 1 we see that $\csc x \to \infty$. Thus the lines $x = 0$ and $x = \pi$ are vertical asymptotes for the graph of csc. The graph in the interval $\pi < x < 2\pi$ is drawn in the same way. Indeed, the values of csc in that interval are the same as those in the interval $0 < x < \pi$ except for the sign.

The complete graph is now drawn in Figure 4 from the fact that the function csc is periodic with period 2π.

To sketch the graph of sec let us recall a reduction formula from Section 5.5:

$$\cos x = \sin\left(x + \frac{\pi}{2}\right)$$

Taking the reciprocals of both sides of this equation, we get the reduction formula

$$\sec x = \csc\left(x + \frac{\pi}{2}\right)$$

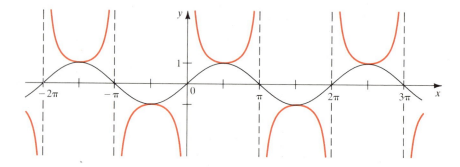

Figure 4
Graph of $y = \csc x$

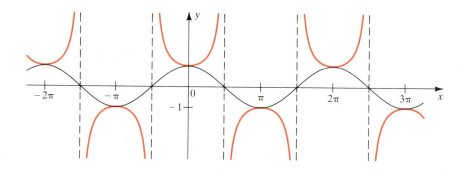

Figure 5
Graph of $y = \sec x$

It follows that the graph of $\sec x$ (see Figure 5) is the same as the graph of $\csc x$ shifted $\pi/2$ units to the left.

Trigonometric Graphs

We now consider graphs of functions that are transformations of the trigonometric functions. The techniques for sketching graphs that were given in Section 2.3 are very useful here. In fact, those graphing techniques can help us see the validity of some of the trigonometric identities we derived in the preceding section. For example, the graph of the function $\sin x$ can be obtained from the graph of the function $\cos x$ by shifting it $\pi/2$ units to the right. Thus $\cos(x - \pi/2) = \sin x$. We can also obtain the graph of $\sin x$ from the graph of $\cos x$ as follows: First reflect the graph of $\cos x$ through the x-axis. The resulting graph is the graph of $y = -\cos x$. By shifting this graph $\pi/2$ units to the left we get the graph of $\sin x$. Thus $\sin x = -\cos(x + \pi/2)$.

It is an instructive exercise to obtain other relations between the trigonometric functions by considering their graphs as we have done here. But the graphing techniques allow us to do much more, as later examples in this section will show. We begin, however, with some simple examples.

Example 1

Sketch the graphs of (a) $f(x) = 2 + \cos x$ and (b) $g(x) = -\cos x$.

Solution

(a) The graph of $f(x)$ is the same as the graph of $\cos x$ shifted up two units (see Figure 6).

(b) The graph of $g(x)$ in Figure 7 is the reflection of the graph of $\cos x$ through the x-axis.

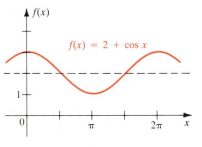

Figure 6

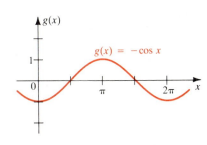

Figure 7 ●

Example 2

Sketch the graph of $y = 2 \sin x$, where $-\pi \le x \le 4\pi$.

Solution

We begin with the graph of $y = \sin x$, where $-\pi \le x \le 4\pi$, and then multiply each y value by 2 (see Figure 8). The zeros of $y = 2 \sin x$ are the same as the zeros of $y = \sin x$. The factor 2 in the function $y = 2 \sin x$ stretches the graph vertically by a factor of 2.

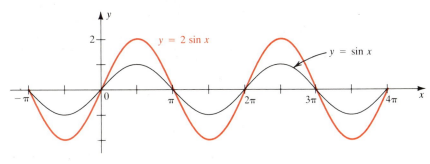

Figure 8 ●

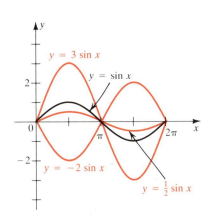

Figure 9

Because of the factor 2 in the preceding example, the maximum value the function assumes is 2 and the minimum value is -2. We say that the function $y = 2 \sin x$ has an amplitude of 2. In general, the **amplitude** of the functions $y = a \sin x$ and $y = a \cos x$ is $|a|$. Figure 9 shows the graphs of $y = a \sin x$ for several values of a.

Example 3

Draw the graphs of (a) $y = \sin 2x$ and (b) $y = \sin\frac{1}{2}x$.

Solution

(a)   Since $y = \sin x$ is periodic of period 2π, we first need to sketch the graph of $y = \sin 2x$ for

$$0 \le 2x \le 2\pi \quad \text{or} \quad 0 \le x \le \pi$$

This means that $y = \sin 2x$ completes one period as x varies from 0 to π, so that its period is π. See Figure 10.

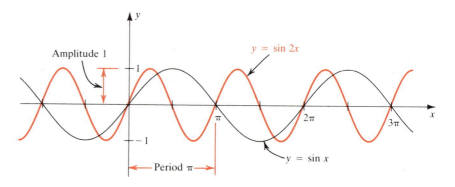

Figure 10

(b)   To sketch one period of $y = \sin\frac{1}{2}x$ we need

$$0 \le \tfrac{1}{2}x \le 2\pi \quad \text{or} \quad 0 \le x \le 4\pi$$

So $y = \sin\frac{1}{2}x$ completes one period as x increases from 0 to 4π. Thus its period is 4π. See Figure 11.

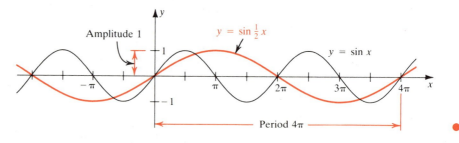

Figure 11

In general, since the functions $y = \sin x$ and $y = \cos x$ have period 2π, the functions

(2) $$y = a \sin kx$$

(3) $$y = a \cos kx$$

where $k > 0$, complete one period as kx varies from 0 to 2π. Thus

$$0 \le kx \le 2\pi \quad \text{or} \quad 0 \le x \le \frac{2\pi}{k}$$

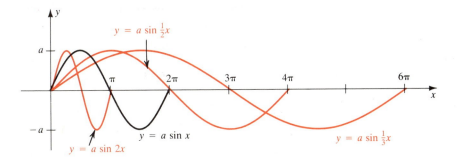

Figure 12

So these functions complete one period as x varies between 0 and $2\pi/k$. Thus the functions of Equations 2 and 3 have period $2\pi/k$. The graphs of these functions are called **sine curves** and **cosine curves**, respectively. Notice the effect of the number k on the graphs of Equations 2 and 3. The effect is to stretch the graph if $k < 1$ and to compress the graph if $k > 1$. For comparison, we sketch in Figure 12 the graphs of one period of the sine curve $y = a \sin kx$ for several values of k.

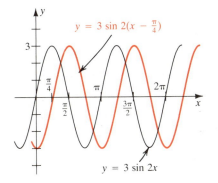

Figure 13

Example 4

Sketch the graph of $y = 3 \sin 2\left(x - \dfrac{\pi}{4}\right)$.

Solution

The graph we want is the same as the graph of $y = 3 \sin 2x$ shifted to the right $\pi/4$ units. The graph of $y = 3 \sin 2x$ is a sine curve with amplitude 3 and period $2\pi/2 = \pi$ (see Figure 13).   ●

The functions we have been considering have the general form, with $k > 0$,

(4) $$y = a \sin k(x - b)$$
(5) $$y = a \cos k(x - b)$$

The graph of Equation 4 is a sine curve and the graph of Equation 5 is a cosine curve that are shifted to the right or left by an amount $|b|$. This number b is called the **phase shift**. (Note that phase shift is not uniquely defined, since adding any multiple of 2π to x will give the same graph.) We summarize the properties of the functions in (4) and (5) in the following box:

Amplitude	$|a|$
Period	$\dfrac{2\pi}{k}$
Phase shift	b

Example 5

Find the amplitude, period, and phase shift of the function $y = \frac{3}{4}\sin\left(2x - \frac{2\pi}{3}\right)$.
Sketch the graph of one period of this function.

Solution

We first write this function in the form of Equation 4:

$$y = a \sin k(x - b)$$

To do this we factor 2 from the expression $\left(2x - \frac{2\pi}{3}\right)$ to get

$$y = \frac{3}{4}\sin 2\left(x - \frac{\pi}{3}\right)$$

Thus we have

$$\text{Amplitude:} \quad |a| = \frac{3}{4}$$

$$\text{Period:} \quad \frac{2\pi}{k} = \frac{2\pi}{2} = \pi$$

$$\text{Phase shift:} \quad b = \frac{\pi}{3}$$

From this information, it follows that one period of this sine curve begins at $\pi/3$ and ends at $\pi + (\pi/3) = 4\pi/3$. To sketch the graph over the interval

$$\frac{\pi}{3} \le x \le \frac{4\pi}{3}$$

we divide this interval into four equal parts and draw a sine curve as shown in Figure 14.

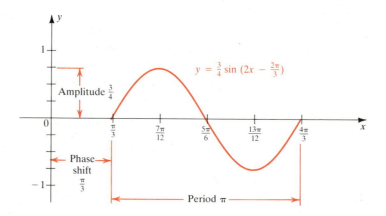

Figure 14

Example 6

Sketch the graph of $y = 2\cos x + \sin 2x$.

Solution

We use graphical addition (see Section 2.5) to sketch this graph in Figure 15.

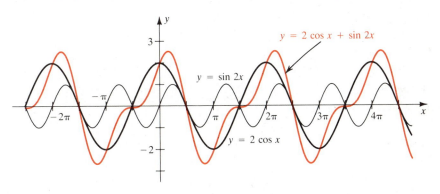

Figure 15

The function in Example 6 is periodic and its period is 2π. In general, functions that are sums of functions from the list

$$\sin x,\ \cos x,\ \sin 2x,\ \cos 2x,\ \sin 3x,\ \cos 3x, \ldots$$

are periodic. Although these functions appear to be very special kinds of periodic functions, they play a basic role in describing all periodic functions.* It is a remarkable fact, discovered by the French mathematician J. B. J. Fourier, that every periodic function can be written as a sum (sometimes an infinite sum) of these functions. This is remarkable because it means that any situation in which periodic variation occurs can be mathematically described using these functions.

Example 7

Sketch the graph of $y = x + \sin x$.

Solution

Using graphical addition, we add each value of x to the corresponding value of $\sin x$ to get the graph of $y = x + \sin x$ in Figure 16.

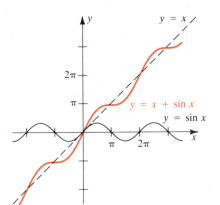

Figure 16

Example 8

Sketch the graph of $y = x^2 \sin x$.

Solution

The values of $\sin x$ are between -1 and 1. In other words,

$$-1 \le \sin x \le 1$$

---

* Some very minimal restrictions are required, but all the periodic functions studied in this course and in calculus courses satisfy these restrictions.

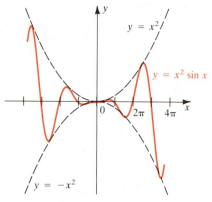

Figure 17

Multiplying this inequality through by x^2 and noting that $x^2 \geq 0$, we get

$$-x^2 \leq x^2\sin x \leq x^2$$

So the values of the function lie between $-x^2$ and x^2. In fact, $y = x^2$ for those values of x for which $\sin x = 1$, and $y = -x^2$ for those values of x for which $\sin x = -1$. This suggests that the functions $y = x^2$ and $y = -x^2$ form a boundary for the graph of $y = x^2\sin x$. We sketch the graphs of these two functions together with the graph of $y = x^2\sin x$ in Figure 17 so we can see how these graphs are related. ●

Intuitively, Example 8 shows that the function $y = x^2$ controls the amplitude of the graph of $y = x^2\sin x$. In general, if $f(x) = a(x)\sin x$, the function a determines how the amplitude of f varies, and the graph of f will lie between the graphs of $y = -a(x)$ and $y = a(x)$. Here is another example.

Example 9

Sketch the graph of $f(x) = e^{-x}\cos x$.

Solution

The graph of f lies between the graphs of $y = -e^{-x}$ and $y = e^{-x}$. We sketch the graphs of both these functions and the graph of f in Figure 18. Notice how the graph of $y = e^{-x}$ controls the amplitude of f.

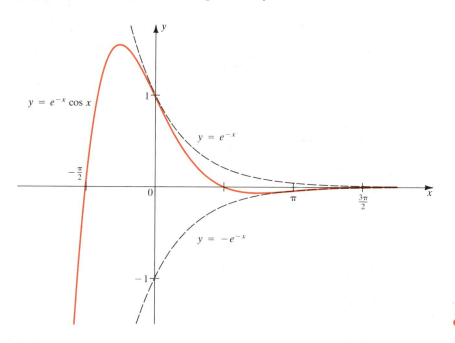

Figure 18

Example 10

Sketch the graph of $f(x) = \cos 2\pi x \cos 16\pi x$.

Solution

Again we can think of the function $\cos 2\pi x$ as determining the amplitude of the function $f(x)$. We first sketch the graphs of $y = -\cos 2\pi x$ and $y = \cos 2\pi x$. The graph of f is a cosine curve between the graphs of these two functions (see Figure 19).

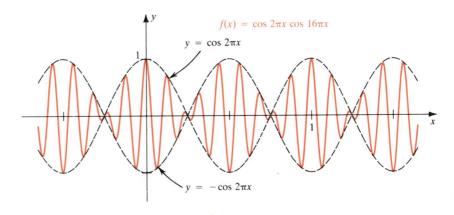

Figure 19

Exercises 5.6

In Exercises 1–10 find the period and amplitude of each function and sketch the graph over one period.

1. $f(x) = 3 \sin 3x$

2. $f(x) = -2 \sin 2\pi x$

3. $f(x) = 10 \sin \frac{1}{2}x$

4. $f(x) = \cos 10\pi x$

5. $g(x) = -\cos \frac{1}{3}x$

6. $g(x) = 2 + \sin(-2x)$

7. $h(x) = 3 + 3 \cos 3\pi x$

8. $h(x) = 5 - 2 \sin 2x$

9. $h(x) = \sin \frac{2}{3}x$

10. $g(x) = -1 - \cos(-x)$

11. Find the function whose graph is the graph of $f(x) = 3 \sin 3x$ shifted two units to the right.

12. Find the function whose graph is the graph of $f(x) = -2 \sin 2\pi x$ shifted π units to the left.

13. Find the function whose graph is the graph of $f(x) = 10 \sin \frac{1}{2}x$ shifted $\pi/2$ units to the right.

14. Find the function whose graph is the graph of $f(x) = \cos 10\pi x$ shifted $\frac{1}{2}$ unit to the left.

In Exercises 15–30 find the amplitude, period, and phase shift of the function and sketch the graph.

15. $f(x) = \cos\left(x - \frac{\pi}{2}\right)$

16. $f(x) = \sin(x + 1)$

17. $f(x) = 2 \sin\left(\pi x - \frac{\pi}{3}\right)$

18. $f(x) = -2 \cos\left(x - \frac{\pi}{12}\right)$

19. $f(x) = 2 \sin\left(\frac{\pi}{2} - x\right)$

20. $f(x) = -4 \sin 2\left(x + \frac{\pi}{2}\right)$

21. $f(x) = 5 \cos\left(3x - \frac{\pi}{4}\right)$

22. $f(x) = \frac{1}{2}\cos\left(\frac{\pi}{2} - 2\pi x\right)$

23. $f(x) = 2 + 2 \sin\left(\frac{2}{3}x - \frac{\pi}{6}\right)$

24. $f(x) = \sin \frac{1}{2}\left(x + \frac{\pi}{4}\right)$

25. $f(x) = 3 \cos \pi\left(x + \frac{1}{2}\right)$

26. $f(x) = 1 + \cos\left(3x + \frac{\pi}{2}\right)$

27. $f(x) = -2 \cos\left(2x - \frac{\pi}{3}\right)$

28. $f(x) = 3 + 2\sin 3(x + 1)$

29. $f(x) = \sin(3x + \pi)$ **30.** $f(x) = \cos\left(\dfrac{\pi}{2} - x\right)$

In Exercises 31–40 find the period and sketch the graph of the given function.

31. $f(x) = \tan(2x)$ **32.** $f(x) = -\tan(\tfrac{1}{2}x)$

33. $f(x) = 2\sec \pi x$ **34.** $f(x) = \csc(3x + \pi)$

35. $f(x) = \sec\left(x + \dfrac{\pi}{2}\right)$ **36.** $f(x) = \csc 2(x + 1)$

37. $f(x) = -\cot\left(2x - \dfrac{\pi}{2}\right)$ **38.** $f(x) = \tfrac{1}{2}\tan(2\pi x)$

39. $f(x) = 2\tan(-x)$ **40.** $f(x) = \sec(-2x)$

In Exercises 41–46 sketch the graph of the given function using graphical addition.

41. $f(x) = x + \cos x$ **42.** $f(x) = \sin x + \cos x$

43. $f(x) = \sin x + \sin 2x$

44. $f(x) = \ln x + \sin 2\pi x \quad (x \geq 1)$

45. $f(x) = e^{-x} + \sin x$ **46.** $f(x) = x^2 + \sin 2x$

In Exercises 47–64 sketch the graph of the given function.

47. $f(x) = x^2\cos x$ **48.** $f(x) = e^{-x}\sin x$

49. $f(x) = x\sin x$ **50.** $f(x) = x\cos x$

51. $f(x) = \ln x \sin \pi x \quad (x \geq 1)$

52. $f(x) = x^3\sin 2x$

53. $f(x) = \cos 3\pi x \cos 21\pi x$

54. $f(x) = \sin 2\pi x \sin 10\pi x$

55. $f(x) = |\sin x|$ **56.** $f(x) = |\cos x| + 1$

57. $f(x) = |\cos x + 1|$ **58.** $f(x) = |\sec x|$

59. $f(x) = |\tan 2x|$ **60.** $f(x) = e^{|x|}\sin x$

61. $f(x) = 2^x\cos x$ **62.** $f(x) = 3^{-x}\sin x$

63. $f(x) = \dfrac{1}{x}\sin \pi x \quad (x \geq 1)$

64. $f(x) = \dfrac{1}{x^2}\cos 2\pi x \quad (x \geq 1)$

In Exercises 65–72:
(a) *Determine whether the function is periodic and if so, find the period.*
(b) *Determine whether the function is even, odd, or neither.*
(c) *Use your answers to parts (a) and (b) to sketch the graph.*

65. $f(x) = \cos|x|$ **66.** $f(x) = \sin|x|$

67. $f(x) = e^{\cos x}$ **68.** $f(x) = e^{\sin 2x}$

69. $f(x) = (\sin x)^2$ **70.** $f(x) = (\cos x)^2$

71. $f(x) = (\sin 3x)^3$ **72.** $f(x) = 2^{-\sin 2\pi x}$

Section 5.7

Applications: Harmonic Motion

Motion that is caused by vibration or oscillation is common in nature. A weight suspended from a spring that has been compressed and then allowed to vibrate vertically is a simple example of such behavior. But this same "back and forth" motion occurs in such diverse phenomena as sound waves, light waves, alternating electrical currents, pulsating stars, and nuclear magnetic resonance, to name a few. All these phenomena exhibit a kind of motion called **harmonic motion**.

The trigonometric functions provide us with a powerful tool for describing such oscillatory motion. It is not hard to see why this is so. A glance at the graphs of the trigonometric functions shows that these functions themselves exhibit this kind of oscillatory behavior. Figure 1 is the graph of the sine function. Let us think of t as time. Then, since $y = \sin t$ is the y-coordinate of the point $P(t)$ on the unit circle in Figure 1, we see that as time increases, $y = \sin t$ moves up and down through the values between -1 and 1. It is this property that makes the functions sine (and cosine) so valuable in describing oscillatory motion.

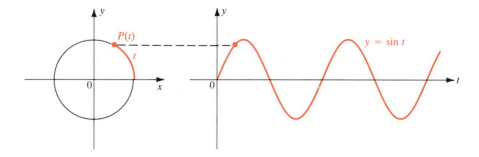

Figure 1

For example, if the weight in Figure 2 completes one cycle of its vibration every $\frac{1}{2}$ s, then its motion is described by

$$y = 10 \sin 4\pi t$$

where y is the distance the weight is displaced at time t from its rest position at the origin of the xy-plane. (We assume here that there is no friction so the spring and weight will oscillate in the same way indefinitely.) Notice that the *amplitude* (the maximum displacement) of the motion is 10. Since the weight completes two cycles of its motion in 1 s, we say that the *frequency* is 2. In this case the *period*, or the time required to complete one cycle, is $\frac{1}{2}$.

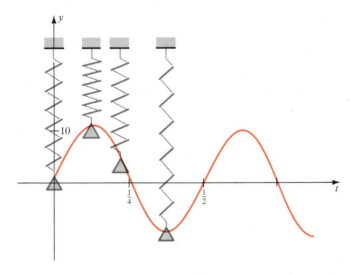

Figure 2

The equation that describes the displacement of an object at time t is called its **equation of motion**. If the equation of motion of an object is one of the following:

(1) $$y = a \sin \omega t$$

(2) $$y = a \cos \omega t$$

then that object is said to move in **simple harmonic motion**. In this case $|a|$ is called the **amplitude** of the motion and is the maximum displacement. The **period** of the motion is $2\pi/\omega$ and is the time required for one complete cycle. The **frequency** $\omega/2\pi$ is the number of complete cycles per unit of time.

(3)

(4)

(5)

Amplitude	$\lvert a \rvert$
Period	$\dfrac{2\pi}{\omega}$
Frequency	$\dfrac{\omega}{2\pi}$

Notice that the functions

(6) $$f(t) = a \sin 2\pi v t \qquad \text{and} \qquad f(t) = a \cos 2\pi v t$$

have frequency v. Since we can see immediately from these equations what the frequency is without doing any calculations, we often write equations of simple harmonic motion in this form.

The main difference between Equations 1 and 2 is the starting point of the motion. In Equation 1, $y = 0$ when $t = 0$, whereas in Equation 2, $y = a$ when $t = 0$. In other words, in Equation 1 the motion "starts" with zero displacement, whereas in Equation 2 the motion "starts" with the displacement at maximum (at the amplitude a). This observation is useful in setting up equations that describe a given motion.

Of course, it is possible for the motion to start at any point in its cycle. In the example of the spring, the displacement of the weight at $t = 0$ was $y = 0$. If at time $t = 0$ the weight was at a different displacement in its cycle, say at $y = 5$, then the equation of motion would be

$$y = 10 \sin\left(4\pi t + \frac{\pi}{6}\right) \qquad \text{or} \qquad y = 10 \sin 4\pi\left(t + \frac{1}{24}\right)$$

(Notice that at $t = 0$, the displacement is $y = 10 \sin \frac{\pi}{6} = 10 \cdot \frac{1}{2} = 5$.) This has the effect of simply shifting the graph in Figure 2 by an amount $1/24$ to the left (see Figure 3).

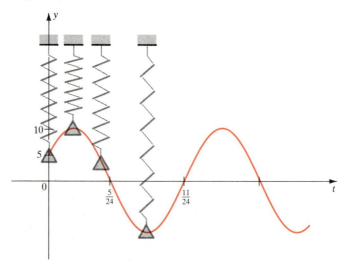

Figure 3

In general, the equations of simple harmonic motion that involve a shift are

(7)
$$y = a \sin \omega(t - \theta)$$

(8)
$$y = a \cos \omega(t - \theta)$$

Notice that the graph of Equation 7 is the same as the graph of Equation 1 shifted by θ units. For this reason θ is called the **phase shift**.

Example 1

A mass is suspended from a spring. The spring is compressed a distance of 4 cm and then released. It is observed that the mass returns to the compressed position after $\frac{1}{3}$ s. Describe the motion of the mass.

Solution

The mass will oscillate in simple harmonic motion. So its motion will be described by Equation 1 or 2. The amplitude of the motion is 4 cm. Since this amplitude is reached at time $t = 0$ when the mass is released, the appropriate equation to use here is Equation 2. The period $p = 1/3$, so the frequency is $1/p = 3$. The motion of the mass is described by the equation

$$f(t) = 4\cos(2\pi)3t \qquad \text{or} \qquad f(t) = 4\cos 6\pi t$$

where $f(t)$ is the displacement from the rest position at time t. Notice that when $t = 0$, $y = 4$, as we expect. ●

An important situation in which simple harmonic motion occurs is in the production of sound. Sound is produced by a regular variation in the air pressure from the normal air pressure. If the pressure varies in simple harmonic motion, then a pure sound is produced. The tone of the sound is determined by the frequency of the motion and the loudness is determined by the amplitude.

Example 2

A tuba player plays the note E and sustains the sound for some time. Assuming that the sound is a pure E, the equation of the sound is given by

(9)
$$V(t) = 0.2 \sin 80\pi t$$

where $V(t)$ is the variation in pressure from the normal pressure at time t. Find the frequency, amplitude, and period of the equation. If the tuba player increases the loudness of her note, how does Equation 9 change? If the player is playing the note incorrectly and it is a little flat, how does this affect Equation 9?

Solution

The frequency is $80\pi/2\pi = 40$, the period is $1/40$, and the amplitude is 0.2. If the player increases the loudness, the amplitude of the wave increases. In other words, the number 0.2 is replaced by a larger number. If the note is flat, then the frequency is less than 40. In this case the coefficient of t is less than 80π. ●

In general the sounds produced by musical instruments are not pure sounds but are composed of a combination of pure sounds, each with a frequency that is a multiple of the lowest, or *fundamental*, frequency. The higher frequencies are called *overtones*. When the note A is played on a musical instrument, the fundamental frequency 440 is produced, along with overtones with frequencies of, for example, 880 (double the frequency) and 1320 (triple the frequency). The equation of this sound may have the form

(10) $$V(t) = 0.5 \sin(440)2\pi t + 0.2 \sin(880)2\pi t + 0.3 \sin(1320)2\pi t$$

The function $V(t)$ in Equation 10 still has frequency 440. Thus we hear the sound as the musical note A. But it is the presence or absence and the intensity of the various overtones that make the sound of the note A on a violin different from the sound of the same note on an oboe. In other words, the overtones determine the character of different instruments and make it possible to distinguish between their sounds.

Example 3

A Cepheid variable is a star whose brightness alternately increases and decreases. The most easily visible such star is Delta Cephei, hence the name. For this star the time between periods of maximum brightness is 5.4 days. The average brightness (or magnitude) of the star is 4.0 and its brightness changes by ± 0.35. Assuming that the brightness of the star changes in simple harmonic motion, express its brightness as a function of time.

Solution

Since the average magnitude of the star is 4.0, the brightness of the star increases and decreases from this value in simple harmonic motion. Let us measure time in days, and take $t = 0$ to be a time when the star is at its average brightness. Since the period of the star is 5.4 days, from Equation 1 we see that the brightness $B(t)$ of the star at time t, where t is measured in days, is given by

$$B(t) = 4.0 + 0.35 \sin 2\pi \left(\frac{1}{5.4}\right)t \qquad \bullet$$

Another situation in which simple harmonic motion occurs is in alternating current (AC) generators. Alternating current is produced when an armature rotates about its axis in a magnetic field. Figure 4 shows a simple version of such a generator. As the wire passes through the magnetic field, a voltage is generated in the wire. If the armature rotates at the rate of ω rad/s, then the voltage generated is proportional to the speed at which the armature crosses the lines of force of the

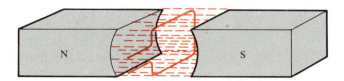

Figure 4

magnetic field. It turns out that this is proportional to $\cos \omega t$, provided the armature is parallel to the lines of force of the magnetic field at time $t = 0$. Thus the voltage generated is given by the function

(11) $E(t) = E_0 \cos \omega t$

where E_0 is the maximum voltage produced and depends on the strength of the magnetic field.

Example 4

Ordinary 110-V household alternating current varies from $+155$ V to -155 V with a frequency of 60 Hz (cycles per second). Find an equation that describes this variation in voltage.

Solution

The variation in voltage is simple harmonic. Since the frequency is 60 Hz, from Equation 5 we have

$$\frac{\omega}{2\pi} = 60$$

or $\omega = 120\pi$

Let us take $t = 0$ to be a time when the voltage is $+155$ V. Then using Equation 11, we get

$$E(t) = 155 \cos 120\pi t \qquad \bullet$$

Why do we say that household current is 110 V when the maximum voltage produced is 155 V? Because of the symmetry of the cosine function, we easily see that the average current produced is zero. This average value would be the same for all AC generators and so gives no information about the voltage generated. To avoid the value zero, engineers use a different method of averaging alternating voltage. It is called the root-mean-square (RMS) method. It can be shown that the RMS value is $1/\sqrt{2}$ times the maximum voltage:

$$155 \times \frac{1}{\sqrt{2}} \approx 110 \text{ V}$$

There are numerous other applications of simple harmonic motion. Some examples are in the exercises at the end of this section.

Damped Harmonic Motion

The spring in Figure 1 was assumed to oscillate in a frictionless environment. In this hypothetical case the amplitude of the oscillation will not change. In the presence of friction, however, the motion of the spring will eventually die down; that is, the amplitude of the motion will decrease with time. Motion of this type is called **damped harmonic motion** and is governed by the equations .

(12) $f(t) = ke^{-ct}\sin \omega t$

(13) $f(t) = ke^{-ct}\cos \omega t$

where k, c, and ω are positive numbers. In other words, this is simple harmonic motion where the amplitude is now modulated by the function $a(t) = ke^{-ct}$. (For the sake of clarity we have taken the phase shift to be zero here.) Figure 5 shows the difference between damped motion and motion that is not damped.

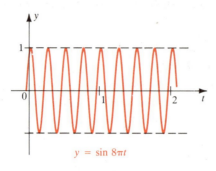

$$y = \sin 8\pi t$$

Simple harmonic motion

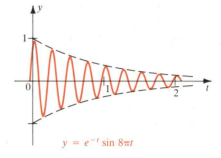

$$y = e^{-t} \sin 8\pi t$$

Damped harmonic motion

Figure 5

The number c is called the **damping constant**. The larger c is, the sooner the oscillation dies down. When a guitar string is plucked and then allowed to vibrate freely, a point on that string undergoes damped harmonic motion. We can hear the damping of the motion as the sound produced by the vibration of the string dies down. How fast the damping of the string occurs (as measured by the size of the constant c) is a property of the size of the string and the material it is made of. Another example of damped harmonic motion is the motion that a shock absorber on a car undergoes when the car hits a bump in the road. In this case the shock absorber is specifically engineered to have the motion damped as quickly as possible (large c) and to have the frequency as small as possible (small ω). On the other hand, the sound produced by a tuba player playing the note C, say, is undamped as long as the player can maintain the loudness of the note. The electromagnetic waves that produce light move in simple harmonic motion that is not damped.

Example 5

The G string on a violin is pulled a distance of 0.5 cm from its rest position. It is then released and allowed to vibrate. The damping constant c for this string is determined to be 1.4. Suppose that the note produced is a pure G (frequency = 200). Find the equation that describes the motion of the point at which the string was plucked.

Solution

Let P be the point at which the string was plucked. We will find a function $f(t)$ that gives the distance at time t of the point P from its original rest position. Equation 13 is the appropriate equation to use here because in this equation the maximum value of $f(t)$ is achieved at $t = 0$, and this is the case with our string.

So the equation we are looking for has the form

$$f(t) = ke^{-ct}\cos \omega t$$

From this equation, $f(0) = k$. But we know that the original displacement of the string is 0.5 cm. Thus $k = 0.5$. Since the frequency of the vibration is 200, we have $\omega/2\pi = 200$ or $\omega = (200)(2\pi)$. Finally, since we know that the damping constant is 1.4, we get

$$f(t) = 0.5e^{-1.4t}\cos 400\pi t$$ ●

Example 6

A stone is dropped in a calm lake, causing waves to form. The up and down motion of a point on the surface of the water is damped harmonic motion. At some time the amplitude of the wave is measured, and 20 s later it is found that the amplitude had dropped to $\frac{1}{10}$ of its value. Use this information to find the value of c.

Solution

In Equation 13 the amplitude is governed by the coefficient of $\cos \omega t$. Thus the amplitude at time t is ke^{-ct}, and 20 s later it is $ke^{-c(t+20)}$. So from the information we are given,

$$\frac{ke^{-ct}}{ke^{-c(t+20)}} = 10$$

We now solve this equation for c. Canceling k and using the Laws of Exponents, we get

$$e^{-ct+c(t+20)} = 10$$

or

$$e^{-ct+ct+20} = e^{20c} = 10$$

Taking the natural logarithms of both sides of this last equation, we get

$$20c = \ln(10)$$

or

$$c = \tfrac{1}{20}\ln(10) \approx \tfrac{1}{20}(2.30) \approx 0.12$$ ●

Exercises 5.7

In Exercises 1–10 the given function specifies the position of an object that is moving along an axis in simple harmonic motion.
(a) Find the amplitude, period, and frequency of the motion.
(b) What is the position of the object at time $t = 0$?
(c) Sketch a graph of the function along one complete period.

1. $f(t) = 4\sin 8\pi t$

2. $f(t) = 10\sin \tfrac{1}{2}t$

3. $f(t) = 0.3\cos 2t$

4. $f(t) = -2\cos\left(\dfrac{3\pi}{2}t\right)$

5. $f(t) = 1000\sin 2\pi t$

6. $f(t) = \dfrac{1}{20}\cos\left(\dfrac{\pi}{3}t\right)$

7. $f(t) = -\cos \pi\left(t + \dfrac{1}{2}\right)$

8. $f(t) = 2\sin\left(\dfrac{\pi}{12}t\right)$

9. $f(t) = \pi\sin\left(3\pi t + \dfrac{\pi}{6}\right)$

10. $f(t) = \dfrac{\pi}{6}\cos\left(2t + \dfrac{\pi}{3}\right)$

11. A point P moves in simple harmonic motion. If the point completes two cycles of its motion every minute and if the amplitude of the motion is 6 ft, find a function in the form of Equation 1 that describes the motion of P.

12. Rework Exercise 11 using Equation 2.

13. A point P moves in simple harmonic motion with an amplitude of 10 cm and a frequency of 3 cycles per second. Find a function in the form of Equation 1 that describes the motion of the point P.

14. Rework Exercise 13 using Equation 2.

15. A point P moves in simple harmonic motion with an amplitude of 2 m and a period of 1/3 s. Find a function in the form of Equation 7 if the phase shift is $\pi/3$.

16. Rework Exercise 15 using Equation 8.

17. A mass suspended from a spring is oscillating with a frequency of 5 cycles per second. If its maximum displacement from rest position is 2 cm, find an equation that describes the motion of this mass. Assume the mass is at maximum displacement when $t = 0$.

18. A mass oscillating on a spring is observed to complete one cycle of its motion in $\frac{1}{10}$ sec. It is also observed that the amplitude of the motion is 8 cm. Find an equation that describes the motion of this mass. Assume the mass has zero displacement when $t = 0$.

19. A mass suspended from a spring is pulled down a distance of 2 ft from its rest position. The mass is released at time $t = 0$ and allowed to oscillate. If the mass returns to this position after 1 s, find an equation that describes its motion.

20. A mass is suspended vertically on a spring. The spring is compressed so that the mass is 5 cm above its rest position. The mass is released at time $t = 0$ and allowed to oscillate. It is observed that the mass reaches its lowest point $\frac{1}{2}$ s after it is released. Find an equation that describes the motion of the mass.

*The frequency of oscillation of an object suspended on a spring depends on the stiffness k of the spring (called the **spring constant**) and on the mass m of the object. If the spring is compressed a distance a and then allowed to oscillate, the equation that describes its motion is*

(14) $$f(t) = a \cos \sqrt{\frac{k}{m}}\, t$$

Use this information in Exercises 21–24.

21. Find the equation that describes the motion of a mass m suspended on a spring with spring constant k that has

been compressed a distance a and released. Use the following values for m, k, and a.
(a) $m = 4$ kg, $k = 3$, $a = 10$ m
(b) $m = 10$ lb, $k = 10$, $a = 100$ in.
(c) $m = 10$ g, $k = 1$, $a = 1$ cm
(d) $m = 100$ g, $k = 2$, $a = 10$ cm

22. Find the period and frequency of the motion described in Equation 14.

23. How is the frequency affected when the mass is increased? Is the oscillation faster or slower? How does Equation 14 tell us this?

24. What is the frequency of the motion described in Exercise 21(a)? Exercise 21(b)? Exercise 21(c)? Exercise 21(d)?

25. A ferris wheel in an amusement park has a radius of 10 m, and the bottom of the wheel is 1 m above the ground. Suppose the ferris wheel makes one complete revolution every 20 s. If a person is riding on the ferris wheel, find an equation that describes how his height above the ground varies with time. Does his height vary in simple harmonic fashion? Assume the person is closest to the ground when $t = 0$.

26. The pistons in the engine of a car (see the figure) move up and down to turn the crankshaft at a constant angular velocity.
(a) If this angular velocity is ω rad/min, find an equation that gives the height of the point P above the center of the crankshaft O at any time t.
(b) Is the motion of the piston simple harmonic motion?

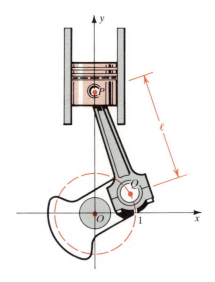

27. The Bay of Fundy in Nova Scotia has the highest tides in the world. In one section of the bay the water level rises to 21 ft above mean sea level and drops back to 21 ft below mean sea level in every 12-hr period. Assuming that the motion of the tides is simple harmonic, find an equation that describes the height of the tide in the Bay of Fundy above mean sea level. Sketch a graph that shows the level of the tides over a 24-hr period. Take $t = 0$ to be a time when the tide is at mean sea level and is rising.

28. Normal 110-V household alternating current actually varies from -155 V to $+155$ V with a frequency of 60 Hz (cycles per second). Suppose that at time $t = 0$ the voltage produced is zero. Write an equation that describes the variation in the voltage of household current with respect to time.

29. The armature in an electric generator is rotating at the rate of 100 revolutions per second. If the constant associated with the magnet is $E_0 = 310$ (see Example 4), find an equation that describes the variation in the voltage of the current produced by this generator. What is the maximum voltage produced? What is the RMS voltage?

30. The variable star Zeta Gemini has a period of 10 days. The average brightness of the star is 3.8 and the maximum variation from the average is 0.2. Assuming that the variation in brightness is simple harmonic, find an equation that describes this variation as a function of time.

31. Astronomers believe that the radius of a variable star increases and decreases as the brightness of the star does. The variable star Delta Cephei (see Example 3) has an average radius of 20 million miles and changes by a maximum of 1.5 million miles from this average in the course of a single pulsation. Find an equation that describes the radius of this star as a function of time.

32. The pendulum in a grandfather clock makes one complete swing every 2 s. The maximum angle that the pendulum makes with respect to its rest position is 10°. We know from physical principles that the angle θ between the pendulum and its rest position changes in simple harmonic fashion. Find an equation that describes the size of the angle θ as a function of time. (Take $t = 0$ to be a time when the pendulum is vertical.)

33. A train whistle produces a sharp pure tone. The maximum change in pressure from the normal air pressure that is produced by this whistle is 10 N/m². The frequency in this variation is 20 cycles per second. Find a function $P(t)$ that describes this variation in pressure (in other words, an equation that describes the sound). Sketch a graph of $P(t)$.

34. As a train speeds toward us, the pitch of the whistle increases, and as the train speeds past us, the pitch of the whistle decreases. This phenomenon is called the *Doppler effect*.
  (a) If the train in Exercise 33 is speeding toward us, modify the graph you sketched in that exercise to reflect the Doppler effect.
  (b) Rework part (a) if the train is speeding away from us.

35. The number of hours of daylight varies through the course of a year, the longest day being June 21 and the shortest December 21. The variation in the length of daylight is simple harmonic. In Philadelphia there are 14 hr, 50 min of daylight on June 21 and 9 hr, 10 min of daylight on December 21.
  (a) Find a function that represents the number of hours of daylight in Philadelphia as a function of the number of days after March 21.
  (b) Find a function that represents the number of hours of daylight in Philadelphia as a function of the number of days after June 21.
  (c) Find a function that represents the number of hours of daylight in Philadelphia as a function of the number of days after January 1.
  (d) The graph in the figure shows the number of hours of daylight for different latitudes. Find an equation that gives the number of hours of daylight in your city as a function of the number of days after March 21. (Graph taken from *Daylight, Twilight, Darkness and Time*, by Lucia Carolyn Harrison, New York: Silver, Burdett and Company, 1935, p. 40.)

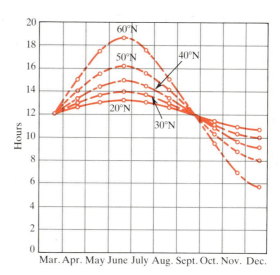

Graph of the relative length of day from March 21 to December 21, inclusive, at the latitudes indicated.

36. A mass is attached to a spring and the spring is compressed a distance of 10 cm from its rest position and then released. The spring moves in damped harmonic motion. The damping constant for this spring is $c = 2.1$ and the spring oscillates with a frequency of 5 cycles per second. Find the equation of motion of this mass if $t = 0$ is taken to be the instant that the spring is released.

37. A strong gust of wind strikes a tall building causing it to sway back and forth in damped harmonic motion. The frequency of the oscillation is 0.5 cycles per second and the damping constant is $c = 0.9$. Find an equation that describes the motion of the building. Assume $k = 1$ and take $t = 0$ to be the instant when the gust of wind strikes the building.

38. A car hits a bump on the road. As a result the shock absorber on the car is compressed a distance of 6 in. and then released. The shock absorber vibrates in damped harmonic motion with a frequency of 2 cycles per second. The damping constant for this particular shock absorber is 2.8.
   (a) Find an equation that describes the displacement of the shock absorber from its rest position as a

function of time. Take $t = 0$ to be the instant that the shock absorber is released.
   (b) How long does it take for the amplitude of the vibration to decrease to 0.5 in.?

39. A tuning fork is struck and oscillates in damped harmonic motion. The amplitude of the motion is measured and 3 s later it is found that the amplitude had dropped to 1/4 of its value. Find the damping constant c for this tuning fork.

40. A guitar string is pulled from its center a distance of 3 cm. It is then released and vibrates in damped harmonic motion with a frequency of 165 cycles per second. After 2 s it is observed that the amplitude of the vibration at the center of the string is 0.6 cm.
   (a) Find the damping constant c.
   (b) Find an equation that describes the position of the point at the center of the string above its rest position as a function of time. Take $t = 0$ to be the instant that the string is released.

CHAPTER 5 REVIEW

Define, state, or discuss the following.

1. Angle measure
2. Positive angle measure, negative angle measure
3. Degree
4. Radian
5. Angles in standard position
6. Coterminal angles
7. Length of an arc of a circle
8. Area of a sector of a circle
9. The trigonometric ratios: sin, cos, tan, csc, sec, cot
10. Solving right triangles
11. Trigonometric ratios of any angle
12. Reference angle
13. Trigonometric identities
14. Reciprocal identities
15. Pythagorean identities
16. Odd-even identities
17. Reduction formulas
18. The Law of Sines
19. The Law of Cosines
20. The ambiguous case
21. Trigonometric functions
22. Circular functions
23. Unit circle
24. Periodic function
25. Trigonometric graphs
26. Sine curve, cosine curve
27. Amplitude
28. Period
29. Phase shift
30. Frequency
31. Simple harmonic motion
32. Damped harmonic motion
33. Damping constant

Review Exercises

1. Find the radian measure that corresponds to the given degree measure.
   - **(a)** 70°
   - **(b)** 420°
   - **(c)** −240°
   - **(d)** 24°
   - **(e)** −330°
   - **(f)** 750°

2. Find the degree measure that corresponds to the given radian measure.
   - **(a)** $\dfrac{7\pi}{2}$
   - **(b)** $-\dfrac{\pi}{3}$
   - **(c)** $\dfrac{7\pi}{4}$
   - **(d)** 3
   - **(e)** $-\dfrac{5}{2}$
   - **(f)** $\dfrac{11\pi}{6}$

3. Find the length of an arc of a circle of radius 8 m if the arc subtends a central angle of 1 rad.

4. Find the measure of a central angle θ in a circle of radius 5 ft if θ is subtended by an arc of length 7 ft.

5. A circular arc of length 100 ft subtends a central angle of 70°. Find the radius of the circle.

6. Find the area of a sector with a central angle of 2 rad in a circle of radius 5 m.

7. The area of a sector of a circle of radius 25 ft is 125 ft². Find the central angle of the sector.

8. Find the values of the six trigonometric ratios of the angle θ in the triangles shown.

   (a)
   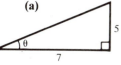

   (b)
   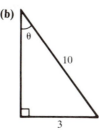

   (c)
   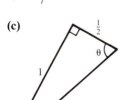

9. Find the exact values of the trigonometric ratios for the given angles.
   - **(a)** $\sin 315°$
   - **(b)** $\csc 765°$
   - **(c)** $\tan(-135°)$
   - **(d)** $\cos\left(\dfrac{5\pi}{6}\right)$
   - **(e)** $\cot\left(-\dfrac{22\pi}{3}\right)$
   - **(f)** $\sin 405°$

   - **(g)** $\cos(585°)$
   - **(h)** $\sec\left(\dfrac{11\pi}{3}\right)$

10. Find the values of the six trigonometric ratios of the angle θ if θ is in standard position and the point $(-5, 13)$ is on the terminal side of θ.

11. Find the acute angle that is formed by the line $y - \sqrt{3}x + 1 = 0$ and the x-axis.

12. Find the six trigonometric ratios of θ in standard position if its terminal side is in quadrant III and is parallel to the line $4y - 2x - 1 = 0$.

13. If $\tan \theta = \frac{1}{4}$ and θ is in quadrant III, find $\sec \theta + \cot \theta$.

14. If $\sin \theta = -\frac{\sqrt{3}}{2}$ and θ is in quadrant IV, find $\cos 2\theta + 2\cos \theta$.

15. If $\cos \theta = \frac{1}{2}$ and θ is in quadrant I, find $\tan \theta + \sec \theta$.

16. If $\sec \theta = -\frac{3}{2}$ and θ is in quadrant II, find $\sin^2\theta + \cos^2\theta$.

17. Write the first trigonometric ratio in terms of the second for θ in the given quadrant.
   - **(a)** $\sin \theta$, $\cos \theta$; θ in quadrant II
   - **(b)** $\tan \theta$, $\sin \theta$; θ in quadrant III

18. Find the side labeled x.

   (a)
   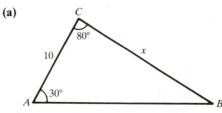

   (b)
   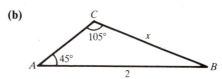

   22·180

   (c)
   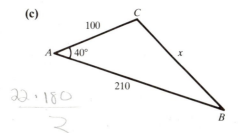

   3

(d)

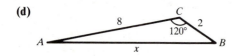

(e)

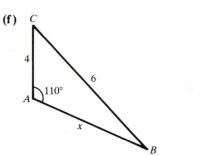

(f)

28 mi/h in a direction S42°E. How far apart are the two ships after 2 h?

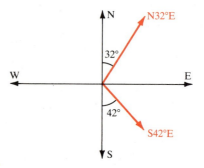

19. The angle of elevation to the top of the highest tower in the world, the CN tower in Toronto, is 28.81° from a distance of 1 km from the base of the tower. Find the height of the tower.

20. Find the perimeter of a regular hexagon that is inscribed in a circle of radius 8 m.

21. A pilot measures the angles of depression to two ships to be 40° and 52°. (See the figure.) If the pilot is flying at an elevation of 35,000 ft, find the distance between the two ships.

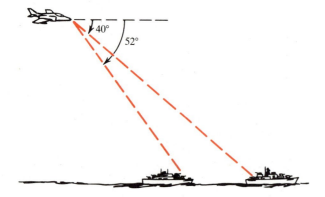

22. Two ships leave a port at the same time. One travels at 20 mi/h in a direction N32°E and the other travels at

23. From a point A on the ground the angle of elevation to a tall building is 24.1°. From a point B that is 600 ft closer to the building, the angle of elevation is measured to be 30.2°. Find the height of the building.

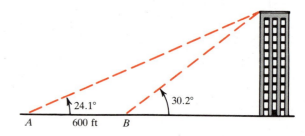

24. Find the distance between the points A and B on the opposite side of a lake from the information given in the figure.

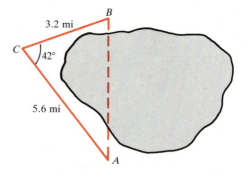

25. A boat is in the ocean off a straight shoreline. Points A and B are 120 mi apart on the shore as in the figure. It is found that $\angle A = 42.3°$ and $\angle B = 68.9°$. Find the shortest distance from the boat to the shore.

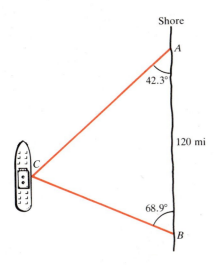

Shore

A

42.3°

120 mi

C

68.9°

B

26. In order to measure the height of an inaccessible cliff on the opposite side of a river, a surveyor makes the measurements shown in the figure. Find the height of the cliff using this information.

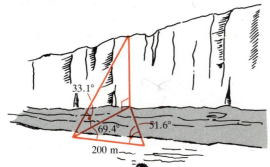

33.1°

69.4°   51.6°

200 m

27. Find the values of the six trigonometric functions at the point t for the given point $P(t)$ on the unit circle.

(a) $\left(\dfrac{\sqrt{3}}{2}, \dfrac{1}{2}\right)$ (b) $\left(-\dfrac{5}{13}, \dfrac{12}{13}\right)$

(c) $\left(-\dfrac{2}{7}, -\dfrac{3\sqrt{5}}{7}\right)$ (d) $(0, -1)$

28. Use the trigonometric identities to find the values of the remaining trigonometric functions.

(a) $\sin t = \dfrac{5}{13}, \cos t = -\dfrac{12}{13}$

(b) $\tan t = \dfrac{\sqrt{7}}{3}, \sec t = \dfrac{4}{3}$

(c) $\cot t = -\dfrac{1}{2}, \csc t = \dfrac{\sqrt{5}}{2}$

(d) $\sec t = -\dfrac{41}{40}, \csc t = -\dfrac{41}{9}$

In Exercises 29–40 sketch the graph of the given function.

29. $f(x) = 10\cos\frac{1}{2}x$ **30.** $f(x) = 2 + 4\sin 2\pi x$

31. $f(x) = 10\sin\left(2x - \dfrac{\pi}{2}\right)$

32. $f(x) = \tan\left(x + \dfrac{\pi}{6}\right)$

33. $f(x) = 3 + 3\cos\left(\dfrac{1}{3}x - \dfrac{\pi}{6}\right)$

34. $f(x) = \tan \pi x$

35. $f(x) = 1 - \cos\left(\dfrac{\pi}{2}x + \dfrac{\pi}{6}\right)$

36. $f(x) = -4\sec 4\pi x$

37. $f(x) = x + \sin 2x$

38. $f(x) = x^2\sin 2x$

39. $f(x) = e^x\cos x$

40. $f(x) = 2^{-x}\sin 4x$

41. The following are graphs of sine and cosine curves. Determine the function whose graph is shown.

(a)

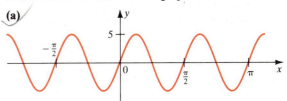

(b)

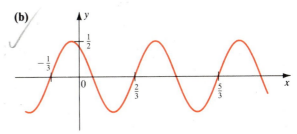

(c)

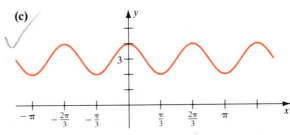

(d)

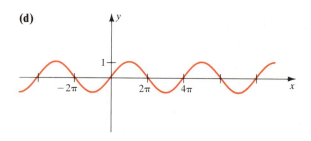

42. A point P moving in simple harmonic motion completes 8 cycles of its motion every second. If the amplitude of the motion is 50 cm, find an equation that describes the motion of P as a function of time. Assume the point P is at its maximum displacement when $t = 0$.

43. A mass that is suspended from a spring is oscillating in simple harmonic motion at a frequency of 4 cycles per second. The distance from the high point to the low point of the oscillation is 100 cm. Find an equation that describes the distance of the mass from its rest position as a function of time. Assume the mass is at its lowest point when $t = 0$.

44. The top of a flagpole is displaced a distance of 5 in. from its rest position by a sharp gust of wind and then released. It then vibrates approximately in damped harmonic motion with a frequency of 2 cycles per second. If the damping constant for this flagpole is $c = 0.3$, write an equation that describes the displacement of the top of the flagpole from its rest position as a function of time. Take $t = 0$ to be the instant the flagpole is released.

45. The flagpole in Exercise 44 is replaced by another pole to avoid excessive vibration. To test this new flagpole, its top is pulled 5 in. from its rest position and then released. It is observed that the flagpole vibrates at a frequency of 2 cycles per second. Two seconds later it is found that the amplitude of the vibration had dropped to 1 in.
(a) Find the damping constant for this flagpole.
(b) Write an equation that describes the displacement of the top of the flagpole as a function of time. Take $t = 0$ to be the instant the flagpole is released.

Chapter 5 Test

1. (a) Find the radian measures that correspond to the following degree measures: $300°$, $-18°$, $225°$.
　(b) Find the degree measures that correspond to the following radian measures: $\dfrac{5\pi}{6}$, $-\dfrac{11\pi}{4}$, 2.4.

2. Find the exact values of the following:
　(a) $\sin 405°$　　　　　**(b)** $\tan(-150°)$
　(c) $\sec\left(\dfrac{5\pi}{3}\right)$　　　　**(d)** $\csc\left(\dfrac{5\pi}{2}\right)$

3. Find $\tan\theta + \sin\theta$ for the angle θ shown in the figure.

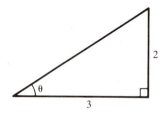

4. If $\cos\theta = -\frac{1}{3}$ and θ is in quadrant III, find $\tan\theta \cot\theta + \csc\theta$.

5. Find the area of the shaded region in the figure.

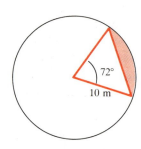

6. Express $\tan\theta$ in terms of $\sec\theta$ for θ in quadrant II.

7. Find the lengths a and b shown in the figure in terms of θ.

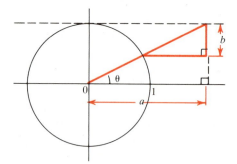

8. Find the side labeled x.

(a)

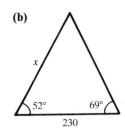

(b)

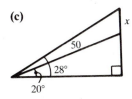

(c)

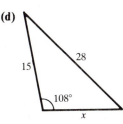

(d)

9. Find the values of the functions $\sin t$ and $\cos t$ at the point $(2/\sqrt{13}, 3/\sqrt{13})$ on the unit circle.

10. Sketch the graphs of the following functions:

(a) $f(x) = 1 - \sin 2x$ **(b)** $f(x) = \cos\left(\dfrac{1}{2}x - \dfrac{\pi}{6}\right)$

(c) $f(x) = 3 + \cos 2\pi x$

(d) $f(x) = \tan\left(2\pi x - \dfrac{\pi}{2}\right)$

11. Sketch the graphs of the following functions:
(a) $f(x) = x + \sin 4x$ **(b)** $f(x) = e^{-x}\cos 2\pi x$

12. The figure shows the variation of the water level above a certain reference point in the Long Beach harbor for a particular day. Assuming that this variation is simple harmonic, find an equation that describes the variation in water level as a function of the number of hours after midnight. Take $t = 0$ to be as shown in the graph.

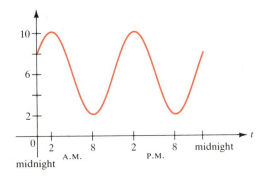

Problems Plus

1. A ball of radius a is inside a cone so that the sides of the cone are tangent to the ball. A larger ball of radius b fits inside the cone in such a way that it is tangent to both the ball of radius a and the sides of the cone. Express b in terms of a and the angle θ shown in the figure.

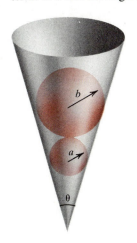

2. Sketch the graphs of the following functions:
(a) $f(x) = \sin \pi \sqrt{x}$ **(b)** $f(x) = x^2 \sin \pi x^2$

3. Find the number of solutions to the equation

$$\sin x = \frac{x}{100}.$$

4. A person starts at a point P on the earth's surface and walks 1 mi south, then 1 mi east, then 1 mi north and finds herself back at P where she started. Describe all points P for which this is possible. (There are infinitely many.)

6

Analytic Trigonometry

The greatest mathematicians, as Archimedes, Newton, and Gauss, always united theory and applications in equal measure.

Felix Klein

There are many relations among the trigonometric functions. The identities we encountered in Chapter 5 are basic examples of these. In this chapter we study some of the deeper properties that relate the trigonometric functions to one another. We will find identities for trigonometric functions of sums and differences of real numbers, multiple-angle formulas, and other related identities.

Because of these properties the trigonometric functions are very useful for certain mathematical applications. In this chapter we will see how they are used in the study of complex numbers and vectors. In Chapter 8 the trigonometric functions will be applied to the study of rotations of graphs in the plane, polar coordinates, and parametric equations.

Section 6.1

Trigonometric Identities

We have been working with equations and identities throughout this book. Let us take a closer look at what these concepts mean.

Equations and Identities

An **equation** is a statement that two mathematical expressions are equal. Thus

$$1 + 1 = 2$$

is an equation. An equation may contain a variable; for example,

(1) $$3x + 2 = 14$$

is an equation. This kind of equation is valid for only certain values of x and so it is sometimes called a **conditional equation**. Such an equation implies a question: For what values of the variable x is the statement $3x + 2 = 14$ true? In this case the value $x = 4$ makes Equation 1 true and it is false for all other values of x. In general, each value of x that makes an equation true is called a **solution** of the equation.

Here are some more examples of equations:

(A) $x + 2 = 5$ (B) $x + 1 = 1 + x$

$2x^2 + 3x = x - 3$ $(x + 1)^2 = x^2 + 2x + 1$

$y + 2x + 1 = y + x$ $(x + y)^2 = x^2 + 2xy + y^2$

$e^{2x} + 5e^x + 6 = 0$ $e^{x+y} = e^x e^y$

$\ln x = 0$ $\ln(xy) = \ln x + \ln y$

$\sin^2 x - 1 = 0$ $\sin^2 x + \cos^2 x = 1$

What is the difference between the equations in (A) and those in (B)? The important difference is that the equations in (B) are true for *every* value of the variables for which both sides are defined. Equations with this property are called **identities**. It is easy to see that some equations are identities. For example, the equation

$$x + 1 = x + 1$$

is clearly an identity, whereas it takes more work to show that

$$\sin^2 x + \cos^2 x = 1$$

is also an identity.

In this section we will be interested in equations and identities that involve the trigonometric functions. We review some of the trigonometric identities from Chapter 5.

Fundamental Trigonometric Identities

Reciprocal identities:

$$\csc\theta = \frac{1}{\sin\theta} \qquad \sec\theta = \frac{1}{\cos\theta} \qquad \cot\theta = \frac{1}{\tan\theta}$$

$$\tan\theta = \frac{\sin\theta}{\cos\theta} \qquad \cot\theta = \frac{\cos\theta}{\sin\theta}$$

Pythagorean identities:

$$\sin^2\theta + \cos^2\theta = 1$$
$$\tan^2\theta + 1 = \sec^2\theta$$
$$1 + \cot^2\theta = \csc^2\theta$$

Even–odd identities:

$$\sin(-t) = -\sin t$$
$$\cos(-t) = \cos t$$
$$\tan(-t) = -\tan t$$

Reduction formulas:

$$\sin(t + \pi) = -\sin t \qquad \sin\left(t + \frac{\pi}{2}\right) = \cos t$$

$$\cos(t + \pi) = -\cos t \qquad \cos\left(t + \frac{\pi}{2}\right) = -\sin t$$

$$\tan(t + \pi) = \tan t \qquad \tan\left(t + \frac{\pi}{2}\right) = -\cot t$$

Identities help us to write the same expression in different ways. It is often possible to use these identities to write a complicated looking expression as a much simpler one. In Example 1 a trigonometric expression involving the sine, secant, and tangent functions turns out to be equal to the cosine function.

Example 1

Simplify the expression $(1 + \sin x)(\sec x - \tan x)$.

Solution

Each of the following steps uses one of the fundamental identities. Provide reasons for each step.

$$(1 + \sin x)(\sec x - \tan x) = (1 + \sin x)\left(\frac{1}{\cos x} - \frac{\sin x}{\cos x}\right)$$

$$= (1 + \sin x)\left(\frac{1 - \sin x}{\cos x}\right)$$

$$= \frac{1 - \sin^2 x}{\cos x}$$

$$= \frac{\cos^2 x}{\cos x} = \cos x \qquad \bullet$$

We see that it is often helpful to rewrite everything in terms of sine and cosine. But this is not always the case, as the next example shows.

Example 2

Simplify the expression

$$\frac{\sin^2 x \tan^2 x}{\tan x + \sin x}$$

Solution

Supply reasons for the following steps:

$$\frac{\sin^2 x \tan^2 x}{\tan x + \sin x} = \frac{(1 - \cos^2 x)\tan^2 x}{\tan x + \sin x}$$

$$= \frac{\tan^2 x - \cos^2 x \tan^2 x}{\tan x + \sin x}$$

$$= \frac{\tan^2 x - \cos^2 x\left(\dfrac{\sin^2 x}{\cos^2 x}\right)}{\tan x + \sin x}$$

$$= \frac{\tan^2 x - \sin^2 x}{\tan x + \sin x}$$

$$= \frac{(\tan x + \sin x)(\tan x - \sin x)}{\tan x + \sin x}$$

$$= \tan x - \sin x \qquad \bullet$$

Proving Trigonometric Identities

Many identities follow from the fundamental identities. In the following examples we learn how to prove that a given trigonometric equation is an identity, and in the process we see how to discover new identities.

First, it is easy to decide when a given equation is *not* an identity. All that needs to be done is to show that the equation does not hold for some value of the variable (or variables). Thus, the equation

$$\sin x + \cos x = 1$$

is not an identity because when $x = \pi/4$ we have

$$\sin\frac{\pi}{4} + \cos\frac{\pi}{4} = \frac{\sqrt{2}}{2} + \frac{\sqrt{2}}{2} = \sqrt{2} \neq 1$$

To verify that a trigonometric equation is an identity we transform one side of the equation into the other side by a series of steps, each of which is itself an identity. The next three examples illustrate this procedure.

Example 3

Verify the identity $\tan x \sin x + \cos x = \sec x$.

Solution

We start with the left side of this equation and transform it into the right side. In each step a trigonometric identity or an algebraic identity is used. Supply the reasons for each step.

$$\tan x \sin x + \cos x = \left(\frac{\sin x}{\cos x}\right)\sin x + \cos x$$

$$= \frac{\sin^2 x}{\cos x} + \cos x$$

$$= \frac{\sin^2 x + \cos^2 x}{\cos x}$$

$$= \frac{1}{\cos x} = \sec x$$

In Example 3 it is not so easy to see how to change the right side into the left side, but it is possible to do so. Just notice that each of the steps is reversible. In other words, if we start with the last expression in the proof (this is the right-hand side of the equation) and read backward through the steps, we see that each step is an identity and so the right-hand side is transformed into the left-hand side. You will agree, however, that in this case it is harder to see the steps necessary to arrive at the left-hand side from the right-hand side.

In general it is better to change the more complicated side of the equation to the simpler side.

Example 4

Prove that the equation

$$\frac{\sin x(\sin x - \csc x) + \cot^2 x}{\cos^2 x} = \cot^2 x$$

is an identity.

Solution

Since the left side appears more complicated, we transform it to the other side. Supply reasons for the following steps:

$$\frac{\sin x(\sin x - \csc x) + \cot^2 x}{\cos^2 x} = \frac{\sin^2 x - \sin x \csc x + \cot^2 x}{\cos^2 x}$$

$$= \frac{\sin^2 x - \sin x\left(\dfrac{1}{\sin x}\right) + \cot^2 x}{\cos^2 x}$$

$$= \frac{\sin^2 x - 1 + \cot^2 x}{\cos^2 x}$$

$$= \frac{\sin^2 x - 1}{\cos^2 x} + \frac{\cot^2 x}{\cos^2 x}$$

$$= \frac{-\cos^2 x}{\cos^2 x} + \left(\frac{\cos^2 x}{\sin^2 x}\right)\frac{1}{\cos^2 x}$$

$$= -1 + \frac{1}{\sin^2 x} = \csc^2 x - 1 = \cot^2 x \qquad \bullet$$

Example 5

Verify the identity

$$2 \tan x \sec x = \frac{1}{1 - \sin x} - \frac{1}{1 + \sin x}$$

Solution

Finding a common denominator and combining the fractions on the right side of this equation, we get

$$\frac{1}{1 - \sin x} - \frac{1}{1 + \sin x} = \frac{(1 + \sin x) - (1 - \sin x)}{(1 - \sin x)(1 + \sin x)} = \frac{2 \sin x}{1 - \sin^2 x}$$

$$= \frac{2 \sin x}{\cos^2 x} = 2\frac{\sin x}{\cos x}\left(\frac{1}{\cos x}\right) = 2 \tan x \sec x \qquad \bullet$$

In Example 6 we introduce something extra to the problem by multiplying the numerator and the denominator by a trigonometric expression, chosen so that we can simplify the result.

Example 6

Verify the identity

$$\frac{\cot x}{1 - \sin x} = \sec x(\csc x + 1)$$

Solution

We multiply the numerator and denominator of the left side of this equation by $1 + \sin x$.

$$\frac{\cot x}{1 - \sin x} = \frac{\cot x}{1 - \sin x}\frac{1 + \sin x}{1 + \sin x} = \frac{\cot x(1 + \sin x)}{1 - \sin^2 x}$$

$$= \frac{\cot x(1 + \sin x)}{\cos^2 x} = \frac{\cos x}{\sin x}\frac{(1 + \sin x)}{\cos^2 x}$$

$$= \frac{1 + \sin x}{\sin x \cos x} = \frac{1}{\sin x \cos x} + \frac{\sin x}{\sin x \cos x}$$

$$= \frac{1}{\sin x \cos x} + \frac{1}{\cos x} = \csc x \sec x + \sec x = \sec x(\csc x + 1) \quad \bullet$$

Here is another method for proving that an equation is an identity. If we can transform each side of the equation separately by way of identities to arrive at the same result, then the equation is an identity. Example 7 illustrates this procedure.

Example 7

Prove that the equation

$$\frac{1}{1 - \sin z} = \sec^2 z + \tan z \sec z$$

is an identity.

Solution

We transform the left-hand side to get

$$\frac{1}{1 - \sin z} = \frac{1}{1 - \sin z}\frac{1 + \sin z}{1 + \sin z} = \frac{1 + \sin z}{1 - \sin^2 z} = \frac{1 + \sin z}{\cos^2 z}$$

From the right side we get

$$\sec^2 z + \tan z \sec z = \frac{1}{\cos^2 z} + \frac{\sin z}{\cos z}\frac{1}{\cos z} = \frac{1}{\cos^2 z} + \frac{\sin z}{\cos^2 z} = \frac{1 + \sin z}{\cos^2 z}$$

It follows that
$$\frac{1}{1 - \sin z} = \sec^2 z + \tan z \sec z \quad \bullet$$

There are often several ways to prove that a given identity is true. For instance, in Example 7 we could have continued to manipulate the left-hand side to arrive at the other side. In all the other examples we could have started with either side and transformed it to the other side by way of identities. Indeed, it is not possible to give rules for proving identities. You will find, however, that with the suggestions made in this section and some practice it will in most cases be easy to see what to do to prove a particular identity.

 As a final warning, notice that to prove an identity we do *not* perform the same operation on both sides of the equation. Remember that we do not know at first that the equation we are given is an identity. It is our task to prove that it is. Indeed, if we start with an equation that is not an identity such as

(2) $$\sin x = -\sin x$$

and square both sides, we get the equation

(3) $\sin^2 x = \sin^2 x$

which is clearly an identity. Does this suggest that the original equation is an identity? Of course not. The problem here is that the operation of squaring is not **reversible** in the sense that we cannot arrive back to (2) from (3) by taking square roots (reversing the procedure). Only operations that are reversible will necessarily transform an identity to an identity.

Exercises 6.1

In Exercises 1–8 show that the given equation is not an identity.

1. $\dfrac{1}{x+1} = \dfrac{1}{x} + 1$ **2.** $\sin 2x = 2 \sin x$

3. $\sec^2\theta + \csc^2\theta = 1$

4. $\dfrac{1}{\sin x + \cos x} = \csc x + \sec x$

5. $\sin(x - \pi) = \sin x$ **6.** $\sqrt{x^2 + 1} = x + 1$

7. $\tan t + \cot t = 1$

8. $\sin(x + y) = \sin x + \sin y$

In Exercises 9–20 determine whether the given equation is an identity. If the equation is not an identity, find all its solutions.

9. $2x + 1 = 3$

10. $(u + 2)^2 = u^2 + 4u + 4$

11. $\ln(3x) = \ln 3 + \ln x$ **12.** $\sqrt{x + 1} = \sqrt{x} + 1$

13. $e^{2x} - 2e^x + 1 = 0$

14. $x^3 + 2x = x(x^2 + 1) + x$

15. $\dfrac{x + 1}{x} = 1 + \dfrac{1}{x}$ **16.** $\tan\theta \cot\theta = 1$

17. $\sin x - 1 = 0$ **18.** $xe^{\ln x^2} = x^3$

19. $\sin t \csc t + \cos t \sec t = 2$

20. $e^{\sin^2 x} e^{\cos^2 x} = e$

In Exercises 21–28 write the given trigonometric expression in terms of sine and cosine and simplify.

21. $\cos x \tan x$ **22.** $\sec\alpha + \tan\alpha$

23. $\sec^2 x - \tan^2 x$

24. $(\tan w - \csc w)(\sin w \cos w)$

25. $\tan A + \cot A$ **26.** $\dfrac{\tan x + \cot x}{\sec x \csc x}$

27. $\cos u + \tan u \sin u$ **28.** $\cos^2 x(1 + \tan^2 x)$

In Exercises 29–50 simplify the given trigonometric expression.

29. $\dfrac{\cos x \sec x}{\cot x}$ **30.** $\cos^3 x + \sin^2 x \cos x$

31. $\dfrac{1 + \sin y}{1 + \csc y}$ **32.** $\dfrac{\tan x}{\sec(-x)}$

33. $\dfrac{\sec^2 x - 1}{\sec^2 x}$ **34.** $\dfrac{\sec x - \cos x}{\tan x}$

35. $\dfrac{1 + \csc x}{\cos x + \cot x}$

36. $(1 + \tan r)(\sin^2 r \cos r + \sin r \cos^2 r)$

37. $\dfrac{1 + \sin u}{\cos u} + \dfrac{\cos u}{1 + \sin u}$ **38.** $\tan x \cos x \csc x$

39. $\dfrac{\cot x - 1}{1 + \tan(-x)}$ **40.** $\dfrac{1 + \cot A}{\csc A}$

41. $\tan\theta + \cos(-\theta) + \tan(-\theta)$

42. $\dfrac{\sin x + \tan x}{\tan x}$

43. $\dfrac{2 + \tan^2 x}{\sec^2 x} - 1$ **44.** $\sin^4 B - \cos^4 B + \cos^2 B$

45. $\dfrac{1}{1 - \sin\alpha} + \dfrac{1}{1 + \sin\alpha}$ **46.** $\dfrac{\cos x}{\sec x + \tan x}$

47. $(\sec x - \tan x)^2(1 + \sin x)$

48. $\dfrac{\sin t}{1 - \cos t} - \csc t$

49. $\dfrac{1}{\sec^2 \beta} + \dfrac{1}{\csc^2 \beta}$

50. $\dfrac{\sin x}{\csc x} + \dfrac{\cos x}{\sec x}$

In Exercises 51–136 verify the identity.

51. $\tan x \cot x = 1$

52. $\sin x \csc x = 1$

53. $\sin \theta \cot \theta = \cos \theta$

54. $\dfrac{\tan x}{\sec x} = \sin x$

55. $\dfrac{\cos u \sec u}{\tan u} = \cot u$

56. $\dfrac{\cot x \sec x}{\csc x} = 1$

57. $\dfrac{\tan y}{\csc y} = \sec y - \cos y$

58. $\dfrac{\cos v}{\sec v \sin v} = \csc v - \sin v$

59. $\sin B + \cos B \cot B = \csc B$

60. $\cos x + \sin x \tan x = \sec x$

61. $\sin(-x)\tan(-x) = \sin^2 x \sec x$

62. $\cos(-x) - \sin(-x) = \cos x + \sin x$

63. $\cot(-\alpha)\cos(-\alpha) + \sin(-\alpha) = -\csc \alpha$

64. $\csc x(\csc x + \sin(-x)) = \cot^2 x$

65. $(1 - \sin x)(1 + \sin x) = \cos^2 x$

66. $(\sin x + \cos x)^2 = 1 + 2 \sin x \cos x$

67. $(1 - \cos \beta)(1 + \cos \beta) = \dfrac{1}{\csc^2 \beta}$

68. $\dfrac{\cos x}{\sec x} + \dfrac{\sin x}{\csc x} = 1$

69. $\dfrac{(\sin x + \cos x)^2}{\sin^2 x - \cos^2 x} = \dfrac{\sin^2 x - \cos^2 x}{(\sin x - \cos x)^2}$

70. $(\sin x + \cos x)^4 = (1 + 2 \sin x \cos x)^2$

71. $\dfrac{\sec t - \cos t}{\sec t} = \sin^2 t$

72. $\dfrac{1 - \sin x}{1 + \sin x} = (\sec x - \tan x)^2$

73. $\dfrac{1}{1 - \sin^2 y} = 1 + \tan^2 y$

74. $\csc x - \sin x = \cos x \cot x$

75. $(\cot x - \csc x)(\cos x + 1) = -\sin x$

76. $\sin^4 \theta - \cos^4 \theta = \sin^2 \theta - \cos^2 \theta$

77. $(1 - \cos^2 x)(1 + \cot^2 x) = 1$

78. $\cos^2 x - \sin^2 x = 2 \cos^2 x - 1$

79. $2 \cos^2 x - 1 = 1 - 2 \sin^2 x$

80. $\tan y + \cot y = \sec y \csc y$

81. $\dfrac{1 - \cos \alpha}{\sin \alpha} = \dfrac{\sin \alpha}{1 + \cos \alpha}$

82. $\csc x \cos^2 x + \sin x = \csc x$

83. $\dfrac{1}{1 - \sin x} = \sec x(\sec x + \tan x)$

84. $\sin^2 \alpha + \cos^2 \alpha + \tan^2 \alpha = \sec^2 \alpha$

85. $\dfrac{\sin x - 1}{\sin x + 1} = \dfrac{-\cos^2 x}{(\sin x + 1)^2}$

86. $\dfrac{\sin w}{\sin w + \cos w} = \dfrac{\tan w}{1 + \tan w}$

87. $\dfrac{(\sin t + \cos t)^2}{\sin t \cos t} = 2 + \sec t \csc t$

88. $\sec t \csc t(\tan t + \cot t) = \sec^2 t + \csc^2 t$

89. $\dfrac{1 + \tan^2 u}{1 - \tan^2 u} = \dfrac{1}{\cos^2 u - \sin^2 u}$

90. $\dfrac{1 + \sec^2 x}{1 + \tan^2 x} = 1 + \cos^2 x$

91. $\dfrac{\sec x}{\sec x - \tan x} = \sec x(\sec x + \tan x)$

92. $\dfrac{\sec x + \csc x}{\tan x + \cot x} = \sin x + \cos x$

93. $\sec v - \tan v = \dfrac{1}{\sec v + \tan v}$

94. $\csc x - \cot x = \dfrac{1}{\csc x + \cot x}$

95. $\dfrac{1 + \sin x}{1 - \sin x} = 2 \sec x(\sec x + \tan x) - 1$

96. $\dfrac{\sin A}{1 - \cos A} - \cot A = \csc A$

97. $\dfrac{\sin x + \cos x}{\sec x + \csc x} = \sin x \cos x$

98. $\dfrac{1 - \cos x}{\sin x} + \dfrac{\sin x}{1 - \cos x} = 2 \csc x$

99. $\dfrac{\csc x - \cot x}{\sec x - 1} = \cot x$

100. $\dfrac{\csc^2 x - \cot^2 x}{\sec^2 x} = \cos^2 x$

101. $\tan^2 u - \sin^2 u = \tan^2 u \sin^2 u$

102. $\cot^2 u - \cos^2 u = \cot^2 u \cos^2 u$

103. $\dfrac{\tan v \sin v}{\tan v + \sin v} = \dfrac{\tan v - \sin v}{\tan v \sin v}$

104. $\dfrac{\cot v \cos v}{\cot v + \cos v} = \dfrac{\cot v - \cos v}{\cot v \cos v}$

105. $\sec^4 x - \tan^4 x = \sec^2 x + \tan^2 x$

106. $\dfrac{\cos \theta}{1 - \sin \theta} = \sec \theta + \tan \theta$

107. $\dfrac{\cos \theta}{1 - \sin \theta} = \dfrac{\sin \theta - \csc \theta}{\cos \theta - \cot \theta}$

108. $\dfrac{1 + \tan x}{1 - \tan x} = \dfrac{\cos x + \sin x}{\cos x - \sin x}$

109. $\dfrac{\cos^2 t + \tan^2 t - 1}{\sin^2 t} = \tan^2 t$

110. $\dfrac{1}{1 - \sin x} - \dfrac{1}{1 + \sin x} = 2 \sec x \tan x$

111. $\dfrac{1}{\sec x + \tan x} + \dfrac{1}{\sec x - \tan x} = 2 \sec x$

112. $\dfrac{1 + \sin x}{1 - \sin x} - \dfrac{1 - \sin x}{1 + \sin x} = 4 \tan x \sec x$

113. $(\tan x + \cot x)^2 = \sec^2 x + \csc^2 x$

114. $\tan^2 x - \cot^2 x = \sec^2 x - \csc^2 x$

115. $\dfrac{\sec u - 1}{\sec u + 1} = \dfrac{1 - \cos u}{1 + \cos u}$

116. $\dfrac{\sec u - 1}{\sec u + 1} = \dfrac{\tan u - \sin u}{\tan u + \sin u}$

117. $\dfrac{\cot x + 1}{\cot x - 1} = \dfrac{1 + \tan x}{1 - \tan x}$

118. $\dfrac{\sin^3 x + \cos^3 x}{\sin x + \cos x} = 1 - \sin x \cos x$

119. $\dfrac{\sin^3 x - \csc^3 x}{\sin x - \csc x} = \sin^2 x + \csc^2 x + 1$

120. $\dfrac{\tan v - \cot v}{\tan^2 v - \cot^2 v} = \sin v \cos v$

121. $\dfrac{1 + \sin x}{1 - \sin x} = (\tan x + \sec x)^2$

122. $\dfrac{1 + \cos x}{1 - \cos x} = (\cot x + \csc x)^2$

123. $\dfrac{(\sin x + \cos x)^2}{(\sin x - \cos x)^2} = \dfrac{\sec x + 2 \sin x}{\sec x - 2 \sin x}$

124. $\dfrac{\tan x + \tan y}{\cot x + \cot y} = \tan x \tan y$

125. $(\tan x + \cot x)^4 = \csc^4 x \sec^4 x$

126. $\dfrac{\sin^2 x - \tan^2 x}{\cos^2 x - \cot^2 x} = \tan^6 x$

127. $(\sin \alpha - \tan \alpha)(\cos \alpha - \cot \alpha) = (\cos \alpha - 1)(\sin \alpha - 1)$

128. $\sin^6 \beta + \cos^6 \beta + 3 \sin^2 \beta \cos^2 \beta = 1$

129. $\dfrac{1 + \cos x + \sin x}{1 + \cos x - \sin x} = \dfrac{1 + \sin x}{\cos x}$

130. $\ln|\tan x \sin x| = 2 \ln|\sin x| + \ln|\sec x|$

131. $\ln|\tan x + \cot x| = 2 \ln|\sec x| - \ln|\tan x|$

132. $\ln|\sec x + \tan x| = -\ln|\sec x - \tan x|$

133. $\ln|\csc x - \cot x| = -\ln|\csc x + \cot x|$

134. $\ln|\tan x| + \ln|\cot x| = 0$

135. $e^{\sin^2 x} e^{\tan^2 x} = e^{\sec^2 x} e^{-\cos^2 x}$

136. $e^{x + 2 \ln|\sin x|} = e^x \sin^2 x$

137. For $-\dfrac{\pi}{2} \le \theta \le \dfrac{\pi}{2}$ and $x = \sin \theta$, show the following:

$$\cos \theta = \sqrt{1 - x^2} \qquad \tan \theta = \dfrac{x}{\sqrt{1 - x^2}}$$

$$\cot \theta = \dfrac{1 - x^2}{x\sqrt{1 - x^2}}$$

138. For $-\dfrac{\pi}{2} \le \theta \le \dfrac{\pi}{2}$ and $x = a \sin \theta$, $a > 0$, show that

$$\sqrt{a^2 - x^2} = a \cos \theta$$

139. For $-\dfrac{\pi}{2} < \theta < \dfrac{\pi}{2}$ and $x = a\tan\theta,\ a > 0$, show that

$$\sqrt{x^2 + a^2} = a\sec\theta$$

140. For $-\dfrac{\pi}{2} \le \theta \le \dfrac{\pi}{2}$ and $x + 1 = 2\sin\theta$, show that

$$\sqrt{3 - 2x - x^2} = 2\cos\theta$$

141. For $0 < \theta \le \dfrac{\pi}{2}$ and $e^x = 3\sin\theta$, show that

$$\sqrt{9 - e^{2x}} = 3\cos\theta$$

Section 6.2

Trigonometric Equations

A **trigonometric equation** is an equation that contains trigonometric functions. The equations

$$2\sin x - 1 = 0 \qquad 4\cos^2 x - 2\cos x + 3 = 0 \qquad (\sin x - 1)(\tan x + 2) = 0$$

are trigonometric equations. We are interested in solving such equations; that is, we would like to find all the values of the variable x that make the equation true. The first of these equations

(1) $$2\sin x - 1 = 0$$

or $$\sin x = \tfrac{1}{2}$$

is the simplest kind of trigonometric equation. The solutions to this equation in the interval $[0, 2\pi]$ are $x = \pi/6$ and $x = 5\pi/6$. But since the sine function is periodic with period 2π, any integer multiple of 2π added to one of these solutions gives another solution. Thus all the solutions of Equation 1 are

$$x = \frac{\pi}{6} + 2k\pi, \qquad k = 0, \pm1, \pm2, \pm3, \ldots$$

$$x = \frac{5\pi}{6} + 2k\pi \qquad k = 0, \pm1, \pm2, \pm3, \ldots$$

Figure 1 shows a graphical representation of the solutions of this equation.

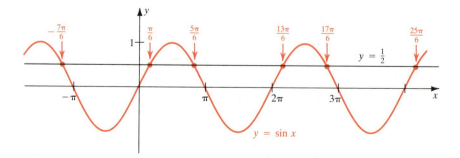

Figure 1

In general, as in this example, if a trigonometric equation has one solution, then it has infinitely many solutions. To find all the solutions of such an equation we need to find only the solutions in an appropriate interval and then use the fact that the trigonometric functions are periodic.

To solve more complicated trigonometric equations we use algebraic techniques to reduce the equation to simple equations like Equation 1. The following examples illustrate this technique.

Example 1

Solve the equation $4\cos^2 x - 8\cos x + 3 = 0$.

Solution

Factoring the left side of this equation, we get

$$(2\cos x - 1)(2\cos x - 3) = 0$$

so $2\cos x - 1 = 0$ or $2\cos x - 3 = 0$

$$\cos x = \tfrac{1}{2} \qquad\qquad \cos x = \tfrac{3}{2}$$

Since $\cos x$ is never greater than 1, we see that $\cos x = \tfrac{3}{2}$ has no solution. In the interval $[0, 2\pi]$, $\cos x = \tfrac{1}{2}$ has the solutions $x = \pi/3$ and $x = 5\pi/3$. Because the cosine function is periodic with period 2π, all the solutions are of the form

$$x = \frac{\pi}{3} + 2k\pi \qquad x = \frac{5\pi}{3} + 2k\pi$$

where $k = 0, \pm 1, \pm 2, \pm 3, \dots$. ●

Example 2

Find all the solutions of the equation $\tan x \sin x - \tan x - \sin x + 1 = 0$.

Solution

Factoring $\tan x$ from the first two terms of the left side of the equation, we get

$$\tan x(\sin x - 1) - \sin x + 1 = 0$$
$$\tan x(\sin x - 1) - (\sin x - 1) = 0$$
$$(\sin x - 1)(\tan x - 1) = 0$$

so $\sin x - 1 = 0$ or $\tan x - 1 = 0$

$$\sin x = 1 \qquad\qquad \tan x = 1$$

The equation $\sin x = 1$ has the solution $x = \pi/2$ in the interval $[0, 2\pi]$. Since sine is periodic with period 2π, we get all the solutions by adding integer multiples of 2π to $x = \pi/2$:

$$x = \frac{\pi}{2} + 2k\pi, \qquad k = 0, \pm 1, \pm 2, \pm 3, \dots$$

Since the function tan is periodic with period π, we need to find the solutions to $\tan x = 1$ in any interval of length π. A convenient such interval is $(-\pi/2, \pi/2)$ in which $\tan x = 1$ has the solution $x = \pi/4$. This gives

$$x = \frac{\pi}{4} + k\pi, \qquad k = 0, \pm 1, \pm 2, \pm 3, \ldots$$

It follows that the solutions to the original equation in this example are

$$x = \frac{\pi}{2} + 2k\pi \qquad x = \frac{\pi}{4} + k\pi$$

where k is an integer.   ●

Example 3

Solve the trigonometric equation $\tan^4 2x - 9 = 0$.

Solution

Adding 9 to both sides of this equation, we get

$$\tan^4 2x = 9$$

Taking the fourth root gives

$$\tan 2x = \sqrt{3} \qquad \text{or} \qquad \tan 2x = -\sqrt{3}$$

The solutions of these equations in the interval $(-\pi/2, \pi/2)$ are

$$2x = \frac{\pi}{3} \qquad \text{or} \qquad 2x = -\frac{\pi}{3}$$

Since the tangent function is periodic with period π, all the solutions are given by

$$2x = \frac{\pi}{3} + k\pi \qquad \text{or} \qquad 2x = -\frac{\pi}{3} + k\pi$$

Dividing both sides of these equations by 2 gives us the solution to the equation of this example:

$$x = \frac{\pi}{6} + \frac{k}{2}\pi \qquad x = -\frac{\pi}{6} + \frac{k}{2}\pi$$

where k is any integer.   ●

Trigonometric identities are useful tools for solving trigonometric equations. They can be used to transform an equation into an equivalent equation that is simpler to solve. The next example illustrates this.

Example 4

Solve the equation $3\sin x = 2\cos^2 x$ in the interval $[0, 2\pi]$.

Solution

Using the identity $\cos^2 x = 1 - \sin^2 x$, we get an equivalent equation that involves only the sine function:

$$3 \sin x = 2(1 - \sin^2 x)$$

or $$2 \sin^2 x + 3 \sin x - 2 = 0$$

Factoring gives

$$(2 \sin x - 1)(\sin x + 2) = 0$$

Thus $$2 \sin x - 1 = 0 \qquad \text{or} \qquad \sin x + 2 = 0$$

$$\sin x = \tfrac{1}{2} \qquad\qquad\qquad \sin x = -2$$

Since $-1 \leq \sin x \leq 1$, the equation $\sin x = -2$ has no solutions. The solutions of the given equation are thus the solutions of $\sin x = \tfrac{1}{2}$. These are $x = \pi/6$, $5\pi/6$.   ●

Example 5

Find the points of intersection of the graphs of $f(x) = \sin x$ and $g(x) = \cos x$.

Solution

The graphs of f and g are shown in Figure 2. The graphs intersect at the points where $f(x) = g(x)$. So we need to find the solutions of the equation

$$\sin x = \cos x$$

Notice that the numbers x for which $\cos x = 0$ are not solutions of this equation. For $\cos x \neq 0$ we can divide both sides of the equation by $\cos x$ to get

$$\frac{\sin x}{\cos x} = 1 \qquad \text{or} \qquad \tan x = 1$$

The solution of this last equation on the interval $(-\pi/2, \pi/2)$ is $x = \pi/4$. Since the tangent function is periodic with period π, all the solutions are

$$x = \frac{\pi}{4} + k\pi, \qquad k = 0, \pm 1, \pm 2, \pm 3, \ldots$$

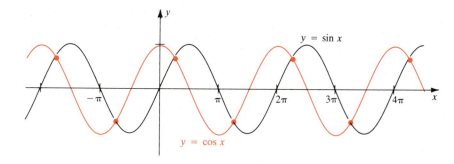

Figure 2

Example 6

Find the solutions of $\cos 2x \sec x = 2 \cos 2x$ in the interval $[0, 2\pi)$.

Solution

We subtract $2\cos 2x$ from both sides of the equation and factor out $\cos 2x$ to get

$$\cos 2x(\sec x - 2) = 0$$

Thus $\qquad\qquad\qquad \cos 2x = 0 \qquad \text{or} \qquad \sec x = 2$

We begin by solving $\cos 2x = 0$. Since we are seeking solutions in the interval $0 \leq x < 2\pi$, $2x$ is in the interval $0 \leq 2x < 4\pi$. In this interval $\cos 2x = 0$ has the solutions

$$2x = \frac{\pi}{2}, \frac{3\pi}{2}, \frac{5\pi}{2}, \frac{7\pi}{2}$$

or $\qquad\qquad\qquad x = \dfrac{\pi}{4}, \dfrac{3\pi}{4}, \dfrac{5\pi}{4}, \dfrac{7\pi}{4}$

Now we solve $\sec x = 2$. In the interval $[0, 2\pi]$, $\sec x = 2$ for $x = \pi/3$ and $x = 5\pi/3$. Thus all the solutions of the given equation are

$$x = \frac{\pi}{4}, \frac{\pi}{3}, \frac{3\pi}{4}, \frac{5\pi}{4}, \frac{5\pi}{3}, \frac{7\pi}{4} \qquad\qquad \bullet$$

Notice that in Example 6 we did not divide both sides by $\cos 2x$. It would have been wrong to do so, since we would be dividing by zero. Indeed, if we divide both sides by $\cos 2x$, then we lose all solutions of the given equation that are solutions of $\cos 2x = 0$.

When solving equations it is necessary that any operation performed on the equation produce an *equivalent* equation in the sense that the new equation has the same roots as the original equation. An example of an operation that may *not* give an equivalent equation is "squaring both sides." Let us again consider the equation $\sin x = -\sin x$. It is easy to see that the solutions to this equation are $x = k\pi$, where k is an integer. But if we square both sides of this equation, we introduce many new solutions. In fact, the equation $\sin^2 x = \sin^2 x$ is obviously an identity and so has every real number x as a solution. Another operation that may introduce new roots to an equation is multiplying both sides by an expression that itself may be zero. For instance,

$$\sin x = 0$$

has the solutions $x = k\pi$, where k is any integer. Multiplying both sides by $\cos x$ gives the equation

$$\cos x \sin x = 0$$

which has as additional solutions the roots of $\cos x = 0$.

If we perform an operation that may introduce new roots of an equation, then it is necessary to check that the solutions obtained are not extraneous; that is, we must verify that they satisfy the original equation.

Example 7

Solve the equation $\cos x + 1 = \sin x$ in the interval $[0, 2\pi)$.

Solution

To get an equation that involves either sine or cosine only, we square both sides and use the Pythagorean identities.

$$(\cos x + 1)^2 = \sin^2 x$$
$$\cos^2 x + 2\cos x + 1 = \sin^2 x$$
$$\cos^2 x + 2\cos x + 1 = 1 - \cos^2 x$$
$$2\cos^2 x + 2\cos x = 0$$
$$2\cos x(\cos x + 1) = 0$$

Thus $\qquad \cos x = 0 \qquad$ or $\qquad \cos x + 1 = 0$

From these we get the possible solutions

$$x = \frac{\pi}{2}, \frac{3\pi}{2}, \pi$$

Since we may have introduced extraneous roots by squaring both sides of the equation, we must check to see whether each of these values for x satisfies the original equation:

$x = \dfrac{\pi}{2}$	$x = \dfrac{3\pi}{2}$	$x = \pi$
$\cos\dfrac{\pi}{2} + 1 = \sin\dfrac{\pi}{2}$	$\cos\dfrac{3\pi}{2} + 1 = \sin\dfrac{3\pi}{2}$	$\cos\pi + 1 = \sin\pi$
$0 + 1 = 1$	$0 + 1 = -1$	$-1 + 1 = 0$
True	False	True

Thus the solutions of the given equation in the interval $[0, 2\pi)$ are $\pi/2$ and π.

●

Example 8

Solve the equation $2\cos x + \tan x - \sec x = 0$ in the interval $[0, 2\pi)$.

Solution

Writing the equation in terms of sine and cosine gives

$$2\cos x + \frac{\sin x}{\cos x} - \frac{1}{\cos x} = 0$$

Multiplying both sides by $\cos x$, we have

$$2\cos^2 x + \sin x - 1 = 0$$

To get an equation that involves the sine function only, we use the identity $\cos^2 x = 1 - \sin^2 x$ to get

$$2(1 - \sin^2 x) + \sin x - 1 = 0$$
$$2 - 2\sin^2 x + \sin x - 1 = 0$$
$$2\sin^2 x - \sin x - 1 = 0$$
$$(2\sin x + 1)(\sin x - 1) = 0$$

Thus $2\sin x + 1 = 0$ or $\sin x - 1 = 0$

$\sin x = -\frac{1}{2}$ $\sin x = 1$

In the interval $[0, 2\pi)$ the first of these equations has the solutions $x = 7\pi/6$, $11\pi/6$ and the other has the solution $x = \pi/2$. Since we have multiplied both sides of the equation by $\cos x$, we must check to see whether any of these solutions are extraneous. Notice that neither $\tan x$ nor $\sec x$ is defined for $x = \pi/2$ and so this value is not a solution of the given equation. It is easy to check that the other two values do satisfy the equation. Thus the solutions of the original equation are

$$x = \frac{7\pi}{6}, \frac{11\pi}{6}$$

We will return to the topic of trigonometric equations in later sections of this chapter.

Exercises 6.2

In Exercises 1–16 find all the solutions of the given equation.

1. $2\sin x - \sqrt{3} = 0$ **2.** $\tan x + 1 = 0$

3. $(\tan x - 1)(2\cos x - \sqrt{3}) = 0$

4. $\sqrt{3}\sin 2x = \cos 2x$ **5.** $4\cos^2 x - 3 = 0$

6. $\cos\dfrac{x}{2} - 1 = 0$

7. $\cos x \sin x - 2\cos x = 0$

8. $\sin x + \cos x = 3$

9. $\sin(x + \pi) + 3\sin x = 1$

10. $\sin(-x) = \sin x + 1$ **11.** $\sin^2 x = 4 - 2\cos^2 x$

12. $\dfrac{1 - \sin x}{1 + \sin x} = -3$ **13.** $\csc 3x = \sin 3x$

14. $\sin^2 x + \frac{1}{2} = \sqrt{2}\sin x$

15. $4\cos^2 x - 4\cos x + 1 = 0$

16. $2\cos^2 x + \sin x = 1$

In Exercises 17–48 find the solution of the given equation in the interval $[0, 2\pi)$.

17. $2\sin x \tan x - \tan x = 1 - 2\sin x$

18. $\tan^2 3x + \cot^2 3x = 2$

19. $\sec x \tan x - \cos x \cot x = \sin x$

20. $\tan^5 x - 9\tan x = 0$

21. $3\tan^3 x - 3\tan^2 x - \tan x + 1 = 0$

22. $4\sin x \cos x + 2\sin x - 2\cos x - 1 = 0$

23. $\tan x - 3\cot x = 0$

24. $2\sin x - 2\sqrt{3}\cos x - \sqrt{3}\tan x + 3 = 0$

25. $\sec x - \csc x - 2\tan x + 2 = 0$

26. $2\sin^2 x - \cos x = 1$ **27.** $4\cos^2 x + 2\sin^2 x = 3$

28. $\cos^2 x - \frac{7}{2}\sin x = -1$ **29.** $\cos^2 \pi x - \sin^2 \pi x = 0$

30. $8\sin^4 x - 10\sin^2 x + 3 = 0$

31. $\sec x - \tan x = \cos x$ **32.** $\tan^2 x + \sec^2 x = 3$

33. $\cos x - 2 = \sin x$ **34.** $\tan 3x + 1 = \sec 3x$

35. $\cot x - \csc x = 1$

36. $\sin x + 2\cot x = 2\csc x$

37. $\tan^3 x - \tan^2 x - 3\tan x + 3 = 0$

38. $x\sin x + 1 = x + \sin x$ **39.** $\sin 2x = 2\tan 2x$

40. $3\sec^2 x + 4\cos^2 x = 7$ **41.** $\csc(x + 1) = \sin(x + 1)$

42. $2\sin\left(\dfrac{\pi}{2} + x\right) = 1$

43. $\sqrt{3} + \cos(t + \pi) = \sin\left(t + \dfrac{\pi}{2}\right)$

44. $\sin(\cos x) = 0$ **45.** $\ln(2 - \sin^2 x) = 0$

46. $\ln|1 - \tan^2 x| = 2\ln|1 + \tan x|$

47. $\log_2 2\cos x = 0$ **48.** $\log_3 2\sin x = \dfrac{1}{2}$

In Exercises 49–52 sketch the graphs of f and g on the same axes and find their points of intersection.

49. $f(x) = 3\cos x + 1; g(x) = \cos x - 1$

50. $f(x) = \sin 2x; g(x) = 2\sin 2x + 1$

51. $f(x) = \tan x; g(x) = \sqrt{3}$

52. $f(x) = \sin x - 1; g(x) = \cos x$

In Exercises 53 and 54 find the points of intersection of the graphs of the given pairs of functions.

53. $f(x) = \tan x \sin x; g(x) = 2 - \cos x$

54. $f(x) = \ln(e + \sin^2 x); g(x) = 1$

55. Find $\dfrac{1}{\tan x} + \dfrac{1}{\cot x}$ if $\tan x + \cot x = \dfrac{9}{2}$.

56. Find $\dfrac{1}{\sin x} - \dfrac{1}{\csc x}$ if $\sin x - \csc x = 3$.

Section 6.3

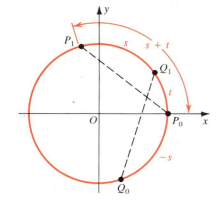

Figure 1

Addition and Subtraction Formulas

In this section we derive identities for the trigonometric functions of sums and differences. The two main identities are

(1) $$\sin(x + y) = \sin x \cos y + \cos x \sin y$$

(2) $$\cos(x + y) = \cos x \cos y - \sin x \sin y$$

These will be of primary importance in deriving the other identities of this section.

To prove Identity 2, let s and t be two real numbers. On the unit circle we mark the distances $t, s + t$, and $-s$ starting at the point $(1, 0)$ and label the points P_0, P_1, Q_0, and Q_1 as shown in Figure 1. The coordinates of these points are

$$P_0(1, 0) \qquad\qquad Q_0(\cos(-s), \sin(-s))$$
$$P_1(\cos(s + t), \sin(s + t)) \qquad Q_1(\cos t, \sin t)$$

Since $\cos(-s) = \cos(s)$ and $\sin(-s) = -\sin s$, it follows that the point Q_0 has the coordinates $Q_0(\cos s, -\sin s)$. Notice that the distances between P_0 and P_1 and between Q_0 and Q_1 measured along the arc of the circle are equal. Since equal arcs subtend equal chords, it follows that $P_0 P_1 = Q_0 Q_1$. Using the Distance Formula from Section 1.7, we get

$$\sqrt{[\cos(s + t) - 1]^2 + [\sin(s + t) - 0]^2}$$
$$= \sqrt{[\cos t - \cos s]^2 + [\sin t + \sin s]^2}$$

Squaring both sides and expanding, we have

$$\cos^2(s + t) - 2\cos(s + t) + 1 + \sin^2(s + t)$$
$$= \cos^2 t - 2\cos s \cos t + \cos^2 s + \sin^2 t + 2\sin s \sin t + \sin^2 s$$

Using the Pythagorean identity $\sin^2 z + \cos^2 z = 1$ three times gives

$$2 - 2\cos(s + t) = 2 - 2\cos s \cos t + 2\sin s \sin t$$

Finally, subtracting 2 from each side and dividing both sides by -2, we get

$$\cos(s + t) = \cos s \cos t - \sin s \sin t$$

which is Equation 2.

Example 1

Find the exact value of $\cos 75°$.

Solution

Notice that $75° = 45° + 30°$. Since we know the exact values of sin and cos at $45°$ and $30°$, we use Equation 2 to get

$$\cos 75° = \cos(45° + 30°)$$
$$= \cos 45° \cos 30° - \sin 45° \sin 30°$$
$$= \frac{\sqrt{2}}{2} \frac{\sqrt{3}}{2} - \frac{\sqrt{2}}{2} \frac{1}{2} = \frac{\sqrt{2}\sqrt{3} - \sqrt{2}}{4} = \frac{\sqrt{6} - \sqrt{2}}{4} \qquad \bullet$$

A formula for $\cos(s - t)$ is easily derived from Equation 2. We have

$$\cos(s - t) = \cos(s + (-t))$$
$$= \cos s \cos(-t) - \sin s \sin(-t)$$
$$= \cos s \cos t + \sin s \sin t$$

We summarize these formulas.

Addition and Subtraction Formulas for Cosine

$$\cos(s + t) = \cos s \cos t - \sin s \sin t$$
$$\cos(s - t) = \cos s \cos t + \sin s \sin t$$

Equation 1, the addition formula for sine, can be derived in a way similar to that used to derive Equation 2, but the following derivation is easier:

$$\sin(s + t) = \cos\left(\frac{\pi}{2} - (s + t)\right) = \cos\left(\left(\frac{\pi}{2} - s\right) - t\right)$$
$$= \cos\left(\frac{\pi}{2} - s\right)\cos t + \sin\left(\frac{\pi}{2} - s\right)\sin t$$
$$= \sin s \cos t + \cos s \sin t$$

So $\sin(s + t) = \sin s \cos t + \cos s \sin t$. This is the addition formula for sine. The subtraction formula for sine can be obtained from this formula in the same way that the subtraction formula for cosine was obtained from the addition formula for cosine. We summarize these formulas.

Addition and Subtraction Formulas for Sine

$$\sin(s + t) = \sin s \cos t + \cos s \sin t$$
$$\sin(s - t) = \sin s \cos t - \cos s \sin t$$

Example 2

Find the exact values of (a) $\sin 105°$ and (b) $\cos \dfrac{\pi}{12}$.

Solution

(a) Since $105° = 60° + 45°$, the addition formula for sine gives

$$\sin 105° = \sin(60° + 45°)$$
$$= \sin 60° \cos 45° + \cos 60° \sin 45°$$
$$= \frac{\sqrt{3}}{2} \frac{\sqrt{2}}{2} + \frac{1}{2} \frac{\sqrt{2}}{2}$$
$$= \frac{\sqrt{6} + \sqrt{2}}{4}$$

(b) Since $\dfrac{\pi}{12} = \dfrac{\pi}{4} - \dfrac{\pi}{6}$, the subtraction formula for cosine gives

$$\cos \frac{\pi}{12} = \cos\left(\frac{\pi}{4} - \frac{\pi}{6}\right)$$
$$= \cos \frac{\pi}{4} \cos \frac{\pi}{6} + \sin \frac{\pi}{4} \sin \frac{\pi}{6}$$
$$= \frac{\sqrt{2}}{2} \frac{\sqrt{3}}{2} + \frac{\sqrt{2}}{2} \frac{1}{2} = \frac{\sqrt{6} + \sqrt{2}}{4}$$

Example 3

If $\sin x = \frac{3}{5}$, x in quadrant I, and $\cos y = -\frac{2}{3}$, y in quadrant II, find $\sin(x + y)$.

Solution

To apply the addition formula for sine we need to find $\cos x$ and $\sin y$. A simple way to do this is to use the definitions of the trigonometric functions and sketch a graph to represent the angles x and y as shown in Figure 2. Thus $\cos x = 4/5$ and $\sin y = \sqrt{5}/3$. The addition formula for sine now gives

$$\sin(x + y) = \sin x \cos y + \cos x \sin y$$
$$= \frac{3}{5}\left(-\frac{2}{3}\right) + \frac{4}{5} \frac{\sqrt{5}}{3} = \frac{-6 + 4\sqrt{5}}{15} = \frac{2(2\sqrt{5} - 3)}{15}$$

Figure 2

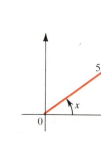
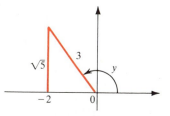

We now obtain a formula for $\tan(s + t)$ in terms of $\tan s$ and $\tan t$. We use the addition formulas for the sine and cosine functions to get

$$\tan(s + t) = \frac{\sin(s + t)}{\cos(s + t)}$$

$$= \frac{\sin s \cos t + \cos s \sin t}{\cos s \cos t - \sin s \sin t}$$

$$= \frac{\dfrac{\sin s \cos t}{\cos s \cos t} + \dfrac{\cos s \sin t}{\cos s \cos t}}{\dfrac{\cos s \cos t}{\cos s \cos t} - \dfrac{\sin s \sin t}{\cos s \cos t}}$$

$$= \frac{\tan s + \tan t}{1 - \tan s \tan t}$$

Of course a formula for $\tan(s - t)$ can easily be derived from this.

Addition and Subtraction Formulas for Tangent

$$\tan(s + t) = \frac{\tan s + \tan t}{1 - \tan s \tan t}$$

$$\tan(s - t) = \frac{\tan s - \tan t}{1 + \tan s \tan t}$$

Example 4

Verify the identity

$$\frac{1 + \tan x}{1 - \tan x} = \tan\left(\frac{\pi}{4} + x\right)$$

Solution

Starting with the right-hand side and using the addition formula for tangent, we get

$$\tan\left(\frac{\pi}{4} + x\right) = \frac{\tan\dfrac{\pi}{4} + \tan x}{1 - \tan\dfrac{\pi}{4}\tan x}$$

$$= \frac{1 + \tan x}{1 - \tan x} \qquad\bullet$$

The addition formulas for sine and cosine can be used to write expressions of the form

(3) $a\cos x + b\sin x$

as a single trigonometric function. For example, consider the expression

(4) $\dfrac{1}{2}\cos x + \dfrac{\sqrt{3}}{2}\sin x$

If we set $\phi = \pi/3$, then $\cos \phi = 1/2$ and $\sin \phi = \sqrt{3}/2$ and we can write

$$\frac{1}{2}\cos x + \frac{\sqrt{3}}{2}\sin x = \cos \phi \cos x + \sin \phi \sin x = \cos(x - \phi) = \cos\left(x - \frac{\pi}{3}\right)$$

The reason we are able to write this expression in terms of cosine only is that the coefficients in (4) are precisely the cosine and sine of a particular number: in this case, $\pi/3$. It is possible to do this in general for expressions of the form of (3). To see this, multiply the numerator and denominator in (3) by $\sqrt{a^2 + b^2}$ to get

(5) $a \cos x + b \sin x = \sqrt{a^2 + b^2}\left(\dfrac{a}{\sqrt{a^2 + b^2}}\cos x + \dfrac{b}{\sqrt{a^2 + b^2}}\sin x\right)$

We need a number ϕ with the property that

(6) $\cos \phi = \dfrac{a}{\sqrt{a^2 + b^2}}$ and $\sin \phi = \dfrac{b}{\sqrt{a^2 + b^2}}$

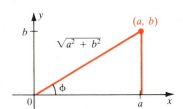

Figure 3

Figure 3 shows that the point (a, b) in the plane determines a number ϕ with precisely this property. Note also that

$$\tan \phi = \frac{b}{a}$$

(We can use this fact to find ϕ.) Equation 5 now becomes

$$a \cos x + b \sin x = \sqrt{a^2 + b^2}(\cos \phi \cos x + \sin \phi \sin x)$$

or

(7)
$$\boxed{a \cos x + b \sin x = \sqrt{a^2 + b^2}\cos(x - \phi)}$$

Example 5

Solve the equation $-\cos 2t + \sqrt{3}\sin 2t = 1$ in the interval $[0, \pi)$.

Solution

We first rewrite the left side of the equation in the form of Equation 7. Since $a = -1$ and $b = \sqrt{3}$, we have $\sqrt{a^2 + b^2} = 2$ and so

$$-\cos 2t + \sqrt{3}\sin 2t = 2\left(-\frac{1}{2}\cos 2t + \frac{\sqrt{3}}{2}\sin 2t\right)$$

Thus $\cos \phi = -1/2$ and $\sin \phi = \sqrt{3}/2$. It follows that ϕ is in quadrant II with $\tan \phi = -\sqrt{3}$, so $\phi = 2\pi/3$. This gives

$$-\cos 2t + \sqrt{3}\sin 2t = 2\cos\left(2t - \frac{2\pi}{3}\right)$$

Our equation becomes

$$2\cos\left(2t - \frac{2\pi}{3}\right) = 1 \quad \text{or} \quad \cos\left(2t - \frac{2\pi}{3}\right) = \frac{1}{2}$$

In the interval $[0, 2\pi)$ we have

$$2t - \frac{2\pi}{3} = \frac{\pi}{3} + 2k\pi \qquad \text{or} \qquad 2t - \frac{2\pi}{3} = \frac{5\pi}{3} + 2k\pi$$

Thus the solutions in the interval $[0, \pi)$ are $t = \pi/6,\ \pi/2,\ 7\pi/6$.

Exercises 6.3

In Exercises 1–6 find the exact value of the given expression.

1. $\sin 15°$ **2.** $\cos 135°$ **3.** $\tan 105°$

4. $\cot \dfrac{\pi}{12}$ **5.** $\sin\left(-\dfrac{11\pi}{12}\right)$

6. $\dfrac{\sqrt{2}}{2} \cos \dfrac{\pi}{12} - \dfrac{\sqrt{2}}{2} \sin \dfrac{\pi}{12}$

In Exercises 7–10 write each expression as a trigonometric function of one number and find its exact value.

7. $\sin 18° \cos 27° + \cos 18° \sin 27°$

8. $\cos \dfrac{3\pi}{7} \cos \dfrac{2\pi}{21} + \sin \dfrac{3\pi}{7} \sin \dfrac{2\pi}{21}$

9. $\dfrac{\tan 73° - \tan 13°}{1 + \tan 73° \tan 13°}$

10. $\cos \dfrac{13\pi}{15} \cos\left(-\dfrac{\pi}{5}\right) - \sin \dfrac{13\pi}{15} \sin\left(-\dfrac{\pi}{5}\right)$

11. If α and β are two angles such that α is in quadrant I and β is in quadrant II and

$$\cos \alpha = \frac{5}{13} \qquad \text{and} \qquad \cos \beta = -\frac{3}{5}$$

find $\sin(\alpha - \beta)$ and $\sin(\alpha + \beta)$.

12. If α and β are two angles such that α is in quadrant II and

$$\sin \alpha = \frac{3}{5} \qquad \text{and} \qquad \tan \beta = \frac{4}{3}$$

find $\tan(\alpha + \beta)$ and $\tan(\alpha - \beta)$.

13. If α and β are two angles such that α and β are in quadrants III and II, respectively, and

$$\cos \alpha = -\frac{3}{5} \qquad \text{and} \qquad \csc \beta = \frac{5}{4}$$

find $\sin(\alpha + \beta)$ and $\cos(\alpha + \beta)$.

14. If α and β are two acute angles such that

$$\cos \beta = \frac{24}{25} \qquad \text{and} \qquad \sin(\alpha + \beta) = \frac{4}{5}$$

find $\sin \alpha$ and $\cos \alpha$.

In Exercises 15–26 prove the given identity.

15. $\sin\left(x - \dfrac{\pi}{2}\right) = -\cos x$ **16.** $\cos\left(x - \dfrac{\pi}{2}\right) = \sin x$

17. $\tan\left(x - \dfrac{\pi}{2}\right) = -\cot x$ **18.** $\sin(x - \pi) = -\sin x$

19. $\cos(x - \pi) = -\cos x$ **20.** $\tan(x - \pi) = \tan x$

21. $\sec\left(x - \dfrac{\pi}{2}\right) = \csc x$ **22.** $\csc(x - \pi) = -\csc x$

23. $\csc\left(x - \dfrac{\pi}{2}\right) = -\sec x$

24. $\sin\left(\dfrac{\pi}{2} - x\right) = \sin\left(\dfrac{\pi}{2} + x\right)$

25. $\cos\left(x - \dfrac{\pi}{2}\right) + \cos\left(x + \dfrac{\pi}{2}\right) = 0$

26. $\cos\left(x + \dfrac{\pi}{6}\right) + \sin\left(x - \dfrac{\pi}{3}\right) = 0$

In Exercises 27–45 verify the identity.

27. $\cos\left(x - \dfrac{\pi}{3}\right) = \dfrac{1}{2}(\cos x + \sqrt{3} \sin x)$

28. $\cos\left(x + \dfrac{\pi}{4}\right) = -\sin\left(x - \dfrac{\pi}{4}\right)$

29. $\tan\left(x + \dfrac{\pi}{4}\right) = \dfrac{1 + \tan x}{1 - \tan x}$

30. $\tan\left(x - \dfrac{\pi}{4}\right) = \dfrac{\tan x - 1}{\tan x + 1}$

31. $\sin(x + y) + \sin(x - y) = 2\sin x \cos y$

32. $\sin(x + y) - \sin(x - y) = 2\cos x \sin y$

33. $\cos(x + y) + \cos(x - y) = 2\cos x \cos y$

34. $\cos(x + y) - \cos(x - y) = -2\sin x \sin y$

35. $\cot(x - y) = \dfrac{\cot x \cot y + 1}{\cot y - \cot x}$

36. $\cot(x + y) = \dfrac{\cot x \cot y - 1}{\cot x + \cot y}$

37. $\tan x - \tan y = \dfrac{\sin(x - y)}{\cos x \cos y}$

38. $1 - \tan x \tan y = \dfrac{\cos(x + y)}{\cos x \cos y}$

39. $\dfrac{\sin(x - y)}{\cos(x + y)} = \dfrac{\tan x - \tan y}{1 - \tan x \tan y}$

40. $\dfrac{\sin(x + y) - \sin(x - y)}{\cos(x + y) + \cos(x - y)} = \tan y$

41. $\dfrac{\sin \pi x}{\sin x} - \dfrac{\cos \pi x}{\cos x} = \dfrac{\sin(\pi - 1)x}{\sin x \cos x}$

42. $\cos(x + y)\cos(x - y) = \cos^2 x - \sin^2 y$

43. $\sin(x - y)\cos y + \cos(x - y)\sin y = \sin x$

44. $\cos(x + y)\cos y + \sin(x + y)\sin y = \cos x$

45. $\cot x - \tan y = \dfrac{\cos(x + y)}{\sin x \cos y}$

In Exercises 46–49 solve the given trigonometric equation in the interval $[0, 2\pi)$.

46. $\cos x \cos 3x = \sin x \sin 3x$

47. $\cos x \cos 2x + \sin x \sin 2x = \dfrac{1}{2}$

48. $\sin 2x \cos x + \cos 2x \sin x = \dfrac{\sqrt{3}}{2}$

49. $\sin 3x \cos x = \cos 3x \sin x$

In Exercises 50–52 prove the given identity.

50. $\sin(x + y + z) = \sin x \cos y \cos z + \cos x \sin y \cos z$
$ + \cos x \cos y \sin z - \sin x \sin y \sin z$

51. $\tan(x - y) + \tan(y - z) + \tan(z - x) =$
$\tan(x - y)\tan(y - z)\tan(z - x)$

52. $\cot(y - z)\cot(z - x) + \cot(x - y)\cot(z - x) +$
$\cot(x - y)\cot(y - z) = 1$

53. Show that if $\beta - \alpha = \pi/2$, then

$$\sin(x + \alpha) + \cos(x + \beta) = 0$$

54. Prove that for any positive integer n,
 (a) $\cos(x - n\pi) = (-1)^n \cos x$
 (b) $\sin(x - n\pi) = (-1)^n \sin x$

55. Prove that for any positive integer n,
 (a) $\cos\left(x - \dfrac{2n + 1}{2}\pi\right) = (-1)^n \sin x$
 (b) $\sin\left(x + \dfrac{2n + 1}{2}\pi\right) = (-1)^n \cos x$

56. If $f(x) = \sin x$ and $g(x) = \cos x$, show that
 (a) $\dfrac{f(x + h) - f(x)}{h} = \left(\dfrac{\sin h}{h}\right)\cos x - \sin x\left(\dfrac{1 - \cos h}{h}\right)$
 (b) $\dfrac{g(x + h) - g(x)}{h} =$
$-\cos x\left(\dfrac{1 - \cos h}{h}\right) - \sin x\left(\dfrac{\sin h}{h}\right)$

In Exercises 57–60 write the given expression as a function of cosine only.

57. $\cos 2x - \sqrt{3}\sin 2x$

58. $\sin 3x + \cos 3x$

59. $5(\sin 7x - \cos 7x)$

60. $3\sqrt{3}\cos \pi x + 3\sin \pi x$

In Exercises 61 and 62 graph the given function. (Hint: First express the function in terms of cosine only.)

61. $f(x) = \sin x + \cos x$

62. $g(x) = \cos 2x + \sqrt{3}\sin 2x$

In Exercises 63–66 solve the equation in the interval $[0, 2\pi)$ *by the method of Example 5.*

63. $\sin x - \cos x = \sqrt{2}$ **64.** $\sin 3x + \cos 3x = 1$

65. $\sin 2x - \sqrt{3}\cos 2x = 2$

66. $3\sqrt{3}\cos \pi x + 3\sin \pi x = 0$

67. Refer to the figure. Show that $\alpha + \beta = \gamma$ and find $\tan \gamma$.

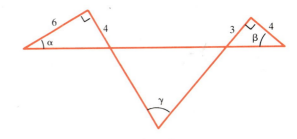

68. **(a)** If L is a line in the plane and θ is the angle formed by the line and the x-axis as shown in the figure, show that the slope m of the line is given by

$$m = \tan \theta$$

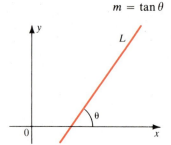

(b) Let L_1 and L_2 be two nonparallel lines in the plane with slopes m_1 and m_2, respectively. Let ψ be the acute angle formed by the two lines as shown in the figure. Show that

$$\tan \psi = \frac{m_2 - m_1}{1 + m_1 m_2}$$

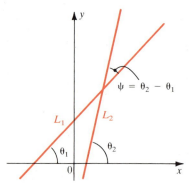

(c) Find the acute angle formed by the two lines $y - \frac{1}{3}x + 1$ and $y = -\frac{1}{2}x - 3$.

(d) Show that if two lines are perpendicular, then the slope of one is the negative reciprocal of the slope of the other. (*Hint:*   First find an expression for $\cot \psi$.)

Section 6.4

Double-Angle and Half-Angle Formulas

The identities we consider in this section are consequences of the addition formulas. The first set of identities we derive are called **double-angle formulas** because they allow us to find the value of the trigonometric functions at $2u$ from the value at u. The **half-angle formulas** relate the values of the trigonometric functions at $\frac{1}{2}u$ to their values at u.

Double-Angle Formulas

We begin with some of the most useful formulas.

Double Angle Formulas

(1) $$\sin 2x = 2 \sin x \cos x$$

(2) $$\cos 2x = \cos^2 x - \sin^2 x$$

(3) $$\cos 2x = 1 - 2 \sin^2 x$$

(4) $$\cos 2x = 2 \cos^2 x - 1$$

(5) $$\tan 2x = \frac{2 \tan x}{1 - \tan^2 x}$$

Formula 1 is derived by letting $x = y$ in the addition formula for sine. We get

$$\sin 2x = \sin(x + x)$$
$$= \sin x \cos x + \cos x \sin x$$
$$= 2 \sin x \cos x$$

Formulas 2, 3, and 4 are double-angle formulas for the cosine functions. The first of these can be derived in a way similar to that used to derive Formula 1. Setting $x = y$ in the addition formula for cosine, we have

$$\begin{aligned} \cos 2x &= \cos(x + x) \\ &= \cos x \cos x - \sin x \sin x \\ &= \cos^2 x - \sin^2 x \end{aligned}$$

Formulas 3 and 4 are obtained from Formula 2 together with the identity $\sin^2 x + \cos^2 x = 1$. To obtain Formula 3 we substitute $\cos^2 x = 1 - \sin^2 x$ in (2) to get

$$\begin{aligned} \cos 2x &= \cos^2 x - \sin^2 x \\ &= (1 - \sin^2 x) - \sin^2 x \\ &= 1 - 2\sin^2 x \end{aligned}$$

Similarly, we get Formula 4 by substituting $\sin^2 x = 1 - \cos^2 x$ in (2).

The formula for $\tan 2x$ follows from the addition formula for tangent. Its derivation is left as an exercise.

Example 1

If $\cos x = -2/3$ and $\pi/2 \le x \le \pi$, find $\cos 2x$ and $\sin 2x$.

Solution

Using the double-angle formula for cosine, we get

$$\begin{aligned} \cos 2x &= 2\cos^2 x - 1 \\ &= 2\left(-\frac{2}{3}\right)^2 - 1 = \frac{8}{9} - 1 = -\frac{1}{9} \end{aligned}$$

To use the formula $\sin 2x = 2\sin x \cos x$ we need to find $\sin x$ first. We have

$$\sin x = \sqrt{1 - \cos^2 x} = \sqrt{1 - \left(\frac{2}{3}\right)^2} = \frac{\sqrt{5}}{3}$$

where we have used the positive square root because $\sin x$ is positive in quadrant II. From Formula 1 we have

$$\begin{aligned} \sin 2x &= 2\sin x \cos x \\ &= 2\left(\frac{\sqrt{5}}{3}\right)\left(-\frac{2}{3}\right) = -\frac{4\sqrt{5}}{9} \end{aligned}$$

Example 2

Find an identity for $\cos 3x$ in terms of $\cos x$.

Solution

Supply the reasons for the following steps:

$$\cos 3x = \cos(2x + x)$$
$$= \cos 2x \cos x - \sin 2x \sin x$$
$$= (2\cos^2 x - 1)\cos x - (2\sin x \cos x)\sin x$$
$$= 2\cos^3 x - \cos x - 2\sin^2 x \cos x$$
$$= 2\cos^3 x - \cos x - 2\cos x(1 - \cos^2 x)$$
$$= 2\cos^3 x - \cos x - 2\cos x + 2\cos^3 x$$
$$= 4\cos^3 x - 3\cos x$$

    Example 2 shows that $\cos 3x$ can be written as a polynomial of degree 3 in $\cos x$. The identity $\cos 2x = 2\cos^2 x - 1$ shows that $\cos 2x$ is a polynomial of degree 2 in $\cos x$. In fact, for any natural number n, $\cos nx$ can be written as a polynomial in $\cos x$ of degree n (see Exercise 83). The analogous result for $\sin nx$ is not true in general. It can easily be seen that $\sin 2x$ is not a polynomial of degree 2 in $\sin x$.

Example 3

Prove the identity

$$\frac{\sin 3x}{\sin x \cos x} = 4\cos x - \sec x$$

Solution

Supply reasons for the following steps:

$$\frac{\sin 3x}{\sin x \cos x} = \frac{\sin(x + 2x)}{\sin x \cos x}$$

$$= \frac{\sin x \cos 2x + \cos x \sin 2x}{\sin x \cos x}$$

$$= \frac{\sin x(2\cos^2 x - 1) + \cos x(2\sin x \cos x)}{\sin x \cos x}$$

$$= \frac{\sin x(2\cos^2 x - 1)}{\sin x \cos x} + \frac{\cos x(2\sin x \cos x)}{\sin x \cos x}$$

$$= \frac{2\cos^2 x - 1}{\cos x} + 2\cos x$$

$$= 2\cos x - \frac{1}{\cos x} + 2\cos x$$

$$= 4\cos x - \sec x$$

Half-Angle Formulas

Solving Formula 3 for $\sin^2 x$, we get

$$\sin^2 x = \frac{1 - \cos 2x}{2}$$

Similarly, solving Formula 4 for $\cos^2 x$ gives

$$\cos^2 x = \frac{1 + \cos 2x}{2}$$

From these two identities and the fact that $\tan^2 x = \sin^2 x / \cos^2 x$ we get

$$\tan^2 x = \frac{1 - \cos 2x}{1 + \cos 2x}$$

We summarize these important identities:

(6)
$$\sin^2 x = \frac{1 - \cos 2x}{2}$$

(7)
$$\cos^2 x = \frac{1 + \cos 2x}{2}$$

(8)
$$\tan^2 x = \frac{1 - \cos 2x}{1 + \cos 2x}$$

Formulas 6–8 are useful because they allow us to write any trigonometric expression that involves even powers of sine and cosine in terms of the first power of cosine only. This is important in calculus.

Example 4

Express $\sin^2 x \cos^2 x$ in terms of the first power of cosine.

Solution

Using Formulas 6 and 7 repeatedly gives

$$\sin^2 x \cos^2 x = \left(\frac{1 - \cos 2x}{2}\right)\left(\frac{1 + \cos 2x}{2}\right)$$

$$= \left(\frac{1 - \cos^2 2x}{4}\right) = \frac{1}{4} - \frac{1}{4}\cos^2 2x$$

$$= \frac{1}{4} - \frac{1}{4}\left(\frac{1 + \cos 4x}{2}\right) = \frac{1}{4} - \frac{1}{8} - \frac{\cos 4x}{8}$$

$$= \tfrac{1}{8} - \tfrac{1}{8}\cos 4x = \tfrac{1}{8}(1 - \cos 4x)$$

Example 5

Find the exact value of $\cos 15°$.

Solution

Using Formula 7 with $x = 15°$, we have

$$\cos^2 15° = \frac{1 + \cos 2(15)}{2} = \frac{1 + \cos 30°}{2}$$

$$= \frac{1 + \frac{\sqrt{3}}{2}}{2} = \frac{2 + \sqrt{3}}{4}$$

Thus
$$\cos 15° = \sqrt{\frac{2 + \sqrt{3}}{4}} = \frac{1}{2}\sqrt{2 + \sqrt{3}}$$

where we have chosen the positive square root because $15°$ is in quadrant I.

Notice that in Example 4 we could find $\cos 15°$ exactly because the right-hand side of Formula 7 required us to find $\cos 2(15°) = \cos 30°$, a quantity whose exact value we know. Thus Formulas 6–8 allow us to find the values of the trigonometric functions of half an angle if we know the values of the trigonometric functions at that angle. To see this more clearly let us substitute $x = u/2$ in Formulas 6–8 and take the square root of both sides of each of these equations to get the following half-angle identities:

Half-Angle Formulas

(9) $\sin\dfrac{u}{2} = \pm\sqrt{\dfrac{1 - \cos u}{2}}$

(10) $\cos\dfrac{u}{2} = \pm\sqrt{\dfrac{1 + \cos u}{2}}$

(11) $\tan\dfrac{u}{2} = \pm\sqrt{\dfrac{1 - \cos u}{1 + \cos u}}$

In these equations the choice of the $+$ or $-$ sign depends on the quadrant in which $u/2$ lies.

Example 6

Find the exact value of $\sin 7.5°$.

Solution

In Example 5 we found the exact value of $\cos 15°$. Since $7.5°$ is half of $15°$, we can use Formula 9 with $u = 15°$ to get

$$\sin 7.5° = \sin\left(\frac{15}{2}\right)° = \sqrt{\frac{1 - \cos 15°}{2}}$$

$$= \sqrt{\frac{1 - \frac{1}{2}\sqrt{2 + \sqrt{3}}}{2}}$$

$$= \sqrt{\frac{2 - \sqrt{2 + \sqrt{3}}}{4}}$$

$$= \frac{1}{2}\sqrt{2 - \sqrt{2 + \sqrt{3}}}$$

We chose the positive sign in Formula 9 because 7.5° is in quadrant I.   ●

Of course it is possible to approximate the square roots in the last expression to find a numerical value for sin 7.5° as close as desired to the exact value. (Try this on your calculator.) By using the half-angle formulas again with $u = 7.5°$ we can find the values of the trigonometric functions at $u = 3.75°$. When this process is repeated it is possible to find approximate values of the trigonometric functions for arbitrarily small angles. Combining this knowledge with the addition formulas, we can find the values of the trigonometric functions at angles that are close to any angle for which the trigonometric functions have already been computed. If we continue in this manner it is possible to construct tables of the trigonometric functions that are as accurate as we wish. In practice, this is not the way in which trigonometric tables are made, but it is nice to know that we can construct our own tables with what we know about the trigonometric functions so far.

Half-angle formulas for the tangent function can be derived that do not involve radicals. To see this, suppose that $u/2$ is in quadrant I. Multiplying the numerator and denominator of the fraction in Formula 11 by $1 - \cos u$, we get

$$\tan\frac{u}{2} = \sqrt{\frac{1 - \cos u}{1 + \cos u}\frac{1 - \cos u}{1 - \cos u}}$$

$$= \sqrt{\frac{(1 - \cos u)^2}{1 - \cos^2 u}}$$

$$= \frac{|1 - \cos u|}{|\sin u|}$$

Since $|\cos u| \leq 1$, $1 - \cos u$ is nonnegative for all values of u. It is also true that $\sin u$ and $\tan(u/2)$ always have the same sign. (Verify this.) It follows that

(12)

$$\tan\frac{u}{2} = \frac{1 - \cos u}{\sin u}$$

for all values of u for which $\sin u \neq 0$. Another half-angle identity for tangent can be derived from this by multiplying the numerator and denominator of Formula 12 by $1 + \cos u$.

Half-Angle Formulas for Tangent

(13)
$$\tan\frac{u}{2} = \frac{1 - \cos u}{\sin u}$$

(14)
$$\tan\frac{u}{2} = \frac{\sin u}{1 + \cos u}$$

Example 7

Find $\tan(u/2)$ if $\sin u = 2/5$ and u is in quadrant II.

Solution

To use Formula 13 or 14 we first need to find $\cos u$. Since cosine is negative in quadrant II, we have

$$\cos u = -\sqrt{1 - \sin^2 u} = -\sqrt{1 - \left(\frac{2}{5}\right)^2} = -\frac{\sqrt{21}}{5}$$

Thus, by Formula 13,

$$\tan\frac{u}{2} = \frac{1 - \cos u}{\sin u} = \frac{1 + (\sqrt{21}/5)}{2/5} = \frac{5 + \sqrt{21}}{2}$$ ●

Example 8

Solve the equation $\cos 3x + \cos x = 0$ for $0 \leq x \leq 2\pi$.

Solution

First we use identities to simplify the equation. In Example 2 we found that $\cos 3x = 4\cos^3 x - 3\cos x$. Thus

$$\cos 3x + \cos x = 0$$
$$4\cos^3 x - 3\cos x + \cos x = 0$$
$$4\cos^3 x - 2\cos x = 0$$
$$2\cos^3 x - \cos x = 0$$
$$\cos x(2\cos^2 x - 1) = 0$$

Thus $\cos x = 0$ or $2\cos^2 x - 1 = 0$

From $\cos x = 0$ we get $x = \pi/2$ or $x = 3\pi/2$. The other equation gives

$$\cos^2 x = \frac{1}{2} \quad \text{or} \quad \cos x = \pm\frac{1}{\sqrt{2}} = \pm\frac{\sqrt{2}}{2}$$

So $x = \pi/4$, $3\pi/4$, $5\pi/4$, $7\pi/4$. The solutions to the equation in the interval

$0 \le x \le 2\pi$ are thus

$$x = \frac{\pi}{2}, \frac{3\pi}{2}, \frac{\pi}{4}, \frac{3\pi}{4}, \frac{5\pi}{4}, \frac{7\pi}{4}$$

Exercises 6.4

In Exercises 1–8 use a double-angle formula to write each expression in terms of a trigonometric expression that involves half the given angle.

1. $\sin 88°$ **2.** $\cos 48°$ **3.** $\tan 68°$

4. $\cos \frac{\pi}{7}$ **5.** $\tan \frac{\pi}{6}$ **6.** $\cos 4\theta$

7. $\sin 5\theta$ **8.** $\tan 8\theta$

In Exercises 9–16 use a double-angle formula to write each expression in terms of a single trigonometric function of twice the given angle.

9. $2\sin 18° \cos 18°$ **10.** $\cos^2 33° - \sin^2 33°$

11. $\dfrac{2\tan 7°}{1 - \tan^2 7°}$ **12.** $2\sin \frac{\pi}{12}\cos \frac{\pi}{12}$

13. $\cos^2 \frac{\pi}{5} - \sin^2 \frac{\pi}{5}$ **14.** $\dfrac{2\tan 7\theta}{1 - \tan^2 7\theta}$

15. $2\sin 3\theta \cos 3\theta$ **16.** $\cos^2 5\theta - \sin^2 5\theta$

In Exercises 17–24 use a half-angle formula to write each expression in terms of a trigonometric expression that involves twice the given angle.

17. $\tan 13°$ **18.** $\sin 28°$ **19.** $\cos 7°$

20. $\cos \frac{\pi}{12}$ **21.** $\tan \frac{\pi}{30}$ **22.** $\sin \frac{\theta}{4}$

23. $\cos \frac{\theta}{6}$ **24.** $\tan \frac{\theta}{8}$

In Exercises 25–32 use a half-angle formula to write each expression in terms of a single trigonometric function of half the given angle.

25. $\sqrt{\dfrac{1 - \cos 30°}{2}}$ **26.** $\sqrt{\dfrac{1 + \cos 20°}{2}}$

27. $\sqrt{\dfrac{1 - \cos \frac{\pi}{6}}{2}}$ **28.** $\sqrt{\dfrac{1 - \cos 16°}{1 + \cos 16°}}$

29. $\dfrac{\sin 8°}{1 + \cos 8°}$ **30.** $\dfrac{1 - \cos 4\theta}{\sin 4\theta}$

31. $\sqrt{\dfrac{1 - \cos 8\theta}{2}}$ **32.** $\sqrt{\dfrac{1 + \cos 20\theta}{2}}$

In Exercises 33–40 use an appropriate half-angle formula to find the exact value of the given expression.

33. $\sin 15°$ **34.** $\tan 15°$ **35.** $\sin 7.5°$

36. $\cos 22.5°$ **37.** $\cos 11.25°$ **38.** $\tan \frac{\pi}{8}$

39. $\sin \frac{\pi}{12}$ **40.** $\cos \frac{5\pi}{12}$

In Exercises 41–46 find $\sin 2x$, $\cos 2x$, and $\tan 2x$ from the given information.

41. $\sin x = \frac{5}{13}$ and x is in quadrant I

42. $\cos x = \frac{4}{5}$ and $\csc x < 0$

43. $\tan x = -\frac{4}{3}$ and x is in quadrant II

44. $\csc x = 4$ and $\tan x < 0$

45. $\sin x = -\frac{3}{5}$ and x is in quadrant III

46. $\cot x = \frac{2}{3}$ and $\sin x > 0$

In Exercises 47–52 find $\sin \frac{x}{2}$, $\cos \frac{x}{2}$, and $\tan \frac{x}{2}$ from the given information where $0 \le x \le 2\pi$.

47. $\sin x = \frac{3}{5}$ and x is in quadrant I

48. $\cos x = -\frac{4}{5}$ and x is in quadrant III

49. $\csc x = 3$ and x is in quadrant II

50. $\tan x = 1$ and x is in quadrant I

51. $\sec x = \frac{3}{2}$ and x is in quadrant IV

52. $\cot x = 5$ and $\csc x < 0$

In Exercises 53–68 prove that the given equation is an identity.

53. $\cos^2 5x - \sin^2 5x = \cos 10x$

54. $\sin 8x = 2 \sin 4x \cos 4x$ **55.** $2 \sin \dfrac{x}{2} \cos \dfrac{x}{2} = \sin x$

56. $\dfrac{1 - \cos 2x}{\sin 2x} = \tan x$

57. $(\sin x + \cos x)^2 = 1 + \sin 2x$

58. $\dfrac{2 \tan \dfrac{x}{2}}{1 + \tan^2 \dfrac{x}{2}} = \sin x$ **59.** $\dfrac{\sin 4x}{\sin x} = 4 \cos x \cos 2x$

60. $\dfrac{1 + \sin 2x}{\sin 2x} = 1 + \frac{1}{2} \sec x \csc x$

61. $\dfrac{2(\tan x - \cot x)}{\tan^2 x - \cot^2 x} = \sin 2x$

62. $\cot 2x = \dfrac{1 - \tan^2 x}{2 \tan x}$ **63.** $\tan x = \dfrac{\sin 2x}{1 + \cos 2x}$

64. $\tan 3x = \dfrac{3 \tan x - \tan^3 x}{1 - 3 \tan^2 x}$

65. $4(\sin^6 x + \cos^6 x) = 4 - 3 \sin^2 2x$

66. $(1 - \cos 4x)(2 + \tan^2 x + \cot^2 x) = 8$

67. $\dfrac{\sin 3x + \cos 3x}{\cos x - \sin x} = 1 + 4 \sin x \cos x$

68. $\cos^4 x - \sin^4 x = \cos 2x$

In Exercises 69–76 solve the given equation in the interval $[0, 2\pi)$.

69. $\sin 2x - \cos x = 0$ **70.** $\sin 2\pi x + \sin \pi x = 0$

71. $\cos 2x + \sin x = 0$ **72.** $\cos 3x - \cos x = 0$

73. $\tan \dfrac{x}{2} - \sin x = 0$

74. $\sin x \cos 2x + 2 \sin^2 x = 1$

75. $\cos 2x + \cos x = 2$

76. $\tan x + \cot x = 4 \sin 2x$

77. Let $z = \tan \dfrac{x}{2}$.

   (a) Show that $\sin x = \dfrac{2z}{1 + z^2}$, $\cos x = \dfrac{1 - z^2}{1 + z^2}$, and

   $\tan x = \dfrac{2z}{1 - z^2}$.

   (b) Deduce that if f is a rational function in $\sin x$, $\cos x$, and $\tan x$, then f can be written to depend on z only.

   (c) Solve the equation

   $$\sqrt{3} \sin x - 3 \cos x + \sqrt{3} \tan x = 3$$

78. Show that the function $f(x) = \dfrac{\sin 3x}{\sin x} - \dfrac{\cos 3x}{\cos x}$ is constant.

79. Use the double-angle identity $\sin 2x = 2 \sin x \cos x$ n times to show that

   $$\sin(2^n x) = 2^n \sin x \cos x \cos 2x \cos 4x \cdots \cos 2^{n-1} x$$

80. In triangle ABC shown in the figure the line segment s bisects angle C. Show that the length of s is given by

   $$s = \dfrac{2ab \cos x}{a + b}$$

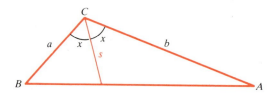

81. Prove the identity

   $$\tan^2 \left(\dfrac{x}{2} + \dfrac{\pi}{4} \right) = \dfrac{1 + \sin x}{1 - \sin x}$$

82. If $\sin x + \cos x = \alpha$, show the following:
   (a) $\sin 2x = \alpha^2 - 1$
   (b) $\sin^3 x + \cos^3 x = \frac{1}{2}(3\alpha - \alpha^3)$ [*Hint:* Find $(\sin x + \cos x)^3$ and use part (a).]

83. **(a)** Show that there is a polynomial $p(t)$ of degree 4 such that $\cos 4x = p(\cos x)$.
   (b) Show that there is a polynomial $p(t)$ of degree 5 such that $\cos 5x = p(\cos x)$.
   [*Note:* In general, there is a polynomial $p(t)$ of degree n such that $\cos nx = p(\cos x)$. These polynomials are named after the Russian mathematician Tchebycheff.]

84. Let $3x = \frac{1}{3}\pi$ and let $y = \cos x$. Use the result of Example 2 to show that y satisfies the equation

$$8y^3 - 6y - 1 = 0$$

(*Note:*   This equation has roots of a certain kind that are used in showing that the angle $\frac{1}{3}\pi$ cannot be trisected using ruler and compass only.)

85. Show that $y = 2\cos\left(\dfrac{2\pi}{7}\right)$ satisfies the equation $y^3 + y^2 - 2y - 1 = 0.$

86. The lower right-hand corner of a long piece of paper that is 6 in. wide is folded over to the right-hand edge as shown in the figure. The length L of the fold depends on the angle θ. Show that

$$L = \frac{3}{\sin\theta\cos^2\theta}$$

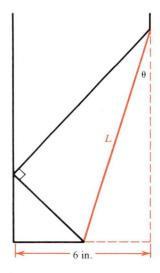

6 in.

Further Identities

It is possible to write the product $\sin u \cos v$ as a sum of trigonometric functions. To see this, consider the addition and subtraction formulas for the sine function:

$$\sin(u + v) = \sin u \cos v + \cos u \sin v$$
$$\sin(u - v) = \sin u \cos v - \cos u \sin v$$

Adding the left and right sides of these identities gives

$$\sin(u + v) + \sin(u - v) = 2\sin u \cos v$$

or $\sin u \cos v = \frac{1}{2}[\sin(u + v) + \sin(u - v)]$

The other three **product-to-sum identities** follow from the addition formulas in a similar way.

Product-to-Sum Identities

(1) $\sin u \cos v = \frac{1}{2}[\sin(u + v) + \sin(u - v)]$

(2) $\cos u \sin v = \frac{1}{2}[\sin(u + v) - \sin(u - v)]$

(3) $\cos u \cos v = \frac{1}{2}[\cos(u + v) + \cos(u - v)]$

(4) $\sin u \sin v = \frac{1}{2}[\cos(u - v) - \cos(u + v)]$

Example 1

Express $\sin 3x \sin 5x$ as a sum of trigonometric functions.

Solution

Using Formula 4 with $u = 3x$ and $v = 5x$ and the fact that cosine is an even function, we get

$$\sin 3x \sin 5x = \tfrac{1}{2}[\cos(3x - 5x) - \cos(3x + 5x)]$$
$$= \tfrac{1}{2}\cos(-2x) - \tfrac{1}{2}\cos 8x$$
$$= \tfrac{1}{2}\cos 2x - \tfrac{1}{2}\cos 8x \qquad\qquad \bullet$$

Example 2

Find the value of the product $\sin 37.5° \cos 7.5°$.

Solution

Using Formula 1 with $u = 37.5°$ and $v = 7.5°$ gives

$$\sin 37.5° \cos 7.5° = \tfrac{1}{2}[\sin(37.5° + 7.5°) + \sin(37.5° - 7.5°)]$$
$$= \tfrac{1}{2}[\sin 45° + \sin 30°]$$
$$= \frac{1}{2}\left(\frac{\sqrt{2}}{2} + \frac{1}{2}\right) = \frac{\sqrt{2} + 1}{4}$$

We could have also obtained this value by using Formula 2. $\bullet$

As we have mentioned, Formulas 1–4 show that certain products of trigonometric functions can be written as sums. These same identities allow us to write the sums of trigonometric functions as products. Indeed, the right side of each of these equations is a sum and the left side is a product. For example, in order to write

$$\sin x + \sin y$$

as a product we use Formula 1 with the appropriate choice for u and v. To find out what u and v should be, notice that we want

$$\begin{cases} x = u + v \\ y = u - v \end{cases}$$

Solving these two equations simultaneously for u and v gives

$$u = \frac{x + y}{2} \qquad \text{and} \qquad v = \frac{x - y}{2}$$

Substituting for $u, v, u + v$, and $u - v$ in Formula 1 gives

$$\sin\frac{x + y}{2}\cos\frac{x - y}{2} = \frac{1}{2}[\sin x + \sin y]$$

or
$$\sin x + \sin y = 2\sin\frac{x + y}{2}\cos\frac{x - y}{2}$$

The remaining three of the following **sum-to-product identities** are obtained in a similar manner.

Sum-to-Product Identities

(5) $$\sin x + \sin y = 2 \sin \frac{x+y}{2} \cos \frac{x-y}{2}$$

(6) $$\sin x - \sin y = 2 \cos \frac{x+y}{2} \sin \frac{x-y}{2}$$

(7) $$\cos x + \cos y = 2 \cos \frac{x+y}{2} \cos \frac{x-y}{2}$$

(8) $$\cos x - \cos y = -2 \sin \frac{x+y}{2} \sin \frac{x-y}{2}$$

Example 3

Write $\sin 7x + \sin 3x$ as a product.

Solution

Formula 5 gives

$$\sin 7x + \sin 3x = 2 \sin \frac{7x+3x}{2} \cos \frac{7x-3x}{2}$$

$$= 2 \sin 5x \cos 2x$$

Example 4

Verify the identity

$$\frac{\sin 3x - \sin x}{\cos 3x + \cos x} = \tan x$$

Solution

We apply Formula 6 to the numerator and Formula 7 to the denominator.

$$\frac{\sin 3x - \sin x}{\cos 3x + \cos x} = \frac{2 \cos \dfrac{3x+x}{2} \sin \dfrac{3x-x}{2}}{2 \cos \dfrac{3x+x}{2} \cos \dfrac{3x-x}{2}}$$

$$= \frac{2 \cos 2x \sin x}{2 \cos 2x \cos x}$$

$$= \frac{\sin x}{\cos x} = \tan x$$

Example 5

Solve the equation $\sin 3x - \sin x - \cos 2x = 0$.

Solution

By Formula 6 we have

$$\sin 3x - \sin x = 2 \cos \frac{3x + x}{2} \sin \frac{3x - x}{2}$$

$$= 2 \cos 2x \sin x$$

So the equation is equivalent to

$$2 \cos 2x \sin x - \cos 2x = 0$$

$$\cos 2x(2 \sin x - 1) = 0$$

Thus $\cos 2x = 0$ or $2 \sin x - 1 = 0$

The equation $\cos 2x = 0$ has the solutions

$$2x = \frac{\pi}{2} + k\pi = \frac{(2k + 1)\pi}{2} \qquad \text{or} \qquad x = \frac{(2k + 1)\pi}{4}$$

where k is any integer. The solutions of the equation $\sin x = \frac{1}{2}$ in the interval $[0, 2\pi)$ are $x = \pi/6, 5\pi/6$, and so all its solutions are given by

$$x = \frac{\pi}{6} + 2k\pi \qquad \text{and} \qquad x = \frac{5\pi}{6} + 2k\pi$$

where k is any integer. Thus the solutions of the given equation are

$$x = \frac{(2k + 1)\pi}{4} \qquad x = \frac{\pi}{6} + 2k\pi \qquad x = \frac{5\pi}{6} + 2k\pi$$

where k is any integer.

Exercises 6.5

In Exercises 1–4 write the given product as a sum.

1. $\sin 2x \cos 3x$

2. $\sin x \sin 5x$

3. $3 \cos 4x \cos 7x$

4. $11 \sin \frac{x}{2} \cos \frac{x}{4}$

In Exercises 5–8 find the value of the product.

5. $2 \sin 52.5° \sin 97.5°$

6. $3 \cos 37.5° \cos 7.5°$

7. $\tan 52.5° \tan 97.5°$

8. $\cos 37.5° \sin 7.5°$

In Exercises 9–16 write the given sum as a product.

9. $\sin 5x + \sin 3x$

10. $\sin x - \sin 4x$

11. $\cos 4x - \cos 6x$

12. $\cos 9x + \cos 2x$

13. $\sin 2x - \sin 7x$

14. $\sin 3x + \sin 4x$

15. $\cos 11\pi x + \cos 9\pi x$

16. $\cos \frac{x}{2} - \cos \frac{5x}{2}$

In Exercises 17–20 find the value of the sum.

17. $\sin 75° + \sin 15°$

18. $\sin 105° - \sin 15°$

19. $\cos 255° - \cos 195°$

20. $\cos \frac{\pi}{12} + \cos \frac{5\pi}{12}$

In Exercises 21–28 verify the given identity.

21. $\dfrac{\sin x + \sin 5x}{\cos x + \cos 5x} = \tan 3x$

22. $\dfrac{\sin 3x + \sin 7x}{\cos 3x - \cos 7x} = \cot 2x$

23. $\dfrac{\sin 10x}{\sin 9x + \sin x} = \dfrac{\cos 5x}{\cos 4x}$

24. $\dfrac{\sin x + \sin 3x + \sin 5x}{\cos x + \cos 3x + \cos 5x} = \tan 3x$

25. $\dfrac{\sin x + \sin y}{\cos x + \cos y} = \tan\left(\dfrac{x + y}{2}\right)$

26. $\dfrac{\cos x - \cos y}{\sin x + \sin y} = -\tan\left(\dfrac{x - y}{2}\right)$

27. $\tan y = \dfrac{\sin(x + y) - \sin(x - y)}{\cos(x + y) + \cos(x - y)}$

28. $\cot x = \dfrac{\sin(x + y) - \sin(x - y)}{\cos(x - y) - \cos(x + y)}$

In Exercises 29–36 solve the given equation.

29. $\sin x + \sin 3x = 0$ **30.** $\cos 5x - \cos 7x = 0$

31. $\cos 4x + \cos 2x = \cos x$

32. $\sin x + \sin 3x + \sin 5x = 0$

33. $\sin 5x - \sin 3x = \cos 4x$

34. $\cos 3x - \cos x - \sin 2x = 0$

35. $\sin 3x - \sin x - \cos 2x = 0$

36. $\sin 4x - \sin 2x - \sqrt{2}\cos 3x = 0$

37. Show that $\sin 45° + \sin 15° = \sin 75°$.

38. Show that $\cos 87° + \cos 33° = \sin 63°$.

39. If $A + B + C = \pi/2$, show that
$$\sin 2A + \sin 2B + \sin 2C = 4\cos A \cos B \cos C$$

40. If $A + B + C = \pi$, show that
$$\tan A + \tan B + \tan C = \tan A \tan B \tan C$$

41. Prove the identity
$$\frac{\sin x + \sin 2x + \sin 3x + \sin 4x + \sin 5x}{\cos x + \cos 2x + \cos 3x + \cos 4x + \cos 5x} = \tan 3x$$

Section 6.6

Inverse Trigonometric Functions

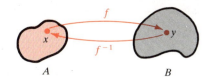

Figure 1

If f is a one-to-one function with domain A and range B, then its inverse f^{-1} is the function with domain B and range A defined by

$$f^{-1}(y) = x \quad \Leftrightarrow \quad f(x) = y$$

(See Section 2.6.) In other words, f^{-1} is the rule that reverses the action of f. Figure 1 shows a graphical representation of the actions of f and f^{-1}.

In order for a function to have an inverse it must be one-to-one. Since the trigonometric functions are not one-to-one, they do not have inverses. It is possible, however, to restrict the domains of the trigonometric functions in such a way that the resulting functions are one-to-one.

The Inverse Sine Function

Let us first consider the sine function. There are many ways to restrict the domain of sine so that the new function is one-to-one. A natural way to do this is to restrict the domain to the interval $[-\pi/2, \pi/2]$. The reason for this choice is that sine attains each of its values exactly once on this interval. We write $\sin x$ (with a capital S) for the new function that has the domain $[-\pi/2, \pi/2]$ and has the same values as $\sin x$ on this interval. The graphs of $\sin x$ and $\sin x$ are shown in Figure 2. Notice that the function $\sin x$ is one-to-one (by the Horizontal Line Test) and so has an inverse.

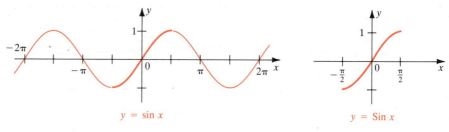

Figure 2

$y = \sin x$ $y = \text{Sin } x$

The inverse of the function Sin x is the function $\text{Sin}^{-1}x$ defined by

$$\text{Sin}^{-1}x = y \quad \Leftrightarrow \quad \text{Sin } y = x$$

where $-1 \leq x \leq 1$ and $-\pi/2 \leq y \leq \pi/2$. This function is also denoted by **arcsin** x. The graph of $\text{Sin}^{-1}x$ is shown in Figure 3; it is obtained by reflecting the graph of Sin x in the line $y = x$.

It is customary to write $\text{Sin}^{-1}x$ without the capital S, as $\sin^{-1}x$, and we will follow this convention. We give a formal definition:

> The **inverse sine** function is the function **sin^{-1}** with domain $[-1, 1]$ and range $[-\pi/2, \pi/2]$ given by
>
> $$\sin^{-1}x = y \quad \Leftrightarrow \quad \sin y = x$$
>
> This function is also called **arcsin**.

Thus $\sin^{-1}x$ *is the number in the interval* $[-\pi/2, \pi/2]$ *whose sine is* x. In other words, $\sin(\sin^{-1}x) = x$. In fact, from the general properties of inverse functions studied in Section 2.6 we have the following relations:

> $$\sin(\sin^{-1}x) = x \qquad \text{for } -1 \leq x \leq 1$$
> $$\sin^{-1}(\sin x) = x \qquad \text{for } -\frac{\pi}{2} \leq x \leq \frac{\pi}{2}$$

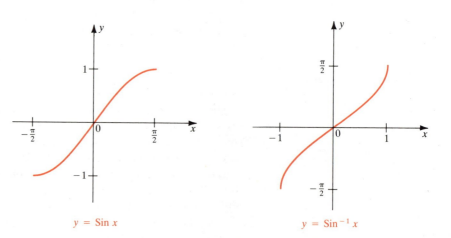

Figure 3

$y = \text{Sin } x$ $y = \text{Sin}^{-1}x$

Example 1

Find (a) $\sin^{-1}\frac{1}{2}$, (b) $\sin^{-1}(-\frac{1}{2})$, (c) $\sin^{-1}(\sqrt{2}/2)$, and (d) $\sin^{-1}\frac{3}{2}$.

Solution

(a)   The number in the interval $[-\pi/2, \pi/2]$ whose sine is $\frac{1}{2}$ is $\pi/6$. Thus $\sin^{-1}\frac{1}{2} = \pi/6$.

(b)   Again, $\sin^{-1}(-\frac{1}{2})$ is the number in the interval $[-\pi/2, \pi/2]$ whose sine is $-\frac{1}{2}$. Since $\sin(-\pi/6) = -\frac{1}{2}$, we have $\sin^{-1}(-\frac{1}{2}) = -\pi/6$.

(c)   Since $\sin(\pi/4) = \sqrt{2}/2$ and $\pi/4$ is in the interval $[-\pi/2, \pi/2]$, we have $\sin^{-1}(\sqrt{2}/2) = \pi/4$.

(d)   Since $\frac{3}{2} > 1$, it is not in the domain of $\sin^{-1}x$, so $\sin^{-1}(\frac{3}{2})$ is not defined.

Example 2

Find approximate values for (a) $\sin^{-1}(0.82)$ and (b) $\sin^{-1}\frac{1}{3}$.

Solution

Since no rational multiple of π has a sine of 0.82 or $\frac{1}{3}$, we use a calculator to approximate these values. Using the INV SIN or SIN^{-1} or ARCSIN key on the calculator (making sure the calculator is in radian mode) gives

(a) $\sin^{-1}(0.82) \approx 0.96141$ (b) $\sin^{-1}\frac{1}{3} \approx 0.33984$

Example 3

Find $\cos(\sin^{-1}\frac{3}{5})$.

Solution 1

It is easy to find $\sin(\sin^{-1}\frac{3}{5})$. In fact, by the properties of inverse functions this value is exactly $\frac{3}{5}$. To find $\cos(\sin^{-1}\frac{3}{5})$ we reduce the problem to this easy problem by writing the cosine function in terms of the sine function. Let $u = \sin^{-1}\frac{3}{5}$. Since $-\pi/2 \leq u \leq \pi/2$, $\cos x$ is positive and we can write

$$\cos u = \sqrt{1 - \sin^2 u}$$

Thus $$\cos\left(\sin^{-1}\frac{3}{5}\right) = \sqrt{1 - \sin^2\left(\sin^{-1}\frac{3}{5}\right)}$$

$$= \sqrt{1 - \left(\frac{3}{5}\right)^2} = \sqrt{1 - \frac{9}{25}} = \sqrt{\frac{16}{25}} = \frac{4}{5}$$

Solution 2

Again let $u = \sin^{-1}\frac{3}{5}$. Then u is the number in the interval $[-\pi/2, \pi/2]$ whose sine is $\frac{3}{5}$. Let us interpret u as an angle and draw a right triangle with u as one of its acute angles, with opposite side 3 and hypotenuse 5 (see Figure 4). The remaining leg of the triangle is found by the Pythagorean Theorem to be 4. From the

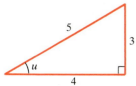

Figure 4

diagram we get

$$\cos\left(\sin^{-1}\frac{3}{5}\right) = \cos u = \frac{4}{5}$$ ●

Solution 2 of Example 3 has the advantage that we can immediately find the values of the other trigonometric functions of $\sin^{-1}\frac{3}{5}$ from the triangle. Thus $\tan(\sin^{-1}\frac{3}{5}) = \frac{3}{4}$, $\sec(\sin^{-1}\frac{3}{5}) = \frac{5}{4}$, and $\csc(\sin^{-1}\frac{3}{5}) = \frac{5}{3}$.

The Inverse Cosine Function

If the domain of the cosine function is restricted to the interval $[0, \pi]$, the resulting function is one-to-one and so has an inverse. (Again, we choose this interval because on it cosine attains each of its values exactly once.) The inverse of this function is denoted by $\cos^{-1}x$ or arccos x.

The **inverse cosine** function is the function **$\cos^{-1}$** with domain $[-1, 1]$ and range $[0, \pi]$ given by

$$\cos^{-1}x = y \iff \cos y = x$$

This function is also called **arccos**.

Thus, $\cos^{-1}x$ *is the number in the interval* $[0, \pi]$ *whose cosine is x.* The following relations hold:

$$\cos(\cos^{-1}x) = x \qquad \text{for } -1 \le x \le 1$$
$$\cos^{-1}(\cos x) = x \qquad \text{for } 0 \le x \le \pi$$

The graph of cos with its restricted domain and the graph of $\cos^{-1}$ are shown in Figure 5.

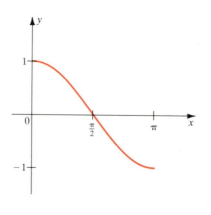

$y = \cos x, \quad 0 \le x \le \pi$

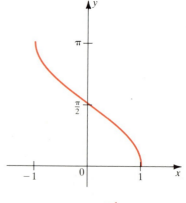

$y = \cos^{-1} x$

Figure 5

Example 4

Find (a) $\cos^{-1}(\sqrt{3}/2)$, (b) $\cos^{-1}1$, (c) $\cos^{-1}0$, and (d) $\cos^{-1}(5/7)$.

Solution

(a)   The number in the interval $[0, \pi]$ whose cosine is $\sqrt{3}/2$ is $\pi/6$. Thus $\cos^{-1}(\sqrt{3}/2) = \pi/6$.

(b)   Since 0 is the number in the interval $[0, \pi]$ whose cosine is 1, it follows that $\cos^{-1}1 = 0$.

(c)   Since $\cos(\pi/2) = 0$, it follows that $\cos^{-1}0 = \pi/2$.

(d)   Since no rational multiple of π has cosine 5/7, we use a calculator to find this value approximately: $\cos^{-1}(5/7) \approx 0.77519$.   ●

Example 5

Find $\sin(\cos^{-1}x)$ and $\tan(\cos^{-1}x)$ for $-1 \le x \le 1$.

Solution

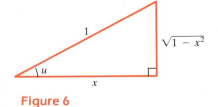

Figure 6

Let $u = \cos^{-1}x$. Then u is the number in the interval $[0, \pi]$ with $\cos u = x$. Let us interpret u as radian measure and sketch a triangle with the property that $\cos u = x$ as shown in Figure 6. Notice that the relations shown in the triangle hold for all values of u in the interval $[0, \pi]$. From the figure we see that $\sin u = \sqrt{1 - x^2}$ and $\tan u = \sqrt{1 - x^2}/x$. Thus for all x, where $-1 \le x \le 1$, we have

$$\sin(\cos^{-1}x) = \sqrt{1 - x^2} \qquad \text{and} \qquad \tan(\cos^{-1}x) = \frac{\sqrt{1 - x^2}}{x} \qquad ●$$

Example 6

Find (a) $\sin(\tfrac{1}{2}\cos^{-1}x)$ and (b) $\sin(\tfrac{1}{2}\cos^{-1}\tfrac{7}{8})$.

Solution

(a)   The half-angle identity for sine is useful here. This identity says that

$$\sin \tfrac{1}{2}u = \pm\sqrt{\frac{1 - \cos u}{2}}$$

To determine the appropriate sign notice that

$$0 \le \cos^{-1}x \le \pi$$

Multiplying this inequality through by $\tfrac{1}{2}$, we get

$$0 \le \tfrac{1}{2}\cos^{-1}x \le \frac{\pi}{2}$$

Since sine is positive on this interval, we use the $+$ sign to get

$$\sin(\tfrac{1}{2}\cos^{-1}x) = \sqrt{\frac{1 - \cos(\cos^{-1}x)}{2}} = \sqrt{\frac{1 - x}{2}}$$

(b)   Setting $x = \frac{7}{8}$ in part (a) gives

$$\sin(\tfrac{1}{2}\cos^{-1}\tfrac{7}{8}) = \sqrt{\frac{1 - \frac{7}{8}}{2}} = \sqrt{\frac{1}{16}} = \frac{1}{4}$$ ●

The Inverse Tangent Function

We restrict the domain of the tangent function to the interval $[-\pi/2, \pi/2]$. The inverse of the resulting one-to-one function is called $\tan^{-1}$ or arctan.

The **inverse tangent** function is the function **$\tan^{-1}$** with domain R and range $(-\pi/2, \pi/2)$ defined by

$$\tan^{-1}x = y \quad \Leftrightarrow \quad \tan y = x$$

This function is also called **arctan**.

Thus $\tan^{-1}x$ *is the number in the interval* $(-\pi/2, \pi/2)$ *whose tangent is x.* The inverse function relations give

---

$$\tan(\tan^{-1}x) = x \qquad \text{for } x \in R$$

$$\tan^{-1}(\tan x) = x \qquad \text{for } -\frac{\pi}{2} < x < \frac{\pi}{2}$$

Figure 7 shows the graph of $\tan x$ on the interval $[-\pi/2, \pi/2]$ and the graph of its inverse function $\tan^{-1}x$.

A similar procedure can be used to define inverses for the other trigonometric functions. Although there is general agreement on how to restrict the domains of

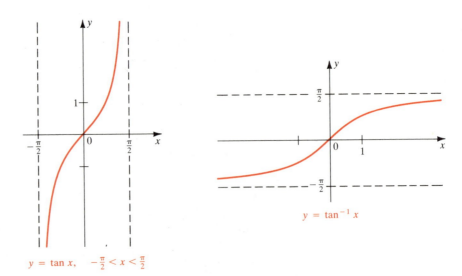

$$y = \tan x, \quad -\tfrac{\pi}{2} < x < \tfrac{\pi}{2}$$

$$y = \tan^{-1} x$$

Figure 7

the sine, cosine, and tangent functions, there is no such agreement for the other trigonometric functions. Of course, any interval on which these functions are one-to-one and on which they attain all their values is appropriate for the restricted domain. Exercises 84–86 outline how this can be done.

Example 7

Find (a) $\tan^{-1}1$, (b) $\tan^{-1}\sqrt{3}$, and (c) $\tan^{-1}(-20)$.

Solution

(a) The number in the interval $(\pi/2, \pi/2)$ with tangent 1 is $\pi/4$. Thus $\tan^{-1}1 = \pi/4$.

(b) Since $\tan \pi/3 = \sqrt{3}$, $\tan^{-1}\sqrt{3} = \pi/3$.

(c) We use a calculator to find that $\tan^{-1}(-20) \approx -1.52084$. ●

Example 8

Find $\sin(\cos^{-1}x + \tan^{-1}y)$, where $-1 \leq x \leq 1$ and y is any real number.

Solution

Let $u = \cos^{-1}x$ and $v = \tan^{-1}y$. Then $0 \leq u \leq \pi$ and $-\pi/2 < v < \pi/2$. Using the method of Example 5, we sketch triangles with angles u and v such that $\cos u = x$ and $\tan v = y$ (see Figure 8). Notice that the relations expressed in these triangles hold for all values of u and v. From the addition formula for sine we get

$$\sin(\cos^{-1}x + \tan^{-1}y) = \sin(u + v)$$

$$= \sin u \cos v + \cos u \sin v$$

$$= \sqrt{1 - x^2}\,\frac{1}{\sqrt{1 + y^2}} + x\,\frac{y}{\sqrt{1 + y^2}}$$

$$= \frac{1}{\sqrt{1 + y^2}}(\sqrt{1 - x^2} + xy) \qquad ●$$

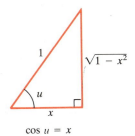

$\cos u = x$

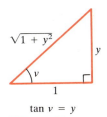

$\tan v = y$

Figure 8

Identities Involving Inverse Trigonometric Functions

There are identities that relate the inverse trigonometric functions to one another. These are consequences of the corresponding identities for the trigonometric functions themselves. Here are some examples.

Example 9

Prove the identity

$$2\tan^{-1}y = \sin^{-1}\!\left(\frac{2y}{1 + y^2}\right)$$

Solution

Let $v = \tan^{-1}y$. From the double-angle formula for sine and the triangle in Figure 8(b) we have

$$\sin 2v = 2 \sin v \cos v$$

$$= 2 \frac{y}{\sqrt{1 + y^2}} \frac{1}{\sqrt{1 + y^2}}$$

$$= \frac{2y}{1 + y^2}$$

So

$$2v = \sin^{-1}\left(\frac{2y}{1 + y^2}\right)$$

Thus

$$2 \tan^{-1}y = \sin^{-1}\left(\frac{2y}{1 + y^2}\right) \qquad \bullet$$

Example 10

Prove that for $-1 \le x \le 1$, $\sin^{-1}x + \cos^{-1}x = \pi/2$.

Solution

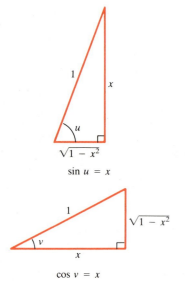

$\sin u = x$

$\cos v = x$

Figure 9

Let $u = \sin^{-1}x$ and $v = \cos^{-1}x$, so that $\sin u = x$ and $\cos v = x$. Since $-\pi/2 \le u \le \pi/2$ and $0 \le v \le \pi$, it follows that $-\pi/2 \le u + v \le 3\pi/2$. From the triangles in Figure 9 and from the addition formula for sine we get

$$\sin(u + v) = \sin u \cos v + \cos u \sin v$$

$$= xx + \sqrt{1 - x^2}\sqrt{1 - x^2} = x^2 + (1 - x^2) = 1$$

Thus

$$u + v = \sin^{-1}1 = \frac{\pi}{2}$$

and so

$$\sin^{-1}x + \cos^{-1}x = \frac{\pi}{2}$$

Notice that this identity states the simple fact that when $\sin u = \cos v$ (they are both equal to x), u and v are complementary ($u + v = \pi/2$). $\qquad \bullet$

Trigonometric Equations

The inverse trigonometric functions are needed to solve trigonometric equations. When we solved trigonometric equations in the preceding sections we implicitly used inverse trigonometric functions without specifically calling them that. The reason is that we dealt with equations for which the solutions were numbers whose trigonometric values we knew exactly. In general to solve any equation involving x, our goal is to isolate x. The inverse trigonometric functions allow us to do this easily. For instance, let us solve the equation

$$\cos x - \tfrac{2}{3} = 0$$

in the interval $[0, \pi]$. First we write the equation as

$$\cos x = \tfrac{2}{3}$$

To isolate x, we apply the inverse cosine function to both sides to get

$$\cos^{-1}(\cos x) = \cos^{-1}\tfrac{2}{3}$$

or
$$x = \cos^{-1}\tfrac{2}{3}$$

Using a calculator, we find that $x \approx 0.84107$.

Example 11

Solve the equation $3\cos^2 x - \cos x - 1 = 0$ in the interval $[0, \pi]$.

Solution

Setting $Y = \cos x$, we get $3Y^2 - Y - 1 = 0$, which is a quadratic equation. By the quadratic formula we have

$$Y = \frac{1 \pm \sqrt{13}}{6} \qquad \text{or} \qquad \cos x = \frac{1 \pm \sqrt{13}}{6}$$

Thus the solutions to this equation in the interval $[0, \pi]$ are

$$x = \cos^{-1}\left(\frac{1 + \sqrt{13}}{6}\right) \qquad \text{and} \qquad x = \cos^{-1}\left(\frac{1 - \sqrt{13}}{6}\right)$$

To find approximations for these solutions we use a calculator to get

$$x = \cos^{-1}\left(\frac{1 + \sqrt{13}}{6}\right) \approx 0.69572$$

and
$$x = \cos^{-1}\left(\frac{1 - \sqrt{13}}{6}\right) \approx 2.02001 \qquad \bullet$$

Example 12

Solve the equation $\tan^{-1}x + \cos^{-1}\left(\dfrac{x}{2}\right) = \dfrac{\pi}{2}$.

Solution

Let $u = \tan^{-1}x$ and $v = \cos^{-1}(x/2)$ and sketch triangles representing u and v as in Figure 10. The equation we want to solve can be written as

$$u + v = \frac{\pi}{2}$$

We take the cosine of both sides, apply the addition formula for cosine, and use the triangles in Figure 10 to get

$$\cos(u + v) = 0$$
$$\cos u \cos v - \sin u \sin v = 0$$
$$\frac{1}{\sqrt{1 + x^2}}\frac{x}{2} - \frac{x}{\sqrt{1 + x^2}}\frac{\sqrt{4 - x^2}}{2} = 0$$

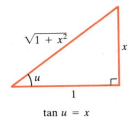

$\tan u = x$

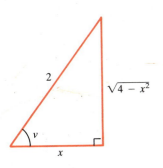

$\cos v = \tfrac{1}{2}x$

Figure 10

Multiplying both sides by $2\sqrt{1 + x^2}$ gives

$$x - x\sqrt{4 - x^2} = 0$$

or
$$x(1 - \sqrt{4 - x^2}) = 0$$

so
$$x = 0 \quad \text{or} \quad 1 - \sqrt{4 - x^2} = 0$$

To solve $1 - \sqrt{4 - x^2} = 0$, we write it as $1 = \sqrt{4 - x^2}$ and square both sides to get

$$1 = 4 - x^2 \quad \text{or} \quad x^2 = 3$$

So,
$$x = \pm\sqrt{3}$$

Thus the possible solutions to the original equation are $x = 0, \pm\sqrt{3}$. It is easy to check that each of these values satisfies the original equation and so the solutions are $x = 0, \sqrt{3}, -\sqrt{3}$.

Exercises 6.6

In Exercises 1–8 find the exact value of the given expression if it is defined.

1. (a) $\sin^{-1}\frac{1}{2}$ (b) $\cos^{-1}\frac{1}{2}$ (c) $\cos^{-1}2$

2. (a) $\sin^{-1}(\sqrt{3}/2)$ (b) $\cos^{-1}(\sqrt{3}/2)$
   (c) $\cos^{-1}(-\sqrt{3}/2)$

3. (a) $\sin^{-1}(\sqrt{2}/2)$ (b) $\cos^{-1}(\sqrt{2}/2)$
   (c) $\sin^{-1}(-\sqrt{2}/2)$

4. (a) $\tan^{-1}\sqrt{3}$ (b) $\tan^{-1}(-\sqrt{3})$
   (c) $\sin^{-1}\sqrt{3}$

5. (a) $\sin^{-1}1$ (b) $\cos^{-1}1$ (c) $\cos^{-1}(-1)$

6. (a) $\tan^{-1}1$ (b) $\tan^{-1}(-1)$ (c) $\tan^{-1}0$

7. (a) $\tan^{-1}(\sqrt{3}/3)$ (b) $\tan^{-1}(-\sqrt{3}/3)$
   (c) $\sin^{-1}(-2)$

8. (a) $\sin^{-1}0$ (b) $\cos^{-1}0$ (c) $\cos^{-1}(-\frac{1}{2})$

In Exercises 9–10 find an approximate value of the given expression to five decimal places using a calculator.

9. (a) $\sin^{-1}(0.7688)$ (b) $\cos^{-1}(-0.5014)$
   (c) $\tan^{-1}(15.2000)$

10. (a) $\cos^{-1}(0.3388)$ (b) $\tan^{-1}(1.0000)$
    (c) $\cos^{-1}(0.9800)$

In Exercises 11–24 find the exact value of the given expression.

11. $\sin(\sin^{-1}\frac{1}{2})$ 12. $\cos(\sin^{-1}\frac{1}{2})$ 13. $\tan(\sin^{-1}\frac{1}{2})$

14. $\sin(\sin^{-1}0)$ 15. $\cos[\sin^{-1}(\sqrt{3}/2)]$

16. $\tan[\sin^{-1}(\sqrt{2}/2)]$ 17. $\sin(\tan^{-1}\sqrt{3})$

18. $\tan^{-1}[\tan(3\pi/4)]$ 19. $\cos^{-1}[\cos(\pi/3)]$

20. $\cos^{-1}[\cos(4\pi/3)]$ 21. $\tan^{-1}[2\sin(\pi/3)]$

22. $\cos^{-1}[\sqrt{3}\sin(\pi/6)]$

23. $\tan^{-1}[\sin(\pi/3) + \cos(\pi/6)]$

24. $\cos^{-1}[\frac{1}{2}\tan(\pi/3) - \sin(\pi/3)]$

In Exercises 25–38 evaluate the given expression. Do not use a calculator.

25. $\sin[\cos^{-1}(3/5)]$ 26. $\tan[\sin^{-1}(4/5)]$

27. $\sin[\tan^{-1}(12/5)]$ 28. $\cos(\tan^{-1}5)$

29. $\sin[2\cos^{-1}(3/5)]$ 30. $\tan[2\tan^{-1}(5/13)]$

31. $\cos[\frac{1}{2}\sin^{-1}(\sqrt{3}/2)]$ 32. $\tan(\frac{1}{2}\sin^{-1}\frac{15}{17})$

33. $\sin(\sin^{-1}\frac{1}{2} + \cos^{-1}\frac{1}{2})$ 34. $\cos(\sin^{-1}\frac{3}{5} - \cos^{-1}\frac{3}{5})$

35. $\sin(\sin^{-1}\frac{5}{13} - \cos^{-1}\frac{12}{13})$

36. $\tan(\sin^{-1}\frac{1}{2} + \cos^{-1}\frac{1}{3})$ 37. $\cos(\tan^{-1}3 + \cos^{-1}\frac{1}{2})$

38. $\cos(2\sin^{-1}\frac{3}{5} + \tan^{-1}\frac{3}{4})$

In Exercises 39–48 rewrite each expression as an algebraic expression in x.

39. $\cos(\sin^{-1}x)$ 40. $\sin(\tan^{-1}x)$ 41. $\tan(\sin^{-1}x)$

42. $\sin(2\cos^{-1}x)$

43. $\cos(2\tan^{-1}x)$

44. $\sin(2\sin^{-1}x)$

45. $\tan(\tfrac{1}{2}\sin^{-1}x)$

46. $\cos(\cos^{-1}x + \sin^{-1}x)$

47. $\sin(\tan^{-1}x - \sin^{-1}x)$

48. $\sin(2\sin^{-1}x + \cos^{-1}x)$

In Exercises 49 and 50 write each expression as an algebraic expression in x and y.

49. $\tan(\tan^{-1}x + \sin^{-1}y)$

50. $\tan(\tan^{-1}x - \tan^{-1}2y)$

51. A painting that is 2 m high hangs in a museum with its bottom edge 3 m above the floor. A person whose eye level is h meters above the floor stands at a distance of x meters directly in front of the painting. The size that the painting appears to the viewer is determined by the size of the angle θ that the painting subtends at the viewer's eyes (see the figure). The larger θ is, the larger the painting appears to the viewer. The angle θ depends on the distance x; in other words, the angle θ is a function of x. Show that

$$\theta = \tan^{-1}\left(\frac{2x}{x^2 + (3-h)(5-h)}\right)$$

(*Hint:* Use the subtraction formula for tangent and the fact that $\theta = \alpha - \beta$.)

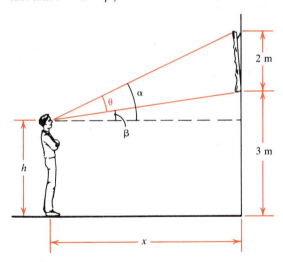

In Exercises 52–65 verify the given identity.

52. $\cos^{-1}x = \sin^{-1}\sqrt{1 - x^2}, |x| \le 1$

53. $\tan^{-1}x = \cos^{-1}\dfrac{1}{\sqrt{1 + x^2}}$

54. $\tan^{-1}x = \sin^{-1}\dfrac{x}{\sqrt{1 + x^2}}$

55. $\sin^{-1}x = \tan^{-1}\dfrac{x}{\sqrt{1 - x^2}}, |x| \le 1$

56. $\tfrac{1}{2}\cos^{-1}x = \cos^{-1}\sqrt{\dfrac{x + 1}{2}}, |x| \le 1$

57. $\sin^{-1}(-x) = -\sin^{-1}x, |x| \le 1$

58. $\cos^{-1}(-x) = \pi - \cos^{-1}x, |x| \le 1$

59. $\tan^{-1}(-x) = -\tan^{-1}x$

60. $2\sin^{-1}x = \cos^{-1}(1 - 2x^2), 0 \le x \le 1$

61. $2\tan^{-1}\tfrac{1}{x} = \cos^{-1}\dfrac{x^2 - 1}{x^2 + 1}$

62. $\tan^{-1}x + \tan^{-1}\tfrac{1}{x} = \tfrac{\pi}{2}, x > 0$

63. $\tan^{-1}x + \tan^{-1}y = \tan^{-1}\dfrac{x + y}{1 - xy}, |x| < 1 \text{ and } |y| < 1$

64. $\sin^{-1}x + \sin^{-1}y = \cos^{-1}(\sqrt{1 - x^2}\sqrt{1 - y^2} - xy),$
$0 \le x, y \le 1$

65. $\sin^{-1}x - \sin^{-1}y = \sin^{-1}(x\sqrt{1 - y^2} - y\sqrt{1 - x^2}),$
$0 \le x, y \le 1$

In Exercises 66–75, (a) solve the trigonometric equation on the given interval and (b) use a calculator to approximate the solutions you found in part (a) to five decimal places.

66. $\tan^2 x - \tan x - 2 = 0$ on $(-\pi/2, \pi/2)$

67. $3\sin^2 x - 7\sin x + 2 = 0$ on $[0, 2\pi)$

68. $6\cos^2 x - 7\cos x + 2 = 0$ on $[0, 2\pi)$

69. $2\sin 2x - \cos x = 0$ on $[0, \pi/2)$

70. $\sin 2x(\sec^2 x - 2) = 0$ on $[0, \pi)$

71. $\cos 2x - 5\cos x + 4 = 0$ on $[0, \pi)$

72. $8\sin^4 x - 6\sin^2 x + 1 = 0$ on $[0, \pi)$

73. $\tan^4 x - 13\tan^2 x + 36 = 0$ on $[0, \pi/2]$

74. $\cos^3 x - 2\cos^2 x + 1 = 0$ on $[0, 2\pi)$

75. $\sec x - \csc x - 3\tan x + 3 = 0$ on $[0, 2\pi)$

In Exercises 76–83 solve the given equation.

76. $2\sin^{-1}x + \cos^{-1}x = \pi$ **77.** $\sin^{-1}x - \cos^{-1}x = 0$

78. $\sin^{-1}2x = \cos^{-1}x$

79. $\sin^{-1}x - \cos^{-1}\sqrt{1 - x^2} = 0$

80. $\tan^{-1}x + \tan^{-1}2x = \dfrac{\pi}{4}$

81. $\tan^{-1}x + \tan^{-1}2x + \tan^{-1}3x = \pi$

82. $\tan^{-1}x - \sin^{-1}x = 0$

83. $2\tan^{-1}\frac{1}{x} - \sin^{-1}\frac{1}{x} = 0$

Exercises 84–86 give possible restricted domains for the csc, sec, *and* cot *functions on which inverses of these functions can be defined.*

84. **(a)** Verify that the function csc is one-to-one on $(0, \pi/2] \cup (\pi, 3\pi/2]$.

(b) Define an inverse, $\csc^{-1}$, for the function csc by restricting its domain as in part (a).

85. **(a)** Verify that the function sec is one-to-one on $[0, \pi/2) \cup [\pi, 3\pi/2)$.
(b) Define an inverse, $\sec^{-1}$, for the function sec by restricting its domain as in part (a).

86. **(a)** Verify that the function cot is one-to-one on $(0, \pi)$.
(b) Define an inverse, $\cot^{-1}$, for the function cot by restricting its domain as in part (a).

Section 6.7

Application: Trigonometric Form of Complex Numbers; DeMoivre's Theorem

Complex numbers were introduced in Chapter 3 in order to solve certain algebraic equations. The applications of complex numbers go far beyond this initial use, however. Complex numbers are now used routinely in physics, electrical engineering, aerospace engineering, and many other fields. In Chapter 3 we learned to perform arithmetic operations on complex numbers and to represent them graphically. In this section we represent complex numbers using the trigonometric functions sine and cosine. This will enable us, among other things, to find the *n*th roots of complex numbers.

Trigonometric Form of Complex Numbers

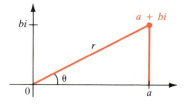

Figure 1

Let $z = a + bi$ be a complex number and let us draw in the complex plane the line segment that joins the origin to the point $a + bi$ (see Figure 1). Let $r = |z| = \sqrt{a^2 + b^2}$ be the length of this line segment. If θ is an angle in standard position whose terminal side coincides with this line segment, then

$$a = r\cos\theta \qquad \text{and} \qquad b = r\sin\theta$$

so $z = r\cos\theta + ir\sin\theta = r(\cos\theta + i\sin\theta)$. It follows that any complex number z can be represented in the **trigonometric form**

$$z = r(\cos\theta + i\sin\theta)$$

where $r = |z|$ and $\tan\theta = b/a$. The angle θ is called the **argument** of the complex number z. We will write $\arg(z)$ for the argument of z. Notice that the argument of z is not unique so that the representation of a complex number in trigonometric form is not unique.

Example 1

Write the following complex numbers in trigonometric form: (a) $z = 1 + i$, (b) $w = 3 - 4i$, (c) $v = \sqrt{3} - i$.

Solution

These complex numbers are graphed in Figure 2.

(a)   The argument of z is $\pi/4$ and $r = |z| = \sqrt{1 + 1} = \sqrt{2}$. Thus

$$z = \sqrt{2}\left(\cos\frac{\pi}{4} + i\sin\frac{\pi}{4}\right)$$

(b)   The argument of w is $\arg(w) = \tan^{-1}(-4/3)$ and $|w| = \sqrt{3^2 + 4^2} = 5$. So

$$w = 5[\cos(\tan^{-1}(-4/3)) + i\sin(\tan^{-1}(-4/3))]$$

Since $\tan^{-1}(-4/3) \approx -0.92730$, we can write z approximately as

$$z \approx 5[\cos(-0.92730) + i\sin(-0.92730)]$$

(c)   Since $|v| = \sqrt{(\sqrt{3})^2 + (-1)^2} = 2$ it follows that $\cos\theta = \sqrt{3}/2$ and $\sin\theta = -1/2$ and so $\theta = -\pi/6$. Thus

$$v = 2[\cos(-\pi/6) + i\sin(-\pi/6)]$$

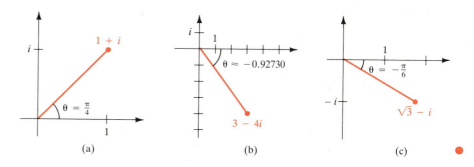

Figure 2

(a)   (b)   (c)

The addition formulas for sine and cosine discussed in Section 6.3 greatly simplify the multiplication of complex numbers in trigonometric form as the following calculations show. Let

$$z_1 = r_1(\cos\theta_1 + i\sin\theta_1)$$
$$z_2 = r_2(\cos\theta_2 + i\sin\theta_2)$$

be two complex numbers written in trigonometric form. Then

$$z_1 z_2 = r_1 r_2(\cos\theta_1 + i\sin\theta_1)(\cos\theta_2 + i\sin\theta_2)$$
$$= r_1 r_2[(\cos\theta_1\cos\theta_2 - \sin\theta_1\sin\theta_2) + i(\sin\theta_1\cos\theta_2 + \cos\theta_1\sin\theta_2)]$$
$$= r_1 r_2[\cos(\theta_1 + \theta_2) + i\sin(\theta_1 + \theta_2)]$$

This formula says that *to multiply two complex numbers we multiply the moduli and add the arguments.*

A similar argument using the subtraction formulas for sine and cosine shows that *to divide two complex numbers we divide the moduli and subtract the arguments.*

We state this precisely:

If the two complex numbers z_1 and z_2 have the trigonometric forms

$$z_1 = r_1(\cos\theta_1 + i\sin\theta_1) \qquad \text{and} \qquad z_2 = r_2(\cos\theta_2 + i\sin\theta_2)$$

then

(1) $$z_1 z_2 = r_1 r_2 [\cos(\theta_1 + \theta_2) + i\sin(\theta_1 + \theta_2)]$$

(2) $$\frac{z_1}{z_2} = \frac{r_1}{r_2}[\cos(\theta_1 - \theta_2) + i\sin(\theta_1 - \theta_2)], \qquad z_2 \neq 0$$

As an important special case of the division formula, we have

---

For the nonzero complex number

$$z = r(\cos\theta + i\sin\theta)$$

we have

(3) $$\frac{1}{z} = \frac{1}{r}(\cos\theta - i\sin\theta)$$

Example 2

Let $z_1 = 2\left(\cos\dfrac{\pi}{4} + i\sin\dfrac{\pi}{4}\right)$ and $z_2 = 5\left(\cos\dfrac{\pi}{3} + i\sin\dfrac{\pi}{3}\right)$. Find the complex numbers $z_1 z_2$, z_1/z_2, and $1/z_1$ in trigonometric form.

Solution

By Equation 1,

$$z_1 z_2 = (2)(5)\left[\cos\left(\frac{\pi}{4} + \frac{\pi}{3}\right) + i\sin\left(\frac{\pi}{4} + \frac{\pi}{3}\right)\right]$$

$$= 10\left(\cos\frac{7\pi}{12} + i\sin\frac{7\pi}{12}\right)$$

By Equation 2 we have

$$\frac{z_1}{z_2} = \frac{2}{5}\left[\cos\left(\frac{\pi}{4} - \frac{\pi}{3}\right) + i\sin\left(\frac{\pi}{4} - \frac{\pi}{3}\right)\right]$$

$$= \frac{2}{5}\left[\cos\left(-\frac{\pi}{12}\right) + i\sin\left(-\frac{\pi}{12}\right)\right]$$

$$= \frac{2}{5}\left(\cos\frac{\pi}{12} - i\sin\frac{\pi}{12}\right)$$

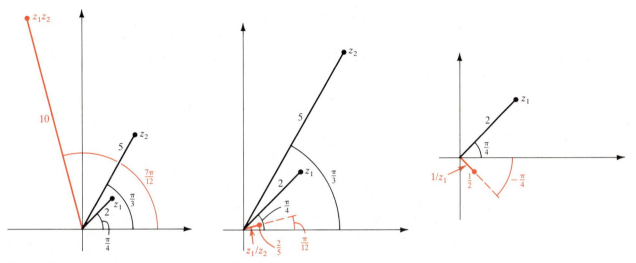

Figure 3

From Equation 3 it follows that

$$\frac{1}{z_1} = \frac{1}{2}\left(\cos\frac{\pi}{4} - i\sin\frac{\pi}{4}\right)$$

The graphs of z_1, z_2, z_1z_2, z_1/z_2, and $1/z_1$ are shown in Figure 3.   ●

Example 3

Find the product of the complex numbers $1 + i$ and $\sqrt{3} - i$ in trigonometric form.

Solution

From Example 1 we have

$$1 + i = \sqrt{2}\left(\cos\frac{\pi}{4} + i\sin\frac{\pi}{4}\right) \quad\text{and}\quad \sqrt{3} - i = 2\left[\cos\left(-\frac{\pi}{6}\right) + i\sin\left(-\frac{\pi}{6}\right)\right]$$

So by Equation 1,

$$(1 + i)(\sqrt{3} - i) = 2\sqrt{2}\left[\cos\left(\frac{\pi}{4} - \frac{\pi}{6}\right) + i\sin\left(\frac{\pi}{4} - \frac{\pi}{6}\right)\right]$$

$$= 2\sqrt{2}\left(\cos\frac{\pi}{12} + i\sin\frac{\pi}{12}\right)$$ ●

DeMoivre's Theorem

Repeated use of the multiplication formula (1) gives a useful formula for raising a complex number to a power n, where n is any positive integer. Let z be a complex number written in trigonometric form

$$z = r(\cos\theta + i\sin\theta)$$

Then by Equation 1,

$$z^2 = zz = r^2[\cos(\theta + \theta) + i\sin(\theta + \theta)]$$
$$= r^2(\cos 2\theta + i\sin 2\theta)$$

By Equation 1 again applied to z and z^2 we get

$$z^3 = z^2 z = r^3[\cos(2\theta + \theta) + i\sin(2\theta + \theta)]$$
$$= r^3(\cos 3\theta + i\sin 3\theta)$$

Similarly, $$z^4 = r^4(\cos 4\theta + i\sin 4\theta)$$

Repeating this argument, we see that for any positive integer n,

(4) $$z^n = r^n(\cos n\theta + i\sin n\theta)$$

A similar argument using the division formula (2) shows that Equation 4 holds for negative integers also. Thus we have the following theorem:

DeMoivre's Theorem

If $z = r(\cos\theta + i\sin\theta)$, then for any integer n,

$$z^n = r^n(\cos n\theta + i\sin n\theta)$$

This says that *to take the nth power of a complex number we take the nth power of the modulus and multiply the argument by n.*

Example 4
Find $(\frac{1}{2} + \frac{1}{2}i)^{10}$.

Solution

Since $(\frac{1}{2} + \frac{1}{2}i) = \frac{1}{2}(1 + i)$, it follows from Example 1(a) that $\frac{1}{2} + \frac{1}{2}i$ has the trigonometric form

$$\frac{1}{2} + \frac{1}{2}i = \frac{\sqrt{2}}{2}\left(\cos\frac{\pi}{4} + i\sin\frac{\pi}{4}\right)$$

So by DeMoivre's Theorem,

$$\left(\frac{1}{2} + \frac{1}{2}i\right)^{10} = \left(\frac{\sqrt{2}}{2}\right)^{10}\left(\cos\frac{10\pi}{4} + i\sin\frac{10\pi}{4}\right)$$

$$= \frac{2^5}{2^{10}}\left(\cos\frac{5\pi}{2} + i\sin\frac{5\pi}{2}\right) = \frac{1}{32}i$$ ●

*n*th Roots of Complex Numbers

DeMoivre's Theorem can also be used to find the *n*th roots of complex numbers. An *n*th root of the complex number z is a complex number w such that

$$w^n = z$$

Writing these two numbers in trigonometric form as

$$w = s(\cos \phi + i \sin \phi) \quad \text{and} \quad z = r(\cos \theta + i \sin \theta)$$

and using DeMoivre's Theorem, we get

$$s^n(\cos n\phi + i \sin n\phi) = r(\cos \theta + i \sin \theta)$$

The equality of these two complex numbers shows that

$$s^n = r \quad \text{or} \quad s = r^{1/n}$$

and $\qquad\qquad \cos n\phi = \cos \theta \quad \text{and} \quad \sin n\phi = \sin \theta$

From the fact that sine and cosine have period 2π it follows that

$$n\phi = \theta + 2k\pi \quad \text{or} \quad \phi = \frac{\theta + 2k\pi}{n}$$

Thus

(5) $\qquad w = r^{1/n}\left[\cos\left(\frac{\theta + 2k\pi}{n}\right) + i \sin\left(\frac{\theta + 2k\pi}{n}\right)\right]$

Since the expression in Equation 5 gives a different value of w for $k = 0, 1, 2, \ldots, n - 1$, we have the following:

Let $z = r(\cos \theta + i \sin \theta)$ and let n be a positive integer. Then z has the n distinct nth roots

(6) $\qquad w_k = r^{1/n}\left[\cos\left(\frac{\theta + 2k\pi}{n}\right) + i \sin\left(\frac{\theta + 2k\pi}{n}\right)\right]$

where $k = 0, 1, 2, \ldots, n - 1$.

Notice that each of the nth roots of z has modulus $|w_k| = r^{1/n}$. Thus all the nth roots of z lie on the circle of radius $r^{1/n}$ in the complex plane. Also, since the argument of each successive nth root exceeds the argument of the previous root by $2\pi/n$, we see that the nth roots of z are equally spaced on this circle.

Example 5

Find the six sixth roots of $z = -8$ and graph these roots in the complex plane.

Solution

In trigonometric form $z = 8(\cos \pi + i \sin \pi)$. Applying Equation 6 with $n = 6$, we get

$$w_k = 8^{1/6}\left(\cos\frac{\pi + 2k\pi}{6} + i \sin\frac{\pi + 2k\pi}{6}\right)$$

We get six roots of -8 by taking $k = 0, 1, 2, 3, 4, 5$ in this formula:

$$w_0 = 8^{1/6}\left(\cos\frac{\pi}{6} + i \sin\frac{\pi}{6}\right) = \sqrt{2}\left(\frac{\sqrt{3}}{2} + \frac{1}{2}i\right)$$

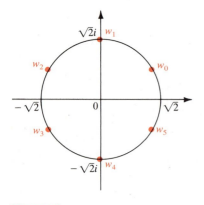

Figure 4
The six sixth roots of $z = -8$

$$w_1 = 8^{1/6}\left(\cos\frac{\pi}{2} + i\sin\frac{\pi}{2}\right) = \sqrt{2}i$$

$$w_2 = 8^{1/6}\left(\cos\frac{5\pi}{6} + i\sin\frac{5\pi}{6}\right) = \sqrt{2}\left(-\frac{\sqrt{3}}{2} + \frac{1}{2}i\right)$$

$$w_3 = 8^{1/6}\left(\cos\frac{7\pi}{6} + i\sin\frac{7\pi}{6}\right) = \sqrt{2}\left(-\frac{\sqrt{3}}{2} - \frac{1}{2}i\right)$$

$$w_4 = 8^{1/6}\left(\cos\frac{3\pi}{2} + i\sin\frac{3\pi}{2}\right) = -\sqrt{2}i$$

$$w_5 = 8^{1/6}\left(\cos\frac{11\pi}{6} + i\sin\frac{11\pi}{6}\right) = \sqrt{2}\left(\frac{\sqrt{3}}{2} - \frac{1}{2}i\right)$$

All these points lie on the circle of radius $\sqrt{2}$ as shown in Figure 4. ●

When finding roots of complex numbers we will sometimes write the argument θ of the complex number in degrees. In this case Equation 6 becomes

(7) $$w_k = r^{1/n}\left[\cos\left(\frac{\theta + 360°k}{n}\right) + i\sin\left(\frac{\theta + 360°k}{n}\right)\right]$$

for $k = 0, 1, 2, \ldots, n - 1$.

___

Example 6

Find the three cube roots of $z = 4\sqrt{2} + 4\sqrt{2}i$ and graph these roots in the complex plane.

Solution

First we write z in trigonometric form using degrees. Thus

$$z = 8(\cos 45° + i\sin 45°)$$

Applying Equation 7 with $n = 3$, we get that the cube roots of z are of the form

$$w_k = 8^{1/3}\left[\cos\left(\frac{45° + 360°k}{3}\right) + i\sin\left(\frac{45° + 360°k}{3}\right)\right]$$

Assigning to k the values $k = 0, 1, 2$ gives

$$w_0 = 2(\cos 15° + i\sin 15°)$$
$$w_1 = 2(\cos 135° + i\sin 135°)$$
$$w_2 = 2(\cos 255° + i\sin 255°)$$

The three cube roots of z are graphed in Figure 5. These roots are equally spaced on the circle $|z| = 2$. ●

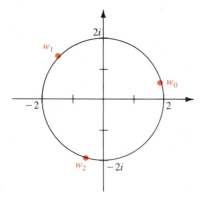

Figure 5
The three cube roots of
$z = 4\sqrt{2} + 4\sqrt{2}i$

___

Example 7

Solve the equation $z^6 + 8 = 0$.

Solution

This equation can be written $z^6 = -8$. Thus the solutions are the sixth roots of -8. These were found in Example 5.   ●

Exercises 6.7

In Exercises 1–24 write each complex number in trigonometric form with argument θ between 0 and 2π.

1. $1 + i$

2. $1 - \sqrt{3}i$

3. $\sqrt{2} - \sqrt{2}i$

4. $1 - i$

5. $2\sqrt{3} - 2i$

6. $-1 + i$

7. $-\sqrt{2}i$

8. $-3 - 3\sqrt{3}i$

9. $5 + 5i$

10. 4

11. $4\sqrt{3} - 4i$

12. $8i$

13. -20

14. $\sqrt{3} + i$

15. $3 + 4i$

16. $i(2 - 2i)$

17. $3i(1 + i)$

18. $2(1 - i)$

19. $4(\sqrt{3} + i)$

20. $-3 - 3i$

21. $2 + i$

22. $3 + \sqrt{3}i$

23. $\sqrt{2} + \sqrt{2}i$

24. $-\pi i$

In Exercises 25–34 write z_1 and z_2 in trigonometric form and then use Equations 1 and 2 to find the product $z_1 z_2$ and the quotients z_1/z_2 and $1/z_1$. Sketch $z_1, z_2, z_1 z_2, z_1/z_2$, and $1/z_1$.

25. $z_1 = \sqrt{3} + i, z_2 = 1 + \sqrt{3}i$

26. $z_1 = \sqrt{2} - \sqrt{2}i, z_2 = 1 - i$

27. $z_1 = 2\sqrt{3} - 2i, z_2 = -1 + i$

28. $z_1 = -\sqrt{2}i, z_2 = -3 - 3\sqrt{3}i$

29. $z_1 = 5 + 5i, z_2 = 4$

30. $z_1 = 4\sqrt{3} - 4i, z_2 = 8i$

31. $z_1 = -20, z_2 = \sqrt{3} + i$

32. $z_1 = 3 + 4i, z_2 = 2 - 2i$

33. $z_1 = 3i(1 + i), z_2 = 2(1 - i)$

34. $z_1 = 4(\sqrt{3} + i), z_2 = -3 - 3i$

In Exercises 35–46 find the indicated power using DeMoivre's Theorem.

35. $(1 + i)^{20}$

36. $(1 - \sqrt{3}i)^5$

37. $(2\sqrt{3} + 2i)^5$

38. $(1 - i)^8$

39. $\left(\dfrac{\sqrt{2}}{2} + \dfrac{\sqrt{2}}{2}i \right)^{12}$

40. $(\sqrt{3} - i)^{-10}$

41. $(2 - 2i)^8$

42. $\left(-\dfrac{1}{2} - \dfrac{\sqrt{3}}{2}i \right)^{15}$

43. $(-1 - i)^7$

44. $(3 + \sqrt{3}i)^4$

45. $(2\sqrt{3} + 2i)^{-5}$

46. $(1 - i)^{-8}$

In Exercises 47–56 find the indicated roots. Sketch the roots in the complex plane.

47. The square roots of $4\sqrt{3} + 4i$

48. The cube roots of $4\sqrt{3} + 4i$

49. The fourth roots of $-81i$

50. The fifth roots of 32

51. The eighth roots of 1

52. The cube roots of $1 + i$

53. The cube roots of i

54. The fifth roots of i

55. The fourth roots of -1

56. The fifth roots of $-16 - 16\sqrt{3}i$

In Exercises 57–64 solve the given equation.

57. $z^4 + 1 = 0$

58. $z^8 - i = 0$

59. $z^3 - 4\sqrt{3} - 4i = 0$

60. $z^2 + z + 1 = 0$

61. $z^6 - 1 = 0$

62. $z^3 + 1 = -i$

63. $iz^2 - 4z - 3i = 0$

64. $z^3 - 1 = 0$

65. Show that the sum of the four fourth roots of 1 is zero. (It is true in general that the sum of the n nth roots of z is zero.)

66. Find the product of the four fourth roots of 1.

67. **(a)** Let n be a positive integer and
$$w = \left(\cos\frac{2\pi}{n} + i\sin\frac{2\pi}{n} \right).$$ Show that
$1, w, w^2, w^3, \ldots, w^{n-1}$ are the n nth roots of 1.
   (b) If $z \neq 0$ is any complex number and $s^n = z$, show that the n nth roots of z are
$s, sw, sw^2, sw^3, \ldots, sw^{n-1}$.

Figure 1

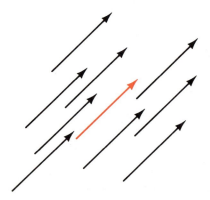

Figure 2

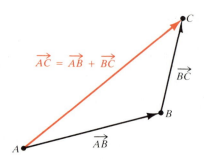

Figure 3

Application: Vectors

In applications of mathematics certain quantities are determined completely by their magnitude—for example, length, mass, area, temperature, and energy. We speak of a length of 5 m or a mass of 3 kg; only one number is needed to describe these quantities. Such quantities are called **scalars**.

On the other hand, to describe the displacement of an object two numbers are required: the *magnitude* and the *direction* of the displacement. Thus to describe a flight on a plane one must give the distance traveled as well as the direction of travel. Another example is velocity. To describe the velocity of a moving object we must specify both the *speed* and the *direction* of travel. Quantities such as displacement, velocity, acceleration, and force that involve magnitude as well as direction are called *directed* quantities. One way to represent such quantities mathematically is through the use of vectors.

Geometric Description of Vectors

A **vector** in the plane is a line segment with an assigned direction. We sketch a vector as in Figure 1 with an arrowhead at one end to specify the direction. We denote this vector by $\overrightarrow{AB}$. The point A is called the **initial point** and B is called the **terminal point** of the vector $\overrightarrow{AB}$. The length of the line segment AB is called the **magnitude** or **length** of the vector and is denoted by $|\mathbf{AB}|$. We use boldface letters to denote vectors. Thus we write $\mathbf{u} = \overrightarrow{AB}$.

Two vectors are considered **equal** if they have equal magnitudes and the same direction. Thus all the vectors in Figure 2 are equal.

This definition of equality makes sense if we think of a vector as representing a displacement. Two such displacements are the same if they have equal magnitudes and the same direction. The initial and terminal points of the vector depend on the initial position of the object to which this displacement is applied. So the vectors in Figure 2 can be thought of as the *same* displacement applied to objects in different locations in the plane. If the displacement $\mathbf{u} = \overrightarrow{AB}$ is followed by the displacement $\mathbf{v} = \overrightarrow{BC}$, then the resulting displacement is $\overrightarrow{AC}$ as shown in Figure 3. In other words, the single displacement represented by the vector $\overrightarrow{AC}$ has the same effect as the other two displacements. We will call the vector $\overrightarrow{AC}$ the **sum** of the vectors $\overrightarrow{AB}$ and $\overrightarrow{BC}$ and we write $\overrightarrow{AC} = \overrightarrow{AB} + \overrightarrow{BC}$. The **zero vector**, denoted by $\mathbf{0}$, represents no displacement.

To find the sum of two vectors $\mathbf{u}$ and $\mathbf{v}$ we sketch vectors equal to $\mathbf{u}$ and $\mathbf{v}$ with the initial point of one at the terminal point of the other [see Figure 4(a)]. Notice that if we draw $\mathbf{u}$ and $\mathbf{v}$ starting at the same point, then $\mathbf{u} + \mathbf{v}$ is the vector that is the diagonal of the parallelogram formed by $\mathbf{u}$ and $\mathbf{v}$, as shown in Figure 4(b).

If a is a real number and $\mathbf{v}$ is a vector, we define a new vector $a\mathbf{v}$ as follows: The vector $a\mathbf{v}$ has magnitude $|a||\mathbf{v}|$ and has the same direction as $\mathbf{v}$ if $a > 0$ and the opposite direction to $\mathbf{v}$ if $a < 0$. If $a = 0$, then $a\mathbf{v} = \mathbf{0}$, the zero vector. This process is called **multiplication of a vector by a scalar**. Notice that multiplication of a vector by a scalar $a > 0$ has the effect of stretching or shrinking a vector while at the same time preserving its direction. Multiplication by $a < 0$ has a

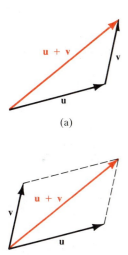

(a)

(b)

Figure 4

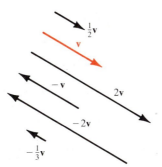

Figure 5

Multiplication of a vector by a scalar

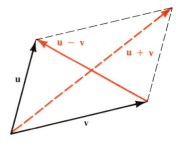

Figure 6

Subtraction of vectors

similar effect on the length of a vector but reverses the direction. Figure 5 shows graphs of the vector $a\mathbf{v}$ for different values of a.

We write the vector $(-1)\mathbf{v}$ as $-\mathbf{v}$. Thus $-\mathbf{v}$ is the vector with the same length as $\mathbf{v}$ but the opposite direction. **Subtraction** of two vectors $\mathbf{u}$ and $\mathbf{v}$ is defined by $\mathbf{u} - \mathbf{v} = \mathbf{u} + (-\mathbf{v})$. Figure 6 shows that the vector $\mathbf{u} - \mathbf{v}$ is the other diagonal of the parallelogram formed by $\mathbf{u}$ and $\mathbf{v}$.

Analytical Description of Vectors

So far we have been discussing vectors geometrically. We now give a method for describing vectors analytically. Consider a vector $\mathbf{v}$ in the coordinate plane as in Figure 7(a). Notice that to go from the initial point to the terminal point of $\mathbf{v}$ we move a units to the right and b units upward. We will represent this vector as an ordered pair of real numbers $\langle a, b \rangle$. The real number a is called the **horizontal component of v** and b is called the **vertical component of v**. We must remember that vectors represent a magnitude and a direction and not a particular arrow in the plane. Thus again there are many different representations of the vector $\langle a, b \rangle$ depending on the initial point [see Figure 7(b)].

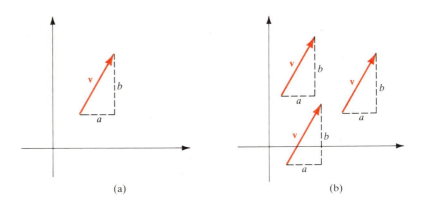

(a)

(b)

Figure 7

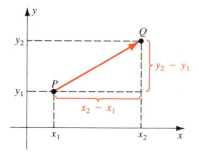

Figure 8

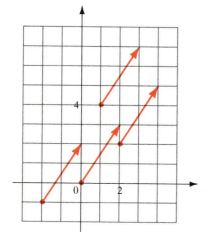

Figure 9

With this in mind we have the following (see Figure 8):

> If a vector **v** in the plane has initial point $P(x_1, y_1)$ and terminal point $Q(x_2, y_2)$, then
>
> **(1)**
> $$\mathbf{v} = \langle x_2 - x_1, y_2 - y_1 \rangle$$

Example 1

(a) Sketch representations of the vector $\mathbf{u} = \langle 2, 3 \rangle$ with initial points at $(0,0)$, $(2,2)$, $(-2,-1)$, and $(1,4)$.

Solution

These vectors are sketched in Figure 9.   ●

Example 2

(a) Find the vector with initial point $(-2, 5)$ and terminal point $(3, 7)$.

(b) Let $\mathbf{v} = \langle 3, 7 \rangle$. If a representation of **v** has initial point $(2, 4)$, what is the terminal point?

Solution

(a) From Equation 1 we see that the desired vector is

$$\langle 3 - (-2), 7 - 5 \rangle = \langle 5, 2 \rangle$$

This vector is sketched in Figure 10(a).

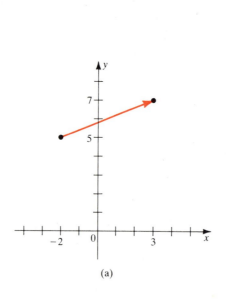

(a)

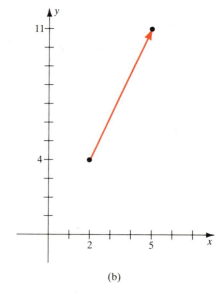

(b)

Figure 10

(b)   Let the terminal point be (x, y). Then by Equation 1 we have

$$\langle x - 2, y - 4 \rangle = \langle 3, 7 \rangle$$

So $x - 2 = 3$ and $y - 4 = 7$, or $x = 5$ and $y = 11$. The terminal point is $(5, 11)$. This representation of **v** is sketched in Figure 10(b).   ●

Two vectors are equal if and only if their corresponding components are equal. If $\mathbf{u} = \langle a, b \rangle$ and $\mathbf{v} = \langle c, d \rangle$, then addition, scalar multiplication, and length of vectors have the following meanings in terms of components:

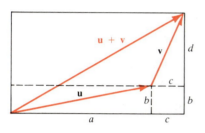

Figure 11

(2)

(3)

(4)

(5)

$$\mathbf{u} + \mathbf{v} = \langle a + c, b + d \rangle$$

$$\mathbf{u} - \mathbf{v} = \langle a - c, b - d \rangle$$

$$k\mathbf{u} = \langle ka, kb \rangle, \qquad k \in R$$

$$|\mathbf{u}| = \sqrt{a^2 + b^2}$$

The zero vector is the vector $\mathbf{0} = \langle 0, 0 \rangle$. The proofs of these properties are left as exercises. Figure 11 shows why (2) is true.

The following rules for manipulating vectors follow easily from Equations 2–5:

Vector addition	**Multiplication by a scalar**
$\mathbf{u} + \mathbf{v} = \mathbf{v} + \mathbf{u}$	$a(\mathbf{u} + \mathbf{v}) = a\mathbf{u} + a\mathbf{v}$
$\mathbf{u} + (\mathbf{v} + \mathbf{w}) = (\mathbf{u} + \mathbf{v}) + \mathbf{w}$	$(a + b)\mathbf{u} = a\mathbf{u} + b\mathbf{u}$
$\mathbf{u} + \mathbf{0} = \mathbf{u}$	$(ab)\mathbf{u} = a(b\mathbf{u}) = b(a\mathbf{u})$
$\mathbf{u} + (-\mathbf{u}) = \mathbf{0}$	$1\mathbf{u} = \mathbf{u}$
Length of a vector	$0\mathbf{u} = \mathbf{0}$
$|a\mathbf{u}| = |a|\,|\mathbf{u}|$	$a\mathbf{0} = \mathbf{0}$

Example 3

If $\mathbf{u} = \langle 2, -3 \rangle$ and $\mathbf{v} = \langle -1, 2 \rangle$, find $2\mathbf{u}$, $3\mathbf{v}$, $2\mathbf{u} - \mathbf{v}$, $2\mathbf{u} + 3\mathbf{v}$, $|\mathbf{v}|$, and $|-3\mathbf{v}|$.

Solution

Using Equations 2–5, we have

$$2\mathbf{u} = 2\langle 2, -3 \rangle = \langle 4, -6 \rangle \qquad 3\mathbf{v} = 3\langle -1, 2 \rangle = \langle -3, 6 \rangle$$

$$2\mathbf{u} - \mathbf{v} = 2\langle 2, -3 \rangle - \langle -1, 2 \rangle = \langle 4, -6 \rangle - \langle -1, 2 \rangle = \langle 5, -8 \rangle$$

$$2\mathbf{u} + 3\mathbf{v} = 2\langle 2, -3 \rangle + 3\langle -1, 2 \rangle = \langle 4, -6 \rangle + \langle -3, 6 \rangle = \langle 1, 0 \rangle$$

$$|\mathbf{v}| = \sqrt{(-1)^2 + 2^2} = \sqrt{5}$$

$$|-3\mathbf{v}| = |-3|\,|\mathbf{v}| = 3\sqrt{5}$$

●

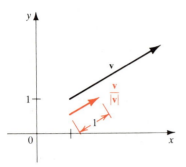

Figure 12

Unit vector in the direction of **v**

A vector of length 1 is called a **unit vector**. For example, the vector $\mathbf{v} = \langle \frac{3}{5}, \frac{4}{5} \rangle$ is a unit vector, since $|\mathbf{v}| = \sqrt{(\frac{3}{5})^2 + (\frac{4}{5})^2} = 1$. If **v** is any nonzero vector, then the vector

(6)
$$\mathbf{u} = \frac{1}{|\mathbf{v}|}\mathbf{v}$$

is a unit vector in the direction of **v** (see Figure 12).

Example 4

Find a unit vector in the same direction as $\mathbf{v} = \langle 2, -3 \rangle$.

Solution

Since $|\mathbf{v}| = \sqrt{2^2 + (-3)^2} = \sqrt{13}$, it follows from Equation 6 that

$$\mathbf{u} = \frac{1}{|\mathbf{v}|}\mathbf{v} = \frac{1}{\sqrt{13}}\langle 2, -3 \rangle = \left\langle \frac{2}{\sqrt{13}}, -\frac{3}{\sqrt{13}} \right\rangle$$

is the desired unit vector. It is a good exercise to check that this vector has length 1. ●

The Vectors i and j

The vectors **i** and **j** are defined as follows:

$$\mathbf{i} = \langle 1, 0 \rangle \qquad \mathbf{j} = \langle 0, 1 \rangle$$

(see Figure 13). These two unit vectors are special because every vector $\mathbf{u} = \langle a, b \rangle$ can be written in terms of them. In fact,

(7)
$$\mathbf{u} = a\mathbf{i} + b\mathbf{j}$$

The expression $a\mathbf{i} + b\mathbf{j}$ is called a **linear combination** of **i** and **j**. In other words, every vector can be written as a linear combination of these two vectors. (We note that **i** and **j** are not the only vectors with this property. See Exercise 58.)

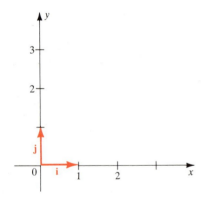

Figure 13

Example 5

(a) Write the vector $\mathbf{u} = \langle 5, -8 \rangle$ as a linear combination of **i** and **j**.

(b) If $\mathbf{u} = 3\mathbf{i} + 2\mathbf{j}$ and $\mathbf{v} = -\mathbf{i} + 6\mathbf{j}$, write $2\mathbf{u} + 5\mathbf{v}$ as a linear combination of **i** and **j**.

Solution

(a) By Equation 7, $\mathbf{u} = 5\mathbf{i} + (-8)\mathbf{j} = 5\mathbf{i} - 8\mathbf{j}$.

(b) The properties of addition and scalar multiplication of vectors show that we can manipulate vectors in the same way as algebraic expressions. Thus

$$2\mathbf{u} + 5\mathbf{v} = 2(3\mathbf{i} + 2\mathbf{j}) + 5(-\mathbf{i} + 6\mathbf{j})$$
$$= (6\mathbf{i} + 4\mathbf{j}) + (-5\mathbf{i} + 30\mathbf{j})$$
$$= \mathbf{i} + 34\mathbf{j}$$

●

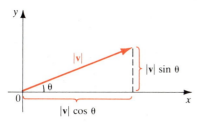

Figure 14

The Direction of a Vector

If the vector **v** is drawn with its initial point at the origin and if θ is the smallest positive angle in standard position formed by the positive x-axis and **v**, then we say that θ is the **direction** of **v** (see Figure 14).

Now suppose that the vector **v** has direction θ. If $\mathbf{v} = \langle a, b \rangle$, then from Figure 14 we have

(8)
$$a = |\mathbf{v}|\cos\theta \quad \text{and} \quad b = |\mathbf{v}|\sin\theta$$

On the other hand, if we know the components of $\mathbf{v} = \langle a, b \rangle$, then **v** has direction θ that satisfies

(9)
$$\tan\theta = \frac{b}{a}$$

Example 6

(a)   Find the components of the vector **v** with length 8 and direction $\pi/3$ and write **v** as a linear combination of **i** and **j**.

(b)   Find the direction of the vector $\mathbf{u} = -\sqrt{3}\mathbf{i} + \mathbf{j}$.

Solution

(a)   From Equation 8, $a = 8\cos(\pi/3) = 4$ and $b = 8\sin(\pi/3) = 4\sqrt{3}$. Thus $\mathbf{v} = \langle 4, 4\sqrt{3} \rangle$. In terms of the vectors **i** and **j** we can write $\mathbf{v} = 4\mathbf{i} + 4\sqrt{3}\mathbf{j}$.

(b)   The direction θ has the property that

$$\tan\theta = \frac{1}{-\sqrt{3}} = -\frac{\sqrt{3}}{3}$$

Thus the reference angle for θ is $\pi/6$. Since the terminal point of the vector **u** is in quadrant II, it follows that $\theta = 5\pi/6$.   ●

Exercises 6.8

*In Exercises 1 and 2 sketch vectors that represent $2\mathbf{v}$, $3\mathbf{v}$, $\frac{1}{2}\mathbf{v}$, $-\mathbf{v}$, $-2\mathbf{v}$, and $-\frac{1}{4}\mathbf{v}$ for the vector **v** shown.*

*In Exercises 3–6 two vectors **u** and **v** are shown. Sketch $\mathbf{u} + \mathbf{v}$, $\mathbf{u} - \mathbf{v}$, $\mathbf{v} - \mathbf{u}$, $2\mathbf{v} + \mathbf{u}$, and $\mathbf{v} - 2\mathbf{u}$.*

1.

2.

3.

4.

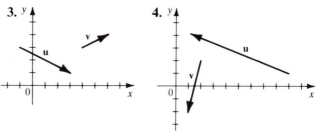

5.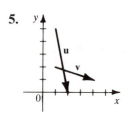

6.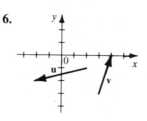

In Exercises 7–16 find $2\mathbf{u}$, $-3\mathbf{v}$, $\mathbf{u} + \mathbf{v}$, and $3\mathbf{u} - 4\mathbf{v}$ for the given vectors $\mathbf{u}$ and $\mathbf{v}$.

7. $\mathbf{u} = \langle 2, 7 \rangle$, $\mathbf{v} = \langle 3, 1 \rangle$

8. $\mathbf{u} = \langle -2, 5 \rangle$, $\mathbf{v} = \langle 2, -8 \rangle$

9. $\mathbf{u} = \langle 0, -1 \rangle$, $\mathbf{v} = \langle -2, 0 \rangle$

10. $\mathbf{u} = \langle 6, 0 \rangle$, $\mathbf{v} = \langle -2, 0 \rangle$

11. $\mathbf{u} = 0$, $\mathbf{v} = \mathbf{i} - 5\mathbf{j}$

12. $\mathbf{u} = -(2\mathbf{i} + \mathbf{j})$, $\mathbf{v} = 2(\mathbf{i} - \mathbf{j})$

13. $\mathbf{u} = 2\mathbf{i}$, $\mathbf{v} = 3\mathbf{i} - 2\mathbf{j}$ **14.** $\mathbf{u} = \mathbf{i}$, $\mathbf{v} = -2\mathbf{j}$

15. $\mathbf{u} = \mathbf{i} + \mathbf{j}$, $\mathbf{v} = \mathbf{i} - \mathbf{j}$

16. $\mathbf{u} = 2\mathbf{i} + \mathbf{j}$, $\mathbf{v} = -2\mathbf{i} - \mathbf{j}$

In Exercises 17–20 find $|\mathbf{u}|$, $|\mathbf{v}|$, $|2\mathbf{u}|$, $|\frac{1}{2}\mathbf{v}|$, $|\mathbf{u} + \mathbf{v}|$, $|2\mathbf{u} - 3\mathbf{v}|$, $|\mathbf{u} - \mathbf{v}|$, and $|\mathbf{u}| - |\mathbf{v}|$.

17. $\mathbf{u} = 2\mathbf{i} + \mathbf{j}$, $\mathbf{v} = 3\mathbf{i} - 2\mathbf{j}$

18. $\mathbf{u} = -2\mathbf{i} + 3\mathbf{j}$, $\mathbf{v} = \mathbf{i} - 2\mathbf{j}$

19. $\mathbf{u} = \langle 10, -1 \rangle$, $\mathbf{v} = \langle -2, -2 \rangle$

20. $\mathbf{u} = \langle -6, 6 \rangle$, $\mathbf{v} = \langle -2, -1 \rangle$

In Exercises 21–28 find the vector with initial point P and terminal point Q.

21. $P(3, 2)$, $Q(8, 9)$ **22.** $P(1, 1)$, $Q(9, 9)$

23. $P(5, 3)$, $Q(1, 0)$ **24.** $P(-1, 3)$, $Q(-6, -1)$

25. $P(6, 4)$, $Q(0, 0)$ **26.** $P(0, 0)$, $Q(6, 4)$

27. $P(-1, -1)$, $Q(-1, 1)$ **28.** $P(-8, -6)$, $Q(-1, -1)$

In Exercises 29–36, (a) find the magnitude and direction (in degrees) of the vector and (b) find a unit vector in the same direction as the given vector.

29. $\mathbf{v} = \langle 3, 4 \rangle$ **30.** $\mathbf{v} = \langle -2, 2 \rangle$

31. $\mathbf{v} = \langle \sqrt{3}, -3 \rangle$ **32.** $\mathbf{v} = \left\langle -\dfrac{\sqrt{2}}{2}, -\dfrac{\sqrt{2}}{2} \right\rangle$

33. $\mathbf{v} = \langle -12, 5 \rangle$ **34.** $\mathbf{v} = \langle 40, 9 \rangle$

35. $\mathbf{v} = \mathbf{i} + \sqrt{3}\mathbf{j}$ **36.** $\mathbf{v} = \mathbf{i} + \mathbf{j}$

In Exercises 37–44 find the horizontal and vertical components of the vectors with given length and direction, and write the vector as a linear combination of the vectors $\mathbf{i}$ and $\mathbf{j}$.

37. $|\mathbf{v}| = 40$, $\theta = 30°$ **38.** $|\mathbf{v}| = 50$, $\theta = 120°$

39. $|\mathbf{v}| = 1$, $\theta = 225°$ **40.** $|\mathbf{v}| = \sqrt{2}$, $\theta = 45°$

41. $|\mathbf{v}| = 800$, $\theta = 125°$ **42.** $|\mathbf{v}| = 22$, $\theta = 15°$

43. $|\mathbf{v}| = 4$, $\theta = 10°$ **44.** $|\mathbf{v}| = \sqrt{3}$, $\theta = 300°$

45. A man pushes a lawn mower with a force of 30 lb exerted at an angle of 30° to the ground. Find the horizontal and vertical components of the force.

46. A jet is flying in a direction N20°E with a speed of 500 mi/h. Find the north and east components of the velocity.

In Exercises 47–57 prove the stated property.

47. $\mathbf{u} + \mathbf{v} = \mathbf{v} + \mathbf{u}$

48. $\mathbf{u} + (\mathbf{v} + \mathbf{w}) = (\mathbf{u} + \mathbf{v}) + \mathbf{w}$

49. $\mathbf{u} + \mathbf{0} = \mathbf{u}$ **50.** $\mathbf{u} + (-\mathbf{u}) = \mathbf{0}$

51. $a(\mathbf{u} + \mathbf{v}) = a\mathbf{u} + a\mathbf{v}$ **52.** $(a + b)\mathbf{u} = a\mathbf{u} + b\mathbf{u}$

53. $(ab)\mathbf{u} = a(b\mathbf{u}) = b(a\mathbf{u})$ **54.** $1\mathbf{u} = \mathbf{u}$

55. $0\mathbf{u} = \mathbf{0}$ **56.** $a\mathbf{0} = \mathbf{0}$ **57.** $|a\mathbf{u}| = |a||\mathbf{u}|$

58. Show that every vector in the plane is a linear combination of the vectors $\mathbf{m} = \langle 1, 1 \rangle$ and $\mathbf{n} = \langle 0, 1 \rangle$.

Section 6.9

Application: Velocity, Force, and Work

Vectors are ideal for describing mathematically such quantities as velocity and force. In this section we consider how vectors are used in the study of these quantities.

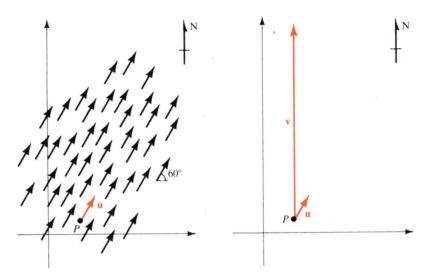

Figure 1

Velocity

Figure 1 shows some vectors that represent wind flowing with a constant velocity of 40 mi/h in the direction N30°E. Suppose that an airplane is flying through this wind at P, the pilot heads his plane straight north, and the speed of the airplane (in still air) is 300 mi/h. It is obvious from experience that the wind will affect the speed and the direction of the airplane. To find out exactly what this effect will be, we draw the vector $\mathbf{u}$ that represents the flow of the wind with its initial point at P. The velocity of the plane is represented by the vector $\mathbf{v}$. The resulting velocity of the plane (relative to the ground) is given by the vector $\mathbf{u} + \mathbf{v}$.

Example 1

Assume the airplane and the wind are as described above.

(a)   Express the velocities of the wind and the airplane as vectors.

(b)   Find the true course of the airplane.

Solution

We choose a coordinate system as shown in Figure 2.

(a)   We write the vector $\mathbf{u}$ in terms of components. We have

$$x\text{-component of } \mathbf{u} \text{ is } 40 \cos 60° = 20$$

$$y\text{-component of } \mathbf{u} \text{ is } 40 \sin 60° = 40\frac{\sqrt{3}}{2} = 20\sqrt{3}$$

Thus $\mathbf{u} = 20\mathbf{i} + 20\sqrt{3}\,\mathbf{j} \approx 20\mathbf{i} + 34.64\mathbf{j}$. Clearly, $\mathbf{v} = 0\mathbf{i} + 300\mathbf{j} = 300\mathbf{j}$.

(b)   The true course of the airplane is given by the vector $\mathbf{w} = \mathbf{u} + \mathbf{v}$:

$$\mathbf{w} = \mathbf{u} + \mathbf{v} = (20\mathbf{i} + 20\sqrt{3}\,\mathbf{j}) + (300\mathbf{j})$$
$$= 20\mathbf{i} + (300 + 20\sqrt{3})\mathbf{j}$$
$$\approx 20\mathbf{i} + 334.64\mathbf{j}$$

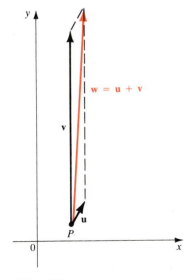

Figure 2

The true speed of the airplane is $|\mathbf{w}| \approx \sqrt{(20)^2 + (334.64)^2} \approx 335.238$ mi/h. The direction of the airplane is the direction of the vector $\mathbf{w}$. By Equation 9 in Section 6.8 the angle θ has the property that $\tan \theta = 334.64/20 = 16.732$. Since θ is clearly between 0 and $\pi/2$, it follows that $\theta \approx 86.6°$. In other words, the plane is heading in the direction N3.4°E.   ●

Example 2

A woman in a boat starts at one shore of a straight river that flows from west to east and she wants to reach the point directly on the opposite side. If the speed of the boat (in still water) is 10 mi/h and the river is flowing east at the rate of 5 mi/h, in what direction should she head her boat in order to arrive at her desired destination?

Solution

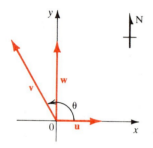

Figure 3

We choose a coordinate system with the origin at the initial position of the boat as shown in Figure 3. Let the vectors $\mathbf{u}$ and $\mathbf{v}$ represent the velocities of the river and the boat, respectively. Clearly,

$$\mathbf{u} = 5\mathbf{i}$$

and since $|\mathbf{v}| = 10$,

$$\mathbf{v} = (10 \cos \theta)\mathbf{i} + (10 \sin \theta)\mathbf{j}$$

where the angle θ is as shown in Figure 3. The true course of the boat is then given by the vector $\mathbf{w} = \mathbf{u} + \mathbf{v}$. So

$$\mathbf{w} = \mathbf{u} + \mathbf{v} = 5\mathbf{i} + (10 \cos \theta)\mathbf{i} + (10 \sin \theta)\mathbf{j} = (5 + 10 \cos \theta)\mathbf{i} + (10 \sin \theta)\mathbf{j}$$

Since the woman wants to reach a point directly on the other side of the river, her direction should have horizontal component zero. In other words, she should choose θ in such a way that

$$5 + 10 \cos \theta = 0$$
$$\cos \theta = -\tfrac{1}{2}$$

so

$$\theta = 120°$$

Thus she should head her boat in the direction $\theta = 120°$ (or N30°W) as shown in Figure 3.   ●

Force

Force is also represented by a vector. Intuitively one can think of force as describing a push or a pull on an object—for example, a horizontal push of a book across a table or the downward pull of the earth's gravity on a ball. Force is measured in pounds (or newtons in the metric system). For instance, a man who weighs 200 lb exerts a force of 200 lb down on the ground. If several forces are acting on an object, the resultant force experienced by the object is the vector sum of these forces.

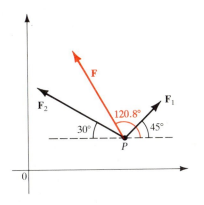

Figure 4

Example 3

Two forces $\mathbf{F}_1$ and $\mathbf{F}_2$ with magnitudes 10 lb and 20 lb act on an object at a point P as shown in Figure 4. Find the resultant force acting at P.

Solution

We write $\mathbf{F}_1$ and $\mathbf{F}_2$ in terms of components:

$$\mathbf{F}_1 = 10\cos 45°\mathbf{i} + 10\sin 45°\mathbf{j} = 10\frac{\sqrt{2}}{2}\mathbf{i} + 10\frac{\sqrt{2}}{2}\mathbf{j} = 5\sqrt{2}\mathbf{i} + 5\sqrt{2}\mathbf{j}$$

$$\mathbf{F}_2 = -20\cos 30°\mathbf{i} + 20\sin 30°\mathbf{j} = -20\frac{\sqrt{3}}{2}\mathbf{i} + 20\frac{1}{2}\mathbf{j} = -10\sqrt{3}\mathbf{i} + 10\mathbf{j}$$

So the resultant force $\mathbf{F}$ is

$$\begin{aligned}
\mathbf{F} &= \mathbf{F}_1 + \mathbf{F}_2 \\
&= (5\sqrt{2}\mathbf{i} + 5\sqrt{2}\mathbf{j}) + (-10\sqrt{3}\mathbf{i} + 10\mathbf{j}) \\
&= (5\sqrt{2} - 10\sqrt{3})\mathbf{i} + (5\sqrt{2} + 10)\mathbf{j} \\
&\approx -10.2\mathbf{i} + 17.1\mathbf{j}
\end{aligned}$$

The length of $\mathbf{F}$ is $|\mathbf{F}| \approx 19.9$ and the direction of $\mathbf{F}$ is θ, where $\tan\theta \approx (17.1)/(-10.2) \approx -1.67647$. Thus the reference angle for θ is approximately $59.2°$. Since the terminal point of $\mathbf{F}$ is in quadrant II, it follows that $\theta \approx 180° - 59.2° = 120.8°$. Thus the resultant force has magnitude 19.9 lb and is in the direction $120.8°$ from the positive x direction.   ●

Suppose that a car weighing 3000 lb sits on a driveway that is inclined 15° to the horizontal. The car exerts a force directly downward of 3000 lb. But what is the actual force caused by the weight of the car on the driveway? The answer is not 3000 lb. To see this notice that part of the 3000-lb force causes the car to roll down the driveway. In other words, part of this force acts parallel to the driveway. The remaining portion of the force acts perpendicular to the driveway. These forces are called $\mathbf{u}$ and $\mathbf{v}$ in Figure 5 and $\mathbf{w}$ is the force of 3000 lb acting directly downward. We say that the vector $\mathbf{w}$ is **resolved** into the vectors $\mathbf{u}$ and $\mathbf{v}$ and these vectors are called **components** of $\mathbf{w}$.

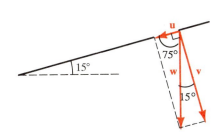

Figure 5

Example 4

Assume the car and driveway described above.

(a)   Find the magnitude of the force required to push up the driveway in order to keep the car from rolling.

(b)   Find the actual force experienced by the driveway due to the weight of the car.

Solution

(a)   In order to keep the car from rolling, the force required is equal to the force exerted by the car down the driveway but in the opposite direction. In other words, we need to find the length of **u** in Figure 5:

$$|\mathbf{u}| = 3000 \cos 75° \approx 776.5$$

(b)   The magnitude of this force is the length of the vector **v**:

$$|\mathbf{v}| = 3000 \sin 75° \approx 2897.8$$
●

The Dot Product and Work

If $\mathbf{u} = \langle a_1, b_1 \rangle$ and $\mathbf{v} = \langle a_2, b_2 \rangle$ the **dot product** of **u** and **v** is defined by

(1) $$\mathbf{u} \cdot \mathbf{v} = a_1 a_2 + b_1 b_2$$

Thus to find the dot product of **u** and **v** we multiply corresponding components and add. The result is *not* a vector; it is a real number, or scalar.

Example 5

Find the dot product of the given pairs of vectors.

(a) $\mathbf{u} = \langle 3, -2 \rangle, \mathbf{v} = \langle 4, 5 \rangle$

(b) $\mathbf{u} = 2\mathbf{i} + 5\mathbf{j}, \mathbf{v} = 8\mathbf{i} - 7\mathbf{j}$

Solution

(a) $\mathbf{u} \cdot \mathbf{v} = \langle 3, -2 \rangle \cdot \langle 4, 5 \rangle = (3)(4) + (-2)(5) = 2$

(b) $\mathbf{u} \cdot \mathbf{v} = (2\mathbf{i} + 5\mathbf{j}) \cdot (8\mathbf{i} - 7\mathbf{j}) = (2)(8) + (5)(-7) = -19$
●

We define the **angle θ between two vectors u** and **v** to be the angle between the representations of **u** and **v** with initial points at the origin, where $0 \leq \theta \leq \pi$. The usefulness of the dot product is apparent from the following property:

If θ is the angle between two nonzero vectors **u** and **v**, then

(2) $$\mathbf{u} \cdot \mathbf{v} = |\mathbf{u}| |\mathbf{v}| \cos \theta$$

(This fact is a consequence of the Law of Cosines. A proof is outlined in Exercise 53.) It follows that the cosine of the angle θ between the vectors **u** and **v** is

(3) $$\cos \theta = \frac{\mathbf{u} \cdot \mathbf{v}}{|\mathbf{u}| |\mathbf{v}|}$$

Example 6

Find the angle between the vectors $\mathbf{u} = \langle 2, 5 \rangle$ and $\mathbf{v} = \langle 4, -3 \rangle$.

Solution

Since

$$|\mathbf{u}| = \sqrt{2^2 + 5^2} = \sqrt{29} \quad \text{and} \quad |\mathbf{v}| = \sqrt{4^2 + (-3)^2} = 5$$

and since $$\mathbf{u} \cdot \mathbf{v} = (2)(4) + (5)(-3) = -7$$

it follows from Equation 3 that

$$\cos \theta = \frac{\mathbf{u} \cdot \mathbf{v}}{|\mathbf{u}| \, |\mathbf{v}|} = \frac{-7}{5\sqrt{29}}$$

Thus the angle between $\mathbf{u}$ and $\mathbf{v}$ is

$$\theta = \cos^{-1}\left(\frac{-7}{5\sqrt{29}}\right) \approx 105.1°$$

●

Two nonzero vectors $\mathbf{u}$ and $\mathbf{v}$ are called **perpendicular** or **orthogonal** if the angle between them is $\pi/2$. In this case it follows from Equation 2 that

$$\mathbf{u} \cdot \mathbf{v} = |\mathbf{u}| \, |\mathbf{v}| \cos \frac{\pi}{2} = 0$$

and conversely, if $\mathbf{u} \cdot \mathbf{v} = 0$, then $\cos \theta = 0$, so $\theta = \pi/2$. Therefore

---

$\mathbf{u}$ and $\mathbf{v}$ are orthogonal if and only if $\mathbf{u} \cdot \mathbf{v} = 0$

Example 7
Determine whether the following pairs of vectors are perpendicular:

(a) $\mathbf{u} = \langle 3, 5 \rangle$ and $\mathbf{v} = \langle 2, -8 \rangle$,

(b) $\mathbf{u} = \langle 2, 1 \rangle$ and $\mathbf{v} = \langle -1, 2 \rangle$,

(c) $\mathbf{u} = \langle -4, 8 \rangle$ and $\mathbf{v} = \langle -2, -1 \rangle$.

Solution

(a) $\mathbf{u} \cdot \mathbf{v} = (3)(2) + (5)(-8) = -34$, so $\mathbf{u}$ and $\mathbf{v}$ are not perpendicular.

(b) $\mathbf{u} \cdot \mathbf{v} = (2)(-1) + (1)(2) = 0$, so $\mathbf{u}$ and $\mathbf{v}$ are perpendicular.

(c) $\mathbf{u} \cdot \mathbf{v} = (-4)(-2) + (8)(-1) = 0$, so $\mathbf{u}$ and $\mathbf{v}$ are perpendicular.   ●

The rules for manipulating the dot product of vectors follow easily from the definition. We list them here and leave their proofs as exercises.

Rules for Dot Products

$$\mathbf{u} \cdot \mathbf{v} = \mathbf{v} \cdot \mathbf{u}$$
$$(a\mathbf{u}) \cdot \mathbf{v} = a(\mathbf{u} \cdot \mathbf{v}) = \mathbf{u} \cdot (a\mathbf{v})$$
$$(\mathbf{u} + \mathbf{v}) \cdot \mathbf{w} = \mathbf{u} \cdot \mathbf{w} + \mathbf{v} \cdot \mathbf{w}$$

Notice that if $\mathbf{u} = \langle a, b \rangle$, then

$$\mathbf{u} \cdot \mathbf{u} = \langle a, b \rangle \cdot \langle a, b \rangle = a^2 + b^2 = |\mathbf{u}|^2$$

Taking the square root of both sides of this equation, we get the following connection between the length of a vector and the dot product:

$$|\mathbf{u}| = \sqrt{\mathbf{u} \cdot \mathbf{u}}$$

One use of the dot product occurs in physics in calculating **work**. The term *work* is used in everyday language to mean the total amount of effort required to perform a task. In physics work has a technical meaning that conforms with this intuitive meaning. If a constant force of magnitude F moves an object through a distance d along a straight line, then the work done is

(4) $W = Fd$ or   work = force × distance

If F is measured in pounds and d in feet, then the unit of work is a foot-pound (ft-lb). For example, how much work is done in lifting a 20-lb weight 6 ft off the ground? Since a force of 20 lb is required to lift this weight and since the weight moves through a distance of 6 ft, the amount of work done is

$$W = Fd = (20)(6) = 120 \text{ ft-lb}$$

Equation 4 applies only when the force is directed along the line of motion of the object. Suppose, however, that the constant force $\mathbf{F} = \overrightarrow{PR}$ is a vector that points in some other direction as in Figure 6. If the force moves the object from P to Q, then the displacement vector is $\mathbf{D} = \overrightarrow{PQ}$. The portion of the force that acts in the direction of motion of the object is $|\mathbf{F}|\cos\theta$. Since this force acts through a distance $|\mathbf{D}|$, the work done is $W = Fd$ or

$$W = (|\mathbf{F}|\cos\theta)|\mathbf{D}| = |\mathbf{F}||\mathbf{D}|\cos\theta$$

From Equation 2 we get

(5) $$\boxed{W = \mathbf{F} \cdot \mathbf{D}}$$

Thus the work done by a constant force $\mathbf{F}$ is the dot product $\mathbf{F} \cdot \mathbf{D}$, where $\mathbf{D}$ is the displacement vector.

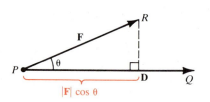

Figure 6

Example 8

A man pulls a wagon horizontally by exerting a force of 20 lb on the handle. If the handle makes an angle of 60° with the horizontal, find the work done in moving the wagon 100 ft.

Solution

Choose a coordinate system with the origin at the initial position of the wagon (see Figure 7). Thus the wagon moves from the point $P(0,0)$ to the point $Q(100,0)$. The vector that represents this displacement is

$$\mathbf{D} = 100\mathbf{i}$$

The force on the handle is given by

$$\mathbf{F} = (20\cos 60°)\mathbf{i} + (20\sin 60°)\mathbf{j} = 10\mathbf{i} + 10\sqrt{3}\mathbf{j}$$

Thus the work done is

$$W = \mathbf{F} \cdot \mathbf{D} = (10\mathbf{i} + 10\sqrt{3}\mathbf{j}) \cdot (100\mathbf{i}) = 1000 \text{ ft-lb}$$

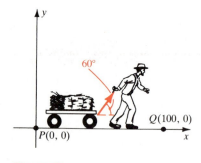

Figure 7

Exercises 6.9

1. A pilot heads his jet due east. The jet has a speed of 425 mi/h in still air. If the wind is blowing due north with a speed of 40 mi/h, find the true speed and direction of the jet.

2. A jet is flying through a wind that is blowing with a speed of 55 mi/h in the direction N30°E. The jet has a speed of 765 mi/h (in still air) and the pilot heads the jet in the direction N45°E. Find the true speed and direction of the jet.

3. Find the true speed and direction of the jet in Exercise 2 if the pilot heads the plane in the direction N30°W.

4. In what direction should the pilot in Exercise 2 head the plane in order for the true course to be due north?

5. A straight river flows east at a speed of 10 mi/h. A boater starts at the south shore of the river and heads in a direction 60° from the shore (see the figure). If the motorboat has a speed of 20 mi/h (in still water), find the true speed and direction of the boat.

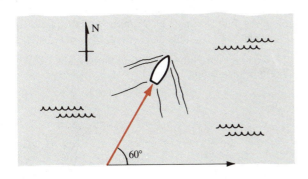

6. The boater in Exercise 5 wants to arrive at a point on the north shore of the river directly opposite the point where she started. In what direction should she head her boat?

7. A boat heads in the direction N72°E. The speed of the boat in still water is 24 mi/h. The water is flowing directly south. It is observed that the true direction of the boat is directly east. Find the speed of the water and the true speed of the boat.

8. A woman walks due west on the deck of an ocean liner at 2 mi/h. The ocean liner is moving due north at a speed of 25 mi/h. Find the speed and direction of the woman relative to the surface of the water.

9. A sailboat has its sail inclined in the direction N20°E.

The wind is blowing into the sail in the direction S45°W with a force of 220 lb (see the figure).

(a) Find the effective force of the wind on the sail. (*Hint:* Resolve the force of the wind into two components, one parallel to the sail and one perpendicular to the sail. The component of the wind parallel to the sail slips by and does not push on the sail.)

(b) If the keel of the ship is aligned directly to the north, find the effective force of the wind that drives the boat forward. (*Hint:* Only the portion of the force found in part (a) that is parallel to the keel will drive the boat forward.)

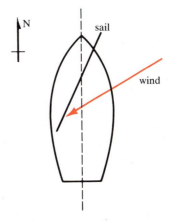

The forces $\mathbf{F}_1, \mathbf{F}_2, \ldots, \mathbf{F}_n$ acting at the same point P are said to be in equilibrium if the resultant force is zero—that is, if $\mathbf{F}_1 + \mathbf{F}_2 + \cdots + \mathbf{F}_n = 0$. In Exercises 10–19 find (a) the resultant sum of the forces acting at 0, and (b) the additional force required (if any) in order for the forces to be in equilibrium.

10. $\mathbf{F}_1 = \langle 2, 5 \rangle, \mathbf{F}_2 = \langle 3, -8 \rangle$

11. $\mathbf{F}_1 = \langle 3, -7 \rangle, \mathbf{F}_2 = \langle 4, -2 \rangle, \mathbf{F}_3 = \langle -7, 9 \rangle$

12. $\mathbf{F}_1 = \langle 2, -5 \rangle, \mathbf{F}_2 = \langle 3, 8 \rangle, \mathbf{F}_3 = \langle 1, 1 \rangle$

13. $\mathbf{F}_1 = \langle 4, -4 \rangle, \mathbf{F}_2 = \langle 2, -7 \rangle, \mathbf{F}_3 = \langle -8, 3 \rangle, \mathbf{F}_4 = \langle -5, -5 \rangle$

14. $\mathbf{F}_1 = 4\mathbf{i} - \mathbf{j}, \mathbf{F}_2 = 3\mathbf{i} - 7\mathbf{j}, \mathbf{F}_3 = -8\mathbf{i} + 3\mathbf{j}, \mathbf{F}_4 = \mathbf{i} + \mathbf{j}$

15. $\mathbf{F}_1 = \mathbf{i} - \mathbf{j}, \mathbf{F}_2 = \mathbf{i} + \mathbf{j}, \mathbf{F}_3 = -2\mathbf{i} + \mathbf{j}$

16.

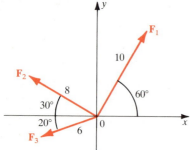

17.

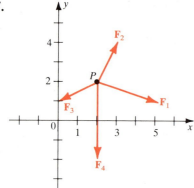

18.

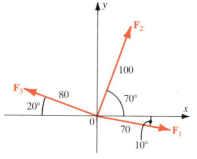

19.

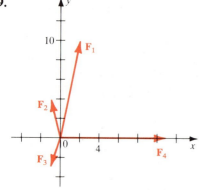

20. A 100-lb weight hangs from a string as shown in the figure. Find the tensions T_1 and T_2 in the string.

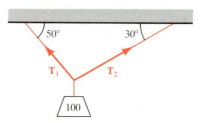

21. A car is on a driveway that is inclined 25° to the horizontal. If the car weighs 2755 lb, find the force exerted by the car on the driveway. Find the force required to push the car up the driveway in order to keep it from rolling down.

22. A car is on a driveway that is inclined 10° to the horizontal. A force of 490 lb is required just to keep the car from rolling down the driveway.
   (a) Find the weight of the car.
   (b) Find the force the car exerts against the driveway.

23. A package that weighs 200 lb is on an inclined plane. If a force of 80 lb is just sufficient to keep the package from sliding, find the angle of inclination of the plane. (Ignore the effects of friction.)

24. Two cranes are lifting an object that weighs 18,278 lb. Find the tensions T_1 and T_2.

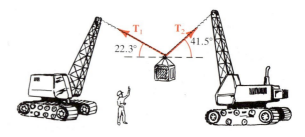

In Exercises 25–28 find $\mathbf{u} \cdot \mathbf{v}$, $\mathbf{u} \cdot (\mathbf{v} + \mathbf{u})$, $2\mathbf{u} \cdot 3\mathbf{v}$, $(\mathbf{u} \cdot \mathbf{v})\mathbf{u}$, *and* $(\mathbf{u} \cdot \mathbf{u})\mathbf{u}$.

25. $\mathbf{u} = \langle 3, -2 \rangle$, $\mathbf{v} = \langle 1, 2 \rangle$

26. $\mathbf{u} = \langle 1, 5 \rangle$, $\mathbf{v} = \langle 1, 1 \rangle$ **27.** $\mathbf{u} = \mathbf{i} - 2\mathbf{j}$, $\mathbf{v} = 3\mathbf{j}$

28. $\mathbf{u} = 2\mathbf{i} + \mathbf{j}$, $\mathbf{v} = \mathbf{i} + 2\mathbf{j}$

In Exercises 29–34 find the angle between the vectors $\mathbf{u}$ *and* $\mathbf{v}$ *to the nearest degree.*

29. $\mathbf{u} = \langle 2, 7 \rangle$, $\mathbf{v} = \langle 3, 1 \rangle$

30. $\mathbf{u} = \langle -2, 5 \rangle$, $\mathbf{v} = \langle 5, 2 \rangle$

31. $\mathbf{u} = \langle -6, 6 \rangle, \mathbf{v} = \langle 1, -1 \rangle$

32. $\mathbf{u} = 2\mathbf{i} + \mathbf{j}, \mathbf{v} = 3\mathbf{i} - 2\mathbf{j}$ **33.** $\mathbf{u} = \mathbf{i}, \mathbf{v} = -5\mathbf{j}$

34. $\mathbf{u} = \mathbf{i} + \mathbf{j}, \mathbf{v} = \mathbf{i} - \mathbf{j}$

35. A constant force given by the vector $\mathbf{F} = \langle 2, 8 \rangle$ moves an object along a straight line from the point $(2, 5)$ to the point $(11, 13)$. Find the work done if the distance is measured in feet and the force is measured in pounds.

36. A lawn mower is pushed a distance of 200 ft along a horizontal path by a constant force of 50 lb. The handle of the lawn mower is at an angle of $30°$ to the horizontal. Find the work done.

37. In this problem we prove that $\mathbf{u} \cdot \mathbf{v} = |\mathbf{u}||\mathbf{v}|\cos \theta$, where θ is the angle between $\mathbf{u}$ and $\mathbf{v}$.
   (a) Apply the Law of Cosines to the triangle in the figure to obtain
$$|\mathbf{u} - \mathbf{v}|^2 = |\mathbf{u}|^2 + |\mathbf{v}|^2 - 2|\mathbf{u}||\mathbf{v}|\cos \theta$$
   (b) Show that $|\mathbf{u} - \mathbf{v}|^2 = |\mathbf{u}|^2 - 2\mathbf{u} \cdot \mathbf{v} + |\mathbf{v}|^2$.
   (c) Conclude from parts (a) and (b) that $\mathbf{u} \cdot \mathbf{v} = |\mathbf{u}||\mathbf{v}|\cos \theta$.

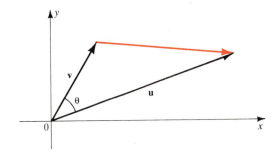

CHAPTER 6 REVIEW

Define, state, or discuss each of the following.

1. Equation
2. Identity
3. Trigonometric equation
4. Trigonometric identity
5. Fundamental trigonometric identities
6. Proving identities
7. Solving trigonometric equations
8. Addition formulas for sine, cosine, and tangent
9. Subtraction formulas for sine, cosine, and tangent
10. Double-angle formulas for sine, cosine, and tangent
11. Half-angle formulas for sine, cosine, and tangent
12. Product-to-sum identities
13. Sum-to-product identities
14. Inverse of a function
15. The inverse trigonometric function
16. Complex numbers
17. The complex plane
18. Graphing complex numbers
19. Trigonometric form of complex numbers
20. Multiplying complex numbers in trigonometric form
21. DeMoivre's Theorem
22. nth roots of complex numbers
23. Scalar and vector quantities
24. Vectors
25. Components of a vector
26. Addition of vectors
27. Scalar multiplication
28. Length of a vector
29. Direction of a vector
30. The vectors $\mathbf{i}$ and $\mathbf{j}$
31. Dot product of vectors
32. Angle between two vectors
33. Orthogonal vectors
34. Work

Review Exercises

In Exercises 1–30 verify the given identity.

1. $\cos^2 x \csc x - \csc x = -\sin x$

2. $\dfrac{1}{1 - \sin^2 x} = 1 + \tan^2 x$

3. $\dfrac{\cos^2 x - \tan^2 x}{\sin^2 x} = \cot^2 x - \sec^2 x$

4. $\dfrac{1 + \sec x}{\sec x} = \dfrac{\sin^2 x}{1 - \cos x}$

5. $\dfrac{\cos^2 x}{1 - \sin x} = \dfrac{\cos x}{\sec x - \tan x}$

6. $(1 - \tan x)(1 - \cot x) = 2 - \sec x \csc x$

7. $\sin^2 x \cot^2 x + \cos^2 x \tan^2 x = 1$

8. $(\tan x + \cot x)^2 = \csc^2 x \sec^2 x$

9. $\dfrac{\sin 2x}{1 + \cos 2x} = \tan x$

10. $\dfrac{\cos(x + y)}{\cos x \sin y} = \cot y - \tan x$

11. $\tan\left(\dfrac{x}{2}\right) = \csc x - \cot x$

12. $\dfrac{\sin(x + y) + \sin(x - y)}{\cos(x + y) + \cos(x - y)} = \tan x$

13. $\sin(x + y)\sin(x - y) = \sin^2 x - \sin^2 y$

14. $\csc x - \tan\dfrac{x}{2} = \cot x$

15. $1 + \tan x \tan\dfrac{x}{2} = \sec x$

16. $\dfrac{\sin 3x + \cos 3x}{\cos x - \sin x} = 1 + 2 \sin 2x$

17. $\left(\cos\dfrac{x}{2} - \sin\dfrac{x}{2}\right)^2 = 1 - \sin x$

18. $\dfrac{\cos 3x - \cos 7x}{\sin 3x + \sin 7x} = \tan 2x$

19. $\dfrac{\sin 2x}{\sin x} - \dfrac{\cos 2x}{\cos x} = \sec x$

20. $\cos 4x = 1 - 8 \sin^2 x + 8 \sin^4 x$

21. $(\cos x + \cos y)^2 + (\sin x - \sin y)^2 = 2 + 2\cos(x + y)$

22. $2 \sin^2(3x) = 1 - \cos^2 6x$

23. $1 + \cos x = \sin x \cot\dfrac{x}{2}$

24. $\tan\left(x + \dfrac{\pi}{4}\right) = \dfrac{1 + \tan x}{1 - \tan x}$

25. $\dfrac{\sec x - 1}{\sin x \sec x} = \tan\dfrac{x}{2}$

26. $(\sin\frac{1}{2}x + \cos\frac{1}{2}x)^2 = 1 + \sin x$

27. $\tan^{-1} x - \tan^{-1} y = \tan^{-1}\left(\dfrac{x - y}{1 + xy}\right)$

28. $2 \tan^{-1} x = \tan^{-1}\left(\dfrac{2x}{1 + x^2}\right)$

29. $2 \cos^{-1} x = \cos^{-1}(2x^2 - 1)$

30. $\cos^{-1} x - \cos^{-1} y = \cos^{-1}(xy + \sqrt{1 - x^2}\sqrt{1 - y^2})$

In Exercises 31–46 solve the given equation on the interval $[0, 2\pi)$.

31. $\cos x \sin x - \sin x = 0$ **32.** $\sin x - 2 \sin^2 x = 0$

33. $2 \sin^2 x - 5 \sin x + 2 = 0$

34. $\sin x - \cos x - \tan x = -1$

35. $2 \cos^2 x - 7 \cos x + 3 = 0$

36. $4 \sin^2 x + 2 \cos^2 x = 3$

37. $\dfrac{1 - \cos x}{1 + \cos x} = 3$ **38.** $\sin x = \cos 2x$

39. $\tan^3 x + \tan^2 x - 3 \tan x - 3 = 0$

40. $\cos 2x \csc^2 x = 2 \cos 2x$

41. $\tan\frac{1}{2}x + 2 \sin 2x = \csc x$

42. $\cos 3x + \cos 2x + \cos x = 0$

43. $2 \cos x + \tan x - \sec x = 0$

44. $2 \cos^2 2x - \cos 2x - 1 = 0$

45. $\tan x + \sec x = \sqrt{3}$ **46.** $2 \cos x - 3 \tan x = 0$

In Exercises 47–54 find the exact value of the given expression.

47. $\cos 15°$ **48.** $\tan 75°$ **49.** $\cos\left(\dfrac{11}{12}\pi\right)$

50. $\dfrac{1}{2}\cos\dfrac{\pi}{12} + \dfrac{\sqrt{3}}{2}\sin\dfrac{\pi}{12}$ **51.** $\tan\dfrac{\pi}{8}$

52. $\sin\left(\dfrac{5\pi}{12}\right)$ **53.** $\cos 7.5°$

54. $2 \sin\dfrac{\pi}{12}\cos\dfrac{\pi}{12}$

In Exercises 55–60 use the appropriate trigonometric identity to simplify the given expression and find its exact value.

55. $\sin 5° \cos 40° + \cos 5° \sin 40°$

56. $\dfrac{\tan 66° - \tan 6°}{1 + \tan 66° \tan 6°}$

57. $\cos^2\left(\dfrac{\pi}{8}\right) - \sin^2\left(\dfrac{\pi}{8}\right)$

58. $\dfrac{2 \tan 22.5°}{1 - \tan^2 22.5°}$

59. $\cos 37.5° \cos 7.5°$

60. $\cos 67.5° + \cos 22.5°$

In Exercises 61–72 find the exact value of the expression given that $\sec x = \frac{3}{2}$ and $\csc y = 3$ and x and y are in quadrant I.

61. $\sin(x + y)$

62. $\sin(x - y)$

63. $\cos(x + y)$

64. $\cos(x - y)$

65. $\tan(x + y)$

66. $\sin 2x$

67. $\cos 2x$

68. $\sin\left(\dfrac{y}{2}\right)$

69. $\cos\left(\dfrac{y}{2}\right)$

70. $\tan\left(\dfrac{y}{2}\right)$

71. $\sin 4x$ (*Hint:*   Use Exercises 66 and 67 and a trigonometric identity.)

72. $\tan\left(\dfrac{y}{4}\right)$ (*Hint:*   Use Exercises 68 and 69 and a trigonometric identity.)

In Exercises 73–82 find the exact value of the given expression.

73. $\sin^{-1}(\sqrt{3}/2)$

74. $\tan^{-1}(\sqrt{3}/3)$

75. $\cos(\tan^{-1}\sqrt{3})$

76. $\sin[\cos^{-1}(\sqrt{3}/2)]$

77. $\tan(\sin^{-1}\frac{2}{5})$

78. $\sin(\cos^{-1}\frac{3}{8})$

79. $\cos(2\sin^{-1}\frac{1}{3})$

80. $\sin(\frac{1}{2}\cos^{-1}\frac{3}{5})$

81. $\cos(\sin^{-1}\frac{5}{13} - \cos^{-1}\frac{4}{5})$

82. $\tan(\cos^{-1}\frac{1}{2} + \cos^{-1}\frac{1}{3})$

In Exercises 83–90 write each complex number in trigonometric form with argument between 0 and 2π.

83. $4 + 4i$

84. $-10i$

85. $5 + 3i$

86. $1 + \sqrt{3}i$

87. $4(-1 + i)$

88. $-\sqrt{2} - \sqrt{2}i$

89. $i(1 + i)$

90. -20

In Exercises 91–96 find the indicated power.

91. $(1 - \sqrt{3}i)^4$

92. $(1 + i)^8$

93. $(2 - 2i)^5$

94. $(\sqrt{3} + i)^{-4}$

95. $(2\sqrt{3} - 2i)^{-8}$

96. $\left(\dfrac{1}{2} + \dfrac{\sqrt{3}}{2}i\right)^{20}$

In Exercises 97–102 find the indicated roots.

97.   The square roots of $-16i$

98.   The square roots of $5 + 5\sqrt{3}i$

99.   The cube roots of $4 + 4\sqrt{3}i$

100.   The fifth roots of 1

101.   The sixth roots of 1

102.   The eighth roots of i

In Exercises 103–106 find all the solutions of the given equation.

103. $z^3 + 1 = 0$

104. $z^2 + z + 2 = 0$

105. $z^6 + 1 = 0$

106. $iz^4 - 1 = 0$

In Exercises 107–110, (a) find the vectors $\mathbf{u} + \mathbf{v}, \mathbf{u} - \mathbf{v}, 2\mathbf{u}$, and $3\mathbf{u} - 2\mathbf{v}$ for the given vectors $\mathbf{u}$ and $\mathbf{v}$, and (b) sketch the vectors you found in part (a).

107. $\mathbf{u} = \langle -2, 3\rangle, \mathbf{v} = \langle 8, 1\rangle$

108. $\mathbf{u} = \langle 1, 1\rangle, \mathbf{v} = \langle -2, -2\rangle$

109. $\mathbf{u} = 2\mathbf{i} + \mathbf{j}, \mathbf{v} = \mathbf{i} - 2\mathbf{j}$

110. $\mathbf{u} = -\mathbf{i} - \mathbf{j}, \mathbf{v} = \mathbf{i} + \mathbf{j}$

In Exercises 111–118 $\mathbf{u} = \mathbf{i} + 2\mathbf{j}$ and $\mathbf{v} = 2\mathbf{i} - 2\mathbf{j}$. Find the indicated vector or scalar.

111. $|\mathbf{u}|$

112. $|\mathbf{u}|\mathbf{v}$

113. $|\sqrt{3}\mathbf{u}|$

114. $|\mathbf{u} + \mathbf{v}|$

115. $|\mathbf{u} - \mathbf{v}|$

116. $\mathbf{u} \cdot \mathbf{v}$

117. $\mathbf{u} \cdot (\mathbf{u} + \mathbf{v})$

118. $|\mathbf{u} \cdot \mathbf{v}|\mathbf{u}$

In Exercises 119–122 find the horizontal and vertical components of the vector $\mathbf{u}$ described and write the vector as a linear combination of the vectors $\mathbf{i}$ and $\mathbf{j}$.

119. $|\mathbf{u}| = 20$, direction $\theta = 60°$

120. $|\mathbf{u}| = 30$, direction $\theta = 315°$

121. $\mathbf{u}$ has terminal point $Q(3, -1)$ and initial point $P(0, 3)$.

122. $\mathbf{u}$ has terminal point $Q(0, -2)$ and initial point $P(1, 1)$.

123.   Find the vector $\mathbf{u}$ if $\mathbf{u} \cdot \mathbf{i} = 3$ and $\mathbf{u} \cdot \mathbf{j} = 4$.

124.   Find the vector $\mathbf{u}$ if $\mathbf{u} \cdot (\mathbf{i} + \mathbf{j}) = 4$ and $\mathbf{u} \cdot (\mathbf{i} - \mathbf{j}) = 3$.

In Exercises 125–128 find the angle between **u** and **v**.

125. $\mathbf{u} = \langle -5, 2 \rangle, \mathbf{v} = \langle -3, 4 \rangle$

126. $\mathbf{u} = \langle 5, 3 \rangle, \mathbf{v} = \langle -2, 6 \rangle$

127. $\mathbf{u} = 2\mathbf{i} + \mathbf{j}, \mathbf{v} = \mathbf{i} + 3\mathbf{j}$

128. $\mathbf{u} = \mathbf{i} - \mathbf{j}, \mathbf{v} = \mathbf{i} + \mathbf{j}$

In Exercises 129 and 130 find the work done by the given force.

129. The force $\mathbf{F} = \langle 2, 9 \rangle$ acting from the point $(1, 1)$ to the point $(7, -1)$

130. The force $\mathbf{F} = \mathbf{i} - 3\mathbf{j}$ acting horizontally through a distance of 10 ft to the right

Chapter 6 Test

1. Verify the identities:

(a) $\dfrac{\tan x}{1 - \cos x} = \csc x(1 + \sec x)$

(b) $\dfrac{2 \tan x}{1 + \tan^2 x} = \sin 2x$

2. Solve the trigonometric equations on the interval $0 \leq x \leq 2\pi$.

(a) $2\cos^2 x + 5\cos x + 2 = 0$

(b) $\sin 2x + 2\sin^2 \frac{1}{2} x = 1$

3. Let $x = 2\sin\theta$, where $-\pi/2 \leq \theta \leq \pi/2$. Simplify the expression $x/\sqrt{4 - x^2}$.

4. Simplify the expression

$$\sin 8° \cos 22° + \cos 8° \sin 22°$$

5. If $\sin x = \frac{3}{5}$, x in quadrant I, and $\cos y = \frac{5}{13}$, y in quadrant IV, find $\sin(x + y)$.

6. (a) Write $\sin 3x \cos 5x$ as a sum of trigonometric functions.

(b) Write $\sin 2x - \sin 5x$ as a product of trigonometric functions.

7. If $\sin\theta = -\frac{4}{5}$ and θ is in quadrant III, find $\tan(\theta/2)$.

8. Sketch the graphs of $y = \sin x$ and $y = \sin^{-1} x$ and specify the domain of each function.

9. Evaluate the following expressions. Do not use a calculator.

(a) $\cos\left(\tan^{-1} \dfrac{9}{40}\right)$ (b) $\tan^{-1}\left(\sin\dfrac{\pi}{3} + \cos\dfrac{\pi}{6}\right)$

10. Write the complex number $1 + \sqrt{3}\,i$ in trigonometric form.

11. Let $z_1 = 4\left(\cos\dfrac{7\pi}{12} + i\sin\dfrac{7\pi}{12}\right)$ and $z_2 = 2\left(\cos\dfrac{5\pi}{12} + i\sin\dfrac{5\pi}{12}\right)$. Find $z_1 z_2$ and $\dfrac{z_1}{z_2}$.

12. Find the cube roots of $27i$ and sketch these roots in the complex plane.

13. (a) Find the vector with initial point $P(3, -1)$ and terminal point $Q(-3, 9)$.

(b) Find a unit vector in the same direction as the vector in part (a).

14. Let $\mathbf{u} = \langle 1, 3 \rangle$ and $\mathbf{v} = \langle -6, 2 \rangle$.

(a) Find $\mathbf{u} - 3\mathbf{v}$.   (b) Find $|\mathbf{u} + \mathbf{v}|$.

(c) Find $\mathbf{u} \cdot \mathbf{v}$.

(d) Are $\mathbf{u}$ and $\mathbf{v}$ perpendicular?

15. A river is flowing due east at 8 mi/h. A man heads his motorboat in a direction N30°E. If the speed of the motorboat in still water is 12 mi/h, find the true speed and direction of the boat.

16. Find the work done by the force $\mathbf{F} = 3\mathbf{i} - 5\mathbf{j}$ in moving an object from the point $(2, 2)$ to the point $(7, -13)$.

Problems Plus

1. Sketch the graphs of the following functions:
  (a) $f(x) = \sin(\sin^{-1}x)$
  (b) $f(x) = \sin^{-1}(\sin x)$

2. Find $\angle A + \angle B + \angle C$ in the figure.

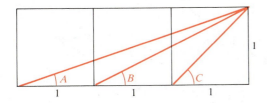

3. If $0 < \theta < \pi/2$ and $\sin 2\theta = a$, find $\sin \theta + \cos \theta$.

4. Find $\cos 36° - \cos 72°$ exactly.

5. (a) Show that for any integers a and b one can find integers c and d such that $c^2 + d^2 = 5(a^2 + b^2)$.
  (*Hint:* Multiply $a + bi$ by an appropriate complex number of modulus $\sqrt{5}$.)
  (b) Can the number 5 in part (a) be replaced by some other integer p? If so, what are the possible values of p for which the statement in part (a) is still valid?
  (*Remark:* This problem shows how complex numbers can be used to find properties of integers.)

6. Show that the sum of the vectors whose initial points are at the center of a regular polygon and whose terminal points are at the vertices of the polygon is zero.

7. Solve for x

$$\cos^2 x + \cos^2 2x + \cos^2 3x = 1 + \cos x \cos 2x \cos 3x.$$

8. Show that $\dfrac{\pi}{4} = \tan^{-1}\left(\dfrac{1}{3}\right) + \tan^{-1}\left(\dfrac{1}{5}\right) + \tan^{-1}\left(\dfrac{1}{8}\right)$.

7

Systems of Equations and Inequalities

A heap and its seventh make nineteen, how large is the heap?
Egyptian Papyrus, 2200 B.C.

As the sun eclipses the stars by his brilliancy, so the man of knowledge will eclipse the fame of others in assemblies of the people if he proposes algebraic problems, and still more if he solves them.
Brahmagupta, 628 A.D.

Many of the problems to which we can apply the techniques of algebra give rise to sets of equations with several unknowns, rather than to just a single equation in a single variable. A set of equations with common variables is called a **system** of equations, and in this chapter we develop techniques for finding simultaneous solutions of systems. We first consider pairs of linear equations in two unknowns, the simplest case of this situation. To help us solve linear equations with an arbitrary number of variables, we study the algebra of matrices and determinants. We also study systems of inequalities and linear programming, which is an optimization technique widely used in business and the social sciences.

Pairs of Lines

In Section 1.7 we saw that the graph of any equation of the form

$$Ax + By = C$$

is a line. Let us consider a **system** of two such equations:

(1)
$$\begin{cases} ax + by = c \\ dx + ey = f \end{cases}$$

A **solution** of this system is an ordered pair of numbers (x_0, y_0) that simultaneously makes each equation a true statement when x is replaced by x_0 and y by y_0. This means that the point (x_0, y_0) lies on both of the lines in the system, and so it must be a point at which they intersect. For example, $(2, 6)$ is a solution of the system

(2)
$$\begin{cases} 3x - y = 0 \\ 5x + 2y = 22 \end{cases}$$

because
$$3(2) - (6) = 0$$
and
$$5(2) + 2(6) = 22$$

Graphing the lines given by Equations 2, we see in Figure 1 that $(2, 6)$ is their point of intersection. The graph also shows that there can be no other solutions of the system because the lines do not intersect anywhere else.

In general, three possible situations can occur when we graph two linear equations. The graphs may intersect at a single point (Figure 2), they may be parallel with no intersection points (Figure 3), or the two equations may just be different equations for the same line (Figure 4). This means that the system can have one solution, no solution, or infinitely many solutions, since each solution corresponds to an intersection point of the lines.

There are two basic methods for solving systems of two linear equations. The first, called the **method of substitution**, is perhaps the more obvious and we use it in Example 1. The second method is easier to extend to situations where we have more equations and more variables. We use it in the remaining examples.

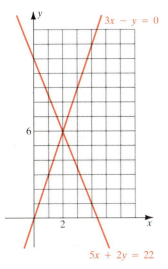

Figure 1

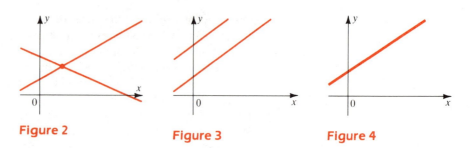

Figure 2

Figure 3

Figure 4

Example 1

Solve the system

$$\begin{cases} 4x - 3y = 11 \\ 6x + 2y = -3 \end{cases}$$

and graph the lines.

Solution

Solving the second equation for y in terms of x, we get

$$y = -3x - \tfrac{3}{2}$$

Now we can substitute this expression for y into the first equation, which gives us an equation that involves only the variable x:

$$4x - 3(-3x - \tfrac{3}{2}) = 11$$
$$13x + \tfrac{9}{2} = 11$$
$$13x = \tfrac{13}{2}$$
$$x = \tfrac{1}{2}$$

We now substitute this value for x back into the original expression for y:

$$y = -3(\tfrac{1}{2}) - \tfrac{3}{2} = -3$$

The solution of the system is $(\tfrac{1}{2}, -3)$, which is also the intersection point of the lines in the system (see Figure 5).   ●

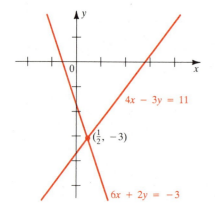

$4x - 3y = 11$

$(\tfrac{1}{2}, -3)$

$6x + 2y = -3$

Figure 5

Example 2

Solve the system

$$\begin{cases} x - 3y = 6 \\ -2x + 5y = -5 \end{cases}$$

Solution

Instead of using the method of substitution, we will try to eliminate either the x or the y from the equations by adding a suitable multiple of one to the other. This method of solving the system is called the **elimination method**. Notice that if we multiply the first equation by 2, the coefficients of x in the two equations are negatives of each other:

$$\begin{cases} 2x - 6y = 12 \\ -2x + 5y = -5 \end{cases}$$

Adding the two equations eliminates the variable x, and we can solve for y:

$$-y = 7 \qquad \text{or} \qquad y = -7$$

At this point we could substitute this value for y into either of the original equations and solve for x. We use the first one because it looks a little easier.

$$x - 3(-7) = 6$$
$$x + 21 = 6$$
$$x = -15$$

The solution of the system is $(-15, -7)$. As a check on our answer, we make sure that the point satisfies the second equation as well:

$$-2(-15) + 5(-7) = -5$$
$$30 - 35 = -5$$

This is true, so our answer satisfies both equations.   ●

Example 3

Solve the system

$$\begin{cases} 8x - 2y = 5 \\ -12x + 3y = 7 \end{cases}$$

and graph the lines.

Solution

This time we try to find a suitable combination of the two equations to eliminate the variable y. Multiplying the first equation by 3 and the second by 2 gives

$$\begin{cases} 24x - 6y = 15 \\ -24x + 6y = 14 \end{cases}$$

Adding the two equations eliminates *both* x and y in this case, and we end up with $0 = 29$, which is obviously false. This means that the given system is **inconsistent**; there is no solution, since no matter what values we assign to x and y, we end up with the false statement $0 = 29$. In slope-intercept form the equations in the system are

$$y = 4x - \tfrac{5}{2}$$

and

$$y = 4x + \tfrac{7}{3}$$

These lines are parallel, with different y-intercepts (see Figure 6), so there is no intersection point.   ●

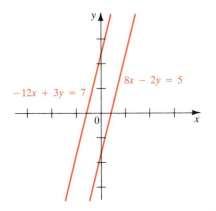

Figure 6

Example 4

Solve the system

$$\begin{cases} 3x - 6y = 12 \\ 4x - 8y = 16 \end{cases}$$

Solution

Multiplying the first equation by 4 and the second by 3, in preparation for subtracting the equations to eliminate the x, gives

$$\begin{cases} 12x - 24y = 48 \\ 12x - 24y = 48 \end{cases}$$

We see that the two equations in the original system are just different ways of expressing the equation of one single line. The coordinates of any point on this line give a solution of the system. Writing the equation in slope-intercept form, we have $y = \frac{1}{2}x - 2$, so any pair of the form

$$(x, \tfrac{1}{2}x - 2)$$

where x can be any real number, is a solution of the system. There are infinitely many solutions. ●

Frequently when we use equations to solve problems in the sciences or in other areas we obtain systems like the ones we have been considering. The next two examples illustrate such situations.

Example 5

A woman rows a boat upstream from one point on a river to another point 4 mi away in $1\frac{1}{2}$ h. The return trip, traveling with the current, takes only 45 min. How fast does she row relative to the water, and at what speed is the current flowing?

Solution

In this and any other problem that involves distance, time, and speed, we make use of the fundamental relationship between these quantities:

$$\text{speed} = \frac{\text{distance}}{\text{time}}$$

or the equivalent formulations

$$\text{distance} = \text{speed} \times \text{time}$$

and

$$\text{time} = \frac{\text{distance}}{\text{speed}}$$

We use these equations to put the English sentences of the problem into mathematical form. Since we are asked to find the rowing speed and the speed of the current, we give names to these quantities. Let

$$x = \text{rowing speed in miles per hour}$$

and

$$y = \text{speed of the current in miles per hour}$$

When she is traveling with the current (downstream), she will therefore be moving at a total of $x + y$ miles per hour, but upstream she moves at $x - y$ miles per hour, since the current decreases her net speed. The distance both upstream and downstream is 4 mi, so using the fact that distance = speed × time for both parts of the trip, we get the equations

$$4 = (x - y) \cdot \tfrac{3}{2}$$

and

$$4 = (x + y) \cdot \tfrac{3}{4}$$

(*Note:*   All times have been converted to hours, since we are expressing the speeds in miles per *hour*.) If we multiply the equations by 2 and 4, respectively, to clear the denominators, we get the system

$$\begin{cases} 3x - 3y = 8 \\ 3x + 3y = 16 \end{cases}$$

Adding the equations eliminates the variable y:

$$6x = 24$$
$$x = 4$$

Substituting this into the first equation in the system (although the second works just as well) and solving for y gives

$$3(4) - 3y = 8$$
$$-3y = 8 - 12$$
$$y = \tfrac{4}{3}$$

The woman rows at 4 mi/h and the current flows at $1\tfrac{1}{3}$ mi/h.   ●

   Many students find that the hardest part of solving a word problem is getting started. Remember the following two key steps:

1.   Assign letters to denote the variable quantities in the problem. Usually the last sentence of the problem tells you what is being asked for, so this is what the variable names will represent.

2.   Translate the given information from English sentences into equivalent mathematical equations involving the variables. Note that an equation is just a sentence written using mathematical notation.

Example 6

A vintner wishes to fortify wine that contains 10% alcohol by adding to it some 70%-alcohol solution. The resulting mixture is to have an alcoholic strength of 16% and is to fill a thousand one-liter bottles. How many liters of the wine and of the alcohol solution should he use?

Solution

Let x = number of liters of wine to be used

            y = number of liters of alcohol solution to be used

To help us translate the information in the problem into equations, we organize the given data in a table:

	Wine	Alcohol solution	Resulting mixture
Volume	x	y	1000
Percent alcohol	10%	70%	16%
Amount of alcohol	$(0.10)x$	$(0.70)y$	$(0.16)1000$

The volume of the mixture must be the total of the two volumes the vintner is adding together, so

$$x + y = 1000$$

Similarly, the amount of alcohol in the mixture must be the total of the alcohol contributed by the wine and by the alcohol solution, which means that

$$(0.10)x + (0.70)y = (0.16)1000$$
$$(0.10)x + (0.70)y = 160$$
$$x + 7y = 1600$$

Thus we must solve the system

$$\begin{cases} x + y = 1000 \\ x + 7y = 1600 \end{cases}$$

Subtracting the first equation from the second eliminates the x, and we get

$$6y = 600$$
$$y = 100$$

We now substitute this into the first equation and solve:

$$x + 100 = 1000$$
$$x = 900$$

The vintner should use 900 L of wine and 100 L of the alcohol solution.　●

Exercises 7.1

In Exercises 1–6 graph each pair of lines on a single set of axes. Determine whether the lines are parallel or not, and if they are not parallel, estimate the coordinates of their point of intersection from your graph.

1. $\begin{cases} 2x + y = 8 \\ 3x - 2y = 12 \end{cases}$

2. $\begin{cases} 3x + 2y = 3 \\ -x + 5y = 16 \end{cases}$

3. $\begin{cases} 6x - 3y = 6 \\ -10x + 5y = 0 \end{cases}$

4. $\begin{cases} 3x + 5y = 15 \\ x + \frac{5}{3}y = 5 \end{cases}$

5. $\begin{cases} 2x + 5y = 15 \\ 4x + 8y = 22 \end{cases}$

6. $\begin{cases} -4x + 14y = 28 \\ 10x - 35y = 70 \end{cases}$

Solve the systems in Exercises 7–12 using the substitution method.

7. $\begin{cases} 3x - 5y = 1 \\ -x + 4y = 2 \end{cases}$

8. $\begin{cases} 4x - 3y = 28 \\ 9x - y = -6 \end{cases}$

9. $\begin{cases} 5x + 2y = 11 \\ 3x + 6y = 9 \end{cases}$

10. $\begin{cases} -4x + 12y = 0 \\ 12x + 4y = 160 \end{cases}$

11. $\begin{cases} \frac{1}{2}x + \frac{1}{3}y = 2 \\ \frac{1}{5}x - \frac{2}{3}y = 8 \end{cases}$

12. $\begin{cases} 0.2x - 0.2y = -1.8 \\ -0.3x + 0.5y = 3.3 \end{cases}$

Solve the systems in Exercises 13–26 using the elimination method. If a system has infinitely many solutions, express them in the form given in Example 4.

13. $\begin{cases} x - 3y = 6 \\ -2x + 5y = -10 \end{cases}$

14. $\begin{cases} 4x + 2y = 16 \\ x - 5y = 70 \end{cases}$

15. $\begin{cases} 2x - 30y = -12 \\ -3x + 45y = 18 \end{cases}$

16. $\begin{cases} 18x + y = 30 \\ 12x - 3y = -24 \end{cases}$

17. $\begin{cases} 3x + 5y = 17 \\ 7x + 9y = 29 \end{cases}$

18. $\begin{cases} 2x - 3y = -8 \\ 14x - 21y = 3 \end{cases}$

19. $\begin{cases} 8s - 3t = -3 \\ 5s - 2t = -1 \end{cases}$

20. $\begin{cases} u - 30v = -5 \\ -3u + 80v = 5 \end{cases}$

21. $\begin{cases} \frac{1}{2}x + \frac{3}{5}y = 3 \\ \frac{1}{3}x + 2y = -6 \end{cases}$

22. $\begin{cases} \frac{3}{2}x - \frac{1}{3}y = \frac{1}{2} \\ 2x - \frac{1}{2}y = -\frac{1}{2} \end{cases}$

23. $\begin{cases} 0.2r + 0.3s = 0.16 \\ -1.2r + 4s = 1.36 \end{cases}$

24. $\begin{cases} 4.8x - 1.6y = 8 \\ 3.6x - 1.2y = 6 \end{cases}$

25. $\begin{cases} \sqrt{3}x + \sqrt{2}y = 5 \\ 2\sqrt{6}x + 4y = \sqrt{5} \end{cases}$

26. $\begin{cases} \sqrt{10}w - \sqrt{2}z = -2 + 5\sqrt{2} \\ \sqrt{2}w + 2\sqrt{5}z = 3\sqrt{10} \end{cases}$

In Exercises 27–30 solve the systems by first making a substitution that will turn the equations into equations of lines and then using the methods discussed in the text. The appropriate substitutions are given in Exercises 27 and 28, but you must determine them for yourself in Exercises 29 and 30.

27. $\begin{cases} \dfrac{2}{u} + \dfrac{1}{v} = 1 \\ \dfrac{3}{u} - \dfrac{2}{v} = 1 \end{cases}$

$\left(\text{Let } x = \dfrac{1}{u}, y = \dfrac{1}{v}. \right)$

28. $\begin{cases} 2r^2 + 3s^2 = 11 \\ 6r^2 - s^2 = 23 \end{cases}$

(Let $x = r^2$, $y = s^2$.)

29. $\begin{cases} 2z^3 + \frac{1}{2}w^3 = 2 \\ -3z^3 + \frac{3}{2}w^3 = 15 \end{cases}$

30. $\begin{cases} \dfrac{2}{x} - \dfrac{4}{y^2} = 8 \\ \dfrac{1}{x} - \dfrac{3}{y^2} = 6 \end{cases}$

Solve the systems in Exercises 31–34.

31. $\begin{cases} \dfrac{2x - 5}{3} + \dfrac{y - 1}{6} = \dfrac{1}{2} \\ \dfrac{x}{5} + \dfrac{3y - 6}{12} = 1 \end{cases}$

32. $\begin{cases} x - 3y = 4x - 6y - 10 \\ 2x = 12y + 10 \end{cases}$

33. $x - 2y = 2x + 2y = 1$ 34. $x = 2x + y = 2y + 1$

In Exercises 35–38 find x and y in terms of a, b, and/or θ.

35. $\begin{cases} ax + by = 1 \\ bx + ay = 1 \end{cases}$ $(a^2 - b^2 \neq 0)$

36. $\begin{cases} ax + by = 0 \\ a^2x + b^2y = 1 \end{cases}$ $(a \neq 0, b \neq 0, a \neq b)$

37. $\begin{cases} x\sin\theta + y\cos\theta = 1 \\ x\cos\theta - y\sin\theta = 1 \end{cases}$

38. $\begin{cases} x + y\tan\theta = \cos\theta \\ -x\tan\theta + y = 0 \end{cases}$

39. Find two numbers whose sum is 34 and whose difference is 10.

40. The sum of two numbers is twice their difference. The larger number is 6 more than twice the smaller. Find the numbers.

41. A man has 14 coins in his pocket, all of which are dimes and quarters. If the total value of his change is $2.75, how many dimes and how many quarters does he have?

42. The admission fee at an amusement park is $1.50 for children and $4.00 for adults. On a certain day, 2200 people entered the park and the admission fees collected totaled $5050. How many children and how many adults were admitted?

43. A man flies a small airplane from Fargo to Bismarck, North Dakota—a distance of 180 mi. Because he is flying into a headwind, the trip takes him 2 h. On the way back, the wind is still blowing at the same speed, so the return trip takes only 1 h and 12 min. What is his speed in still air, and how fast is the wind blowing?

44. A boat travels downstream between two points on a river 20 mi apart in 1 h. The return trip against the current takes $2\frac{1}{2}$ h. What is the boat's speed, and how fast does the current in the river flow?

45. A woman keeps fit by bicycling and running every day. On Monday she spends $\frac{1}{2}$ h at each activity, covering a total of $12\frac{1}{2}$ mi. On Tuesday she runs for 12 min and cycles for 45 min and covers a total of 16 mi. Assuming her running and cycling speeds do not change from day to day, find these speeds.

46. A biologist has two brine solutions, one containing 5% salt and another containing 20% salt. How many milliliters of each solution should he mix to obtain 1 L of a solution that contains 14% salt?

47. A researcher performs an experiment to test a hypothesis that involves the nutrients niacin and retinol. She wishes to feed one group of laboratory rats a diet that contains precisely 32 units of niacin and 22,000 units of retinol per day. She has two types of commercial pellet foods available. Food A contains 0.12 unit of niacin and 100 units of retinol per gram. Food B contains 0.20 unit of niacin and 50 units of retinol per gram. How many grams of each food should she feed this group of rats each day?

48. A customer in a coffee shop wishes to purchase a blend of two coffees: Kenyan, costing $3.50 a pound, and Sri Lankan, costing $5.60 a pound. He ends up buying 3 lb of such a blend, which costs him $11.55. How many pounds of each kind went into the mixture?

49. A chemist has two large containers of sulfuric acid solution, with different concentrations of acid in each container. Blending 300 mL of the first solution and 600 mL of the second gives a mixture that is 15% acid, whereas 100 mL of the first mixed with 500 mL of the second gives a $12\frac{1}{2}$% acid mixture. What is the concentration of sulfuric acid in each of the original containers?

50. John and Mary leave their house at the same time and drive off in opposite directions. John drives at 60 mi/h and travels 35 mi farther than Mary, who drives at 40 mi/h. Mary's trip takes 15 min longer than John's. For what length of time does each of them drive?

51. A business executive normally leaves her office at 5:00 P.M. to drive home. On Monday she is able to leave the office at 4:45 P.M. She is able to drive at 30 mi/h and arrives home 19 min earlier than usual. On Tuesday she leaves the office at 5:20 P.M. Because the traffic is so heavy, she drives at an average speed of only 15 mi/h and arrives home 36 min later than usual. What is her usual speed on the way home, and at what time does she usually get there?

52. The sum of the digits of a two-digit number is 7. When the digits are reversed, the number is increased by 27. Find the number.

53. The sum of the digits of a two-digit number is 9. When the digits are reversed, the value of the number is decreased to $\frac{3}{8}$ of its original value. What is the number?

54. Find the area of the triangle that lies in the first quadrant (with its base on the x-axis) and that is bounded by the lines $y = 2x - 4$ and $y = -4x + 20$.

55. Find the equation of the parabola that passes through the origin and through the points $(1, 12)$ and $(3, 6)$. (*Hint:* Recall that the general equation of a parabola of this type is $y = ax^2 + bx + c$.)

Systems of Linear Equations

A **linear equation in n variables** is an equation that can be put in the form

(1) $$a_1 x_1 + a_2 x_2 + \cdots + a_n x_n = c$$

where $a_1, a_2, \ldots, a_n$ and c are real numbers, and $x_1, x_2, \ldots, x_n$ are the variables. If there are no more than three or four variables, we generally use x, y, z, and w instead of x_1, x_2, x_3, and x_4. Such equations are called linear because if $n = 2$, Equation 1 is just

$$a_1 x + a_2 y = c$$

which is the equation of a line. The equations

$$6x_1 - 3x_2 + \sqrt{5}x_3 = 1000$$

and

$$x + y + z = 2w - \tfrac{1}{2}$$

are thus linear, but the equations

$$x^2 + 3y - \sqrt{z} = 5$$

and

$$x_1 x_2 + 6x_3 = -6$$

are not. Each term of a linear equation is either a constant or a constant multiple of one of the variables.

We are going to adapt the method of elimination introduced in Section 7.1 to solve systems of linear equations in any number of variables. We begin with an example to show how the method works before formally describing the technique.

Example 1

Solve the system

$$\begin{cases} x + 2y + 4z = 7 \\ -x + y + 2z = 5 \\ 2x + 3y + 3z = 7 \end{cases}$$

Solution

We first eliminate the x from the second and third equations. If we add the first equation to the second, we get

$$\begin{array}{r} x + 2y + 4z = 7 \\ -x + y + 2z = 5 \\ \hline 3y + 6z = 12 \end{array}$$

which does not contain the variable x. Similarly, if we add -2 times the first equation to the third, we get

$$\begin{array}{r} -2x - 4y - 8z = -14 \\ 2x + 3y + 3z = 7 \\ \hline -y - 5z = -7 \end{array}$$

which also does not contain x. This means that the original system is equivalent to the simpler system

$$\begin{cases} x + 2y + 4z = 7 \\ 3y + 6z = 12 \\ -y - 5z = -7 \end{cases}$$

Since each term in the second equation has a common factor of 3, we multiply both sides by $\frac{1}{3}$:

$$\begin{cases} x + 2y + 4z = 7 \\ y + 2z = 4 \\ -y - 5z = -7 \end{cases}$$

We now eliminate the y from the last equation by adding the second equation to it. This gives

$$\begin{cases} x + 2y + 4z = & 7 \\ y + 2z = & 4 \\ -3z = & -3 \end{cases}$$

Finally, multiplying the last equation by $-\frac{1}{3}$, we get

$$\begin{cases} x + 2y + 4z = 7 \\ y + 2z = 4 \\ z = 1 \end{cases}$$

We now know from the third equation that $z = 1$. Substituting this into the second equation allows us to solve for y:

$$y + 2(1) = 4$$
$$y = 4 - 2 = 2$$

Putting these values for y and z into the first equation gives us x:

$$x + 2(2) + 4(1) = 7$$
$$x = 7 - 4 - 4 = -1$$

Thus the simultaneous solution to the system is the ordered triple $(-1, 2, 1)$; that is, $x = -1$, $y = 2$, and $z = 1$.   ●

The technique we have been using here is called **Gaussian elimination** in honor of the German mathematician C. F. Gauss (whom we previously encountered in Section 3.5 in connection with the Fundamental Theorem of Algebra). The method consists of using algebraic operations to change the linear system we are solving into an equivalent system in **triangular form**. A system of three equations in three variables is in triangular form if the second equation does not have x in it and the third has neither x nor y. The last system in the solution to Example 1 is in triangular form. As we have seen, it is easy to solve a system in this form. The algebraic operations we are permitted to use to change the system to triangular form are the following:

1. Add a multiple of one equation to another.
2. Multiply an equation by a constant.
3. Rearrange the order of the equations.

These are called the **elementary row operations**. None of these operations will change the solutions of an equation, so after performing them we will always end up with an equivalent system—that is, one that has the same solution.

Example 2

Solve the following system using Gaussian elimination:

$$\begin{cases} x - y + 3z = 4 \\ x + 2y - 2z = 10 \\ 3x - y + 5z = 14 \end{cases}$$

Solution

We first eliminate the x from the second and third equations by subtracting the first from the second, and subtracting three times the first from the third:

$$\begin{cases} x - y + 3z = 4 \\ 3y - 5z = 6 \\ 2y - 4z = 2 \end{cases}$$

Multiplying the last equation by $\frac{1}{2}$ to simplify, we get

$$\begin{cases} x - y + 3z = 4 \\ 3y - 5z = 6 \\ y - 2z = 1 \end{cases}$$

Subtracting three times the third equation from the second (to eliminate y in the second equation) gives us

$$\begin{cases} x - y + 3z = 4 \\ z = 3 \\ y - 2z = 1 \end{cases}$$

Finally, we interchange the second and third equations to put the system into triangular form:

(2)
$$\begin{cases} x - y + 3z = 4 \\ y - 2z = 1 \\ z = 3 \end{cases}$$

Working backward from the last equation, we get

$$y - 2(3) = 1$$
$$y = 7$$

and, from the first equation,

$$x - (7) + 3(3) = 4$$
$$x = 2$$

The solution of the system is $(2, 7, 3)$. ●

If we examine the solution to Example 2, we see that the variables x, y, and z act simply as place-holders in our computations. It is only the coefficients of the variables and the constants that actually enter into the calculations. We will use this fact to simplify our notation for the Gaussian elimination process. Instead of writing the equations in a system out in full, we write only the coefficients and constants in a rectangular array. The system of Example 2 is thus written as

$$\begin{bmatrix} 1 & -1 & 3 & 4 \\ 1 & 2 & -2 & 10 \\ 3 & -1 & 5 & 14 \end{bmatrix}$$

where each row represents an equation. We will also use the following symbols to represent the elementary row operations:

Symbol	Meaning
$R_i + kR_j$	Change the ith row by adding k times row j to it
kR_i	Multiply the ith row by k
$R_i \leftrightarrow R_j$	Interchange the ith and jth rows

With this notation, the calculations in the solution to Example 2 are written as follows:

$$\begin{bmatrix} 1 & -1 & 3 & 4 \\ 1 & 2 & -2 & 10 \\ 3 & -1 & 5 & 14 \end{bmatrix} \xrightarrow[R_3 - 3R_1]{R_2 - R_1} \begin{bmatrix} 1 & -1 & 3 & 4 \\ 0 & 3 & -5 & 6 \\ 0 & 2 & -4 & 2 \end{bmatrix}$$

$$\xrightarrow{\frac{1}{2}R_3} \begin{bmatrix} 1 & -1 & 3 & 4 \\ 0 & 3 & -5 & 6 \\ 0 & 1 & -2 & 1 \end{bmatrix} \xrightarrow{R_2 - 3R_3} \begin{bmatrix} 1 & -1 & 3 & 4 \\ 0 & 0 & 1 & 3 \\ 0 & 1 & -2 & 1 \end{bmatrix}$$

$$\xrightarrow{R_2 \leftrightarrow R_3} \begin{bmatrix} 1 & -1 & 3 & 4 \\ 0 & 1 & -2 & 1 \\ 0 & 0 & 1 & 3 \end{bmatrix}$$

At this point we would rewrite this as a system of equations and continue as in Example 2.

A rectangular array of numbers like the ones we have been using is called a **matrix**. We will study the algebra of matrices in Sections 7.5 and 7.6, but for now we use them as a shorthand device when solving systems of equations.

Note that the goal in Gaussian elimination is to end up with a system in triangular form, one that can be easily solved. Each elementary row operation we perform should take us one step closer to that goal. At each stage in the process there will be several possible operations to choose from, so there is no single "right" way to solve such a problem. Of course, no matter what route we decide to take to the triangular form, the final answer will be the same.

Example 3

Solve the system

$$\begin{cases} 3x + y + 4z + w = 6 \\ 2x + 3z + 4w = 13 \\ y - 2z - w = 0 \\ x - y + z + w = 3 \end{cases}$$

Solution

In matrix form the system is

$$\begin{bmatrix} 3 & 1 & 4 & 1 & 6 \\ 2 & 0 & 3 & 4 & 13 \\ 0 & 1 & -2 & -1 & 0 \\ 1 & -1 & 1 & 1 & 3 \end{bmatrix}$$

Since it is advantageous to have a 1 in the upper left corner, we first re-arrange the order of the rows and then create as many zeros as possible in the first column.

$$\begin{bmatrix} 1 & -1 & 1 & 1 & 3 \\ 0 & 1 & -2 & -1 & 0 \\ 3 & 1 & 4 & 1 & 6 \\ 2 & 0 & 3 & 4 & 13 \end{bmatrix} \xrightarrow[R_4 - 2R_1]{R_3 - 3R_1} \begin{bmatrix} 1 & -1 & 1 & 1 & 3 \\ 0 & 1 & -2 & -1 & 0 \\ 0 & 4 & 1 & -2 & -3 \\ 0 & 2 & 1 & 2 & 7 \end{bmatrix}$$

To get this into triangular form, we must change the 4 and the 2 in the second column to zeros, so we continue as follows:

$$\xrightarrow[R_4 - 2R_2]{R_3 - 4R_2} \begin{bmatrix} 1 & -1 & 1 & 1 & 3 \\ 0 & 1 & -2 & -1 & 0 \\ 0 & 0 & 9 & 2 & -3 \\ 0 & 0 & 5 & 4 & 7 \end{bmatrix}$$

We would now like to change the 5 to a 0. Although we might be tempted to do this by subtracting five times row 1 from row 4, this will not work because that would eliminate the first two zeros in that row, which we worked so hard to get. So instead we perform the following operations on the last two rows:

$$\xrightarrow{R_3 - 2R_4} \begin{bmatrix} 1 & -1 & 1 & 1 & 3 \\ 0 & 1 & -2 & -1 & 0 \\ 0 & 0 & -1 & -6 & -17 \\ 0 & 0 & 5 & 4 & 7 \end{bmatrix} \xrightarrow{R_4 + 5R_3}$$

$$\begin{bmatrix} 1 & -1 & 1 & 1 & 3 \\ 0 & 1 & -2 & -1 & 0 \\ 0 & 0 & -1 & -6 & -17 \\ 0 & 0 & 0 & -26 & -78 \end{bmatrix} \xrightarrow[-\frac{1}{26}R_4]{-R_3} \begin{bmatrix} 1 & -1 & 1 & 1 & 3 \\ 0 & 1 & -2 & -1 & 0 \\ 0 & 0 & 1 & 6 & 17 \\ 0 & 0 & 0 & 1 & 3 \end{bmatrix}$$

The last equation tells us that $w = 3$, so working backward through the equations as before, we get

$$\begin{aligned} z + 6w &= 17 & y - 2z - w &= 0 & x - y + z + w &= 3 \\ z + 6(3) &= 17 & y - 2(-1) - 3 &= 0 & x - 1 + (-1) + 3 &= 3 \\ z &= -1 & y &= 1 & x &= 2 \end{aligned}$$

The solution is $(2, 1, -1, 3)$.

Linear equations, often containing hundreds or even thousands of variables, occur frequently in the applications of algebra to the sciences and to other fields. For now, we consider an example that involves only three variables.

Example 4

A nutritionist is performing an experiment on student volunteers. He wishes to feed one of his subjects a daily diet that consists of a combination of three commercial diet foods: MiniCal, SloStarve, and SlimQuick. For the experiment it is important that the subject consume exactly 500 mg of potassium, 75 g of protein, and 1150 units of vitamin D every day. The amounts of these nutrients in one ounce of each food are given in the following table:

	mg Potassium	g Protein	Units vitamin D
MiniCal	50	5	90
SloStarve	75	10	100
SlimQuick	10	3	50

How many ounces of each food should the subject eat every day to satisfy the nutrient requirements exactly?

Solution

Let x, y, and z represent the number of ounces of MiniCal, SloStarve, and SlimQuick, respectively, that the subject should eat every day. This means that he will get $50x$ mg of potassium from MiniCal, $75y$ from SloStarve, and $10z$ mg from SlimQuick, for a total of $50x + 75y + 10z$ mg of potassium. Since the potassium requirement is 500 mg, we get the equation

$$50x + 75y + 10z = 500$$

Similar reasoning for the protein and vitamin D requirements leads to

$$5x + 10y + 3z = 75$$

and

$$90x + 100y + 50z = 1150$$

Dividing the first equation by 5 and the third by 10 gives the system

$$\begin{cases} 10x + 15y + 2z = 100 \\ 5x + 10y + 3z = 75 \\ 9x + 10y + 5z = 115 \end{cases}$$

We solve this using Gaussian elimination:

$$\begin{bmatrix} 10 & 15 & 2 & 100 \\ 5 & 10 & 3 & 75 \\ 9 & 10 & 5 & 115 \end{bmatrix} \xrightarrow{R_1 - R_3} \begin{bmatrix} 1 & 5 & -3 & -15 \\ 5 & 10 & 3 & 75 \\ 9 & 10 & 5 & 115 \end{bmatrix}$$

$$\xrightarrow[R_3 - 9R_1]{R_2 - 5R_1} \begin{bmatrix} 1 & 5 & -3 & -15 \\ 0 & -15 & 18 & 150 \\ 0 & -35 & 32 & 250 \end{bmatrix} \xrightarrow{-\frac{1}{3}R_2} \begin{bmatrix} 1 & 5 & -3 & -15 \\ 0 & 5 & -6 & -50 \\ 0 & -35 & 32 & 250 \end{bmatrix}$$

$$\xrightarrow{R_3 + 7R_2} \begin{bmatrix} 1 & 5 & -3 & -15 \\ 0 & 5 & -6 & -50 \\ 0 & 0 & -10 & -100 \end{bmatrix} \xrightarrow{-\frac{1}{10}R_3} \begin{bmatrix} 1 & 5 & -3 & -15 \\ 0 & 5 & -6 & -50 \\ 0 & 0 & 1 & 10 \end{bmatrix}$$

Now we work backward through the equations to get $z = 10$, $y = 2$, and $x = 5$. The subject should be fed 5 oz of MiniCal, 2 oz of SloStarve, and 10 oz of SlimQuick every day.   ●

A more practical application might involve dozens of foods and nutrients rather than just three. As you may imagine, such a problem would be almost impossible to solve without the assistance of a computer.

Exercises 7.2

State whether or not the equations or systems of equations in Exercises 1–6 are linear.

1. $6x - 3y + 1000z - w = \sqrt{13}$

2. $6xy - 3yz + 15zx = 0$ **3.** $x_1^2 + x_2^2 + x_3^2 = 36$

4. $e^2 x_1 + \pi x_2 - \sqrt{5} = x_3 - \frac{1}{2}x_4$

5. $\begin{cases} x - 3xy + 5y = 0 \\ 12x + 321y = 123 \end{cases}$ **6.** $\begin{cases} x - 3y = 15z + \dfrac{1}{\sqrt{3}} \\ x = 3z \\ y - z = \dfrac{x}{\sqrt{47}} \end{cases}$

Write systems of equations that correspond to the matrices in Exercises 7–10.

7. $\begin{bmatrix} 2 & 3 & 1 \\ 4 & 2 & 3 \end{bmatrix}$ **8.** $\begin{bmatrix} 1 & 2 & 4 & 6 \\ 3 & -1 & 2 & 4 \\ -1 & -1 & 0 & 7 \end{bmatrix}$

9. $\begin{bmatrix} 0 & 1 & 0 & 0 \\ 1 & 0 & 1 & 0 \\ 0 & -2 & 2 & 7 \end{bmatrix}$

10. $\begin{bmatrix} 1 & 2 & 3 & 4 & 5 \\ -1 & 0 & 1 & 0 & 6 \\ 2 & 3 & 5 & 0 & 0 \\ 0 & 1 & 1 & 0 & -2 \end{bmatrix}$

Use Gaussian elimination to solve the systems in Exercises 11–30.

11. $\begin{cases} x + y - z = 2 \\ 2y + z = 8 \\ 2x + 3y - 5z = 1 \end{cases}$ **12.** $\begin{cases} 3x - y = -9 \\ -x + 2y + 3z = 17 \\ x + y + z = 4 \end{cases}$

13. $\begin{cases} 3x + y + 3z = 11 \\ 4x - y + z = 8 \\ 5x + 6y + 2z = 5 \end{cases}$ **14.** $\begin{cases} 2x - 3y + 5z = 15 \\ 4x + 2y - 3z = -6 \\ 6x + y + z = 9 \end{cases}$

15. $\begin{cases} x_1 + 2x_2 - x_3 = 9 \\ 2x_1 - x_3 = -2 \\ 3x_1 + 5x_2 + 2x_3 = 22 \end{cases}$

16. $\begin{cases} 2x_1 + x_2 = 7 \\ 2x_1 - x_2 + x_3 = 6 \\ 3x_1 - 2x_2 + 4x_3 = 11 \end{cases}$

17. $\begin{cases} 2x - 3y - z = 13 \\ -x + 2y - 5z = 6 \\ 5x - y - z = 49 \end{cases}$

18. $\begin{cases} 10x + 10y - 20z = 60 \\ 15x + 20y + 30z = -25 \\ -5x + 30y - 10z = 45 \end{cases}$

19. $\begin{cases} 0.1x + y - 0.2z = 0.6 \\ -0.2x + 1.1y + 0.6z = -1.6 \\ 0.3x + 0.2y + z = -1.4 \end{cases}$

20. $\begin{cases} \frac{1}{2}x + \frac{1}{3}y - \frac{1}{6}z = 8 \\ x - \frac{2}{3}y + \frac{1}{3}z = -4 \\ -\frac{1}{3}x + \frac{1}{2}y - z = 10 \end{cases}$

21. $\begin{cases} 3x + y + z = \frac{3}{2} \\ 3x + 12z = -5 \\ 2y - 4z = 4 \end{cases}$

22. $\begin{cases} x - y + 2z = 1.9 \\ 5x - 6y + z = 8 \\ 7x + y - 2z = -1.1 \end{cases}$

23. $\begin{cases} x + y - z - w = 6 \\ 2x + z - 3w = 8 \\ x - y + 4w = -10 \\ 3x + 5y - z - w = 20 \end{cases}$

24. $\begin{cases} -x + 2y + z - 3w = 3 \\ 3x - 4y + z + w = 9 \\ -x - y + z + w = 0 \\ 2x + y + 4z - 2w = 3 \end{cases}$

25. $\begin{cases} x_1 - x_2 + x_3 + 2x_4 + 3x_5 = 0 \\ -x_1 - 2x_2 + x_3 - 2x_4 + x_5 = 7 \\ -x_1 + x_2 + x_4 - x_5 = -4 \\ 2x_1 - 2x_2 + 3x_3 - x_4 = 12 \\ x_1 + x_3 - x_4 - 5x_5 = 5 \end{cases}$

26. $\begin{cases} x + y + z + w + u + v = 12 \\ y - z + u - v = -1 \\ 2x - 2z + 4w - 4v = -6 \\ 3y - z + v = 4 \\ x - y + z - w + u - v = 0 \\ -x - y + z + w = 2 \end{cases}$

27. $\begin{cases} x + y = -2 \\ 2y + z = 1 \\ x - 3z = -20 \end{cases}$

28. $\begin{cases} 3x + 5z = 56 \\ 4y - 2z = 14 \\ 7x + 4y = 77 \end{cases}$

29. $\begin{cases} x_1 + 7x_3 = -20 \\ 2x_1 - 5x_2 = 7 \\ -3x_2 + x_3 = 0 \end{cases}$

30. $\begin{cases} x_1 + x_2 - x_3 = 0 \\ x_1 + 3x_4 = 13 \\ 3x_2 - 2x_3 = 0 \\ 2x_1 + 5x_3 = 17 \end{cases}$

31. A doctor recommends that one of her patients take 50 mg each of niacin, riboflavin, and thiamin daily to help alleviate a deficiency. Looking into his medicine chest at home, the patient finds three brands of vitamin pills. The amounts of the relevant vitamins per pill are given in the table:

	Niacin	Riboflavin	Thiamin
VitaMax	5 mg	15 mg	10 mg
Vitron	10 mg	20 mg	10 mg
VitaPlus	15 mg	none	10 mg

How many pills of each type should he take every day to fulfill the doctor's prescription?

32. A chemist has three containers of acid solution at various concentrations. The first is 10% acid, the second is 20%, and the third is 40%. How many milliliters of each should he mix together to make 100 mL of acid at 18% concentration, if he has to use four times as much of the 10% solution as the 40% solution?

33. The drawer of a cash register contains 30 coins (pennies, nickels, dimes, and quarters). The total value of the coins is $3.31. The total number of pennies and nickels combined is the same as the total number of dimes and quarters combined. The total value of the quarters is five times the total value of the dimes. How many coins of each type are there?

34. A small school has 100 students who occupy three classrooms: rooms A, B, and C. After the first class period of the school day, half the students in room A move to room B, one fifth of the students in room B move to room C, and a third of the students in room C move to room A. Nevertheless, the total number of students in each room remains the same after this shift. How many students are there in each room?

35. A hotel offers three classes of accommodation: standard, deluxe, and first-class rooms. A group of ten employees of a manufacturing company attend a trade convention and stay in this hotel. If six of them take a standard room, two a deluxe, and two a first-class room, the total hotel bill would be $530 per day. If five stay in a standard room, four in a deluxe, and only one in a first-class room, the bill would decrease to $510 per day. If they splurge and have three use a standard room, three a deluxe, and four a first-class room, the bill would be $645 per day. How much is the daily rate for each type of room?

36. Amanda, Bryce, and Corey enter a race in which they will have to run, swim, and cycle over a marked course. Amanda runs at an average speed of 10 mi/h, swims at 4 mi/h, and cycles at 20 mi/h during this race. Bryce runs at $7\frac{1}{2}$ mi/h, swims at 6 mi/h, and cycles at 15 mi/h. Corey runs at 15 mi/h, swims at 3 mi/h, and cycles at 40 mi/h. Corey finishes first with a total time of 1 h and

45 min. Amanda comes in second with a time of 2 h and 30 min. Bryce finishes last with a time of 3 h. How many miles long is each part of the race?

37. Determine a, b, and c so that the graph of the parabola $y = ax^2 + bx + c$ passes through the points $(-2, 24)$, $(1, 3)$, and $(3, 9)$.

38. Determine a, b, c, and d so that the points $(1, 1)$, $(2, 45)$, $(-1, -3)$, and $(-3, 225)$ all lie on the graph of the function $f(x) = ax^4 + bx^2 + cx + d$.

Solve the equations in Exercises 39 and 40.

39. $5x + 2y = 4x - z = 4y + 3z = 1$

40. $6x + 2y = 2z - 2y = 3w - 7y = x + z = 2$

Inconsistent and Dependent Systems

All the systems of linear equations that we considered in the last section had one unique solution for each of the unknowns. But as we saw in Section 7.1, a system of two linear equations in two variables can have one solution, no solutions, or infinitely many solutions. The same cases arise when we study linear systems with more equations and more variables. We use the Gaussian elimination process, described in the preceding section, to determine how many solutions any given linear system may have.

In Example 1 we see how to determine when a system has no solutions.

Example 1

Use Gaussian elimination to solve the following system:

$$\begin{cases} x - 3y + 2z = 12 \\ 2x \qquad\; + 5z = 14 \\ x + 9y + 4z = 20 \end{cases}$$

Solution

$$\begin{bmatrix} 1 & -3 & 2 & 12 \\ 2 & 0 & 5 & 14 \\ 1 & 9 & 4 & 20 \end{bmatrix} \xrightarrow[\;R_3 - R_1\;]{R_2 - 2R_1} \begin{bmatrix} 1 & -3 & 2 & 12 \\ 0 & 6 & 1 & -10 \\ 0 & 12 & 2 & 8 \end{bmatrix}$$

$$\xrightarrow{\;R_3 - 2R_2\;} \begin{bmatrix} 1 & -3 & 2 & 12 \\ 0 & 6 & 1 & -10 \\ 0 & 0 & 0 & 28 \end{bmatrix}$$

This is in triangular form, but if we translate the last row back into equation form, we get

$$0x + 0y + 0z = 28$$

or

$$0 = 28$$

which is false. No matter what values we pick for x, y, and z, the last equation will never be a true statement. This means the system *has no solution*. ●

A system that has no solution is said to be **inconsistent**. The procedure we used to show that the system in Example 1 is inconsistent works in general. If we use Gaussian elimination to change a system to triangular form, and if one of the equations we end up with is false, then the system is inconsistent. (The false equation will always have the form $0 = N$, where N is nonzero.)

The next example shows what happens when we apply Gaussian elimination to a system with infinitely many solutions.

Example 2

Find the complete solution of the following system:

$$\begin{cases} -3x - 5y + 36z = 10 \\ -x \qquad\quad + 7z = 5 \\ x + y - 10z = -4 \end{cases}$$

Solution

$$\begin{bmatrix} -3 & -5 & 36 & 10 \\ -1 & 0 & 7 & 5 \\ 1 & 1 & -10 & -4 \end{bmatrix} \xrightarrow{R_1 \leftrightarrow R_3} \begin{bmatrix} 1 & 1 & -10 & -4 \\ -1 & 0 & 7 & 5 \\ -3 & -5 & 36 & 10 \end{bmatrix}$$

$$\xrightarrow[\substack{R_2 + R_1 \\ R_3 + 3R_1}]{} \begin{bmatrix} 1 & 1 & -10 & -4 \\ 0 & 1 & -3 & 1 \\ 0 & -2 & 6 & -2 \end{bmatrix} \xrightarrow{R_3 + 2R_2} \begin{bmatrix} 1 & 1 & -10 & -4 \\ 0 & 1 & -3 & 1 \\ 0 & 0 & 0 & 0 \end{bmatrix}$$

The system is now in triangular form. Translating the last row back into an equation, we get

$$0x + 0y + 0z = 0$$

or

$$0 = 0$$

This equation is always true, no matter what x, y, and z are. Since the equation adds no new information about the variables, we can drop it from the system, which we now write in the form

$$\begin{bmatrix} 1 & 1 & -10 & -4 \\ 0 & 1 & -3 & 1 \end{bmatrix}$$

This is equivalent to the system

$$\begin{cases} x + y - 10z = -4 \\ \quad\ y - 3z = 1 \end{cases}$$

Neither of these equations determines a value for z, but we can use them to express x and y in terms of z. From the last equation we get

$$y = 3z + 1$$

Substituting this value for y into the first equation gives us

$$x + (3z + 1) - 10z = -4$$
$$x - 7z + 1 = -4$$
$$x = 7z - 5$$

Since no value is determined for z, we can get a solution to the system by letting z be any real number and then using the above equations to calculate x and y. For example, if $z = 1$, then

$$x = 7z - 5 = 7(1) - 5 = 2$$
and
$$y = 3z + 1 = 3(1) + 1 = 4$$

Thus $(2, 4, 1)$ is a solution to the system. We would get a different solution if we let $z = 2$ because then

$$x = 7z - 5 = 7(2) - 5 = 9$$
and
$$y = 3z + 1 = 3(2) + 1 = 7$$

So $(9, 7, 2)$ is also a solution. There are infinitely many solutions because z can be given any value. We write the complete solution as follows:

$$\begin{cases} x = 7z - 5 \\ y = 3z + 1 \\ z = \text{any real number} \end{cases}$$

●

A system with infinitely many solutions is called **dependent**. In the complete solution to such a system, one or more of the variables will be arbitrary, and the values of the others will *depend* on the arbitrary one(s). If we use Gaussian elimination to convert a dependent system to triangular form and then discard any equations of the form $0 = 0$, we end up with a system that has fewer equations than variables. In Example 2, we ended up with only two equations in the three variables x, y, and z. In general, if we end up with n equations in m variables after this process, the complete solution will have $m - n$ arbitrary variables, and the values of the others will be expressed in terms of these.

Example 3

Find the complete solution of the system

$$\begin{cases} x + 2y - 3z - 4w = 10 \\ x + 3y - 3z - 4w = 15 \\ 2x + 2y - 6z - 8w = 10 \end{cases}$$

Solution

$$
\begin{bmatrix}
1 & 2 & -3 & -4 & 10 \\
1 & 3 & -3 & -4 & 15 \\
2 & 2 & -6 & -8 & 10
\end{bmatrix}
\xrightarrow[R_3 - 2R_1]{R_2 - R_1}
\begin{bmatrix}
1 & 2 & -3 & -4 & 10 \\
0 & 1 & 0 & 0 & 5 \\
0 & -2 & 0 & 0 & -10
\end{bmatrix}
$$

$$
\xrightarrow{R_3 + 2R_2}
\begin{bmatrix}
1 & 2 & -3 & -4 & 10 \\
0 & 1 & 0 & 0 & 5 \\
0 & 0 & 0 & 0 & 0
\end{bmatrix}
$$

$$
\xrightarrow{\text{discard } R_3}
\begin{bmatrix}
1 & 2 & -3 & -4 & 10 \\
0 & 1 & 0 & 0 & 5
\end{bmatrix}
$$

At this stage we have two equations in four unknowns, so the system is dependent and the solution will have two arbitrary variables. From the second equation, $y = 5$, so substituting this into the first, we get

$$x + 2y - 3z - 4w = 10$$
$$x + 2(5) - 3z - 4w = 10$$
$$x = 3z + 4w$$

The complete solution is therefore

$$
\begin{cases}
x = 3z + 4w \\
y = 5 \\
z = \text{any real number} \\
w = \text{any real number}
\end{cases}
$$

 Note that z and w do *not* necessarily have to be the *same* real number in the solution for Example 3. We can choose arbitrary values for each if we wish to construct a specific solution to the system. For example, if we let $z = 1$ and $w = 2$, we get the solution $(11, 5, 1, 2)$. You should check that this does indeed satisfy all three of the original equations in Example 3.

Many of the linear systems that arise in practical problems turn out to be inconsistent or dependent. Both situations arise in the next two examples.

Example 4

A biologist is performing an experiment on the effects of various combinations of vitamins. She wishes to feed each of her laboratory rabbits a diet that contains exactly 9 mg of niacin, 14 mg of thiamine, and 32 mg of riboflavin. She has available three different types of commercial rabbit pellets whose content per ounce of the relevant vitamins is given in the following table:

	mg Niacin	mg Thiamine	mg Riboflavin
Type A	2	3	8
Type B	3	1	5
Type C	1	3	7

How many ounces of each type of food should she give each rabbit daily to satisfy her requirements?

Solution

If we let x represent the amount of type A to be fed to each rabbit, y the amount of type B, and z the amount of type C, then the daily requirements she has established lead to the linear equations

$$\begin{cases} 2x + 3y + z = 9 \\ 3x + y + 3z = 14 \\ 8x + 5y + 7z = 32 \end{cases}$$

We solve this system as follows:

$$\begin{bmatrix} 2 & 3 & 1 & 9 \\ 3 & 1 & 3 & 14 \\ 8 & 5 & 7 & 32 \end{bmatrix} \xrightarrow{R_2 - R_1} \begin{bmatrix} 2 & 3 & 1 & 9 \\ 1 & -2 & 2 & 5 \\ 8 & 5 & 7 & 32 \end{bmatrix} \xrightarrow[R_3 - 8R_2]{R_1 - 2R_2}$$

$$\begin{bmatrix} 0 & 7 & -3 & -1 \\ 1 & -2 & 2 & 5 \\ 0 & 21 & -9 & -8 \end{bmatrix} \xrightarrow[R_1 \leftrightarrow R_2]{R_3 - 3R_1} \begin{bmatrix} 1 & -2 & 2 & 5 \\ 0 & 7 & -3 & -1 \\ 0 & 0 & 0 & -5 \end{bmatrix}$$

The last row translates into the equation $0 = -5$, which is always false. The system has no solution, so no combination of the three foods will give the required vitamin combination.   ●

Example 5

Suppose the biologist in Example 4 had specified 37 mg instead of 32 mg as the riboflavin requirement, but all other aspects of the experiment remained unchanged. Would there now be a combination of the three foods that would satisfy the requirements?

Solution

The only change we need to make in the solution to Example 4 is to replace the 32 in the original system of equations by 37. If we then carry out the same row operations as before, we arrive at the matrix

$$\begin{bmatrix} 1 & -2 & 2 & 5 \\ 0 & 7 & -3 & -1 \\ 0 & 0 & 0 & 0 \end{bmatrix}$$

The last equation now simply states that $0 = 0$ and so can be eliminated. The system has infinitely many solutions. If we add $\frac{2}{7}$ of the second row to the first (to eliminate the variable y from the first equation), we get

$$\begin{bmatrix} 1 & 0 & \frac{8}{7} & \frac{33}{7} \\ 0 & 7 & -3 & -1 \end{bmatrix}$$

so that the solution is

$$\begin{cases} x = (-8z + 33)/7 \\ y = (3z - 1)/7 \\ z = \text{any real number} \end{cases}$$

Because an amount of food cannot be negative, not every solution to the system provides a practical solution to the problem. Since $y \geq 0$, we have

$$3z - 1 \geq 0$$
$$z \geq \tfrac{1}{3}$$

and since $x \geq 0$,

$$-8z + 33 \geq 0$$
$$\tfrac{33}{8} \geq z$$

This means that the solution to the problem would have to include the condition that the amount z of type C rabbit food used should be between $\tfrac{1}{3}$ oz and $\tfrac{33}{8}$ oz.

Exercises 7.3

In Exercises 1–22 find the complete solution to each system of equations, or show that none exists.

1. (a) $\begin{cases} x - y - 2z = 2 \\ 2x - 3y + 6z = 5 \\ 3x - 4y + 5z = 12 \end{cases}$

(b) $\begin{cases} x - y - 2z = 2 \\ 2x - 3y + 6z = 5 \\ 3x - 4y + 4z = 12 \end{cases}$

(c) $\begin{cases} x - y - 2z = 2 \\ 2x - 3y + 6z = 5 \\ 3x - 4y + 4z = 7 \end{cases}$

2. (a) $\begin{cases} 3x - 2y + 5z = 1 \\ 4x + y - 2z = -16 \\ 4x - 10y + 24z = 36 \end{cases}$

(b) $\begin{cases} 3x - 2y + 5z = 1 \\ 4x + y - 2z = -16 \\ 4x - 11y + 24z = 36 \end{cases}$

(c) $\begin{cases} 3x - 2y + 5z = 1 \\ 4x + y - 2z = -16 \\ 4x - 10y + 24z = 37 \end{cases}$

3. $\begin{cases} x - y + 3z = 3 \\ 4x - 8y + 32z = 24 \\ 2x - 3y + 11z = 4 \end{cases}$

4. $\begin{cases} -2x + 6y - 2z = -12 \\ x - 3y + 2z = 10 \\ -x + 3y + 2z = 6 \end{cases}$

5. $\begin{cases} x + 5y = 12 \\ 3x - 7y = 14 \\ 2x - 4y = 10 \end{cases}$

6. $\begin{cases} 12x - 7y = 11 \\ -3x - 14y = 12 \\ 15x + 8y = 13 \end{cases}$

7. $\begin{cases} x - 3y = 1 \\ 3x - y = 5 \\ 4x - 8y = 3 \end{cases}$

8. $\begin{cases} x + 2y + 3z = 7 \\ 3x + 2y + z = 21 \end{cases}$

9. $\begin{cases} 5x - 4y + 7z = 12 \\ 4x - 3y + 8z = 12 \end{cases}$

10. $\begin{cases} 3x - 6y - 12z = 0 \\ -4x + 8y + 16z = 0 \end{cases}$

11. $\begin{cases} 2x - y + 5z = 12 \\ x + 4y - 2z = -3 \\ 8x + 5y + 11z = 30 \end{cases}$

12. $\begin{cases} 3r + 2s - 3t = 10 \\ r - s - t = -5 \\ r + 4s - t = 20 \end{cases}$

13. $\begin{cases} 2x + y - 2z = 12 \\ -x - \tfrac{1}{2}y + z = -6 \\ 3x + \tfrac{3}{2}y - 3z = 18 \end{cases}$

14. $\begin{cases} y - 5z = 7 \\ 3x + 2y = 12 \\ 3x + 10z = 80 \end{cases}$

15. $\begin{cases} x + y + z + w = 8 \\ y - w = 0 \\ 3x + 2y + z = 12 \\ -3x - 2y + z + 4w = 0 \end{cases}$

16. $\begin{cases} y - z + 2w = 0 \\ 3x + 2y + w = 0 \\ 2x + 4w = 12 \\ -2x - 2z + 5w = 6 \end{cases}$

17. $\begin{cases} 2x - y + 2z + w = 5 \\ -x + y + 4z - w = 3 \\ 3x - 2y - z = 0 \end{cases}$

18. $\begin{cases} 3t - u + v + 2w = 5 \\ t + u - v - w = 7 \\ 4t - 4u + 4v + 6w = 3 \end{cases}$

19. $\begin{cases} x - y + w = 0 \\ 3x - z + 2w = 0 \\ x - 4y + z + 2w = 0 \end{cases}$

20. $\begin{cases} 3x_1 - 2x_2 + 4x_3 = -2 \\ x_1 - 2x_2 + x_3 = 0 \\ 4x_1 - 4x_2 + 5x_3 = -2 \\ -4x_2 - x_3 = 2 \end{cases}$

21. $\begin{cases} 2x - y + z = 5 \\ 3x - 4y - 2z = 1 \\ x - 2y + 4z = 9 \\ 2x - 3y + 5z = 0 \end{cases}$

22. $\begin{cases} a + b + c + d + e = 2 \\ a - c + e = 2 \\ -2a + b - d = 0 \\ 2b + 2e = 4 \end{cases}$

23. A nutritionist wishes to make a milk substitute by combining soya powder, ground millet, and nonfat dried milk powder with enough water to make 1 qt. She wants the mixture to contain 1.1 mg of thiamin, 3.1 mg of riboflavin, and 3.5 mg of niacin. The amounts of these nutrients per ounce of each substance are given in the following table:

	mg Thiamine	mg Riboflavin	mg Niacin
Soya powder	0.2	0.2	1.0
Ground millet	0.5	2.0	1.0
Dried milk	0.4	1.4	1.0

How many ounces of each food should she combine to satisfy her requirements for these nutrients? (Give all possible combinations.)

24. If the nutritionist of Exercise 23 decides she wants her product to have 1.2 mg of thiamin instead of 1.1 mg (without changing the other requirements), what combination of the three foods could she use?

25. A furniture factory makes wooden tables, chairs, and armoires. Each piece of furniture requires three production steps: cutting the wood, assembling, and finishing. The number of hours of each operation required to make a piece of furniture is given in the following table:

	Cutting	Assembling	Finishing
Tables	$\frac{1}{2}$	$\frac{1}{2}$	1
Chairs	1	$1\frac{1}{2}$	$1\frac{1}{2}$
Armoires	1	1	2

The workers in the plant can provide 300 labor-hours of cutting, 400 h of assembling, and 590 h of finishing each week. How many tables, chairs, and armoires should be produced so that all available labor-hours are used? Or is this impossible?

26. Do Exercise 25 assuming that one worker has been laid off, so that only 550 h of finishing labor are available each week.

27. I have some pennies, nickels, and dimes in my pocket. The total value of the coins is 72 cents, and the number of dimes is one third of the total number of nickels and pennies. How many coins of each denomination do I have? (*Hint:* The number of each type of coin must be a nonnegative integer.)

28. A diagram of part of the network of streets in a city is shown in the figure, where the arrows indicate one-way streets. The numbers on the diagram show how many cars enter or leave this section of the city via the indicated street in a certain one-hour period. The variables x, y, z, and w represent the number of cars that travel along the portions of First, Second, Avocado, and Birch Streets shown in the figure during this 1-h period. Find x, y, z, and w, assuming that none of the cars involved in this problem stopped or parked on any of the streets shown in the diagram.

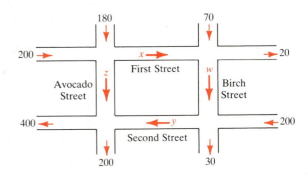

29. (a) Suppose that (x_0, y_0, z_0) and (x_1, y_1, z_1) are solutions of the system

$$\begin{cases} a_1 x + b_1 y + c_1 z = d_1 \\ a_2 x + b_2 y + c_2 z = d_2 \\ a_3 x + b_3 y + c_3 z = d_3 \end{cases}$$

Show that $\left(\dfrac{x_0 + x_1}{2}, \dfrac{y_0 + y_1}{2}, \dfrac{z_0 + z_1}{2} \right)$ is also a solution.

(b) Use the result of part (a) to prove that if the system has two different solutions, it has infinitely many.

Application: Partial Fractions

The process of adding and subtracting fractions by first writing them with a common denominator is familiar for both numerical fractions and rational functions. Thus it is easy to add

$$\frac{1}{3} + \frac{1}{4} = \frac{4}{12} + \frac{3}{12} = \frac{7}{12}$$

or to subtract

$$\frac{1}{x-1} - \frac{1}{x+1} = \frac{x+1}{(x+1)(x-1)} - \frac{x-1}{(x+1)(x-1)} = \frac{2}{x^2-1}$$

The ancient Egyptians used a number notation that required them to write all fractions as sums of reciprocals of whole numbers. For them it was therefore important to know how to reverse this process—for example, to be able to write $7/12$ as the sum of the more elementary fractions $1/3$ and $1/4$. For us this skill is of little use in the context of numerical fractions. For rational functions, however, the process opposite to "bringing to a common denominator" will turn out to be very important in your study of calculus. You will have to know, for example, how to split up $2/(x^2 - 1)$ into the difference of the simpler functions $1/(x - 1)$ and $1/(x + 1)$. These simpler functions are called **partial fractions**, and since the process of finding them involves solving linear equations, we study it here.

Let r be the rational function

$$r(x) = \frac{P(x)}{Q(x)}$$

where the degree of P is less than the degree of Q. It can be shown, using advanced algebra techniques, that every polynomial with real coefficients can be factored completely into linear and irreducible quadratic factors—that is, factors of the form $ax + b$ and $ax^2 + bx + c$, which do not factor any further. For instance,

$$x^4 - 1 = (x^2 - 1)(x^2 + 1) = (x - 1)(x + 1)(x^2 + 1)$$

After we have factored the denominator Q of r completely, we will be able to express $r(x)$ as a sum of **partial fractions** of the form

$$\frac{A}{(ax + b)^i} \quad \text{and} \quad \frac{Ax + B}{(ax^2 + bx + c)^j}$$

This sum is called the **partial fraction decomposition** of r. We now explain the details in the four cases that occur.

Case 1: The Denominator $Q(x)$ Is a Product of Distinct Linear Factors.
This means that we can write

$$Q(x) = (a_1 x + b_1)(a_2 x + b_2) \cdots (a_n x + b_n)$$

where no factor is repeated. In this case the partial fraction decomposition of r takes the form

$$r(x) = \frac{P(x)}{Q(x)} = \frac{A_1}{a_1 x + b_1} + \frac{A_2}{a_2 x + b_2} + \cdots + \frac{A_n}{a_n x + b_n}$$

where the constants $A_1, A_2, \ldots, A_n$ are determined as in the following example.

Example 1

Find the partial fraction decomposition of

$$\frac{5x + 7}{x^3 + 2x^2 - x - 2}$$

Solution

The denominator factors as follows:

$$\begin{aligned} x^3 + 2x^2 - x - 2 &= x^2(x + 2) - (x + 2) = (x^2 - 1)(x + 2) \\ &= (x - 1)(x + 1)(x + 2) \end{aligned}$$

This gives us the partial fraction decomposition

$$\frac{5x + 7}{x^3 + 2x^2 - x - 2} = \frac{A}{x - 1} + \frac{B}{x + 1} + \frac{C}{x + 2}$$

Multiplying both sides by the common denominator $(x - 1)(x + 1)(x + 2)$, we get

$$\begin{aligned} 5x + 7 &= A(x + 1)(x + 2) + B(x - 1)(x + 2) + C(x - 1)(x + 1) \\ &= A(x^2 + 3x + 2) + B(x^2 + x - 2) + C(x^2 - 1) \\ &= (A + B + C)x^2 + (3A + B)x + (2A - 2B - C) \end{aligned}$$

If two polynomials are equal, their coefficients are equal. Thus, since $5x + 7$ has no x^2 term, we have that $A + B + C = 0$. Similarly, by comparing the coefficients of x, we see that $3A + B = 5$, and by comparing constant terms, we get that $2A - 2B - C = 7$. This leads to the following system of linear equations for A, B, and C:

$$\begin{aligned} A + B + C &= 0 \\ 3A + B &= 5 \\ 2A - 2B - C &= 7 \end{aligned}$$

We solve this system using the methods developed in Section 7.2.

$$\begin{bmatrix} 1 & 1 & 1 & 0 \\ 3 & 1 & 0 & 5 \\ 2 & -2 & -1 & 7 \end{bmatrix} \xrightarrow[\substack{R_2 - 3R_1 \\ R_3 - 2R_1}]{} \begin{bmatrix} 1 & 1 & 1 & 0 \\ 0 & -2 & -3 & 5 \\ 0 & -4 & -3 & 7 \end{bmatrix} \xrightarrow[\substack{R_3 - 2R_2 \\ -R_2}]{}$$

$$\begin{bmatrix} 1 & 1 & 1 & 0 \\ 0 & 2 & 3 & -5 \\ 0 & 0 & 3 & -3 \end{bmatrix} \xrightarrow[\substack{R_2 - R_3 \\ \frac{1}{3}R_3}]{} \begin{bmatrix} 1 & 1 & 1 & 0 \\ 0 & 2 & 0 & -2 \\ 0 & 0 & 1 & -1 \end{bmatrix}$$

Thus we see that $C = -1$, $B = -1$, and $A = 2$, so the required partial fraction decomposition is

$$\frac{5x + 7}{x^3 + 2x^2 - x - 2} = \frac{2}{x - 1} + \frac{-1}{x + 1} + \frac{-1}{x + 2} \qquad \bullet$$

   The same method of attack works in each of the remaining cases. We set up the partial fraction decomposition with the unknown constants $A, B, C, \ldots$. We then multiply both sides of the resulting equation by the common denominator, simplify the right-hand side of the equation, and equate coefficients. This gives a set of linear equations that will always have a unique solution (provided the partial fraction decomposition has been set up correctly).

Case 2: The Denominator $Q(x)$ Is a Product of Linear Factors, Some of Which Are Repeated. Suppose the complete factorization of $Q(x)$ contains the linear factor $ax + b$ repeated k times; that is, $(ax + b)^k$ is a factor of $Q(x)$. Then corresponding to each such factor, the partial fraction decomposition for $P(x)/Q(x)$ will contain

$$\frac{A_1}{ax + b} + \frac{A_2}{(ax + b)^2} + \cdots + \frac{A_k}{(ax + b)^k}$$

Example 2

Find the partial fraction decomposition of

$$\frac{x^2 + 1}{x(x - 1)^3}$$

Solution

Because the factor $x - 1$ is repeated three times in the denominator, the partial fraction decomposition is

$$\frac{x^2 + 1}{x(x - 1)^3} = \frac{A}{x} + \frac{B}{x - 1} + \frac{C}{(x - 1)^2} + \frac{D}{(x - 1)^3}$$

Multiplying both sides by the common denominator $x(x - 1)^3$ gives

$$x^2 + 1 = A(x - 1)^3 + Bx(x - 1)^2 + Cx(x - 1) + Dx$$
$$= A(x^3 - 3x^2 + 3x - 1) + B(x^3 - 2x^2 + x) + C(x^2 - x) + Dx$$
$$= (A + B)x^3 + (-3A - 2B + C)x^2 + (3A + B - C + D)x - A$$

Equating coefficients, we get the equations

$$
\begin{aligned}
A + B & = 0 \\
-3A - 2B + C & = 1 \\
3A + B - C + D & = 0 \\
-A & = 1
\end{aligned}
$$

If we rearrange these equations by putting the last one in first position, we can easily see (without having to use matrix techniques) that the solution to the system is $A = -1$, $B = 1$, $C = 0$, and $D = 2$, so

$$
\frac{x^2 + 1}{x(x - 1)^3} = \frac{-1}{x} + \frac{1}{x - 1} + \frac{2}{(x - 1)^3}
$$

●

Case 3: The Denominator $Q(x)$ Has Irreducible Quadratic Factors, None of Which Is Repeated.   If the complete factorization of $Q(x)$ contains the quadratic factor $ax^2 + bx + c$ (which cannot be factored further), then corresponding to this the partial fraction decomposition of $P(x)/Q(x)$ will have a term of the form

$$
\frac{Ax + B}{ax^2 + bx + c}
$$

Example 3

Find the partial fraction decomposition of

$$
\frac{2x^2 - x + 4}{x^3 + 4x}
$$

Solution

Since $x^3 + 4x = x(x^2 + 4)$, which cannot be factored further, we write

$$
\frac{2x^2 - x + 4}{x^3 + 4x} = \frac{A}{x} + \frac{Bx + C}{x^2 + 4}
$$

Multiplying by $x(x^2 + 4)$, we get

$$
\begin{aligned}
2x^2 - x + 4 &= A(x^2 + 4) + (Bx + C)x \\
&= (A + B)x^2 + Cx + 4A
\end{aligned}
$$

Equating coefficients gives the equations

$$
\begin{aligned}
A + B &= 2 \\
C &= -1 \\
4A &= 4
\end{aligned}
$$

and so $A = 1$, $B = 1$, and $C = -1$. The required partial fraction decomposition is

$$
\frac{2x^2 - x + 4}{x^3 + 4x} = \frac{1}{x} + \frac{x - 1}{x^2 + 4}
$$

●

Case 4: The Denominator $Q(x)$ Has a Repeated Irreducible Quadratic Factor. If the complete factorization of $Q(x)$ contains the factor $(ax^2 + bx + c)^k$, where $ax^2 + bx + c$ cannot be factored further, then corresponding to this the partial fraction decomposition of $P(x)/Q(x)$ will have the terms

$$\frac{A_1 x + B_1}{ax^2 + bx + c} + \frac{A_2 x + B_2}{(ax^2 + bx + c)^2} + \cdots + \frac{A_k x + B_k}{(ax^2 + bx + c)^k}$$

Example 4

Write out the form of the partial fraction decomposition of

$$\frac{x^5 - 3x^2 + 12x - 1}{x^3(x^2 + x + 1)(x^2 + 2)^3}$$

Solution

$$\frac{x^5 - 3x^2 + 12x - 1}{x^3(x^2 + x + 1)(x^2 + 2)^3} = \frac{A}{x} + \frac{B}{x^2} + \frac{C}{x^3} + \frac{Dx + E}{x^2 + x + 1} + \frac{Fx + G}{x^2 + 2}$$

$$+ \frac{Hx + I}{(x^2 + 2)^2} + \frac{Jx + K}{(x^2 + 2)^3} \qquad \bullet$$

In order to find the values of $A, B, C, D, E, F, G, H, I, J$, and K in Example 4, we would have to solve a system of 11 linear equations. Although certainly possible, this would involve a great deal of work!

It is important to note that the techniques we have described in this section apply only to rational functions $P(x)/Q(x)$ in which the degree of P is less than the degree of Q. If this is not the case, we must first use long division to divide Q into P.

Example 5

Find the partial fraction decomposition of

$$\frac{2x^4 + 4x^3 - 2x^2 + x + 7}{x^3 + 2x^2 - x - 2}$$

Solution

Since the degree of the numerator is larger than the degree of the denominator, we use long division to obtain

$$\frac{2x^4 + 4x^3 - 2x^2 + x + 7}{x^3 + 2x^2 - x - 2} = 2x + \frac{5x + 7}{x^3 + 2x^2 - x - 2}$$

The remainder term now satisfies the requirement that the degree of the numerator is less than the degree of the denominator. At this point we would proceed as in Example 1 to obtain the decomposition

$$\frac{2x^4 + 4x^3 - 2x^2 + x + 7}{x^3 + 2x^2 - x - 2} = 2x + \frac{2}{x - 1} + \frac{-1}{x + 1} + \frac{-1}{x + 2} \qquad \bullet$$

Exercises 7.4

In Exercises 1–26 find the partial fraction decomposition of the rational functions given.

1. $\dfrac{4}{x^2 - 4}$

2. $\dfrac{2x + 1}{x^2 + x - 2}$

3. $\dfrac{x + 14}{x^2 - 2x - 8}$

4. $\dfrac{8x - 3}{2x^2 - x}$

5. $\dfrac{x}{8x^2 - 10x + 3}$

6. $\dfrac{7x - 3}{x^3 + 2x^2 - 3x}$

7. $\dfrac{9x^2 - 9x + 6}{2x^3 - x^2 - 8x + 4}$

8. $\dfrac{-3x^2 - 3x + 27}{(x + 2)(2x^2 + 3x - 9)}$

9. $\dfrac{x^2 + 1}{x^3 + x^2}$

10. $\dfrac{3x^2 + 5x - 13}{(3x + 2)(x^2 - 4x + 4)}$

11. $\dfrac{2x}{4x^2 + 12x + 9}$

12. $\dfrac{x - 4}{(2x - 5)^2}$

13. $\dfrac{4x^2 - x - 2}{x^4 + 2x^3}$

14. $\dfrac{x^3 - 2x^2 - 4x + 3}{x^4}$

15. $\dfrac{-10x^2 + 27x - 14}{(x - 1)^3(x + 2)}$

16. $\dfrac{-2x^2 + 5x - 1}{x^4 - 2x^3 + 2x - 1}$

17. $\dfrac{3x^3 + 22x^2 + 53x + 41}{(x + 2)^2(x + 3)^2}$

18. $\dfrac{3x^2 + 12x - 20}{x^4 - 8x^2 + 16}$

19. $\dfrac{x - 3}{x^3 + 3x}$

20. $\dfrac{3x^2 - 2x + 8}{x^3 - x^2 + 2x - 2}$

21. $\dfrac{2x^3 + 7x + 5}{(x^2 + x + 2)(x^2 + 1)}$

22. $\dfrac{x^2 + x + 1}{2x^4 + 3x^2 + 1}$

23. $\dfrac{x^4 + x^3 + x^2 - x + 1}{x(x^2 + 1)^2}$

24. $\dfrac{2x^2 - x + 8}{(x^2 + 4)^2}$

25. $\dfrac{x^5 - 2x^4 + x^3 + x + 5}{x^3 - 2x^2 + x - 2}$

26. $\dfrac{x^5 - 3x^4 + 3x^3 - 4x^2 + 4x + 12}{(x - 2)^2(x^2 + 2)}$

In Exercises 27 and 28 give the form of the partial fraction decomposition of the given rational function (as in Example 4 in the text).

27. $\dfrac{x^3 + x + 1}{x(2x - 5)^3(x^2 + 2x + 5)^2}$

28. $\dfrac{1}{(x^6 - 1)(x^4 - 1)}$

29. Determine A and B in terms of a and b:

$$\frac{ax + b}{x^2 - 1} = \frac{A}{x - 1} + \frac{B}{x + 1}$$

30. Determine A, B, C, and D in terms of a and b:

$$\frac{ax^3 + bx^2}{(x^2 + 1)^2} = \frac{Ax + B}{x^2 + 1} + \frac{Cx + D}{(x^2 + 1)^2}$$

Algebra of Matrices

Up to this point we have been using matrices simply as a notational convenience, to make our work in solving linear equations a little easier. Matrices have many other uses in mathematics and the sciences, and for most of these applications a knowledge of matrix algebra is essential. Like numbers, matrices can be added, subtracted, multiplied, and divided under certain circumstances, and in this section we learn how to perform these algebraic operations on matrices.

Recall that a matrix is simply a rectangular array of numbers enclosed between brackets. For example, let A be the matrix

$$A = \begin{bmatrix} -1 & 4 & 7 & 0 \\ 0 & 2 & 13 & 14 \\ \frac{1}{2} & 22 & 8 & -2 \end{bmatrix}$$

The **dimension** of a matrix is a pair of numbers that indicates how many rows and columns a matrix has. The matrix A is a 3×4 matrix because it has 3 rows and 4 columns. The individual numbers that make up a matrix are called its **entries**, and they are specified by their row and column position. In the above matrix A, the number 13 is the $(2, 3)$ entry, since it is in the second row and the third column. If the name of a matrix is A, we will often use the symbol a_{ij} to denote the (i, j) entry of the matrix. Thus for the above matrix, $a_{24} = 14$ and $a_{32} = 22$.

Two matrices are **equal** if they have the same dimension and their corresponding entries are equal. So

> $A = B$ if and only if both A and B have dimension $m \times n$ and $a_{ij} = b_{ij}$ for $i = 1, 2, \ldots, m$ and $j = 1, 2, \ldots, n$.

For example,

$$\begin{bmatrix} \sqrt{4} & 2^2 & 0 & e^0 \\ 0.5 & 1 & 0 & 1-1 \end{bmatrix} = \begin{bmatrix} 2 & 4 & 0 & 1 \\ \frac{1}{2} & \frac{2}{2} & 0 & 0 \end{bmatrix}$$

but

$$\begin{bmatrix} 1 & 2 \\ 3 & 4 \\ 5 & 6 \end{bmatrix} \neq \begin{bmatrix} 1 & 3 & 5 \\ 2 & 4 & 6 \end{bmatrix}$$

Two matrices can be added or subtracted whenever they have the same dimension. (Otherwise their sum or difference is undefined.) We add or subtract the matrices by adding or subtracting corresponding entries. Thus we have the sum

$$\begin{bmatrix} 2 & -3 \\ 0 & 5 \\ 7 & -\frac{1}{2} \end{bmatrix} + \begin{bmatrix} 1 & 0 \\ -3 & 1 \\ 2 & 2 \end{bmatrix} = \begin{bmatrix} 3 & -3 \\ -3 & 6 \\ 9 & \frac{3}{2} \end{bmatrix}$$

because both the matrices being added have dimension 3×2. The difference

$$\begin{bmatrix} 7 & -3 & 0 \\ 0 & 1 & 5 \end{bmatrix} - \begin{bmatrix} 6 & 0 & -6 \\ 8 & 1 & 9 \end{bmatrix} = \begin{bmatrix} 1 & -3 & 6 \\ -8 & 0 & -4 \end{bmatrix}$$

is also defined, since the matrices being subtracted are both 2×3. But the result of the operation

$$\begin{bmatrix} 7 & -3 & 0 \\ 0 & 1 & 5 \end{bmatrix} + \begin{bmatrix} 2 & -3 \\ 0 & 5 \\ 7 & -\frac{1}{2} \end{bmatrix}$$

is undefined, since you cannot take the sum of a 2×3 and a 3×2 matrix.

We can multiply a number times a matrix by multiplying every entry in the matrix by that number. For example,

$$5 \begin{bmatrix} 2 & -3 \\ 0 & 5 \\ 7 & -\frac{1}{2} \end{bmatrix} = \begin{bmatrix} 10 & -15 \\ 0 & 25 \\ 35 & -\frac{5}{2} \end{bmatrix}$$

Multiplication of two matrices is not quite so easy to describe. We will see in later examples why taking the matrix product involves the following rather complex procedure.

First of all, the product AB (or $A \cdot B$) of two matrices A and B is defined only when the number of columns in A is equal to the number of rows in B. This means that if we write their dimensions side by side, the two inner numbers must match:

$$\begin{array}{ccc} \text{matrices} & A & B \\ \text{dimensions} & m \times n & n \times k \end{array}$$

columns in A rows in B

If the dimensions of A and B match in this fashion, then the product AB will have dimension $m \times k$. Before describing the procedure for obtaining the elements of AB, we define the **inner product** of a row of A and a column of B.

If $[a_1 \quad a_2 \quad \cdots \quad a_n]$ is a row of A, and if $\begin{bmatrix} b_1 \\ b_2 \\ \vdots \\ b_n \end{bmatrix}$ is a column of B, then

their **inner product** is the number $a_1 b_1 + a_2 b_2 + \cdots + a_n b_n$.

For example,

$$[2 \quad -1 \quad 0 \quad 4] \cdot \begin{bmatrix} 5 \\ 4 \\ -3 \\ \frac{1}{2} \end{bmatrix} = 2 \cdot 5 + (-1) \cdot 4 + 0 \cdot (-3) + 4 \cdot \tfrac{1}{2} = 8$$

We now define the **product** AB of two matrices as follows:

Suppose that A is an $m \times n$ matrix and B an $n \times k$ matrix. Then $C = AB$ is an $m \times k$ matrix, where c_{ij} is the inner product of the ith row of A and the jth column of B.

Example 1

Let

$$A = \begin{bmatrix} 1 & 3 \\ -1 & 0 \end{bmatrix} \quad \text{and} \quad B = \begin{bmatrix} -1 & 5 & 2 \\ 0 & 4 & 7 \end{bmatrix}$$

Calculate, if possible, the products AB and BA.

Solution

Since A has dimension 2×2 and B has dimension 2×3, the product AB will have dimension 2×3. We can thus write

$$AB = \begin{bmatrix} 1 & 3 \\ -1 & 0 \end{bmatrix} \begin{bmatrix} -1 & 5 & 2 \\ 0 & 4 & 7 \end{bmatrix} = \begin{bmatrix} ? & ? & ? \\ ? & ? & ? \end{bmatrix}$$

where the question marks must be filled in using the rule that defines the entries of a matrix product. The $(1, 1)$ entry will be the inner product of the first row of A and the first column of B:

$$\begin{bmatrix} 1 & 3 \end{bmatrix} \cdot \begin{bmatrix} -1 \\ 0 \end{bmatrix} = 1 \cdot (-1) + 3 \cdot 0 = -1$$

Similarly, we calculate the remaining entries as follows:

Entry	Inner product of	Value
$(1, 2)$	$\begin{bmatrix} 1 & 3 \\ -1 & 0 \end{bmatrix} \begin{bmatrix} -1 & 5 & 2 \\ 0 & 4 & 7 \end{bmatrix}$	$1 \cdot 5 + 3 \cdot 4 = 17$
$(1, 3)$	$\begin{bmatrix} 1 & 3 \\ -1 & 0 \end{bmatrix} \begin{bmatrix} -1 & 5 & 2 \\ 0 & 4 & 7 \end{bmatrix}$	$1 \cdot 2 + 3 \cdot 7 = 23$
$(2, 1)$	$\begin{bmatrix} 1 & 3 \\ -1 & 0 \end{bmatrix} \begin{bmatrix} -1 & 5 & 2 \\ 0 & 4 & 7 \end{bmatrix}$	$(-1) \cdot (-1) + 0 \cdot 0 = 1$
$(2, 2)$	$\begin{bmatrix} 1 & 3 \\ -1 & 0 \end{bmatrix} \begin{bmatrix} -1 & 5 & 2 \\ 0 & 4 & 7 \end{bmatrix}$	$(-1) \cdot 5 + 0 \cdot 4 = -5$
$(2, 3)$	$\begin{bmatrix} 1 & 3 \\ -1 & 0 \end{bmatrix} \begin{bmatrix} -1 & 5 & 2 \\ 0 & 4 & 7 \end{bmatrix}$	$(-1) \cdot 2 + 0 \cdot 7 = -2$

Thus we have

$$AB = \begin{bmatrix} -1 & 17 & 23 \\ 1 & -5 & -2 \end{bmatrix}$$

The product BA is not defined, however, because the dimensions are

$$2 \times 3 \quad \text{and} \quad 2 \times 2$$

The inner two numbers are not the same, so the rows and columns will not match up when we try to calculate the product. ●

 The next example shows that even when both AB and BA are defined, they are not necessarily equal. This will prove that matrix multiplication is *not* commutative.

Example 2

Let

$$A = \begin{bmatrix} 5 & 7 \\ -3 & 0 \end{bmatrix} \quad \text{and} \quad B = \begin{bmatrix} 1 & 2 \\ 9 & -1 \end{bmatrix}$$

Calculate the products AB and BA.

Solution

Since both A and B are 2×2 matrices, both AB and BA are defined and are also 2×2 matrices.

$$AB = \begin{bmatrix} 5 & 7 \\ -3 & 0 \end{bmatrix} \begin{bmatrix} 1 & 2 \\ 9 & -1 \end{bmatrix} = \begin{bmatrix} 5 \cdot 1 + 7 \cdot 9 & 5 \cdot 2 + 7 \cdot (-1) \\ (-3) \cdot 1 + 0 \cdot 9 & (-3) \cdot 2 + 0 \cdot (-1) \end{bmatrix}$$

$$= \begin{bmatrix} 68 & 3 \\ -3 & -6 \end{bmatrix}$$

$$BA = \begin{bmatrix} 1 & 2 \\ 9 & -1 \end{bmatrix} \begin{bmatrix} 5 & 7 \\ -3 & 0 \end{bmatrix} = \begin{bmatrix} 1 \cdot 5 + 2 \cdot (-3) & 1 \cdot 7 + 2 \cdot 0 \\ 9 \cdot 5 + (-1) \cdot (-3) & 9 \cdot 7 + (-1) \cdot 0 \end{bmatrix}$$

$$= \begin{bmatrix} -1 & 7 \\ 48 & 63 \end{bmatrix}$$

This shows that, in general, $AB \neq BA$. In fact, in this example, AB and BA do not even have any entries in common.   ●

Although matrix multiplication is not commutative, it does obey the associative and distributive laws. That is, if $A, B, C,$ and D are matrices for which the products below are defined, we have

$$A(BC) = (AB)C \quad \text{(associativity)}$$
$$A(B + C) = AB + AC, \quad (B + C)D = BD + CD \quad \text{(distributivity)}$$

The next two examples give some indication of why mathematicians chose to define the matrix product in such an apparently bizarre fashion.

Example 3

Show that the matrix equation

$$\begin{bmatrix} 1 & 2 & 4 \\ -1 & 1 & 2 \\ 2 & 3 & 3 \end{bmatrix} \begin{bmatrix} x \\ y \\ z \end{bmatrix} = \begin{bmatrix} 7 \\ 5 \\ 7 \end{bmatrix}$$

is equivalent to the system of equations in Example 1 of Section 7.2.

Solution

If we perform the matrix multiplication on the left-hand side of the given equation, we get

$$\begin{bmatrix} x + 2y + 4z \\ -x + y + 2z \\ 2x + 3y + 3z \end{bmatrix} = \begin{bmatrix} 7 \\ 5 \\ 7 \end{bmatrix}$$

Since two matrices are equal if their corresponding entries are equal, this means that

$$\begin{aligned} x + 2y + 4z &= 7 \\ -x + y + 2z &= 5 \\ 2x + 3y + 3z &= 7 \end{aligned}$$

This is exactly the system of equations we had in Example 1 of Section 7.2.   ●

The preceding example shows that our definition of matrix product allows us to express a system of linear equations as a single matrix equation in a natural way.

Example 4

In a certain city the proportion of voters in each age group who registered as Democrats, Republicans, or Independents is given by the following matrix:

$$\begin{array}{c} \\ \\ \text{Democrats} \\ \text{Republicans} \\ \text{Independents} \end{array} \begin{array}{c} \qquad\qquad Age \\ \begin{array}{ccc} 18\text{--}30 & 31\text{--}50 & \text{Over } 50 \end{array} \\ \left[\begin{array}{ccc} 0.30 & 0.60 & 0.50 \\ 0.50 & 0.35 & 0.25 \\ 0.20 & 0.05 & 0.25 \end{array} \right] = A \end{array}$$

The next matrix gives the distribution, by age and gender, of the voting population of this city.

$$\begin{array}{cc} & \begin{array}{cc} \text{Male} & \text{Female} \end{array} \\ \begin{array}{cc} & 18\text{--}30 \\ Age & 31\text{--}50 \\ & \text{Over } 50 \end{array} & \left[\begin{array}{cc} 5{,}000 & 6{,}000 \\ 10{,}000 & 12{,}000 \\ 12{,}000 & 15{,}000 \end{array} \right] = B \end{array}$$

For the purposes of this problem, let us make the (highly unrealistic) assumption that within each age group, political preference is not related to gender. That is, the percentage of Democrat males in the 18–30 group, for example, is the same as the percentage of Democrat females in this group.

(a)   Calculate the product AB.

(b)   How many male Democrats are there in this city?

(c)   How many female Republicans are there?

Solution

(a)

$$AB = \left[\begin{array}{ccc} 0.30 & 0.60 & 0.50 \\ 0.50 & 0.35 & 0.25 \\ 0.20 & 0.05 & 0.25 \end{array} \right] \left[\begin{array}{cc} 5{,}000 & 6{,}000 \\ 10{,}000 & 12{,}000 \\ 12{,}000 & 15{,}000 \end{array} \right] = \left[\begin{array}{cc} 13{,}500 & 16{,}500 \\ 9{,}000 & 10{,}950 \\ 4{,}500 & 5{,}550 \end{array} \right]$$

(b)   When we take the inner product of a row from A with a column from B, we are adding the number of people in each of the three age groups who belong to the category in question. For example, the $(2, 1)$ entry of AB (9,000) was obtained by taking the inner product of the Republican row from A with the male column from B. This number is therefore the total number of male Republicans in this city. We can label the rows and columns of AB as follows:

$$
\begin{array}{c}
 \\
\text{Democratic} \\
\text{Republican} \\
\text{Independent}
\end{array}
\begin{array}{cc}
\text{Male} & \text{Female} \\
\begin{bmatrix} 13{,}500 & 16{,}500 \\ 9{,}000 & 10{,}950 \\ 4{,}500 & 5{,}550 \end{bmatrix} = AB
\end{array}
$$

There are 13,500 male Democrats in this city.

(c)   There are 10,950 female Republicans.   ●

If you add the entries in the columns of matrix A in Example 4, you will see that in each case the sum is 1. (Can you see why this has to be true, given what the matrix is describing?) A matrix with this property is called **stochastic**. Stochastic matrices are studied extensively in statistics, where they arise frequently in situations like the one described in Example 4.

Exercises 7.5

In Exercises 1–21 the matrices A, B, C, D, E, F, and G are defined as follows:

$$A = \begin{bmatrix} 2 & -5 \\ 0 & 7 \end{bmatrix} \qquad B = \begin{bmatrix} 3 & \frac{1}{2} & 5 \\ 1 & -1 & 3 \end{bmatrix}$$

$$C = \begin{bmatrix} 2 & -\frac{5}{2} & 0 \\ 0 & 2 & -3 \end{bmatrix} \qquad D = \begin{bmatrix} 7 & 3 \end{bmatrix}$$

$$E = \begin{bmatrix} 0 & 0 & 0 & 0 & 0 \\ 0 & 0 & 0 & 0 & 0 \\ 0 & 0 & 0 & 0 & 0 \end{bmatrix} \qquad F = \begin{bmatrix} 1 & 0 & 0 \\ 0 & 1 & 0 \\ 0 & 0 & 1 \end{bmatrix}$$

$$G = \begin{bmatrix} 5 & -3 & 10 \\ 6 & 1 & 0 \\ -5 & 2 & 2 \\ 0 & 0 & 0 \end{bmatrix}$$

Carry out the algebraic operation indicated in each exercise, or explain why it cannot be performed.

1. $B + C$

2. $B + F$

3. $C - B$

4. $5A$

5. $3B + 2C$

6. $C - 5A$

7. $2C - 6B$

8. DA

9. AD

10. BC

11. BF

12. GF

13. $(DA)B$

14. $D(AB)$

15. GE

16. A^2

17. A^3

18. $DB + DC$

19. B^2

20. F^2

21. $BF + FE$

22. What must be true about the dimensions of the matrices A and B if both products AB and BA are defined?

Write the systems of equations in Exercises 23–26 as matrix equations. (See Example 3.)

23. $\begin{cases} 2x - 5y = 7 \\ 3x + 2y = 4 \end{cases}$

24. $\begin{cases} 6x - y + z = 12 \\ 2x + z = 7 \\ y - 2z = 4 \end{cases}$

25. $\begin{cases} 3x_1 + 2x_2 - x_3 + x_4 = 0 \\ x_1 \qquad\;\; - x_3 \qquad\;\; = 5 \\ 3x_2 + x_3 - x_4 = 4 \end{cases}$

26. $\begin{cases} x - y + z = 2 \\ 4x - 2y - z = 2 \\ x + y + 5z = 2 \\ -x - y - z = 2 \end{cases}$

Let

$$A = \begin{bmatrix} 4 & 6 \\ 1 & 3 \end{bmatrix} \quad B = \begin{bmatrix} 2 & 5 \\ 3 & 7 \end{bmatrix} \quad C = \begin{bmatrix} 2 & 3 \\ 1 & 0 \\ 0 & 2 \end{bmatrix}$$

$$D = \begin{bmatrix} 10 & 20 \\ 30 & 20 \\ 10 & 0 \end{bmatrix}$$

Solve the matrix equations in Exercises 27–30 for the unknown matrix X, or explain why there is no solution.

27. $2X - A = B$ **28.** $5(X - C) = D$

29. $3X + B = C$ **30.** $A + D = 3X$

31. Let O represent the 2×2 **zero matrix**:

$$O = \begin{bmatrix} 0 & 0 \\ 0 & 0 \end{bmatrix}$$

If A and B are 2×2 matrices with $AB = O$, is it necessarily true that $A = O$ or $B = O$?

32. Prove that if A and B are 2×2 matrices, then

$$(A + B)^2 = A^2 + AB + BA + B^2$$

33. If A and B are 2×2 matrices, is it necessarily true that

$$(A + B)^2 \overset{?}{=} A^2 + 2AB + B^2$$

34. Let

$$A = \begin{bmatrix} 1 & 1 \\ 0 & 1 \end{bmatrix}$$

(a) Calculate A^2, A^3, and A^4.
(b) Find a general formula for A^n.

35. Let

$$A = \begin{bmatrix} 1 & 1 \\ 1 & 1 \end{bmatrix}$$

(a) Calculate A^2, A^3, and A^4.
(b) Find a general formula for A^n.

36. A small fast-food chain has restaurants in Santa Monica, Long Beach, and Anaheim. Only hamburgers, hot dogs, and milk shakes are sold by this chain. On a certain day, sales were distributed according to the following matrix:

Number of items sold:

	Santa Monica	Long Beach	Anaheim	
Hamburgers	4000	1000	3500	= A
Hot dogs	400	300	200	
Milk shakes	700	500	900	

The price of each item is given by the matrix

Hamburger	Hot dog	Milk shake	
[$0.90	$0.80	$1.10] = B

(a) Calculate the product BA.
(b) Interpret the entries in the product matrix BA.

37. A specialty car manufacturer has plants in Auburn, Biloxi, and Chattanooga. Three models are produced, with daily production given in the following matrix:

Cars produced each day:

	Model K	Model R	Model W	
Auburn	12	10	0	
Biloxi	4	4	20	= A
Chattanooga	8	9	12	

Because of a wage increase, February profits are lower than January profits. The profits per car are tabulated in the matrix below:

	January	February	
Model K	$1000	$500	
Model R	$2000	$1200	= B
Model W	$1500	$1000	

(a) Calculate AB.
(b) Assuming that all cars produced were sold, what was the daily profit in January from the Biloxi plant?
(c) What was the total daily profit (from all three plants) in February?

38. Let

$$A = \begin{bmatrix} 1 & 0 & 6 & -1 \\ 2 & \frac{1}{2} & 4 & 0 \end{bmatrix}$$

$$B = \begin{bmatrix} 1 & 7 & -9 & 2 \end{bmatrix}$$

$$C = \begin{bmatrix} 1 \\ 0 \\ -1 \\ -2 \end{bmatrix}$$

Determine which of the following products are defined, and calculate the ones that are: ABC, ACB, BAC, BCA, CAB, CBA.

Inverses of Matrices and Matrix Equations

We have seen in the preceding section that matrices can, when the dimensions are appropriate, be added, subtracted, and multiplied. In this section we will investigate division of matrices, which will allow us to solve equations that involve matrices.

First, we define **identity matrices**, which play the same role for matrix multiplication that the number 1 does for ordinary multiplication of numbers; that is, $1 \cdot a = a \cdot 1 = a$ for all numbers a. The term **main diagonal** in this definition refers to the entries of a square matrix whose row and column numbers are the same. (Note that these entries stretch diagonally down the matrix, from top left to bottom right.)

> The **identity matrix** I_n is the $n \times n$ matrix for which each main diagonal entry is a 1 and for which all other entries are 0.

Thus the 2×2, 3×3, and 4×4 identity matrices are, respectively,

$$I_2 = \begin{bmatrix} 1 & 0 \\ 0 & 1 \end{bmatrix} \qquad I_3 = \begin{bmatrix} 1 & 0 & 0 \\ 0 & 1 & 0 \\ 0 & 0 & 1 \end{bmatrix} \qquad I_4 = \begin{bmatrix} 1 & 0 & 0 & 0 \\ 0 & 1 & 0 & 0 \\ 0 & 0 & 1 & 0 \\ 0 & 0 & 0 & 1 \end{bmatrix}$$

Identity matrices behave like the number 1 in the sense that

$$A \cdot I_n = A \qquad \text{and} \qquad I_n \cdot B = B$$

whenever these products are defined. Thus multiplication by an identity of the appropriate size leaves a matrix unchanged. For example, one can verify by direct calculation that

$$\begin{bmatrix} 1 & 0 \\ 0 & 1 \end{bmatrix} \begin{bmatrix} 3 & 5 & 6 \\ -1 & 2 & 7 \end{bmatrix} = \begin{bmatrix} 3 & 5 & 6 \\ -1 & 2 & 7 \end{bmatrix}$$

or that

$$\begin{bmatrix} -1 & 7 & \frac{1}{2} \\ 12 & 1 & 3 \\ -2 & 0 & 7 \end{bmatrix} \begin{bmatrix} 1 & 0 & 0 \\ 0 & 1 & 0 \\ 0 & 0 & 1 \end{bmatrix} = \begin{bmatrix} -1 & 7 & \frac{1}{2} \\ 12 & 1 & 3 \\ -2 & 0 & 7 \end{bmatrix}$$

If A and B are $n \times n$ matrices, and if $AB = BA = I_n$, then we say that B is the **inverse** of A and we write $B = A^{-1}$. The concept of the inverse of a matrix is analogous to that of the reciprocal of a real number. The following rule allows us to calculate the inverse of a 2×2 matrix:

> If $A = \begin{bmatrix} a & b \\ c & d \end{bmatrix}$, then $A^{-1} = \dfrac{1}{ad - bc} \begin{bmatrix} d & -b \\ -c & a \end{bmatrix}$.

Example 1

Let

$$A = \begin{bmatrix} 4 & 5 \\ 2 & 3 \end{bmatrix}$$

Find A^{-1} and verify that $AA^{-1} = A^{-1}A = I_2$.

Solution

Using the rule, we get

$$A^{-1} = \frac{1}{4\cdot 3 - 5\cdot 2}\begin{bmatrix} 3 & -5 \\ -2 & 4 \end{bmatrix} = \frac{1}{2}\begin{bmatrix} 3 & -5 \\ -2 & 4 \end{bmatrix} = \begin{bmatrix} \frac{3}{2} & -\frac{5}{2} \\ -1 & 2 \end{bmatrix}$$

To verify that this is indeed the inverse of A, we calculate

$$AA^{-1} = \begin{bmatrix} 4 & 5 \\ 2 & 3 \end{bmatrix}\begin{bmatrix} \frac{3}{2} & -\frac{5}{2} \\ -1 & 2 \end{bmatrix} = \begin{bmatrix} 4\cdot\frac{3}{2} + 5(-1) & 4(-\frac{5}{2}) + 5\cdot 2 \\ 2\cdot\frac{3}{2} + 3(-1) & 2(-\frac{5}{2}) + 3\cdot 2 \end{bmatrix} = \begin{bmatrix} 1 & 0 \\ 0 & 1 \end{bmatrix}$$

and

$$A^{-1}A = \begin{bmatrix} \frac{3}{2} & -\frac{5}{2} \\ -1 & 2 \end{bmatrix}\begin{bmatrix} 4 & 5 \\ 2 & 3 \end{bmatrix} = \begin{bmatrix} \frac{3}{2}\cdot 4 + (-\frac{5}{2})2 & \frac{3}{2}\cdot 5 + (-\frac{5}{2})3 \\ (-1)4 + 2\cdot 2 & (-1)5 + 2\cdot 3 \end{bmatrix} = \begin{bmatrix} 1 & 0 \\ 0 & 1 \end{bmatrix}$$

The quantity $ad - bc$ that appears in the rule for calculating the inverse is called the **determinant** of the matrix. If the determinant is 0, then the matrix will not have an inverse (since we cannot divide by 0). In the next section we will learn how to calculate the determinants of square matrices of any size, and how to use them to solve systems of equations.

For 3×3 and larger square matrices, the following technique provides the most efficient way to calculate the inverse. If A is an $n \times n$ matrix, we begin by constructing the $n \times 2n$ matrix, which has the entries of A on the left and of the identity matrix I_n on the right:

$$\begin{bmatrix} a_{11} & a_{12} & \cdots & a_{1n} & 1 & 0 & \cdots & 0 \\ a_{21} & a_{22} & \cdots & a_{2n} & 0 & 1 & \cdots & 0 \\ \vdots & \vdots & \ddots & \vdots & \vdots & \vdots & \ddots & \vdots \\ a_{n1} & a_{n2} & \cdots & a_{nn} & 0 & 0 & \cdots & 1 \end{bmatrix}$$

We then use the elementary row operations on this new large matrix to change the left side into the identity matrix. The right side will be automatically transformed into A^{-1}. (We omit the proof of this fact.)

Example 2

Find the inverse of the matrix

$$A = \begin{bmatrix} 1 & -2 & -4 \\ 2 & -3 & -6 \\ -3 & 6 & 15 \end{bmatrix}$$

and verify that $AA^{-1} = A^{-1}A = I_3$.

Solution

We begin with the 3×6 matrix whose left half is A and whose right half is the identity matrix:

$$\begin{bmatrix} 1 & -2 & -4 & 1 & 0 & 0 \\ 2 & -3 & -6 & 0 & 1 & 0 \\ -3 & 6 & 15 & 0 & 0 & 1 \end{bmatrix}$$

We then transform the left half of this new matrix into the identity matrix by performing the following sequence of elementary row operations on the *entire* new matrix:

$$\xrightarrow[\substack{R_2 - 2R_1 \\ R_3 + 3R_1}]{} \begin{bmatrix} 1 & -2 & -4 & 1 & 0 & 0 \\ 0 & 1 & 2 & -2 & 1 & 0 \\ 0 & 0 & 3 & 3 & 0 & 1 \end{bmatrix}$$

$$\xrightarrow{\frac{1}{3}R_3} \begin{bmatrix} 1 & -2 & -4 & 1 & 0 & 0 \\ 0 & 1 & 2 & -2 & 1 & 0 \\ 0 & 0 & 1 & 1 & 0 & \frac{1}{3} \end{bmatrix}$$

$$\xrightarrow{R_1 + 2R_2} \begin{bmatrix} 1 & 0 & 0 & -3 & 2 & 0 \\ 0 & 1 & 2 & -2 & 1 & 0 \\ 0 & 0 & 1 & 1 & 0 & \frac{1}{3} \end{bmatrix}$$

$$\xrightarrow{R_2 - 2R_3} \begin{bmatrix} 1 & 0 & 0 & -3 & 2 & 0 \\ 0 & 1 & 0 & -4 & 1 & -\frac{2}{3} \\ 0 & 0 & 1 & 1 & 0 & \frac{1}{3} \end{bmatrix}$$

We have now transformed the left half of this matrix into the identity matrix. Note that to do this in as systematic a fashion as possible, we first changed the elements below the main diagonal to zeros, just as we would if we were doing Gaussian elimination. We then changed the main diagonal elements to ones by multiplying by the appropriate constant(s). Finally, we completed the process by changing the remaining entries on the left side to zeros. The right half will now be A^{-1}.

$$A^{-1} = \begin{bmatrix} -3 & 2 & 0 \\ -4 & 1 & -\frac{2}{3} \\ 1 & 0 & \frac{1}{3} \end{bmatrix}$$

To verify this, we multiply

$$AA^{-1} = \begin{bmatrix} 1 & -2 & -4 \\ 2 & -3 & -6 \\ -3 & 6 & 15 \end{bmatrix} \begin{bmatrix} -3 & 2 & 0 \\ -4 & 1 & -\frac{2}{3} \\ 1 & 0 & \frac{1}{3} \end{bmatrix} = \begin{bmatrix} 1 & 0 & 0 \\ 0 & 1 & 0 \\ 0 & 0 & 1 \end{bmatrix}$$

and $\quad A^{-1}A = \begin{bmatrix} -3 & 2 & 0 \\ -4 & 1 & -\frac{2}{3} \\ 1 & 0 & \frac{1}{3} \end{bmatrix} \begin{bmatrix} 1 & -2 & -4 \\ 2 & -3 & -6 \\ -3 & 6 & 15 \end{bmatrix} = \begin{bmatrix} 1 & 0 & 0 \\ 0 & 1 & 0 \\ 0 & 0 & 1 \end{bmatrix}$ ●

The next example shows that not all square matrices have inverses.

Example 3

Try to find the inverse of the matrix

$$\begin{bmatrix} 2 & -3 & -7 \\ 1 & 2 & 7 \\ 1 & 1 & 4 \end{bmatrix}$$

Solution

We proceed as follows:

$$\begin{bmatrix} 2 & -3 & -7 & 1 & 0 & 0 \\ 1 & 2 & 7 & 0 & 1 & 0 \\ 1 & 1 & 4 & 0 & 0 & 1 \end{bmatrix} \xrightarrow{R_1 \leftrightarrow R_2} \begin{bmatrix} 1 & 2 & 7 & 0 & 1 & 0 \\ 2 & -3 & -7 & 1 & 0 & 0 \\ 1 & 1 & 4 & 0 & 0 & 1 \end{bmatrix}$$

$$\xrightarrow[R_3 - R_1]{R_2 - 2R_1} \begin{bmatrix} 1 & 2 & 7 & 0 & 1 & 0 \\ 0 & -7 & -21 & 1 & -2 & 0 \\ 0 & -1 & -3 & 0 & -1 & 1 \end{bmatrix}$$

$$\xrightarrow{-\frac{1}{7}R_2} \begin{bmatrix} 1 & 2 & 7 & 0 & 1 & 0 \\ 0 & 1 & 3 & -\frac{1}{7} & \frac{2}{7} & 0 \\ 0 & -1 & -3 & 0 & -1 & 1 \end{bmatrix}$$

$$\xrightarrow[R_1 - 2R_2]{R_3 + R_2} \begin{bmatrix} 1 & 0 & 1 & \frac{2}{7} & \frac{3}{7} & 0 \\ 0 & 1 & 3 & -\frac{1}{7} & \frac{2}{7} & 0 \\ 0 & 0 & 0 & -\frac{1}{7} & -\frac{5}{7} & 1 \end{bmatrix}$$

At this point we would like to change the 0 in the (3, 3) position of this matrix to a 1 without changing the zeros in the (3, 1) and (3, 2) positions. But there is no way to accomplish this because no matter what multiple of rows 1 and/or 2 we add to row 3, we cannot change the third zero in row 3 without changing the first or second as well. Thus we cannot change the left half to the identity matrix. The original matrix does not have an inverse.   ●

If we encounter a row of zeros on the left when trying to find an inverse, then the original matrix does not have an inverse.

Matrix Equations

We saw in the preceding section that a system of linear equations can be written as a single matrix equation. For example, the system

$$\begin{cases} x + 2y + 4z = 7 \\ -x + y + 2z = 5 \\ 2x + 3y + 3z = 7 \end{cases}$$

is equivalent to the matrix equation

$$\begin{bmatrix} 1 & 2 & 4 \\ -1 & 1 & 2 \\ 2 & 3 & 3 \end{bmatrix} \begin{bmatrix} x \\ y \\ z \end{bmatrix} = \begin{bmatrix} 7 \\ 5 \\ 7 \end{bmatrix}$$

(See Example 3 in Section 7.5.)

If we let

$$A = \begin{bmatrix} 1 & 2 & 4 \\ -1 & 1 & 2 \\ 2 & 3 & 3 \end{bmatrix} \qquad X = \begin{bmatrix} x \\ y \\ z \end{bmatrix} \qquad B = \begin{bmatrix} 7 \\ 5 \\ 7 \end{bmatrix}$$

then this matrix equation can be written

(1)
$$AX = B$$

This has the same form as, for example, the following simple real number equation:

$$3x = 12$$

We solve this latter equation by multiplying both sides by the reciprocal (or inverse) of 3:

$$\tfrac{1}{3}(3x) = \tfrac{1}{3}(12)$$

$$x = 4$$

Similarly, we can solve the matrix equation (1) by multiplying both sides by the inverse of A (provided this inverse exists):

$$AX = B$$
$$A^{-1}(AX) = A^{-1}B$$
$$(A^{-1}A)X = A^{-1}B$$
$$I_3 X = A^{-1}B$$

(2)
$$X = A^{-1}B$$

In this example,

$$A^{-1} = \frac{1}{9} \begin{bmatrix} 3 & -6 & 0 \\ -7 & 5 & 6 \\ 5 & -1 & -3 \end{bmatrix}$$

(verify!), so that Equation 2 becomes

$$\begin{bmatrix} x \\ y \\ z \end{bmatrix} = \frac{1}{9} \begin{bmatrix} 3 & -6 & 0 \\ -7 & 5 & 6 \\ 5 & -1 & -3 \end{bmatrix} \begin{bmatrix} 7 \\ 5 \\ 7 \end{bmatrix}$$

$$= \frac{1}{9} \begin{bmatrix} -9 \\ 18 \\ 9 \end{bmatrix} = \begin{bmatrix} -1 \\ 2 \\ 1 \end{bmatrix}$$

Thus $x = -1$, $y = 2$, and $z = 1$ is the solution to the original system. (Compare this with the solution to Example 1 in Section 7.2.)

Example 4

Solve the following matrix equation:

$$\begin{bmatrix} 2 & -5 \\ 3 & -6 \end{bmatrix} \begin{bmatrix} x \\ y \end{bmatrix} = \begin{bmatrix} 15 \\ 36 \end{bmatrix}$$

Solution

Using the rule for calculating the inverse of a 2×2 matrix, we get

$$\begin{bmatrix} 2 & -5 \\ 3 & -6 \end{bmatrix}^{-1} = \frac{1}{2(-6) - (-5)3} \begin{bmatrix} -6 & -(-5) \\ -3 & 2 \end{bmatrix} = \frac{1}{3} \begin{bmatrix} -6 & 5 \\ -3 & 2 \end{bmatrix}$$

Multiplying both sides of the equation by this inverse matrix, we get

$$\begin{bmatrix} x \\ y \end{bmatrix} = \frac{1}{3} \begin{bmatrix} -6 & 5 \\ -3 & 2 \end{bmatrix} \begin{bmatrix} 15 \\ 36 \end{bmatrix} = \begin{bmatrix} 30 \\ 9 \end{bmatrix}$$

So $x = 30$ and $y = 9$.   ●

Example 5

A pet store owner feeds his hamsters and gerbils different mixtures of three types of rodent food pellets, which we will call brands A, B, and C. He wishes to be sure to feed his animals the correct amount of each brand to satisfy exactly their optimal daily requirements for protein, fat, and carbohydrates. Suppose that hamsters need 340 mg of protein, 280 mg of fat, and 440 mg of carbohydrates, and gerbils need 480 mg of protein, 360 mg of fat, and 680 mg of carbohydrates each day. How many grams of each food should the store owner feed his hamsters and gerbils daily if the amounts of these nutrients in one gram of each brand are given in the following table:

	Protein	Fat	Carbohydrates
Brand A	10 mg	10 mg	5 mg
Brand B	none	20 mg	10 mg
Brand C	20 mg	10 mg	30 mg

Solution

If we let x_1, x_2, and x_3 be the grams of brands A, B, and C, respectively, that the hamsters should eat, and if we let y_1, y_2, and y_3 be the corresponding amounts for the gerbils, then we want to solve the matrix equations

(3)
$$\begin{bmatrix} 10 & 0 & 20 \\ 10 & 20 & 10 \\ 5 & 10 & 30 \end{bmatrix} \begin{bmatrix} x_1 \\ x_2 \\ x_3 \end{bmatrix} = \begin{bmatrix} 340 \\ 280 \\ 440 \end{bmatrix}$$

and

(4)
$$\begin{bmatrix} 10 & 0 & 20 \\ 10 & 20 & 10 \\ 5 & 10 & 30 \end{bmatrix} \begin{bmatrix} y_1 \\ y_2 \\ y_3 \end{bmatrix} = \begin{bmatrix} 480 \\ 360 \\ 680 \end{bmatrix}$$

Since the coefficient matrix on the left is the same in both of these equations, we can solve each one by multiplying both sides by the inverse of this matrix. We therefore begin by finding this inverse.

$$\begin{bmatrix} 10 & 0 & 20 & 1 & 0 & 0 \\ 10 & 20 & 10 & 0 & 1 & 0 \\ 5 & 10 & 30 & 0 & 0 & 1 \end{bmatrix} \xrightarrow{\;2\cdot R_3\;} \begin{bmatrix} 10 & 0 & 20 & 1 & 0 & 0 \\ 10 & 20 & 10 & 0 & 1 & 0 \\ 10 & 20 & 60 & 0 & 0 & 2 \end{bmatrix}$$

$$\xrightarrow[R_3 - R_1]{R_2 - R_1} \begin{bmatrix} 10 & 0 & 20 & 1 & 0 & 0 \\ 0 & 20 & -10 & -1 & 1 & 0 \\ 0 & 20 & 40 & -1 & 0 & 2 \end{bmatrix}$$

$$\xrightarrow{\;R_3 - R_2\;} \begin{bmatrix} 10 & 0 & 20 & 1 & 0 & 0 \\ 0 & 20 & -10 & -1 & 1 & 0 \\ 0 & 0 & 50 & 0 & -1 & 2 \end{bmatrix}$$

$$\xrightarrow{\;\frac{1}{5}R_3\;} \begin{bmatrix} 10 & 0 & 20 & 1 & 0 & 0 \\ 0 & 20 & -10 & -1 & 1 & 0 \\ 0 & 0 & 10 & 0 & -\frac{1}{5} & \frac{2}{5} \end{bmatrix}$$

$$\xrightarrow[R_1 - 2R_3]{R_2 + R_3} \begin{bmatrix} 10 & 0 & 0 & 1 & \frac{2}{5} & -\frac{4}{5} \\ 0 & 20 & 0 & -1 & \frac{4}{5} & \frac{2}{5} \\ 0 & 0 & 10 & 0 & -\frac{1}{5} & \frac{2}{5} \end{bmatrix}$$

$$\xrightarrow[\frac{1}{10}R_3]{\frac{1}{10}R_1, \frac{1}{20}R_2} \begin{bmatrix} 1 & 0 & 0 & 0.10 & 0.04 & -0.08 \\ 0 & 1 & 0 & -0.05 & 0.04 & 0.02 \\ 0 & 0 & 1 & 0 & -0.02 & 0.04 \end{bmatrix}$$

So
$$\begin{bmatrix} 10 & 0 & 20 \\ 10 & 20 & 10 \\ 5 & 10 & 30 \end{bmatrix}^{-1} = \frac{1}{100} \begin{bmatrix} 10 & 4 & -8 \\ -5 & 4 & 2 \\ 0 & -2 & 4 \end{bmatrix}$$

and if we now multiply both sides of Equations 3 and 4 by this inverse matrix, we get

$$\begin{bmatrix} x_1 \\ x_2 \\ x_3 \end{bmatrix} = \frac{1}{100} \begin{bmatrix} 10 & 4 & -8 \\ -5 & 4 & 2 \\ 0 & -2 & 4 \end{bmatrix} \begin{bmatrix} 340 \\ 280 \\ 440 \end{bmatrix} = \begin{bmatrix} 10 \\ 3 \\ 12 \end{bmatrix}$$

and

$$\begin{bmatrix} y_1 \\ y_2 \\ y_3 \end{bmatrix} = \frac{1}{100} \begin{bmatrix} 10 & 4 & -8 \\ -5 & 4 & 2 \\ 0 & -2 & 4 \end{bmatrix} \begin{bmatrix} 480 \\ 360 \\ 680 \end{bmatrix} = \begin{bmatrix} 8 \\ 4 \\ 20 \end{bmatrix}$$

This means that each hamster should be fed 10 g of brand A, 3 g of brand B, and 12 g of brand C, whereas each gerbil should be fed 8 g of brand A, 4 g of brand B, and 20 g of brand C daily.   ●

Since there is usually a lot of work involved in finding the inverse of a 3 × 3 or larger matrix, the method used in Example 5 is really useful only when we are solving several systems of equations with the same coefficient matrix.

Exercises 7.6

In Exercises 1 and 2 find the inverses of the given matrices, and verify that $A^{-1}A = AA^{-1} = I_2$ and $B^{-1}B = BB^{-1} = I_3$.

1. $A = \begin{bmatrix} 7 & 4 \\ 3 & 2 \end{bmatrix}$

2. $B = \begin{bmatrix} 1 & 3 & 2 \\ 0 & 2 & 2 \\ -2 & -1 & 0 \end{bmatrix}$

Find the inverses of the matrices in Exercises 3–18, if they exist.

3. $\begin{bmatrix} 3 & 7 \\ 2 & 5 \end{bmatrix}$

4. $\begin{bmatrix} 3 & 5 \\ 4 & 7 \end{bmatrix}$

5. $\begin{bmatrix} 2 & 5 \\ -5 & -13 \end{bmatrix}$

6. $\begin{bmatrix} -7 & 4 \\ 8 & -5 \end{bmatrix}$

7. $\begin{bmatrix} 6 & -3 \\ -8 & 4 \end{bmatrix}$

8. $\begin{bmatrix} \frac{1}{2} & \frac{1}{3} \\ 5 & 4 \end{bmatrix}$

9. $\begin{bmatrix} 0.4 & -1.2 \\ 0.3 & 0.6 \end{bmatrix}$

10. $\begin{bmatrix} 4 & 2 & 3 \\ 3 & 3 & 2 \\ 1 & 0 & 1 \end{bmatrix}$

11. $\begin{bmatrix} 2 & 4 & 1 \\ -1 & 1 & -1 \\ 1 & 4 & 0 \end{bmatrix}$

12. $\begin{bmatrix} 5 & 7 & 4 \\ 3 & -1 & 3 \\ 6 & 7 & 5 \end{bmatrix}$

13. $\begin{bmatrix} 1 & 2 & 3 \\ 4 & 5 & -1 \\ 1 & -1 & -10 \end{bmatrix}$

14. $\begin{bmatrix} 2 & 1 & 0 \\ 1 & 1 & 4 \\ 2 & 1 & 2 \end{bmatrix}$

15. $\begin{bmatrix} 0 & -2 & 2 \\ 3 & 1 & 3 \\ 1 & -2 & 3 \end{bmatrix}$

16. $\begin{bmatrix} 3 & -2 & 0 \\ 5 & 1 & 1 \\ 2 & -2 & 0 \end{bmatrix}$

17. $\begin{bmatrix} 1 & 2 & 0 & 3 \\ 0 & 1 & 1 & 1 \\ 0 & 1 & 0 & 1 \\ 1 & 2 & 0 & 2 \end{bmatrix}$

18. $\begin{bmatrix} 1 & 0 & 1 & 0 \\ 0 & 1 & 0 & 1 \\ 1 & 1 & 1 & 0 \\ 1 & 1 & 1 & 1 \end{bmatrix}$

Solve the systems of equations in Exercises 19–26 by converting to a matrix equation and using the inverse of the coefficient matrix, as in Example 4. Use the inverses from Exercises 3–6, 11, 12, 16, and 17.

19. $\begin{cases} 3x + 7y = 4 \\ 2x + 5y = 0 \end{cases}$

20. $\begin{cases} 3x + 5y = 10 \\ 4x + 7y = 20 \end{cases}$

21. $\begin{cases} 2x + 5y = 2 \\ -5x - 13y = 20 \end{cases}$

22. $\begin{cases} -7x + 4y = 0 \\ 8x - 5y = 100 \end{cases}$

23. $\begin{cases} 2x + 4y + z = 7 \\ -x + y - z = 0 \\ x + 4y = -2 \end{cases}$

24. $\begin{cases} 5x + 7y + 4z = 1 \\ 3x - y + 3z = 1 \\ 6x + 7y + 5z = 1 \end{cases}$

25. $\begin{cases} 3x - 2y = 6 \\ 5x + y + z = 12 \\ 2x - 2y = 18 \end{cases}$

26. $\begin{cases} x + 2y + 3w = 0 \\ y + z + w = 1 \\ y + w = 2 \\ x + 2y + 2w = 3 \end{cases}$

Solve the matrix equations in Exercises 27 and 28 by multiplying both sides by the appropriate inverse matrix.

27. $\begin{bmatrix} 3 & -2 \\ -4 & 3 \end{bmatrix} \begin{bmatrix} x & y & z \\ u & v & w \end{bmatrix} = \begin{bmatrix} 1 & 0 & -1 \\ 2 & 1 & 3 \end{bmatrix}$

28. $\begin{bmatrix} 0 & -2 & 2 \\ 3 & 1 & 3 \\ 1 & -2 & 3 \end{bmatrix} \begin{bmatrix} x & u \\ y & v \\ z & w \end{bmatrix} = \begin{bmatrix} 3 & 6 \\ 6 & 12 \\ 0 & 0 \end{bmatrix}$

29. A nutritionist is studying the effects of the nutrients folic acid, choline, and inositol. He has three different types of food available, which contain the following amounts of these nutrients per ounce:

	Folic acid	Choline	Inositol
Type A	3 mg	4 mg	3 mg
Type B	1 mg	2 mg	2 mg
Type C	3 mg	4 mg	4 mg

(a) Find the inverse of the matrix

$$\begin{bmatrix} 3 & 1 & 3 \\ 4 & 2 & 4 \\ 3 & 2 & 4 \end{bmatrix}$$

and use it to solve the remaining parts of this problem.

(b) How many ounces of each food should the nutritionist feed his laboratory rats if he wants their diet to contain 10 mg of folic acid, 14 mg of choline, and 13 mg of inositol?

(c) How much of each food should be given to supply 9 mg of folic acid, 12 mg of choline, and 10 mg of inositol?

(d) Is there any combination of these foods that will supply 2 mg of folic acid, 4 mg of choline, and 11 mg of inositol?

30. Refer to Exercise 29. Suppose it is found that food C has been improperly labeled and actually contains 4 mg of folic acid, 6 mg of choline, and 5 mg of inositol per ounce. Would it still be possible to use matrix inversion to solve parts (b), (c), and (d) of Exercise 29? Why or why not?

Find the inverses of the matrices in Exercises 31–34.

31. $\begin{bmatrix} \sin\theta & -\cos\theta \\ \cos\theta & \sin\theta \end{bmatrix}$

32. $\begin{bmatrix} \sec\theta & \tan\theta \\ \tan\theta & \sec\theta \end{bmatrix}$

33. $\begin{bmatrix} e^x & -e^{2x} \\ e^{2x} & e^{3x} \end{bmatrix}$

34. $\begin{bmatrix} 1 & e^x & 0 \\ e^x & -e^{2x} & 0 \\ 0 & 0 & 2 \end{bmatrix}$

35. A matrix that has an inverse is called **invertible**. Find two 2×2 invertible matrices whose sum is not invertible.

36. Find the inverse of the matrix

$$\begin{bmatrix} a & 0 & 0 & 0 \\ 0 & b & 0 & 0 \\ 0 & 0 & c & 0 \\ 0 & 0 & 0 & d \end{bmatrix}$$

where $abcd \neq 0$.

Section 7.7

Determinants and Cramer's Rule

If a matrix is **square**—that is, if it has the same number of rows as columns—then we can assign to it a number called its **determinant**. Determinants can be used to solve matrix equations, as we will see later in this section. They are also useful in determining whether a matrix has an inverse or not, without actually going through the process of trying to find its inverse.

We denote the determinant of a square matrix A by the symbol $|A|$, and we begin by defining $|A|$ for the simplest cases. If A is a 1×1 matrix, then it has only one entry and we define the determinant to be the value of that entry; that is, if $A = [a]$, then $|A| = a$. If A is a 2×2 matrix, then

$$A = \begin{bmatrix} a & b \\ c & d \end{bmatrix}$$

and we define the determinant of A to be

$$|A| = \begin{vmatrix} a & b \\ c & d \end{vmatrix} = ad - bc$$

Example 1

Evaluate $|A|$ for $A = \begin{bmatrix} 6 & -3 \\ 2 & 3 \end{bmatrix}$.

Solution

$$\begin{vmatrix} 6 & -3 \\ 2 & 3 \end{vmatrix} = 6 \cdot 3 - (-3) \cdot 2 = 18 - (-6) = 24 \qquad \bullet$$

Note that we can think of the evaluation of a 2×2 determinant as a "cross-product" operation. We take the product of the diagonal from top left to bottom right, and subtract the product from top right to bottom left.

To define the concept of determinant for an arbitrary $n \times n$ matrix, we must first introduce the following terminology:

Let A be an $n \times n$ matrix.

1. The **minor** M_{ij} of the element a_{ij} is the determinant of the matrix obtained by deleting the ith row and jth column of A.

2. The **cofactor** A_{ij} of the element a_{ij} is

$$A_{ij} = (-1)^{i+j} M_{ij}$$

For example, if A is the matrix

$$\begin{bmatrix} 2 & 3 & -1 \\ 0 & 2 & 4 \\ -2 & 5 & 6 \end{bmatrix}$$

then M_{12} is the determinant of the matrix obtained by deleting the first row and second column from A. Thus

$$M_{12} = \begin{vmatrix} 2 & 3 & 1 \\ 0 & 2 & 4 \\ -2 & 5 & 6 \end{vmatrix} = \begin{vmatrix} 0 & 4 \\ -2 & 6 \end{vmatrix} = 0(6) - 4(-2) = 8$$

so

$$A_{12} = (-1)^{1+2} M_{12} = -8$$

Similarly,

$$M_{33} = \begin{vmatrix} 2 & 3 & -1 \\ 0 & 2 & 4 \\ -2 & 5 & 6 \end{vmatrix} = \begin{vmatrix} 2 & 3 \\ 0 & 2 \end{vmatrix} = 2 \cdot 2 - 3 \cdot 0 = 4$$

so

$$A_{33} = (-1)^{3+3} M_{33} = 4$$

Note that the cofactor of a_{ij} is just the minor of a_{ij} multiplied by either 1 or -1, depending on whether $i + j$ is even or odd. Thus in a 3×3 matrix we obtain the cofactor of any element by prefixing its minor with the sign obtained from the following checkerboard pattern:

$$\begin{bmatrix} + & - & + \\ - & + & - \\ + & - & + \end{bmatrix}$$

We are now ready to define the determinant of any square matrix.

If A is an $n \times n$ matrix, then the **determinant** of A is obtained by multiplying each element of the first row by its cofactor and then adding the results. In symbols,

$$|A| = \begin{vmatrix} a_{11} & a_{12} & \cdots & a_{1n} \\ a_{21} & a_{22} & \cdots & a_{2n} \\ \vdots & \vdots & \ddots & \vdots \\ a_{n1} & a_{n2} & \cdots & a_{nn} \end{vmatrix} = a_{11}A_{11} + a_{12}A_{12} + \cdots + a_{1n}A_{1n}$$

Example 2

Evaluate the determinant of the matrix

$$A = \begin{bmatrix} 2 & 3 & -1 \\ 0 & 2 & 4 \\ -2 & 5 & 6 \end{bmatrix}$$

Solution

$$\begin{aligned} |A| &= 2\begin{vmatrix} 2 & 4 \\ 5 & 6 \end{vmatrix} - 3\begin{vmatrix} 0 & 4 \\ -2 & 6 \end{vmatrix} + (-1)\begin{vmatrix} 0 & 2 \\ -2 & 5 \end{vmatrix} \\ &= 2(2 \cdot 6 - 4 \cdot 5) - 3[0 \cdot 6 - 4(-2)] - [0 \cdot 5 - 2(-2)] \\ &= -16 - 24 - 4 \\ &= -44 \end{aligned}$$

●

In our definition of the determinant, we used the cofactors of elements in the first row only. This is sometimes called **expanding the determinant by the first row**. In fact, we can expand the determinant by any row or column in the same way and obtain the same result. Although we will not prove this, the next example illustrates this principle.

Example 3

Expand the determinant of the matrix A in Example 2 about the second row and about the third column, and show that the value obtained is the same in each case.

Solution

Expanding by the second row, we get

$$|A| = \begin{vmatrix} 2 & 3 & -1 \\ 0 & 2 & 4 \\ -2 & 5 & 6 \end{vmatrix} = -0\begin{vmatrix} 3 & -1 \\ 5 & 6 \end{vmatrix} + 2\begin{vmatrix} 2 & -1 \\ -2 & 6 \end{vmatrix} - 4\begin{vmatrix} 2 & 3 \\ -2 & 5 \end{vmatrix}$$
$$= 0 + 2[2 \cdot 6 - (-1)(-2)] - 4[2 \cdot 5 - 3(-2)]$$
$$= 0 + 20 - 64$$
$$= -44$$

Expanding by the third column gives

$$|A| = -1\begin{vmatrix} 0 & 2 \\ -2 & 5 \end{vmatrix} - 4\begin{vmatrix} 2 & 3 \\ -2 & 5 \end{vmatrix} + 6\begin{vmatrix} 2 & 3 \\ 0 & 2 \end{vmatrix}$$
$$= -[0 \cdot 5 - 2(-2)] - 4[2 \cdot 5 - 3(-2)] + 6(2 \cdot 2 - 2 \cdot 0)$$
$$= -4 - 64 + 24$$
$$= -44$$

In both cases we obtained the same value for the determinant as when we expanded by the first row in Example 2.   ●

The following principle allows us to determine whether a square matrix has an inverse or not, without actually calculating the inverse. This is one of the most important uses of the determinant in matrix algebra and is the reason for the name *determinant*.

Invertibility Criterion

If A is a square matrix, then A has an inverse if and only if $|A| \neq 0$.

Although we will not prove this fact, we have already seen (in the preceding section) why it is true in the case of 2×2 matrices.

Example 4

Show that the matrix A has no inverse, where

$$A = \begin{bmatrix} 1 & 2 & 0 & 4 \\ 0 & 0 & 0 & 3 \\ 5 & 6 & 2 & 6 \\ 2 & 4 & 0 & 9 \end{bmatrix}$$

Solution

We begin by calculating the determinant of A. Since all but one of the elements of the second row is zero, we need to calculate only the cofactor of the 3 if we expand the determinant by the second row.

$$|A| = -0 \cdot A_{21} + 0 \cdot A_{22} - 0 \cdot A_{23} + 3A_{24} = 3A_{24}$$

$$= 3\begin{vmatrix} 1 & 2 & 0 \\ 5 & 6 & 2 \\ 2 & 4 & 0 \end{vmatrix} \quad \text{(expand this by the third column)}$$

$$= 3(-2)\begin{vmatrix} 1 & 2 \\ 2 & 4 \end{vmatrix}$$

$$= 3(-2)(1 \cdot 4 - 2 \cdot 2) = 0$$

Since the determinant of A is zero, A cannot have an inverse, by the Invertibility Criterion.   ●

   The preceding example shows that if we expand a determinant about a row or column that contains many zeros, our work is considerably reduced because we do not have to evaluate the cofactors of the elements that are zero. The following principle enables us in many cases to simplify the process of finding a determinant by introducing zeros into it without changing its value.

Row and Column Transformations of a Determinant

If A is a square matrix, and if the matrix B is obtained from A by adding a multiple of one row to another, or a multiple of one column to another, then $|A| = |B|$.

Example 5

Find the determinant of the matrix A. Does it have an inverse?

$$A = \begin{bmatrix} 8 & 2 & -1 & -4 \\ 3 & 5 & -3 & 11 \\ 24 & 6 & 1 & -12 \\ 2 & 2 & 7 & -1 \end{bmatrix}$$

Solution

If we subtract three times row 1 from row 3, that will change all but one of the elements of row 3 to zeros:

$$\begin{bmatrix} 8 & 2 & -1 & -4 \\ 3 & 5 & -3 & 11 \\ 0 & 0 & 4 & 0 \\ 2 & 2 & 7 & -1 \end{bmatrix}$$

This new matrix will have the same determinant as A, and if we expand its determinant by the third row, we get

$$|A| = 4\begin{vmatrix} 8 & 2 & -4 \\ 3 & 5 & 11 \\ 2 & 2 & -1 \end{vmatrix}$$

Now adding two times column 3 to column 1 in this determinant gives

$$|A| = 4 \begin{vmatrix} 0 & 2 & -4 \\ 25 & 5 & 11 \\ 0 & 2 & -1 \end{vmatrix} \quad \textit{(expand this by the first column)}$$

$$= 4(-25) \begin{vmatrix} 2 & -4 \\ 2 & -1 \end{vmatrix}$$

$$= 4(-25)[2(-1) - (-4)2] = -600$$

Since the determinant of A is not zero, A does have an inverse.   ●

Cramer's Rule

The solutions of linear equations can sometimes be expressed using determinants. To illustrate, let us try to solve the following pair of linear equations for the variable x:

$$\begin{cases} ax + by = r \\ cx + dy = s \end{cases}$$

If $d \neq 0$, then we can eliminate the variable y from the first equation by multiplying the second equation by b/d and then subtracting it from the first, which gives

$$ax - \left(\frac{b}{d}\right)cx = r - \left(\frac{b}{d}\right)s$$

If we now multiply both sides of the equation by d and factor x from the left, we get

$$(ad - bc)x = rd - bs$$

Assuming that $ad - bc \neq 0$, we can now solve this equation for x, obtaining

$$x = \frac{rd - bs}{ad - bc}$$

The numerator and denominator of this fraction look like the determinants of 2×2 matrices. In fact, we can write the solution for x as

$$x = \frac{\begin{vmatrix} r & b \\ s & d \end{vmatrix}}{\begin{vmatrix} a & b \\ c & d \end{vmatrix}}$$

Using the same sort of technique, we can solve the original pair of equations for y to get

$$y = \frac{\begin{vmatrix} a & r \\ b & s \end{vmatrix}}{\begin{vmatrix} a & b \\ c & d \end{vmatrix}}$$

Notice that the denominator in each case is the determinant of the coefficient matrix, which we will call D. The numerator in the solution for x is the determinant of the matrix obtained from D by replacing the coefficients of x by r and s, respectively. Similarly, in the solution for y the numerator is the determinant of the matrix obtained from D by replacing the coefficients of y by r and s. Thus if we define

$$D = \begin{bmatrix} a & b \\ c & d \end{bmatrix} \qquad D_x = \begin{bmatrix} r & b \\ s & d \end{bmatrix} \qquad D_y = \begin{bmatrix} a & r \\ c & s \end{bmatrix}$$

we can write the solution of the system as

$$x = \frac{|D_x|}{|D|} \quad \text{and} \quad y = \frac{|D_y|}{|D|}$$

This pair of formulas is known as **Cramer's Rule**, and it can be used to solve any pair of linear equations in two unknowns in which the determinant of the coefficient matrix is not zero.

Example 6

Use Cramer's Rule to solve the system

$$\begin{cases} 2x + 6y = -1 \\ x + 8y = 2 \end{cases}$$

Solution

For this system, we have

$$|D| = \begin{vmatrix} 2 & 6 \\ 1 & 8 \end{vmatrix} = 2 \cdot 8 - 6 \cdot 1 = 10$$

$$|D_x| = \begin{vmatrix} -1 & 6 \\ 2 & 8 \end{vmatrix} = (-1)8 - 6 \cdot 2 = -20$$

and
$$|D_y| = \begin{vmatrix} 2 & -1 \\ 1 & 2 \end{vmatrix} = 2 \cdot 2 - (-1)1 = 5$$

The solution is

$$x = \frac{|D_x|}{|D|} = \frac{-20}{10} = -2$$

and
$$y = \frac{|D_y|}{|D|} = \frac{5}{10} = \frac{1}{2} \qquad \bullet$$

Cramer's Rule can be extended to apply to any system of n linear equations in n variables in which the determinant of the coefficient matrix is not zero. As we saw in the preceding section, any such system can be written in matrix form as

$$\begin{bmatrix} a_{11} & a_{12} & \cdots & a_{1n} \\ a_{21} & a_{22} & \cdots & a_{2n} \\ \vdots & \vdots & \ddots & \vdots \\ a_{n1} & a_{n2} & \cdots & a_{nn} \end{bmatrix} \begin{bmatrix} x_1 \\ x_2 \\ \vdots \\ x_n \end{bmatrix} = \begin{bmatrix} b_1 \\ b_2 \\ \vdots \\ b_n \end{bmatrix}$$

By analogy with what we did in the case of two equations in two unknowns, we let D be the coefficient matrix in the above system, and we let D_{x_i} be the matrix obtained by replacing the ith column of D by the numbers $b_1, b_2, \ldots, b_n$ that appear to the right of the equal sign in the system. The solution of the system is then given by the following rule:

Cramer's Rule

$$x_1 = \frac{|D_{x_1}|}{|D|}, \quad x_2 = \frac{|D_{x_2}|}{|D|}, \quad \ldots, \quad x_n = \frac{|D_{x_n}|}{|D|}$$

Example 7

Use Cramer's Rule to solve the system

$$\begin{cases} 2x - 3y + 4z = 1 \\ x \qquad + 6z = 0 \\ 3x - 2y \qquad = 5 \end{cases}$$

Solution

First we evaluate the determinants that appear in Cramer's Rule.

$$|D| = \begin{vmatrix} 2 & -3 & 4 \\ 1 & 0 & 6 \\ 3 & -2 & 0 \end{vmatrix} = -38 \qquad |D_x| = \begin{vmatrix} 1 & -3 & 4 \\ 0 & 0 & 6 \\ 5 & -2 & 0 \end{vmatrix} = -78$$

$$|D_y| = \begin{vmatrix} 2 & 1 & 4 \\ 1 & 0 & 6 \\ 3 & 5 & 0 \end{vmatrix} = -22 \qquad |D_z| = \begin{vmatrix} 2 & -3 & 1 \\ 1 & 0 & 0 \\ 3 & -2 & 5 \end{vmatrix} = 13$$

Now we use Cramer's Rule to get the solution:

$$x = \frac{|D_x|}{|D|} = \frac{-78}{-38} = \frac{39}{19}$$

$$y = \frac{|D_y|}{|D|} = \frac{-22}{-38} = \frac{11}{19}$$

$$z = \frac{|D_z|}{|D|} = \frac{13}{-38} = -\frac{13}{38}$$

If we had solved the system in Example 7 using Gaussian elimination, our work would have involved matrices whose elements are fractions with fairly large denominators. Thus in cases like Examples 6 and 7, Cramer's Rule provides an efficient method for solving systems of linear equations. But in systems with more than three equations, evaluating the various determinants involved is usually a long and tedious task. Moreover, the rule does not apply if $|D| = 0$ or if D is not a square matrix. So Cramer's Rule is a useful alternative to Gaussian elimination in only some situations.

Exercises 7.7

In Exercises 1–8 find the determinant of the given matrix, if it exists.

1. $\begin{bmatrix} 2 & 3 \\ -1 & 0 \end{bmatrix}$ **2.** $\begin{bmatrix} -4 & 6 \\ 2 & -5 \end{bmatrix}$ **3.** $[6]$

4. $\begin{bmatrix} 4 \\ 2 \end{bmatrix}$ **5.** $[3 \quad -1]$ **6.** $[0]$

7. $\begin{bmatrix} \frac{1}{2} & \frac{1}{3} \\ 1 & \frac{2}{3} \end{bmatrix}$ **8.** $\begin{bmatrix} \frac{3}{5} & -0.6 \\ \frac{1}{3} & 4 \end{bmatrix}$

In Exercises 9–14 evaluate the given minor and cofactor using the matrix

$$A = \begin{bmatrix} 1 & 0 & \frac{1}{2} \\ -3 & 5 & 2 \\ 0 & 0 & 4 \end{bmatrix}$$

9. M_{11}, A_{11} **10.** M_{33}, A_{33} **11.** M_{12}, A_{12}

12. M_{13}, A_{13} **13.** M_{23}, A_{23} **14.** M_{32}, A_{32}

In Exercises 15–20 find the determinant of the matrix. Determine whether the matrix has an inverse or not, but do not try to find the inverse.

15. $\begin{bmatrix} 1 & 3 & 7 \\ 2 & 0 & -1 \\ 0 & 2 & 6 \end{bmatrix}$ **16.** $\begin{bmatrix} -2 & -\frac{3}{2} & \frac{1}{2} \\ 2 & 4 & 0 \\ \frac{1}{2} & 2 & 1 \end{bmatrix}$

17. $\begin{bmatrix} 30 & 0 & 20 \\ 0 & -10 & -20 \\ 40 & 0 & 10 \end{bmatrix}$ **18.** $\begin{bmatrix} 1 & 2 & 5 \\ -2 & -3 & 2 \\ 3 & 5 & 3 \end{bmatrix}$

19. $\begin{bmatrix} 1 & 3 & 3 & 0 \\ 0 & 2 & 0 & 1 \\ -1 & 0 & 0 & 2 \\ 1 & 6 & 4 & 1 \end{bmatrix}$ **20.** $\begin{bmatrix} 1 & 2 & 0 & 2 \\ 3 & -4 & 0 & 4 \\ 0 & 1 & 6 & 0 \\ 1 & 0 & 2 & 0 \end{bmatrix}$

Evaluate the determinants in Exercises 21–24. Use row or column operations whenever possible to simplify your work.

21. $\begin{bmatrix} 0 & 0 & 4 & 6 \\ 2 & 1 & 1 & 3 \\ 2 & 1 & 2 & 3 \\ 3 & 0 & 1 & 7 \end{bmatrix}$ **22.** $\begin{bmatrix} -2 & 3 & -1 & 7 \\ 4 & 6 & -2 & 3 \\ 7 & 7 & 0 & 5 \\ 3 & -12 & 4 & 0 \end{bmatrix}$

23. $\begin{bmatrix} 1 & 2 & 3 & 4 & 5 \\ 0 & 2 & 4 & 6 & 8 \\ 0 & 0 & 3 & 6 & 9 \\ 0 & 0 & 0 & 4 & 8 \\ 0 & 0 & 0 & 0 & 5 \end{bmatrix}$ **24.** $\begin{bmatrix} 2 & -1 & 6 & 4 \\ 7 & 2 & -2 & 5 \\ 4 & -2 & 10 & 8 \\ 6 & 1 & 1 & 4 \end{bmatrix}$

25. Let

$$B = \begin{bmatrix} 4 & 1 & 0 \\ -2 & -1 & 1 \\ 4 & 0 & 3 \end{bmatrix}$$

(a) Evaluate $|B|$ by expanding by the second row.
(b) Evaluate $|B|$ by expanding by the third column.
(c) Do your results in parts (a) and (b) agree?

26. Consider the system

$$\begin{cases} x + 2y + 6z = 5 \\ -3x - 6y + 5z = 8 \\ 2x + 6y + 9z = 7 \end{cases}$$

(a) Verify that $x = -1, y = 0, z = 1$ is a solution of the system.
(b) Find the determinant of the coefficient matrix.
(c) Without solving the system, determine whether or not there are any other solutions.
(d) Can Cramer's Rule be used to solve this system? Why or why not?

Use Cramer's Rule to solve the systems in Exercises 27–42.

27. $\begin{cases} 2x - y = -9 \\ x + 2y = 8 \end{cases}$ **28.** $\begin{cases} 6x + 12y = 33 \\ 4x + 7y = 20 \end{cases}$

29. $\begin{cases} x - 6y = 3 \\ 3x + 2y = 1 \end{cases}$

30. $\begin{cases} \frac{1}{2}x + \frac{1}{3}y = 1 \\ \frac{1}{4}x - \frac{1}{6}y = -\frac{3}{2} \end{cases}$

31. $\begin{cases} 0.4x + 1.2y = 0.4 \\ 1.2x + 1.6y = 3.2 \end{cases}$

32. $\begin{cases} 10x - 17y = 21 \\ 20x - 31y = 39 \end{cases}$

33. $\begin{cases} x - y + 2z = 0 \\ 3x \qquad + z = 11 \\ -x + 2y \qquad = 0 \end{cases}$

34. $\begin{cases} 5x - 3y + z = 6 \\ \quad 4y - 6z = 22 \\ 7x + 10y \qquad = -13 \end{cases}$

35. $\begin{cases} 2x_1 + 3x_2 - 5x_3 = 1 \\ x_1 + x_2 - x_3 = 2 \\ \quad 2x_2 + x_3 = 8 \end{cases}$

36. $\begin{cases} -2a \qquad + c = 2 \\ a + 2b - c = 9 \\ 3a + 5b + 2c = 22 \end{cases}$

37. $\begin{cases} \frac{1}{3}x - \frac{1}{5}y + \frac{1}{2}z = \frac{7}{10} \\ -\frac{2}{3}x + \frac{2}{5}y + \frac{3}{2}z = \frac{11}{10} \\ x - \frac{4}{5}y + z = \frac{9}{5} \end{cases}$

38. $\begin{cases} 2x - y = 5 \\ 5x + 3z = 19 \\ 4y + 7z = 17 \end{cases}$

39. $\begin{cases} 3y + 5z = 4 \\ 2x - z = 10 \\ 4x + 7y = 0 \end{cases}$

40. $\begin{cases} 2x - 5y \qquad = 4 \\ x + y - z = 8 \\ 3x \qquad + 5z = 0 \end{cases}$

41. $\begin{cases} 3r - s + 3t = 7 \\ 4r + 5s - 2t = 0 \\ 9r + s + t = 0 \end{cases}$

42. $\begin{cases} \theta + \phi + \psi = 2 \\ 2\theta - \phi + \psi = 4 \\ \theta - 3\phi + 2\psi = 0 \end{cases}$

43. **(a)** Show that the equation

$$\begin{vmatrix} x_1 & y_1 & 1 \\ x_2 & y_2 & 1 \\ x & y & 1 \end{vmatrix} = 0$$

is an equation for the line that passes through the points (x_1, y_1) and (x_2, y_2).

(b) Use the result of part (a) to find an equation for the line that passes through the points $(20, 50)$ and $(-10, 25)$.

44. Evaluate the determinant

$$\begin{vmatrix} a & a & a & a & a \\ 0 & a & a & a & a \\ 0 & 0 & a & a & a \\ 0 & 0 & 0 & a & a \\ 0 & 0 & 0 & 0 & a \end{vmatrix}$$

Solve for x in Exercises 45–48.

45. $\begin{vmatrix} x & 12 & 13 \\ 0 & x-1 & 23 \\ 0 & 0 & x-2 \end{vmatrix} = 0$

46. $\begin{vmatrix} x & 1 & 1 \\ 1 & 1 & x \\ x & 1 & x \end{vmatrix} = 0$

47. $\begin{vmatrix} 1 & 0 & x \\ x^2 & 1 & 0 \\ x & 0 & 1 \end{vmatrix} = 0$

48. $\begin{vmatrix} a & b & x-a \\ x & x+b & x \\ 0 & 1 & 1 \end{vmatrix} = 0$

Nonlinear Systems

Up to this point we have been studying systems of *linear* equations. As we have seen, mathematicians have developed several techniques for handling such systems. In calculus and the sciences, however, one often encounters systems of nonlinear equations as well, so we study them in this section. Unfortunately, there are no general techniques for nonlinear systems like the ones we have been applying to linear systems. We have to approach each problem on an individual basis and solve it using whatever ad hoc method or combination of methods happens to work in that particular situation.

The technique we use most often is simple substitution. If the system we are dealing with consists of a linear equation and a quadratic polynomial in two variables (as in Example 1), then this method always gives us the complete solution.

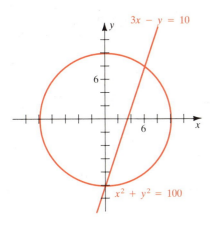

$3x - y = 10$

6

6

$x^2 + y^2 = 100$

Figure 1

Example 1

Find all solutions of the following system:

$$\begin{cases} x^2 + y^2 = 100 \\ 3x - y = 10 \end{cases}$$

Solution

The graph of the first equation is a circle and the graph of the second is a line (see Figure 1). The figure shows that the graphs intersect in two points, so the system has two solutions. We solve the system by solving for y in the second equation and then substituting into the first.

$$y = 3x - 10$$
$$x^2 + (3x - 10)^2 = 100$$
$$x^2 + (9x^2 - 60x + 100) = 100$$
$$10x^2 - 60x = 0$$
$$10x(x - 6) = 0$$
$$x = 0 \quad \text{or} \quad x = 6$$

If $x = 0$, then $y = 3(0) - 10 = -10$, and if $x = 6$, then $y = 3(6) - 10 = 8$. Thus the solutions are $(0, -10)$ and $(6, 8)$.   ●

Example 2

Solve the system

$$\begin{cases} x^2 + 2y^2 = 11 \\ 3x^2 + 4y = 23 \end{cases}$$

Solution

Here we could solve for y in the second equation and substitute into the first, just as in Example 1. But this would lead to a fourth-degree equation in x involving fractions, which might be difficult to solve. Instead, we multiply the first equation by 3 and subtract the second from it to eliminate the x:

$$3x^2 + 6y^2 = 33$$
$$\underline{3x^2 + 4y = 23}$$
$$6y^2 - 4y = 10$$

We now solve this new equation by factoring:

$$6y^2 - 4y - 10 = 0$$
$$2(3y^2 - 2y - 5) = 0$$
$$2(y + 1)(3y - 5) = 0$$
$$y = -1 \quad \text{or} \quad y = \tfrac{5}{3}$$

We can now solve for the corresponding values of x by substituting into either of the two original equations. If $y = -1$, then using the second equation, we get

$$3x^2 + 4(-1) = 23$$
$$3x^2 = 27$$
$$x^2 = 9$$
$$x = 3 \quad \text{or} \quad x = -3$$

If $y = \frac{5}{3}$, then

$$3x^2 + 4(\tfrac{5}{3}) = 23$$
$$3x^2 = 23 - \tfrac{20}{3} = \tfrac{49}{3}$$
$$x^2 = \tfrac{49}{9}$$
$$x = \tfrac{7}{3} \quad \text{or} \quad x = -\tfrac{7}{3}$$

This means that there are four solutions:

$$(3, -1), (-3, -1), (\tfrac{7}{3}, \tfrac{5}{3}), (-\tfrac{7}{3}, \tfrac{5}{3})$$

Some nonlinear systems are really just disguised versions of linear ones, like the system in the next example. We can solve these using any of the available linear methods.

Example 3

Solve the system

$$\begin{cases} 4x^3 + 6y^2 = 22 \\ 5x^3 + 8y^2 = 32 \end{cases}$$

Solution

If we let $u = x^3$ and $v = y^2$, then the system becomes linear:

$$\begin{cases} 4u + 6v = 22 \\ 5u + 8v = 32 \end{cases}$$

We can solve this using Gaussian elimination, matrix inversion, Cramer's Rule, or any method that works on linear systems. Omitting the details of the calculations, we have

$$u = -8 \quad \text{and} \quad v = 9$$

Since $u = x^3$ and $v = y^2$, this gives us

$$x = -2 \quad \text{and} \quad y = \pm 3$$

so the solutions of the original system are

$$(-2, 3) \quad \text{and} \quad (-2, -3)$$

Example 4

Solve the system

$$\begin{cases} 3x^2 + 2y^2 + 15x = 0 \\ xy + y^2 = 0 \end{cases}$$

Solution

We begin by tackling the second equation, since it looks more manageable. Factoring, we get

$$(x + y)y = 0$$

so $x = -y$ or $y = 0$

We now substitute each of these possibilities into the first equation to see what its consequences are. If $x = -y$, then

$$3(-y)^2 + 2y^2 + 15(-y) = 0$$
$$5y^2 - 15y = 0$$
$$5y(y - 3) = 0$$
$$y = 0 \quad \text{or} \quad y = 3$$

Since these values for y were obtained from the assumption that $x = -y$, we get the following solutions of the original system:

$$(0, 0) \quad \text{and} \quad (-3, 3)$$

Now we check what happens in the first equation if $y = 0$:

$$3x^2 + 2(0)^2 + 15x = 0$$
$$3x^2 + 15x = 0$$
$$3x(x + 5) = 0$$
$$x = 0 \quad \text{or} \quad x = -5$$

This leads to the solutions $(0, 0)$ and $(-5, 0)$. We have already found the first of these, so the solutions are

$$(0, 0), \quad (-3, 3), \quad \text{and} \quad (-5, 0) \qquad \bullet$$

The next example involves three equations and three variables.

Example 5

Solve the system

$$\begin{cases} x + yz = 0 \\ y + 4xz = 0 \\ x^2 + y^2 = 20 \end{cases}$$

Solution

If we solve for x in the first equation and substitute into the second, we get

$$y + 4(-yz)z = 0$$
$$y(1 - 4z^2) = 0$$
$$y = 0 \quad \text{or} \quad z^2 = \tfrac{1}{4}$$

Substituting $y = 0$ into the first equation leads to $x = 0$, but this does not satisfy the third equation. So it must be true that $z^2 = \tfrac{1}{4}$ or $z = \pm\tfrac{1}{2}$. Thus $y = \mp 2x$ (from the second equation) and the third equation becomes

$$x^2 + 4x^2 = 20$$
$$5x^2 = 20$$
$$x^2 = 4$$
$$x = \pm 2$$

This leads to four possible solutions:

$$(2, 4, -\tfrac{1}{2}), (2, -4, \tfrac{1}{2}), (-2, 4, \tfrac{1}{2}), (-2, -4, -\tfrac{1}{2})$$ ●

We conclude this section with an example that arises from a problem in geometry.

Example 6

A right triangle has an area of 120 ft^2 and a perimeter of 60 ft. Find the lengths of its sides.

Solution

Let x and y be the lengths of the sides adjacent to the right angle. Then by the Pythagorean Theorem, the hypotenuse has length $\sqrt{x^2 + y^2}$. (See Figure 2.) Since the area is 120 ft^2, we have $\tfrac{1}{2}xy = 120$, so that

(1) $$xy = 240$$

Also, since the perimeter is 60 ft,

(2) $$x + y + \sqrt{x^2 + y^2} = 60$$

We simplify Equation 2 as follows:

$$\sqrt{x^2 + y^2} = 60 - x - y$$
$$(\sqrt{x^2 + y^2})^2 = (60 - x - y)^2$$
$$x^2 + y^2 = 3600 + x^2 + y^2 - 120x - 120y + 2xy$$
$$120x + 120y = 3600 + 2xy$$
(3) $$60x + 60y = 1800 + xy$$

Thus we must solve the system that combines Equations 1 and 3:

$$\begin{cases} xy = 240 \\ 60x + 60y = 1800 + xy \end{cases}$$

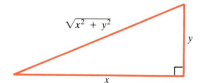

Figure 2

Adding these equations (to eliminate the xy term) and simplifying, we get

$$60x + 60y = 2040$$
$$x + y = 34$$
$$y = 34 - x$$

Substituting this into Equation 1 gives

$$x(34 - x) = 240$$
$$x^2 - 34x + 240 = 0$$
$$(x - 24)(x - 10) = 0$$

So either $x = 24$ and $y = 10$, or $x = 10$ and $y = 24$. In either case, the hypotenuse is

$$\sqrt{(10)^2 + (24)^2} = \sqrt{676} = 26$$

The sides of the triangle are 10 ft, 24 ft, and 26 ft long.

Exercises 7.8

Find all real solutions (x, y) of the systems of equations in Exercises 1–24.

1. $\begin{cases} x^2 + y^2 = 8 \\ x + y = 0 \end{cases}$

2. $\begin{cases} x^2 + y = 9 \\ x - y + 3 = 0 \end{cases}$

3. $\begin{cases} y + x^2 = 4x \\ y + 4x = 16 \end{cases}$

4. $\begin{cases} x - y^2 = 0 \\ y - x^2 = 0 \end{cases}$

5. $\begin{cases} x - 2y = 2 \\ y^2 - x^2 = 2x + 4 \end{cases}$

6. $\begin{cases} y = 4 - x^2 \\ y = x^2 - 4 \end{cases}$

7. $\begin{cases} x - y = 4 \\ xy = 12 \end{cases}$

8. $\begin{cases} x^3 - y = 0 \\ -2x + y = 4 \end{cases}$

9. $\begin{cases} 3x^2 - y^2 = 11 \\ x^2 + 4y^2 = 8 \end{cases}$

10. $\begin{cases} xy = 24 \\ 2x^2 - y^2 + 4 = 0 \end{cases}$

11. $\begin{cases} x^2 y = 16 \\ x^2 + 4y + 16 = 0 \end{cases}$

12. $\begin{cases} 2x^2 + 4y = 13 \\ x^2 - y^2 = \frac{7}{2} \end{cases}$

13. $\begin{cases} x + \sqrt{y} = 0 \\ y^2 - 4x^2 = 12 \end{cases}$

14. $\begin{cases} \sqrt{x} - 2\sqrt{y} = 1 \\ 2x - 4(y + \sqrt{y}) = 2 \end{cases}$

15. $\begin{cases} x^2 + y^2 = 9 \\ x^2 - y^2 = 1 \end{cases}$

16. $\begin{cases} x^2 + 2y^2 = 2 \\ 2x^2 - 3y = 15 \end{cases}$

17. $\begin{cases} 2x^2 - 8y^3 = 19 \\ 4x^2 + 16y^3 = 34 \end{cases}$

18. $\begin{cases} x^4 - y^3 = 15 \\ 3x^4 + 5y^3 = 53 \end{cases}$

19. $\begin{cases} \dfrac{2}{x} - \dfrac{3}{y} = 1 \\ \dfrac{4}{x} + \dfrac{7}{y} = 1 \end{cases}$

20. $\begin{cases} \dfrac{4}{x^2} + \dfrac{6}{y^4} = \dfrac{7}{2} \\ \dfrac{1}{x^2} - \dfrac{2}{y^4} = 0 \end{cases}$

21. $\begin{cases} 3\sqrt{x} + 5\sqrt{y} = 19 \\ 2\sqrt{x} + 7\sqrt{y} = 20 \end{cases}$

22. $\begin{cases} 2\sqrt{x} - \sqrt{y} = 3 \\ 4\sqrt{x} + 3\sqrt{y} = 1 \end{cases}$

23. $\begin{cases} x^2 - xy + 2y^2 = 8 \\ x^3 - xy^2 = 0 \end{cases}$

24. $\begin{cases} xy - 3x = 0 \\ x^3 - y + 11 = 0 \end{cases}$

Find all real solutions (x, y, z) of the systems of equations in Exercises 25–30.

25. $\begin{cases} x - y = 2 \\ y + z = 0 \\ x^2 + y^2 + z^2 = 4 \end{cases}$

26. $\begin{cases} xy + z = 0 \\ yz + x = 0 \\ x^2 + y^2 = 2 \end{cases}$

27. $\begin{cases} x^2 + yz = 0 \\ y + xz = 2 \\ xyz = 1 \end{cases}$

28. $\begin{cases} xy - xz = 0 \\ y + yz = 2 \\ x^2 + y^2 = 5 \end{cases}$

29. $\begin{cases} x^2 + y + z = 0 \\ 2x^2 - y + 3z = -4 \\ y^2 + yz = 0 \end{cases}$

30. $\begin{cases} x^2 + y^3 + z^4 = 4 \\ 3x^2 - y^3 - z^4 = 12 \\ 2x^2 - 3y^3 + 2z^4 = 13 \end{cases}$

31. A right triangle has a perimeter of 40 cm and an area of 60 cm^2. What are the lengths of its sides?

32. A rectangle has an area of 180 cm^2 and a perimeter of 54 cm. What are its dimensions?

33. A right triangle has an area of 54 in.^{2}. The product of the lengths of the three sides is 1620 in.^{3}. What are the lengths of its sides?

34. A right triangle has an area of 84 ft^2 and a hypotenuse 25 ft long. What are the lengths of its other two sides?

35. The perimeter of a rectangle is 70 and its diagonal is 25. Find its length and width.

36. A circular piece of sheet metal has a diameter of 20 in. The edges are to be cut off to form a rectangle of area 160 in.^{2} (see the figure). What are the dimensions of the rectangle?

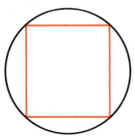

37. Find an equation for the line that passes through the points of intersection of the circles $x^2 + y^2 = 25$ and $x^2 - 3x + y^2 + y = 30$.

38. **(a)** For what value of k does the system

$$\begin{cases} y = x^2 \\ y = x + k \end{cases}$$

have exactly one solution?

(b) Graph both equations in the system on the same set of axes, using the value of k you chose in part (a).

(c) Based on your graph, how many solutions will the system have if k is smaller than your value from part (a)? How many will there be if k is larger?

In Exercises 39–42 find all real solutions of the system.

39. $\begin{cases} x - y = 3 \\ x^3 - y^3 = 387 \end{cases}$ (*Hint:* Factor the left side of the second equation.)

40. $\begin{cases} x^2 + xy + xz = 1 \\ xy + y^2 + yz = 3 \\ xz + yz + z^2 = 5 \end{cases}$ (*Hint:* Add the equations and factor the result.)

41. $\begin{cases} 2^x + 2^y = 10 \\ 4^x + 4^y = 68 \end{cases}$

42. $\begin{cases} \sin x + \cos y = \frac{3}{2} \\ 2\sin x - \cos y = 0 \end{cases}$ (*x* and *y* in radians)

Section 7.9

Systems of Inequalities

In this section we study systems of inequalities in two variables from a graphical point of view. First we consider the graph of a single inequality. We already know that the graph of $y = x^2$, for example, is the parabola in Figure 1. If we replace the equal sign by the symbol $\geq$, we obtain the **inequality**

$$y \geq x^2$$

Its graph consists of not just the parabola in Figure 1, but also every point whose y-coordinate is *larger* than x^2; that is, its graph in Figure 2 includes all the points *above* the parabola as well.

Similarly, the graph of $y \leq x^2$ in Figure 3 consists of all points on and *below* the parabola, whereas the graphs of $y > x^2$ and $y < x^2$ do not include the points on the parabola itself, as indicated by the broken curves in Figures 4 and 5.

The graph of an inequality, in general, consists of a region in the plane whose boundary is the graph of the equation obtained by replacing the inequality sign

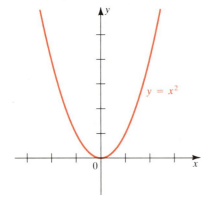

Figure 1

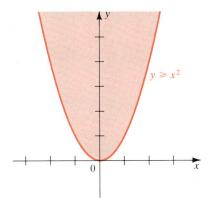

Figure 2

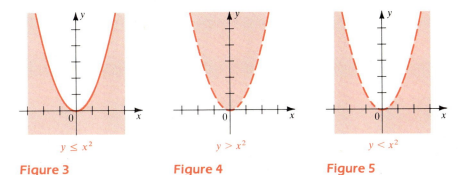

Figure 3   **Figure 4**   **Figure 5**

($\geq$, $\leq$, $>$, or $<$) by an equal sign. To determine which side of this graph gives the solution set of the inequality, we need to check only **test points**, as illustrated in the next example.

Example 1

Graph the following inequalities:

(a) $x^2 + y^2 < 25$ (b) $x + 2y \geq 5$

Solution

(a)   The graph of $x^2 + y^2 = 25$ is a circle of radius 5 centered at the origin. The points on the circle itself do not satisfy the inequality, since it is of the form $<$, so we graph the circle with a broken line in Figure 6.

To determine whether the inside or the outside of the circle satisfies the inequality, we use the test points $(0, 0)$ on the inside and $(6, 0)$ on the outside. (Note that *any* points inside and outside the circle can serve as test points. We chose these for simplicity.)

Check: Does $(0, 0)$ satisfy $x^2 + y^2 < 25$?

$$0^2 + 0^2 \overset{?}{<} 25$$
$$0 \overset{?}{<} 25 \qquad \textit{yes}$$

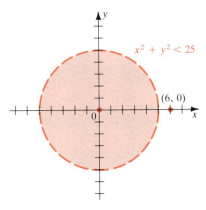

Figure 6

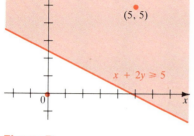

Figure 7

Check: Does $(6, 0)$ satisfy $x^2 + y^2 < 25$?

$$6^2 + 0^2 \overset{?}{<} 25$$
$$36 \overset{?}{<} 25 \qquad no$$

Thus the graph of $x^2 + y^2 < 25$ consists of the points inside the circle only (see Figure 6).

(b)   The graph of $x + 2y = 5$ is the line shown in Figure 7. We use the test points $(0, 0)$ and $(5, 5)$ on opposite sides of the line.

Check: Does $(0, 0)$ satisfy $x + 2y \geq 5$?

$$(0) + 2(0) \overset{?}{\geq} 5$$
$$0 \overset{?}{\geq} 5 \qquad no$$

Check: Does $(5, 5)$ satisfy $x + 2y \geq 5$?

$$(5) + 2(5) \overset{?}{\geq} 5$$
$$15 \overset{?}{\geq} 5 \qquad yes$$

Our check shows that the points *above* the line satisfy the inequality.

Alternatively, we could put the inequality into slope-intercept form and graph it directly:

$$x + 2y \geq 5$$
$$2y \geq -x + 5$$
$$y \geq -\tfrac{1}{2}x + \tfrac{5}{2}$$

From this form, we see that the graph includes the points whose y-coordinates are *larger* than those on the line $y = -\tfrac{1}{2}x + \tfrac{5}{2}$; that is, the graph consists of the points on or *above* this line as in Figure 7.   ●

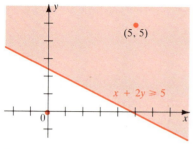

Figure 8

Example 2

Graph the solution set of the pair of inequalities

$$\begin{cases} x^2 + y^2 < 25 \\ x + 2y \geq 5 \end{cases}$$

Solution

These are the two inequalities of Example 1. In this example we wish to graph those points that simultaneously satisfy both inequalities. The solution thus consists of the intersection of the graphs in Example 1 (see Figure 8).   ●

The points $(-3, 4)$ and $(5, 0)$ in Figure 8 are called the **vertices** of the solution set. They are obtained by simultaneously solving the *equations*

$$\begin{cases} x^2 + y^2 = 25 \\ x + 2y = 5 \end{cases}$$

Note that in this case the vertices are not part of the solution set, since they do not satisfy the inequality $x^2 + y^2 < 25$. They simply show where the "corners" of the solution set lie.

An inequality is **linear** if it can be put into one of the following forms:

$$ax + by \geq c \qquad ax + by \leq c \qquad ax + by > c \qquad ax + by < c$$

In the next example we graph the solution set of a system of linear inequalities.

Example 3

Graph the system

$$\begin{cases} x + 3y \leq 12 \\ x + y \leq 8 \\ x \geq 3 \\ y \geq 0 \end{cases}$$

Solution

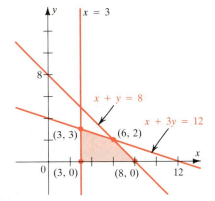

Figure 9

In Figure 9 we first graph the lines given by the equations that correspond to each of the inequalities. The shaded region is the set of points that satisfy all four inequalities simultaneously. The coordinates of the vertices are obtained by simultaneously solving the equations of the lines that intersect at that vertex. For example, the vertex $(6, 2)$ lies on both the lines

$$x + 3y = 12$$

and

$$x + y = 8$$

In this case the vertices *are* part of the solution set.   ●

Example 4

A manufacturer of insulating materials produces two different brands: Foamboard and Plastiflex. Each cubic yard of Foamboard weighs 8 lb, and each cubic yard of Plastiflex weighs 24 lb. The products are moved from the plant to the loading dock on carts that have a maximum capacity of 25 yd^3 and 432 lb. Find a system of inequalities that describes all possible combinations of Foamboard and Plastiflex that can be carried on such a cart. Graph the solution set of this system.

Solution

First we let

$$x = \text{number of cubic yards of Foamboard on a cart}$$
$$y = \text{number of cubic yards of Plastiflex on a cart}$$

Since a cart can carry no more than 25 yd^3, we have

$$x + y \leq 25$$

In addition, the total weight cannot exceed 432 lb. Since there are $8x$ lb of Foamboard and $24y$ lb of Plastiflex on the cart, this means that

$$8x + 24y \leq 432$$

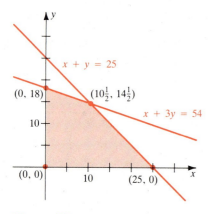

Figure 10

Dividing both sides of this inequality by 8 simplifies it to

$$x + 3y \leq 54$$

Finally, negative amounts would be meaningless in this context, so we get

$$x \geq 0 \quad \text{and} \quad y \geq 0$$

Thus the possible amounts of material that a cart can hold are given by the system

$$\begin{cases} x + y \leq 25 \\ x + 3y \leq 54 \\ \quad\quad x \geq 0 \\ \quad\quad y \geq 0 \end{cases}$$

The graph is shown in Figure 10.   ●

When a region in the plane can be covered by a (sufficiently large) circle, it is said to be **bounded**. A region that is not bounded is called **unbounded**. For example, the regions graphed in Figures 6, 8, 9, and 10 are bounded, whereas those in Figures 2, 3, 4, 5, and 7 are unbounded. Unbounded regions cannot be "fenced in"; they extend infinitely far in some direction.

Exercises 7.9

Graph the inequalities in Exercises 1–12.

1. $y > -1$ **2.** $x \leq 4$ **3.** $y \geq 2x + 2$

4. $y < -x + 5$ **5.** $2x - y \leq 8$

6. $3x + 4y + 12 > 0$ **7.** $4x + 5y < 25$

8. $-x^2 + y \geq 10$ **9.** $y > x^3 + 1$

10. $x - y^2 \leq 0$ **11.** $x^2 + y^2 \geq 5$ **12.** $xy < 4$

Graph the solutions of the systems of inequalities in Exercises 13–34. In each case, find the coordinates of all vertices and determine whether the solution set is bounded or not.

13. $\begin{cases} x + y \leq 4 \\ \quad\quad y \geq x \end{cases}$ **14.** $\begin{cases} 2x + 3y > 12 \\ 3x - y < 21 \end{cases}$

15. $\begin{cases} y < \frac{1}{4}x + 2 \\ y \geq 2x - 5 \end{cases}$ **16.** $\begin{cases} x - y > 0 \\ 4 + y \leq 2x \end{cases}$

17. $\begin{cases} x \geq 0, \quad y \geq 0 \\ 3x + 5y \leq 15 \\ 3x + 2y \leq 9 \end{cases}$ **18.** $\begin{cases} \quad\quad x > 2 \\ \quad\quad y < 12 \\ 2x - 4y > 8 \end{cases}$

19. $\begin{cases} y < 9 - x^2 \\ y \geq x + 3 \end{cases}$ **20.** $\begin{cases} x \geq y^2 \\ x + y \geq 6 \end{cases}$

21. $\begin{cases} x^2 + y^2 \leq 4 \\ x - y > 0 \end{cases}$ **22.** $\begin{cases} x > 0, \quad y > 0 \\ x + y < 10 \\ x^2 + y^2 > 9 \end{cases}$

23. $\begin{cases} x^2 - y \leq 0 \\ 2x^2 + y \leq 12 \end{cases}$ **24.** $\begin{cases} x^2 + y^2 < 9 \\ 2x + y^2 \geq 1 \end{cases}$

25. $\begin{cases} x + 2y \leq 14 \\ 3x - y \geq 0 \\ x - y \geq 2 \end{cases}$ **26.** $\begin{cases} y < x + 6 \\ 3x + 2y \geq 12 \\ x - 2y \leq 2 \end{cases}$

27. $\begin{cases} x \geq 0, \quad y \geq 0 \\ \quad\quad x \leq 5 \\ x + y \leq 7 \\ x + 2y \geq 4 \end{cases}$ **28.** $\begin{cases} x \geq 0, \quad y \geq 0 \\ \quad\quad y \leq 4 \\ 2x + y \leq 8 \\ 20x + 3y \leq 66 \end{cases}$

29. $\begin{cases} y > x + 1 \\ x + 2y \leq 12 \\ x + 1 > 0 \end{cases}$ **30.** $\begin{cases} x + y > 12 \\ y < \frac{1}{2}x - 6 \\ 3x + y < 6 \end{cases}$

31. $\begin{cases} x^2 + y^2 \le 8 \\ \quad\quad x \ge 2 \\ \quad\quad y \ge 0 \end{cases}$ **32.** $\begin{cases} x^2 - y \ge 0 \\ x + y < 6 \\ x - y < 6 \end{cases}$

33. $\begin{cases} x^2 + y^2 < 9 \\ \quad x + y > 0 \\ \quad\quad x \le 0 \end{cases}$ **34.** $\begin{cases} \quad\quad y \ge x^3 \\ \quad\quad y \le 2x + 4 \\ x + y \ge 0 \end{cases}$

35. A publishing company publishes a total of no more than 100 books every year. At least 20 of these are nonfiction, but the company always puts out at least as much fiction as nonfiction. Find a system of inequalities that describes the possible number of fiction and nonfiction books the company can produce each year consistent with these policies. Graph the solution set.

36. A man and his daughter manufacture unfinished tables and chairs. Each table requires 3 h of sawing and 1 h of

assembly. Each chair requires 2 h of sawing and 2 h of assembly. The two of them can do a total of up to 12 h of sawing and 8 h of assembly work each day. Find a system of inequalities that describes all possible combinations of tables and chairs that they can make daily. Graph the solution set.

In Exercises 37–40 graph the solution set of the system of inequalities.

37. $\begin{cases} \quad\quad xy > 0 \\ |x| + |y| < 1 \end{cases}$ **38.** $\begin{cases} y \le \ln x \\ x \le e \\ y \ge 0 \end{cases}$

39. $\begin{cases} x \ge 0, \quad y \ge 0 \\ \quad\quad y \le \sin x \\ \pi y - 2x \ge 0 \end{cases}$ **40.** $\begin{cases} \quad\quad y \ge e^x \\ 2y + x \le e^2 x + 2 \end{cases}$

Application: Linear Programming

Linear programming is a mathematical technique used to determine the optimal allocation of resources in business, the military, and other areas of human endeavor. For example, a manufacturer who makes several different products from the same raw materials can use linear programming to tell how much of each he or she should produce to maximize profits. This technique is probably the most important practical application of systems of linear inequalities. In 1975 Leonid Kantorovich and T. C. Koopmans won the Nobel Prize in economics for their work in the development of this subject.

Although linear programming can be applied to very complex problems with hundreds or even thousands of variables, we will consider only a few simple examples to which the graphical methods of the preceding section can be applied. We introduce the technique with a typical problem.

Example 1

A small shoe manufacturer makes two different styles of shoes: oxfords and loafers. Two machines are used in the process: a cutting machine and a sewing machine. Each type of shoe requires 15 min per pair on the cutting machine, but oxfords require 10 min of sewing per pair, whereas loafers require 20 min of sewing per pair. Because the manufacturer can hire only one operator for each machine, each is available for just 8 h per day. If the profit on each pair of oxfords is $15 and on each pair of loafers is $20, how many pairs of each type should be produced per day for maximum profit?

Solution

First we organize the information given into a table. To be consistent, we convert all times to hours.

	Oxfords	Loafers	Time available
Time on cutting machine	$\frac{1}{4}$	$\frac{1}{4}$	8
Time on sewing machine	$\frac{1}{6}$	$\frac{1}{3}$	8
Profit	$15	$20	

Let

$$x = \text{number of pairs of oxfords made daily}$$
$$y = \text{number of pairs of loafers made daily}$$

The total number of cutting hours needed is then $\frac{1}{4}x + \frac{1}{4}y$. Since only 8 h are available on the cutting machine, we have

$$\tfrac{1}{4}x + \tfrac{1}{4}y \le 8$$

Similarly, by considering the amount of time needed and available on the sewing machine, we get

$$\tfrac{1}{6}x + \tfrac{1}{3}y \le 8$$

Since we cannot produce a negative number of shoes, we also have

$$x \ge 0 \qquad \text{and} \qquad y \ge 0$$

Thus x and y must satisfy the system of inequalities

$$\begin{cases} \tfrac{1}{4}x + \tfrac{1}{4}y \le 8 \\ \tfrac{1}{6}x + \tfrac{1}{3}y \le 8 \\ \qquad\quad x \ge 0 \\ \qquad\quad y \ge 0 \end{cases}$$

If we multiply the first inequality by 4 and the second by 6, we obtain the simplified system

$$\begin{cases} x + y \le 32 \\ x + 2y \le 48 \\ \quad\; x \ge 0 \\ \quad\; y \ge 0 \end{cases}$$

The graph of this system (with vertices labeled) is shown in Figure 1.

We wish to determine which values for x and y give maximum profit. The only values that satisfy the restrictions of the problem are the ones that correspond to points of the shaded region in Figure 1. This is called the **feasible region** for the problem. Since each pair of oxfords provides $15 profit and each pair of loafers $20, the total profit will be

$$P = 15x + 20y$$

As x or y increases, profit will increase as well. Thus it seems reasonable that the maximum profit will occur at a point on one of the outside edges of the feasible

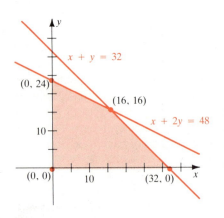

Figure 1

region, where it is impossible to increase x or y without going outside the region. In fact, it can be shown that the maximum value will occur at a vertex. This means that we need to check the profit only at the vertices.

Vertex	$P = 15x + 20y$
$(0,0)$	0
$(0,24)$	$15(0) + 20(24) = \$480$
$(16,16)$	$15(16) + 20(16) = \$560$ ← maximum profit
$(32,0)$	$15(32) + 20(0) = \$480$

The largest value of P occurs at the point $(16, 16)$, where $P = \$560$. Thus the manufacturer should make 16 pairs of oxfords and 16 pairs of loafers, for a maximum daily profit of $\$560$. ●

All the linear programming problems that we consider follow the pattern of this example. Each problem involves two variables, and certain restrictions described in the problem lead to a system of linear inequalities that involve these variables. The graph of this system is called the **feasible region**. We consider only bounded feasible regions. The function we are trying to maximize or minimize is called the **objective function**. This function will always attain its largest and smallest values at the **vertices** of the feasible region, so checking its value at all vertices gives the solution to the problem.

Example 2

A car dealer has warehouses in Millville and Trenton and has dealerships in Camden and Atlantic City. Every car sold at the dealerships has to be delivered from one of the warehouses. On a certain day the Camden dealers sell 10 cars and the Atlantic City dealers sell 12. The Millville warehouse has 15 cars available and the Trenton warehouse has 10. It costs $\$50$ to ship a car from Millville to Camden, $\$40$ from Millville to Atlantic City, $\$60$ from Trenton to Camden, and $\$55$ from Trenton to Atlantic City. How many cars should be moved from each warehouse to each dealership to fill the orders at minimum cost?

Solution

The first step is to organize the given information. Rather than construct a table, we draw a diagram to show the flow of cars from the warehouses to the dealerships (see Figure 2). The diagram shows the number of cars available or required at each location, and the cost of shipping between locations.

Since there are four possible routes, there seem to be four variables here. But if we let x be the number of cars to be shipped from Millville to Camden, then $10 - x$ cars would have to be shipped from Trenton to Camden, since the Camden dealership needs 10 cars in all. Similarly, if y cars are shipped from Millville to Atlantic City, $12 - y$ would be shipped from Trenton to Atlantic City.

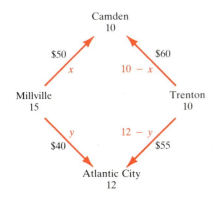

Figure 2

We now derive the inequalities that define the feasible region. First of all, the number of cars shipped on each route cannot be negative, so

$$x \geq 0$$
$$10 - x \geq 0$$
$$y \geq 0$$
$$12 - y \geq 0$$

Second, the total number of cars shipped from each warehouse cannot exceed the number of cars available there, so

$$x + y \leq 15$$
$$(10 - x) + (12 - y) \leq 10$$

Simplifying the latter inequality, we get

$$22 - x - y \leq 10$$
$$-x - y \leq -12$$
$$x + y \geq 12$$

The inequalities $10 - x \geq 0$ and $12 - y \geq 0$ can be rewritten as $x \leq 10$ and $y \leq 12$, respectively. Thus the feasible region is described by the system of inequalities

$$\begin{cases} x \geq 0, \quad y \geq 0 \\ x \leq 10, \quad y \leq 12 \\ x + y \leq 15 \\ x + y \geq 12 \end{cases}$$

The feasible region is graphed in Figure 3.

From Figure 2 we see that the total cost of shipping the cars is

$$\begin{aligned} C &= 50x + 40y + 60(10 - x) + 55(12 - y) \\ &= 50x + 40y + 600 - 60x + 660 - 55y \\ &= 1260 - 10x - 15y \end{aligned}$$

This is the objective function. We check its value at each vertex.

Vertex	$C = 1260 - 10x - 15y$	
$(0, 12)$	$1260 - 10(0) - 15(12) = \$1080$	
$(3, 12)$	$1260 - 10(3) - 15(12) = \$1050$	← minimum cost
$(10, 5)$	$1260 - 10(10) - 15(5) = \$1085$	
$(10, 2)$	$1260 - 10(10) - 15(2) = \$1130$	

The lowest cost is incurred at the point $(3, 12)$. Thus the dealer should ship

3 cars from Millville to Camden

12 cars from Millville to Atlantic City

7 cars from Trenton to Camden

0 cars from Trenton to Atlantic City

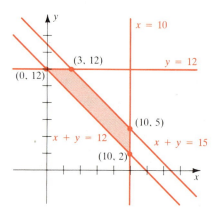

Figure 3

In the 1940s mathematicians developed matrix methods for solving linear programming problems that involve more than two variables. These methods were first used by the Allied military in World War II to solve supply problems similar to (but of course much more complicated than) Example 2. Improving these matrix methods is an active and exciting area of current mathematical research.

Exercises 7.10

In Exercises 1–4 you are given a set of inequalities describing a feasible region, and an objective function. Graph the feasible region, determine the coordinates of its vertices, and then find the maximum and minimum values of the objective function on the feasible region by checking its values at the vertices.

1. $\begin{cases} x \geq 0, \quad y \geq 0 \\ 2x + y \leq 10 \\ 2x + 4y \leq 28 \end{cases}$
$P = 140 - x + 3y$

2. $\begin{cases} x \geq 0, \quad y \geq 0 \\ x \leq 10, \quad y \leq 20 \\ x + y \geq 5 \\ x + 2y \leq 18 \end{cases}$
$Q = 70x + 82y$

3. $\begin{cases} x \geq 3, \quad y \geq 4 \\ 2x + y \leq 24 \\ 2x + 3y \leq 36 \end{cases}$
$R = 12 + x + 4y$

4. $\begin{cases} x \leq 36, \quad y \leq 40 \\ x \leq 2y \\ 2x + y \geq 60 \end{cases}$
$S = 200 + 3x - y$

5. A furniture manufacturer makes wooden tables and chairs. The production process involves two basic types of labor: carpentry and finishing. Making a table requires 2 h of carpentry and 1 h of finishing, whereas making a chair requires 3 h of carpentry and $\frac{1}{2}$ h of finishing. The profit per table is $35 and per chair is $20. The manufacturer's employees can supply a maximum of 108 h of carpentry work and 20 h of finishing work per day. How many tables and chairs should be made each day to maximize profits?

6. A housing contractor has subdivided a farm into 100 building lots. He has designed two types of homes for these lots: colonial and ranch style. A colonial requires $30,000 of capital and provides a profit of $4000. A ranch style house requires $40,000 of capital and provides an $8000 profit. If he has $3.6 million of capital at hand, how many houses of each type should he build for maximum profit? Will any of the lots be left vacant?

7. A trucker is planning to carry citrus fruit from Florida to Montreal. Each crate of oranges is 4 ft^3 in volume and weighs 80 lb. Each crate of grapefruit has a volume of 6 ft^3 and weighs 100 lb. Her truck has a maximum capacity of 300 ft^3 and can carry no more that 5600 lb. Moreover, she is not permitted to carry more crates of grapefruit than crates of oranges. If she makes a profit

of $2.50 on each crate of oranges and $4 on each crate of grapefruit, how many crates of each type of fruit should she carry for maximum profit?

8. A manufacturer of calculators produces two types: a standard type and a scientific model. Long-term demand for the two types mandates that the company manufacture at least 100 standard and 80 scientific calculators each day. However, because of limitations on production capacity, no more than 200 standard and 170 scientific calculators can be made daily. To satisfy a shipping contract, a total of at least 200 calculators must be shipped every day.
(a) If it costs $5 to produce a standard calculator and $7 for a scientific one, how many of each type should be made each day to minimize costs?
(b) If each standard calculator results in a $2 loss, but each scientific one produces a $5 profit, how many of each type should be made each day to maximize net profits?

9. An electronics discount chain has a sale on a certain brand of stereo. The chain has stores in Santa Monica and El Toro and warehouses in Long Beach and Pasadena. To satisfy rush orders, 15 sets must be shipped from the warehouses to the Santa Monica store, and 19 must be shipped to the El Toro store. The cost of shipping a set from Long Beach to Santa Monica is $5; from Long Beach to El Toro, $6; from Pasadena to Santa Monica, $4; and from Pasadena to El Toro, $5.50. If the Long Beach warehouse has 24 sets and the Pasadena warehouse has 18, how many sets should be shipped from each warehouse to each store to fill the orders at a minimum shipping cost?

10. A man owns two building supply stores on the east and west sides of a city. Two customers order some $\frac{1}{2}$-inch plywood. Customer A needs 50 sheets and customer B needs 70 sheets. The east-side store has 80 sheets and the west-side store has 45 sheets of this plywood in stock. Delivery costs per sheet are: $0.50 from the east-side store to customer A, $0.60 from the east-side store to customer B, $0.40 from the west-side store to A, and $0.55 from the west-side store to B. How many sheets should be shipped from each store to each customer to minimize delivery costs?

11. Refer to Example 4 of Section 7.9. Suppose that a worker can load 12 yd^3 of Foamboard per minute and 10 yd^3 of Plastiflex per minute. If each cubic yard of Foamboard brings $0.50 profit and each cubic yard of Plastiflex $0.70, how should the carts be loaded to provide maximum profit per minute for each cartload?

12. A confectioner sells two different types of nut mixtures. Each package of the standard mixture contains 100 g of cashews and 200 g of peanuts and sells for $1.95. Each package of the deluxe mixture contains 150 g of cashews and 50 g of peanuts and sells for $2.25. The confectioner has 15 kg of cashews and 20 kg of peanuts available. Based on past sales statistics, he wishes to have at least as many standard as deluxe packages available. How many bags of each type of mixture should he package to maximize his revenue?

13. A furniture manufacturer has factories in two locations. The Vancouver factory can produce 20 sofas, 20 chairs, and 35 ottomans each day, whereas the Seattle factory can produce 50 sofas, 25 chairs, and 25 ottomans each day. It costs $3000 per day to operate the Vancouver factory and $4000 per day to operate the Seattle factory. An order for 400 sofas, 300 chairs, and 375 ottomans is received. The order is to be filled in no more than 30 days. How many days should each factory be operated to fill the order at minimum cost?

14. A biologist wishes to feed laboratory rabbits a mixture of two different types of foods. Type I contains 8 g of fat, 12 g of carbohydrate, and 2 g of protein per ounce, whereas type II contains 12 g of fat, 12 g of carbohydrate, and 1 g of protein per ounce. Type I costs $0.20 per ounce and type II costs $0.30 per ounce. Each rabbit is to receive a daily minimum of 24 g of fat, 36 g of carbohydrate, and 4 g of protein, but should get no more than 5 oz of food per day. How many ounces of each type of food should be fed to each rabbit daily to satisfy the requirements at minimum cost?

15. A woman wishes to invest $12,000 in three types of bonds: municipal bonds paying 7% interest per annum, bank investment certificates paying 8%, and high-risk bonds paying 12%. For tax reasons, she wants the amount invested in municipal bonds to be at least three times as much as the amount invested in bank certificates. To keep her level of risk manageable, she will invest no more than $2000 in high-risk bonds. How much should she invest in each type of bond to maximize her annual interest yield? (*Hint:* Let x = amount in municipal bonds and y = amount in bank certificates. Then the amount in high-risk bonds would be $12,000 - x - y$.)

16. Refer to Exercise 15. Suppose the investor decides to increase to $3000 the maximum amount she will allow herself to invest in high-risk bonds but leaves the other conditions unchanged. By how much will her maximum possible interest yield increase?

17. A small software company publishes computer games and educational and utility software. Their policy is to market a total of 36 new programs each year, with at least 4 of these being games. The number of utility programs published is never more than twice the number of educational programs. On the average, the company can expect to make an annual profit of $5000 on each computer game, $8000 on each educational program, and $6000 on each utility program. How many of each type of program should they publish annually for maximum profit?

18. All parts of this problem refer to the following feasible region and objective function:

$$\begin{cases} y \geq 0 \\ x \geq y \\ x + 2y \leq 12 \\ x + y \leq 10 \end{cases}$$
$$P = x + 4y$$

(a) Graph the feasible region.
(b) On your graph from part (a), sketch the graphs of the linear equations obtained by setting P equal to 40, 36, 32, and 28.
(c) If we continue to decrease the value of P, at which vertex of the feasible region will these lines first touch the feasible region?
(d) Verify that the maximum value of P on the feasible region occurs at the vertex you chose in part (c).

CHAPTER 7 REVIEW

Define, state, or discuss the following.

1. System of equations
2. Linear equation
3. Gaussian elimination
4. Triangular form
5. Elementary row operations

6. Matrix
7. Inconsistent system of equations
8. Dependent system of equations
9. Partial fraction decomposition
10. Addition and subtraction of matrices

11. Product of matrices

12. Identity matrix

13. Inverse of a matrix

14. Minor

15. Cofactor

16. Determinant

17. Invertibility Criterion

18. Row and column transformations of a determinant

19. Cramer's Rule

20. Nonlinear system of equations

21. System of inequalities

22. Bounded and unbounded regions

23. Vertex

24. Linear programming

25. Feasible region

26. Objective function

Review Exercises

In Exercises 1–6 solve the systems of equations and graph the lines.

1. $\begin{cases} 2x + 4y = 16 \\ 4x - y = 5 \end{cases}$

2. $\begin{cases} y = 4x + 4 \\ x = 3y + 10 \end{cases}$

3. $\begin{cases} 2x - 7y = 28 \\ y = \frac{2}{7}x - 4 \end{cases}$

4. $\begin{cases} 6x - 8y = 15 \\ -\frac{3}{4}x + 2y = -4 \end{cases}$

5. $\begin{cases} 2x - y = -1 \\ x + 3y = 10 \\ 3x + 4y = 15 \end{cases}$

6. $\begin{cases} 2x + 5y = 9 \\ -x + 3y = 1 \\ 7x - 2y = 14 \end{cases}$

In Exercises 7–14 find the complete solution of the system using Gaussian elimination, or show that there is no solution.

7. $\begin{cases} x - 2y + z = 0 \\ 3x - y + 2z = 0 \\ 4x - 9y = 21 \end{cases}$

8. $\begin{cases} 2x - y + 3z = 2 \\ 4x - 9z = 2 \\ 3x + 2y + 6z = 18 \end{cases}$

9. $\begin{cases} x - 2y + 3z = 1 \\ 2x - y + z = 3 \\ 2x - 7y + 11z = 2 \end{cases}$

10. $\begin{cases} x + y + z + w = 2 \\ 2x - 3z = 5 \\ x - 2y + 4w = 9 \\ x + y + 2z + 3w = 5 \end{cases}$

11. $\begin{cases} x - 3y + z = 4 \\ 4x - y + 15z = 5 \end{cases}$

12. $\begin{cases} 2x - 3y + 4z = 3 \\ 4x - 5y + 9z = 13 \\ 2x + 7z = 0 \end{cases}$

13. $\begin{cases} -x + 4y + z = 8 \\ 2x - 6y + z = -9 \\ x - 6y - 4z = -15 \end{cases}$

14. $\begin{cases} x - z + w = 2 \\ 2x + y - 2w = 12 \\ 3y + z + w = 4 \\ x + y - z = 10 \end{cases}$

15. A man invests his savings in two accounts, one paying 6% interest per annum and the other paying 7%. He has twice as much invested in the 7% account as in the 6% account, and his annual interest income is $600. How much does he have invested in each account?

16. Find the values of a, b, and c if the parabola

$$y = ax^2 + bx + c$$

is to pass through the points $(1, 0)$, $(-1, -4)$, and $(2, 11)$.

Find the partial fraction decompositions of the rational functions in Exercises 17–20.

17. $\dfrac{3x + 1}{x^2 - 2x - 15}$

18. $\dfrac{8}{x^3 - 4x}$

19. $\dfrac{2x - 4}{x(x - 1)^2}$

20. $\dfrac{x + 6}{x^3 - 2x^2 + 4x - 8}$

In Exercises 21–32, let

$$A = \begin{bmatrix} 2 & 0 & -1 \end{bmatrix} \quad B = \begin{bmatrix} 1 & 2 & 4 \\ -2 & 1 & 0 \end{bmatrix}$$

$$C = \begin{bmatrix} \frac{1}{2} & 3 \\ 2 & \frac{3}{2} \\ -2 & 1 \end{bmatrix} \quad D = \begin{bmatrix} 1 & 4 \\ 0 & -1 \\ 2 & 0 \end{bmatrix}$$

$$E = \begin{bmatrix} 2 & -1 \\ -\frac{1}{2} & 1 \end{bmatrix} \quad F = \begin{bmatrix} 4 & 0 & 2 \\ -1 & 1 & 0 \\ 7 & 5 & 0 \end{bmatrix}$$

$$G = \begin{bmatrix} 5 \end{bmatrix}$$

Carry out the operation indicated in each exercise, or explain why it cannot be performed.

21. $A + B$ **22.** $C - D$ **23.** $2C + 3D$

24. $5B - 2C$ **25.** GA **26.** AG

27. BC **28.** CB **29.** BF

30. FC **31.** $(C + D)E$ **32.** $F(2C - D)$

In Exercises 33–38 find the determinant and, if possible, the inverse of the matrix.

33. $\begin{bmatrix} 1 & 4 \\ 2 & 9 \end{bmatrix}$

34. $\begin{bmatrix} 2 & 2 \\ 1 & -3 \end{bmatrix}$

35. $\begin{bmatrix} 4 & -12 \\ -2 & 6 \end{bmatrix}$

36. $\begin{bmatrix} 2 & 4 & 0 \\ -1 & 1 & 2 \\ 0 & 3 & 2 \end{bmatrix}$

37. $\begin{bmatrix} 3 & 0 & 1 \\ 2 & -3 & 0 \\ 4 & -2 & 1 \end{bmatrix}$

38. $\begin{bmatrix} 1 & 0 & 0 & 1 \\ 0 & 2 & 0 & 2 \\ 0 & 0 & 3 & 3 \\ 0 & 0 & 0 & 4 \end{bmatrix}$

In Exercises 39 and 40 express the system of linear equations as a matrix equation. Then solve the matrix equation by multiplying both sides by the inverse of the coefficient matrix.

39. $\begin{cases} 12x - 5y = 10 \\ 5x - 2y = 17 \end{cases}$

40. $\begin{cases} 2x + y + 5z = \frac{1}{3} \\ x + 2y + 2z = \frac{1}{4} \\ x + 3z = \frac{1}{6} \end{cases}$

Solve the systems in Exercises 41–44 using Cramer's Rule.

41. $\begin{cases} 2x + 7y = 13 \\ 6x + 16y = 30 \end{cases}$

42. $\begin{cases} 12x - 11y = 140 \\ 7x + 9y = 20 \end{cases}$

43. $\begin{cases} 2x - y + 5z = 0 \\ -x + 7y = 9 \\ 5x + 4y + 3z = -9 \end{cases}$

44. $\begin{cases} 3x + 4y - z = 10 \\ x - 4z = 20 \\ 2x + y + 5z = 30 \end{cases}$

Find all solutions of the systems of equations in Exercises 45–48.

45. $\begin{cases} x^2 + y^2 + 6y = 0 \\ x - 2y = 3 \end{cases}$

46. $\begin{cases} x^2 + y^2 = 10 \\ x^2 + 2y^2 - 7y = 0 \end{cases}$

47. $\begin{cases} 3x^4 + \frac{4}{y} = 50 \\ x^4 - \frac{8}{y} = 12 \end{cases}$

48. $\begin{cases} x^2 + yz = 0 \\ y^2 + xz = 0 \\ x^2 + xy + y^2 = 3 \end{cases}$

Graph the solution sets of the systems of inequalities in Exercises 49–52. Find the coordinates of all vertices, and determine whether the solution set is bounded or unbounded.

49. $\begin{cases} x^2 + y^2 < 9 \\ x + y < 0 \end{cases}$

50. $\begin{cases} y - x^2 \geq 4 \\ y < 20 \end{cases}$

51. $\begin{cases} x \geq 0, \quad y \geq 1 \\ x + 2y \leq 12 \\ y \leq x + 4 \end{cases}$

52. $\begin{cases} x \geq 4 \\ x + y \geq 24 \\ x \leq 2y + 12 \end{cases}$

53. Find the maximum and minimum values of the function $P = 3x + 4y$ on the region described by the inequalities in Exercise 51.

54. (a) Find the minimum value of the function $Q = 60 + 3x + 5y$ on the region described by the inequalities in Exercise 52.
(b) Explain why Q has no maximum value on this region.

55. A farmer wishes to plant oats and barley on 400 acres of land. The land can produce 40 bushels of oats or 50 bushels of barley per acre. After harvest, the farmer will have to store the grain for several months in order to get the best price for it, and he has facilities to store no more than 18,000 bushels of grain.
(a) If he can get $2.05 per bushel for oats and $1.80 for barley, how many acres of each grain should he plant for maximum revenue?
(b) If instead the price for oats is $1.20 and for barley is $1.60 per bushel, how many acres of each should he plant to maximize his revenue?

56. A woman wishes to invest $12,000—some in a high-risk stock with an expected annual dividend of 15%, some in long-term bonds yielding 10% interest per annum, and the remainder in a bank money market account paying an annual yield of 6% interest. She wishes to put at least $4000 into the money market account. Also, the amount invested in the high-risk stock should be no more than half the total of her other two investments. How much should she put in each investment to maximize her annual interest and dividend yield?

In Exercises 57 and 58 solve for x, y, and z in terms of a, b, and c.

57. $\begin{cases} -x + y + z = a \\ x - y + z = b \\ x + y - z = c \end{cases}$

58. $\begin{cases} ax + by + cz = a - b + c \\ bx + by + cz = c \\ cx + cy + cz = c \end{cases}$
$(a \neq b, b \neq c, c \neq 0)$

59. For what values of k do the three lines

$$x + y = 12$$
$$kx - y = 0$$
$$y - x = 2k$$

have a common point of intersection?

60. For what value of k does the system

$$\begin{cases} kx + y + z = 0 \\ x + 2y + kz = 0 \\ -x + 3z = 0 \end{cases}$$

have infinitely many solutions?

Chapter 7 Test

1. An airplane takes $2\frac{1}{2}$ h to make a trip of 600 km against the wind. It takes 50 min to travel 300 km with the wind. Find the speed of the wind and the speed of the airplane in still air.

In Problems 2–5 find all solutions of the system. Determine whether the system is linear or nonlinear. If it is linear, state whether it is inconsistent, dependent, or neither.

2. $\begin{cases} 2x - 5y = 9 \\ 7x + 6y = 8 \end{cases}$

3. $\begin{cases} 3x - y + z = 5 \\ x - 4z = 7 \\ x - y + 9z = -8 \end{cases}$

4. $\begin{cases} 2x - y + z = 0 \\ 3x + 2y - 3z = 1 \\ x - 4y + 5z = -1 \end{cases}$

5. $\begin{cases} 2x^2 + y^2 = 6 \\ 3x^2 - 4y = 11 \end{cases}$

In Problems 6–13 let

$$A = \begin{bmatrix} 2 & 3 \\ 2 & 4 \end{bmatrix} \quad B = \begin{bmatrix} 1 & 6 \\ 0 & 1 \\ -1 & 5 \end{bmatrix}$$

$$C = \begin{bmatrix} 1 & 0 & 4 \\ -1 & 1 & 2 \\ 0 & 1 & 3 \end{bmatrix}$$

Find the following, or explain why the indicated operation cannot be performed.

6. $A + B$

7. AB

8. $BA - 3B$

9. CBA

10. A^{-1}

11. B^{-1}

12. $|B|$

13. $|C|$

14. Write a matrix equation equivalent to the given system of linear equations. Find the inverse of the coefficient matrix and use it to solve the system.

$$\begin{cases} 3x - 5y = 51 \\ 2x + 3y = 64 \end{cases}$$

15. Solve using Cramer's Rule:

$$\begin{cases} 2x - z = 14 \\ 3x - y + 5z = 0 \\ 4x + 2y + 3z = -2 \end{cases}$$

16. Only one of the following matrices has an inverse. Find the determinant of each matrix, and use the determinants to decide which one has an inverse. Then find the inverse of the invertible one.

$$A = \begin{bmatrix} 1 & 4 & 1 \\ 0 & 2 & 0 \\ 1 & 0 & 1 \end{bmatrix}$$

$$B = \begin{bmatrix} 1 & 4 & 0 \\ 0 & 2 & 0 \\ -3 & 0 & 1 \end{bmatrix}$$

17. Graph the following system of inequalities, indicating the coordinates of the vertices:

$$\begin{cases} x^2 - 2x - y + 5 \le 0 \\ y \le 5 + 2x \end{cases}$$

18. Find the partial fraction decomposition of the rational function

$$\frac{4x - 1}{(x - 1)^2(x + 2)}$$

19. A farmer grows wheat and barley on 200 acres of land. He can borrow no more than $10,000 at the beginning of the season for production costs, and he has no other resources available to him. It costs $60 per acre to grow wheat and $40 per acre to grow barley, and the land yields 50 bushels of wheat or 40 bushels of barley per acre. The market predictions for the fall harvest indicate that he can make a profit of $2.50 per bushel of wheat and $2.00 per bushel of barley. How many acres of each grain should he plant to maximize his profits?

Problems Plus

1. Justin, Sasha, and Vanessa each have a bag of marbles. First Justin gives Sasha and Vanessa each as many marbles as they already have. Then Sasha gives Justin and Vanessa as many marbles as they now have. Finally, Vanessa gives Justin and Sasha as many marbles as they now have. Everyone ends up with 16 marbles. How many did each person have to begin with?

2. Find the complete solution of the following system of equations:

$$\begin{cases} x + y + z = 2 \\ x^2 + y^2 + z^2 = 2 \\ \qquad\qquad xy = z^2 \end{cases}$$

3. (a) By inspection, find one solution of the system

$$\begin{cases} ax - by + cz = 0 \\ dx + ey - fz = 0 \\ \qquad\quad gy + hz = 0 \end{cases}$$

where a, b, c, d, e, f, g, and h are all positive real numbers.

(b) Prove that the system has no other solutions.

4. Find the largest and smallest values of the function $P = 100 - x - y$ on the feasible region described by the following system of inequalities:

$$\begin{cases} x^2 + y^2 \le 25 \\ x + 3y \ge 5 \end{cases}$$

8

Analytic Geometry

In studying the procedures of geometric thought we may hope to reach what is most essential in the human mind.

Henri Poincaré

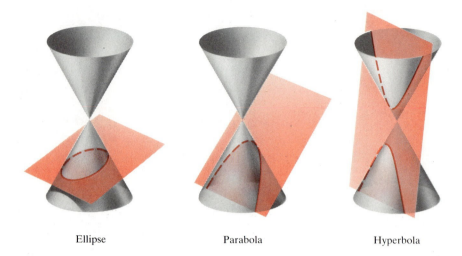

Figure 1

Ellipse   Parabola   Hyperbola

In this chapter we study the geometry of **conic sections** (or simply **conics**). Conic sections are the curves formed by the intersection of a plane with a pair of circular cones. These curves can have three basic shapes, called **ellipses**, **parabolas**, and **hyperbolas**, as illustrated in Figure 1.

The ancient Greeks studied these curves because they considered the geometry of conic sections to be very beautiful. The mathematician Apollonius (262–190 B.C.) wrote a definitive eight-volume work on the subject. In more modern times, conics were found to be useful as well as beautiful. Galileo discovered in 1590 that the path of a missile shot upward at an angle is a parabola. Kepler (1609) found that the planets move in elliptical orbits around the sun. Newton (1668) was the first to build a reflecting telescope, whose principle is based on the properties of parabolas and hyperbolas. In this century many more applications of conic sections have been developed. One important application is the LORAN radio navigation system, which uses the intersection points of hyperbolas to pinpoint the location of ships and aircraft.

In addition to studying conics, we learn in this chapter about two other ways of describing points and curves in the Cartesian plane: polar coordinates and parametric equations. Both of these topics require a thorough understanding of trigonometry.

Section 8.1

Parabolas

We have seen that the graph of the equation $y = ax^2 + bx + c$ is a U-shaped curve called a **parabola** that opens either upward or downward depending on whether the sign of a is positive or negative (see Figure 1). The lowest or highest point of the parabola is called the **vertex**, and the parabola is symmetric about its **axis**.

In this section we study parabolas from a more geometric (rather than algebraic) point of view. We begin with the geometric definition of a parabola.

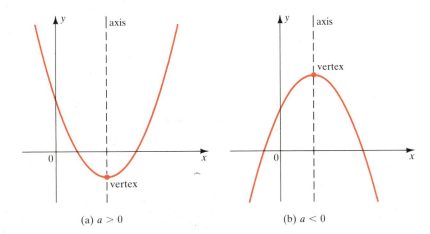

(a) $a > 0$ (b) $a < 0$

Figure 1
$y = ax^2 + bx + c$

> A **parabola** is the set of points in a plane equidistant from a fixed point F (called the **focus**) and a fixed line l (called the **directrix**).

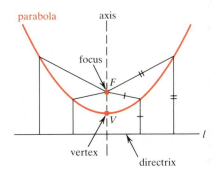

Figure 2

This definition is illustrated in Figure 2. Note that the vertex V of the parabola lies halfway between the focus and the directrix, and that the axis of symmetry is the line that runs through the focus perpendicular to the directrix.

In this section we restrict our attention to parabolas that are situated with the vertex at the origin and that have a vertical or horizontal axis of symmetry. (Parabolas in more general positions will be considered in Sections 8.4 and 8.5.) If the focus of such a parabola is the point $F(0, p)$, then the axis of symmetry must be vertical and the directrix has the equation $y = -p$. (See Figure 3, which illustrates the case where $p > 0$.)

If $P(x, y)$ is any point on the parabola, then the distance from P to the focus F (using the Distance Formula) is

$$\sqrt{x^2 + (y - p)^2}$$

and the distance from P to the directrix is

$$|y - (-p)| = |y + p|$$

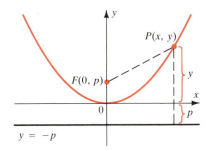

Figure 3

By the definition of a parabola, these two distances must be equal:

$$\sqrt{x^2 + (y - p)^2} = |y + p|$$

Squaring both sides and simplifying, we get

$$x^2 + (y - p)^2 = |y + p|^2 = (y + p)^2$$
$$x^2 + y^2 - 2py + p^2 = y^2 + 2py + p^2$$
$$x^2 - 2py = 2py$$
$$x^2 = 4py$$

> The equation of a parabola with focus $F(0, p)$ and directrix $y = -p$ is
>
> **(1)** $\qquad\qquad x^2 = 4py$

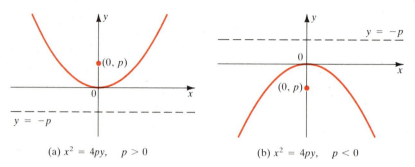

Figure 4

(a) $x^2 = 4py$, $p > 0$

(b) $x^2 = 4py$, $p < 0$

If $p > 0$, then the parabola opens upward, but if $p < 0$, it opens downward (see Figure 4). If x is replaced by $-x$, the equation remains unchanged, so the graph is symmetric about the y-axis.

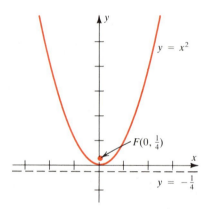

Figure 5

Example 1

Find the focus and directrix of the parabola $y = x^2$ and sketch the graph.

Solution

Comparing the equation $y = x^2$ with Equation 1, we see that $4p = 1$, so $p = \frac{1}{4}$. Thus the focus is $(0, \frac{1}{4})$ and the directrix is $y = -\frac{1}{4}$. The graph is shown in Figure 5.   ●

Reflecting the graph in Figure 3 about the diagonal line $y = x$ has the effect of interchanging the roles of x and y. Thus using the same sort of argument as in the derivation of Equation 1, we can show the following:

> The equation of a parabola with focus $F(p, 0)$ and directrix $x = -p$ is
>
> (2) $$y^2 = 4px$$

In this case the x-axis is the axis of symmetry, and if $p > 0$, the parabola opens to the right, whereas if $p < 0$, it opens to the left (see Figure 6).

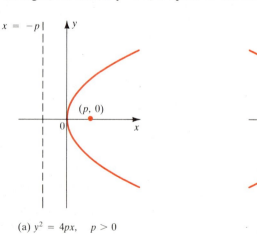

Figure 6

(a) $y^2 = 4px$, $p > 0$

(b) $y^2 = 4px$, $p < 0$

Example 2

Find the focus and directrix of the parabola $6x + y^2 = 0$ and sketch the graph.

Solution

We first write the equation as $y^2 = -6x$. Comparing this with Equation 2, we see that $-6 = 4p$, so $p = -\frac{3}{2}$. Thus the focus is $(-\frac{3}{2}, 0)$ and the directrix is $x = \frac{3}{2}$. The graph is shown in Figure 7.

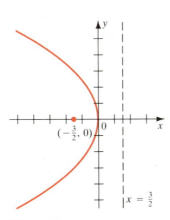

Figure 7
$6x + y^2 = 0$

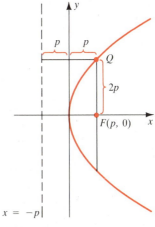

Figure 8

The coordinates of the focus can be used to help us estimate the "width" of a parabola when sketching its graph. The line segment that runs through the focus perpendicular to the axis, with endpoints on the parabola, is called the **latus rectum**, and its length is the **focal diameter** of the parabola (see Figure 8). From the figure we can see that the distance from an endpoint Q of the latus rectum to the directrix is $2p$. Thus the distance from Q to the focus must be $2p$ as well (by the definition of a parabola), and so the focal diameter is $4p$.

Example 3

Find the focus, directrix, and focal diameter of the parabola $y = \frac{1}{2}x^2$ and sketch the graph.

Solution

First we put the equation in the form of Equation 1:

$$x^2 = 2y$$

Thus $4p = 2$, so the focal diameter is 2. Also $p = \frac{1}{2}$, so the focus is $(0, \frac{1}{2})$ and the directrix is $y = -\frac{1}{2}$. The latus rectum extends one unit to the left and to the right of the focus, since the focal diameter is 2 (see Figure 9).

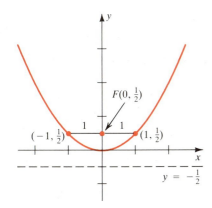

Figure 9
$y = \frac{1}{2}x^2$

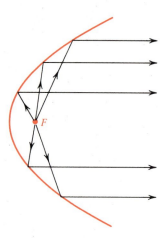

Figure 10
Parabolic reflector

Parabolas have an important property that makes them useful as reflectors for lamps and telescopes. Light from a source placed at the focus of a surface with a parabolic cross-section is reflected in such a way as to travel parallel to the axis of the parabola (see Figure 10). Thus a parabolic reflector concentrates the light into a beam of parallel rays. Conversely, light that approaches the reflector in rays parallel to its axis of symmetry is reflected to the focus. This principle is used in the construction of reflecting telescopes.

Exercises 8.1

Find the focus, directrix, and focal diameter for each parabola in Exercises 1–12 and sketch the graph.

1. $y^2 = 4x$ **2.** $x^2 = y$

3. $x^2 = 9y$ **4.** $y^2 = 3x$

5. $y = 5x^2$ **6.** $y = -2x^2$

7. $x = -8y^2$ **8.** $x = \frac{1}{2}y^2$

9. $x^2 + 6y = 0$ **10.** $x - 7y^2 = 0$

11. $5x + 3y^2 = 0$ **12.** $8x^2 + 12y = 0$

Find equations for the parabolas sketched in Exercises 13–20.

13.

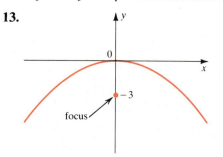

14.

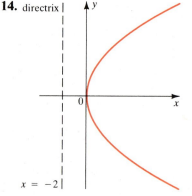

15.

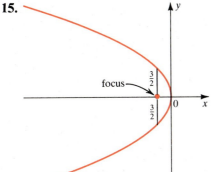

focus — $\dfrac{3}{2}$

$\dfrac{3}{2}$

16.

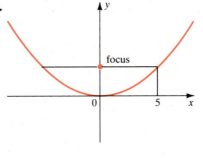

focus

5

17.

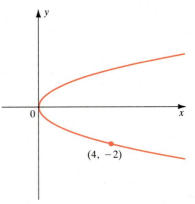

$(4, -2)$

18.

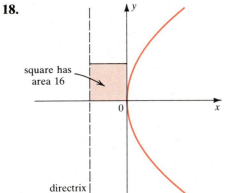

square has
area 16

directrix

19.

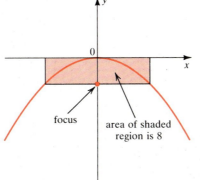

0

focus

area of shaded
region is 8

20.

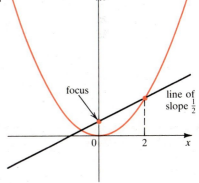

focus

line of
slope $\frac{1}{2}$

2

*In Exercises 21–26 find equations for the parabolas described.
In each exercise the vertex of the parabola is to be the origin.*

21. Its directrix is $x = 3$.

22. Its focus is $(0, -\frac{1}{2})$.

23. Its focus is on the positive x-axis, two units away from the directrix.

24. Its focal diameter is 7 and its focus is on the negative y-axis.

25. Its directrix has y-intercept 6.

26. It opens upward with the focus five units from the vertex.

27. A lamp with a parabolic reflector is shown in the figure. The bulb is at the focus and the focal diameter is 12 cm.
   (a) Find an equation of the parabola.
   (b) Find the diameter $|CD|$ of the opening 20 cm from the vertex.

28. A reflector for a radiotelescope is parabolic in cross-section, with the receiver at the focus. The reflector is 20 ft wide from rim to rim and 1 ft deep (see the figure). How far is the receiver from the vertex of the parabolic reflector?

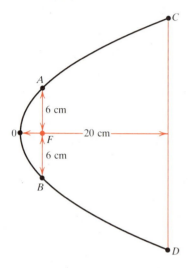

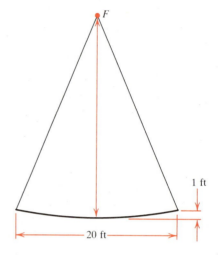

Section 8.2

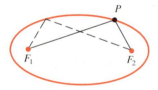

Figure 1

Ellipses

An ellipse is an oval curve that looks like an elongated circle. More precisely, we have the following definition:

> An **ellipse** is the set of all points in the plane the sum of whose distances from two fixed points F_1 and F_2 is a constant (see Figure 1). These two fixed points are called the **foci** (plural of **focus**) of the ellipse.

The definition suggests a simple method for drawing an ellipse. Place a sheet of paper on a drawing board and insert thumbtacks at the two points that are to be the foci of the ellipse. Attach the ends of a string to the thumbtacks, as shown in Figure 2. With the point of a pencil, hold the string taut. Then carefully move the pencil around the foci, keeping the string taut at all times. The pencil will trace out an ellipse, since the sum of the distances from its point to the foci will always equal the length of the string, which is constant.

If the string is only slightly longer than the distance between the foci, the ellipse traced out will be elongated as in Figure 3(a), but if the foci are close together relative to the length of the string, the ellipse will be almost circular as in Figure 3(b).

To obtain the simplest possible equation for an ellipse, we place the foci on the x-axis at $F_1(-c, 0)$ and $F_2(c, 0)$, so that the origin is halfway between them (see

Figure 2

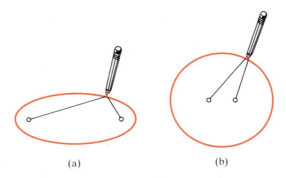

(a)   (b)

Figure 3

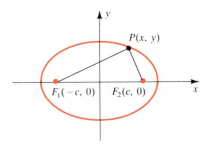

Figure 4

Figure 4). For later convenience, we let the sum of the distances from a point on the ellipse to the foci be $2a$. Then if $P(x, y)$ is any point on the ellipse, we have

$$|PF_1| + |PF_2| = 2a$$

so, from the Distance Formula,

$$\sqrt{(x + c)^2 + y^2} + \sqrt{(x - c)^2 + y^2} = 2a$$

or

$$\sqrt{(x - c)^2 + y^2} = 2a - \sqrt{(x + c)^2 + y^2}$$

Squaring both sides and multiplying out, we get

$$x^2 - 2cx + c^2 + y^2 = 4a^2 - 4a\sqrt{(x + c)^2 + y^2} + (x^2 + 2cx + c^2 + y^2)$$

which simplifies to

$$4a\sqrt{(x + c)^2 + y^2} = 4a^2 + 4cx$$

Dividing both sides by 4 and squaring again, we get

$$a^2[(x + c)^2 + y^2] = (a^2 + cx)^2$$
$$a^2x^2 + 2a^2cx + a^2c^2 + a^2y^2 = a^4 + 2a^2cx + c^2x^2$$

(1)
$$(a^2 - c^2)x^2 + a^2y^2 = a^2(a^2 - c^2)$$

Since the sum of the distances from P to the foci must be greater than the distance between the foci, we have that $2a > 2c$, or $a > c$. Thus $a^2 - c^2 > 0$, and we can divide both sides of Equation 1 by $a^2(a^2 - c^2)$ to get

$$\frac{x^2}{a^2} + \frac{y^2}{a^2 - c^2} = 1$$

For convenience, let $b^2 = a^2 - c^2$, where $b > 0$. Since $b^2 < a^2$, it follows that $b < a$. The preceding equation then becomes

(2)
$$\frac{x^2}{a^2} + \frac{y^2}{b^2} = 1, \qquad a > b$$

To graph the ellipse represented by Equation 2, we need to know the x- and y-intercepts. Setting $y = 0$, we get

$$\frac{x^2}{a^2} = 1$$

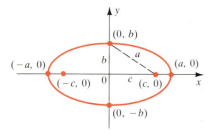

Figure 5
$$\frac{x^2}{a^2} + \frac{y^2}{b^2} = 1 \qquad (a > b)$$

so $x^2 = a^2$ and $x = \pm a$. Thus the ellipse crosses the x-axis at $(a, 0)$ and $(-a, 0)$. These points are called the **vertices** of the ellipse, and the segment that joins them is called the **major axis**. Its length is $2a$.

Similarly, if we set $x = 0$ in Equation 2, we get $y = \pm b$, so the ellipse crosses the y-axis at $(0, b)$ and $(0, -b)$. The segment that joins these points is called the **minor axis** and has length $2b$. Note that $2a > 2b$, so the major axis is longer than the minor axis.

In Section 1.7 we studied several tests that detect symmetry in a graph. If we replace x by $-x$ or y by $-y$ in Equation 2, the equation remains unchanged. Thus the ellipse is symmetric about both the x- and y-axes and hence about the origin as well. For this reason, the origin is called the **center** of the ellipse. The complete graph is shown in Figure 5.

We saw earlier in this section (Figure 3) that if $2a$ is only slightly greater than $2c$, the ellipse is long and thin, whereas if $2c$ is much less than $2a$, the ellipse is almost circular. We define the **eccentricity** of the ellipse to be the ratio

$$e = \frac{c}{a}$$

Thus if e is close to 1, c is almost equal to a and the ellipse is elongated, but if e is close to 0, the ellipse is close to a circle. (Note that $0 < e < 1$ for any ellipse.) The eccentricity is a measure of how "stretched out" the ellipse is.

We summarize the preceding discussion as follows:

The graph of the equation

$$\frac{x^2}{a^2} + \frac{y^2}{b^2} = 1, \qquad a > b$$

is an ellipse with foci $(\pm c, 0)$, where $c^2 = a^2 - b^2$. The center is $(0, 0)$ and the vertices are $(\pm a, 0)$. The major axis is horizontal and has length $2a$, and the minor axis is vertical and has length $2b$. The eccentricity is $e = c/a$. (See Figure 5.)

Example 1

Find the foci, vertices, eccentricity, and the lengths of the major and minor axes for the ellipse

$$\frac{x^2}{9} + \frac{y^2}{4} = 1$$

and sketch the graph.

Solution

Here $a^2 = 9$ and $b^2 = 4$, so $c^2 = a^2 - b^2 = 9 - 4 = 5$. Thus we have

foci: $(\pm\sqrt{5}, 0)$

vertices: $(\pm 3, 0)$

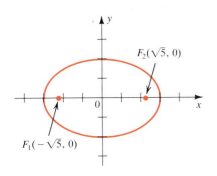

Figure 6
$$\frac{x^2}{9} + \frac{y^2}{4} = 1$$

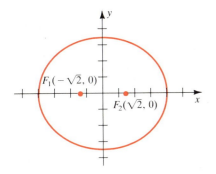

Figure 7
$$\frac{x^2}{16} + \frac{y^2}{12} = 1$$

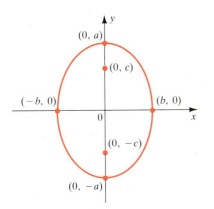

Figure 8
$$\frac{x^2}{b^2} + \frac{y^2}{a^2} = 1 \quad (a > b)$$

eccentricity: $\sqrt{5}/3 \approx 0.745$
length of major axis: 6
length of minor axis: 4

The graph is shown in Figure 6.   ●

Example 2
The vertices of an ellipse are $(\pm 4, 0)$ and the eccentricity is $\frac{1}{2}$. Find its equation and sketch the graph.

Solution
Since the vertices are $(\pm 4, 0)$, we have that $a = 4$. Also $e = c/a = 1/2$, so

$$\frac{c}{4} = \frac{1}{2}$$
$$c = 2$$

Finally, $c^2 = a^2 - b^2$, so $2^2 = 4^2 - b^2$, and hence

$$b = \sqrt{4^2 - 2^2} = 2\sqrt{3}$$

Thus the equation of the ellipse is

$$\frac{x^2}{16} + \frac{y^2}{12} = 1$$

Its graph is shown in Figure 7.   ●

If the foci of the ellipse are placed on the y-axis at $(0, \pm c)$ rather than on the x-axis, then the roles of x and y are reversed in the discussion leading up to Equation 2. Thus we have the following description of such ellipses:

> The graph of the equation
> $$\frac{x^2}{b^2} + \frac{y^2}{a^2} = 1, \qquad a > b$$
> is an ellipse with foci $(0, \pm c)$, where $c^2 = a^2 - b^2$. The center is $(0, 0)$ and the vertices are $(0, \pm a)$. The major axis is vertical and has length $2a$, and the minor axis is horizontal and has length $2b$. The eccentricity is $e = c/a$. (See Figure 8.)

Example 3
Find the foci of the ellipse $16x^2 + 9y^2 = 144$ and sketch the graph.

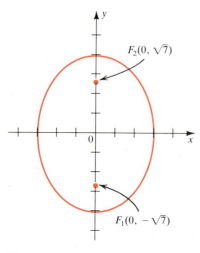

Figure 9
$16x^2 + 9y^2 = 144$

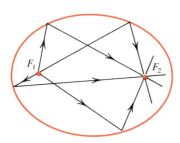

Figure 10

Solution

Dividing through by 144, we get

$$\frac{x^2}{9} + \frac{y^2}{16} = 1$$

Since $16 > 9$, this is an ellipse with its foci on the y-axis, and with $a = 4$ and $b = 3$. We have

$$c^2 = a^2 - b^2 = 16 - 9 = 7$$
$$c = \sqrt{7}$$

Thus the foci are $(0, \pm\sqrt{7})$. The graph is shown in Figure 9.

Example 4

Find an equation for the ellipse with foci $(0, \pm 2)$ and vertices $(0, \pm 3)$.

Solution

For this ellipse, $c = 2$ and $a = 3$. Therefore $b^2 = a^2 - c^2 = 9 - 4 = 5$, so the equation of the ellipse is

$$\frac{x^2}{5} + \frac{y^2}{9} = 1$$

Like parabolas, ellipses have an interesting reflection property that has a number of practical consequences. If a light source is placed at one focus of a reflecting surface with elliptical cross-sections, then all the light is reflected off the surface to the other focus, as shown in Figure 10. This principle, which works for sound waves as well as light, is used in *lithotripsy*, the new treatment for kidney stones. A reflector with elliptical cross-section is placed in such a way that the kidney stone is at one focus. High intensity sound waves generated at the other focus are reflected to the stone and destroy it without damaging surrounding tissue. The patient is spared the trauma of surgery and recovers within a few days.

Exercises 8.2

In Exercises 1–14 find the vertices, foci, and eccentricity of the ellipse. Determine the lengths of the major and minor axes and sketch the graph.

1. $\dfrac{x^2}{25} + \dfrac{y^2}{9} = 1$

2. $\dfrac{x^2}{16} + \dfrac{y^2}{25} = 1$

3. $9x^2 + 4y^2 = 36$

4. $4x^2 + 25y^2 = 100$

5. $x^2 + 4y^2 = 16$

6. $4x^2 + y^2 = 16$

7. $2x^2 + y^2 = 3$

8. $5x^2 + 6y^2 = 30$

9. $x^2 + 4y^2 = 1$

10. $9x^2 + 4y^2 = 1$

11. $\frac{1}{2}x^2 + \frac{1}{8}y^2 = \frac{1}{4}$

12. $x^2 = 4 - 2y^2$

13. $y^2 = 1 - 2x^2$

14. $20x^2 + 4y^2 = 5$

In Exercises 15–26 find an equation for the ellipse that satisfies the given conditions.

15. Foci $(\pm 4, 0)$, vertices $(\pm 5, 0)$

16. Foci $(0, \pm 3)$, vertices $(0, \pm 5)$

17. Length of major axis 4, length of minor axis 2, foci on y-axis

18. Length of major axis 6, length of minor axis 4, foci on x-axis

19. Foci $(0, \pm 2)$, length of minor axis 6

20. Foci $(\pm 5, 0)$, length of major axis 12

21. Endpoints of major axis $(\pm 10, 0)$, distance between foci 6

22. Endpoints of minor axis $(0, \pm 3)$, distance between foci 8

23. Length of major axis 10, foci on x-axis, ellipse passes through point $(\sqrt{5}, 2)$

24. Eccentricity $\frac{1}{9}$, foci $(0, \pm 2)$

25. Eccentricity 0.8, foci $(\pm 1.5, 0)$

26. Eccentricity $\sqrt{3}/2$, foci on y-axis, length of major axis 4

In Exercises 27 and 28, find the intersection points of the pairs of ellipses. Sketch the graphs of each pair of equations on the same coordinate axes and show the points of intersection.

27. $\begin{cases} 4x^2 + y^2 = 4 \\ 4x^2 + 9y^2 = 36 \end{cases}$

28. $\begin{cases} \dfrac{x^2}{16} + \dfrac{y^2}{9} = 1 \\ \dfrac{x^2}{9} + \dfrac{y^2}{16} = 1 \end{cases}$

29. The planets move around the sun in elliptical orbits with the sun at one focus. The point in the orbit at which the planet is closest to the sun is called **perihelion**, and the point at which it is farthest is called **aphelion**. These points are the vertices of the orbit. The earth is 147,000,000 km from the sun at perihelion and 153,000,000 km from the sun at aphelion. Find an equation for the earth's orbit. (Place the origin at the center of the orbit, with the sun on the x-axis.)

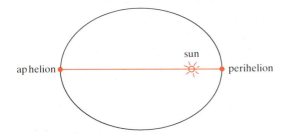

30. With an eccentricity of 0.25, Pluto's orbit is the most eccentric in the solar system. The length of the minor axis of its orbit is approximately 10,000,000,000 km. Find an equation for Pluto's orbit. (See Exercise 29.)

31. For an object in an elliptical orbit around the moon, the points in the orbit that are closest to and farthest from the center of the moon are called **perilune** and **apolune**, respectively. These are the vertices of the orbit. The center of the moon will be at one of the foci of the orbit. The *Apollo 11* spacecraft was placed in a lunar orbit with perilune at 68 mi and apolune at 195 mi above the surface of the moon. Assuming the moon is a sphere with a radius of 1074 mi, find an equation for the orbit of *Apollo 11*.

32. If $k > 0$, the equation

$$\frac{x^2}{k} + \frac{y^2}{4 + k} = 1$$

represents an ellipse. Show that all the ellipses represented by this equation have the same foci, no matter what the value of k is.

33. A "sunburst" window above a doorway is in the shape of the top half of an ellipse. The window is 20 in. tall at its highest point and 80 in. wide at the bottom. How tall is the window at a point 25 in. from the center of the base?

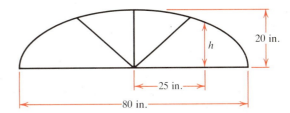

34. The **ancillary circle** of an ellipse is the circle with radius equal to half the length of the minor axis and with the same center as the ellipse (see the figure). Note that the ancillary circle is the largest circle that can fit inside an ellipse.

(a) Find an equation for the ancillary circle of the ellipse

$$x^2 + 4y^2 = 16$$

(b) Show that if (s, t) is a point on the ancillary circle in part (a), then $(2s, t)$ is a point on the ellipse.

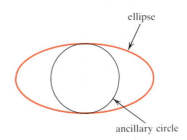

ellipse

ancillary circle

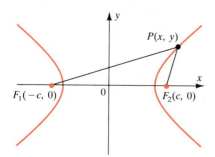

Figure 1

P is on the hyperbola if
$\||PF_1| - |PF_2|\| = 2a$

Hyperbolas

Although ellipses and hyperbolas have completely different shapes, their definitions and equations are similar. Instead of using a *sum* of distances from two fixed foci, as in the case of an ellipse, we use the *difference* to define a hyperbola.

> A **hyperbola** is the set of all points in the plane the difference of whose distances from two fixed points F_1 and F_2 is a constant (see Figure 1). The points F_1 and F_2 are called the **foci** of the hyperbola.

As in the case of the ellipse, we get the simplest equation for the hyperbola by placing the foci on the x-axis at $(\pm c, 0)$, as shown in Figure 1. From the definition, if $P(x, y)$ lies on the hyperbola, then either $|PF_1| - |PF_2|$ or $|PF_2| - |PF_1|$ must equal some positive constant, which we call $2a$. Thus we have

$$|PF_1| - |PF_2| = \pm 2a$$

or

$$\sqrt{(x + c)^2 + y^2} - \sqrt{(x - c)^2 + y^2} = \pm 2a$$

Proceeding as we did in the case of the ellipse in Section 8.2, we simplify this to

(1)
$$(c^2 - a^2)x^2 - a^2 y^2 = a^2(c^2 - a^2)$$

If we consider the triangle PF_1F_2 in Figure 1, we see that

$$\||PF_1| - |PF_2|\| < 2c$$

so

$$2a < 2c \quad \text{or} \quad a < c$$

Thus $c^2 - a^2 > 0$, so we can write $b^2 = c^2 - a^2$. We then simplify Equation 1 to

(2)
$$\frac{x^2}{a^2} - \frac{y^2}{b^2} = 1$$

If we replace x by $-x$ or y by $-y$ in Equation 2, the equation remains unchanged, so the hyperbola is symmetric about both the x- and y-axes and about the origin. The x-intercepts are $\pm a$, and the points $(a, 0)$ and $(-a, 0)$ are the **vertices** of the hyperbola. But setting $x = 0$ in Equation 2 leads to $-y^2 = b^2$, which is impossible, so there is no y-intercept. Furthermore, from Equation 2 we obtain

$$\frac{x^2}{a^2} = \frac{y^2}{b^2} + 1 \geq 1$$

so $x^2 \geq a^2$ and hence $x \geq a$ or $x \leq -a$. This means that the hyperbola consists of two parts, called its **branches**. The segment that joins the two vertices of the branches is called the **transverse axis** of the hyperbola, and the origin is called its **center**.

As a further guide to graphing the hyperbola, we solve for y in Equation 2:

$$y = \pm \frac{b}{a} \sqrt{x^2 - a^2}$$

Considering only the portion of the hyperbola in the first quadrant (so that $x > 0$ and $y \geq 0$), we have

$$y = \frac{b}{a} \sqrt{x^2 - a^2}$$

(3)
$$y = \frac{b}{a} x \sqrt{1 - \frac{a^2}{x^2}}$$

As x becomes larger and larger, a^2/x^2 becomes smaller and smaller, so that $\sqrt{1 - a^2/x^2}$ approaches $\sqrt{1} = 1$ in value. Thus from Equation 3 we see that as x gets larger, the line $y = (b/a)x$ and the hyperbola get closer together. By symmetry, the same sort of situation exists in the other quadrants, and we see that the hyperbola approaches the lines

$$y = \pm \frac{b}{a} x$$

These lines are called the **asymptotes** of the hyperbola and are a useful guide in graphing it. (Note that asymptotes were discussed for rational functions in Section 3.7.)

A convenient way to locate the asymptotes and graph the hyperbola is to first plot the vertices $(a, 0)$ and $(-a, 0)$ and the points $(0, b)$ and $(0, -b)$. Then draw horizontal and vertical segments through these points to construct a rectangle as in Figure 2. We call this rectangle the **central box** of the hyperbola. The slopes of the diagonals of the central box are $\pm b/a$, so by extending them we obtain the asymptotes $y = \pm(b/a)x$. Finally, we use the asymptotes as a guide in sketching the hyperbola.

We summarize the main results of our discussion as follows:

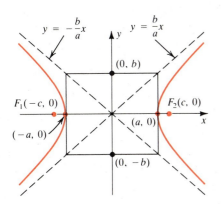

Figure 2

Hyperbola $\dfrac{x^2}{a^2} - \dfrac{y^2}{b^2} = 1$ with

asymptotes $y = \pm \dfrac{b}{a} x$

The hyperbola

(4)
$$\frac{x^2}{a^2} - \frac{y^2}{b^2} = 1$$

has vertices $(\pm a, 0)$, asymptotes $y = \pm(b/a)x$, and foci $(\pm c, 0)$, where $c^2 = a^2 + b^2$. (See Figure 2.)

Example 1

Find the foci, vertices, and asymptotes of the hyperbola

$$9x^2 - 16y^2 = 144$$

and sketch the graph.

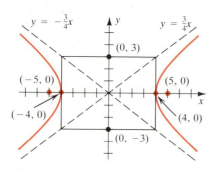

Figure 3

$9x^2 - 16y^2 = 144$

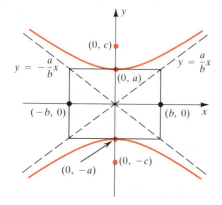

Figure 4

Hyperbola $\dfrac{y^2}{a^2} - \dfrac{x^2}{b^2} = 1$ with

asymptotes $y = \pm\dfrac{a}{b}x$

Figure 5

$\dfrac{y^2}{4} - x^2 = 1$

Solution

First we divide both sides of the equation by 144 to put it into the form of Equation 4:

$$\frac{x^2}{16} - \frac{y^2}{9} = 1$$

This means that $a = 4$, $b = 3$, and $c = \sqrt{16 + 9} = 5$. The vertices are $(\pm 4, 0)$, the foci are $(\pm 5, 0)$, and the asymptotes are $y = \pm(3/4)x$. After drawing the central box and asymptotes, we complete the sketch of the hyperbola in Figure 3.

●

Placing the foci on the y-axis rather than on the x-axis has the effect of reversing the roles of x and y in the derivation of Equation 4. This leads to the following information, which is illustrated in Figure 4:

The hyperbola

(5) $$\frac{y^2}{a^2} - \frac{x^2}{b^2} = 1$$

has vertices $(0, \pm a)$, asymptotes $y = \pm(a/b)x$, and foci $(0, \pm c)$, where $c^2 = a^2 + b^2$.

Example 2

Find the foci and equation of the hyperbola with vertices $(0, \pm 2)$ and asymptotes $y = \pm 2x$. Sketch the graph.

Solution

Since the foci are on the y-axis, the equation will take the form of Equation 5, with $a = 2$. From the asymptote equation, we see that $a/b = 2$. Thus $b = a/2 = 1$,

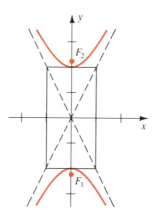

and $c^2 = a^2 + b^2 = 2^2 + 1^2 = 5$. The foci are $(0, \pm\sqrt{5})$ and the equation of the hyperbola is

$$\frac{y^2}{4} - x^2 = 1$$

The graph is shown in Figure 5. ●

Example 3

Find the foci, vertices, and asymptotes of the hyperbola

$$x^2 - 9y^2 + 9 = 0$$

and sketch the graph.

Solution

We begin by writing the equation in the standard form for hyperbolas:

$$x^2 - 9y^2 = -9 \qquad \text{or} \qquad y^2 - \frac{x^2}{9} = 1$$

This is in the form of Equation 5, so the hyperbola has its foci and vertices on the y-axis. We have $a^2 = 1$ and $b^2 = 9$, so $c = \sqrt{a^2 + b^2} = \sqrt{10}$. Thus the foci are $(0, \pm\sqrt{10})$ and the vertices are $(0, \pm 1)$. Since $a = 1$ and $b = 3$, the asymptotes are $y = \pm(1/3)x$. The graph is shown in Figure 6. ●

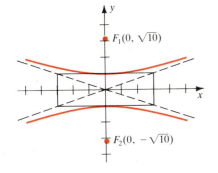

Figure 6
$x^2 - 9y^2 + 9 = 0$

In the LORAN (LOng RAnge Navigation) system, hyperbolas are used on board ships and aircraft to determine their location. In Figure 7 radio stations at A and B transmit simultaneous signals, which are received by the ship at P. The onboard computer converts the time difference in receiving these signals into a distance difference $|PA| - |PB|$. By the definition of a hyperbola, this locates the ship on one branch of a hyperbola with foci at A and B (sketched in black in the

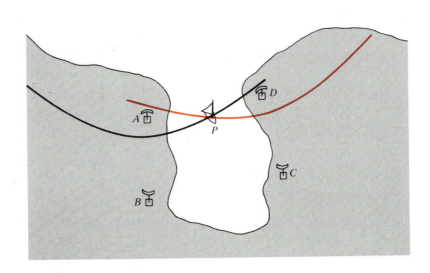

Figure 7

figure). The same procedure is carried out with two other radio stations at C and D, and this locates the ship on a second hyperbola (in red in Figure 7). The coordinates of the intersection point of these two hyperbolas, which can be calculated precisely by the computer, give the location of P.

Exercises 8.3

In Exercises 1–12 find the vertices, foci, and asymptotes of the given hyperbola and sketch the graph.

1. $\dfrac{x^2}{4} - \dfrac{y^2}{16} = 1$ **2.** $\dfrac{y^2}{9} - \dfrac{x^2}{16} = 1$

3. $y^2 - \dfrac{x^2}{25} = 1$ **4.** $\dfrac{x^2}{2} - y^2 = 1$

5. $x^2 - y^2 = 1$ **6.** $9x^2 - 4y^2 = 36$

7. $25y^2 - 9x^2 = 225$ **8.** $x^2 - y^2 + 4 = 0$

9. $x^2 - 4y^2 - 8 = 0$ **10.** $x^2 - 2y^2 = 3$

11. $4y^2 - x^2 = 1$ **12.** $9x^2 - 16y^2 = 1$

In Exercises 13–24 find an equation for the hyperbola that satisfies the given conditions.

13. Foci $(\pm 5, 0)$, vertices $(\pm 3, 0)$

14. Foci $(0, \pm 10)$, vertices $(0, \pm 8)$

15. Foci $(0, \pm 2)$, vertices $(0, \pm 1)$

16. Foci $(\pm 6, 0)$, vertices $(\pm 2, 0)$

17. Vertices $(\pm 1, 0)$, asymptotes $y = \pm 5x$

18. Vertices $(0, \pm 6)$, asymptotes $y = \pm \frac{1}{3}x$

19. Foci $(0, \pm 8)$, asymptotes $y = \pm \frac{1}{2}x$

20. Vertices $(0, \pm 6)$, hyperbola passes through $(-5, 9)$

21. Asymptotes $y = \pm x$, hyperbola passes through $(5, 3)$

22. Foci $(\pm 3, 0)$, hyperbola passes through $(4, 1)$

23. Foci $(\pm 5, 0)$, length of transverse axis 6

24. Foci $(0, \pm 1)$, length of transverse axis 1

25. (a) Show that the asymptotes of the hyperbola $x^2 - y^2 = 5$ are perpendicular to each other.

(b) Find the equation for a hyperbola with foci $(\pm c, 0)$ and with asymptotes perpendicular to each other.

26. The hyperbolas

$$\frac{x^2}{a^2} - \frac{y^2}{b^2} = 1 \quad \text{and} \quad \frac{x^2}{a^2} - \frac{y^2}{b^2} = -1$$

are said to be **conjugate** to each other.
(a) Show that the hyperbolas

$$x^2 - 4y^2 + 16 = 0 \quad \text{and} \quad 4y^2 - x^2 + 16 = 0$$

are conjugate to each other, and graph them on the same coordinate axes.
(b) What do the hyperbolas of part (a) have in common?
(c) Show that any pair of conjugate hyperbolas have the relationship you discovered in part (b).

27. Derive Equation 1 in the text from the equation that precedes it.

28. (a) For the hyperbola

$$\frac{x^2}{9} - \frac{y^2}{16} = 1$$

determine the values of a, b, and c and find the coordinates of the foci F_1 and F_2.
(b) Show that the point $P(5, \frac{16}{3})$ lies on this hyperbola.
(c) Find $|PF_1|$ and $|PF_2|$, using the point P of part (b).
(d) Verify that the difference between $|PF_1|$ and $|PF_2|$ is $2a$.

29. Refer to Figure 7 in the text. Suppose that the radio stations at A and B are 500 mi apart and that the ship at P receives the signal from A 2640 microseconds (μs) before it receives the signal from B.
(a) Assuming that radio signals travel at 980 ft/μs, find $|PA| - |PB|$.
(b) Find an equation for the branch of the hyperbola indicated in black in the figure. (Place A and B on

the y-axis with the origin halfway between them. Use miles as the unit of distance.)

(c) If A is due north of B and P is due east of A, how far is P from A?

30. Some comets, such as Halley's comet, are a permanent part of the solar system, traveling in elliptical orbits around the sun. Others pass through the solar system only once, following a hyperbolic path with the sun at a focus. The figure shows the path of such a comet. Find an equation for the path, assuming that the closest the comet gets to the sun is 2×10^9 mi and that the path the comet was taking before it neared the solar system is at right angles to the path it continues on after leaving the solar system.

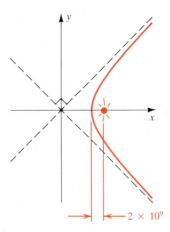

$\leftarrow 2 \times 10^9$

Shifted Conics

In the preceding sections we studied parabolas with vertices at the origin, and ellipses and hyperbolas with centers at the origin. We restricted ourselves to these cases because the equations then have the simplest forms. In this section we consider conics whose vertices and centers are not necessarily at the origin, and we determine how this affects their equations.

If we replace x by $x - h$ in any equation, the graph of the new equation is simply the old graph shifted to the right by h units if h is positive, or to the left by $-h$ units if h is negative. Similarly, the effect of replacing y by $y - k$ is to shift the graph upward by k units (or downward by $-k$ units if k is negative). In Section 2.3 we applied these principles to the graphs of functions. Here we apply them to the conics.

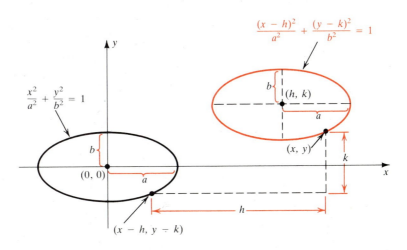

Figure 1
Shifted ellipse

For example, if we take the ellipse with equation

$$\frac{x^2}{a^2} + \frac{y^2}{b^2} = 1$$

and shift it so that its center is at the point (h, k) instead of at the origin, then its equation becomes

$$\frac{(x - h)^2}{a^2} + \frac{(y - k)^2}{b^2} = 1$$

(See Figure 1.)

Example 1

Sketch the graph of the ellipse

(1)
$$\frac{(x + 1)^2}{4} + \frac{(y - 2)^2}{9} = 1$$

and determine the coordinates of the foci.

Solution

From the preceding discussion, we see that this ellipse has the same shape as the ellipse

(2)
$$\frac{x^2}{4} + \frac{y^2}{9} = 1$$

but that its center is at the point $(-1, 2)$ instead of at the origin. In fact, the graph of Equation 1 is the graph of Equation 2 shifted to the left by one unit and upward by two units.

Since $a^2 = 9$ and $b^2 = 4$, we have $c^2 = 9 - 4 = 5$, so $c = \sqrt{5}$. Thus the foci of the ellipse of Equation 2 are $(0, \pm\sqrt{5})$. This means we obtain the foci of the ellipse of Equation 1 by shifting these points one unit to the left and two units upward to get

$$(-1, 2 + \sqrt{5}) \qquad \text{and} \qquad (-1, 2 - \sqrt{5})$$

The same shifts are used to find the location of the vertices as shown in Figure 2.

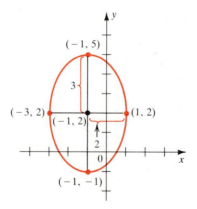

Figure 2

$$\frac{(x + 1)^2}{4} + \frac{(y - 2)^2}{9} = 1$$

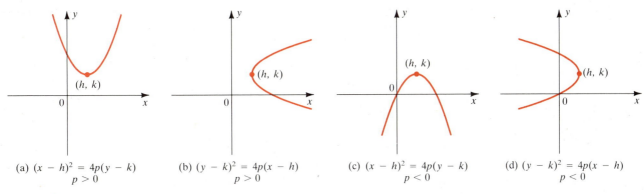

(a) $(x - h)^2 = 4p(y - k)$
$p > 0$

(b) $(y - k)^2 = 4p(x - h)$
$p > 0$

(c) $(x - h)^2 = 4p(y - k)$
$p < 0$

(d) $(y - k)^2 = 4p(x - h)$
$p < 0$

Figure 3
Shifted parabolas

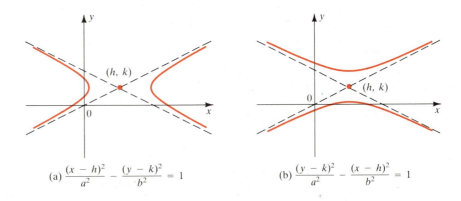

(a) $\dfrac{(x - h)^2}{a^2} - \dfrac{(y - k)^2}{b^2} = 1$

(b) $\dfrac{(y - k)^2}{a^2} - \dfrac{(x - h)^2}{b^2} = 1$

Figure 4
Shifted hyperbolas

This shifting technique can be applied to parabolas and hyperbolas as well. The results are summarized in Figures 3 and 4.

Example 2

Determine the vertex, focus, and directrix and sketch the graph of the parabola

$$x^2 - 4x = 8y - 28$$

Solution

We complete the square in x to put this equation into one of the forms in Figure 3.

$$x^2 - 4x + 4 = 8y - 28 + 4$$
$$(x - 2)^2 = 8y - 24$$
$$(x - 2)^2 = 8(y - 3)$$

This is a parabola that opens upward, with vertex at $(2, 3)$. Since $4p = 8$, we have $p = 2$, so the focus is two units above the vertex and the directrix is two units below the vertex. Thus the focus is $(2, 5)$ and the directrix is $y = 1$. The graph is shown in Figure 5.

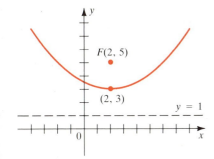

Figure 5
$x^2 - 4x = 8y - 28$

Example 3

Show that the equation

$$9x^2 - 72x - 16y^2 - 32y = 16$$

represents a hyperbola. Find its center, vertices, foci, and asymptotes and sketch its graph.

Solution

We first complete the square in both x and y:

$$9(x^2 - 8x) - 16(y^2 + 2y) = 16$$
$$9(x^2 - 8x + 16) - 16(y^2 + 2y + 1) = 16 + 9 \cdot 16 - 16 \cdot 1$$
$$9(x - 4)^2 - 16(y + 1)^2 = 144$$

Now we divide both sides of the equation by 144 to get

(3)
$$\frac{(x - 4)^2}{16} - \frac{(y + 1)^2}{9} = 1$$

This is a hyperbola with center at $(4, -1)$ and with a horizontal transverse axis. Its graph will have the same shape as the unshifted hyperbola

(4)
$$\frac{x^2}{16} - \frac{y^2}{9} = 1$$

Since $a^2 = 16$ and $b^2 = 9$, we have $a = 4$, $b = 3$, and $c = \sqrt{a^2 + b^2} = \sqrt{16 + 9} = 5$. Thus the foci lie five units to the left and to the right of the center, and the vertices lie four units on either side of the center.

$$\text{foci:} \quad (9, -1) \text{ and } (-1, -1)$$
$$\text{vertices:} \quad (8, -1) \text{ and } (0, -1)$$

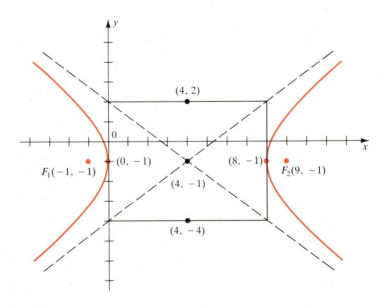

Figure 6
$9x^2 - 72x - 16y^2 - 32y = 16$

The asymptotes of the unshifted hyperbola of Equation 4 are $y = \pm(3/4)x$, so the asymptotes of the hyperbola of Equation 3 are

$$y + 1 = \pm\tfrac{3}{4}(x - 4)$$
$$y + 1 = \pm\tfrac{3}{4}x \mp 3$$
$$y = \tfrac{3}{4}x - 4 \quad \text{and} \quad y = -\tfrac{3}{4}x + 2$$

The graph is shown in Figure 6.   ●

If we multiply out and simplify the equations of any of the shifted conics illustrated in Figures 1, 3, and 4, we will always obtain an equation of the form

(5) $$Ax^2 + Cy^2 + Dx + Ey + F = 0$$

where A and C are not both 0. Conversely, if we begin with an equation of this form, we can complete the square in x and y to see which type of conic section the equation represents. In some cases, called the **degenerate cases**, the graph of the equation turns out to be just a pair of lines or a single point, or there may be no graph at all. The next example illustrates such a case.

Example 4

Sketch the graph of the equation

$$9x^2 - y^2 + 18x + 6y = 0$$

Solution

Because the coefficients of x^2 and y^2 are of opposite sign, this equation looks as if it should represent a hyperbola, like the equation of Example 3. To see if this is in fact the case, we complete the square:

$$9(x^2 + 2x) - (y^2 - 6y) = 0$$
$$9(x^2 + 2x + 1) - (y^2 - 6y + 9) = 9 - 9 = 0$$
$$9(x + 1)^2 - (y - 3)^2 = 0$$

For this to fit the form of the equation of a hyperbola, we would need a nonzero constant to the right of the equal sign. In fact, further analysis shows that this is the equation of a pair of intersecting lines:

$$(y - 3)^2 = 9(x + 1)^2$$
$$y - 3 = \pm 3(x + 1)$$
$$y = 3x + 6 \quad \text{or} \quad y = -3x$$

These lines are graphed in Figure 7.   ●

Because the equation in Example 4 looked at first glance like the equation of a hyperbola but in fact turned out to represent simply a pair of lines, we refer to its graph as a **degenerate hyperbola**. Degenerate ellipses and parabolas can also arise when we complete the square in an equation of the form of Equation 5. For example, the equation

$$4x^2 + y^2 - 8x + 2y + 6 = 0$$

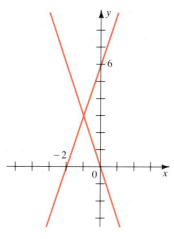

Figure 7
$9x^2 - y^2 + 18x + 6y = 0$

looks as if it should represent an ellipse because the coefficients of x^2 and y^2 are of the same sign. But completing the square leads to

$$4(x - 1)^2 + (y + 1)^2 = -1$$

which has no solutions at all (since the sum of two squares cannot be negative). This is therefore degenerate.

To sum up, we have the following theorem:

The graph of the equation

$$Ax^2 + Cy^2 + Dx + Ey + F = 0$$

where not both A and C are 0, is a conic or a degenerate conic. In the nondegenerate cases, the graph is:

1.   a parabola if A or C is 0
2.   an ellipse if A and C are of the same sign (or a circle if $A = C$)
3.   a hyperbola if A and C are of opposite sign

Exercises 8.4

In Exercises 1–4 find the center, foci, and vertices of the given ellipse, and determine the lengths of the major and minor axes. Then sketch the graph.

1. $\dfrac{(x - 2)^2}{9} + \dfrac{(y - 1)^2}{4} = 1$

2. $\dfrac{(x - 3)^2}{16} + (y + 3)^2 = 1$

3. $\dfrac{x^2}{9} + \dfrac{(y + 5)^2}{25} = 1$ **4.** $\dfrac{(x + 2)^2}{4} + y^2 = 1$

In Exercises 5–8 find the vertex, focus, and directrix of the given parabola. Then sketch the graph.

5. $(x - 3)^2 = 8(y + 1)$ **6.** $(y + 5)^2 = -6x + 12$

7. $-4(x + \frac{1}{2})^2 = y$ **8.** $y^2 = 16x - 8$

In Exercises 9–12 find the center, foci, vertices, and asymptotes of the given hyperbola. Then sketch the graph.

9. $\dfrac{(x + 1)^2}{9} - \dfrac{(y - 3)^2}{16} = 1$

10. $(x - 8)^2 - (y + 6)^2 = 1$

11. $y^2 - \dfrac{(x + 1)^2}{4} = 1$

12. $\dfrac{(y - 1)^2}{25} - (x + 3)^2 = 1$

In Exercises 13–24 complete the square to determine whether the equation represents an ellipse, a parabola, a hyperbola, or a degenerate conic, and sketch a graph of the equation. If the graph is an ellipse, find the center, foci, vertices, and lengths of the major and minor axes. If it is a parabola, find the vertex, focus, and directrix. If it is a hyperbola, find the center, foci, vertices, and asymptotes. If there is no graph, explain why.

13. $9x^2 - 36x + 4y^2 = 0$ **14.** $y^2 = 4(x + 2y)$

15. $x^2 - 4y^2 - 2x + 16y = 20$

16. $x^2 + 6x + 12y + 9 = 0$

17. $4x^2 + 25y^2 - 24x + 250y + 561 = 0$

18. $2x^2 + y^2 = 2y + 1$

19. $16x^2 - 9y^2 - 96x + 288 = 0$

20. $4x^2 - 4x - 8y + 9 = 0$

21. $x^2 + 16 = 4(y^2 + 2x)$

22. $x^2 - y^2 = 10(x - y) + 1$

23. $3x^2 + 4y^2 - 6x - 24y + 39 = 0$

24. $x^2 + 4y^2 + 20x - 40y + 300 = 0$

25. What must the value of F be if the graph of the equation

$$4x^2 + y^2 + 4(x - 2y) + F = 0$$

is **(a)** an ellipse? **(b)** a single point? **(c)** the empty set?

26. Find an equation for the ellipse that shares a vertex and a focus with the parabola $x^2 + y = 100$ and that has its other focus at the origin.

Rotation of Axes

In Section 8.4 we studied conics with equations of the form

$$Ax^2 + Cy^2 + Dx + Ey + F = 0$$

We saw that the graph is always an ellipse, parabola, or hyperbola with horizontal or vertical axes (except in the degenerate cases). In this section we study the most general second-degree equation

(1) $$Ax^2 + Bxy + Cy^2 + Dx + Ey + F = 0$$

We will see that by rotating the coordinate axes through an appropriate angle, we can eliminate the term Bxy and then use our knowledge of conic sections to analyze the graph.

In Figure 1 the x- and y-axes have been rotated through an acute angle ϕ about the origin to produce a new pair of axes, which we call the X- and Y-axes. A point P that has coordinates (x, y) in the old system has coordinates (X, Y) in the new system. If we let r denote the distance of P from the origin and let θ be the angle that the segment OP makes with the new X-axis, then we can see from Figure 2 (by considering the two right triangles in the figure) that

$$X = r\cos\theta \qquad Y = r\sin\theta$$
$$x = r\cos(\theta + \phi) \qquad y = r\sin(\theta + \phi)$$

Using the addition formulas for cosine, we see that

$$\begin{aligned} x &= r\cos(\theta + \phi) \\ &= r(\cos\theta\cos\phi - \sin\theta\sin\phi) \\ &= (r\cos\theta)\cos\phi - (r\sin\theta)\sin\phi \\ &= X\cos\phi - Y\sin\phi \end{aligned}$$

Similarly, we can apply the addition formula for sine to the expression for y, and so we have the following relations between the xy- and the XY-coordinate

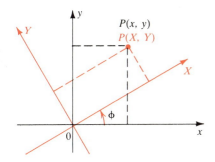

Figure 1

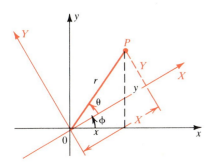

Figure 2

systems:

(2)

$$
\begin{aligned}
x &= X \cos \phi - Y \sin \phi \\
y &= X \sin \phi + Y \cos \phi
\end{aligned}
$$

By solving this system of equations for X and Y in terms of x and y (see Exercise 25), we obtain the following:

(3)

$$
\begin{aligned}
X &= x \cos \phi + y \sin \phi \\
Y &= -x \sin \phi + y \cos \phi
\end{aligned}
$$

Example 1

If the axes are rotated through $30°$, find the XY-coordinates of the point with xy-coordinates $(2, -4)$.

Solution

Using Equations 3 with $x = 2$, $y = -4$, and $\phi = 30°$, we get

$$
X = 2 \cos 30° + (-4)\sin 30° = 2\left(\frac{\sqrt{3}}{2}\right) - 4\left(\frac{1}{2}\right) = \sqrt{3} - 2
$$

$$
Y = -2 \sin 30° + (-4)\cos 30° = -2\left(\frac{1}{2}\right) - 4\left(\frac{\sqrt{3}}{2}\right) = -1 - 2\sqrt{3}
$$

The XY-coordinates are $(-2 + \sqrt{3}, -1 - 2\sqrt{3})$.   ●

Example 2

Show by rotating the axes through $45°$ that the graph of the equation $xy = 2$ is a hyperbola.

Solution

We use Equations 2 with $\phi = 45°$ to obtain

$$
x = X \cos 45° - Y \sin 45° = \frac{X}{\sqrt{2}} - \frac{Y}{\sqrt{2}}
$$

$$
y = X \sin 45° + Y \cos 45° = \frac{X}{\sqrt{2}} + \frac{Y}{\sqrt{2}}
$$

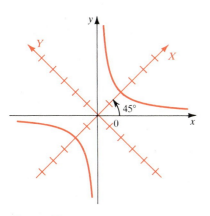

Figure 3
$xy = 2$

Substituting these expressions into the original equation gives

$$\left(\frac{X}{\sqrt{2}} - \frac{Y}{\sqrt{2}}\right)\left(\frac{X}{\sqrt{2}} + \frac{Y}{\sqrt{2}}\right) = 2$$

$$\frac{(X - Y)(X + Y)}{(\sqrt{2})(\sqrt{2})} = 2$$

$$\frac{X^2}{4} - \frac{Y^2}{4} = 1$$

We recognize this as a hyperbola with vertices $(\pm 2, 0)$ in the XY-coordinate system. Its asymptotes are $Y = \pm X$, which correspond to the coordinate axes in the xy-system (see Figure 3).   ●

The method illustrated in Example 2 can be used in any equation of the form of Equation 1 to eliminate the term that involves the product xy. We can always choose an angle ϕ so that when the coordinate axes are rotated through this angle, the equation in the new coordinate system does not involve XY. In Example 2 the angle ϕ was given to us, but to see what angle works in any given situation, we perform the following analysis.

After rotating the coordinate axes through an angle ϕ, we substitute the values of x and y from Equations 2 into Equation 1 to see what it becomes in XY-coordinates:

$$A(X \cos\phi - Y \sin\phi)^2 + B(X \cos\phi - Y \sin\phi)(X \sin\phi + Y \cos\phi)$$
$$+ C(X \sin\phi + Y \cos\phi)^2 + D(X \cos\phi - Y \sin\phi)$$
$$+ E(X \sin\phi + Y \cos\phi) + F = 0$$

If we multiply this out and collect like terms, we obtain an equation of the form

(4) $$A'X^2 + B'XY + C'Y^2 + D'X + E'Y + F' = 0$$

where the coefficient B' of XY is

$$B' = 2(C - A)\sin\phi \cos\phi + B(\cos^2\phi - \sin^2\phi)$$

(See Exercise 26.) To eliminate the XY term we would like to choose ϕ so that $B' = 0$; that is,

$$2(C - A)\sin\phi \cos\phi + B(\cos^2\phi - \sin^2\phi) = 0$$
$$B(\cos^2\phi - \sin^2\phi) = 2(A - C)\sin\phi \cos\phi$$
$$\frac{\cos^2\phi - \sin^2\phi}{2\sin\phi \cos\phi} = \frac{A - C}{B}$$

Using the double-angle formulas for cosine and sine, we can rewrite this as

$$\frac{\cos 2\phi}{\sin 2\phi} = \frac{A - C}{B}$$

or

(5)

$$\cot 2\phi = \frac{A - C}{B}$$

Example 3

Identify and sketch the curve

$$6\sqrt{3}x^2 + 6xy + 4\sqrt{3}y^2 = 21\sqrt{3}$$

Solution

To eliminate the xy term, we rotate the axes through an angle ϕ that satisfies Equation 5.

$$\cot 2\phi = \frac{A - C}{B} = \frac{6\sqrt{3} - 4\sqrt{3}}{6} = \frac{1}{\sqrt{3}}$$

Thus $2\phi = 60°$ and hence $\phi = 30°$. With this value of ϕ, Equations 2 become

$$x = X\left(\frac{\sqrt{3}}{2}\right) - Y\left(\frac{1}{2}\right)$$

$$y = X\left(\frac{1}{2}\right) + Y\left(\frac{\sqrt{3}}{2}\right)$$

Substituting these values for x and y into the given equation leads to

$$6\sqrt{3}\left(\frac{X\sqrt{3}}{2} - \frac{Y}{2}\right)^2 + 6\left(\frac{X\sqrt{3}}{2} - \frac{Y}{2}\right)\left(\frac{X}{2} + \frac{Y\sqrt{3}}{2}\right) + 4\sqrt{3}\left(\frac{X}{2} + \frac{Y\sqrt{3}}{2}\right)^2 = 21\sqrt{3}$$

Multiplying this out and simplifying, we get

$$7\sqrt{3}X^2 + 3\sqrt{3}Y^2 = 21\sqrt{3}$$

or, after dividing through by $21\sqrt{3}$,

$$\frac{X^2}{3} + \frac{Y^2}{7} = 1$$

This is the equation of an ellipse in the XY-coordinate system. The foci lie on the Y-axis. Because $a^2 = 7$ and $b^2 = 3$, the length of the major axis is $2\sqrt{7}$ and the length of the minor axis is $2\sqrt{3}$. The ellipse is sketched in Figure 4. ●

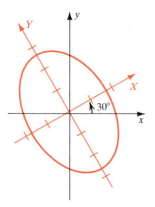

Figure 4
$6\sqrt{3}x^2 + 6xy + 4\sqrt{3}y^2 = 21\sqrt{3}$

In the preceding example we were able to determine ϕ without difficulty, since we remembered that $\cot 60° = 1/\sqrt{3}$. In general, finding ϕ is not quite so easy. As the next example illustrates, the following half-angle formulas, which are valid for $0 < \phi < \pi/2$, turn out to be useful in determining ϕ (see Section 6.4):

$$\cos\phi = \sqrt{\frac{1 + \cos 2\phi}{2}} \qquad \sin\phi = \sqrt{\frac{1 - \cos 2\phi}{2}}$$

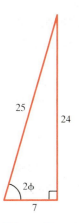

Figure 5

Example 4

Identify and sketch the curve

$$64x^2 + 96xy + 36y^2 - 15x + 20y - 25 = 0$$

Solution

This is in the form of Equation 1, with $A = 64$, $B = 96$, and $C = 36$. Thus

$$\cot 2\phi = \frac{A - C}{B} = \frac{64 - 36}{96} = \frac{7}{24}$$

In Figure 5 we sketch a triangle with $\cot 2\phi = 7/24$. We see that

$$\cos 2\phi = \frac{7}{25}$$

so, using the half-angle formulas, we get

$$\cos \phi = \sqrt{\frac{1 + \frac{7}{25}}{2}} = \sqrt{\frac{16}{25}} = \frac{4}{5}$$

and

$$\sin \phi = \sqrt{\frac{1 - \frac{7}{25}}{2}} = \sqrt{\frac{9}{25}} = \frac{3}{5}$$

The rotation Equations 2 then become

$$x = \tfrac{4}{5}X - \tfrac{3}{5}Y \qquad \text{and} \qquad y = \tfrac{3}{5}X + \tfrac{4}{5}Y$$

Substituting into the given equation, we have

$$64(\tfrac{4}{5}X - \tfrac{3}{5}Y)^2 + 96(\tfrac{4}{5}X - \tfrac{3}{5}Y)(\tfrac{3}{5}X + \tfrac{4}{5}Y)$$
$$+ 36(\tfrac{3}{5}X + \tfrac{4}{5}Y)^2 - 15(\tfrac{4}{5}X - \tfrac{3}{5}Y) + 20(\tfrac{3}{5}X + \tfrac{4}{5}Y) - 25 = 0$$

Multiplying out and collecting common terms, we get

$$100X^2 + 25Y - 25 = 0$$

which simplifies to

$$-4X^2 = Y - 1$$

We recognize this as being a parabola that opens along the negative Y-axis, with vertex $(0, 1)$ in XY-coordinates. Since $4p = -4$, we have $p = -1$, so the focus is $(0, 0)$ and the directrix is $Y = 2$. Using

$$\phi = \cos^{-1}(\tfrac{4}{5}) \approx 37°$$

we sketch the graph in Figure 6.

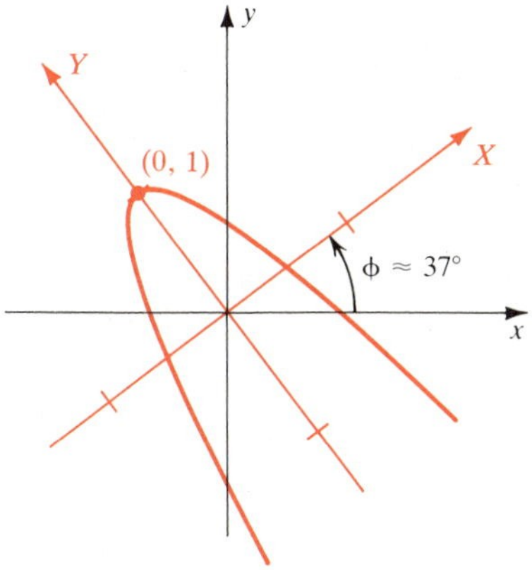

Figure 6
$64x^2 + 96xy + 36y^2 - 15x + 20y - 25 = 0$

Exercises 8.5

In Exercises 1–6 determine the XY-coordinates of the given point if the coordinate axes are rotated through the indicated angle.

1. $(1, 1)$, $\phi = 45°$

2. $(-2, 1)$, $\phi = 30°$

3. $(3, -\sqrt{3})$, $\phi = 60°$

4. $(2, 0)$, $\phi = 15°$

5. $(0, 2)$, $\phi = 55°$

6. $(\sqrt{2}, 4\sqrt{2})$, $\phi = 45°$

In Exercises 7 and 8 determine the equation of the given conic in XY-coordinates when the coordinate axes are rotated through the indicated angle.

7. $y = (x - 1)^2, \phi = 45°$

8. $x^2 - y^2 = 2y, \phi = \cos^{-1}(\frac{3}{5})$

Use rotation of axes to identify and sketch the graph of each of the equations in Exercises 9–22.

9. $xy = 8$

10. $xy + 4 = 0$

11. $x^2 + 2xy + y^2 + x - y = 0$

12. $13x^2 + 6\sqrt{3}xy + 7y^2 = 16$

13. $x^2 + 2\sqrt{3}xy - y^2 + 2 = 0$

14. $21x^2 + 10\sqrt{3}xy + 31y^2 = 144$

15. $11x^2 - 24xy + 4y^2 + 20 = 0$

16. $25x^2 - 120xy + 144y^2 - 156x - 65y = 0$

17. $\sqrt{3}x^2 + 3xy = 3$

18. $153x^2 + 192xy + 97y^2 = 225$

19. $2\sqrt{3}x^2 - 6xy + \sqrt{3}x + 3y = 0$

20. $9x^2 - 24xy + 16y^2 = 100(x - y - 1)$

21. $52x^2 + 72xy + 73y^2 = 40x - 30y + 75$

22. $(7x + 24y)^2 = 600x - 175y + 25$

23. **(a)** Use rotation of axes to show that the equation

$$7x^2 + 48xy - 7y^2 - 200x - 150y + 600 = 0$$

represents a hyperbola.
(b) Find the XY- and xy-coordinates of the center, vertices, and foci.
(c) Find the equations of the asymptotes in XY- and xy-coordinates.

24. **(a)** Use rotation of axes to show that the equation

$$2\sqrt{2}(x + y)^2 = 7x + 9y$$

represents a parabola.

(b) Find the XY- and xy-coordinates of the vertex and focus.
(c) Find the equation of the directrix in XY- and xy-coordinates.

25. Solve Equations 2 in the text for X and Y in terms of x and y. (*Hint:* Begin by multiplying the first equation by $\cos \theta$ and the second by $\sin \theta$, and then add the two equations.)

26. Suppose that a rotation through the angle ϕ changes Equation 1 to Equation 4 (in the text). Show the following:
(a) $B' = 2(C - A)\sin \phi \cos \phi + B(\cos^2\phi - \sin^2\phi)$
(b) $A + C = A' + C'$

27. For an equation of the form of Equation 1, the quantity

$$B^2 - 4AC$$

is called the **discriminant** of the equation. Suppose that a rotation through the angle ϕ changes Equation 1 to Equation 4 (in the text). Show that Equation 1 and Equation 4 have the same discriminant.

28. Use Exercise 27 to show that, except in the degenerate cases, Equation 1 represents an ellipse (or a circle) if the discriminant is negative, a parabola if the discriminant is zero, and a hyperbola if the discriminant is positive.

29. Use Exercise 28 to determine the type of curve in Exercises 9–22.

30. Show that the graph of the equation

$$\sqrt{x} + \sqrt{y} = 1$$

is a part of a parabola by rotating the axes through an angle of 45°. (*Hint:* First convert the equation to one that does not involve radicals.)

31. Let Z, Z', and R be the matrices

$$Z = \begin{bmatrix} x \\ y \end{bmatrix}, \qquad Z' = \begin{bmatrix} X \\ Y \end{bmatrix},$$

$$R = \begin{bmatrix} \cos \phi & -\sin \phi \\ \sin \phi & \cos \phi \end{bmatrix}$$

(a) Show that Equations 2 can be written $Z = RZ'$.
(b) Show that Equations 3 can be written $Z' = R^{-1}Z$.

Section 8.6

Polar Coordinates

Coordinate systems are methods for specifying the location of a point on the plane. Up to now we have been dealing with the rectangular (or Cartesian) coordinate system, which describes locations using a rectangular grid. Using rectangular coordinates is like describing a location in a city by saying that, for example, it is at the corner of 48th Street and 7th Avenue. But we might also

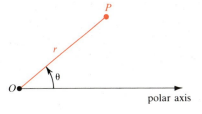

Figure 1

describe this same location by saying that it is three miles northwest of city hall. Instead of specifying the location with respect to a grid of streets and avenues, we can describe it by giving its distance and direction from a fixed reference point.

The polar coordinate system uses distances and directions to specify the location of points in the plane. To set up this system, we first choose a fixed point O called the **origin** (or **pole**). We then draw a ray (half-line) starting at O called the **polar axis**. This axis is usually drawn horizontally to the right and coincides with the x-axis in rectangular coordinates. Now let P be any point in the plane. Let r be the distance from P to the origin, and let θ be the angle between the polar axis and the segment OP, as shown in Figure 1. Then the ordered pair (r, θ) uniquely specifies the location of P. We write $P(r, \theta)$ and refer to r and θ as the **polar coordinates** of P. We use the convention that θ is positive if measured in a counterclockwise direction from the polar axis, and negative if measured in a clockwise direction. It is customary to use radian measure for θ. If $r = 0$, then $P = O$ no matter what value θ has, so $(0, \theta)$ represents the pole for any value of θ.

Because the angles $\theta + 2n\pi$ (for $n = \pm 1, \pm 2, \pm 3, \ldots$) all have the same terminal side as the angle θ, each point has infinitely many representations in polar coordinates. For example, $(2, \pi/3), (2, 7\pi/3)$, and $(2, -5\pi/3)$ all represent the same point, as illustrated in Figure 2. Moreover, we also allow r to take on negative values, with the understanding that if $-r$ is negative, then the point $P(-r, \theta)$ lies r units away from the origin in the direction *opposite* to that given by θ. Thus the point P graphed in Figure 2 can also be described by the coordinates $(-2, 4\pi/3)$ or $(-2, -2\pi/3)$.

Note that by this convention, the coordinates (r, θ) and $(-r, \theta + \pi)$ represent the same point (see Figure 3). In fact, any point $P(r, \theta)$ can also be represented by

$$P(r, \theta + 2n\pi) \qquad \text{and} \qquad P(-r, \theta + (2n + 1)\pi)$$

where n is any integer.

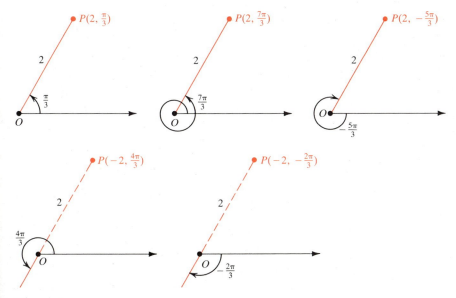

Figure 2

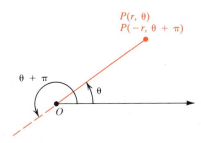

Figure 3

Example 1

Plot the points whose polar coordinates are (a) $(1, 3\pi/4)$, (b) $(3, -\pi/6)$, (c) $(3, 3\pi)$, and (d) $(-4, \pi/4)$.

Solution

The points are plotted in Figure 4. Note that the point in part (d) lies four units from the origin along the angle $5\pi/4$, since the given value of r is negative.

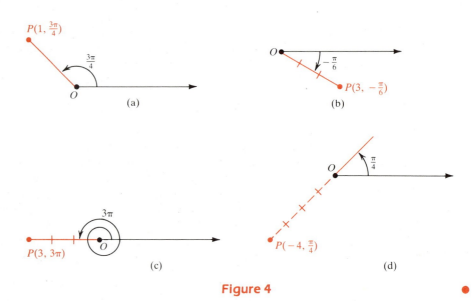

Figure 4

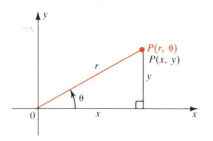

Figure 5

Situations often arise in which we need to consider polar and rectangular coordinates simultaneously. The connection between the two systems is illustrated in Figure 5, where the polar axis coincides with the positive x-axis. Although we have pictured the case where $r > 0$ and θ is acute, the following discussion holds for any angle θ and for any nonzero value of r. From the figure, we see that

$$\cos \theta = \frac{x}{r} \quad \text{and} \quad \sin \theta = \frac{y}{r}$$

Thus we have

(1)
$$x = r \cos \theta$$
$$y = r \sin \theta$$

Note that Equations 1 hold also at the origin, since $r = 0$ and $x = y = 0$ there.

From Figure 5 we also see that the following relations hold:

(2)

$$\tan \theta = \frac{y}{x} \quad (x \neq 0)$$

$$r^2 = x^2 + y^2$$

Equations 1 and 2 can be used to convert between the two coordinate systems, as we see in the next examples.

Example 2
Find rectangular coordinates for the point that has polar coordinates $(4, 2\pi/3)$.

Solution
Using Equations 1 with $r = 4$ and $\theta = 2\pi/3$, we have

$$x = r \cos \theta = 4 \cos\left(\frac{2\pi}{3}\right) = 4 \cdot \left(-\frac{1}{2}\right) = -2$$

$$y = r \sin \theta = 4 \sin\left(\frac{2\pi}{3}\right) = 4 \cdot \frac{\sqrt{3}}{2} = 2\sqrt{3}$$

Thus the point has rectangular coordinates $(-2, 2\sqrt{3})$.   ●

Example 3
Convert the point $(2, -2)$ from rectangular to polar coordinates.

Solution
From Equations 2 we see that

$$r^2 = x^2 + y^2 = 2^2 + (-2)^2 = 8$$

so $r = 2\sqrt{2}$ or $-2\sqrt{2}$. Also

$$\tan \theta = \frac{y}{x} = \frac{-2}{2} = -1$$

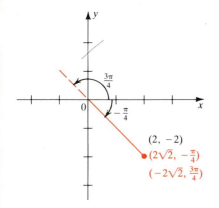

$(2, -2)$
$(2\sqrt{2}, -\frac{\pi}{4})$
$(-2\sqrt{2}, \frac{3\pi}{4})$

Figure 6

so $\theta = 3\pi/4$ or $-\pi/4$. Since the point $(2, -2)$ lies in the fourth quadrant (see Figure 6), we can represent it in polar coordinates as $(2\sqrt{2}, -\pi/4)$ or $(-2\sqrt{2}, 3\pi/4)$.   ●

   Note that Equations 2 do not uniquely determine r or θ. When we use these equations to find the polar coordinates of a point, we must be careful that the values we choose for r and θ give us a point in the correct quadrant, as we saw in Example 3.

Graphs of Polar Equations

The **graph of a polar equation** $r = f(\theta)$ consists of all points P that have at least one polar representation (r, θ) whose coordinates satisfy the equation. In the next two examples, we see that circles centered at the origin and lines that pass through the origin have particularly simple equations in polar coordinates.

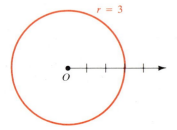

Figure 7

Example 4

Sketch the graph of the equation $r = 3$.

Solution

The graph consists of all points whose r-coordinate is 3; that is, all points that are three units away from the origin. Therefore this is a circle of radius 3 centered at the origin, as shown in Figure 7.   ●

In general the graph of the equation $r = a$ is a circle of radius $|a|$ centered at the origin. By squaring both sides of this equation and using Equation 2, we see that the equivalent equation in rectangular coordinates is $x^2 + y^2 = a^2$.

Example 5

Sketch the graph of the equation $\theta = \pi/3$ and express the equation in rectangular coordinates.

Solution

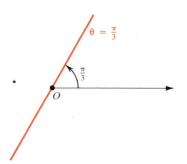

Figure 8

The graph consists of all points whose θ-coordinate is $\pi/3$. This is the straight line that passes through the origin and makes an angle of $\pi/3$ with the polar axis (see Figure 8). Note that the points $(r, \pi/3)$ on the line with $r > 0$ lie in the first quadrant, whereas those with $r < 0$ lie in the third. From (2), we see that if the point (x, y) lies on this line, then

$$\frac{y}{x} = \tan\theta = \tan\left(\frac{\pi}{3}\right) = \sqrt{3}$$

Thus the rectangular equation of this line is $y = \sqrt{3}x$.   ●

To sketch polar curves whose graphs are not so obvious as the ones in the preceding examples, we rely on two techniques. One technique is to plot points calculated for sufficiently many values of θ and then join them in a continuous curve. This is what we did when we first learned to graph functions in rectangular coordinates. The other technique is to convert the equation we are graphing into rectangular coordinates using (1) and (2), in the hope that the resulting equation is one that we recognize from our previous work. Both methods are used in the next example.

Example 6

(a)   Sketch the curve with polar equation $r = 2\sin\theta$.

(b)   Convert this equation to rectangular coordinates.

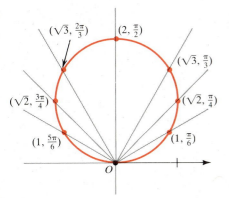

Figure 9
$r = 2\sin\theta$

Solution

(a)   We first use the equation to determine the polar coordinates of several points on the curve. The results are shown in the following table:

θ	0	$\pi/6$	$\pi/4$	$\pi/3$	$\pi/2$	$2\pi/3$	$3\pi/4$	$5\pi/6$	π
$r = 2\sin\theta$	0	1	$\sqrt{2}$	$\sqrt{3}$	2	$\sqrt{3}$	$\sqrt{2}$	1	0

We plot these points in Figure 9 and then join them to sketch the curve. The graph appears to be a circle. We have used values of θ only between 0 and π, since the same points (this time expressed with negative r-coordinates) would be obtained if we allowed θ to range from π to 2π.

(b)   If we multiply both sides of the equation of this curve by r, we get

$$r^2 = 2r\sin\theta$$

Using the identities in (1) and (2), we replace r^2 by $x^2 + y^2$ and $r\sin\theta$ by y to obtain the rectangular equation

$$x^2 + y^2 = 2y$$

Thus $x^2 + y^2 - 2y = 0$, and we complete the square in y to get

$$x^2 + (y - 1)^2 = 1$$

which is a circle of radius 1 centered at the point $(0, 1)$.   ●

Using the method from part (b) of the preceding example, we can show the following:

1.   The equation

$$r = 2a\sin\theta$$

represents a circle of radius $|a|$ centered at the point that has polar coordinates $(a, \pi/2)$.

2.   The equation

$$r = 2a\cos\theta$$

represents a circle of radius $|a|$ centered at the point with polar coordinates $(a, 0)$.

Example 7

Sketch the curve $r = 2 + 2\cos\theta$.

Solution

We tabulate the values of r for some convenient values of θ, as in Example 6.

θ	0	$\pi/6$	$\pi/4$	$\pi/3$	$\pi/2$	$2\pi/3$	$3\pi/4$	$5\pi/6$	π
$r = 2 + 2\cos\theta$	4	$2 + \sqrt{3}$	$2 + \sqrt{2}$	3	2	1	$2 - \sqrt{2}$	$2 - \sqrt{3}$	0

We see that as θ increases from 0 to π, r decreases from 4 to 0. Plotting the points in this table gives us the upper half of the curve shown in Figure 10. We obtain the lower half of this curve by noting that replacing θ by $-\theta$ in the equation leaves it unchanged, since $\cos(-\theta) = \cos\theta$. Thus for every point (r, θ) on the upper half of the curve, there is a corresponding point $(r, -\theta)$ below the polar axis. This means that the lower half is simply the mirror image of the upper half, and we complete the graph using symmetry.

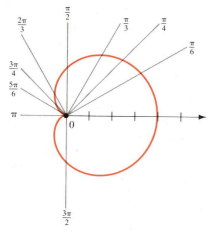

Figure 10

$r = 2 + 2\cos\theta$

The curve in Figure 10 is called a **cardioid** because it is heart-shaped. In general, any equation of the form

$$r = a(1 \pm \cos\theta) \qquad \text{or} \qquad r = a(1 \pm \sin\theta)$$

represents a cardioid.

We saw in Example 7 that exploiting symmetry often saves a lot of work when sketching polar curves. We list three tests for symmetry. Figure 11 shows why these tests work.

Tests for Symmetry

1. If a polar equation is unchanged when we replace θ by $-\theta$, the graph is symmetric about the polar axis [Figure 11(a)].
2. If the equation is unchanged when we replace r by $-r$, the graph is symmetric about the origin [Figure 11(b)].
3. If the equation is unchanged when we replace θ by $\pi - \theta$, the graph is symmetric about the vertical line $\theta = \pi/2$ (the y-axis) [Figure 11(c)].

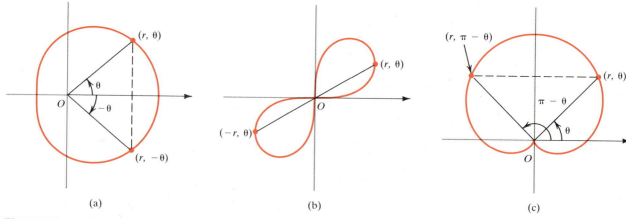

(a)   (b)   (c)

Figure 11

Example 8

Sketch the curve $r = \cos 2\theta$.

Solution

Instead of plotting points as in Examples 6 and 7, we reason as follows. First we sketch the graph of $r = \cos 2\theta$ in *rectangular* coordinates, with θ plotted along the horizontal axis and r on the vertical axis. The graph is shown in Figure 12. From this graph we can read off at a glance the values of r that correspond to increasing values of θ. As θ increases from 0 to $\pi/4$, Figure 12 shows that r decreases from 1 to 0, and so we draw the corresponding portion of the polar curve in Figure 13 (indicated by a single arrow). As θ increases from $\pi/4$ to $\pi/2$, r goes from 0 to -1. This means that the distance from the origin increases from 0 to 1, but instead of being in the second quadrant, this portion of the polar curve (indicated by a double arrow) lies on the opposite side of the origin in the fourth quadrant. The remainder of the curve is drawn in a similar fashion, with the arrows and numbers indicating the order in which the portions are traced out. The resulting curve has four petals and is called a **four-leafed rose**.

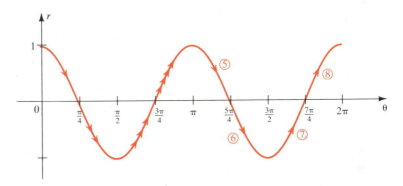

Figure 12
$r = \cos 2\theta$ in rectangular coordinates

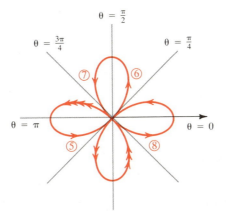

Figure 13
Four-leafed rose $r = \cos 2\theta$

In general, the graphs of equations of the form

$$r = a \cos n\theta \qquad \text{or} \qquad r = a \sin n\theta$$

are n-leafed roses if n is odd and $2n$-leafed roses if n is even (as in Example 8).

Exercises 8.6

In Exercises 1–6 plot the points that have the given polar coordinates. Then give two other polar coordinate representations of these points, one with $r < 0$ and the other with $r > 0$.

1. $(3, \pi/2)$ **2.** $(2, 3\pi/4)$ **3.** $(-1, 7\pi/6)$

4. $(-2, -\pi/3)$ **5.** $(-5, 0)$ **6.** $(3, 1)$

In Exercises 7–12 find the rectangular coordinates for the points whose polar coordinates are given.

7. $(4, \pi/6)$ **8.** $(6, 2\pi/3)$ **9.** $(\sqrt{2}, -\pi/4)$

10. $(-1, 5\pi/2)$ **11.** $(5, 5\pi)$ **12.** $(0, 13\pi)$

In Exercises 13–18 convert the given rectangular coordinates to polar coordinates, with $r > 0$ and $0 \le \theta < 2\pi$.

13. $(-1, 1)$ **14.** $(3\sqrt{3}, -3)$ **15.** $(\sqrt{8}, \sqrt{8})$

16. $(-\sqrt{6}, -\sqrt{2})$ **17.** $(3, 4)$ **18.** $(1, -2)$

In Exercises 19–24 convert the given equation to polar form.

19. $x = y$ **20.** $x^2 + y^2 = 9$ **21.** $y = x^2$

22. $y = 5$ **23.** $x = 4$ **24.** $x^2 - y^2 = 1$

In Exercises 25–34 convert the given polar equation to rectangular coordinates.

25. $r = 7$ **26.** $\theta = \pi$ **27.** $r \cos \theta = 6$

28. $r = 6 \cos \theta$ **29.** $r^2 = \tan \theta$ **30.** $r^2 = \sin 2\theta$

31. $r = \dfrac{1}{\sin \theta - \cos \theta}$ **32.** $r = \dfrac{1}{1 + \sin \theta}$

33. $r = 1 + \cos \theta$ **34.** $r = \dfrac{4}{1 + 2 \sin \theta}$

In Exercises 35–56 sketch the curve whose polar equation is given.

35. $r = 3$ **36.** $r = -1$ **37.** $\theta = -\pi/2$

38. $\theta = 5\pi/6$ **39.** $r = 6 \sin \theta$ **40.** $r = \cos \theta$

41. $r = -2 \cos \theta$ **42.** $r = 2 \sin \theta + 2 \cos \theta$

43. $r = 2 - 2 \cos \theta$ **44.** $r = 1 + \sin \theta$

45. $r = -3(1 + \sin \theta)$ **46.** $r = \cos \theta - 1$

47. $r = \theta, \theta \ge 0$ (spiral)

48. $r\theta = 1, \theta > 0$ (reciprocal spiral)

49. $r = \sin 2\theta$ (four-leafed rose)

50. $r = 2 \cos 3\theta$ (three-leafed rose)

51. $r^2 = \cos 2\theta$ (lemniscate)

52. $r^2 = 4 \sin 2\theta$ (lemniscate)

53. $r = 2 + \sin \theta$ (limaçon)

54. $r = 1 - 2 \cos \theta$ (limaçon)

55. $r = 2 + \sec \theta$ (conchoid)

56. $r = \sin \theta \tan \theta$ (cissoid)

57. **(a)** Show that the distance between the points whose polar coordinates are (r_1, θ_1) and (r_2, θ_2) is

$$\sqrt{r_1^2 + r_2^2 - 2 r_1 r_2 \cos(\theta_2 - \theta_1)}$$

(*Hint:* Use the Law of Cosines.)
    (b) Find the distance between the points whose polar coordinates are $(3, 3\pi/4)$ and $(-1, 7\pi/6)$.

58. Show that the graph of $r = a \cos \theta + b \sin \theta$ is a circle, and find its center and radius.

Parametric Equations

So far we have described curves by giving an equation that the coordinates of all points on the curve must satisfy. For example, we know that the equation $y = x^2$ represents a parabola in rectangular coordinates and that $r = \sin \theta$ represents a circle in polar coordinates. We now study another method for describing curves in the plane, which in some situations turns out to be more useful and natural than rectangular or polar equations. In this method, the x- and y-coordinates of points on the curve are given separately as functions of an additional variable t called the **parameter**:

$$x = f(t) \qquad y = g(t)$$

These are called **parametric equations** for the curve. Substituting a value of t determines the coordinates of a point (x, y). As t varies, the point $(x, y) = (f(t), g(t))$ varies and traces out the curve. If we think of t as representing time, then as t increases, we can imagine a particle at $(x, y) = (f(t), g(t))$ moving along the curve.

Example 1

Sketch and identify the curve defined by the parametric equations

(1) $$x = t^2 - 3t \qquad y = t - 1$$

Solution

For every value of t we get a point on the curve. For example, if $t = 0$, then $x = 0$ and $y = -1$, so the corresponding point is $(0, -1)$. In Figure 1 we plot the points (x, y) determined by the values of t shown in the following table:

t	-2	-1	0	1	2	3	4	5
x	10	4	0	-2	-2	0	4	10
y	-3	-2	-1	0	1	2	3	4

As t increases, a particle whose position is given by the parametric equations will move along the curve in the direction of the arrows. The curve seems to be a parabola. This can be confirmed by eliminating the parameter t from the

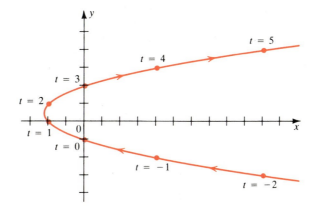

Figure 1

parametric equations and reducing them to a single equation as follows. First we solve for t in the second equation to get $t = y + 1$. Substituting this into the first equation, we get

$$x = (y + 1)^2 - 3(y + 1) = y^2 - y - 2$$

The curve is the parabola $x = y^2 - y - 2$. ●

Notice that we would obtain the same graph as in Example 1 from the parametrization

(2) $x = t^2 - t - 2 \qquad y = t$

because the points on this curve also satisfy the equation $x = y^2 - y - 2$. But the same value of t produces different points on the curve in these two parametrizations. For example, when $t = 0$, the particle that traces out the curve in Figure 1 is at $(0, -1)$, whereas in the parametrization of Equations 2, the particle is already at $(-2, 0)$ when $t = 0$. Thus a parametrization contains more information than just the curve being parametrized. It also indicates *how* that curve is being traced out.

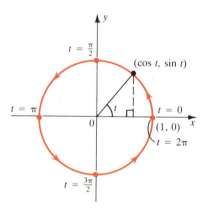

Figure 2

Example 2

Describe and graph the curve represented by the parametric equations

$$x = \cos t \qquad y = \sin t \qquad 0 \le t \le 2\pi$$

Solution

To eliminate the parameter and identify the curve, we will use the trigonometric identity $\cos^2 t + \sin^2 t = 1$. This gives

$$x^2 + y^2 = 1$$

so we see that the graph is a circle of radius 1 centered at the origin. As t increases from 0 to 2π, the point given by the parametric equations starts at $(1, 0)$ and moves counterclockwise once around the circle, as shown in Figure 2. Notice that the parameter t can be interpreted as the angle shown in the figure. ●

Example 3

Find parametric equations for the line of slope 3 that passes through the point $(2, 6)$.

Solution

Let us start at the point $(2, 6)$ and move up and to the right along this line. Because the line has slope 3, for every one unit we move to the right, we must move up three units. In other words, if we increase the x-coordinate by t units, we must correspondingly increase the y-coordinate by $3t$ units. This leads to the parametric equations:

$$x = 2 + t \qquad y = 6 + 3t$$

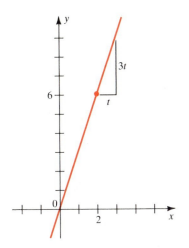

Figure 3

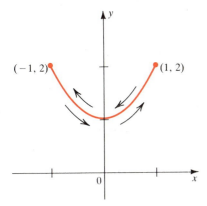

Figure 4

Figure 5
Cycloid

To confirm that these equations give the desired line, we eliminate the parameter by solving for t in the first equation and substituting into the second, to get

$$y = 6 + 3(x - 2) = 3x$$

Thus the slope-intercept form of the equation of this line is $y = 3x$, which is a line of slope 3 that does pass through $(2, 6)$ as required. The graph is shown in Figure 3. ●

Example 4

Sketch the curve with parametric equations

$$x = \sin t \qquad y = 2 - \cos^2 t$$

Solution

To eliminate the parameter, we first use the trigonometric identity $\cos^2 t = 1 - \sin^2 t$ to change the second equation to

$$y = 2 - \cos^2 t = 2 - (1 - \sin^2 t) = 1 + \sin^2 t$$

Now we can substitute $\sin t = x$ from the first equation to get

$$y = 1 + x^2$$

so the point (x, y) moves along the parabola $y = 1 + x^2$. But since $-1 \le \sin t \le 1$, we have $-1 \le x \le 1$, so the parametric equations represent only the part of the parabola between $x = -1$ and $x = 1$. Since $\sin t$ is periodic, the point $(x, y) = (\sin t, 2 - \cos^2 t)$ moves back and forth infinitely often along the parabola between the points $(-1, 2)$ and $(1, 2)$ as shown in Figure 4. ●

Example 5

The curve traced out by a fixed point P on the circumference of a circle as the circle rolls along a straight line is called a **cycloid** (see Figure 5). If the circle has radius a and rolls along the x-axis, with one position of the point P being at the origin, find parametric equations for the cycloid.

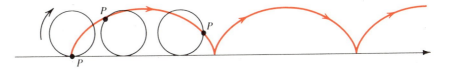

Solution

Figure 6 shows the circle and the point P after the circle has rolled through an angle θ (measured in radians). The distance $|OT|$ that the circle has rolled must be the same as the length of the arc PT, which, by definition of radian measure, is $a\theta$. This means that the center of the circle is $C(a\theta, a)$.

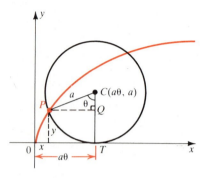

Figure 6

Let the coordinates of P be (x, y). Then from Figure 6 (which illustrates the case where $0 < \theta < \pi/2$) we see that

$$x = |OT| - |PQ| = a\theta - a\sin\theta = a(\theta - \sin\theta)$$
$$y = |TC| - |QC| = a - a\cos\theta = a(1 - \cos\theta)$$

so parametric equations for the cycloid are

$$x = a(\theta - \sin\theta) \qquad y = a(1 - \cos\theta) \qquad \bullet$$

The cycloid has a number of interesting physical properties. It is the "curve of quickest descent" in the following sense. Let us choose two points P and Q not directly above each other and join them with a wire. If we allow a frictionless bead to slide down the wire under the influence of gravity, then of all possible shapes that the wire can be bent into, the bead will slide from P to Q the fastest when the shape is half of an arch of an inverted cycloid (see Figure 7). The cycloid is also the "curve of equal descent" in the sense that no matter where we place a bead B on a cycloid-shaped wire, it takes the same time to slide to the bottom (see Figure 8). These rather surprising properties of the cycloid were proved (using calculus) by several mathematicians and physicists in the 17th century, including Johann Bernoulli, Blaise Pascal, and Christiaan Huygens.

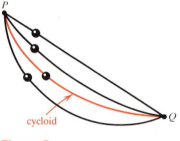

Figure 7

Figure 8

Exercises 8.7

In Exercises 1–22, (a) sketch the curve represented by the parametric equations, and (b) find a rectangular coordinate equation for the curve by eliminating the parameter.

1. $x = 2t, \quad y = t + 6$

2. $x = 6t - 4, \quad y = 3t, \quad t \geq 0$

3. $x = t^2, \quad y = t - 2, \quad 2 \leq t \leq 4$

4. $x = 2t + 1, \quad y = (t + \tfrac{1}{2})^2$

5. $x = \sqrt{t}, \quad y = 1 - t$

6. $x = t^2, \quad y = t^4 + 1$

7. $x = \dfrac{1}{t}, \quad y = t + 1$

8. $x = t + 1, \quad y = \dfrac{t}{t + 1}$

9. $x = 4t^2, \quad y = 8t^3$

10. $x = |t|, \quad y = |1 - |t||$

11. $x = 2\sin t, \quad y = 2\cos t, \quad 0 \leq t \leq \pi$

12. $x = 2\cos t, \quad y = 3\sin t, \quad 0 \leq t \leq 2\pi$

13. $x = \sin^2 t, \quad y = \sin^4 t$

14. $x = \sin^2 t, \quad y = \cos t$

15. $x = \cos t, \quad y = \cos 2t$

16. $x = \cos 2t, \quad y = \sin 2t$

17. $x = \sec t, \quad y = \tan t, \quad 0 \leq t < \pi/2$

18. $x = \cot t, \quad y = \csc t, \quad 0 < t < \pi$

19. $x = e^t, \quad y = e^{-t}$

20. $x = e^{2t}, \quad y = e^t, \quad t \geq 0$

21. $x = \cos^2 t, \quad y = \sin^2 t$

22. $x = \cos^3 t, \quad y = \sin^3 t, \quad 0 \leq t \leq 2\pi$

In Exercises 23–26 find parametric equations for the line with the given properties.

23. Slope $\frac{1}{2}$, passing through $(4, -1)$

24. Slope -2, passing through $(-10, -20)$

25. Passing through $(6, 7)$ and $(7, 8)$

26. Passing through $(12, 7)$ and the origin

27. Find parametric equations for the circle $x^2 + y^2 = a^2$.

28. Find parametric equations for the ellipse

$$\frac{x^2}{a^2} + \frac{y^2}{b^2} = 1$$

29. Show by eliminating the parameter θ that the parametric equations
$$x = a\tan\theta \qquad y = b\sec\theta$$

represent a hyperbola.

30. Show that the parametric equations
$$x = a\sqrt{t} \qquad y = b\sqrt{t + 1}$$

represent a part of the hyperbola of Exercise 29.

In Exercises 31–34 sketch the curve given by the parametric equations.

31. $x = t\cos t, \quad y = t\sin t, \quad t \geq 0$

32. $x = \sin t, \quad y = \sin 2t$

33. $x = \dfrac{3t}{1 + t^3}, \quad y = \dfrac{3t^2}{1 + t^3}$

34. $x = \cot t, \quad y = 2\sin^2 t, \quad 0 < t < \pi$

35. If a projectile is fired with an initial speed of v_0 ft/s at an angle α above the horizontal, then its position after t seconds is given by the parametric equations
$$x = (v_0 \cos\alpha)t \qquad y = (v_0 \sin\alpha)t - 16t^2$$

(where x and y are measured in feet). Show that the path

of the projectile is a parabola by eliminating the parameter t.

36. Referring to Exercise 35, suppose that a gun fires a bullet into the air with an initial speed of 2048 ft/s at an angle of $30°$ to the horizontal.
  (a) After how many seconds will it hit the ground?
  (b) How far from the gun will it hit the ground?
  (c) What is the maximum height attained by the bullet?

37. In Example 5 suppose that the point P that traces out the curve lies not on the edge of the circle but rather at a fixed point inside the rim, at a distance b from the center (where $b < a$). The curve traced out by P is called a **curtate cycloid** (or **trochoid**). Show that parametric equations for the curtate cycloid are

$$x = a\theta - b\sin\theta \qquad y = a - b\cos\theta$$

Sketch the graph.

38. In Exercise 37 if the point P lies *outside* the circle at a distance b from the center (where $b > a$), then the curve traced out by P is called a **prolate cycloid**. Show that parametric equations for the prolate cycloid are the same as the equations for the curtate cycloid, and sketch the graph for the case where $a = 1$ and $b = 2$.

39. A circle C of radius b rolls on the inside of a larger circle of radius a centered at the origin. Let P be a fixed point on the smaller circle, with initial position at the point $(a, 0)$ (see the figure). The curve traced out by P is called a **hypocycloid**.

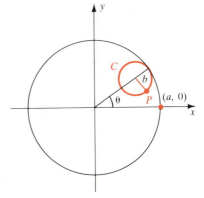

  (a) Show that parametric equations for the hypocycloid are

$$x = (a - b)\cos\theta + b\cos\left(\frac{a - b}{b}\theta\right)$$

$$y = (a - b)\sin\theta - b\sin\left(\frac{a - b}{b}\theta\right)$$

(b) If $a = 4b$, the hypocycloid is called an **astroid**. Show that in this case the parametric equations can be reduced to

$$x = a \cos^3 \theta \qquad y = a \sin^3 \theta$$

Sketch the curve and eliminate the parameter to obtain an equation for the astroid in rectangular coordinates.

40. If the circle C of Exercise 39 rolls on the *outside* of the larger circle, the curve traced out by P is called an **epicycloid**. Find parametric equations for the epicycloid.

41. In the figure the circle of radius a is stationary, and for every θ, the point P is the midpoint of the segment QR. The curve traced out by P for $0 < \theta < \pi$ is called the **longbow curve**. Find parametric equations for this curve.

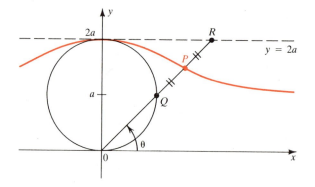

42. A string is wound around a circle and then unwound while being held taut. The curve traced out by the point P at the end of the string is called the **involute** of the circle. If the circle has radius a and is centered at the origin, and if the initial position of P is at $(a, 0)$, show that parametric equations for the involute in terms of the parameter θ (see the figure) are

$$x = a(\cos \theta + \theta \sin \theta) \qquad y = a(\sin \theta - \theta \cos \theta)$$

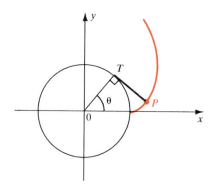

43. Eliminate the parameter θ in the parametric equations for the cycloid (Example 5) to obtain a rectangular coordinate equation for the section of the curve given by $0 \le \theta \le \pi$.

CHAPTER 8 REVIEW

Define, state, or discuss the following.

1. Conic section
2. Parabola
3. Focus, directrix, and vertex of a parabola
4. Reflection property of a parabola
5. Ellipse
6. Foci of an ellipse
7. Major and minor axes of an ellipse
8. Eccentricity of an ellipse
9. Hyperbola
10. Foci of a hyperbola
11. Asymptotes of a hyperbola
12. Shifted conics
13. The equation $Ax^2 + Cy^2 + Dx + Ey + F = 0$
14. Rotation of axes
15. The equation $Ax^2 + Bxy + Cy^2 + Dx + Ey + F = 0$
16. Polar coordinates
17. Parametric equations
18. Cycloid

Review Exercises

Find the vertex, focus, and directrix of the parabolas whose equations are given in Exercises 1–4, and sketch the graphs.

1. $x^2 + 8y = 0$

2. $2x - y^2 = 0$

3. $x - y^2 + 4y - 2 = 0$

4. $2x^2 + 6x + 5y + 10 = 0$

Find the center, vertices, foci, and the lengths of the major and minor axes of the ellipses whose equations are given in Exercises 5–8, and sketch the graphs.

5. $x^2 + 4y^2 = 16$

6. $9x^2 + 4y^2 = 1$

7. $4x^2 + 9y^2 = 36y$

8. $2x^2 + y^2 = 2 + 4(x - y)$

Find the center, vertices, foci, and asymptotes of the hyperbolas whose equations are given in Exercises 9–12, and sketch the graphs.

9. $x^2 - 2y^2 = 16$

10. $x^2 - 4y^2 + 16 = 0$

11. $9y^2 + 18y = x^2 + 6x + 18$

12. $y^2 = x^2 + 6y$

In Exercises 13–24 determine the type of curve represented by the given equation. Find the foci and vertices (if any) and sketch the graph.

13. $\dfrac{x^2}{12} + y = 1$

14. $\dfrac{x^2}{12} + \dfrac{y^2}{144} = \dfrac{y}{12}$

15. $x^2 - y^2 + 144 = 0$

16. $x^2 + 6x = 9y^2$

17. $4x^2 + y^2 = 8(x + y)$

18. $3x^2 - 6(x + y) = 10$

19. $x = y^2 - 16y$

20. $2x^2 + 4 = 4x + y^2$

21. $2x^2 - 12x + y^2 + 6y + 26 = 0$

22. $36x^2 - 4y^2 - 36x - 8y = 31$

23. $9x^2 + 8y^2 - 15x + 8y + 27 = 0$

24. $x^2 + 4y^2 = 4x + 8$

In Exercises 25–32 find an equation for the conic section with the given properties.

25. The parabola with focus $F(0, 1)$ and directrix $y = -1$

26. The ellipse with center $C(0, 4)$, foci $F_1(0, 0)$ and $F_2(0, 8)$, and major axis of length 10

27. The hyperbola with vertices $V(0, \pm 2)$ and asymptotes $y = \pm \frac{1}{2}x$

28. The hyperbola with center $C(2, 4)$, foci $F_1(2, 7)$ and $F_2(2, 1)$, and vertices $V_1(2, 6)$ and $V_2(2, 2)$

29. The ellipse with foci $F_1(1, 1)$ and $F_2(1, 3)$, and with one vertex on the x-axis

30. The parabola with vertex $V(5, 5)$ and directrix the y-axis

31. The ellipse with vertices $V_1(7, 12)$ and $V_2(7, -8)$, and containing the point $P(1, 8)$

32. The parabola with vertex $V(-1, 0)$, horizontal axis of symmetry, and crossing the y-axis where $y = 2$

In Exercises 33–36 use rotation of axes to determine what type of conic is represented by the given equation. Sketch the graph.

33. $x^2 + 4xy + y^2 = 1$

34. $5x^2 - 6xy + 5y^2 - 8x + 8y - 8 = 0$

35. $7x^2 - 6\sqrt{3}xy + 13y^2 - 4\sqrt{3}x - 4y = 0$

36. $9x^2 + 24xy + 16y^2 = 25$

In Exercises 37–44 sketch the curve whose polar equation is given. Express the equation in rectangular coordinates.

37. $r = 3 + 3\cos\theta$

38. $r = 3\sin\theta$

39. $r = 2\sin 2\theta$

40. $r = 4\cos 3\theta$

41. $r^2 = \sec 2\theta$

42. $r^2 = 4\sin 2\theta$

43. $r = \sin\theta + \cos\theta$

44. $r = \dfrac{4}{2 + \cos\theta}$

In Exercises 45–48 sketch the given parametric curve and eliminate the parameter to find an equation in rectangular coordinates that all points on the curve satisfy.

45. $x = 1 - t^2$, $y = 1 + t$

46. $x = t^2 - 1$, $y = t^2 + 1$

47. $x = 1 + \cos t$, $y = 1 - \sin t$, $0 \le t \le \dfrac{\pi}{2}$

48. $x = \dfrac{1}{t} + 2, \quad y = \dfrac{2}{t^2}, \quad 0 < t \le 2$

49. The curves C, D, E, and F are defined parametrically as follows, where the parameter t takes on all real values unless otherwise stated:

$$C: \quad x = t, \quad y = t^2$$
$$D: \quad x = \sqrt{t}, \quad y = t, \quad t \ge 0$$
$$E: \quad x = \sin t, \quad y = 1 - \cos^2 t$$
$$F: \quad x = e^t, \quad y = e^{2t}$$

  (a) Show that the points on all four of these curves satisfy the same rectangular coordinate equation.

  (b) Sketch the graph of each curve and explain how the curves differ from one another.

50. In the figure the point P is the midpoint of the segment $\overline{QR}$ and $0 \le \theta < \pi/2$. Find a parametric representation for the curve traced out by P, using θ as the parameter.

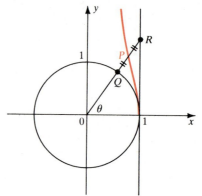

Chapter 8 Test

1. Find the focus and directrix of the parabola $x^2 = -6y$ and sketch the graph.

2. Find the vertices, foci, and the lengths of the major and minor axes for the ellipse

$$\frac{x^2}{8} + \frac{y^2}{12} = 1$$

and sketch the graph.

3. Find the vertices, foci, and asymptotes of the hyperbola

$$\frac{y^2}{49} - \frac{x^2}{36} = 1$$

and sketch the graph.

In Problems 4–6 sketch the graph of the equation.

4. $16x^2 + 36y^2 - 96x + 36y + 9 = 0$

5. $9x^2 - 8y^2 + 36x + 64y = 92$

6. $2x + y^2 + 8y + 8 = 0$

7. Find an equation for the hyperbola with foci $(0, \pm 5)$ and with asymptotes $y = \pm\frac{3}{4}x$.

8. Find an equation for the parabola with focus $(2, 4)$ and with directrix the x-axis.

9. **(a)** Use rotation of axes to eliminate the xy term in the equation

$$5x^2 + 4xy + 2y^2 = 18$$

  (b) Sketch the graph of the equation.

  (c) Find the coordinates of the vertices of this conic (in the xy-coordinate system).

10. Graph the polar equation $r = 2 + \cos\theta$.

11. Convert the polar equation

$$r = 2\cos\theta - 4\sin\theta$$

to rectangular coordinates and identify the graph.

12. **(a)** Sketch the graph of the parametric curve

$$x = 3\sin\theta + 3 \qquad y = 2\cos\theta \qquad 0 \le \theta \le \pi$$

  (b) Eliminate the parameter θ to obtain an equation for this curve in rectangular coordinates.

Problems Plus

1. For what values of the real number k does the circle $(x - k)^2 + y^2 = 4$ intersect the ellipse

$$x^2 + \frac{y^2}{9} = 1$$

in exactly $0, 1, 2, 3, 4,$ or 5 points?

2. A **radius** of the ellipse

$$\frac{x^2}{a^2} + \frac{y^2}{b^2} = 1$$

is a line segment that joins any point on the ellipse to the center. Show that the radius whose endpoint is (x_0, y_0) has length

$$\frac{1}{a}\sqrt{(a^2 - b^2)x_0^2 + a^2b^2}$$

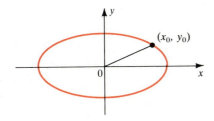

3. On the parabola $x^2 = 4py$, let P and Q be two points with the property that $\angle POQ$ is a right angle. Show that the y-intercept of the segment $\overline{PQ}$ is the same for any such pair of points P and Q.

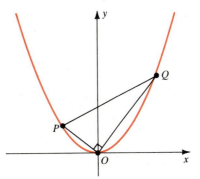

4. In the figure the circle is stationary and the triangle PQR is a right triangle with its legs parallel to the coordinate axes. The curve traced out by P for $0 < \theta < \pi$ is called the **witch of Maria Agnesi**. Show that the curve is the graph of the equation

$$y = \frac{8}{4 + x^2}$$

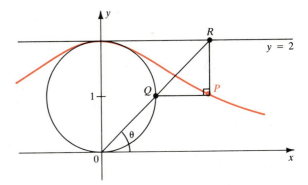

9

Sequences and Series

A mathematician, like a painter or poet, is a maker of patterns.
G. H. Hardy

Certainly let us learn proving, but also let us learn guessing.
George Polya

In this chapter we study sequences and series of numbers. Roughly speaking, a sequence is a list of numbers written in a specific order and a series is what one gets by adding the numbers in a sequence. Sequences and series are important in calculus because of Newton's idea of representing functions as series. They have many theoretical and practical uses. Among other applications, we will consider how series are used to calculate the value of an annuity.

Section 9.6 introduces a special kind of proof called mathematical induction, which is used to prove properties of the sequence of natural numbers. In Section 9.7 we discuss how to find patterns in sequences. In Section 9.8 we use the techniques of Sections 9.6 and 9.7 to find and prove a formula for expanding $(a + b)^n$ for any natural number n.

Section 9.1

Sequences

A **sequence** is a set of numbers written in a specific order:

$$a_1, a_2, a_3, a_4, \ldots, a_n, \ldots$$

The number a_1 is called the **first term**, a_2 is the **second term**, and in general a_n is the **nth term**.

Notice that for every natural number n there is a corresponding number a_n, so *a sequence can be regarded as a function f whose domain is the set of natural numbers*. We usually write a_n instead of the function notation $f(n)$ for the value of the function at the number n.

Here is a simple example of a sequence:

(1) $2, 4, 6, 8, 10, \ldots$

The dots indicate that the sequence continues indefinitely. When we write a sequence in this way, we are saying that it is clear what the subsequent terms of the sequence are. The sequence in (1) consists of even numbers. To be more accurate, however, we need to specify a procedure for finding *all* the terms of the sequence. This can be done by giving a formula for the nth term a_n of the sequence. In this case,

(2) $a_n = 2n$

and the sequence in (1) can be written as

$$2, 4, 6, 8, 10, \ldots, 2n, \ldots$$

Notice how Equation 2 gives all the terms of the sequence. For instance, substituting $n = 1, 2, 3$, and 4 in (2) gives the first four terms:

$$a_1 = 2 \cdot 1 = 2$$
$$a_2 = 2 \cdot 2 = 4$$
$$a_3 = 2 \cdot 3 = 6$$
$$a_4 = 2 \cdot 4 = 8$$

To find the 103rd term of this sequence we use (2) with $n = 103$ to get

$$a_{103} = 2 \cdot 103 = 206$$

Example 1

Find the first five terms and the 100th term of the sequences defined by the following formulas:

(a) $a_n = 2n - 1$

(b) $c_n = n^2 - 1$

(c) $t_n = \dfrac{n}{n + 1}$

(d) $r_n = \dfrac{(-1)^n}{2^n}$

Solution

(a)   Using the formula, we have $a_1 = 2(1) - 1 = 1$, $a_2 = 2(2) - 1 = 3$, $a_3 = 2(3) - 1 = 5$, $a_4 = 2(4) - 1 = 7$, $a_5 = 2(5) - 1 = 9$, and $a_{100} = 2(100) - 1 = 199$. This sequence can be written as

$$1, 3, 5, 7, 9, \ldots, 2n - 1, \ldots$$

Notice that this is the sequence of odd numbers.

(b)   We have $c_1 = 1^2 - 1 = 0$, $c_2 = 2^2 - 1 = 3$, $c_3 = 3^2 - 1 = 8$, $c_4 = 4^2 - 1 = 15$, $c_5 = 5^2 - 1 = 24$, and $c_{100} = 100^2 - 1 = 9999$. This sequence can be written as

$$0, 3, 8, 15, 24, \ldots, n^2 - 1, \ldots$$

(c)   From the formula for t_n we get $t_1 = \frac{1}{2}, t_2 = \frac{2}{3}, t_3 = \frac{3}{4}, t_4 = \frac{4}{5}, t_5 = \frac{5}{6}$, and $t_{100} = \frac{100}{101}$. This sequence can be written as follows:

$$\frac{1}{2}, \frac{2}{3}, \frac{3}{4}, \frac{4}{5}, \frac{5}{6}, \ldots, \frac{n}{n+1}, \ldots$$

(d) $r_1 = -\frac{1}{2}, r_2 = \frac{1}{4}, r_3 = -\frac{1}{8}, r_4 = \frac{1}{16}, r_5 = -\frac{1}{32}$, and $r_{100} = 1/2^{100}$. (The number 2^{100} has 31 digits, so we will not write it here.) This sequence can be written as follows:

$$-\frac{1}{2}, \frac{1}{4}, -\frac{1}{8}, \frac{1}{16}, -\frac{1}{32}, \ldots, \frac{(-1)^n}{2^n}, \ldots$$

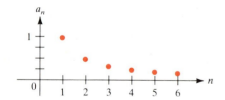

Figure 1

Notice the presence of $(-1)^n$ in this last sequence. It has the effect of making successive terms of the sequence alternately positive and negative.

It is often useful to picture a sequence by sketching its graph. Since a sequence is a function whose domain is the natural numbers, we can draw its graph in the Cartesian plane. For instance, the graph of the sequence

$$1, \frac{1}{2}, \frac{1}{3}, \frac{1}{4}, \frac{1}{5}, \frac{1}{6}, \ldots, \frac{1}{n}, \ldots$$

is shown in Figure 1. Compare this to the graph of

$$1, -\frac{1}{2}, \frac{1}{3}, -\frac{1}{4}, \frac{1}{5}, -\frac{1}{6}, \ldots, \frac{(-1)^{n+1}}{n}, \ldots$$

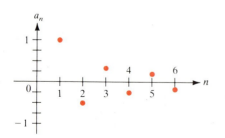

Figure 2

shown in Figure 2.

Some sequences do not have simple defining formulas like those of Example 1. The nth term of a sequence may depend on some or all of the terms that precede it. Sequences defined in this way are called **recursive**. Two examples follow.

Example 2

Find the first five terms of the sequence defined by

$$a_n = 3(a_{n-1} + 2)$$

and $a_1 = 1$.

Solution

The defining formula for this sequence is recursive. It allows us to find the nth term a_n if we know the preceding term a_{n-1}. Thus we can find the second term from the first term, the third term from the second term, the fourth term from the third term, and so on. Since we are given the first term $a_1 = 1$, we can proceed as follows:

$$a_2 = 3(a_1 + 2) = 3(1 + 2) = 9$$
$$a_3 = 3(a_2 + 2) = 3(9 + 2) = 33$$
$$a_4 = 3(a_3 + 2) = 3(33 + 2) = 105$$
$$a_5 = 3(a_4 + 2) = 3(105 + 2) = 321$$

Thus the first five terms of this sequence are

$$1, 9, 33, 105, 321 \qquad \bullet$$

Notice that in order to find the 100th term of the sequence in Example 2 we must first find all the preceding 99 terms.

Example 3

Find the first eleven terms of the sequence defined recursively by

$$F_n = F_{n-1} + F_{n-2}$$

where $F_1 = 1$ and $F_2 = 1$.

Solution

To find F_n we need to find the two preceding terms F_{n-1} and F_{n-2}. Since we are given F_1 and F_2, we proceed as follows:

$$F_3 = F_2 + F_1 = 1 + 1 = 2$$
$$F_4 = F_3 + F_2 = 2 + 1 = 3$$
$$F_5 = F_4 + F_3 = 3 + 2 = 5$$

It is clear what is happening here. Each term is simply the sum of the two terms that precede it, so we can easily write down as many terms as we please. Here are the first eleven terms:

$$1, 1, 2, 3, 5, 8, 13, 21, 34, 55, 89 \qquad \bullet$$

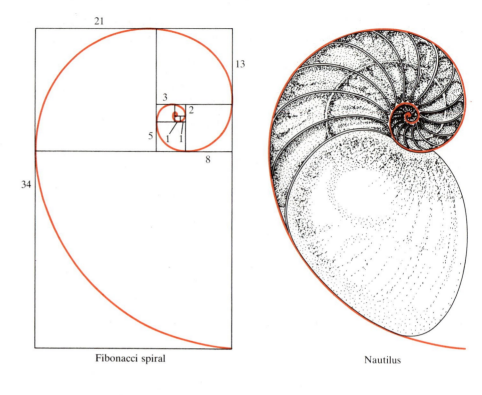

Figure 3

Fibonacci spiral

Nautilus

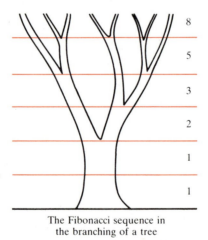

Figure 4

The Fibonacci sequence in
the branching of a tree

The sequence in Example 3 is called the **Fibonacci sequence**, named after the 13th-century Italian mathematician who used this sequence to solve a problem about the breeding of rabbits (see Exercise 37). This sequence has numerous other applications in nature. In fact so many phenomena behave according to this sequence that there is a mathematical journal (the *Fibonacci Quarterly*) devoted entirely to its properties. Figures 3 and 4 show two applications.

The sequences we have considered so far are defined by either a formula or a recursive procedure. But not all sequences can be defined in this way. The sequence

$$2, 3, 5, 7, 11, 13, 17, 19, 23, \ldots$$

is the sequence of prime numbers. No formula is known that will produce all the primes. Also, if we let a_n be the digit in the nth decimal place of the number π, we get the sequence

$$1, 4, 1, 5, 9, 2, 6, 5, 4, \ldots$$

A prime number is a natural number with no factors other than 1 and itself. By convention, 1 is not considered prime.

Again there is no simple formula for finding the terms of this sequence.

Finding patterns is a very important part of mathematics. Consider a sequence

$$1, 4, 9, 16, \ldots$$

Can you detect a pattern in these numbers? In other words, can you define a sequence whose first four terms are these numbers? The answer to this question seems easy; these numbers are the squares of the numbers 1, 2, 3, and 4. Thus the sequence we are looking for is defined by $a_n = n^2$. We point out, however, that this is not the only sequence whose first four terms are 1, 4, 9, and 16. In other words, the answer to our problem is not unique (see Exercises 31 and 32). In the next example we are interested in finding an *obvious* sequence whose first few terms agree with the given ones.

Example 4

Find the nth term of the given sequence:

(a) $\dfrac{1}{2}, \dfrac{3}{4}, \dfrac{5}{6}, \dfrac{7}{8}, \ldots$

(b) $-2, 4, -8, 16, -32, \ldots$

Solution

(a) We notice that the numerators of these fractions are the odd numbers and the denominators are the even numbers. Even numbers are of the form $2n$ and odd numbers are of the form $2n - 1$ (an odd number differs from an even number by 1). So a sequence that has these numbers for its first four terms is given by

$$a_n = \frac{2n - 1}{2n}$$

(b) These numbers are powers of 2 and since they alternate in sign, a sequence that agrees with these terms is given by

$$a_n = (-1)^n 2^n$$

It is a good idea to check that these formulas do indeed generate the given terms. ●

Exercises 9.1

In Exercises 1–12 find the first four terms as well as the 1000th term of each sequence.

1. $a_n = \dfrac{1}{n + 2}$

2. $a_n = n^3 + n^2 + n + 1$

3. $a_n = \dfrac{(-1)^n}{n^2}$

4. $a_n = \cos n\pi$

5. $a_n = 1 + (-1)^n$

6. $a_n = 1 - \dfrac{1}{2^n}$

7. $a_n = \sin \dfrac{n\pi}{2}$

8. $a_n = \dfrac{n^2}{n^3 + 1}$

9. $a_n = \dfrac{(-1)^{n+1} n}{n + 1}$

10. $a_n = 3n - 2$

11. $a_n = (-1)^n \dfrac{2n + 1}{\sqrt{n + 1}}$

12. $a_n = 3$

In Exercises 13–20 a sequence is defined recursively. Find the first five terms of the sequence.

13. $a_n = 2(a_{n-1} - 2)$ and $a_1 = 3$

14. $a_n = \dfrac{a_{n-1}}{2}$ and $a_1 = -8$

15. $a_n = 2a_{n-1} + 1$ and $a_1 = 1$

16. $a_n = (a_{n-1})^2$ and $a_1 = 2$

17. $a_n = a_{n-1} - a_{n-2}$ and $a_1 = 0, a_2 = 1$

18. $a_n = \dfrac{1}{1 + a_{n-1}}$ and $a_1 = 1$

19. $a_n = a_{n-1} + a_{n-2}$ and $a_1 = 1, a_2 = 2$

20. $a_n = a_{n-1} + a_{n-2} + a_{n-3}$ and $a_1 = a_2 = a_3 = 1$

In Exercises 21–30 find the nth term of the given sequence.

21. $2, 4, 8, 16, \ldots$

22. $-\dfrac{1}{3}, \dfrac{1}{9}, -\dfrac{1}{27}, \dfrac{1}{81}, \ldots$

23. $1, 4, 7, 10, \ldots$

24. $5, -25, 125, -625, \ldots$

25. $1, \dfrac{3}{4}, \dfrac{5}{9}, \dfrac{7}{16}, \dfrac{9}{25}, \ldots$

26. $\dfrac{3}{4}, \dfrac{4}{5}, \dfrac{5}{6}, \dfrac{6}{7}, \ldots$

27. $r, r^2, r^3, r^4, \ldots$

28. $a, a + d, a + 2d, a + 3d, \ldots$

29. $0, 2, 0, 2, 0, 2, \ldots$

30. $1, \dfrac{1}{2}, 3, \dfrac{1}{4}, 5, \dfrac{1}{6}, \ldots$

Exercises 31 and 32 explain why a finite number of terms do not uniquely determine a sequence.

31. (a) Show that the first four terms of the sequence $a_n = n^2$ are

$$1, 4, 9, 25$$

(b) Show that the first four terms of the sequence $a_n = n^2 + (n - 1)(n - 2)(n - 3)(n - 4)$ are also

$$1, 4, 9, 25$$

but that the sequences do not agree from the fifth term on.

(c) Find a sequence whose first six terms are the same as those of $a_n = n^2$ but whose succeeding terms differ from this sequence.

32. Find two different sequences that begin

$$2, 4, 8, 16, \ldots$$

33. Find a formula for the nth term of the sequence

$$\sqrt{2}, \sqrt{2\sqrt{2}}, \sqrt{2\sqrt{2\sqrt{2}}}, \sqrt{2\sqrt{2\sqrt{2\sqrt{2}}}}, \ldots$$

(*Hint:* Write each term as a power of 2.)

34. Find the first 100 terms of the sequence defined by

$$a_{n+1} = \begin{cases} \dfrac{a_n}{2} & \text{if } a_n \text{ is an even number} \\ 3a_n + 1 & \text{if } a_n \text{ is an odd number} \end{cases}$$

and $a_1 = 11$.

35. Repeat Exercise 34 with $a_1 = 25$.

36. Find the first ten terms of the sequence defined by

$$a_n = a_{n - a_{n-1}} + a_{n - a_{n-2}}$$

(a) where $a_1 = 1$ and $a_2 = 1$.
(b) where $a_1 = 1$ and $a_2 = 2$.

37. Fibonacci posed the following problem: Suppose that rabbits live forever and that every month each pair produces a new pair that becomes productive at age 2 months. If we start with one newborn pair, how many rabbits will there be in the nth month? Show that the answer is F_n, where F_n is the nth term of the Fibonacci sequence.

Arithmetic and Geometric Sequences

In this section we study two special kinds of sequences: arithmetic sequences, whose terms are generated by successively adding a fixed constant, and geometric sequences, whose terms are generated by successively multiplying by a fixed constant.

Arithmetic Sequences

Perhaps the simplest way to generate a sequence is to start with a number a and add to it a fixed constant d over and over again. This gives the sequence

(1) $$a, a + d, a + 2d, a + 3d, a + 4d, \ldots$$

A sequence of this type is called an **arithmetic sequence** (or an **arithmetic progression**). For example, if $a = 2$ and $d = 3$ we get the arithmetic sequence

(2) $$2, 5, 8, 11, \ldots$$

Any two consecutive terms of this sequence differ by 3. In general, two consecutive terms of (1) differ by d, so d is called the **common difference**.

As another example consider the sequence

(3) $$9, 4, -1, -6, -11, \ldots$$

Here the common difference is $d = -5$. Notice that the terms of an arithmetic sequence decrease if the common difference is negative.

It is easy to find a formula for the nth term of an arithmetic sequence. Observe that in (1)

$$\begin{aligned}
&\text{the 1st term is} \quad a + 0d \\
&\text{the 2nd term is} \quad a + 1d \\
&\text{the 3rd term is} \quad a + 2d \\
&\text{the 4th term is} \quad a + 3d \\
&\qquad\qquad\quad \vdots \qquad\quad \vdots
\end{aligned}$$

Continuing in this manner, we see that the nth term is given by

$$a_n = a + (n - 1)d$$

For example, the nth terms of the sequences in (2) and (3) are

$$a_n = 2 + 3(n - 1)$$

and

$$a_n = 9 - 5(n - 1)$$

An arithmetic sequence is completely determined by the first term a and the common difference d. Thus if we know the first two terms of an arithmetic sequence, we can find a formula for the nth term, as the following example shows.

Example 1

Find the first six terms as well as the 300th term of the arithmetic sequence

$$13, 7, \ldots$$

Solution

Since the first term is 13, we have $a = 13$. The common difference is $d = 7 - 13 = -6$. Thus the nth term of this sequence is

$$a_n = 13 - 6(n - 1)$$

From this we find the first six terms:

$$13, 7, 1, -5, -11, -17$$

The 300th term is $a_{300} = 13 - 6(299) = -1781$.   ●

In fact, an arithmetic sequence is completely determined by any two of its terms. Here is an example.

Example 2

The 11th term of an arithmetic sequence is 52 and the 19th term is 92. Find the 1000th term.

Solution

To find the nth term of this sequence we need to find a and d in the formula

$$a_n = a + (n - 1)d$$

From this formula we get

$$a_{11} = a + (11 - 1)d = a + 10d$$
$$a_{19} = a + (19 - 1)d = a + 18d$$

Since $a_{11} = 52$ and $a_{19} = 92$, we get the two equations:

$$\begin{cases} 52 = a + 10d \\ 92 = a + 18d \end{cases}$$

Solving this system for a and d, we get $a = 2$ and $d = 5$. (Verify this.) Thus the nth term of this sequence is

$$a_n = 2 + 5(n - 1)$$

The 1000th term is $a_{1000} = 2 + 5(999) = 4997$.   ●

Geometric Sequences

Another simple way of generating a sequence is to start with a number a and repeatedly multiply by a fixed constant r, where $r \neq 0$. In this way we arrive at the sequence

(4) $$a, ar, ar^2, ar^3, ar^4, \ldots$$

A sequence of this kind is called a **geometric sequence** (or a **geometric progression**). The ratio of any two consecutive terms of (4) is r. For this reason r is called the **common ratio** of this geometric sequence.

Here are three examples of geometric sequences:

(5) $$3, 6, 12, 24, 48, \ldots$$

(6) $$2, -10, 50, -250, 1250, \ldots$$

(7) $$1, \frac{1}{3}, \frac{1}{9}, \frac{1}{27}, \frac{1}{81}, \ldots$$

In the first sequence $a = 3$ and $r = 2$, in the second sequence $a = 2$ and $r = -5$, and in the third sequence $a = 1$ and $r = \frac{1}{3}$. Notice that when r is negative the terms of the sequence alternate in sign. When $0 < r < 1$ the terms of the sequence decrease, and when $r > 1$ the terms increase. (What happens if $r = 1$?)

From (4) we have the following:

$$\text{the 1st term is} \quad ar^0$$
$$\text{the 2nd term is} \quad ar^1$$
$$\text{the 3rd term is} \quad ar^2$$
$$\text{the 4th term is} \quad ar^3$$
$$\vdots \qquad\qquad \vdots$$

Since this pattern continues, we see that the nth term of a geometric sequence is given by

$$a_n = ar^{n-1}$$

Thus the nth terms of the sequences in (5), (6), and (7) are

$$a_n = 3(2)^{n-1}$$
$$a_n = 2(-5)^{n-1}$$
$$a_n = 1\left(\frac{1}{3}\right)^{n-1}$$

Geometric sequences occur naturally. Here is a simple example. Suppose that when a ball is dropped, its elasticity is such that it bounces up one third of the distance it has fallen. If this ball is dropped from a height of 2 m, it bounces up to a height of $2(\frac{1}{3}) = \frac{2}{3}$ m. On its second bounce it returns to a height of $(\frac{2}{3})(\frac{1}{3}) = \frac{2}{9}$ m, and so on (see Figure 1). Thus the height h_n that the ball reaches on its nth bounce is given by the geometric sequence

$$h_n = \frac{2}{3}\left(\frac{1}{3}\right)^{n-1} = 2\left(\frac{1}{3}\right)^n$$

We can find the nth term of a geometric sequence if we know any two terms, as the following examples show.

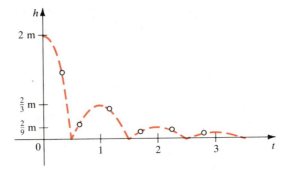

Figure 1

Example 3

Find the eighth term of the geometric sequence $5, 15, 45, \ldots$.

Solution

To find a formula for the nth term of this sequence we need to find a and r. Clearly $a = 5$. To find r we find the ratio of any two consecutive terms. For instance,

$$r = \frac{45}{15} = 3$$

Thus

$$a_n = 5(3)^{n-1}$$

The eighth term is

$$a_8 = 5(3)^{8-1} = 5(3)^7 = 10{,}935 \qquad \bullet$$

Example 4

The third term of a geometric series is $\frac{63}{4}$ and the sixth term is $\frac{1701}{32}$. Find the fifth term.

Solution

Since this series is geometric, its nth term is given by the formula

$$a_n = ar^{n-1}$$

Thus

$$a_3 = ar^{3-1} = ar^2$$

and

$$a_6 = ar^{6-1} = ar^5$$

From the values we are given for these two terms we get the system of equations:

(8)

(9)

$$\begin{cases} \dfrac{63}{4} = ar^2 \\[2mm] \dfrac{1701}{32} = ar^5 \end{cases}$$

One way to solve this system is to divide Equation 9 by Equation 8 to get

$$\frac{ar^5}{ar^2} = \frac{\frac{1701}{32}}{\frac{63}{4}}$$

So $$r^3 = \frac{27}{8} \quad \text{or} \quad r = \frac{3}{2}$$

Substituting for r in Equation 8 gives

$$\frac{63}{4} = a\left(\frac{3}{2}\right)^2 \quad \text{or} \quad a = 7$$

It follows that the nth term of this sequence is

$$a_n = 7\left(\frac{3}{2}\right)^{n-1}$$

and so the fifth term is

$$a_5 = 7\left(\frac{3}{2}\right)^{5-1} = 7\left(\frac{3}{2}\right)^4 = \frac{567}{16}$$

Exercises 9.2

In Exercises 1–8 determine the common difference, the fifth term, the nth term, and the 100th term for the given arithmetic sequence.

1. $4, 9, 14, 19, \ldots$

2. $11, 8, 5, 2, \ldots$

3. $-12, -8, -4, 0, \ldots$

4. $\frac{7}{6}, \frac{5}{3}, \frac{13}{6}, \frac{8}{3}, \ldots$

5. $25, 26.5, 28, 29.5, \ldots$

6. $15, 12.3, 9.6, 6.9, \ldots$

7. $2, 2 + s, 2 + 2s, 2 + 3s, \ldots$

8. $-t, -t + 3, -t + 6, -t + 9, \ldots$

In Exercises 9–18 determine the common ratio, the fifth term, and the nth term for the given geometric sequence.

9. $2, 6, 18, 54, \ldots$

10. $7, \frac{14}{3}, \frac{28}{9}, \frac{56}{27}, \ldots$

11. $0.3, -0.09, 0.027, -0.0081$

12. $1, \sqrt{2}, 2, 2\sqrt{2}, \ldots$

13. $144, -12, 1, -\frac{1}{12}, \ldots$

14. $-8, -2, -\frac{1}{2}, -\frac{1}{8}, \ldots$

15. $3, 3^{5/3}, 3^{7/3}, 27, \ldots$

16. $t, \dfrac{t^2}{2}, \dfrac{t^3}{4}, \dfrac{t^4}{8}, \ldots$

17. $1, s^{2/7}, s^{4/7}, s^{6/7}, \ldots$

18. $5, 5^{c+1}, 5^{2c+1}, 5^{3c+1}, \ldots$

In Exercises 19–30 the first four terms of a sequence are given. In each exercise determine whether the given terms can be the terms of an arithmetic sequence, a geometric sequence, or neither. Find the next term if the sequence is arithmetic or geometric.

19. $5, -3, 5, -3, \ldots$

20. $\frac{1}{3}, 1, \frac{5}{3}, \frac{7}{3}, \ldots$

21. $\sqrt{3}, 3, 3\sqrt{3}, 9, \ldots$

22. $1, -1, 1, -1, \ldots$

23. $2, -1, \frac{1}{2}, 2, \ldots$

24. $-3, 1, 5, 8, \ldots$

25. $x - 1, x, x + 1, x + 2, \ldots$

26. $\dfrac{\sqrt{2}}{\sqrt{2}+1}, \dfrac{2}{\sqrt{2}+1}, \dfrac{4}{\sqrt{2}+2}, \dfrac{4}{\sqrt{2}+1}, \ldots$

27. $16, 8, 4, 1, \ldots$

28. $-3, -\frac{3}{2}, 0, \frac{3}{2}, \ldots$

29. $1, \frac{3}{2}, 2, \frac{5}{2}, \ldots$

30. $\sqrt{5}, \sqrt[3]{5}, \sqrt[6]{5}, 1, \ldots$

31.   The tenth term of an arithmetic sequence is $\frac{55}{2}$ and the second term is $\frac{7}{2}$. Find the first term.

32.   The 12th term of an arithmetic sequence is 32 and the fifth term is 18. Find the 20th term.

33.   The 100th term of an arithmetic sequence is 98 and the common difference is 2. Find the first three terms.

34. The first term of a geometric sequence is 8 and the second term is 4. Find the fifth term.

35. The first term of a geometric sequence is 3 and the third term is $\frac{4}{3}$. Find the fifth term.

36. The first term of a geometric sequence is 1 and the fifth term is 7^4. Find the common ratio, assuming it is positive.

37. The common ratio in a geometric sequence is $\frac{2}{5}$ and the fourth term is $\frac{5}{2}$. Find the third term.

38. The common ratio in a geometric sequence is $\frac{3}{2}$ and the fifth term is 1. Find the first three terms.

39. Which term of the arithmetic sequence $1, 4, 7, \ldots$ is 88?

40. Which term of the geometric sequence $2, 6, 18, \ldots$ is 118,098?

41. The first term of an arithmetic sequence is 1 and the common difference is 4. Is 11,937 a term of this sequence? If so, which term is it?

42. The second term and the fifth term of a geometric sequence are 10 and 1250, respectively. Is 31,250 a term of this sequence? If so, which term is it?

43. A ball is dropped from a height of 80 ft. The elasticity of this ball is such that it rebounds three-fourths of the distance it has fallen. How high does the ball go on the fifth bounce? Find a formula for how high the ball goes on the nth bounce.

44. The size of a bacteria culture increases by 8% every hour and there are 5000 bacteria present initially. How many bacteria will be present at the end of 5 h? Find a formula for the number of bacteria present after n hours.

45. Suppose that the value of a certain machine depreciates 15% each year. What is the value of the machine after 6 yr if its original cost is $12,500?

46. A certain radioactive substance decays so that at the end of each year there is only 0.85% as much as there was at the beginning of the year. If there was originally 1 kg of the substance, find the amount that remains after 5 yr. Find a formula for the amount that remains after n years.

47. A truck radiator holds 5 gal and is filled with water. A gallon of water is removed from the radiator and replaced with a gallon of antifreeze; then a gallon of the mixture is removed from the radiator and again replaced with a gallon of antifreeze. This process is repeated indefinitely. How much water remains in the tank after this process is repeated three times? five times? n times?

48. Show that the sequence defined recursively by

$$a_{n+1} = \frac{3a_n + 1}{3} \quad \text{and} \quad a_1 = c$$

is an arithmetic sequence and find the common difference.

49. Show that the sequence defined recursively by $a_{n+1} = 3a_n$ and $a_1 = c$ is a geometric sequence and find the common ratio.

50. If $a_1, a_2, a_3, \ldots$ is a geometric sequence with common ratio r, show that the sequence

$$\frac{1}{a_1}, \frac{1}{a_2}, \frac{1}{a_3}, \ldots$$

is also a geometric sequence and find the common ratio.

51. If $a_1, a_2, a_3, \ldots$ is a geometric sequence with common ratio r, show that the sequence

$$a_1^2, a_2^2, a_3^2, \ldots$$

is also a geometric sequence and find the common ratio.

52. If $a_1, a_2, a_3, \ldots$ is a geometric sequence with common ratio $r > 0$ and $a_1 > 0$, show that the sequence

$$\log a_1, \log a_2, \log a_3, \ldots$$

is an arithmetic sequence and find the common difference.

53. If $a_1, a_2, a_3, \ldots$ is an arithmetic sequence with common difference d, show that the sequence

$$10^{a_1}, 10^{a_2}, 10^{a_3}, \ldots$$

is a geometric sequence and find the common ratio.

54. Show that a right triangle whose sides are in arithmetic progression is similar to a 3, 4, 5 triangle.

55. If the sum of three consecutive terms in an arithmetic sequence is 15 and their product is 80, find the three terms. (*Hint:* Let x denote the middle term.)

56. The sum of five consecutive terms of an arithmetic sequence is 260. Find the middle term.

57. The first three terms of an arithmetic sequence are $x - 1$, $x + 1$, and $3x + 3$. Find x.

58. If the product of three consecutive terms in a geometric sequence is 216 and their sum is 21, find the three terms. (*Hint:* Let x denote the middle term.)

59. Let $a_1, a_2, a_3, \ldots$ be a geometric sequence with positive terms that satisfy

$$a_n = a_{n+1} + a_{n+2}$$

Find the common ratio.

60. If the numbers $a_1, a_2, \ldots, a_n$ form an arithmetic sequence, then $a_2, a_3, \ldots, a_{n-1}$ are called *arithmetic means* between a_1 and a_n. Insert three arithmetic means between 2 and 14. (Notice that if only one arithmetic mean is inserted between two numbers, then it is their average.)

61. If the numbers $a_1, a_2, \ldots, a_n$ form a geometric sequence, then $a_2, a_3, \ldots, a_{n-1}$ are called *geometric means* between a_1 and a_n. Insert three geometric means between 5 and 80.

62. A sequence is called *harmonic* if the reciprocals of the terms of the sequence form an arithmetic sequence. Determine whether the sequence

$$1, \frac{3}{5}, \frac{3}{7}, 3, \ldots$$

is harmonic.

63. The *harmonic mean* of two numbers is the reciprocal of the arithmetic mean (see Exercise 60) of the reciprocals of the two numbers. Find the harmonic mean of 3 and 5.

Series

In this section we are interested in adding the terms of a sequence. For example, we might want to add the first 100 terms of the sequence

$$1, 2, 3, 4, \ldots$$

to find

$$1 + 2 + 3 + 4 + \cdots + 100$$

Before tackling these problems we need a more compact way of writing such long sums.

Sigma Notation

Given a sequence

$$a_1, a_2, a_3, a_4, \ldots$$

we can write the sum of the first n terms using **sigma notation**. This notation derives its name from the Greek letter Σ (capital sigma, corresponding to our S for sum) and is used as follows:

(1)
$$\sum_{k=1}^{n} a_k = a_1 + a_2 + a_3 + a_4 + \cdots + a_n$$

The left side of (1) is read "The sum of a_k from $k = 1$ to $k = n$." The letter k is called the **index of summation** or the **summation variable**, and the idea is to replace k in the expression after the sigma by the integers $1, 2, 3, \ldots, n$ and add the resulting expressions, arriving at the right side of (1). For example, the sum of the squares of the first five integers can be written as

$$\sum_{k=1}^{5} k^2 = 1^2 + 2^2 + 3^2 + 4^2 + 5^2 = 55$$

Although we often use the letter k for the index of summation, any other letter can be used without affecting the sum. So this last sum can be written as

$$\sum_{j=1}^{5} j^2 = 1^2 + 2^2 + 3^2 + 4^2 + 5^2$$

The index of summation need not start at 1. For example,

$$\sum_{j=3}^{7} j^2 = 3^2 + 4^2 + 5^2 + 6^2 + 7^2$$

The following examples illustrate these points.

Example 1
Find the following sums:

(a) $\displaystyle\sum_{k=1}^{3} k^3(k-1)$ (b) $\displaystyle\sum_{j=3}^{5} \frac{1}{j}$ (c) $\displaystyle\sum_{i=5}^{10} i$ (d) $\displaystyle\sum_{i=1}^{6} 2$

Solution

(a) $\displaystyle\sum_{k=1}^{3} k^3(k-1) = 1^3(1-1) + 2^3(2-1) + 3^3(3-1) = 0 + 8 + 54 = 62$

(b) $\displaystyle\sum_{j=3}^{5} \frac{1}{j} = \frac{1}{3} + \frac{1}{4} + \frac{1}{5} = \frac{47}{60}$

(c) $\displaystyle\sum_{i=5}^{10} i = 5 + 6 + 7 + 8 + 9 + 10 = 45$

(d) $\displaystyle\sum_{i=1}^{6} 2 = 2 + 2 + 2 + 2 + 2 + 2 = 12$ ●

Example 2
Write the following sums using sigma notation:
(a) $1^3 + 2^3 + 3^3 + 4^3 + 5^3 + 6^3 + 7^3$
(b) $\sqrt{3} + \sqrt{4} + \sqrt{5} + \cdots + \sqrt{77}$

Solution

(a)   We can write

$$1^3 + 2^3 + 3^3 + 4^3 + 5^3 + 6^3 + 7^3 = \sum_{k=1}^{7} k^3$$

(b)   A natural way to write this sum is

$$\sqrt{3} + \sqrt{4} + \sqrt{5} + \cdots + \sqrt{77} = \sum_{k=3}^{77} \sqrt{k}$$

There is no unique way of writing a sum in sigma notation. We could also write this last sum as

$$\sqrt{3} + \sqrt{4} + \sqrt{5} + \cdots + \sqrt{77} = \sum_{k=0}^{74} \sqrt{k+3}$$

or

$$\sqrt{3} + \sqrt{4} + \sqrt{5} + \cdots + \sqrt{77} = \sum_{k=1}^{75} \sqrt{k+2}$$ ●

The following properties of sums are natural consequences of properties of the real numbers.

Properties of Sums

Let $a_1, a_2, a_3, a_4, \ldots$ and $b_1, b_2, b_3, b_4, \ldots$ be sequences. Then for every positive integer n,

1. $\displaystyle\sum_{k=1}^{n} (a_k + b_k) = \sum_{k=1}^{n} a_k + \sum_{k=1}^{n} b_k$

2. $\displaystyle\sum_{k=1}^{n} (a_k - b_k) = \sum_{k=1}^{n} a_k - \sum_{k=1}^{n} b_k$

3. $\displaystyle\sum_{k=1}^{n} ca_k = c\left(\sum_{k=1}^{n} a_k\right)$ for any real number c

To prove Property 1 we write out the left side of the equation to get

$$\sum_{k=1}^{n} (a_k + b_k) = (a_1 + b_1) + (a_2 + b_2) + (a_3 + b_3) + \cdots + (a_n + b_n)$$

Since addition is commutative and associative, we can rearrange the terms on the right side to read

$$\sum_{k=1}^{n} (a_k + b_k) = (a_1 + a_2 + a_3 + \cdots + a_n) + (b_1 + b_2 + b_3 + \cdots + b_n)$$

Now rewriting the right side using sigma notation gives Property 1. Property 2 is proved in a similar manner. To prove Property 3 we have

$$\sum_{k=1}^{n} ca_k = ca_1 + ca_2 + ca_3 + \cdots + ca_n$$

$$= c(a_1 + a_2 + a_3 + \cdots + a_n)$$

$$= c\left(\sum_{k=1}^{n} a_k\right)$$

Series

When we add some of the terms of a sequence we get a **series** (or a **finite series**). For example,

$$\sum_{k=1}^{1,000,000} a_k$$

is the series that consists of the first million terms of the sequence $a_1, a_2, a_3, a_4, \ldots$ added together, whereas

(2)
$$\sum_{k=1}^{n} a_k$$

is the series that consists of the first n terms of this sequence added together. The number a_1 is called the **first term** of the series, a_2 is called the **second term**, and in general a_k is called the **kth term**. The number that a series adds to is called the **sum** of the series.

In this and the next section we find the sums of series that consist of many terms. For example, let us find the sum of the series

(3)
$$\sum_{k=1}^{1000} \left(\frac{1}{k} - \frac{1}{k+1} \right)$$

Since this series has 1000 terms, it would take a long time to write down all the terms and add them together. We need a better way of doing this. We will start by adding a few terms of the series and try to detect a pattern as we add more and more terms.

To do this we need some notation. We write S_1 for the first term of a series, S_2 for the sum of the first two terms, S_3 for the sum of the first three terms, and in general S_n for the sum of the first n terms of a series. S_n is called the **nth partial sum** of the series, and the sequence

$$S_1, S_2, S_3, \ldots, S_n, \ldots$$

is called the **sequence of partial sums**. For the series in (2) we have

$$
\begin{aligned}
S_1 &= a_1 \\
S_2 &= a_1 + a_2 \\
S_3 &= a_1 + a_2 + a_3 \\
S_4 &= a_1 + a_2 + a_3 + a_4 \\
&\ \ \vdots \\
S_n &= a_1 + a_2 + a_3 + \cdots + a_n
\end{aligned}
$$

Let us find the partial sums of the series in (3):

$$
\begin{aligned}
S_1 &= \left(1 - \frac{1}{2}\right) & &= 1 - \frac{1}{2} \\
S_2 &= \left(1 - \frac{1}{2}\right) + \left(\frac{1}{2} - \frac{1}{3}\right) & &= 1 - \frac{1}{3} \\
S_3 &= \left(1 - \frac{1}{2}\right) + \left(\frac{1}{2} - \frac{1}{3}\right) + \left(\frac{1}{3} - \frac{1}{4}\right) & &= 1 - \frac{1}{4} \\
S_4 &= \left(1 - \frac{1}{2}\right) + \left(\frac{1}{2} - \frac{1}{3}\right) + \left(\frac{1}{3} - \frac{1}{4}\right) + \left(\frac{1}{4} - \frac{1}{5}\right) &= 1 - \frac{1}{5}
\end{aligned}
$$

Do we detect a pattern here? Of course, we have

$$S_n = 1 - \frac{1}{n + 1}$$

It is now easy to find the sum of as many terms of this series as we please. For instance, the sum of the first 100 terms is

$$S_{100} = 1 - \frac{1}{101} = \frac{100}{101}$$

Notice that the sum of the series is the sum of all 1000 terms. Thus the sum of this series is

$$S_{1000} = 1 - \frac{1}{1001} = \frac{1000}{1001}$$

Example 3

Find the sum of the series

$$\sum_{k=1}^{100} \frac{1}{2^k}$$

Solution

We first find a formula for the nth partial sum of this series. We do this by writing down the first few partial sums and trying to see a pattern:

$$S_1 = \frac{1}{2} \qquad\qquad = \frac{1}{2}$$

$$S_2 = \frac{1}{2} + \frac{1}{4} \qquad\qquad = \frac{3}{4}$$

$$S_3 = \frac{1}{2} + \frac{1}{4} + \frac{1}{8} \qquad = \frac{7}{8}$$

$$S_4 = \frac{1}{2} + \frac{1}{4} + \frac{1}{8} + \frac{1}{16} = \frac{15}{16}$$

$$\vdots$$

Notice that in the value of each partial sum the denominator is a power of 2 and the numerator is one less than the denominator. In general,

$$S_n = \frac{2^n - 1}{2^n} = 1 - \frac{1}{2^n}$$

Now we see that the sum of the given series is

$$S_{100} = 1 - \frac{1}{2^{100}}$$

●

Exercises 9.3

In Exercises 1–12 find the given sum.

1. $\displaystyle\sum_{k=3}^{6} k^2$

2. $\displaystyle\sum_{n=1}^{4} \frac{1}{n}$

3. $\displaystyle\sum_{k=0}^{5} (7 - 2k)$

4. $\displaystyle\sum_{j=1}^{100} (-1)^j$

5. $\displaystyle\sum_{n=1}^{8} [1 + (-1)^n]$

6. $\displaystyle\sum_{n=4}^{12} 10$

7. $\displaystyle\sum_{k=1}^{5} 2^{k-1}$

8. $\displaystyle\sum_{j=1}^{10} \frac{3}{j + 2}$

9. $\displaystyle\sum_{m=0}^{4} (-3)^{m+2}$

10. $\displaystyle\sum_{n=1}^{3} n2^n$

11. $\displaystyle\sum_{m=3}^{5} (2^m + m^2)$

12. $\displaystyle\sum_{k=1}^{1} k^{100}$

In Exercises 13–20 write the given sum without using sigma notation.

13. $\displaystyle\sum_{k=1}^{5} \sqrt{k}$

14. $\displaystyle\sum_{n=0}^{4} \frac{2n - 1}{2n + 1}$

15. $\displaystyle\sum_{k=0}^{6} \sqrt{k + 4}$

16. $\displaystyle\sum_{k=3}^{100} x^k$

17. $\displaystyle\sum_{n=1}^{8} nx^{n+1}$

18. $\displaystyle\sum_{k=6}^{9} k(k + 3)$

19. $\displaystyle\sum_{j=1}^{n} (-1)^{j+1} x^j$

20. $\displaystyle\sum_{j=1}^{8} \frac{x^j}{j^2}$

In Exercises 21–30 write the given sum in sigma notation.

21. $1 + 2 + 3 + 4 + \cdots + 100$

22. $2 + 4 + 6 + \cdots + 2n$

23. $\dfrac{1}{2\ln 2} - \dfrac{1}{3\ln 3} + \dfrac{1}{4\ln 4} - \dfrac{1}{5\ln 5} + \cdots + \dfrac{1}{100\ln 100}$

24. $1 + x + x^2 + x^3 + \cdots + x^{100}$

25. $1 - \dfrac{x}{3} + \dfrac{x^2}{9} - \dfrac{x^3}{27} + \dfrac{x^4}{81} - \dfrac{x^5}{243}$

26. $\dfrac{10}{15} + \dfrac{11}{16} + \dfrac{12}{17} + \cdots + \dfrac{100}{105}$

27. $1 - 2x + 3x^2 - 4x^3 + 5x^4 + \cdots + 100x^{99}$

28. $\dfrac{1}{1 \cdot 2} + \dfrac{1}{2 \cdot 3} + \dfrac{1}{3 \cdot 4} + \cdots + \dfrac{1}{999 \cdot 1000}$

29. $1 \cdot 2 \cdot 3 + 2 \cdot 3 \cdot 4 + 3 \cdot 4 \cdot 5 + \cdots + 97 \cdot 98 \cdot 99$

30. $\dfrac{\sin 1}{1^2} + \dfrac{\sin 2}{2^2} + \dfrac{\sin 3}{3^2} + \cdots + \dfrac{\sin n}{n^2}$

In Exercises 31–39 find a formula for the nth partial sum of the series (as in Example 3) and use this to find the sum of the series.

31. $\displaystyle\sum_{k=1}^{1000} \left(\frac{1}{k + 1} - \frac{1}{k + 2} \right)$

32. $\displaystyle\sum_{k=1}^{100} \left(\frac{1}{2k - 1} - \frac{1}{2k + 1} \right)$

33. $\displaystyle\sum_{k=1}^{20} \frac{2}{3^k}$

34. $\displaystyle\sum_{j=1}^{20} \frac{4}{5^j}$

35. $\displaystyle\sum_{n=1}^{99} (\sqrt{n} - \sqrt{n + 1})$

36. $\displaystyle\sum_{k=0}^{1000} \cos k\pi$

37. $\displaystyle\sum_{m=1}^{101} \left[\tan\left(\frac{m\pi}{3}\right) - \tan\left(\frac{(m + 1)\pi}{3}\right) \right]$

38. $\displaystyle\sum_{k=1}^{20} (2^{k-1} - 2^k)$

39. $\displaystyle\sum_{k=1}^{999999} \log\left(\frac{k}{k + 1}\right)$ (*Hint:* Use a property of logarithms to write the *k*th term as a difference.)

40. Let $a_1, a_2, a_3, \ldots$ be a sequence.
   (a) Show that $\displaystyle\sum_{k=1}^{n} (a_k - a_{k+1}) = a_1 - a_{n+1}$. A series of this form is called a **telescoping series**.
   (b) Which of the series in Exercises 31–39 are telescoping?

Arithmetic and Geometric Series

In this section we find formulas for the sums of series whose terms form an arithmetic or geometric sequence.

Arithmetic Series

An **arithmetic series** is a series whose terms form an arithmetic sequence. We will find a formula for the nth partial sum of an arithmetic series.

Let us begin with a simple example. Suppose we want to find the sum of the numbers $1, 2, 3, 4, \ldots, 100$; that is,

(1)
$$\sum_{k=1}^{100} k$$

When the famous mathematician Carl Friedrich Gauss was a schoolboy, his teacher asked the class this question and expected that it would keep the students busy for a long time. But Gauss answered the question almost immediately. His idea was that, since we are adding numbers that are produced according to a fixed pattern, there must also be a pattern (or formula) for finding the sum. He started by writing the numbers from 1 to 100 and below them the same numbers in reverse order. Writing S for the sum and adding corresponding terms, we get

$$
\begin{aligned}
S &= 1 + 2 + 3 + \cdots + 98 + 99 + 100 \\
S &= 100 + 99 + 98 + \cdots + 3 + 2 + 1 \\
\hline
2S &= 101 + 101 + 101 + \cdots + 101 + 101 + 101
\end{aligned}
$$

It follows that $2S = 100(101) = 10{,}100$ and so $S = 5050$.

Of course the sequence of natural numbers $1, 2, 3, \ldots$ is an arithmetic sequence (with $a = 1$ and $d = 1$) and the method used for summing the first 100 terms of the series in (1) can be used to find a formula for the nth partial sum of any arithmetic series. We want to find the sum of the first n terms of the arithmetic sequence whose terms are $a_k = a + (k-1)d$; that is, we want to find

(2)
$$S_n = \sum_{k=1}^{n} [a + (k-1)d]$$
$$= a + (a+d) + (a+2d) + (a+3d) + \cdots + [a + (n-1)d]$$

Using Gauss's method we write

$$
\begin{aligned}
S_n &= a + (a+d) + \cdots + [a+(n-2)d] + [a+(n-1)d] \\
S_n &= [a+(n-1)d] + [a+(n-2)d] + \cdots + (a+d) + a \\
\hline
2S_n &= [2a+(n-1)d] + [2a+(n-1)d] + \cdots + [2a+(n-1)d] + [2a+(n-1)d]
\end{aligned}
$$

There are n identical terms on the right side of this equation, so

$$2S_n = n[2a + (n-1)d] \qquad \text{or} \qquad S_n = \frac{n}{2}[2a + (n-1)d]$$

Notice that $a_n = a + (n-1)d$ is the last term of the series in (2). So we can write

$$S_n = \frac{n}{2}[a + a + (n-1)d] = n\left(\frac{a + a_n}{2}\right)$$

This last formula says that the sum of the first n terms of an arithmetic series is the average of the first and last terms multiplied by n, the number of terms in the series. We summarize these results:

Sum of an Arithmetic Series

The sum of the first n terms of an arithmetic series

$$a + (a + d) + (a + 2d) + (a + 3d) + \cdots + [a + (n - 1)d]$$

is given by

(3)
$$S_n = \frac{n}{2}[2a + (n - 1)d]$$

or

(4)
$$S_n = n\left(\frac{a + a_n}{2}\right)$$

Example 1

Find the sum of the first 50 odd numbers.

Solution

The odd numbers form an arithmetic sequence whose nth term is given by $a_n = 2n - 1$, so the 50th odd number is $a_{50} = 2(50) - 1 = 99$. Substituting in Formula 4, we get

$$S_{50} = 50\left(\frac{a + a_{50}}{2}\right)$$

$$= 50\left(\frac{1 + 99}{2}\right) = 50 \cdot 50 = 2500$$

●

Example 2

Find the sum of the first 40 terms of the arithmetic sequence $3, 7, 11, 15, \ldots$.

Solution

For this arithmetic sequence $a = 3$ and $d = 4$. Using Formula 3, we get

$$S_{40} = \frac{40}{2}[2(3) + (40 - 1)4]$$

$$= 20[6 + 156] = 3240$$

●

Example 3

An arithmetic series has first term 5 and 50th term 103. How many terms of this series must be added to get 572?

Solution

We first find the common difference. Since for an arithmetic sequence $a_n = a + (n - 1)d$, we get

$$a_{50} = a + (50 - 1)d$$

so
$$103 = 5 + 49d$$

Solving for d gives $d = 2$.

We are asked to find n when $S_n = 572$. Using (3) and substituting for S_n, a, and d give

$$572 = \frac{n}{2}[2 \cdot 5 + (n - 1)2]$$

Solving for n, we have

$$572 = 5n + n(n - 1)$$
$$n^2 + 4n - 572 = 0$$

so
$$(n - 22)(n + 26) = 0$$

This gives $n = 22$ or $n = -26$. But since n is a *number* of terms in a sequence, we must have $n = 22$. ●

Geometric Series

A **geometric series** is a series whose terms form a geometric sequence. So, adding the first n terms of the geometric sequence

$$a, ar, ar^2, ar^3, ar^4, \ldots, ar^{n-1}, \ldots$$

we get the geometric series

$$S_n = \sum_{k=1}^{n} ar^{k-1} = a + ar + ar^2 + ar^3 + ar^4 + \cdots + ar^{n-1}$$

To find a formula for S_n we multiply S_n by r and subtract from S_n to get

$$S_n = a + ar + ar^2 + ar^3 + ar^4 + \cdots + ar^{n-1}$$
$$\underline{rS_n = \quad\quad ar + ar^2 + ar^3 + ar^4 + \cdots + ar^{n-1} + ar^n}$$
$$S_n - rS_n = a - ar^n$$

So
$$S_n(1 - r) = a(1 - r^n)$$

or
$$S_n = \frac{a(1 - r^n)}{1 - r}, \quad r \neq 1$$

Sum of a Geometric Series

The sum of the first n terms of a geometric series is given by

(5)
$$S_n = a\frac{1 - r^n}{1 - r}, \quad r \neq 1$$

Example 4

Find the sum of the first five terms of the geometric sequence $1, 0.7, 0.49, 0.343, \ldots$.

Solution

The sum is a geometric series with $a = 1$ and $r = 0.7$. Using Formula 5 with $n = 5$, we get

$$S_5 = 1 \frac{1 - (0.7)^5}{1 - 0.7} = 2.7731$$

Thus the sum of the first five terms of this sequence is 2.7731.   ●

Example 5

Find the sum of the series

$$\sum_{k=1}^{5} 7\left(-\frac{2}{3}\right)^k$$

Solution

This is a geometric series with first term $a = 7(-\frac{2}{3}) = -\frac{14}{3}$ and common ratio $r = -\frac{2}{3}$, and there are five terms. Thus by Formula 5,

$$S_5 = -\frac{14}{3} \frac{[1 - (-2/3)^5]}{1 - (-2/3)} = -\frac{14}{3} \frac{1 + \frac{32}{243}}{\frac{5}{3}} = -\frac{770}{243}$$ ●

Application to Annuities

An **annuity** is a sum of money that is paid in regular equal payments. Although the word *annuity* suggests annual (or yearly) payments, they can be made semiannually, quarterly, monthly, or at other regular intervals. Payments are usually made at the end of the payment interval. The **amount of an annuity** is the sum of all the individual payments from the time of payment until the last payment is made, together with all the interest.

Example 6

An investor deposits $400 every December 15 and June 15 for ten years into an account that bears interest at the rate of 8% a year compounded semiannually. How much will she have in the account at the time of the last payment?

Solution

We need to find the amount of an annuity that consists of 20 semiannual payments of $400 each. Since the interest rate is 8% a year compounded semiannually, the interest rate per time period is $i = 0.04$. Notice that the first payment is in the account for 19 time periods, the second payment is in for 18 time periods, and so on. The last payment receives no interest. The situation can be illustrated by the time line in Figure 1.

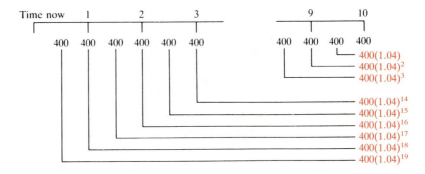

Figure 1

The amount A of the annuity is the sum of these 20 amounts. Thus

$$A = 400 + 400(1.04) + 400(1.04)^2 + \cdots + 400(1.04)^{19}$$

But this is a geometric series with $a = 400$, $r = 1.04$, and $n = 20$, so by Equation 5,

$$A = 400\frac{1 - (1.04)^{20}}{1 - 1.04} \approx 11{,}911.23$$

Thus the amount of the annuity after ten years is $11,911.23.   ●

In general, the regular annuity payment is called the **periodic rent** and is denoted by R. We also let i denote the interest rate per time period and n the number of payments. (We will always assume that the time period in which interest is compounded is equal to the time between payments.) By the same reasoning as in Example 6, we see that the amount A of an annuity is

$$A = R + R(1 + i) + R(1 + i)^2 + \cdots + R(1 + i)^{n-1}$$

Since this is a geometric series with n terms, where $a = R$ and $r = 1 + i$, Equation 5 gives

$$A = R\frac{1 - (1 + i)^n}{1 - (1 + i)} = R\frac{1 - (1 + i)^n}{-i} = R\frac{(1 + i)^n - 1}{i}$$

Amount of an Annuity

(6) $$A = R\frac{(1 + i)^n - 1}{i}$$

Example 7

How much money should be invested every month at 12% a year compounded monthly in order to have $4000 in 18 months?

Solution

In this problem $i = 0.01$, $A = 4000$, and $n = 18$. We need to find the periodic rent R. Using Equation 6, we have

$$4000 = R\frac{(1 + 0.01)^{18} - 1}{0.01}$$

Solving for R, we get

$$R = \frac{4000(0.01)}{(1 + 0.01)^{18} - 1} \approx \frac{40}{1.196147 - 1} \approx 203.928$$

Thus the monthly investment should be $203.93.

Exercises 9.4

In Exercises 1–6 find the sum S_n of the arithmetic series that satisfies the given conditions.

1. $a = 4, d = 2, n = 20$

2. $a = 100, d = -5, n = 8$

3. $a_1 = 55, d = 12, n = 10$

4. $a_2 = 8, a_5 = 9.5, n = 15$

5. $a_3 = 980, a_{10} = 910, n = 5$

6. $a_4 = 21, d = 3, n = 10$

In Exercises 7–12 find the sum S_n of the geometric series that satisfies the given conditions.

7. $a = 5, r = 2, n = 6$

8. $a = \frac{2}{3}, r = \frac{1}{3}, n = 4$

9. $a_3 = 28, a_6 = 224, n = 6$

10. $a_2 = \frac{10}{3}, a_4 = \frac{80}{27}, r < 0, n = 4$

11. $a_3 = 0.18, r = 0.3, n = 5$

12. $a_2 = 0.12, a_5 = 0.00096, n = 4$

In Exercises 13–18 find the sum of the arithmetic series.

13. $1 + 5 + 9 + \cdots + 401$

14. $-3 + (-\frac{3}{2}) + 0 + \frac{3}{2} + 3 + \cdots + 30$

15. $0.7 + 2.7 + 4.7 + \cdots + 56.7$

16. $-10 - 9.9 - 9.8 - \cdots - 0.1$

17. $\displaystyle\sum_{k=0}^{10} (3 + 0.25k)$ **18.** $\displaystyle\sum_{n=0}^{20} (1 - 2n)$

In Exercises 19–24 find the sum of the geometric series.

19. $1 + 3 + 9 + \cdots + 2187$

20. $1 - \frac{1}{2} + \frac{1}{4} - \frac{1}{8} + \cdots - \frac{1}{512}$

21. $0.7 + 0.49 + 0.343 + \cdots + 0.16807$

22. $1 - \sqrt{2} + 2 - 2\sqrt{2} + \cdots + 32$

23. $\displaystyle\sum_{k=0}^{10} 3(\tfrac{1}{2})^k$ **24.** $\displaystyle\sum_{j=0}^{5} 7(\tfrac{3}{2})^j$

In Exercises 25–36 determine whether the series is arithmetic or geometric and find its sum.

25. $4 + 2.4 + 1.44 + \cdots + 0.5184$

26. $2 + 5 + 8 + \cdots + 32$

27. $1 - x + x^2 - x^3 + \cdots + x^{20}$

28. $1 - \sqrt{3} + 3 - 3\sqrt{3} + \cdots + 243$

29. $2 + 4 + 6 + \cdots + 1000$

30. $\sqrt{5} + 2\sqrt{5} + 3\sqrt{5} + \cdots + 100\sqrt{5}$

31. $\frac{1}{2} + 1 + \frac{3}{2} + \cdots + 64$ **32.** $\frac{1}{2} + 1 + 2 + \cdots + 64$

33. $\displaystyle\sum_{i=0}^{8} (1 + \sqrt{2}i)$ **34.** $\displaystyle\sum_{k=0}^{8} 2(\sqrt{3})^k$

35. $\displaystyle\sum_{n=0}^{8} 5^{n/3}$ **36.** $\displaystyle\sum_{i=0}^{8} \sqrt{5}(2)^i$

37. An arithmetic sequence has first term $a = 5$ and common difference $d = 2$. How many terms of this sequence must be added to get 2700?

38. A geometric sequence has first term $a = 1$ and common ratio $r = -\frac{1}{2}$. How many terms of this sequence must be added to get $\frac{341}{512}$?

39. The sum of the first four terms of a geometric series is 50 and the common ratio is $r = \frac{1}{2}$. Find the first term.

40. The sum of the first ten terms of an arithmetic series is 100 and the first term is 1. Find the tenth term.

41. The sum of the first 20 terms of an arithmetic series is 155 and the first term is 3. Find the common difference.

42. The sum of the first and 20th terms of an arithmetic sequence is 182. Find the sum of the first 20 terms.

43. The second term in a geometric series is $\frac{14}{3}$ and the fifth term is $\frac{112}{81}$. Find the sum of the first four terms.

44. The common ratio in a certain geometric sequence is $r = 0.2$ and the sum of the first four terms is 1248. Find the first term.

45. Find the product of the numbers $10^{1/10}, 10^{2/10}, 10^{3/10}, 10^{4/10}, \ldots, 10^{19/10}$.

46. Find the sum of the first ten terms of the sequence
$$a + b, a^2 + 2b, a^3 + 3b, a^4 + 4b, \ldots$$

47. A very patient woman wishes to become a billionaire. She decides on a simple scheme. She puts aside 1 cent the first day, 2 cents the second day, 4 cents the third day, and so on, doubling the number of cents she puts aside each day. How much money will she have at the end of 30 days? How many days will it take for this woman to realize her wish?

48. A ball is dropped from a height of 9 ft. The elasticity of the ball is such that it always bounces back one third of the distance from which it falls.
   (a) Find the total distance the ball has traveled at the instant it hits the ground for the fifth time.
   (b) Find a formula for the total distance the ball has traveled at the instant it hits the ground for the nth time.

49. When an object is allowed to fall freely near the surface of the earth, the gravitational pull is such that it falls 16 ft in the first second, 48 ft in the next second, 80 ft in the next second, and so on.
   (a) Find the total distance the ball falls in 6 s.
   (b) Find a formula for the distance the ball falls in n seconds.

50. Find the amount of an annuity that consists of 20 semi-

annual payments of $500 each into an account that pays 6% interest a year compounded semiannually.

51. Find the amount of an annuity that consists of 16 quarterly payments of $300 each into an account that pays interest of 8% a year compounded quarterly.

52. How much money should be invested every quarter at 10% a year compounded quarterly in order to have $5000 in 2 yr?

53. How much money should be invested monthly at 6% a year compounded monthly in order to have $2000 in 8 mo?

54. The **present value of an annuity (PV)** is the principal that must be invested now at the rate of interest i per time period in order to provide n payments each of amount R. We will always assume that the time period in which interest is compounded is equal to the time between payments.
   (a) Draw a time line as in Example 6 to show that the present value of an annuity is the sum of the present values of each payment (see Exercise 15 of Section 4.2); that is,
$$PV = \frac{R}{1 + i} + \frac{R}{(1 + i)^2} + \frac{R}{(1 + i)^3} + \cdots + \frac{R}{(1 + i)^n}$$
   (b) Show that
$$PV = R\frac{(1 + i)^n - 1}{i(1 + i)^n}$$

55. How much money must be invested now at 9% a year compounded semiannually to provide an annuity of 20 payments of $200 every 6 mo, with the first payment being in 6 mo? (Refer to Exercise 54.)

56. A 55-year-old man deposits $50,000 to set up an annuity with an insurance company at 8% a year compounded semiannually until age 65. If he is to be paid twice a year, how large is each payment he receives? (Refer to Exercise 54.)

57. A **mortgage** is a loan made by a bank for the purpose of purchasing a house. The home buyer repays the loan in equal monthly installments whose size R is determined by the amount of the loan A and the annual rate of interest r. In effect, the bank purchases an annuity from the borrower with present value $PV = A$ and periodic rent R. Thus the formula in Exercise 54(b) can be used to determine R. Solve the formula in Exercise 54(b) for R to show that the monthly payment on a mortgage of amount A that is to be repaid in n monthly payments at an annual interest rate r is given by
$$R = \frac{Ai}{1 - (1 + i)^{-n}}$$
where $i = r/12$ is the interest rate per month.

58. What is the monthly payment on a 30-yr mortgage of $80,000 at 12% interest? What is the monthly payment on this same mortgage if it is to be repaid over a 15-yr period? (Refer to Exercise 57.)

59. What is the monthly payment on a 30-yr mortgage of $100,000 at 8% interest? What is the total amount paid on this loan over the 30-yr period? (Refer to Exercise 57.)

Section 9.5

Infinite Geometric Series

So far we have been discussing series with a finite number of terms. Let us write down a series with an infinite number of terms

$$a_1 + a_2 + a_3 + a_4 + \cdots$$

The dots mean that we are to continue the addition indefinitely. A series of this kind is called an **infinite series**.

What Is an Infinite Series?

What meaning can we attach to the sum of infinitely many numbers? It seems at first that it is not possible to add infinitely many numbers and arrive at a finite number. But consider the following problem. You have a cake and you want to eat it by first eating half the cake, then eating half of what remains, and then again eating half of what remains. This process can continue indefinitely, since at each stage some of the cake remains (see Figure 1).

Does this mean that it is impossible to eat all of the cake? Of course not. Let us write down what you have eaten from this cake:

(1)
$$\frac{1}{2} + \frac{1}{4} + \frac{1}{8} + \frac{1}{16} + \cdots + \frac{1}{2^n} + \cdots$$

This is an infinite series and we notice two things about it. First, from Figure 1 it is clear that no matter how many terms of this series we add, the total will never exceed 1. Second, the more terms of this series that we add, the closer the sum is to 1. This suggests that the number 1 can be written as the sum of infinitely many smaller numbers:

$$1 = \frac{1}{2} + \frac{1}{4} + \frac{1}{8} + \frac{1}{16} + \cdots + \frac{1}{2^n} + \cdots$$

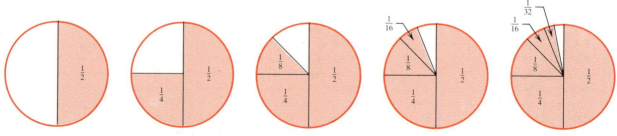

Figure 1

To make this more precise, let us look at the partial sums of this series:

$$S_1 = \frac{1}{2} \qquad\qquad\qquad = \frac{1}{2}$$

$$S_2 = \frac{1}{2} + \frac{1}{4} \qquad\qquad = \frac{3}{4}$$

$$S_3 = \frac{1}{2} + \frac{1}{4} + \frac{1}{8} \qquad = \frac{7}{8}$$

$$S_4 = \frac{1}{2} + \frac{1}{4} + \frac{1}{8} + \frac{1}{16} = \frac{15}{16}$$

$$\vdots$$

and in general (see Example 3 of Section 9.3),

$$S_n = 1 - \frac{1}{2^n}$$

As n gets larger and larger, we are adding more and more of the terms of this series. Intuitively, as n gets larger, S_n gets closer to the sum of the series. Now notice that as n gets large, $1/2^n$ gets closer and closer to 0. Thus S_n gets close to $1 - 0 = 1$. Using the notation of Section 3.7 we can write

$$S_n \to 1 \qquad \text{as} \qquad n \to \infty$$

In general, if S_n gets close to a finite number S as n gets large, we say that S is the **sum of the infinite series.**

Infinite Geometric Series

The terms of the series in (1) form a geometric sequence. We call an infinite series of the form

$$a + ar + ar^2 + ar^3 + ar^4 + \cdots + ar^{n-1} + \cdots$$

an **infinite geometric series**. We can apply the reasoning used earlier to find the sum of an infinite geometric series. The nth partial sum of such a series is given by Equation 5 of Section 9.4:

$$S_n = a\frac{1 - r^n}{1 - r}, \qquad r \neq 1$$

It can be shown that if $|r| < 1$, then r^n gets close to 0 as n gets large (you can easily convince yourself of this using a calculator). It follows that S_n gets close to $a/(1 - r)$ as n gets large, or

$$S_n \to \frac{a}{1 - r} \qquad \text{as} \qquad n \to \infty$$

Thus the sum of this infinite geometric series is $a/(1 - r)$. We summarize this result.

Sum of an Infinite Geometric Series

If $|r| < 1$, then the infinite geometric series

$$a + ar + ar^2 + ar^3 + ar^4 + \cdots + ar^{n-1} + \cdots$$

has the sum

(2)
$$S = \frac{a}{1-r}$$

Example 1

Find the sum of the infinite geometric series

$$2 + \frac{2}{5} + \frac{2}{25} + \frac{2}{125} + \cdots + \frac{2}{5^n} + \cdots$$

Solution

Using Equation 2 with $a = 2$ and $r = \frac{1}{5}$, we find that the sum of this series is

$$S = \frac{2}{1 - \frac{1}{5}} = \frac{5}{2}$$

Example 2

Find the fraction that represents the rational number $2.3\overline{51}$.

Solution

This repeating decimal can be written as a series:

$$\frac{23}{10} + \frac{51}{1000} + \frac{51}{100,000} + \frac{51}{10,000,000} + \frac{51}{1,000,000,000} + \cdots$$

The terms of this series after the first term form an infinite geometric series with

$$a = \frac{51}{1000} \qquad \text{and} \qquad r = \frac{1}{100}$$

By Equation 2 the sum of this part of the series is

$$S = \frac{\frac{51}{1000}}{1 - \frac{1}{100}} = \frac{\frac{51}{1000}}{\frac{99}{100}} = \frac{51}{1000} \cdot \frac{100}{99} = \frac{51}{990}$$

Thus
$$2.3\overline{51} = \frac{23}{10} + \frac{51}{990} = \frac{2328}{990} = \frac{388}{165}$$

Example 3

Show that the series

$$(3) \qquad \frac{1}{1 \cdot 2} + \frac{1}{2 \cdot 3} + \frac{1}{3 \cdot 4} + \cdots + \frac{1}{n(n + 1)} + \cdots$$

has a sum and find its sum.

Solution

We can write the expression for the nth term using the partial fraction decomposition (see Section 7.4) as

$$\frac{1}{n(n + 1)} = \frac{1}{n} - \frac{1}{n + 1}$$

Thus the series in (3) can be written as

$$\left(1 - \frac{1}{2}\right) + \left(\frac{1}{2} - \frac{1}{3}\right) + \left(\frac{1}{3} - \frac{1}{4}\right) + \left(\frac{1}{4} - \frac{1}{5}\right) + \cdots + \left(\frac{1}{n} - \frac{1}{n + 1}\right) + \cdots$$

(This is an example of a telescoping series; see Exercise 40 of Section 9.3). It was shown in Section 9.3 that the partial sums of this series are given by the formula

$$S_n = 1 - \frac{1}{n + 1}$$

As n gets large, $1/(n + 1)$ gets close to 0 so that S_n gets close to $1 - 0 = 1$. In symbols,

$$S_n \to 1 \qquad \text{as} \qquad n \to \infty$$

Thus the series has the sum $S = 1$. ●

Exercises 9.5

In Exercises 1–8 find the sum of the given infinite geometric series.

1. $1 - \frac{1}{3} + \frac{1}{9} - \frac{1}{27} + \cdots$

2. $\frac{2}{5} + \frac{4}{25} + \frac{8}{125} + \cdots$

3. $\frac{1}{3^6} + \frac{1}{3^8} + \frac{1}{3^{10}} + \frac{1}{3^{12}} + \cdots$

4. $3 - \frac{3}{2} + \frac{3}{4} - \frac{3}{8} + \cdots$

5. $-\frac{9}{100} + \frac{3}{10} - 1 + \frac{10}{3} - \cdots$

6. $\frac{1}{\sqrt{2}} + \frac{1}{2} + \frac{1}{2\sqrt{2}} + \frac{1}{4} + \cdots$

7. $5^{4/3} - 5^{5/3} + 5^{6/3} - 5^{7/3} + \cdots$

8. $\frac{1}{1 + \sqrt{2}} - 1 - \frac{1}{1 - \sqrt{2}} - \cdots$

In Exercises 9–14 express each repeating decimal as a fraction.

9. $0.777\ldots$

10. $0.2\overline{53}$

11. $0.030303\ldots$

12. $2.11\overline{25}$

13. $0.\overline{112}$

14. $0.123123123\ldots$

15. The elasticity of a ball is such that it rebounds two-thirds of the distance that it falls. This ball is dropped from a distance of 12 ft from the ground. Use an infinite geometric series to approximate the total distance the ball travels before it stops bouncing.

16. The midpoints of the sides of a square of side 1 are joined to form a new square. This procedure is repeated for each new square. (See the figure.)
(a) Find the sum of the areas of all the squares.
(b) Find the sum of the perimeters of all the squares.

17. A circular disk of radius R is cut out of paper as shown in part (a) of the figure. Two disks of radius $\frac{1}{2}R$ are cut out of paper and placed on top of the first disk [part (b)] and then four disks of radius $\frac{1}{4}R$ are placed on these two disks [part (c)]. Assuming that this process can be repeated indefinitely, find the total area of all the disks.

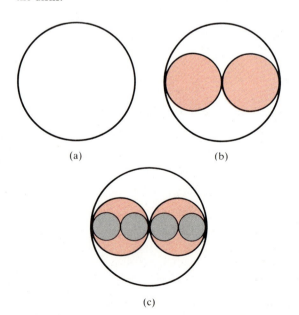

(a)   (b)

(c)

18. A right triangle ABC is given with $\angle A = \theta$ and $AC = b$. CD is drawn perpendicular to AB, DE is drawn perpendicular to BC, EF is perpendicular to AB, and this process is continued indefinitely. Find the total length of all the perpendiculars

$$|CD| + |DE| + |EF| + |FG| + \cdots$$

in terms of b and θ.

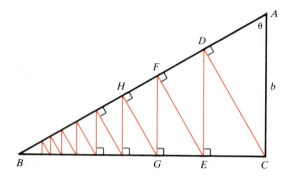

In Exercises 19–22 use the method of Example 3 to find the sum of the given infinite series.

19. $\dfrac{2}{3 \cdot 5} + \dfrac{2}{5 \cdot 7} + \dfrac{2}{7 \cdot 9} + \cdots + \dfrac{2}{(2n + 1)(2n + 3)} + \cdots$

20. $\left(\dfrac{1}{\sqrt{1}} - \dfrac{1}{\sqrt{2}}\right) + \left(\dfrac{1}{\sqrt{2}} - \dfrac{1}{\sqrt{3}}\right) + \left(\dfrac{1}{\sqrt{3}} - \dfrac{1}{\sqrt{4}}\right) + \cdots +$
$\left(\dfrac{1}{\sqrt{n}} - \dfrac{1}{\sqrt{n + 1}}\right) + \cdots$

21. $[\tan^{-1}1 - \tan^{-1}2] + [\tan^{-1}2 - \tan^{-1}3] + \cdots +$
$[\tan^{-1}n - \tan^{-1}(n + 1)] + \cdots$

22. $\dfrac{1}{1 \cdot 3} + \dfrac{1}{3 \cdot 5} + \dfrac{1}{5 \cdot 7} + \cdots + \dfrac{1}{4n^2 - 1} + \cdots$

23. Find the sum of the infinite series

$$\frac{1}{5} + \frac{2}{5^2} + \frac{1}{5^3} + \frac{2}{5^4} + \frac{1}{5^5} + \cdots$$

24. An **annuity in perpetuity** is one that continues forever. Such annuities are useful in setting up scholarship funds to ensure that the award continues.
(a) Draw a time line (as in Example 6 of Section 9.4) to show that the amount of money to be invested now (PV) at interest rate i per time period in order to set up an annuity in perpetuity of amount R per time

period is

$$PV = \frac{R}{1 + i} + \frac{R}{(1 + i)^2} + \frac{R}{(1 + i)^3} + \cdots$$

$$+ \frac{R}{(1 + i)^n} + \cdots$$

(b)   Find the sum of the infinite series in part (a) to show that

$$PV = \frac{R}{i}$$

25.   How much money must be invested now at 10% a year compounded annually to provide an annuity of $5000 every year in perpetuity? The first payment is due in one year. (Refer to Exercise 24.)

26.   How much money must be invested now at 8% a year compounded quarterly in order to provide an annuity of $3000 a year in perpetuity? The first payment is due in one year. (Refer to Exercise 24.)

Section 9.6

Mathematical Induction

There are two aspects to mathematics—discovery and proof—and both are of equal importance. It is necessary to discover something before attempting to prove it, and we can only be certain of its truth once it has been proved. In this section we look more carefully at the relationship between these two parts of mathematics.

Conjecture and Proof

Let us try a simple experiment. We add more and more of the odd numbers as follows:

$$1 = 1$$
$$1 + 3 = 4$$
$$1 + 3 + 5 = 9$$
$$1 + 3 + 5 + 7 = 16$$
$$1 + 3 + 5 + 7 + 9 = 25$$

What do you notice about the numbers on the right side of these equations? They are in fact all perfect squares. These equations say that:

The sum of the first 1 odd number is 1^2.

The sum of the first 2 odd numbers is 2^2.

The sum of the first 3 odd numbers is 3^2.

The sum of the first 4 odd numbers is 4^2.

The sum of the first 5 odd numbers is 5^2.

This leads naturally to the following question: Is it true that for every natural number n, the sum of the first n odd numbers is n^2? Could this remarkable property be true? We could try a few more numbers and find that the pattern persists for the first 6, 7, 8, 9, and 10 odd numbers. At this point we feel quite sure that this is always true, so we make a conjecture:

The sum of the first n odd numbers is n^2.

Since we know that the nth odd number is $2n - 1$, we can write this statement more precisely as:

$$1 + 2 + 3 + \cdots + (2n - 1) = n^2$$

It is important to realize that this is still a question. We cannot conclude that a property is true for all numbers (there are infinitely many) by checking a finite number of cases. To see this more clearly, suppose someone tells us that he has added up the first trillion odd numbers and found out that they do *not* add up to 1 trillion squared. What would you tell this person? It would be silly to tell him that you are sure it is true because you have already checked it out for the first five cases. You could, however, take out paper and pencil and start checking it out yourself, but this task would probably take the rest of your life. The tragedy would be that after completing this task, you would still not be sure of the truth of this conjecture. Do you see why?

Herein lies the power of mathematical proof. A *proof* is a clear argument that demonstrates the truth of a statement beyond doubt. We consider here a special kind of proof called mathematical induction that will help us prove statements like the one we were just considering. But before we do this let us try another experiment.

Consider the polynomial

$$p(n) = n^2 - n + 41$$

Let us find some of the values of $p(n)$ for natural numbers n:

$$p(1) = 41 \qquad p(2) = 43 \qquad p(3) = 47$$
$$p(4) = 53 \qquad p(5) = 61 \qquad p(6) = 71$$
$$p(7) = 83 \qquad p(8) = 97 \qquad p(9) = 113$$

We notice this time that all the values that we have calculated are prime numbers. We might want to conclude at this point that all the values of this polynomial are prime. But let's try a few more values:

$$p(10) = 131 \text{ (prime)} \qquad p(11) = 151 \text{ (prime)} \qquad p(12) = 173 \text{ (prime)}$$
$$p(13) = 197 \text{ (prime)} \qquad p(14) = 223 \text{ (prime)} \qquad p(15) = 251 \text{ (prime)}$$

Now we are getting tired of calculating values. We make the following conjecture:

For every natural number n, $p(n)$ is a prime number.

In fact, if you try values for n from 16 to 40 you will find that $p(n)$ is again prime. But our conjecture is too hasty. It is easily seen that $p(41) = 41^2 - 41 + 41 = 41^2$ is not prime. This is the first n for which $p(n)$ is not prime! So our conjecture is false.

This illustrates clearly that we cannot be certain of the truth of a statement by checking out special cases. We need a convincing argument to determine the truth of a statement—a proof.

Mathematical Induction

We consider a special kind of proof called **mathematical induction**. Here is how it works. Suppose we have a statement that says something about all natural numbers n. Let's call this statement P. For example, we could consider the statement

P: For every natural number n, the sum of the first n odd numbers is n^2.

Since this statement is about *all* natural numbers, it contains infinitely many statements; we will call them $P(1)$, $P(2)$, and so on.

$P(1)$:   The sum of the first 1 odd number is 1^2.
$P(2)$:   The sum of the first 2 odd numbers is 2^2.
$P(3)$:   The sum of the first 3 odd numbers is 3^2.
$\vdots$

How can we prove all of these statements at once? Mathematical induction is a clever way of doing just that.

The crux of the idea is this: suppose we can prove that whenever one of these statements is true, the one following it in the list is also true. In other words,

For every k, if $P(k)$ is true, then $P(k + 1)$ is true.

This is called the **induction step** because it leads us from the truth of one statement to the truth of the next. Now suppose we can also prove that

$P(1)$ is true.

The induction step now leads us through the following chain of statements:

$P(1)$ is true, so $P(2)$ is true.
$P(2)$ is true, so $P(3)$ is true.
$P(3)$ is true, so $P(4)$ is true.
$\vdots$

So we see that if both of these statements are proved, then statement P is proved for all n. We summarize this important method of proof.

Principle of Mathematical Induction

For each natural number n, let $P(n)$ be a statement that depends on n. Suppose that the following two conditions are satisfied:

1. $P(1)$ is true.
2.   For every natural number k, if $P(k)$ is true, then $P(k + 1)$ is true.

Then $P(n)$ is true for all natural numbers n.

FOOTLES AT THE INDUCTION STEP

© 1979 National Council of Teachers of Mathematics. Used by permission. Courtesy of Andrejs Dunkels, Sweden.

To apply this principle there are two steps:

Step 1:   Prove that $P(1)$ is true.

Step 2:   Assume that $P(k)$ is true and use this assumption to prove that $P(k + 1)$ is true.

Notice that in Step 2 we do not prove that $P(k)$ is true. We only show that *if* $P(k)$ is true, *then* $P(k + 1)$ is also true. The assumption that $P(k)$ is true is called the **induction hypothesis**.

We use mathematical induction to prove that the conjecture we made at the beginning of this section is true.

Example 1

Prove that for all natural numbers n, $1 + 3 + 5 + \cdots + (2n - 1) = n^2$.

Solution

Let $P(n)$ denote the statement

(1) $$1 + 3 + 5 + \cdots + (2n - 1) = n^2$$

Step 1:   We need to show that $P(1)$ is true. But $P(1)$ is simply the statement that $1 = 1^2$, which is of course true.

Step 2:   We assume that $P(k)$ is true. Thus our induction hypothesis is that

(2) $$1 + 3 + 5 + \cdots + (2k - 1) = k^2$$

We need to show that $P(k + 1)$ is true; that is,

$$1 + 3 + 5 + \cdots + (2k - 1) + [2(k + 1) - 1] = (k + 1)^2$$

[We get $P(k + 1)$ by substituting $k + 1$ for n in Equation 1.] To show this let us add the quantity $[2(k + 1) - 1]$ to both sides of Equation 2 to get

$$1 + 3 + 5 + \cdots + (2k - 1) + [2(k + 1) - 1] = k^2 + [2(k + 1) - 1]$$
$$= k^2 + [2k + 2 - 1]$$
$$= k^2 + 2k + 1$$
$$= (k + 1)^2$$

Thus $P(k + 1)$ follows from $P(k)$ and this completes the induction step.

Having proved Steps 1 and 2, we conclude from the Principle of Mathematical Induction that $P(n)$ is true for all natural numbers n.   ●

Example 2

Prove that for every natural number n,

$$1 + 2 + 3 + \cdots + n = \frac{n(n + 1)}{2}$$

Solution

We let $P(n)$ denote the statement $1 + 2 + 3 + \cdots + n = n(n + 1)/2$.

  Step 1:   We need to show that $P(1)$ is true. But $P(1)$ says that

$$1 = \frac{1(1 + 1)}{2}$$

which is clearly true.

  Step 2:   Assume that $P(k)$ is true; that is,

(3) $$1 + 2 + 3 + \cdots + k = \frac{k(k + 1)}{2}$$

This is our induction hypothesis. We want to use this to show that $P(k + 1)$ is true. The statement $P(k + 1)$ is

(4) $$1 + 2 + 3 + \cdots + k + (k + 1) = \frac{(k + 1)[(k + 1) + 1]}{2}$$

In other words we want to derive Equation 4 from Equation 3. Since the left sides of these two equations differ by the quantity $k + 1$, let us try adding $k + 1$ to both sides of Equation 3 and manipulating the right side to arrive at

Equation 4. Thus

$$1 + 2 + 3 + \cdots + k + (k + 1) = \frac{k(k + 1)}{2} + (k + 1)$$

$$= (k + 1)\left(\frac{k}{2} + 1\right)$$

$$= (k + 1)\left(\frac{k + 2}{2}\right)$$

$$= \frac{(k + 1)[(k + 1) + 1]}{2}$$

But this is exactly $P(k + 1)$. Thus we have shown that if $P(k)$ is true, then $P(k + 1)$ is true. This completes the induction step.

Steps 1 and 2 together show that the given statement is true for all n by the Principle of Mathematical Induction. ●

It might happen that a statement $P(n)$ is false for the first few natural numbers but true from some number on. For example, we may want to prove that $P(n)$ is true for $n \geq 5$. Notice that if we prove that $P(5)$ is true, then this together with the induction step would imply the truth of $P(5), P(6), P(7), \ldots$ The next example illustrates this point.

Example 3

Prove that $4n < 2^n$ for all $n \geq 5$.

Solution

Let $P(n)$ denote the statement $4n < 2^n$.

Step 1: $P(5)$ is the statement that $4 \cdot 5 < 2^5$ or $20 < 32$, which is true.
Step 2:   Assume that $P(k)$ is true. Thus our induction hypothesis is

(5) $4k < 2^k$

We want to use this to show that $P(k + 1)$ is true; that is,

(6) $4(k + 1) < 2^{k+1}$

So let us start by adding 4 to both sides of (5). (The motivation for this is that we are trying to derive (6), and the left side of (6) is $4k + 4$.) This gives

(7) $4k + 4 < 2^k + 4$

Now since $k \geq 5$, it follows that $4 < 2^k$. From this and (7) we get

$$4(k + 1) = 4k + 4 < 2^k + 4 < 2^k + 2^k = 2 \cdot 2^k = 2^{k+1}$$

or $4(k + 1) < 2^{k+1}$

But this is exactly (6). Thus we have derived $P(k + 1)$ from $P(k)$ and this completes the induction step.

Having proved Steps 1 and 2, we conclude that $P(n)$ is true for all natural numbers $n \geq 5$. ●

In Exercises 1–13 use mathematical induction to prove that the given formula is true for all natural numbers n.

1. $5 + 8 + 11 + \cdots + (3n + 2) = \dfrac{n(3n + 7)}{2}$

2. $1^2 + 2^2 + 3^2 + \cdots + n^2 = \dfrac{n(n + 1)(2n + 1)}{6}$

3. $1^3 + 2^3 + 3^3 + \cdots + n^3 = \dfrac{n^2(n + 1)^2}{4}$

4. $1 \cdot 2 + 2 \cdot 3 + 3 \cdot 4 + \cdots + n(n + 1)$
$= \dfrac{n(n + 1)(n + 2)}{3}$

5. $1 \cdot 3 + 2 \cdot 4 + 3 \cdot 5 + \cdots + n(n + 2)$
$= \dfrac{n(n + 1)(2n + 7)}{6}$

6. $1^3 + 3^3 + 5^3 + \cdots + (2n - 1)^3 = n^2(2n^2 - 1)$

7. $2^3 + 4^3 + 6^3 + \cdots + (2n)^3 = 2n^2(n + 1)^2$

8. $\dfrac{1}{1 \cdot 2} + \dfrac{1}{2 \cdot 3} + \dfrac{1}{3 \cdot 4} + \cdots + \dfrac{1}{n(n + 1)} = \dfrac{n}{n + 1}$

9. $\dfrac{1}{1 \cdot 2 \cdot 3} + \dfrac{1}{2 \cdot 3 \cdot 4} + \dfrac{1}{3 \cdot 4 \cdot 5} + \cdots + \dfrac{1}{n(n + 1)(n + 2)}$
$= \dfrac{n(n + 3)}{4(n + 1)(n + 2)}$

10. $1 \cdot 2 + 2 \cdot 2^2 + 3 \cdot 2^3 + 4 \cdot 2^4 + \cdots + n \cdot 2^n$
$= 2[1 + (n - 1)2^n]$

11. $\sin(\theta + n\pi) = (-1)^n \sin \theta$

12. $\cos(\theta + n\pi) = (-1)^n \cos \theta$

13. $\sin(2^n x) = 2^n \sin x[\cos x \cdot \cos 2x \cdot \cos 4x \cdot \cdots \cdot \cos 2^{n-1} x]$

14. Prove that $n < 2^n$ for all natural numbers n.

15. Prove that if $x > -1$, then $(1 + x)^n \geq 1 + nx$ for all natural numbers n.

16. Prove that $(n + 1)^2 < 2n^2$ for all natural numbers $n \geq 3$.

17. Show that $n^2 - n + 41$ is odd for all natural numbers n.

18. Show that $n^3 - n + 3$ is divisible by 3 for all natural numbers n.

19. Show that $8^n - 3^n$ is divisible by 5 for all natural numbers n.

20. Show that $3^{2n} - 1$ is divisible by 8 for every natural number n.

21. Let $a_{n+1} = 3a_n$ and $a_1 = 5$. Show that $a_n = 5 \cdot 3^{n-1}$ for all natural numbers n.

22. Show that $x - y$ is a factor of $x^n - y^n$ for all natural numbers n.
[*Hint:* $x^{k+1} - y^{k+1} = x^k(x - y) + (x^k - y^k)y$.]

23. Show that $x + y$ is a factor of $x^{2n-1} + y^{2n-1}$ for all natural numbers n.

24. Prove DeMoivre's Theorem: For every natural number n,

$$[r(\cos \theta + i \sin \theta)]^n = r^n(\cos n\theta + i \sin \theta)$$

In Exercises 25–29 F_n denotes the nth term of the Fibonacci sequence discussed in Section 9.1. Use mathematical induction to prove the given statements.

25. F_{3n} is even for all natural numbers n.

26. $F_1 + F_2 + F_3 + \cdots + F_n = F_{n+2} - 1$

27. $F_1^2 + F_2^2 + F_3^2 + \cdots + F_n^2 = F_n F_{n+1}$
[*Hint:* Notice that
$F_n^2 = F_n(F_{n+1} - F_{n-1}) = F_n F_{n+1} - F_n F_{n-1}$.]

28. For all $n \geq 2$,

$$\begin{bmatrix} 1 & 1 \\ 1 & 0 \end{bmatrix}^n = \begin{bmatrix} F_{n+1} & F_n \\ F_n & F_{n-1} \end{bmatrix}$$

29. If $a_{n+2} = a_{n+1} \cdot a_n$ and $a_1 = a_2 = 2$, then $a_n = 2^{F_n}$ for all natural numbers n.

30. What is wrong with the following "proof" by mathematical induction that all girls have blond hair?
    Let $P(n)$ denote the statement: In any group of n girls, if one of them has blond hair, then they all do.
    Step 1: The statement is clearly true for $n = 1$.
    Step 2: Suppose that $P(k)$ is true. We show that $P(k + 1)$ is true. Consider a group of $k + 1$ girls, one of whom has blond hair; we call her Ex. Remove a girl, call her Es, from the group. Now we have a group of k girls, one of whom has blond hair, and by our induction hypothesis all k girls have blond hair. Now put Es back in the group and remove another girl. Again by our induction hypothesis all the girls in this group have blond hair, and so Es also has blond hair. It follows that all $k + 1$ girls in the group have blond hair and this completes the induction step.
    Since everyone knows at least one girl with blond hair, it follows that all girls have blond hair.

Finding Patterns

In the exercises of the preceding section we were given formulas involving natural numbers and we were asked to prove them. But how does one find such formulas in the first place? The answer is through experimentation. In this section we give some examples that will help you find your own formulas and prove them.

We began Section 9.6 by finding a formula for the sum of the first n odd numbers. How did we arrive at this formula? The first step was to look at several cases and *make a conjecture*. The second step was to *state the conjecture precisely* in mathematical terms. The final step was to *prove the conjecture is true*.

In the following examples you are asked to find a pattern or a formula and then prove that your guess is true.

Example 1

Find a formula for the sum

$$S_n = \frac{1}{1 \cdot 3} + \frac{1}{3 \cdot 5} + \frac{1}{5 \cdot 7} + \cdots + \frac{1}{(2n-1)(2n+1)}$$

and prove that your result is valid for all n.

Solution

Let us begin by calculating a few values of S_n:

$$n = 1 \quad S_1 = \frac{1}{1 \cdot 3} = \frac{1}{3}$$

$$n = 2 \quad S_2 = \frac{1}{1 \cdot 3} + \frac{1}{3 \cdot 5} = \frac{2}{5}$$

$$n = 3 \quad S_3 = \frac{1}{1 \cdot 3} + \frac{1}{3 \cdot 5} + \frac{1}{5 \cdot 7} = \frac{3}{7}$$

$$n = 4 \quad S_4 = \frac{1}{1 \cdot 3} + \frac{1}{3 \cdot 5} + \frac{1}{5 \cdot 7} + \frac{1}{7 \cdot 9} = \frac{4}{9}$$

Remember that we want to find a relationship between n and S_n. (We are writing the value of n at the left to emphasize this.) Do you see any simple relationship? Notice that the numerator of S_n is always n. The denominators appear to be the odd numbers $3, 5, 7, \ldots$. Let us see if this pattern persists for the next value of n:

$$n = 5 \quad S_5 = \frac{1}{1 \cdot 3} + \frac{1}{3 \cdot 5} + \frac{1}{5 \cdot 7} + \frac{1}{7 \cdot 9} + \frac{1}{9 \cdot 11} = \frac{5}{11}$$

It does. We now need to state our observation as a formula. Since odd numbers are of the form $2n + 1$, our guess is that

$$S_n = \frac{n}{2n+1}$$

To check that we have written this correctly it is a good idea to see, for example, whether the formula gives S_1 and S_2: $S_1 = 1/(2 \cdot 1 + 1) = 1/3$ (correct), $S_2 = 2/(2 \cdot 2 + 1) = 2/5$ (correct). So we make the conjecture that for all natural numbers n,

$$\frac{1}{1 \cdot 3} + \frac{1}{3 \cdot 5} + \frac{1}{5 \cdot 7} + \cdots + \frac{1}{(2n - 1)(2n + 1)} = \frac{n}{2n + 1}$$

It remains to prove using mathematical induction that this formula holds for all values of n. This is left to Exercise 1. ●

In Section 9.4 we found a formula for the sum of the first n natural numbers. In the next example we give another way of discovering this formula.

Example 2

Find a formula for the sum of the first n natural numbers and prove that the formula you found is valid for all natural numbers n.

Solution

Again we start by making a table of values for the sum of the first few natural numbers and see whether we can detect a pattern for the relationship between n and the sum of the first n natural numbers.

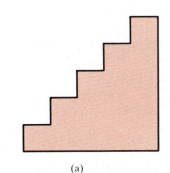

(a)

$n = 1$	$1 = 1$
$n = 2$	$1 + 2 = 3$
$n = 3$	$1 + 2 + 3 = 6$
$n = 4$	$1 + 2 + 3 + 4 = 10$
$n = 5$	$1 + 2 + 3 + 4 + 5 = 15$
$n = 6$	$1 + 2 + 3 + 4 + 5 + 6 = 21$
$n = 7$	$1 + 2 + 3 + 4 + 5 + 6 + 7 = 28$

Do you see a relationship between n and these numbers? Sometimes it is helpful to draw a picture of a special case. In Figure 1(a) we have drawn columns with areas 1, 2, 3, 4, and 5. Thus the total area of this shape is $1 + 2 + 3 + 4 + 5$. Now notice that two such shapes placed together as in Figure 1(b) make a rectangle whose area is $5(5 + 1)$. The area of this rectangle is twice the sum of the numbers from 1 to 5. In other words,

$$2(1 + 2 + 3 + 4 + 5) = 5(5 + 1)$$

Since we are looking for a formula for the sum, we write this as

$$1 + 2 + 3 + 4 + 5 = \frac{5(5 + 1)}{2}$$

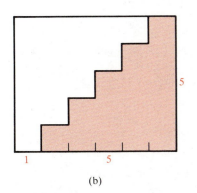

(b)

Figure 1

Of course there is nothing special about the number 5. We could have done the same thing for any natural number n. So our guess is that the sum of the first n natural numbers is the same as half the area of a rectangle with sides n and $n + 1$.

We write the precise statement we want to prove: for every natural number n,

$$1 + 2 + 3 + \cdots + n = \frac{n(n + 1)}{2}$$

This formula was proved in Example 2 of Section 9.6.   ●

Alkarchi (A.D. 1010) used a geometric method similar to that of Example 2 to find a formula for the sum of the cubes of the first n natural numbers. His method is described in Exercise 24.

Example 3

Find and prove an inequality that relates n^2 and 2^n for n a natural number.

Solution

n	n^2	2^n
1	1	2
2	4	4
3	9	8
4	16	16
5	25	32
6	36	64
7	49	128

Let us make a table comparing the values of n^2 and 2^n. We see from the table that for $n = 1$, $2^n > n^2$; for $n = 3$, $n^2 > 2^n$; and $n^2 = 2^n$ for $n = 2$ and 4. But for n larger than 4, it seems that 2^n is always larger than n^2. So our conjecture is

$$2^n > n^2 \text{ for all natural numbers } n \geq 5.$$

To prove this, let $P(n)$ denote the statement

$$2^n > n^2$$

We want to show that $P(n)$ is true for all $n \geq 5$. So we use mathematical induction, starting the induction at $n = 5$.

Step 1: We must show that $P(5)$ is true. This follows from the table.

Step 2: Assume that $P(k)$ is true; that is, $2^k > k^2$, where $k \geq 5$. This is our induction hypothesis. We need to show that $P(k + 1)$ is true; that is, $2^{k+1} > (k + 1)^2$. We proceed as follows:

$$2^{k+1} = 2 \cdot 2^k$$
$$> 2k^2 \qquad \textit{(by the induction hypothesis)}$$
(1) $$= k^2 + k^2$$

We wrote $2k^2$ as $k^2 + k^2$ for a reason. We want to show that this quantity is larger than $(k + 1)^2 = k^2 + 2k + 1$. In other words, we need to show that $k^2 \geq 2k + 1$. This is equivalent to $k^2 - 2k \geq 1$, or $k(k - 2) \geq 1$. But since $k \geq 5$, this last inequality is clearly true. Continuing from (1) we have

$$2^{k+1} > k^2 + k^2$$
$$\geq k^2 + 2k + 1$$
$$= (k + 1)^2$$

This shows that $P(k + 1)$ is true and so completes the induction step.

Having proved Steps 1 and 2, we conclude that $P(n)$ is true for all $n \geq 5$.   ●

Example 4

Find an explicit formula for the sequence defined by the recursion relation $a_{n+1} = 3a_n + 2$ and $a_1 = 2$. Prove that your formula holds for all natural numbers n.

Solution

Let us look at some values of a_n:

$$n = 1 \qquad a_1 = 2$$
$$n = 2 \qquad a_2 = 3a_1 + 2 = 3(2) + 2 = 8$$
$$n = 3 \qquad a_3 = 3a_2 + 2 = 3(8) + 2 = 26$$
$$n = 4 \qquad a_4 = 3a_3 + 2 = 3(26) + 2 = 80$$
$$n = 5 \qquad a_5 = 3a_4 + 2 = 3(80) + 2 = 242$$

Is there a simple relationship between n and a_n? The sequence is neither arithmetic nor geometric because there is no common difference or common ratio. We need to compare this sequence to some sequences that we already know. In this case, let's compare the values of a_n to 3^n because the recursion formula involves multiplying the preceding term by 3. We notice from the table that a_n is always 1 less than the corresponding power of 3. We state our conjecture precisely:

n	a_n	3^n
1	2	3
2	8	9
3	26	27
4	80	81
5	242	243

For every natural number n, $a_n = 3^n - 1$.

To prove this we let $P(n)$ denote the statement $a_n = 3^n - 1$ and we use mathematical induction to show that $P(n)$ is true for all natural numbers n.

Step 1:   We must show that $P(1)$ is true. But $P(1)$ is the statement that

$$a_1 = 3^1 - 1 = 2$$

which is true.

Step 2:   Assuming that $P(k)$ is true, we need to show that $P(k + 1)$ is true. That is, our induction hypothesis is

(2) $$a_k = 3^k - 1$$

and we must show that

(3) $$a_{k+1} = 3^{k+1} - 1$$

To show this notice that

$$
\begin{aligned}
a_{k+1} &= 3a_k + 2 && \textit{(by the recursion formula)}\\
&= 3(3^k - 1) + 2 && \textit{(by the induction hypothesis)}\\
&= 3^{k+1} - 3 + 2\\
&= 3^{k+1} - 1
\end{aligned}
$$

which is exactly (3). This completes the induction step.

Having proved Steps 1 and 2, we conclude from the Principle of Mathematical Induction that $P(n)$ is true for all n. So the formula we found for a_n holds for all natural numbers n.   ●

We summarize three broad steps we have used to find and prove a formula: (1) Look for a pattern (see Section 1.9 for some ideas on how to do this) and make a conjecture, (2) write your conjecture as a mathematical statement, and (3) prove your conjecture. If you can't prove your conjecture, see whether the conjecture is still valid for other cases that you have not yet checked. If it fails, revise your conjecture and try again.

Exercises 9.7

1. Use mathematical induction to prove that the formula found in Example 1 is true for all natural numbers n.

2. Find a formula for the sum of the first n even numbers and prove that your formula holds for all natural numbers n.

3. Find a formula for the nth partial sum of the series $2 + 6 + 10 + \cdots + (4n - 2) + \cdots$ and prove that your formula holds for all natural numbers n.

4. Find a formula for the nth partial sum of the series $3 + 9 + 15 + \cdots + (6n - 3) + \cdots$ and prove that your formula holds for all natural numbers n.

In Exercises 5–7 find a formula for the sums S_n and prove that your formula holds for all natural numbers n.

5. $S_n = \dfrac{1}{1 \cdot 2} + \dfrac{1}{2 \cdot 3} + \dfrac{1}{3 \cdot 4} + \cdots + \dfrac{1}{n(n + 1)}$

6. $S_n = \dfrac{1}{1 \cdot 4} + \dfrac{1}{4 \cdot 7} + \dfrac{1}{7 \cdot 10} + \cdots + \dfrac{1}{(3n - 2)(3n + 1)}$

7. $S_n = \dfrac{1}{1 \cdot 6} + \dfrac{1}{6 \cdot 11} + \dfrac{1}{11 \cdot 16} + \cdots + \dfrac{1}{(5n - 4)(5n + 1)}$

In Exercises 8–10 find a formula for the given product P_n and prove that your formula holds for all natural numbers n.

8. $P_n = \left(1 - \dfrac{1}{2}\right)\left(1 - \dfrac{1}{3}\right)\left(1 - \dfrac{1}{4}\right) \cdots \left(1 - \dfrac{1}{n}\right)$

9. $P_n = \left(1 - \dfrac{1}{4}\right)\left(1 - \dfrac{1}{9}\right)\left(1 - \dfrac{1}{16}\right) \cdots \left(1 - \dfrac{1}{n^2}\right)$

10. $P_n = \left(1 + \dfrac{1}{1}\right)\left(1 + \dfrac{1}{2}\right)\left(1 + \dfrac{1}{3}\right) \cdots \left(1 + \dfrac{1}{n}\right)$

11. Find and prove an inequality relating n and 2^n.

12. Find and prove an inequality relating $1000n$ and n^3.

13. Let F_n be the nth term of the Fibonacci sequence. Find and prove an inequality relating n and F_n for natural numbers n.

In Exercises 14–17 a sequence with nth term a_n is defined recursively. Find an explicit formula for a_n and prove your result.

14. $a_{n+1} = a_n + 2n + 1$ and $a_1 = 1$

15. $a_{n+1} = 5a_n + 4$ and $a_1 = 4$

16. $a_{n+1} = 3a_n - 8$ and $a_1 = 4$

17. $a_{n+1} = \sqrt{5a_n}$ and $a_1 = \sqrt{5}$

18. A sequence is defined recursively by

$$a_{n+2} = a_{n+1} \cdot a_n$$

with $a_2 = a_1 = c$. Find and prove a formula for a_n in terms of the Fibonacci numbers F_n.

19. Let a_n be the nth term of the sequence defined recursively by

$$a_{n+1} = \frac{1}{1 + a_n}$$

and $a_1 = 1$. Find a formula for a_n in terms of the Fibonacci numbers F_n. Prove that the formula you found is valid for all natural numbers n.

20. The product of the first n terms of a sequence is n^3. Find an explicit formula for the nth term a_n of this sequence and prove your result.

21. **(a)** Use long division to divide $x^2 - 1$, $x^3 - 1$, and $x^4 - 1$ by $x - 1$.
  (b) Guess a general formula for factoring $x - 1$ from $x^n - 1$ and prove that your guess is true for all n.

22. Determine whether each of the following statements is true or false. If the statement is true, prove it. If it is false, give an example where it fails.

(a) $p(n) = n^2 - n + 11$ is prime for all n.

(b) $n^2 > n$ for all $n \geq 2$.

(c) $3^n \geq n^3$ for all $n \geq 1$.

(d) $2^{2n+1} + 1$ is divisible by 3 for all $n \geq 1$.

(e) $n^3 \geq (n + 1)^2$ for all $n \geq 2$.

(f) $n^3 - n$ is divisible by 3 for all $n \geq 2$.

(g) $n^3 - 6n^2 + 11n$ is divisible by 6 for all $n \geq 1$.

23. State and prove a formula that is suggested by the following:

$$
\begin{array}{rcl}
1 & = & 1 \\
1 + 2 + 1 & = & 4 \\
1 + 2 + 3 + 2 + 1 & = & 9 \\
1 + 2 + 3 + 4 + 3 + 2 + 1 & = & 16
\end{array}
$$

24. In this exercise we give a method for finding a formula for the sum of the cubes of the first n natural numbers. This method was first published in about A.D. 1010 by Abu Bekr Mohammed ibn Alhusain Alkarchi. The figure shows a square $ABCD$ in which sides AB and AD have been divided into segments of lengths $1, 2, 3, 4, \ldots, n$.

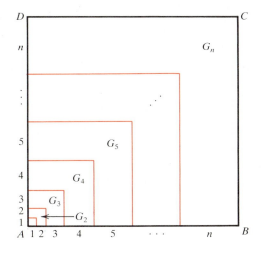

(a) What is the length of the side of the square $ABCD$? Use the result of Example 2 to write a formula for this length.

(b) Use part (a) to find a formula for the area of this square.

(c) Show that the area of each "gnomon" G_k is k^3 and deduce that the area of the square $ABCD$ is the sum of the cubes of the first n natural numbers.

(d) Use parts (b) and (c) to find a formula for the sum of the cubes of the first n natural numbers.

(e) Prove the formula you found in part (d) using mathematical induction.

25. The **Tower of Brahma** consists of three pegs with several rings on the first peg, with each ring smaller than the one below it (see the figure). We are to move the rings from the first peg to the last peg so that they are on the last peg in the same order as they were on the first peg. The rings must be moved according to the following rules: a ring may be moved to any peg, and a ring may not be placed on top of a smaller ring.

(a) What is the smallest number of moves required to accomplish this task if the tower contains two rings? three rings? four rings? five rings?

(b) Make a conjecture as to the smallest number of moves required to accomplish this task if the tower has n rings and prove that your conjecture is true using mathematical induction.

(c) If there are 64 rings in the tower and if each move requires one second, what is the minimum time required to accomplish this task?

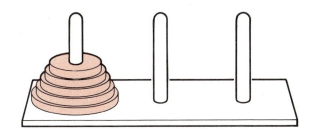

Section 9.8

The Binomial Theorem

An expression of the form $a + b$ is called a **binomial**. Although in principle it is easy to raise $a + b$ to any power, raising it to very high powers would be a tedious task. In this section we find a formula that gives the expansion of $(a + b)^n$ for any natural number n.

We discover this formula by the methods of the preceding section. To find the pattern, let us look at some special cases:

$$(a + b)^1 = a + b$$
$$(a + b)^2 = a^2 + 2ab + b^2$$
$$(a + b)^3 = a^3 + 3a^2b + 3ab^2 + b^3$$
$$(a + b)^4 = a^4 + 4a^3b + 6a^2b^2 + 4ab^3 + b^4$$
$$(a + b)^5 = a^5 + 5a^4b + 10a^3b^2 + 10a^2b^3 + 5ab^4 + b^5$$
$$\vdots$$

The following simple patterns emerge for the expansion of $(a + b)^n$:

1. There are $n + 1$ terms, the first being a^n and the last b^n.
2. The exponents of a decrease by 1 from term to term, while the exponents of b increase by 1.
3. The sum of the exponents of a and b in each term is n.

For instance, notice how the exponents of a and b behave in the expansion of $(a + b)^5$:

The exponents of a decrease:

$$(a + b)^5 = a^{⑤} + 5a^{④}b^1 + 10a^{③}b^2 + 10a^{②}b^3 + 5a^{①}b^4 + b^5$$

The exponents of b increase:

$$(a + b)^5 = a^5 + 5a^4b^{①} + 10a^3b^{②} + 10a^2b^{③} + 5a^1b^{④} + b^{⑤}$$

With these observations we can write the form of the expansion of $(a + b)^n$ for any natural number n. For example, putting a question mark for the missing coefficients, we have

$$(a + b)^8 = a^8 + ?a^7b + ?a^6b^2 + ?a^5b^3 + ?a^4b^4 + ?a^3b^5 + ?a^2b^6 + ?ab^7 + b^8$$

To complete the expansion we need to determine these coefficients. To find a pattern, let us write the coefficients in the expansion of $(a + b)^n$ for the first few values of n in a triangular array as shown below. This array is called **Pascal's triangle**.

$(a + b)^0$			1			
$(a + b)^1$			1	1		
$(a + b)^2$		1	2	1		
$(a + b)^3$	1	3	3	1		
$(a + b)^4$	1	4	6	4	1	
$(a + b)^5$	1	5	10	10	5	1

The row corresponding to $(a + b)^0$ is called the zeroth row and is added for purposes of symmetry. The key observation about Pascal's triangle is the following:

(1)   Every entry (other than a 1) is the sum of the
                two entries diagonally above it.

From this property it is easy to find any row of Pascal's triangle from the row above it. For instance we find the sixth and seventh rows starting with the fifth row:

$(a + b)^5$ 1   5   10   10   5   1

$(a + b)^6$ 1   6   15   20   15   6   1

$(a + b)^7$ 1   7   21   35   35   21   7   1

To see why property (1) holds let us consider the following expansions:

$$(a + b)^4 = \quad a^4 + 4a^3b + 6a^2b^2 + 4ab^3 + b^4$$

$$(a + b)^5 = a^5 + 5a^4b + 10a^3b^2 + 10a^2b^3 + 5ab^4 + b^5$$

We arrive at the expansion of $(a + b)^5$ by multiplying $(a + b)^4$ by $(a + b)$. Now notice, for instance, that the circled term in the expansion of $(a + b)^5$ is obtained via this multiplication from the two circled terms above it. (In fact, we get this term when the two terms above it are multiplied by b and a, respectively.) Thus its coefficient is the sum of the coefficients of these two terms. This is the observation we will use in proving the Binomial Theorem at the end of this section.

Having found these patterns, we can easily obtain the expansion of any binomial, at least to relatively small powers.

Example 1

Find the expansion of $(a + b)^7$ using Pascal's triangle.

Solution

The first term in the expansion is a^7 and the last term is b^7. Using the fact that the exponent of a decreases by 1 from term to term and that of b increases by 1 from term to term, we have

$$(a + b)^7 = a^7 + ?a^6b + ?a^5b^2 + ?a^4b^3 + ?a^3b^4 + ?a^2b^5 + ?ab^6 + b^7$$

The appropriate coefficients appear in the seventh row of Pascal's triangle. Thus

$$(a + b)^7 = a^7 + 7a^6b + 21a^5b^2 + 35a^4b^3 + 35a^3b^4 + 21a^2b^5 + 7ab^6 + b^7$$

Example 2

Use Pascal's triangle to expand $(2 - 3x)^5$.

Solution

We find the expansion of $(a + b)^5$ and then substitute 2 for a and $-3x$ for b. Using Pascal's triangle for the coefficients, we get

$$(a + b)^5 = a^5 + 5a^4b + 10a^3b^2 + 10a^2b^3 + 5ab^4 + b^5$$

Substituting $a = 2$ and $b = -3x$ gives

$$(a + b)^5 = (2)^5 + 5(2)^4(-3x) + 10(2)^3(-3x)^2 + 10(2)^2(-3x)^3$$
$$+ 5(2)(-3x)^4 + (-3x)^5$$
$$= 32 - 240x + 720x^2 - 1080x^3 + 810x^4 - 243x^5 \qquad \bullet$$

Although Pascal's triangle is useful for finding the binomial expansion for reasonably small values of n, it is not practical to use it for finding $(a + b)^n$ for large values of n. The reason is that the method we use for finding the successive rows of Pascal's triangle is recursive. Thus to find the 100th row of this triangle we must first find all the preceding rows.

We need to examine the pattern in the coefficients more carefully to get a formula that will allow us to calculate directly any coefficient in the binomial expansion. Such a formula exists and the rest of this section is devoted to finding and proving it. However, to state this formula we need some notation, which we discuss next.

Factorials and the Binomial Coefficients

The product of the first n natural numbers is called **n factorial** and is denoted by $n!$:

$$\boxed{n! = 1 \cdot 2 \cdot 3 \cdots \cdots (n - 1)n}$$

By convention,

$$\boxed{0! = 1}$$

For example,

$$4! = 1 \cdot 2 \cdot 3 \cdot 4 = 24$$
$$7! = 1 \cdot 2 \cdot 3 \cdot 4 \cdot 5 \cdot 6 \cdot 7 = 5040$$
$$10! = 1 \cdot 2 \cdot 3 \cdot 4 \cdot 5 \cdot 6 \cdot 7 \cdot 8 \cdot 9 \cdot 10 = 3,628,800$$

Example 3

Simplify the following expressions involving factorials:

(a) $\dfrac{7!}{5!}$ (b) $\dfrac{6!100!}{2!102!}$ (c) $\dfrac{n!}{(n - 2)!}$

Solution

(a) $\dfrac{7!}{5!} = \dfrac{\cancel{1} \cdot \cancel{2} \cdot \cancel{3} \cdot \cancel{4} \cdot \cancel{5} \cdot 6 \cdot 7}{\cancel{1} \cdot \cancel{2} \cdot \cancel{3} \cdot \cancel{4} \cdot \cancel{5}} = 6 \cdot 7 = 42$

(b) $\dfrac{6!100!}{2!102!} = \dfrac{(\cancel{1} \cdot \cancel{2} \cdot 3 \cdot 4 \cdot 5 \cdot 6)(\cancel{1} \cdot \cancel{2} \cdot \cancel{3} \cdot \ \cdots \ \cancel{100})}{(\cancel{1} \cdot \cancel{2}) \cdot (\cancel{1} \cdot \cancel{2} \cdot \cancel{3} \cdot \ \cdots \ \cancel{100} \cdot 101 \cdot 102)} = \dfrac{3 \cdot 4 \cdot 5 \cdot 6}{101 \cdot 102} = \dfrac{60}{1717}$

(c) $\dfrac{n!}{(n-2)!} = \dfrac{\cancel{1} \cdot \cancel{2} \cdot \cancel{3} \cdot \ \cdots \ (n \cancel{-2})(n-1)n}{\cancel{1} \cdot \cancel{2} \cdot \cancel{3} \cdot \ \cdots \ (n \cancel{-2})} = (n-1)n = n^2 - n$ ●

Using factorials, we define the symbol $\dbinom{n}{r}$, which is called a **binomial coefficient**. If r and n are nonnegative integers with $r \le n$, then

$$\binom{n}{r} = \frac{n!}{r!(n-r)!}$$

Example 4

Find the following:

(a) $\dbinom{9}{4}$ (b) $\dbinom{20}{16}$ (c) $\dbinom{100}{3}$ (d) $\dbinom{100}{97}$

Solution

(a) $\dbinom{9}{4} = \dfrac{9!}{4!(9-4)!} = \dfrac{9!}{4!5!} = \dfrac{1 \cdot 2 \cdot 3 \cdot 4 \cdot 5 \cdot 6 \cdot 7 \cdot 8 \cdot 9}{(1 \cdot 2 \cdot 3 \cdot 4)(1 \cdot 2 \cdot 3 \cdot 4 \cdot 5)} = \dfrac{6 \cdot 7 \cdot 8 \cdot 9}{1 \cdot 2 \cdot 3 \cdot 4} = 126$

(b) $\dbinom{20}{16} = \dfrac{20!}{16!(20-16)!} = \dfrac{20!}{16!4!} = \dfrac{17 \cdot 18 \cdot 19 \cdot 20}{1 \cdot 2 \cdot 3 \cdot 4} = 4845$

(c) $\dbinom{100}{3} = \dfrac{100!}{3!(100-3)!} = \dfrac{100!}{3!97!} = \dfrac{98 \cdot 99 \cdot 100}{1 \cdot 2 \cdot 3} = 161{,}700$

(d) $\dbinom{100}{97} = \dfrac{100!}{97!(100-97)!} = \dfrac{100!}{97!3!} = \dfrac{98 \cdot 99 \cdot 100}{1 \cdot 2 \cdot 3} = 161{,}700$ ●

We point out that although the binomial coefficient $\dbinom{n}{r}$ is defined in terms of a fraction, all the results of this example are natural numbers. In fact, the binomial coefficient $\dbinom{n}{r}$ is always a natural number. Notice also that the binomial coefficients in parts (c) and (d) of Example 4 are equal. This fact is a special case of the second of the following two properties:

(2)

$$\binom{n}{0} = \binom{n}{n} = 1$$

(3)

$$\binom{n}{r} = \binom{n}{n-r}$$

To see the connection between the binomial coefficients and the binomial expansion of $(a + b)^n$, we calculate the following binomial coefficients:

$$\binom{5}{0} = 1 \qquad \binom{5}{1} = 5 \qquad \binom{5}{2} = 10 \qquad \binom{5}{3} = 10 \qquad \binom{5}{4} = 5 \qquad \binom{5}{5} = 1$$

These are precisely the entries in the fifth row of Pascal's triangle. In fact for all natural numbers n, the nth row of Pascal's triangle has the entries

$$\binom{n}{0} \qquad \binom{n}{1} \qquad \binom{n}{2} \cdots \binom{n}{n-1} \qquad \binom{n}{n}$$

So we can write Pascal's triangle as follows:

$$\binom{0}{0}$$

$$\binom{1}{0} \qquad \binom{1}{1}$$

$$\binom{2}{0} \qquad \binom{2}{1} \qquad \binom{2}{2}$$

$$\binom{3}{0} \qquad \binom{3}{1} \qquad \binom{3}{2} \qquad \binom{3}{3}$$

$$\binom{4}{0} \qquad \binom{4}{1} \qquad \binom{4}{2} \qquad \binom{4}{3} \qquad \binom{4}{4}$$

$$\binom{5}{0} \qquad \binom{5}{1} \qquad \binom{5}{2} \qquad \binom{5}{3} \qquad \binom{5}{4} \qquad \binom{5}{5}$$

$$\binom{n}{0} \qquad \binom{n}{1} \qquad \binom{n}{2} \cdots \cdots \binom{n}{n-1} \qquad \binom{n}{n}$$

The Binomial Theorem

We are now ready to state the Binomial Theorem:

The Binomial Theorem

$$(a + b)^n = \binom{n}{0}a^n + \binom{n}{1}a^{n-1}b + \binom{n}{2}a^{n-2}b^2 + \cdots + \binom{n}{n-1}ab^{n-1} + \binom{n}{n}b^n$$

We prove this thoerem at the end of this section. We first give some examples illustrating the theorem.

Example 5

Use the Binomial Theorem to expand $(x + y)^4$.

Solution

By the Binomial Theorem,

$$(x + y)^4 = \binom{4}{0}x^4 + \binom{4}{1}x^3y + \binom{4}{2}x^2y^2 + \binom{4}{3}xy^3 + \binom{4}{4}y^4$$

Verify that

$$\binom{4}{0} = 1 \quad \binom{4}{1} = 4 \quad \binom{4}{2} = 6 \quad \binom{4}{3} = 4 \quad \binom{4}{4} = 1$$

It follows that

$$(x + y)^4 = x^4 + 4x^3y + 6x^2y^2 + 4xy^3 + y^4$$

Example 6

Use the Binomial Theorem to expand $(\sqrt{x} - 1)^8$.

Solution

We first find the expansion of $(a + b)^8$ and then substitute $\sqrt{x}$ for a and -1 for b. Using the Binomial Theorem, we have

$$(a + b)^8 = \binom{8}{0}a^8 + \binom{8}{1}a^7b + \binom{8}{2}a^6b^2 + \binom{8}{3}a^5b^3 + \binom{8}{4}a^4b^4$$

$$+ \binom{8}{5}a^3b^5 + \binom{8}{6}a^2b^6 + \binom{8}{7}ab^7 + \binom{8}{8}b^8$$

Verify that

$$\binom{8}{0} = 1 \quad \binom{8}{1} = 8 \quad \binom{8}{2} = 28 \quad \binom{8}{3} = 56 \quad \binom{8}{4} = 70$$

$$\binom{8}{5} = 56 \quad \binom{8}{6} = 28 \quad \binom{8}{7} = 8 \quad \binom{8}{8} = 1$$

So

$$(a + b)^8 = a^8 + 8a^7b + 28a^6b^2 + 56a^5b^3 + 70a^4b^4 + 56a^3b^5 \\ + 28a^2b^6 + 8ab^7 + b^8$$

Performing the substitutions $a = x^{1/2}$ and $b = -1$ gives

$$(\sqrt{x} - 1)^8 = (x^{1/2})^8 + 8(x^{1/2})^7(-1) + 28(x^{1/2})^6(-1)^2 + 56(x^{1/2})^5(-1)^3 \\ + 70(x^{1/2})^4(-1)^4 + 56(x^{1/2})^3(-1)^5 \\ + 28(x^{1/2})^2(-1)^6 + 8(x^{1/2})(-1)^7 + (-1)^8$$

This simplifies to

$$(\sqrt{x} - 1)^8 = x^4 - 8x^{7/2} + 28x^3 - 56x^{5/2} + 70x^2 - 56x^{3/2} + 28x - x^{1/2} + 1$$

●

The Binomial Theorem can be used to find particular terms of a binomial expansion without having to find the entire expansion. To do this notice that the term in the expansion of $(a + b)^n$ that contains a^r is

(4)

$$\binom{n}{r} a^r b^{n-r}$$

Example 7

Find the term that contains x^5 in the expansion of $(2x + y)^{20}$.

Solution

The term that contains x^5 is given by (4), where $a = 2x$, $b = y$, $n = 20$, and $r = 5$. So this term is

$$\binom{20}{5} a^5 b^{20-5} = \frac{20!}{5!(20-5)!}(2x)^5 y^{15} = \frac{20!}{5!15!}32x^5y^{15} = 496{,}128x^5y^{15}$$

●

Example 8

Find the coefficient of x^8 in the expansion of $\left(x^2 + \dfrac{1}{x}\right)^{10}$.

Solution

Notice that both x^2 and $1/x$ are powers of x. So the power of x in each term of the expansion is determined by both terms of the binomial. To find the required coefficient we first find the rth term in the expansion. By (4) with $a = x^2, b = 1/x$, and $n = 10$, the rth term is

$$\binom{10}{r}(x^2)^r\left(\frac{1}{x}\right)^{10-r} = \binom{10}{r}(x^2)^r(x^{-1})^{10-r} = \binom{10}{r}x^{3r-10}$$

Thus the term that contains x^8 is the rth term, where

$$3r - 10 = 8$$

or
$$r = 6$$

So the required coefficient is that of the sixth term and is given by

$$\binom{10}{6} = 210$$

●

Proof of the Binomial Theorem

The key property of Pascal's triangle is (1). Stated in terms of the binomial coefficients, it says:

(5)

> For any nonnegative integers r and k with $r \leq k$, we have
> $$\binom{k}{r-1} + \binom{k}{r} = \binom{k+1}{r}$$

Notice that the two terms on the left of this equation are adjacent entries in the kth row of Pascal's triangle, and the term on the right is the entry diagonally below them in the $(k + 1)$st row. Thus this equation is a restatement of property (1) in terms of the binomial coefficients. A proof of this formula is outlined in Exercise 71. We are now ready to prove the Binomial Theorem by mathematical induction.

Proof

Let $P(n)$ denote the statement

$$(a + b)^n = \binom{n}{0}a^n + \binom{n}{1}a^{n-1}b + \binom{n}{2}a^{n-2}b^2 + \cdots + \binom{n}{n-1}ab^{n-1} + \binom{n}{n}b^n$$

Step 1:   We show that $P(1)$ is true. But $P(1)$ is just the statement

$$(a + b)^1 = \binom{1}{0}a^1 + \binom{1}{1}b^1 = 1a + 1b = a + b$$

which is certainly true.

Step 2:   We assume that $P(k)$ is true and show that $P(k + 1)$ is true. The statement $P(k)$ reads

$$(a + b)^k = \binom{k}{0}a^k + \binom{k}{1}a^{k-1}b + \binom{k}{2}a^{k-2}b^2 + \cdots + \binom{k}{k-1}ab^{k-1} + \binom{k}{k}b^k$$

Multiplying both sides of this equation by $(a + b)$ and collecting like terms gives:

$$(a + b)^{k+1} = (a + b)\left[\binom{k}{0}a^k + \binom{k}{1}a^{k-1}b + \binom{k}{2}a^{k-2}b^2 + \cdots + \binom{k}{k-1}ab^{k-1} + \binom{k}{k}b^k\right]$$

$$= a\left[\binom{k}{0}a^k + \binom{k}{1}a^{k-1}b + \binom{k}{2}a^{k-2}b^2 + \cdots + \binom{k}{k-1}ab^{k-1} + \binom{k}{k}b^k\right]$$

$$+ b\left[\binom{k}{0}a^k + \binom{k}{1}a^{k-1}b + \binom{k}{2}a^{k-2}b^2 + \cdots + \binom{k}{k-1}ab^{k-1} + \binom{k}{k}b^k\right]$$

$$= \binom{k}{0}a^{k+1} + \binom{k}{1}a^k b + \binom{k}{2}a^{k-1}b^2 + \cdots + \binom{k}{k-1}a^2 b^{k-1} + \binom{k}{k}ab^k$$

$$+ \binom{k}{0}a^k b + \binom{k}{1}a^{k-1}b^2 + \binom{k}{2}a^{k-2}b^3 + \cdots + \binom{k}{k-1}ab^k + \binom{k}{k}b^{k+1}$$

$$= \binom{k}{0}a^{k+1} + \left[\binom{k}{0} + \binom{k}{1}\right]a^k b + \left[\binom{k}{1} + \binom{k}{2}\right]a^{k-1}b^2 + \cdots + \left[\binom{k}{k-1} + \binom{k}{k}\right]ab^k + \binom{k}{k}b^{k+1}$$

Using Equation 5, we can write each of the expressions in square brackets as a single binomial coefficient. Also, writing the first and last coefficients as $\binom{k+1}{0}$ and $\binom{k+1}{k+1}$ (these are equal to 1 by Equation 2) gives

$$(a + b)^{k+1} = \binom{k+1}{0}a^{k+1} + \binom{k+1}{1}a^k b + \binom{k+2}{2}a^{k-1}b^2 + \cdots + \binom{k+1}{k}ab^k + \binom{k+1}{k+1}b^{k+1}$$

But this last equation is precisely $P(k + 1)$ and this completes the induction step. Having proved Steps 1 and 2, we conclude by the Principle of Mathematical Induction that the theorem is true for all natural numbers n.   ●

Exercises 9.8

In Exercises 1–12 use Pascal's triangle to expand the given expression.

1. $(x + y)^7$

2. $(2x + 1)^4$

3. $\left(x + \dfrac{1}{x}\right)^4$

4. $(x - y)^5$

5. $(3 - \sqrt{3})^5$

6. $(\sqrt{a} + \sqrt{b})^6$

7. $(x^2 y - 1)^5$

8. $(1 + \sqrt{2})^6$

9. $(2x - 3y)^3$

10. $(1 + x^3)^3$

11. $\left(\dfrac{1}{x} - \sqrt{x}\right)^5$

12. $\left(2 + \dfrac{x}{2}\right)^5$

In Exercises 13–20 evaluate the given expression.

13. $10!$

14. $3! \, 4!$

15. $\dfrac{20!}{22!}$

16. $\dfrac{5!}{2! \, 4!}$

17. $\dfrac{100!}{98!}$

18. $\dfrac{12! \, 14!}{11! \, 13!}$

19. $\dfrac{8! + 9!}{8!}$

20. $\dfrac{100! - 99!}{98!}$

In Exercises 21–24 simplify the given expression.

21. $(n - 1)! \, n$

22. $\dfrac{(n - 1)! \, n}{n!}$

23. $\dfrac{(n + 2)!}{(n - 1)!}$

24. $\dfrac{(n + 1)!}{(n - 2)! \, n}$

In Exercises 25–32 evaluate the given expression.

25. $\binom{5}{2}$ **26.** $\binom{10}{6}$ **27.** $\binom{100}{98}$

28. $\binom{10}{5}$ **29.** $\binom{6}{3} + \binom{6}{2} - \binom{7}{3}$

30. $\binom{10}{4} + \binom{10}{3} - \binom{11}{4}$

31. $\binom{5}{0} + \binom{5}{1} + \binom{5}{2} + \binom{5}{3} + \binom{5}{4} + \binom{5}{5}$

32. $\binom{5}{0} - \binom{5}{1} + \binom{5}{2} - \binom{5}{3} + \binom{5}{4} - \binom{5}{5}$

In Exercises 33–36 use the Binomial Theorem to expand the given expression.

33. $(x + 2y)^4$ **34.** $(1 - x)^5$

35. $\left(1 + \dfrac{1}{x}\right)^6$ **36.** $(2A + B^2)^4$

37. Find the first three terms in the expansion of $(x + 2y)^{20}$.

38. Find the first four terms in the expansion of $(x^{1/2} + 1)^{30}$.

39. Find the last two terms in the expansion of $(a^{2/3} + a^{1/3})^{25}$.

40. Find the first three terms in the expansion of $\left(x + \dfrac{1}{x}\right)^{40}$.

41. Find the middle term in the expansion of $(x^2 + 1)^{18}$.

42. Find the fifth term in the expansion of $(ab - 1)^{20}$.

43. Find the 24th term in the expansion of $(a + b)^{25}$.

44. Find the 28th term in the expansion of $(A - B)^{30}$.

45. Find the 100th term in the expansion of $(1 + y)^{100}$.

46. Find the second term in the expansion of $\left(x^2 - \dfrac{1}{x}\right)^{25}$.

47. Find the term that contains x^4 in the expansion of $(x + 2y)^{10}$.

48. Find the term that contains y^3 in the expansion of $(\sqrt{2} + y)^{12}$.

49. Find the term that contains b^8 in the expansion of $(a + b^2)^{12}$.

50. Find the term that contains a in the expansion of $\left(\sqrt{a} + \dfrac{1}{\sqrt{a}}\right)^{10}$.

51. Find the term that does not contain x in the expansion of $\left(8x + \dfrac{1}{2x}\right)^8$.

52. Find the term that does not contain m in the expansion of $(mn + m^{-4})^5$.

53. Find the term that contains c^7 in the expansion of $(2c + \sqrt{c})^8$.

54. Find the coefficient of r^{-5} in the expansion of $\left(\dfrac{r^2}{4} - \dfrac{4}{r^3}\right)^5$.

In Exercises 55–58 find the number of distinct terms in the expansion of the given expression.

55. $\left(x + \dfrac{1}{x}\right)^{20}$ **56.** $\left(x^2 + \dfrac{1}{x}\right)^6$

57. $(a^2 - 2ab + b^2)^5$ **58.** $[(x + y)^2(x - y)^2]^3$

In Exercises 59–62 simplify the given expression.

59. $x^4 + 4x^3y + 6x^2y^2 + 4xy^3 + y^4$

60. $(x - 1)^5 + 5(x - 1)^4 + 10(x - 1)^3 + 10(x - 1)^2 + 5(x - 1) + 1$

61. $8a^3 + 12a^2b + 6ab^2 + b^3$

62. $x^8 + 4x^6y + 6x^4y^2 + 4x^2y^3 + y^4$

63. Expand $(a^2 + a + 1)^4$.
[*Hint:* $a^2 + a + 1 = a^2 + (a + 1)$.]

64. Show that $(1.01)^{100} > 2$. [*Hint:* Note that $(1.01)^{100} = (1 + 0.01)^{100}$ and use the Binomial Theorem.]

65. Show that $(1.001)^{1000} > 2.4$.

66. Show that $\binom{n}{0} = 1$ and $\binom{n}{n} = 1$.

67. Show that $\binom{n}{1} = \binom{n}{n-1} = n$.

68. Show that $\binom{n}{r} = \binom{n}{n-r}$ for $0 \le r \le n$.

69. Show that $\binom{n}{0} + \binom{n}{1} + \binom{n}{2} + \cdots + \binom{n}{n} = 2^n$.
[*Hint:* $2^n = (1 + 1)^n$.]

70. Show that $\binom{n}{0} - \binom{n}{1} + \binom{n}{2} - \cdots +$
$(-1)^k \binom{n}{k} + \cdots + (-1)^n \binom{n}{n} = 0.$

71. In this exercise we prove the identity
$$\binom{n}{r-1} + \binom{n}{r} = \binom{n+1}{r}.$$

    (a) Write out the left side of this equation as the sum of two fractions.

    (b) Show that a common denominator of the expression you found in part (a) is $r!(n-r+1)!$.

    (c) Add the two fractions using the common denominator in part (b), simplify the numerator, and notice that the resulting expression is equal to the right side of the equation.

72. Which is larger: $(100!)^{101}$ or $(101!)^{100}$?

73. A **combination** of n objects taken r at a time is a collection of r objects chosen from the n objects. For example, combinations of the five objects A, B, C, D, and E taken three at a time are ABC, BDE, ACD, and so on. It can be shown that the number of combinations of n objects taken r at a time is $\binom{n}{r}$.

    (a) Find the number of different five-card hands that can be chosen from a deck of 52 cards.

    (b) Find the number of ways a student can choose three classes from a schedule of eight classes.

CHAPTER 9 REVIEW

Define, state, or discuss the following:

1. Sequence
2. Recursive sequence
3. Fibonacci sequence
4. Arithmetic sequence
5. Common difference
6. Geometric sequence
7. Common ratio
8. Sigma notation
9. Series
10. Arithmetic series
11. Sum of an arithmetic series
12. Geometric series
13. Sum of a geometric series
14. Annuity
15. Infinite series
16. Sum of an infinite geometric series
17. Mathematical induction
18. Pascal's triangle
19. Factorial
20. Binomial coefficients
21. The Binomial Theorem

Review Exercises

In Exercises 1–6 find the first four terms as well as the tenth term of the sequence with the given nth term.

1. $a_n = \dfrac{n^2}{n+1}$

2. $a_n = (-1)^n \dfrac{2^n}{n}$

3. $a_n = \dfrac{(-1)^n + 1}{n^3}$

4. $a_n = \dfrac{n(n+1)}{2}$

5. $a_n = \dfrac{(2n)!}{2^n n!}$

6. $a_n = \binom{n+1}{2}$

In Exercises 7–12 a sequence is defined recursively. Find the first seven terms of the sequence.

7. $a_n = a_{n-1} + 2n - 1, \quad a_1 = 1$

8. $a_n = \dfrac{a_{n-1}}{n}, \quad a_1 = 1$

9. $a_n = a_{n-1} + 2a_{n-2}, \quad a_1 = 1, a_2 = 3$

10. $a_n = \sqrt{3a_{n-1}}, \quad a_1 = \sqrt{3}$

11. $a_n = (a_{n-1} - 1)!, \quad a_1 = 3$

12. $a_n = \begin{pmatrix} n+1 \\ a_{n-1} \end{pmatrix}, \quad a_1 = 1$

In Exercises 13–24 the first four terms of a sequence are given. In each case determine whether the given terms can be the terms of an arithmetic sequence, a geometric sequence, or neither. For those sequences that are arithmetic or geometric, find the fifth term.

13. $5, 5.5, 6, 6.5, \ldots$

14. $1, -\frac{3}{2}, 2, -\frac{5}{2}, \ldots$

15. $\sqrt{2}, 2\sqrt{2}, 3\sqrt{2}, 4\sqrt{2}, \ldots$

16. $\sqrt{2}, 2, 2\sqrt{2}, 4, \ldots$

17. $t - 3, t - 2, t - 1, t, \ldots$

18. $t^3, t^2, t, 1, \ldots$

19. $\frac{3}{4}, \frac{1}{2}, \frac{1}{3}, \frac{2}{9}, \ldots$

20. $\frac{a}{c}, 1, \frac{c}{a}, \left(\frac{c}{a}\right)^2, \ldots$

21. $\ln a, \ln 2a, \ln 3a, \ln 4a, \ldots$

22. $\sin^2 x, 1, 1 + \cos^2 x, 1 + 2\cos^2 x, \ldots$

23. $a, abc^3, ab^2c^6, ab^3c^9, \ldots$

24. $a, a + b^2, a + 2b^2, a + 3b^2, \ldots$

25. Show that $3, 6i, -12, -24i, \ldots$ is a geometric sequence and find the common ratio.

26. Find the nth term of the geometric sequence $2, 2 + 2i, 4i, -4 + 4i, -8, \ldots$.

27. The sixth term of an arithmetic sequence is 17 and the fourth term is 11. Find the second term.

28. The 20th term of an arithmetic sequence is 96 and the common difference is 5. Find the nth term.

29. The third term of a geometric sequence is 9 and the common ratio is $\frac{3}{2}$. Find the fifth term.

30. The second term of a geometric sequence is 10 and the fifth term is $\frac{1250}{27}$. Find the nth term.

31. The frequencies of musical notes measured in cycles per second form a geometric sequence. Middle C has a frequency of 256 and the C an octave higher has a frequency of 512. Find the frequency of the C two octaves below middle C.

32. A person has two parents, four grandparents, eight great-grandparents, and so on. How many ancestors does a person have 15 generations back?

33. **(a)** Find a formula for the area A_n of the regular polygon of n sides inscribed in a circle of radius 1 (see the figure).

(b) Find A_3, A_4, A_{100}, and A_{1000}. Notice that the terms of the sequence get closer and closer to π. Why?

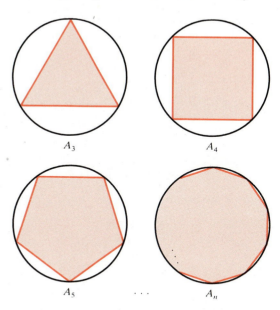

A_3

A_4

A_5

$\cdots$

A_n

34. If $a_1, a_2, a_3, \ldots$ and $b_1, b_2, b_3, \ldots$ are arithmetic sequences, show that $a_1 + b_1, a_2 + b_2, a_3 + b_3, \ldots$ is also an arithmetic sequence.

35. If $a_1, a_2, a_3, \ldots$ and $b_1, b_2, b_3, \ldots$ are geometric sequences, show that $a_1 b_1, a_2 b_2, a_3 b_3, \ldots$ is also a geometric sequence.

36. **(a)** If $a_1, a_2, a_3, \ldots$ is an arithmetic sequence, is the sequence $a_1 + 2, a_2 + 2, a_3 + 2, \ldots$ arithmetic?
(b) If $a_1, a_2, a_3, \ldots$ is a geometric sequence, is the sequence $5a_1, 5a_2, 5a_3, \ldots$ geometric?

37. Find the values of x for which the sequence $6, x, 12, \ldots$ is **(a)** arithmetic; **(b)** geometric.

38. Find the values of x and y for which the sequence $2, x, y, 17, \ldots$ is **(a)** arithmetic; **(b)** geometric.

In Exercises 39–42 find the given sum.

39. $\displaystyle\sum_{k=3}^{6} (k+1)^2$

40. $\displaystyle\sum_{i=1}^{4} \frac{2i}{2i-1}$

41. $\displaystyle\sum_{k=1}^{6} (k+1)2^{k-1}$

42. $\displaystyle\sum_{m=1}^{5} 3^{m-2}$

In Exercises 43–46 write the given sum without using sigma notation. Do not evaluate.

43. $\displaystyle\sum_{k=1}^{10} (k-1)^2$

44. $\displaystyle\sum_{j=2}^{100} \frac{1}{j-1}$

45. $\displaystyle\sum_{k=1}^{50} \frac{3^k}{2^{k+1}}$ 　　　　　**46.** $\displaystyle\sum_{n=1}^{10} n^2 2^n$

In Exercises 47–50 write the given sum using sigma notation. Do not evaluate.

47. $3 + 6 + 9 + 12 + \cdots + 99$

48. $1^2 + 2^2 + 3^2 + \cdots + 100^2$

49. $1 \cdot 2^3 + 2 \cdot 2^4 + 3 \cdot 2^5 + 4 \cdot 2^6 + \cdots + 100 \cdot 2^{102}$

50. $\dfrac{1}{1 \cdot 2} + \dfrac{1}{2 \cdot 3} + \dfrac{1}{3 \cdot 4} + \cdots + \dfrac{1}{999 \cdot 1000}$

In Exercises 51–58 determine whether the series is arithmetic or geometric and find its sum.

51. $1 + 0.9 + (0.9)^2 + \cdots + (0.9)^5$

52. $3 + 3.7 + 4.4 + \cdots + 10$

53. $1 - \sqrt{5} + 5 - 5\sqrt{5} + \cdots + 625$

54. $\sqrt{5} + 2\sqrt{5} + 3\sqrt{5} + \cdots + 100\sqrt{5}$

55. $\dfrac{1}{3} + \dfrac{2}{3} + 1 + \dfrac{4}{3} + \cdots + 33$

56. $a + abc^3 + ab^2c^6 + ab^3c^9 + \cdots + ab^8c^{24}$

57. $\displaystyle\sum_{n=0}^{6} 3(-4)^n$ 　　　　　**58.** $\displaystyle\sum_{k=0}^{8} 7(5)^{k/2}$

59. The first term of an arithmetic sequence is $a = 7$ and the common difference $d = 3$. How many terms of this sequence must be added to get 325?

60. A geometric sequence has first term $a = 81$ and common ratio $r = -\frac{2}{3}$. How many terms of this sequence must be added to get 55?

61. The sum of the first eight terms of an arithmetic series is 100 and the first term is 2. Find the tenth term.

62. The sum of the first three terms of a geometric series is 52 and the common ratio is $r = 3$. Find the first term.

63. A city has a population of 100,000. If the population is increasing at the rate of 10% a year, what will be the population of this city in 10 years? Find a formula for the population of the city after n years.

64. Refer to Exercise 32. What is the total number of ancestors of a person in 15 generations?

65. Find the amount of an annuity that consists of 16 annual payments of $1000 each into an account that pays 8% interest a year compounded annually.

66. How much money should be invested every quarter at 12% a year compounded quarterly in order to have $10,000 in one year?

67. What are the monthly payments on a mortgage of $60,000 at 9% interest if **(a)** the loan is to be repaid in 30 years; **(b)** the loan is to be repaid in 15 years. (Refer to Exercise 57 of Section 9.4.)

In Exercises 68–71 find the sum of the given infinite geometric series.

68. $1 - \dfrac{2}{5} + \dfrac{4}{25} - \dfrac{8}{125} + \cdots$

69. $0.1 + 0.01 + 0.001 + 0.0001 + \cdots$

70. $1 + \dfrac{1}{3^{1/2}} + \dfrac{1}{3} + \dfrac{1}{3^{3/2}} + \cdots$

71. $a + ab^2 + ab^4 + ab^6 + \cdots$

In Exercises 72 and 73 find the sum of the given infinite series.

72. $\left(\dfrac{1}{3} - \dfrac{1}{4}\right) + \left(\dfrac{1}{4} - \dfrac{1}{5}\right) + \left(\dfrac{1}{5} - \dfrac{1}{6}\right) + \cdots$
$$+ \left(\dfrac{1}{n+2} - \dfrac{1}{n+3}\right) + \cdots$$

73. $\dfrac{1}{1 \cdot 4} + \dfrac{1}{4 \cdot 7} + \dfrac{1}{7 \cdot 10} + \cdots + \dfrac{1}{(3n-2)(3n+1)} + \cdots$

74. The nth partial sum of an infinite series is $S_n = 2 - (1/3)^n$. Find the sum of the series.

In Exercises 75–79 use mathematical induction to prove that the given formula is true for all natural numbers n.

75. $1 + 4 + 7 + \cdots + (3n - 2) = \dfrac{n(3n-1)}{2}$

76. $1^3 + 3^3 + 5^3 + \cdots + (2n-1)^3 = n^2(2n^2 - 1)$

77. $1^4 + 2^4 + 3^4 + \cdots + n^4$
$$= \dfrac{n(n+1)(2n+1)(3n^2 + 3n - 1)}{30}$$

78. $\dfrac{1}{1 \cdot 3} + \dfrac{1}{3 \cdot 5} + \dfrac{1}{5 \cdot 7} + \cdots + \dfrac{1}{(2n-1)(2n+1)} = \dfrac{n}{2n+1}$

79. $\left(1 + \dfrac{1}{1}\right)\left(1 + \dfrac{1}{2}\right)\left(1 + \dfrac{1}{3}\right) \cdots \left(1 + \dfrac{1}{n}\right) = n + 1$

80. Show that $7^n - 1$ is divisible by 6 for all natural numbers n.

81. Show that $11^{n+2} + 12^{2n+1}$ is divisible by 133 for every natural number n.

82. Let $a_{n+1} = 3a_n + 4$ and $a_1 = 4$. Show that $a_n = 2 \cdot 3^n - 2$ for all natural numbers n.

83. Prove that the Fibonacci number F_{4n} is divisible by 3 for all natural numbers n.

84. Find and prove an inequality that relates 2^n and $n!$.

85. Find a formula for the sums

$$S_n = 1 \cdot 1! + 2 \cdot 2! + 3 \cdot 3! + \cdots + n \cdot n!$$

and prove that your formula holds for all natural numbers n.

86. Find a formula for the sums

$$S_n = \frac{1}{2!} + \frac{2}{3!} + \frac{3}{4!} + \cdots + \frac{n}{(n+1)!}$$

and prove that your formula holds for all natural numbers n.

87. Let Q_n and S_n be defined by

$$Q_n = \left(1 + \frac{1}{n}\right)^n$$

and $S_n = 1 + \dfrac{1}{1!} + \dfrac{1}{2!} + \dfrac{1}{3!} + \dfrac{1}{4!} + \cdots + \dfrac{1}{n!}$

Find and prove an inequality between Q_n and S_n for all natural numbers n.

In Exercises 88–91 evaluate the given expression.

88. $\dbinom{5}{2}\dbinom{5}{3}$

89. $\dbinom{10}{2} + \dbinom{10}{6}$

90. $\displaystyle\sum_{k=0}^{5} \binom{5}{k}$

91. $\displaystyle\sum_{k=0}^{8} \binom{8}{k}\binom{8}{8-k}$

In Exercises 92 and 93 expand the given expression.

92. $(2x + y)^4$

93. $(1 - x^2)^6$

94. Find the first three terms in the expansion of $(b^{-2/3} + b^{1/3})^{20}$.

95. Find the 20th term in the expansion of $(a + b)^{22}$.

96. Find the term that contains A^6 in the expansion of $(A + 3B)^{10}$.

97. Find the coefficient of s^5 in the expansion of $\left(\dfrac{s^3}{2} - \dfrac{2}{s^2}\right)^5$.

In Exercises 98 and 99 simplify the given expression.

98. $x^5 + 5x^4y + 10x^3y^2 + 10x^2y^3 + 5xy^4 + y^5$

99. $(a + 1)^3 - 3(a + 1)^2 + 3(a + 1) - 1$

Chapter 9 Test

1. Find the tenth term of the sequence whose nth term is

$$a_n = \frac{n}{1 - n^2}.$$

2. A sequence is defined recursively by $a_{n+2} = (a_n)^2 - a_{n+1}$. If $a_1 = 1$ and $a_2 = 1$, find a_5.

3. Find the 30th term in the arithmetic sequence $80, 76, 72, \ldots$.

4. The second term of a geometric sequence is 125 and the fifth term is 1. Is $\frac{1}{5}$ a term of this sequence? If so, which term is it?

5. Determine whether each of the following statements is true or false. If it is true, prove it. If it is false, give an example where it fails.
(a) If $a_1, a_2, a_3, \ldots$ is an arithmetic sequence, then the sequence $a_1^2, a_2^2, a_3^2, \ldots$ is also arithmetic.
(b) If $a_1, a_2, a_3, \ldots$ is a geometric sequence, then the sequence $a_1^2, a_2^2, a_3^2, \ldots$ is also geometric.

6. (a) Write the formula for the sum of a (finite) arithmetic series.
(b) The first term of an arithmetic series is 10 and the tenth term is 2. Find the sum of the first ten terms.
(c) Find the common difference and the 100th term of the series in part (b).

7. (a) Write the formula for the sum of a (finite) geometric series.
(b) Find the sum of the geometric series

$$\frac{1}{3} + \frac{2}{3^2} + \frac{2^2}{3^3} + \frac{2^3}{3^4} + \cdots + \frac{2^9}{3^{10}}$$

8. Find the sum of the infinite geometric series

$$1 + \frac{1}{2^{1/2}} + \frac{1}{2} + \frac{1}{2^{3/2}} + \cdots$$

9. Use mathematical induction to prove that for all natural numbers n,

$$1^2 + 2^2 + 3^2 + \cdots + n^2 = \frac{n(n + 1)(2n + 1)}{6}$$

10. Write the following expressions without using sigma notation; then find the sum:

    (a) $\displaystyle\sum_{n=1}^{5} (1 - n^2)$ **(b)** $\displaystyle\sum_{n=3}^{6} (-1)^n 2^{n-2}$

11. Write the expansion of $(a + b)^n$ using sigma notation.

12. Expand $(2x + y^2)^5$.

13. Find the term that contains a^3 in the expansion of $(2a + b)^{100}$.

14. Find the term that does not contain x in the expansion of $\left(2x + \dfrac{1}{x}\right)^{10}$.

Problems Plus

1. For fixed n and for $0 \le r \le n$, show that the sum of the binomial coefficients $\dbinom{n}{r}$ for r odd is equal to the sum of the binomial coefficients $\dbinom{n}{r}$ for r even.

2. Find the sum of the infinite series

$$\frac{1}{10} + \frac{2}{10^2} + \frac{3}{10^3} + \frac{4}{10^4} + \frac{5}{10^5} + \cdots$$

3. Prove that

$$\frac{1}{\sin 2} + \frac{1}{\sin 4} + \frac{1}{\sin 8} + \cdots + \frac{1}{\sin 2^n} = \cot 1 - \cot 2^n.$$

4. Find and prove a formula suggested by the figure.

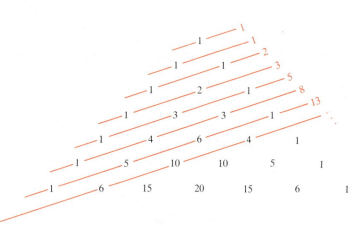

5. To construct the **snowflake curve** start with an equilateral triangle with side of length 1. Step 1 in the construction is to divide each side into three equal parts, construct an equilateral triangle on the middle part, and then delete the middle part (see the figure). Step 2 is to repeat Step 1 for each side of resulting polygon. This process is repeated at each succeeding step. The snowflake curve is the curve that results from repeating this process indefinitely.

    (a) Let s_n, l_n, and p_n represent the number of sides, the length of a side, and the total length of this curve after Step n of the construction, respectively. Find formulas for s_n, l_n, and p_n and prove that your formulas are true for all n.

    (b) Observe that as n becomes large, p_n also becomes large so that the length of the snowflake curve is infinitely large.

    (c) Show that the area enclosed by the snowflake curve is the sum of an infinite series and find the area.

Notice that parts (b) and (c) show that the snowflake curve is infinitely long but encloses only a finite area!

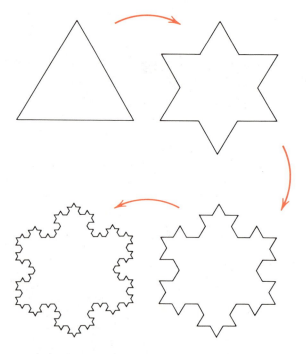

Answers to Odd-Numbered Exercises

Chapter 1

Exercises 1.1

1. false **3.** true **5.** true **7.** true **9.** true **11.** $\{1, 2, 3, 4, 5, 6, 8\}$ **13.** $\{8\}$
15. $\{1, 2, 3, 4, 5, 6, 7, 8, 9, 10\}$ **17.** $-3 < x < 0$ **19.** $2 \le x < 8$
21. $-1 \le x \le 1$ **23.** $x \ge 2$
25. $x \le -2$ **27.** $(-\infty, 1]$ **29.** $[1, 2]$
31. $(-2, 1]$ **33.** $(-1, \infty)$ **35.**
37. **39.** **41.** **43.** 100 **45.** 4
47. π **49.** $3 - \sqrt{3}$ **51.** -1 **53.** 10 **55.** 15 **57.** 26 **59.** 19

Exercises 1.2

1. -243 **3.** $\frac{625}{8}$ **5.** 6 **7.** 25 **9.** -2 **11.** 2^7 **13.** 2^{36} **15.** $2^{-1.5}$ **17.** $2^{5/2}$ **19.** t^5
21. $6x^7y^5$ **23.** $16x^{10}$ **25.** $4/b^2$ **27.** $64r^7s$ **29.** $648y^7$ **31.** x^3/y **33.** y^2z^9/x^5 **35.** $x^{13/15}$
37. $16b^{9/10}$ **39.** $1/c^{2/3}d$ **41.** $y^{1/2}$ **43.** $32x^{12}/y^{16/15}$ **45.** $x^{15}/y^{15/2}$ **47.** $4a^2/3b^{1/3}$ **49.** $3t^{25/6}/s^{1/2}$
51. x^{a+b+c} **53.** $x^{5/3}$ **55.** $x^{5/2}$ **57.** x^{n+5} **59.** $x^{3/4}$ **61.** 0.01 **63.** 3 **65.** $5\sqrt{3}$ **67.** $2\sqrt{5}$
69. $|x|$ **71.** $x\sqrt[3]{y}$ **73.** $a^{6/5}b^{7/5}$ **75.** $|xy^3|$ **77.** $2\sqrt[6]{x}$ **79.** $\sqrt{6}/6$ **81.** $\sqrt{15}/10$ **83.** $\sqrt{2x^5}/2$

85. $x\sqrt[3]{y^2}/y$ **87.** $(3 + \sqrt{5})/4$ **89.** $2(\sqrt{a} - 1)/(a - 1)$ **91.** no   **93.** yes   **95.** no
97. 27   **99.** $7^{1/4}$

Exercises 1.3

1. $-8x - 4$ **3.** $3x^2 - x + 7$ **5.** $x^3 + 3x^2 - 6x + 11$ **7.** $-t^4 + t^3 - t^2 - 10t + 5$ **9.** $x^{3/2} - x$
11. $y^{7/3} - y^{1/3}$ **13.** $2x^2 + 9x - 18$ **15.** $21t^2 - 29t + 10$ **17.** $1 - 4y + 4y^2$ **19.** $x - y$
21. $2x^3 - 7x^2 + 7x - 5$ **23.** $x^3 + x^2 - 2x$ **25.** $30y^4 + y^5 - y^6$ **27.** $4x^4 + 12x^2y^2 + 9y^4$ **29.** $x^4 - a^4$
31. $1 + 3a^3 + 3a^6 + a^9$ **33.** $a - 1/b^2$ **35.** $x^5 + x^4 - 3x^3 + 3x - 2$ **37.** $1 - 2b^2 + b^4$ **39.** $3xy + 7y^2$
43. $2x(1 + 6x^2)$ **45.** $(x + 6)(x + 1)$ **47.** $(x - 4)(x + 2)$ **49.** $9(x - 2)(x + 2)$ **51.** $(3x + 2)(2x - 3)$
53. $(t + 1)(t^2 - t + 1)$ **55.** $(2t - 3)^2$ **57.** $x(x + 1)^2$ **59.** $x^2(x + 3)(x - 1)$ **61.** $(2x - 5)(4x^2 + 10x + 25)$
63. $(x^2 + 2)(x - 1)(x + 1)$ **65.** $(y - 2)(y + 2)(y - 3)$ **67.** $(x - y)(x + y)(x^2 + xy + y^2)(x^2 - xy + y^2)$
69. $x^{1/2}(x - 1)(x + 1)$ **71.** $x^{-3/2}(x + 1)^2$ **73.** $(x^2 + 3)(x^2 + 1)^{-1/2}$ **77.** true   **79.** false   **81.** false

83. $\dfrac{x + 1}{x + 3}$ **85.** $\dfrac{-y}{y + 1}$ **87.** $\dfrac{x(2x + 3)}{2x - 3}$ **89.** $\dfrac{1}{t^2 + 9}$ **91.** $\dfrac{(2x + 1)(2x - 1)}{(x + 5)^2}$ **93.** $\dfrac{x}{yz}$

95. $\dfrac{3x + 7}{(x - 3)(x + 5)}$ **97.** $\dfrac{1}{(x + 1)(x + 2)}$ **99.** $\dfrac{3x + 2}{(x + 1)^2}$ **101.** $\dfrac{u^2 + 3u + 1}{u + 1}$ **103.** $\dfrac{2x + 1}{x^2(x + 1)}$

105. $\dfrac{2x + 7}{(x + 3)(x + 4)}$ **107.** $\dfrac{x - 2}{x^2 - 9}$ **109.** $\dfrac{5x - 6}{x(x - 1)}$ **111.** $\dfrac{-5}{(x + 1)(x + 2)(x - 3)}$ **113.** $-xy$ **115.** $\dfrac{c}{c - 2}$

117. $\dfrac{3x + 7}{x^2 + 2x - 1}$ **119.** $\dfrac{y - x}{xy}$ **121.** $\dfrac{-1}{a(a + h)}$ **123.** $\dfrac{-3}{(2 + x)(2 + x + h)}$ **125.** $\dfrac{1}{\sqrt{1 - x^2}}$

127. $\dfrac{r - s}{t(\sqrt{r} - \sqrt{s})}$ **129.** $\dfrac{-1}{\sqrt{x}\sqrt{x + h}(\sqrt{x} + \sqrt{x + h})}$ **131.** $\dfrac{x + 1}{\sqrt{x^2 + x + 1} - x}$ **133.** $(x^2 - x + 2)(x^2 + x + 2)$

Exercises 1.4

1. 2   **3.** $\frac{32}{9}$ **5.** $-\frac{1}{3}$ **7.** -20 **9.** $-3, -5$ **11.** $-6, 2$ **13.** $\frac{1}{2}, -2$ **15.** $0, -1$ **17.** $\pm\sqrt{5}$
19. $-1 \pm \sqrt{3}$ **21.** $(-1 \pm \sqrt{5})/2$ **23.** $5, -3$ **25.** $-60, 24$ **27.** $(-1 \pm \sqrt{7})/3$ **29.** $(1 \pm \sqrt{5})/4$
31. no real roots   **33.** $(-5 \pm \sqrt{13})/2$ **35.** $(\sqrt{5} \pm 1)/2$ **37.** $100, -50$ **39.** -4 **41.** 4   **43.** 4
45. 21   **47.** no real roots   **49.** $-1, 0, 3$ **51.** $W = \frac{1}{2}(P - 2L)$ **53.** $r = 1 - a/s$ **55.** $b = \pm\sqrt{c^2 - a^2}$
57. 3   **59.** 0   **61.** 1   **63.** 2   **65.** **(a)** ± 20 **(b)** $k > \frac{9}{5}$ **(c)** $k > -8$
67. 48 lb of $3.00 tea, 32 lb of $2.75 tea   **69.** $4500 at 9%, $1500 at 8%   **71.** 112 mi
73. 24 and 26, or -26 and -24 **75.** 13 in. $\times$ 13 in.

Exercises 1.5

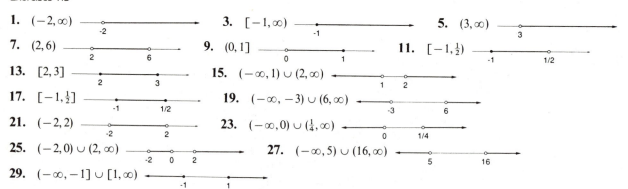

1. $(-2, \infty)$

3. $[-1, \infty)$

5. $(3, \infty)$

7. $(2, 6)$

9. $(0, 1]$

11. $[-1, \frac{1}{2})$

13. $[2, 3]$

15. $(-\infty, 1) \cup (2, \infty)$

17. $[-1, \frac{1}{2}]$

19. $(-\infty, -3) \cup (6, \infty)$

21. $(-2, 2)$

23. $(-\infty, 0) \cup (\frac{1}{4}, \infty)$

25. $(-2, 0) \cup (2, \infty)$

27. $(-\infty, 5) \cup (16, \infty)$

29. $(-\infty, -1] \cup [1, \infty)$

31. $10 \le C \le 35$ **33.** **(a)** $T = 20 - 10h$, h in kilometers   **(b)** $-30°C \le T \le 20°C$ **39.** $x \ge (a + b)c/ab$
41. $x > (c - b)/a$

Exercises 1.6

1. $|x - 3| = \begin{cases} x - 3 & \text{if } x \geq 3 \\ 3 - x & \text{if } x < 3 \end{cases}$ **3.** $|3x - 10| = \begin{cases} 3x - 10 & \text{if } x \geq 10/3 \\ 10 - 3x & \text{if } x < 10/3 \end{cases}$ **5.** $x^2 + 1$ **7.** $3|x + 3|$

9. $\frac{1}{2}|x - 5|$ **11.** $x^2 + 9$ **13.** $\pm\frac{3}{2}$ **15.** $4.01, 3.99$ **17.** $-4, -\frac{2}{5}$ **19.** $-3 < x < 3$ **21.** $3 < x < 5$

23. $x \leq -7$ or $x \geq -3$ **25.** $1.3 \leq x \leq 1.7$ **27.** $-4 < x < 8$ **29.** $-6.001 < x < -5.999$

31. $-4 \leq x \leq -1$ or $1 \leq x \leq 4$ **33.** $x > -1$ **35.** $x > \frac{1}{2}$ **37.** **(a)** **(i)** $-2x + 2$

(ii) 2 **(iii)** $2x - 2$ **(b)** $-\frac{1}{2} < x < \frac{5}{2}$

Exercises 1.7

1. 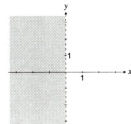   **3.** 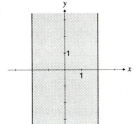   **5.**    **7.**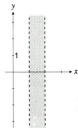

9. 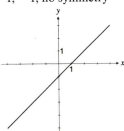   **11.**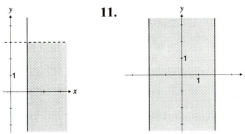

13. **(a)** 5 **(b)** $(\frac{5}{2}, 3)$ **15.** **(a)** $\sqrt{74}$ **(b)** $(\frac{5}{2}, \frac{1}{2})$

17. **(a)** $2\sqrt{37}$ **(b)** $(3, -1)$ **19.** A **25.** $(0, -4)$

27. $1, -1$, no symmetry   **29.** $\frac{5}{3}, -5$, no symmetry   **31.** $\pm 1, 1$, symmetry about y-axis

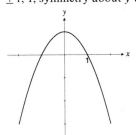

33. $0, 0$, symmetry about y-axis   **35.** no intercepts, symmetry about origin   **37.** $0, 0$, no symmetry

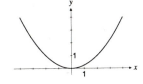   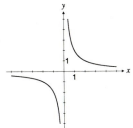   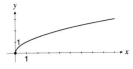

39. $\pm 3, \pm 3$, symmetry about both axes and origin

41. $\pm 2, 2$, symmetry about y-axis

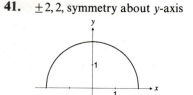

43. $0, 0$, symmetry about y-axis

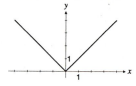

45. $\pm 4, 4$, symmetry about y-axis

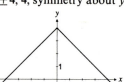

47. $(x - 3)^2 + (y + 1)^2 = 25$

49. $x^2 + y^2 = 65$

51. $(x - 3)^2 + (y + 1)^2 = 32$

53. $(2, -5), 4$

55. $(-\frac{1}{2}, 0), \frac{1}{2}$ **57.** $(\frac{1}{4}, -\frac{1}{4}), \sqrt{5/8}$ **59.**

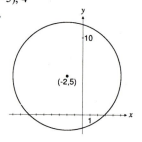

61. $a^2 + b^2 > 4c, (-a/2, -b/2), \frac{1}{2}\sqrt{a^2 + b^2 - 4c}$ **63.**

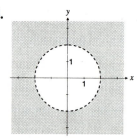

Exercises 1.8

1. 2 **3.** $-\frac{9}{2}$ **5.** $6x - y - 15 = 0$ **7.** $2x - 3y + 19 = 0$ **9.** $5x + y - 11 = 0$ **11.** $3x - y - 2 = 0$
13. $3x - y - 3 = 0$ **15.** $y = 5$ **17.** $x + 2y + 11 = 0$ **19.** $5x - 2y + 1 = 0$
21. **(a)**   **(b)** $3x - 2y + 8 = 0$ **23.** $-\frac{1}{3}, 0$ **25.** $0, -2$

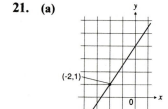

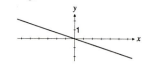

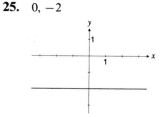

27. $\frac{3}{4}, -3$ **29.** $-\frac{3}{4}, \frac{1}{4}$ **35.** $x - y - 3 = 0$ **37.** **(b)** $4x - 3y - 24 = 0$

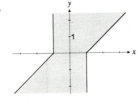

39. **(a)** $d = 48t$ **(b)**   **(c)** 48, speed

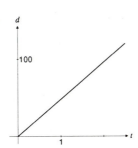

Exercises 1.9

1. 9   **3.** $-7, -1, 3, 9$ **5.** $-\frac{2}{3}, \frac{4}{3}$ **7.** 2   **9.** 52

11. $15 + 3\sqrt{37} \approx 33\frac{1}{4}$ h with Bob's hose, $21 + 3\sqrt{37} \approx 39\frac{1}{4}$ h with Jim's hose   **13.** 37.5 mi/h   **15.** 2.4 cm

19.   **21.**   **23.** 1090   **25.** not true   **29.** $\frac{200}{13} \approx 15.38$ in.

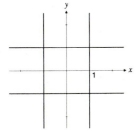

Review Exercises for Chapter 1

1. $-1 < x \le 3$ **3.** $(2, \infty)$ **5.** 6   **7.** $\frac{1}{72}$ **9.** $\frac{1}{6}$ **11.** 11

13. $12x^5y^4$ **15.** $9x^3$ **17.** x^2y^2 **19.** $\sqrt{3x}/3$ **21.** $(x - 2)(x + 5)$ **23.** $(5 - 4t)(5 + 4t)$

25. $(x - 1)(x^2 + x + 1)(x + 1)(x^2 - x + 1)$ **27.** $x^{-1/2}(x - 1)^2$ **29.** $3(2x^2 - 7x + 1)$ **31.** $4a^4 - 4a^2b + b^2$

33. $x^3 - 6x^2 + 11x - 6$ **35.** $2x^{3/2} + x - x^{1/2}$ **37.** $\dfrac{x - 3}{2x + 3}$ **39.** $\dfrac{x + 1}{x - 4}$ **41.** $\dfrac{x + 1}{(x - 1)(x^2 + 1)}$

43. $\dfrac{1}{x + 1}$ **45.** $-1/2x$ **47.** $6x + 3h - 5$ **49.** $\frac{2}{33}$ **51.** $-1, \frac{1}{2}$ **53.** $(-2 \pm \sqrt{7})/3$ **55.** 1

57. 20 lb of raisins, 30 lb of nuts   **59.** $(-3, -1]$

61. $(-\infty, -6) \cup (2, \infty)$ **63.** $[-4, -1)$

65. $(3.98, 4.02)$ **67.** $(-\infty, 2)$ **69.** **(a)** 10   **(b)** $(2, 1)$

71. B **73.** $(x - 2)^2 + (y + 5)^2 = 2$ **75.** $2x - 3y - 16 = 0$ **77.** no symmetry

79. symmetry about y-axis   **81.** no symmetry   **83.** symmetry about both axes and origin

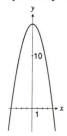      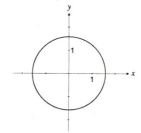

85. no   **87.** yes   **89.** no   **91.** no   **93.** $(x^2 + y^2 + xy)(x^2 + y^2 - xy)$

Chapter 1 Test

1.   **2.** $\frac{1}{8}$ **3.** $x^{b/2}$ **4. (a)** $8\sqrt{2}$ **(b)** $(x + 2)/(x - 2)$ **(c)** $2(x - 1)/(x^2 - 4)$
5. $x(x + 3)(x^2 - 3x + 9)$ **6.** $8 - 12x^2 + 6x^4 - x^6$ **7.** $3 < x < 9$ **8. (a)** 1 **(b)** $(-1 \pm \sqrt{5})/2$
(c) $0, \frac{1}{2}, 1$ **9. (a)** 13 **(b)** $(\frac{3}{2}, 1)$ **10.** $(x - 1)^2 + (y + 2)^2 = 41$ **11.** $2x - y + 9 = 0$
12.   **13.** 75

Chapter 2

Exercises 2.1

1. $0, 12, \frac{3}{4}, 7 - 3\sqrt{5}, a^2 - 3a + 2, a^2 + 3a + 2, a^2 + 2ab + b^2 - 3a - 3b + 2$
3. $-\frac{1}{3}, -3, (1 - \pi)/(1 + \pi), (1 - a)/(1 + a), (2 - a)/a, (1 + a)/(1 - a)$
5. $-4, 10, 3\sqrt{2}, 5 + 7\sqrt{2}, 2x^2 - 3x - 4, 2x^2 + 7x + 1, 4x^2 + 6x - 8, 8x^2 + 6x - 4$
7. $1 + 2a, 2 + 2a + 2h, 1 + 2a + 2h, 2$
9. $3 - 5a + 4a^2, 6 - 5a + 4a^2 - 5h + 4h^2, 3 - 5a - 5h + 4a^2 + 8ah + 4h^2, -5 + 8a + 4h$
11. $15 + 8h, 8x + 8h - 1, 8$ **13.** $1/x, 1/(2 + h), 1/(x + h), -1/x(x + h)$ **15.** $x \rightarrow \boxed{\text{square root}} \rightarrow \sqrt{x}$
17. $[-1, 5], [-2, 10]$ **19.** $[-2, 3], [-6, 14]$ **21.** $(-\infty, \infty), (-\infty, 2]$
23. $[\frac{5}{2}, \infty), [0, \infty)$ **25.** $[-1, 1], [3, 4]$ **27.** $\{x \mid x \neq -4\}$ $4 \rightarrow \boxed{\text{square root}} \rightarrow 2$
29. $\{x \mid x \neq \pm 1\}$ **31.** $(-\infty, \infty)$ **33.** $(-\infty, \frac{5}{2}]$ **35.** $(-\infty, \infty)$
37. $[-2, 3) \cup (3, \infty)$ **39.** $(10, \infty)$ **41.** $[0, 1]$ **43.** $(-\infty, -\frac{1}{2}] \cup [\frac{1}{2}, \infty)$ $2 \rightarrow \boxed{\text{square root}} \rightarrow \sqrt{2}$
45. $(-\infty, 0] \cup [6, \infty)$ **47.** $[0, \pi]$ **49.** $A = 10x - x^2, 0 < x < 10$
51. $A = \sqrt{3}x^2/4, x > 0$ **53.** $r = \sqrt{A/\pi}, A > 0$ **55.** $A = x^2 + 48/x, x > 0$
57. $A = 15x - (\pi + 4)x^2/8, x > 0$ **59.** $A = 2x(1200 - x), 0 < x < 1200$ **61.** $d = 25t, t \geq 0$

Exercises 2.2

1. (a) 1, −1, 3, 4 **(b)** [−3, 4] **(c)** [−1, 4] **(d)** decreasing on [−3, 0], [3, 4], increasing on [0, 3] **3. (a), (c)**
5. function, domain [−3, 2], range [−2, 2] **7.** not a function **9. (a)**

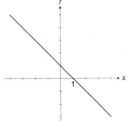

(b) R **(c)** decreasing on R **11. (a)**

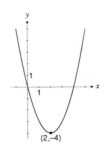

(b) R **(c)** decreasing on (−∞, 2], increasing on [2, ∞)
13. (a)

(b) (−∞, 4] **(c)** decreasing on (−∞, 4]

15. (a)

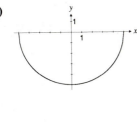

(b) [−5, 5] **(c)** decreasing on [−5, 0], increasing on [0, 5]

17.

19.

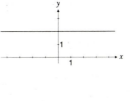

21.

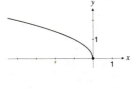

23.

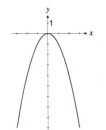

25.

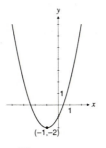

27.

29.

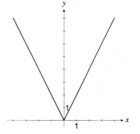

31.

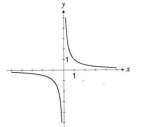

33.

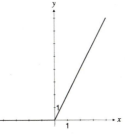

35.

37.

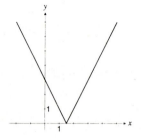

39.

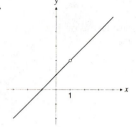

41.

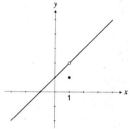

43.

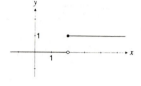

45.

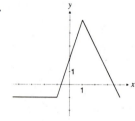

47.

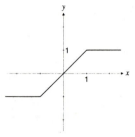

49.

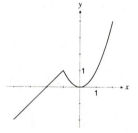

51.

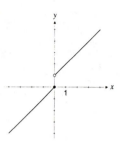

53. $f(x) = -\frac{7}{6}x - \frac{4}{3}, \; -2 \leq x \leq 4$

55. $f(x) = 1 - \sqrt{-x}$

57. even

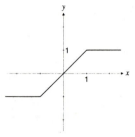

59. neither   **61.** odd

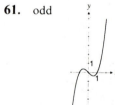

63. **(a)**

(b)

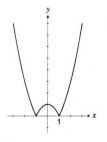

65.

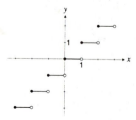

Exercises 2.3

1. Shift 10 units downward   3. Shift 1 unit left   5. Stretch vertically by a factor 8
7. Stretch vertically by a factor 6 and reflect in x-axis   9. Shift 2 units right and 3 units downward
11. Shrink vertically by a factor 2 and shift 9 units upward   13. (a)

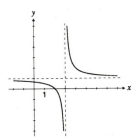

(b) (i) 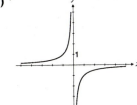   (ii) 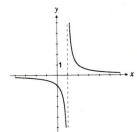   (iii) 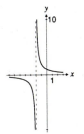   (iv)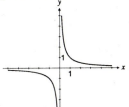

15. 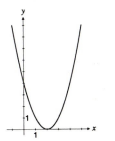   17. 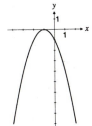   19. 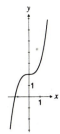   21.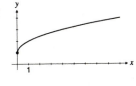

23. 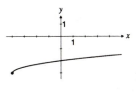   25.    27. 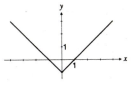   29.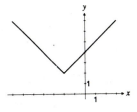

31. (a) (i) 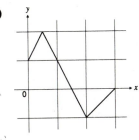   (ii)    (iii)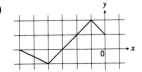

(iv)

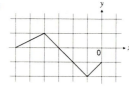

(b) (i) Shrink horizontally by a factor of a
(iii) Reflect in the y-axis

(ii) Stretch horizontally by a factor of a

Exercises 2.4

1.

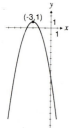

vertex $(0, -1)$, x-intercepts $\pm\sqrt{2}$,
y-intercept -1

3.

vertex $(0, -2)$, no x-intercept,
y-intercept -2

5.

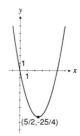

vertex $(\frac{5}{2}, -\frac{25}{4})$, x-intercepts 0,
5, y-intercept 0

7.

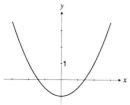

vertex $(-3, 1)$, x-intercepts -4,
-2, y-intercept -8

9.

vertex $(5, 7)$, no x-intercept,
y-intercept 57

11.

maximum $f(1) = 1$

13.

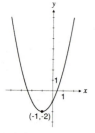

minimum $f(-1) = -2$

15.

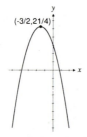

maximum $f(-\frac{3}{2}) = \frac{21}{4}$

17.

minimum $g(2) = 1$

19.

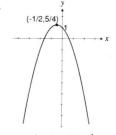

maximum $h(-\frac{1}{2}) = \frac{5}{4}$

21. minimum $f(-\frac{1}{2}) = \frac{3}{4}$ **23.** maximum $f(-\frac{25}{7}) = \frac{1325}{7}$ **25.** $f(x) = 2x^2 - 4x$ **27.** $f(x) = x^2 + x - 3$
29. 25 ft   **31.** ± 50 **33.** square with side 5 ft   **35.** 600 ft by 1200 ft   **37.** 14,062.5 ft

Exercises 2.5

1. $(f + g)(x) = x^2 + 5, (-\infty, \infty); (f - g)(x) = x^2 - 2x - 5, (-\infty, \infty); (fg)(x) = x^3 + 4x^2 - 5x, (-\infty, \infty);$
$(f/g)(x) = (x^2 - x)/(x + 5), (-\infty, -5) \cup (-5, \infty)$
3. $(f + g)(x) = \sqrt{1 + x} + \sqrt{1 - x}, [-1, 1]; (f - g)(x) = \sqrt{1 + x} - \sqrt{1 - x}, [-1, 1]; (fg)(x) = \sqrt{1 - x^2}, [-1, 1];$
$(f/g)(x) = \sqrt{(1 + x)/(1 - x)}, [-1, 1)$

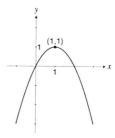

5. $(f + g)(x) = \sqrt{x} + \sqrt[3]{x}, [0, \infty); (f - g)(x) = \sqrt{x} - \sqrt[3]{x}, [0, \infty); (fg)(x) = x^{5/6}, [0, \infty); (f/g)(x) = \sqrt[6]{x}, (0, \infty)$

7. $\{x \mid -3 \le x \le 4, x \neq \pm\sqrt{2}\} = [-3, -\sqrt{2}) \cup (-\sqrt{2}, \sqrt{2}) \cup (\sqrt{2}, 4]$ **9.**

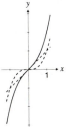

11.

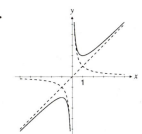

13. 1   **15.** 16   **17.** -11 **19.** -29 **21.** $1 - 3x^2$ **23.** $9x - 20$
25. 4   **27.** 5   **29.** 4

31. $(f \circ g)(x) = 8x + 1, (-\infty, \infty); (g \circ f)(x) = 8x + 11, (-\infty, \infty); (f \circ f)(x) = 4x + 9, (-\infty, \infty); (g \circ g)(x) = 16x - 5, (-\infty, \infty)$
33. $(f \circ g)(x) = 3(6x^2 + 7x + 2), (-\infty, \infty); (g \circ f)(x) = 6x^2 - 3x + 2, (-\infty, \infty);$
$(f \circ f)(x) = 8x^4 - 8x^3 + x, (-\infty, \infty); (g \circ g)(x) = 9x + 8, (-\infty, \infty)$
35. $(f \circ g)(x) = 1/(x^3 + 2x), x \neq 0; (g \circ f)(x) = 1/x^3 + 2/x, x \neq 0; (f \circ f)(x) = x, x \neq 0;$
$(g \circ g)(x) = x^9 + 6x^7 + 12x^5 + 10x^3 + 4x, (-\infty, \infty)$
37. $(f \circ g)(x) = \sqrt[3]{1 - \sqrt{x}}, [0, \infty); (g \circ f)(x) = 1 - \sqrt[6]{x}, [0, \infty); (f \circ f)(x) = \sqrt[9]{x}, (-\infty, \infty); (g \circ g)(x) = 1 - \sqrt{1 - \sqrt{x}}, [0, 1]$
39. $(f \circ g)(x) = (3x - 4)/(3x - 2), x \neq 2, x \neq \frac{2}{3}; (g \circ f)(x) = -(x + 2)/3x, x \neq -\frac{1}{2}, x \neq 0;$
$(f \circ f)(x) = (5x + 4)/(4x + 5), x \neq -\frac{1}{2}, x \neq -\frac{5}{4}; (g \circ g)(x) = x/(4 - x), x \neq 2, x \neq 4$
41. $(f \circ g \circ h)(x) = \sqrt{x - 1} - 1$
43. $(f \circ g \circ h)(x) = (\sqrt{x} - 5)^4 + 1$ **45.** $g(x) = x - 9, f(x) = x^5$ **47.** $g(x) = x^2, f(x) = x/(x + 4)$
49. $f(x) = |x|, g(x) = 1 - x^3$ **51.** $h(x) = x^2, g(x) = x + 1, f(x) = 1/x$ **53.** $f(x) = x^9, g(x) = 4 + x, h(x) = \sqrt[3]{x}$
55. $A(t) = 3600\pi t^2$ **57.** $g(x) = x^2 + x - 1$

Exercises 2.6

1. no   **3.** yes   **5.** no   **7.** yes   **9.** yes   **11.** no   **13.** 2   **15.** 1   **17.** $f^{-1}(x) = \frac{1}{4}(x - 7)$
19. $f^{-1}(x) = (5x - 1)/(2x + 3)$ **21.** $f^{-1}(x) = \frac{1}{5}(x^2 - 2), x \ge 0$ **23.** $f^{-1}(x) = \frac{1}{5}(3 - x)$
25. $f^{-1}(x) = \sqrt{4 - x}$ **27.** $f^{-1}(x) = (x - 4)^3$
29. **(a), (b)**   **(c)** $f^{-1}(x) = \frac{1}{2}(x - 1)$ **31.** **(a), (b)**

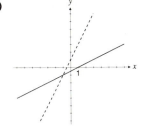

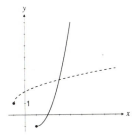

(c) $f^{-1}(x) = x^2 - 2x, x \ge 1$

33. **(a), (b)**

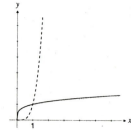

(c) $f^{-1}(x) = \sqrt[4]{x}$

Review Exercises for Chapter 2

1. $3, 1 + 2\sqrt{2}, 1 + \sqrt{a}, 1 + \sqrt{-x-1}, 1 + \sqrt{x^2-1}, x + 2\sqrt{x-1}$ **3.** **(b), (c)** are graphs of functions, **(c)** is one-to-one

5. $(-\infty, \infty), [4, \infty)$ **7.** $(-1, \infty)$ **9.** $\{x \mid x \ne -\frac{1}{2}, x \ne 3\}$

11.

13.

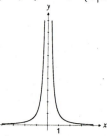

15.

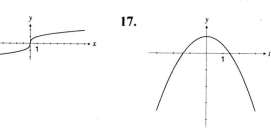

17.

19.

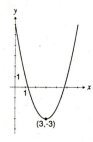

21.

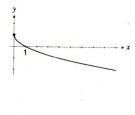

23.

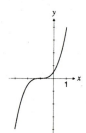

25.

27.

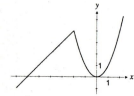

29. **(a)** Shift 8 units upward   **(b)** Shift 8 units left
(c) Stretch vertically by a factor 2, then shift 1 unit upward
(d) Shift 2 units right and 2 units downward   **(e)** Reflect in x-axis
(f) Reflect in line $y = x$ **31.** $f(-\frac{1}{2}) = \frac{5}{4}$ **33.** **(a)** neither   **(b)** odd
(c) even   **(d)** neither

35. $(f \circ g)(x) = -3x^2 + 6x - 1, R, (g \circ f)(x) = -9x^2 + 12x - 3, R, (f \circ f)(x) = 9x - 4, R,$
$(g \circ g)(x) = -x^4 + 4x^3 - 6x^2 + 4x, R$ **37.** $(f \circ g \circ h)(x) = 1 + \sqrt{x}$ **39.** yes   **41.** no
43. $f^{-1}(x) = \frac{1}{3}(x + 2)$ **45.** **(a), (b)**

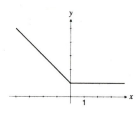

(c) $f^{-1}(x) = \sqrt{x + 4}$ **47.** $A = b\sqrt{4 - b}$
49. **(a)** $A(x) = x^2/16 + (\sqrt{3}/36)(10 - x)^2, 0 \le x \le 10$
(b) $40\sqrt{3}/(9 + 4\sqrt{3}) \approx 4.35$ m

Chapter 2 Test

1. **(a), (b)** are graphs of functions, **(a)** is one-to-one **2.** $(2, \infty)$ **3.** **(a)**  **(b)**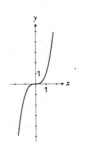

4. **(a)**  **(b)** $f(4) = 12$ **5.** **(a)** $-3, 3$ **(b)**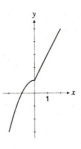

6. Shift 3 units right, reflect in x-axis, then shift 2 units upward

7. $(f \circ g)(x) = 4x^2 - 8x + 2, (g \circ f)(x) = 2x^2 + 4x - 5$

8. **(a)** $f^{-1}(x) = 3 - x^2, x \geq 0$ **(b)**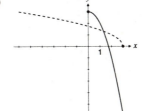

9. **(a)** $A = 400x - 2x^2$ **(b)** 100 ft

Problems Plus

1. $[-1, 2]$ **3.** 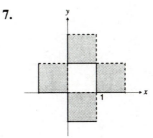 **5.** $\frac{999}{1000}$ **7.**

Chapter 3

Exercises 3.1

In answers 1–25, the first polynomial given is the quotient and the second is the remainder.
1. $4x + 5, 12$ **3.** $x^2 + 5, 13$ **5.** $x^4 - x^3 + 2x^2 - 2x + 3, -3$ **7.** $x^5 + 5x^4 + 10x^3 + 10x^2 + 5x + 1, 0$
9. $x + 2, 8x - 1$ **11.** $3x + 1, 7x - 5$ **13.** $x^3 + 1, 0$ **15.** $2x^2 - 1, 4x^2 + 2x$ **17.** $0, 6x - 10$

19. $2x^2 + 4x, 1$ **21.** $x^{100} + x^{99} + x^{98} + \cdots + x^3 + x^2 + x + 1, 0$ **23.** $\frac{1}{2}x + \frac{1}{2}, -1$
25. $\frac{1}{2}x^3 - \frac{3}{4}x^2 - \frac{3}{8}x + \frac{5}{16}, \frac{37}{16}$ **27.** 17 **29.** 1.92 **31.** 20 **33.** -273 **35.** 100 **37.** $\frac{49}{64}$ **39.** 5

45. $-1 \pm \sqrt{6}$ **47.** $-\frac{3}{2}x^3 + 3x^2 + \frac{15}{2}x - 9$ **49.** $x^4 - 4x^3 - 7x^2 + 22x + 24$ **51.** no   **53.** $\dfrac{-1 \pm \sqrt{13}}{3}$

Exercises 3.2

1. $\pm 1, \pm 7$ **3.** $\pm 1, \pm 3, \pm 5, \pm 15, \pm\frac{1}{2}, \pm\frac{3}{2}, \pm\frac{5}{2}, \pm\frac{15}{2}, \pm\frac{1}{4}, \pm\frac{3}{4}, \pm\frac{5}{4}, \pm\frac{15}{4}$ **5.** 3 or 1 positive, 1 negative; 4 or 2 real
7. 1 positive, 1 negative; 2 real   **9.** 2 or 0 positive, 0 negative; 3 or 1 real
11. 5, 3, or 1 positive, 2 or 0 negative; 7, 5, 3, or 1 real   **15.** $-5, 0$ **17.** $2, -3$ **19.** $5, 2, -3$ **21.** $1, \frac{1}{2}, -\frac{3}{2}$

23. $\pm\frac{3}{2}, \pm 2$ **25.** $-2, -\frac{1}{2}, \dfrac{-5 \pm \sqrt{17}}{2}$ **27.** $1, \frac{2}{5}, 2 \pm \sqrt{2}$ **29.** $4, \frac{3}{4}, -1$ **31.** $1, \frac{3}{4}, -2, \dfrac{2 \pm \sqrt{2}}{2}$

33. $2, \frac{1}{2}, -3$ **43.** $k = 2, -1$

Exercises 3.3

1. 0.26 **3.** 4.18 **5.** 0.62 **7.** $2, -2 \pm \sqrt{2}$ **9.** $\frac{1}{2}, -1, -1.47$ **11.** $-0.88, 1.35, 2.53$

13. $1, -2, \dfrac{-1 \pm \sqrt{5}}{2}$ **15.** $\frac{1}{2}, -0.71$ **17.** 2.626 ft by 3.808 ft   **19.** 2.76 m

Exercises 3.4

1. real part 5, imaginary part -6 **3.** real part 2, imaginary part $\dfrac{\sqrt{3}}{2}$ **5.** real part $-2 + \sqrt{2}$, imaginary part $\sqrt{3} - \sqrt{5}$

7. $10 - i$ **9.** $\frac{11}{2} + \frac{9}{2}i$ **11.** $-19 + 4i$ **13.** $8 + 24i$ **15.** $13 - i$ **17.** $-33 - 56i$
19. $-i$ **21.** $-\frac{1}{2} + \frac{1}{2}i$ **23.** $-5 + 12i$ **25.** $-\frac{8}{13} + \frac{12}{13}i$ **27.** $-i$ **29.** 1

31. $5i$ **33.** -6 **35.** $(3 + 5\sqrt{3}) + (3\sqrt{15} - \sqrt{5})i$ **37.** $\dfrac{2 + 7\sqrt{2}}{15} + \dfrac{\sqrt{7} - 2\sqrt{14}}{15}i$

45. 3

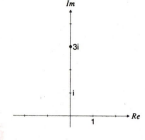

47. $\sqrt{29}$

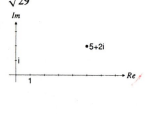

49. 2

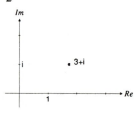

51. 1

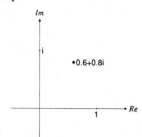

53.

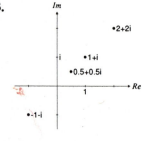

55.

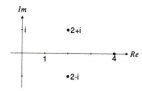

57.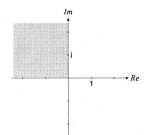

59.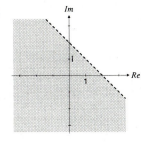

61.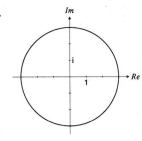

63.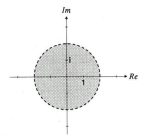

Exercises 3.5

1. $\pm 2i$ **3.** $4 \pm i$ **5.** $\dfrac{5}{6} \pm i\dfrac{\sqrt{23}}{6}$ **7.** $-\dfrac{3}{2} \pm i\dfrac{\sqrt{3}}{2}$ **9.** $0, i$

11. $P(x) = 3x^3 - 2x^2 + 27x - 18$ (or any integer multiple)
13. $R(x) = x^6 + 6x^5 + 15x^4 + 20x^3 + 15x^2 + 6x + 1$ (or any integer multiple)
15. $T(x) = 2x^4 - 8x^3 + 16x^2 - 16x + 8$ **17.** $1 \pm i, \frac{1}{3}, -3$ **19.** $1, \pm 2i, -1 \pm 2i$ **21.** $\pm\frac{3}{2}, \pm\frac{3}{2}i$

23. $\pm 3, \dfrac{\pm 3 \pm 3\sqrt{3}i}{2}$ **25.** $\dfrac{\pm 3 \pm i\sqrt{11}}{2}$ **27.** $1, \dfrac{1 \pm i\sqrt{3}}{2}$ **29.** $\frac{2}{3}, -1, \dfrac{-3 \pm i\sqrt{7}}{2}$ **31.** $1, 2, 4, \dfrac{1 \pm i\sqrt{2}}{3}$

33. $2(x + \frac{3}{2})(x + 1 - i\sqrt{2})(x + 1 + i\sqrt{2})$ **35.** $(x + 3)\left(x - \dfrac{3 + 3\sqrt{3}i}{2}\right)\left(x - \dfrac{3 - 3\sqrt{3}i}{2}\right)$

37. $(x - 2)(x + 2)[x + (1 + i\sqrt{3})][x + (1 - i\sqrt{3})][x - (1 + i\sqrt{3})][x - (1 - i\sqrt{3})]$ **41.** $-22i$
43. **(a)** $x^4 - 2x^3 + 3x^2 - 2x + 2$ **(b)** $x^2 - (1 + 2i)x - 1 + i$

Exercises 3.6

1.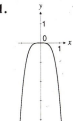

3.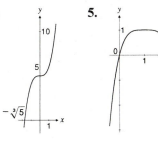

5.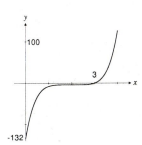

7.

9.

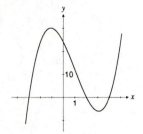

11.

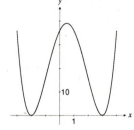

13.

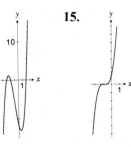

15.

17.

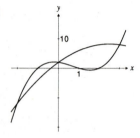

19.

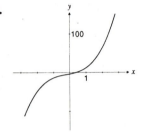

21.

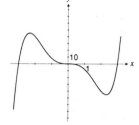

23. (a)

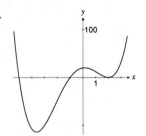

(b) three   **(c)** $(0, 2), (3, 8), (-2, -12)$

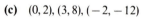

Exercises 3.7

1. x-intercept 6; y-intercept -6 **3.** x-intercepts 5, -3; y-intercept $-\frac{15}{16}$ **5.** x-intercept 3; no y-intercept
7. vertical: $x = -\frac{3}{2}$; horizontal: $y = 0$ **9.** vertical: $x = 3, x = -2$; horizontal: $y = 1$ **11.** horizontal: $y = 0$
13. vertical: $x = 4$; slant: $y = 3x + 5$ **15.** vertical: $x = -3$
17.

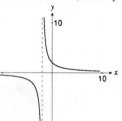

19.

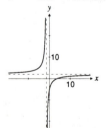

21.

23.

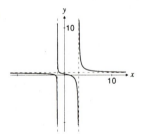

25. 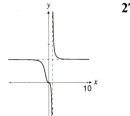 **27.** 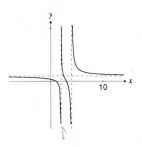 **29.** 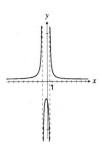 **31.**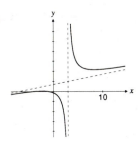

33. 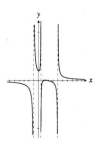 **35.** 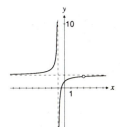 **37.** 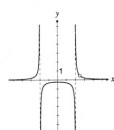 **39.**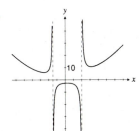

41. 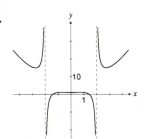 **45.** 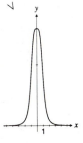 **47.**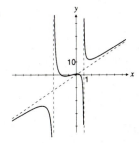

49. $y = \dfrac{x^2 - 5x + 6}{x^2 + 3x - 4}$

51. $y = \dfrac{6x^2 - 15x}{2x - 1}$

Exercises 3.8

1. $(-\infty, -4) \cup (5, \infty)$ **3.** $\left(-\infty, \dfrac{-3 - 3\sqrt{5}}{2}\right] \cup \left[\dfrac{-3 + 3\sqrt{5}}{2}, \infty\right)$ **5.** $[-4, 4]$ **7.** $[-4, \frac{1}{3}]$ **9.** $(-2, 4)$

11. $[-2, \frac{7}{3}]$ **13.** all $x \in R$ **15.** $[-1 - 2\sqrt{2}, -1 + 2\sqrt{2}]$ **17.** $(-\infty, 1) \cup (\frac{7}{2}, \infty)$

19. $(-\infty, -4] \cup [-2, 1] \cup [3, \infty)$ **21.** $[-1, 1]$ **23.** $(-\infty, -3) \cup (\frac{1}{2}, 3)$ **25.** $(-\infty, -3] \cup [1, \infty)$

27. $(-3.53, 3.53)$ **29.** $(0, 2) \cup (3, \infty)$ **31.** $[-\frac{1}{3}, \frac{1}{2}] \cup [1, \infty)$ **33.** $(1, 10)$ **35.** $(-7, \frac{5}{2}] \cup (5, \infty)$

37. $(-\infty, -1 - \sqrt{3}) \cup [0, -1 + \sqrt{3})$ **39.** $(-\infty, -3) \cup (-\frac{2}{3}, 1) \cup (3, \infty)$ **41.** $(-4, 3]$ **43.** $[-8, -\frac{5}{2})$

45. $\left(0, \dfrac{3 - \sqrt{3}}{2}\right] \cup \left(1, \dfrac{3 + \sqrt{3}}{2}\right]$ **47.** $(-\infty, -2) \cup (-1, 1) \cup (1, \infty)$ **49.** $[-2, 0) \cup (1, 3]$ **51.** $(-3, -\frac{1}{2}) \cup (2, \infty)$

53. $(0, 1]$ **55.** $(-\infty, -2) \cup (5, \infty)$ **57.** $(-\frac{1}{2}, 0) \cup (\frac{1}{2}, \infty)$ **59.** $[-2, 3]$ **61.** $(-\infty, -1] \cup [1, \infty)$

63. $(-\infty, -c) \cup [-a, b]$

Review Exercises for Chapter 3

In answers 1–7, the first polynomial given is the quotient and the second is the remainder.

1. $x^2 + 2x + 7, 10$ **3.** $x - 3, -9$ **5.** $x^3 - 5x^2 + 4, -5$ **7.** $x^3 + (\sqrt{3} + 1)x^2 + (\sqrt{3} + 1)x + \sqrt{3}, 2$

9. 3 **13.** 8 **15.** $\pm 1, \pm 2, \pm 4, \pm \frac{1}{2}, \pm \frac{1}{3}, \pm \frac{2}{3}, \pm \frac{4}{3}, \pm \frac{1}{6}$; no positive real roots; 4, 2, or 0 negative real roots

17. $19 + 40i, \sqrt{1961}$ **19.** $\frac{1}{5} + \frac{2}{5}i, 1/\sqrt{5}$ **21.** $-10, 10$

23. $x^4 - 8x^3 + 25x^2 - 72x + 144$ (or any integer multiple) **25.** $\frac{3}{2}, -1, -3$ **27.** $-4, -1, -1 \pm \sqrt{6}$

29. $\pm 3, \pm 3i$ **31.** $\frac{1}{2}, -\frac{1}{3}$ (double root) **33.** $\pm 3i, \pm i\sqrt{6}$ **35.** $-i, -1 - i\sqrt{6}, \dfrac{1 \pm i\sqrt{3}}{2}$

37. **(b)** $0.83, -0.48$ **39.**

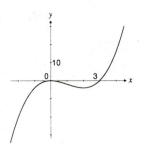

41.

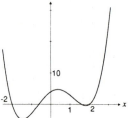

43.

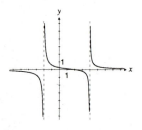

45.

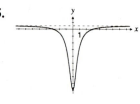

47. $(-\infty, -1] \cup [\frac{3}{2}, \infty)$ **49.** $(-\infty, -2) \cup (1, 2)$
51. $(-2, \frac{1}{7}]$ **53.** **(a)** $[-3, \frac{8}{3}]$ **(b)** $(0, 1)$

Chapter 3 Test

1. **(a)** 19 **(b)** -5 **2.** **(a)** $\pm 1, \pm 2, \pm 3, \pm 4, \pm 6, \pm 9, \pm 12, \pm 18, \pm 36, \pm\frac{1}{2}, \pm\frac{3}{2}, \pm\frac{9}{2}$
(b) $4, 2,$ or 0 positive real roots; 0 negative real roots   **(d)** $\frac{1}{2}, 1, \frac{3}{2}, 2, 3, 4, \frac{9}{2}, 6$ **(e)** $\frac{3}{2}, 2$ (double), 3
(f) $2(x - 2)^2(x - 3)(x - \frac{3}{2})$ **4.** $x^5 + x^4 + 2x^3 + 10x^2 + 13x + 5$
5. **(a)** P and Q: by Rational Roots Theorem; R: by Descartes' Rule   **(b)** no; Descartes' Rule
(c) two (one positive, one negative) by Descartes' Rule   **(d)** only possible rational roots are 1 and -1, neither of which is a root

6. 1.2 **7.** $4, \frac{1}{2}, 2 \pm \sqrt{3}$ **8.** $\pm 2i, \dfrac{-1 \pm i\sqrt{3}}{2}$ **9.**

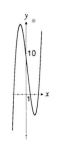

10. **(a)** $r, u,$ **(b)** s **(c)** s
(d)

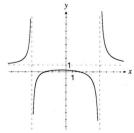

11. **(a)** $\frac{6}{13} - \frac{22}{13}i$ **(b)** $2 - 11i$ **(c)** $-i$ **12.** $(-\infty, -1] \cup (\frac{5}{2}, 3]$ **13.** $(-1 - \sqrt{5}, -1 + \sqrt{5})$

Problems Plus

1. Use the discriminant and the fact that $-p = r_1 + r_2$ **3.** $1, -7$

5. **(a)**

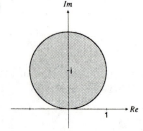

(b)

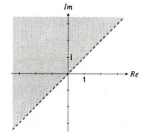

Chapter 4

Exercises 4.1

1.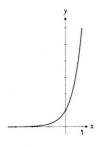

3.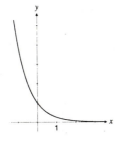

5.

7. $R, (-\infty, 0), y = 0$
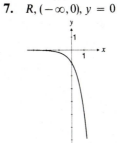

9. $R, (-3, \infty), y = -3$
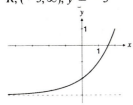

11. $R, (4, \infty), y = 4$
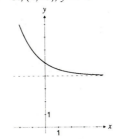

13. $R, (0, \infty), y = 0$
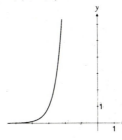

15. $R, (-\infty, 0), y = 0$
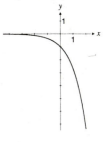

17. $R, (-1, \infty), y = -1$

19. $R, (0, \infty), y = 0$
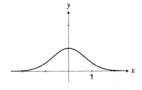

21. $R, (-\infty, 5), y = 5$
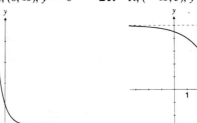

23. $R, [1, \infty)$, no asymptote
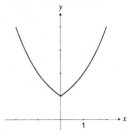

25.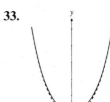

27. ± 1

29. $0, \frac{4}{3}$

31. $-\sqrt{2} < x < \sqrt{2}$

33.

Exercises 4.2

1. **(a)** $(1500)2^{2t}$ **(b)** 2381   **(c)** $(1500)2^{48} \approx 4.22 \times 10^{17}$ **3.** **(a)** $(10,000)3^t$ **(b)** $(10,000)3^{2.5} \approx 1.56 \times 10^5$
5. **(a)** 46.9 billion   **(b)** 46.7 billion   **7.** $(150)2^{-t/1590}$, 97.0 mg   **9.** **(a)** 12 g   **(b)** 7.135 g   **(c)** 0.315 g
11. **(a)** \$16,288.95   **(b)** \$26,532.98   **(c)** \$43,219.42   **13.** **(a)** \$4,615.87   **(b)** \$4,658.91   **(c)** \$4,697.04
(d) \$4,703.11   **(e)** \$4,704.68   **(f)** \$4,704.93   **(g)** \$4,704.94   **15.** \$7,678.96

Exercises 4.3

1. $2^6 = 64$ **3.** $10^{-2} = 0.01$ **5.** $8^3 = 512$ **7.** $a^c = b$ **9.** $\log_2 8 = 3$ **11.** $\log_{10} 0.0001 = -4$
13. $\log_4 0.125 = -\frac{3}{2}$ **15.** $\log_r t = s$ **17.** 4   **19.** 3   **21.** 1   **23.** -3 **25.** $\frac{1}{2}$ **27.** -1 **29.** $-\frac{2}{3}$
31. 37   **33.** 4   **35.** $\sqrt{5}$ **37.** 1024   **39.** $\frac{95}{3}$ **41.** 2   **43.** $1 - \log_2 3$ **45.** e^{10} **47.** $\frac{1}{12}\ln 17$
49. $3, -2$ **51.** 3^{16} **53.**   **55.** $(4, \infty)$, R, $x = 4$ **57.** $(-\infty, 0)$, R, $x = 0$

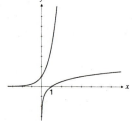 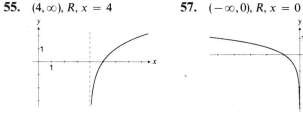

59. $(0, \infty)$, R, $x = 0$ **61.** $(0, \infty)$, R, $x = 1$ **63.** $(0, \infty)$, $[0, \infty)$, $x = 0$

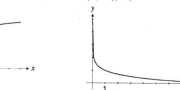 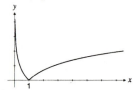

65. $(-\frac{2}{5}, \infty)$ **67.** $(-\infty, -1) \cup (1, \infty)$ **69.** $(0, 2)$ **71.** about 265 mi   **73.** $\log_{10} 2 < x < \log_{10} 5$
75. **(a)** $(1, \infty)$ **(b)** $f^{-1}(x) = 10^{2x}$ **77.** $\log_2 3$

Exercises 4.4

1. $\log_2 x + \log_2(x - 1)$ **3.** $23 \log 7$ **5.** $\log_2 A + 2 \log_2 B$ **7.** $\log_3 x + \frac{1}{2}\log_3 y$ **9.** $\frac{1}{3}\log_5(x^2 + 1)$
11. $\frac{1}{2}(\ln a + \ln b)$ **13.** $3 \log x + 4 \log y - 6 \log z$ **15.** $\log_2 x + \log_2(x^2 + 1) - \frac{1}{2}\log_2(x^2 - 1)$
17. $\ln x + \frac{1}{2}(\ln y - \ln z)$ **19.** $\frac{1}{4}\log(x^2 + y^2)$ **21.** $\frac{1}{3}[\log(x^2 + 4) - \log(x^2 + 1) - 2\log(x^3 - 7)]$
23. $\frac{1}{2}\ln x + 4\ln z - \frac{1}{3}\ln(y^2 + 6y + 17)$ **25.** $\frac{3}{2}$ **27.** 1   **29.** 3   **31.** $\ln 8$ **33.** 16   **35.** $\log_3 160$
37. $\log_2(AB/C^2)$ **39.** $\log[x^4(x - 1)^2/\sqrt[3]{x^2 + 1}]$ **41.** $\ln[5x^2(x^2 + 5)^3]$
43. $\log[\sqrt[3]{2x + 1}\sqrt{(x - 4)/(x^4 - x^2 - 1)}]$ **45.** no   **47.** yes   **49.** no   **51.** yes   **53.** yes   **55.** $\frac{3}{2}$
57. 5   **59.** 5   **61.** $\frac{13}{12}$ **63.** 6   **65.** 6.584963   **67.** 0.943028   **69.** -43.067656 **71.** -2.946865
73. 2.807355   **75.** 2.182658   **81.** $2 < x < 4$ or $7 < x < 9$

Exercises 4.5

1. **(a)** 2.3   **(b)** 3.5   **(c)** 8.3   **3.** **(a)** 10^{-3} M   **(b)** 3.2×10^{-7} M   **5.** $4.8 \leq \text{pH} \leq 6.4$ **7.** 42.5 min
9. 43.5 min   **11.** 2103   **13.** 16 years   **15.** 149 h   **17.** **(a)** 137°F   **(b)** 116 min   **19.** 5 years
21. 8.15 years   **23.** $\log 20 \approx 1.3$ **25.** Twice as intense   **27.** 6.3×10^{-3} watts/m^2 **29.** $t = -\frac{5}{13}\ln(1 - \frac{13}{60}I)$
31. **(b)** 106 dB

Review Exercises for Chapter 4

1. $R, (0, \infty), y = 0$

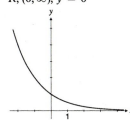

3. $R, (-\infty, 5), y = 5$

5. $(1, \infty), R, x = 1$

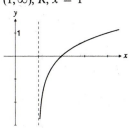

7. $(0, \infty), R, x = 0$

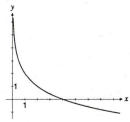

9. $R, (-1, \infty), y = -1$

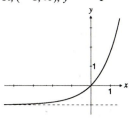

11. $(0, \infty), R, x = 0$

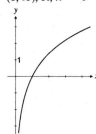

13. $(-\infty, \frac{1}{2})$ **15.** $2^{10} = 1024$ **17.** $10^y = x$
19. $\log_2 64 = 6$ **21.** $\log 74 = x$ **23.** 7 **25.** 45
27. 6 **29.** -3 **31.** $\frac{1}{2}$ **33.** 2 **35.** 92
37. $\frac{2}{3}$ **39.** $\log A + 2 \log B + 3 \log C$
41. $\frac{1}{2}[\ln(x^2 - 1) - \ln(x^2 + 1)]$
43. $2 \log_5 x + \frac{3}{2} \log_5 (1 - 5x) - \frac{1}{2} \log_5 (x^3 - x)$
45. $\log 96$ **47.** $\log[(x - y)^{3/2}/(x^2 + y^2)^2]$
49. $\log[(x^2 - 4)/\sqrt{x^2 + 4}]$ **51.** -15
53. $\frac{1}{3}(5 - \log_5 26)$ **55.** $\frac{4}{3} \ln 10$ **57.** 3 **59.** $-4, 2$
61. 0.430618 **63.** 2.303600 **65.** 1.953445

67. $\log_4 258$ **69.** 7.9, basic **71.** **(a)** 5278 **(b)** 58 min **73.** **(a)** 9.97 mg **(b)** 1.39×10^5 years
75. **(a)** \$16,081.15 **(b)** \$16,178.18 **(c)** \$16,197.65 **(d)** \$16,198.31 **77.** 8.0

Chapter 4 Test

1.

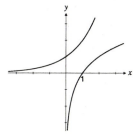

2. $(3, \infty), R, x = 3$ **3.** **(a)** $\frac{2}{3}$ **(b)** 2 **4.** $\frac{1}{2}[\log(x^2 - 1) - 3 \log x - 5 \log(y^2 + 1)]$
5. $\ln[x\sqrt{3 - x^4}/(x^2 + 1)^2]$ **6.** **(a)** $\log_2 10$ **(b)** $3 - e^4$ **7.** **(a)** 20 min
(b) $(1000)2^{3t}$ **(c)** 22,627 **(d)** 1.3 h **(e)** $n(t)$ **8.** $(-4, \frac{8}{5})$

Problems Plus

1. *Hint:* Use proof by contradiction. **5.** $[-1, 1 - \sqrt{3}) \cup (1 + \sqrt{3}, 3]$

Chapter 5

Exercises 5.1

1. $2\pi/9 \approx 0.698$ rad **3.** $2\pi/5 \approx 1.257$ rad **5.** $\pi/4 \approx 0.785$ rad **7.** $17\pi/4 \approx 13.352$ rad **9.** $135°$
11. $150°$ **13.** $270/\pi \approx 85.944°$ **15.** $-15°$ **17.** IV **19.** II **21.** III **23.** III **25.** III **27.** no
29. yes **31.** yes **33.** yes **35.** yes **37.** $13°$ **39.** $63°$ **41.** $280°$ **43.** $359°$ **45.** $2\pi/5$

47. π **49.** $\pi/4$ **51.** $-23 + 8\pi \approx 2.133$ rad   **53.** $55\pi/9 \approx 19.2$ **55.** 4   **57.** 4 mi   **59.** 2 rad $\approx 114.6°$
61. $36/\pi \approx 11.459$ m   **63.** 21,609   **65.** 346 mi   **67.** 3980 mi, 25,000 mi   **69.** 0.619 rad $\approx 35.4°$
71. **(a)** $128\pi/9 \approx 44.7$ **(b)** 25   **73.** $3\pi/2 \approx 4.7$ **75.** 6 ft   **77.** 1 rad $\approx 57.3°$ **79.** 3770 rad/min
81. **(a)** 0.5 rad/s   **(b)** 4.8   **83.** 4712 ft/min   **85.** 16,400 mi/h, 3.59 rad/h   **87.** **(a)** 3 m/s

(b) 20 rad/min   **89.** **(a)** 13   **(b)** $\sqrt{189}$ **91.** **(a)** $\sqrt{63}/4$ **(b)** $\dfrac{b}{4}\sqrt{4a^2 - b^2}$ **93.** $\pi/4$

Exercises 5.2

1. $\sin\theta = 4/5, \cos\theta = 3/5, \tan\theta = 4/3, \csc\theta = 5/4, \sec\theta = 5/3, \cot\theta = 3/4$
3. $\sin\theta = 4/7, \cos\theta = \sqrt{33}/7, \tan\theta = 4\sqrt{33}/33, \csc\theta = 7/4, \sec\theta = 7\sqrt{33}/33, \cot\theta = \sqrt{33}/4$
5. $\sin\theta = 5\sqrt{34}/34, \cos\theta = 3\sqrt{34}/34, \tan\theta = 5/3, \csc\theta = \sqrt{34}/5, \sec\theta = \sqrt{34}/3, \cot\theta = 3/5$
7. **(a)** $2\sqrt{13}/13, 2\sqrt{13}/13$ **(b)** $2/3, 2/3$ **(c)** $\sqrt{13}/3, \sqrt{13}/3$ **9.** **(a)** $5/13, 5/13$ **(b)** $5/12, 5/12$
(c) $13/12, 13/12$ **11.** $\cos\theta = 4/5, \tan\theta = 3/4, \sec\theta = 5/4, \csc\theta = 5/3, \cot\theta = 4/3$
13. $\sin\theta = \sqrt{2}/2, \cos\theta = \sqrt{2}/2, \tan\theta = 1, \sec\theta = \sqrt{2}, \csc\theta = \sqrt{2}$
15. $\sin\theta = 4\sqrt{3}/7, \cos\theta = 1/7, \tan\theta = 4\sqrt{3}, \csc\theta = 7\sqrt{3}/12, \cot\theta = \sqrt{3}/12$
17. $\sin\theta = a/\sqrt{a^2 + b^2}, \cos\theta = b/\sqrt{a^2 + b^2}, \csc\theta = \sqrt{a^2 + b^2}/a, \sec\theta = \sqrt{a^2 + b^2}/b, \cot\theta = b/a$
19. $\sin\theta = \sqrt{1 - d^2}, \tan\theta = \sqrt{1 - d^2}/d, \csc\theta = 1/\sqrt{1 - d^2}, \sec\theta = 1/d, \cot\theta = d/\sqrt{1 - d^2}$
21. $\sin\theta = a/\sqrt{1 + a^2}, \cos\theta = 1/\sqrt{1 + a^2}, \tan\theta = a, \csc\theta = \sqrt{1 + a^2}/a, \sec\theta = \sqrt{1 + a^2}, \cot\theta = 1/a$
23. $\sin\theta = \sqrt{2a + 1}/(a + 1), \cos\theta = a/(1 + a), \tan\theta = \sqrt{2a + 1}/a, \csc\theta = (a + 1)/\sqrt{2a + 1}, \sec\theta = (a + 1)/a,$
$\cot\theta = a/\sqrt{2a + 1}$ **25.** $12\sqrt{2}$ **27.** $4\sqrt{3}$ **29.** 31.30339   **31.** $x = 100\cos\theta, y = 100\sin\theta$
33. $x = 14\cos\theta, y = 14\sin\theta$ **35.** 0.38051 rad   **37.** 0.77519 rad   **39.** 1   **41.** 1   **43.** $-1/2$
45. $(2 - \sqrt{3})/4$ **47.**

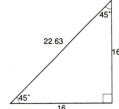

49.

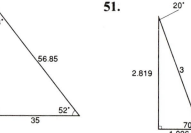

51.

53. 1026 ft   **55.** 2100 mi; no   **57.** 19 ft   **59.** 38.7°   **61.** 345 ft   **63.** 415 ft, 152 ft   **65.** 2570 ft
67. 5800   **69.** **(a)** 89.0544°   **(b)** 236,000 mi   **71.** 1076 ft   **73.** 230.9   **75.** 63.7   **77.** $\sin\theta\cos\theta$
79. $a = \sin\theta, b = \tan\theta, c = \sec\theta, d = \cos\theta$ **81.** $d = 2\sin(\theta/2)$ **83.** **(b)** $\sqrt{2} - 1, 2 - \sqrt{3}$ **85.** $1/(1 + \sqrt{3})$

Exercises 5.3

1. 45°   **3.** 1°   **5.** 30°   **7.** 50°   **9.** 26°   **11.** 85°   **13.** $2\pi/5$ **15.** $\pi/3$ **17.** $\pi/4$ **19.** 1.44
21. 0.7   **23.** 1.05   **25.** negative   **27.** positive   **29.** IV   **31.** $1/2$ **33.** $\sqrt{2}/2$ **35.** $-\sqrt{3}/2$
37. 1   **39.** $-\sqrt{3}/2$ **41.** $\sqrt{3}/3$ **43.** 1   **45.** -2 **47.** $-\sqrt{3}/2$ **49.** $1/2$ **51.** $-\sqrt{3}/3$
53. $-\sqrt{2}$ **55.** $\sqrt{2}/2$ **57.** undefined   **59.** $-1/2$ **61.** $-\sin\theta/\sqrt{1 - \sin^2\theta}$ **63.** $\sqrt{1 - \sin^2\theta}$
65. $-1/\sqrt{1 + \tan^2\theta}$ **67.** $-\sqrt{\sec^2\theta - 1}/\sec\theta$
69. $\cos\theta = -\sqrt{7}/4, \tan\theta = -3\sqrt{7}/7, \csc\theta = 4/3, \sec\theta = -4\sqrt{7}/7, \cot\theta = -\sqrt{7}/3$
71. $\sin\theta = -3/5, \cos\theta = 4/5, \csc\theta = -5/3, \sec\theta = 5/4, \cot\theta = -4/3$
73. $\sin\theta = 1/2, \cos\theta = \sqrt{3}/2, \tan\theta = \sqrt{3}/3, \sec\theta = 2\sqrt{3}/3, \cot\theta = \sqrt{3}$
75. $\sin\theta = 3\sqrt{5}/7, \tan\theta = -3\sqrt{5}/2, \csc\theta = 7\sqrt{5}/15, \sec\theta = -7/2, \cot\theta = -2\sqrt{5}/15$
77. **(a)** $\sqrt{3}/2, \sqrt{3}$ **(b)** $1/2, \sqrt{3}/4$ **(c)** $3/4, 0.88967$ **79.** 19.1   **81.** 43.3   **83.** 2.5
85. **(a)** $\frac{1}{2}\sqrt{2 - \sqrt{3}}, \frac{1}{2}\sqrt{2 + \sqrt{3}}$ **(b)** $\frac{1}{2}\sqrt{2 - \sqrt{2}}, \frac{1}{2}\sqrt{2 + \sqrt{2}}$ **(c)** $\frac{1}{2}\sqrt{2 - \sqrt{2 + \sqrt{3}}}, \frac{1}{2}\sqrt{2 + \sqrt{2 + \sqrt{3}}}$

Exercises 5.4

1. one

3. one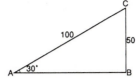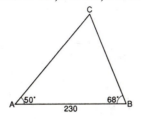

5. none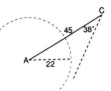

7. 21.5 **9.** 134.6 **11.** 44° **13.** $\angle C = 62°, a \approx 200, b \approx 242$ **15.** $\angle B = 85°, a \approx 5, c \approx 9$

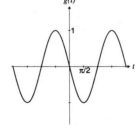

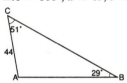

17. $\angle A = 100°, a \approx 89, c \approx 71$ **19.** $\angle B \approx 30°, \angle C \approx 40°, c \approx 19$ **21.** no solution

23. $\angle A_1 \approx 125°, \angle C_1 \approx 30°, a_1 \approx 49; \angle A_2 \approx 5°, \angle C_2 \approx 150°, a_2 \approx 5.6$

25. no solution **27.** 13 **29.** 29.89° **31.** 15 **33.** 2 **35.** 25

37. 84.6° **39.** 24 **41.** $\angle C = 114°, a \approx 51, b \approx 24$

43. $\angle A \approx 63°, \angle B \approx 15°, \angle C \approx 102°$ **45.** no solution

47. $\angle A_1 \approx 83.6°, \angle C_1 \approx 56.4°, a_1 \approx 193; \angle A_2 \approx 16.4°, \angle C_2 \approx 123.6°, a_2 \approx 54.9$

49. $\angle A \approx 48°, \angle B \approx 79°, c \approx 3.2$ **51.** $\angle A \approx 50°, \angle B \approx 73°, \angle C \approx 57°$

53. 2.30 mi **55.** 175 ft **57.** 28 mi **65.** **(a)** 1.14 **(b)** 1.73 **(c)** 36.8 **(d)** 1.06 **(e)** 1.31

(f) 5.76 **67.** **(b)** 12 **(c)** plane

Exercises 5.5

1. 4/5, 3/5, 4/3 **3.** $-\sqrt{13}/7, 6/7, -\sqrt{13}/6$ **5.** 9/41, 40/41, 9/40 **7.** $2\sqrt{2}/3, 1/3, 2\sqrt{2}$

9. $-12/13, -5/13, 12/5$ **11.** 4/5, 3/5 **13.** $-\sqrt{5}/3, 2/3$ **15.** $-\sqrt{7}/3, \sqrt{2}/3$ **17.** $(\sqrt{2}/2, \sqrt{2}/2)$

19. $(1/2, -\sqrt{3}/2)$ **21.** $(-1, 0)$ **27.** $f(g(t)) = \cos t$ **29.** $f(g(t)) = 4 \tan t$ **31.** $f(g(t)) = \cot^2 t \csc t$

33. period 2 **35.** not periodic **39.** 2π **41.** π **43.** 2π **45.** π **47.** odd **49.** odd

51. even **53.** neither **57.** **59.**

61. **63.** **65.** Graph is empty

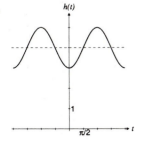

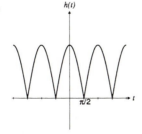

Exercises 5.6

1. $2\pi/3$, 3

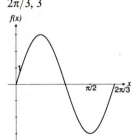

3. 4π, 10

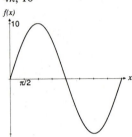

5. 6π, 1

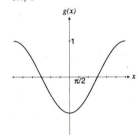

7. 2/3, 3

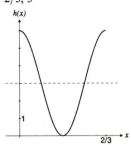

9. 3π, 1

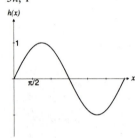

11. $3\sin 3(x-2)$

13. $10\sin\dfrac{1}{2}\left(x-\dfrac{\pi}{2}\right)$

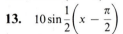

15. 1, 2π, $\pi/2$

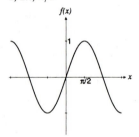

17. 2, 2, 1/3

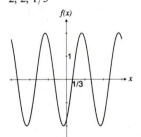

19. 2, 2π, $\pi/2$

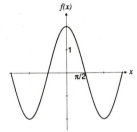

21. 5, $2\pi/3$, $\pi/12$

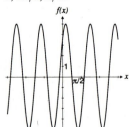

23. 2, 3π, $\pi/4$

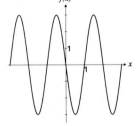

25. 3, 2, $-1/2$

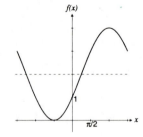

27. $2, \pi, \pi/6$

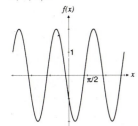

29. $1, 2\pi/3, -\pi/3$

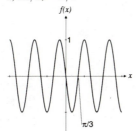

31. $\pi/2$

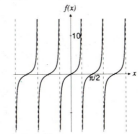

33. 2

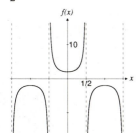

35. 2π

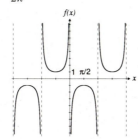

37. $\pi/2$

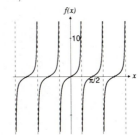

39. π

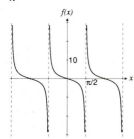

41.

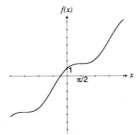

43.

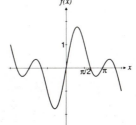

45.

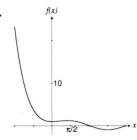

47.

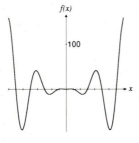

49.

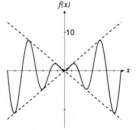

51.

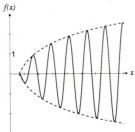

53.

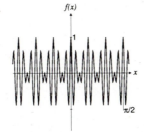

55.

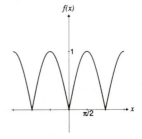

57.

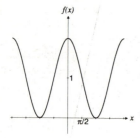

59.

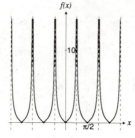

61.

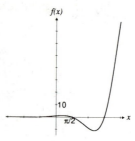

63.

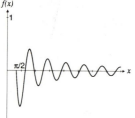

65. **(a)** period π **(b)** even

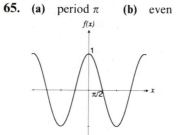

67. **(a)** period 2π **(b)** even

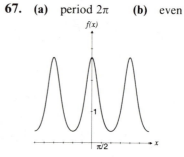

69. **(a)** period π **(b)** even

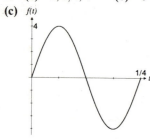

71. **(a)** period $2\pi/3$ **(b)** odd

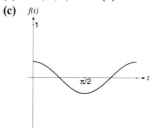

Exercises 5.7

1. **(a)** $4, 1/4, 4$ **(b)** 0
(c)

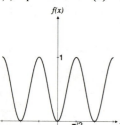

3. **(a)** $0.3, \pi, 1/\pi$ **(b)** 0.3
(c)

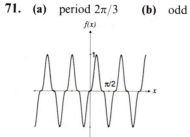

5. **(a)** $1000, 1, 1$ **(b)** 0
(c)

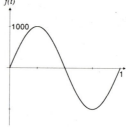

7. (a) 1, 2, 1/2 **(b)** 0 **9. (a)** π, 2/3, 3/2 **(b)** $\pi/2$

(c)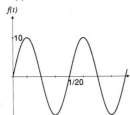

(c)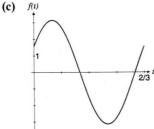

11. $y = 6 \sin 4\pi t$ **13.** $y = 10 \sin 6\pi t$

15. $y = 2 \sin 6\pi[t - (\pi/3)]$ **17.** $y = 2 \cos 10\pi t$

19. $y = -2 \cos 2\pi t$ **21. (a)** $f(t) = 10 \cos(\sqrt{3}\, t/2)$

(b) $f(t) = 100 \cos t$ **(c)** $f(t) = \cos(\sqrt{10}\, t/10)$

(d) $f(t) = 10 \cos(\sqrt{2}\, t/10)$ **23.** decreased, slower

25. $y = 11 - 10 \cos 6\pi t$, t in minutes

27. $y = 21 \sin(\pi t/6)$

29. $E(t) = 310 \cos 200\pi t$, 310 V, 219 V

31. $R = 20 + 1.5 \sin(2\pi t/5.4)$, R in millions of miles

33. $P(t) = 10 \sin 40\pi t$

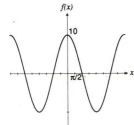

35. (a) $h(t) = 12 + (17/6)\sin(2\pi t/365)$ **(b)** $h(t) = 12 + (17/6)\sin[2\pi(t + 92)/365]$

(c) $h(t) = 12 + (17/6)\sin[2\pi(t - 79)/365]$ **37.** $y = e^{-0.9t} \sin \pi t$ **39.** $\frac{1}{3}\ln 4 \approx 0.46$

Review Exercises for Chapter 5

1. (a) $7\pi/18 \approx 1.22$ rad **(b)** $7\pi/3 \approx 7.33$ rad **(c)** $-4\pi/3 \approx -4.19$ rad **(d)** $2\pi/15 \approx 0.42$ rad

(e) $-11\pi/6 \approx -5.76$ rad **(f)** $25\pi/6 \approx 13.09$ rad **3.** 8 m **5.** 82 ft **7.** 0.4 rad **9. (a)** $-\sqrt{2}/2$

(b) $\sqrt{2}$ **(c)** 1 **(d)** $-\sqrt{3}/2$ **(e)** $-\sqrt{3}/3$ **(f)** $\sqrt{2}/2$ **(g)** $-\sqrt{2}/2$ **(h)** 2 **11.** 60°

13. $(16 - \sqrt{17})/4$ **15.** $2 + \sqrt{3}$ **17. (a)** $\sqrt{1 - \cos^2\theta}$ **(b)** $\sin\theta/-\sqrt{1 - \sin^2\theta}$ **19.** 550 m

21. 14,400 ft **23.** 1160 ft **25.** 80.8 mi

27. (a) $\sin t = 1/2$, $\cos t = \sqrt{3}/2$, $\tan t = \sqrt{3}/3$, $\csc t = 2$, $\sec t = 2\sqrt{3}/3$, $\cot t = \sqrt{3}$

(b) $\sin t = 12/13$, $\cos t = -5/13$, $\tan t = -12/5$, $\csc t = 13/12$, $\sec t = -13/5$, $\cot t = -5/12$

(c) $\sin t = -3\sqrt{5}/7$, $\cos t = -2/7$, $\tan t = 3\sqrt{5}/2$, $\csc t = -7\sqrt{5}/15$, $\sec t = -7/2$, $\cot t = 2\sqrt{5}/15$

(d) $\sin t = -1$, $\cos t = 0$, $\tan t$ undefined, $\csc t = -1$, $\sec t$ undefined, $\cot t = 0$

29. 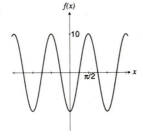 **31.** 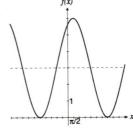 **33.**

35. 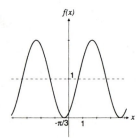 **37.** 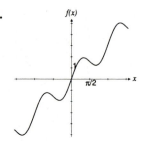 **39.**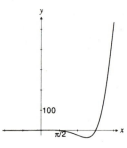

41. (a) $y = 5\sin 4x$ (b) $y = \frac{1}{2}\sin 2\pi(x + \frac{1}{3})$ (c) $y = 3 + \cos 3x$ (d) $y = \sin \frac{1}{2}x$ **43.** $y = -50\cos 8\pi t$
45. (a) $\frac{1}{2}\ln 5 \approx 0.8$ (b) $y = 5e^{-0.8t}\cos 4\pi t$

Chapter 5 Test

1. (a) $5\pi/3, -\pi/10, 5\pi/4$ (b) $150°, -495°, 137.5°$ **2.** (a) $\sqrt{2}/2$ (b) $\sqrt{3}/3$ (c) 2   (d) 1
3. $(26 + 6\sqrt{13})/39$ **4.** $(4 - 3\sqrt{2})/4$ **5.** 15.3 m²   **6.** $-\sqrt{\sec^2\theta - 1}$ **7.** $a = \cot\theta, b = 1 - \sin\theta$
8. (a) 9.1   (b) 250.5   (c) 7.9   (d) 19.5   **9.** $\sin t = 3/\sqrt{13}, \cos t = 2/\sqrt{13}$
10. (a) 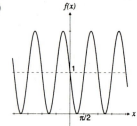   (b) 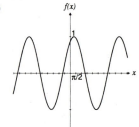   (c)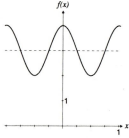

(d) 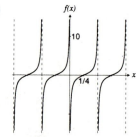   **11.** (a) 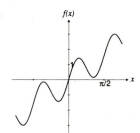   (b)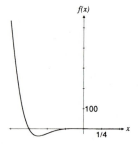

12. $y = 6 + 4\cos\dfrac{\pi}{6}(t - 2)$

Problems Plus

1. $b = a\left(1 + \sin\dfrac{\theta}{2}\right)\bigg/\left(1 - \sin\dfrac{\theta}{2}\right)$ **3.** 63

Chapter 6

Exercises 6.1

9. 1   **11.** identity   **13.** 0   **15.** identity   **17.** $(\pi/2) + 2k\pi$ **19.** identity   **21.** $\sin x$ **23.** 1
25. $\sec A \csc A$ **27.** $\sec u$ **29.** $\tan x$ **31.** $\sin y$ **33.** $\sin^2 x$ **35.** $\sec x$ **37.** $2 \sec u$ **39.** $\cot x$
41. $\cos\theta$ **43.** $\cos^2 x$ **45.** $2\sec^2\alpha$ **47.** $1 - \sin x$ **49.** 1

Exercises 6.2

1. $(\pi/3) + 2k\pi, (2\pi/3) + 2k\pi$ **3.** $(\pi/6) + 2k\pi, (\pi/4) + k\pi, (11\pi/6) + 2k\pi$ **5.** $(\pi/6) + k\pi, (5\pi/6) + k\pi$
7. $(\pi/2) + k\pi$ **9.** $(\pi/6) + 2k\pi, (5\pi/6) + 2k\pi$ **11.** no solutions   **13.** $(\pi/6) + (k\pi/3)$
15. $(\pi/3) + 2k\pi, (5\pi/3) + 2k\pi$ **17.** $\pi/6, 3\pi/4, 5\pi/6, 7\pi/4$ **19.** $\pi/4, 3\pi/4, 5\pi/4, 7\pi/4$
21. $\pi/6, \pi/4, 5\pi/6, 7\pi/6, 5\pi/4, 11\pi/6$ **23.** $\pi/3, 2\pi/3, 4\pi/3, 5\pi/3$ **25.** $\pi/6, \pi/4, 5\pi/6, 5\pi/4$ **27.** $\pi/4, 3\pi/4, 5\pi/4, 7\pi/4$
29. $1/4, 3/4, 5/4, 7/4$ **31.** $0, \pi$ **33.** no solutions   **35.** $3\pi/2$ **37.** $\pi/4, \pi/3, 2\pi/3, 3\pi/4, 4\pi/3, 5\pi/3$
39. $0, \pi/2, \pi, 3\pi/2$ **41.** $(\pi/2) - 1, (3\pi/2) - 1$ **43.** $\pi/6, 11\pi/6$ **45.** $\pi/2, 3\pi/2$ **47.** $\pi/3, 5\pi/3$
49. $((2k + 1)\pi, -2)$ **51.** $(\pi/3 + k\pi, \sqrt{3})$ **53.** $((\pi/3) + 2k\pi, 3/2), ((5\pi/3) + 2k\pi, 3/2)$ **55.** $9/2$

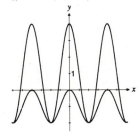
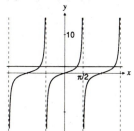

Exercises 6.3

1. $(\sqrt{6} - \sqrt{2})/4$ **3.** $-2 - \sqrt{3}$ **5.** $(\sqrt{2} - \sqrt{6})/4$ **7.** $\sqrt{2}/2$ **9.** $\sqrt{3}$ **11.** $-56/65, -16/65$
13. $0, 1$ **47.** $\pi/3, 5\pi/3$ **49.** $0, \pi/2, \pi, 3\pi/2$ **57.** $2\cos[2x + (\pi/3)]$ **59.** $-5\sqrt{2}\cos[7x + (\pi/4)]$
61. $\sqrt{2}\cos[x - (\pi/4)]$ **63.** $3\pi/4$ **65.** $5\pi/12, 17\pi/12$ **67.** $17/6$

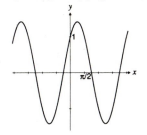

Exercises 6.4

1. $2\sin 44° \cos 44°$ **3.** $(2\tan 34°)/(1 - \tan^2 34°)$ **5.** $\left(2\tan\dfrac{\pi}{12}\right)\Big/\left(1 - \tan^2\dfrac{\pi}{12}\right)$ **7.** $2\sin\frac{5}{2}\theta\cos\frac{5}{2}\theta$ **9.** $\sin 36°$

11. $\tan 14°$ **13.** $\cos(2\pi/5)$ **15.** $\sin 6\theta$ **17.** $(1 - \cos 26°)/\sin 26°$ **19.** $\pm\sqrt{(1 + \cos 14°)/2}$

21. $\left(1 - \cos\dfrac{\pi}{15}\right)\Big/\sin\dfrac{\pi}{15}$ **23.** $\pm\sqrt{\left(1 + \cos\dfrac{\theta}{3}\right)\Big/2}$ **25.** $\sin 15°$ **27.** $\sin(\pi/12)$ **29.** $\tan 4°$ **31.** $\sin 4\theta$

33. $\frac{1}{2}\sqrt{2 - \sqrt{3}}$ **35.** $\frac{1}{2}\sqrt{2 - \sqrt{2 + \sqrt{3}}}$ **37.** $\frac{1}{2}\sqrt{2 + \sqrt{2 + \sqrt{2}}}$ **39.** $\frac{1}{2}\sqrt{2 - \sqrt{3}}$
41. $120/169, 119/169, 120/119$ **43.** $-24/25, -7/25, 24/7$ **45.** $24/25, 7/25, 24/7$ **47.** $\sqrt{10}/10, 3\sqrt{10}/10, 1/3$
49. $\sqrt{(3 + 2\sqrt{2})/6}, \sqrt{(3 - 2\sqrt{2})/6}, 3 + 2\sqrt{2}$ **51.** $\sqrt{6}/6, -\sqrt{30}/6, -\sqrt{5}/5$ **69.** $\pi/6, \pi/2, 5\pi/6, 3\pi/2$
71. $\pi/2, 7\pi/6, 11\pi/6$ **73.** $0, \pi/2, 3\pi/2$ **75.** 0 **77.** **(c)** $(\pi/3) + k\pi$
83. **(a)** $p(t) = 8t^4 - 8t^2 + 1$ **(b)** $p(t) = 16t^5 - 20t^3 + 5t$

Exercises 6.5

1. $\frac{1}{2}(\sin 5x - \sin x)$ **3.** $\frac{3}{2}(\cos 11x + \cos 3x)$ **5.** $(\sqrt{2} + \sqrt{3})/2$ **7.** $-5 - 2\sqrt{6}$ **9.** $2\sin 4x \cos x$
11. $2\sin 5x \sin x$ **13.** $-2\cos\frac{9}{2}x \sin\frac{5}{2}x$ **15.** $2\cos 10\pi x \cos \pi x$ **17.** $\sqrt{6}/2$ **19.** $\sqrt{2}/2$ **29.** $k\pi/2$
31. $(\pi/9) + (2k\pi/3), (\pi/2) + k\pi, (5\pi/9) + (2k\pi/3)$ **33.** $(\pi/8) + (k\pi/4), (\pi/6) + 2k\pi, (5\pi/6) + 2k\pi$
35. $(\pi/4) + (k\pi/2), (\pi/6) + 2k\pi, (5\pi/6) + 2k\pi$

Exercises 6.6

1. (a) $\pi/6$ (b) $\pi/3$ (c) not defined   **3.** (a) $\pi/4$ (b) $\pi/4$ (c) $-\pi/4$ **5.** (a) $\pi/2$ (b) 0
(c) π **7.** (a) $\pi/6$ (b) $-\pi/6$ (c) not defined   **9.** (a) 0.87696   (b) 2.09601   (c) 1.50510
11. $1/2$ **13.** $\sqrt{3}/3$ **15.** $1/2$ **17.** $\sqrt{3}/2$ **19.** $\pi/3$ **21.** $\pi/3$ **23.** $\pi/3$ **25.** $4/5$ **27.** $12/13$
29. $24/25$ **31.** $\sqrt{3}/2$ **33.** 1   **35.** 0   **37.** $(\sqrt{10} - 3\sqrt{30})/20$ **39.** $\sqrt{1 - x^2}$ **41.** $x/\sqrt{1 - x^2}$
43. $(1 - x^2)/(1 + x^2)$ **45.** $(1 - \sqrt{1 - x^2})/x$ **47.** $(x\sqrt{1 - x^2} - 1)/\sqrt{1 + x^2}$
49. $(x\sqrt{1 - y^2} + y)/(\sqrt{1 - y^2} - xy)$ **67.** (a) $\sin^{-1}\frac{1}{3}, \pi - \sin^{-1}\frac{1}{3}$ (b) 0.33984, 2.80176
69. (a) $\pi/2, \sin^{-1}\frac{1}{4}$ (b) 1.57080, 0.25268   **71.** (a) $\cos^{-1}1$ (b) 0   **73.** (a) $\tan^{-1}2, \tan^{-1}3$
(b) 1.10715, 1.24905   **75.** (a) $\pi/4, 5\pi/4, \sin^{-1}\frac{1}{3}, \pi - \sin^{-1}\frac{1}{3}$ (b) 0.78540, 3.92699, 0.33984, 2.80176
77. $\sqrt{2}/2$ **79.** $-1 \le x \le 1$ **81.** 1   **83.** $1, -1$

Exercises 6.7

1. $\sqrt{2}\left(\cos\frac{\pi}{4} + i\sin\frac{\pi}{4}\right)$ **3.** $2\left(\cos\frac{7\pi}{4} + i\sin\frac{7\pi}{4}\right)$ **5.** $4\left(\cos\frac{11\pi}{6} + i\sin\frac{11\pi}{6}\right)$ **7.** $\sqrt{2}\left(\cos\frac{3\pi}{2} + i\sin\frac{3\pi}{2}\right)$

9. $5\sqrt{2}\left(\cos\frac{\pi}{4} + i\sin\frac{\pi}{4}\right)$ **11.** $8\left(\cos\frac{11\pi}{6} + i\sin\frac{11\pi}{6}\right)$ **13.** $20(\cos\pi + i\sin\pi)$ **15.** $5[\cos(\tan^{-1}\frac{4}{3}) + i\sin(\tan^{-1}\frac{4}{3})]$

17. $3\sqrt{2}\left(\cos\frac{3\pi}{4} + i\sin\frac{3\pi}{4}\right)$ **19.** $8\left(\cos\frac{\pi}{6} + i\sin\frac{\pi}{6}\right)$ **21.** $\sqrt{5}[\cos(\tan^{-1}\frac{1}{2}) + i\sin(\tan^{-1}\frac{1}{2})]$

23. $2\left(\cos\frac{\pi}{4} + i\sin\frac{\pi}{4}\right)$

25. $z_1 = 2\left(\cos\frac{\pi}{6} + i\sin\frac{\pi}{6}\right)$
$z_2 = 2\left(\cos\frac{\pi}{3} + i\sin\frac{\pi}{3}\right)$
$z_1 z_2 = 4\left(\cos\frac{\pi}{2} + i\sin\frac{\pi}{2}\right)$
$z_1/z_2 = \cos\frac{\pi}{6} - i\sin\frac{\pi}{6}$
$1/z_1 = \frac{1}{2}\left(\cos\frac{\pi}{6} - i\sin\frac{\pi}{6}\right)$

27. $z_1 = 4\left(\cos\frac{11\pi}{6} + i\sin\frac{11\pi}{6}\right)$
$z_2 = \sqrt{2}\left(\cos\frac{3\pi}{4} + i\sin\frac{3\pi}{4}\right)$
$z_1 z_2 = 4\sqrt{2}\left(\cos\frac{7\pi}{12} + i\sin\frac{7\pi}{12}\right)$
$z_1/z_2 = 2\sqrt{2}\left(\cos\frac{13\pi}{12} + i\sin\frac{13\pi}{12}\right)$
$1/z_1 = \frac{1}{4}\left(\cos\frac{11\pi}{6} - i\sin\frac{11\pi}{6}\right)$

29. $z_1 = 5\sqrt{2}\left(\cos\frac{\pi}{4} + i\sin\frac{\pi}{4}\right)$
$z_2 = 4(\cos 0 + i\sin 0)$
$z_1 z_2 = 20\sqrt{2}\left(\cos\frac{\pi}{4} + i\sin\frac{\pi}{4}\right)$
$z_1/z_2 = \frac{5\sqrt{2}}{4}\left(\cos\frac{\pi}{4} + i\sin\frac{\pi}{4}\right)$
$1/z_1 = \frac{\sqrt{2}}{10}\left(\cos\frac{\pi}{4} - i\sin\frac{\pi}{4}\right)$

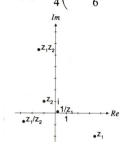

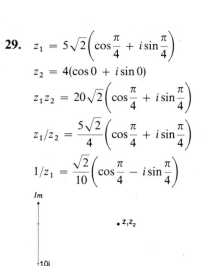

31. $z_1 = 20(\cos \pi + i \sin \pi)$

$z_2 = 2\left(\cos \dfrac{\pi}{6} + i \sin \dfrac{\pi}{6}\right)$

$z_1 z_2 = 40\left(\cos \dfrac{7\pi}{6} + i \sin \dfrac{7\pi}{6}\right)$

$z_1/z_2 = 10\left(\cos \dfrac{5\pi}{6} + i \sin \dfrac{5\pi}{6}\right)$

$1/z_1 = \dfrac{1}{20}(\cos \pi - i \sin \pi)$

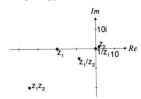

33. $z_1 = 3\sqrt{2}\left(\cos \dfrac{3\pi}{4} + i \sin \dfrac{3\pi}{4}\right)$

$z_2 = 2\sqrt{2}\left(\cos \dfrac{7\pi}{4} + i \sin \dfrac{7\pi}{4}\right)$

$z_1 z_2 = 12\left(\cos \dfrac{\pi}{2} + i \sin \dfrac{\pi}{2}\right)$

$z_1/z_2 = \dfrac{3}{2}(\cos \pi - i \sin \pi)$

$1/z_1 = \dfrac{\sqrt{2}}{6}\left(\cos \dfrac{3\pi}{4} - i \sin \dfrac{3\pi}{4}\right)$

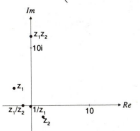

35. -1024 **37.** $512(-\sqrt{3} + i)$

39. -1 **41.** 4096 **43.** $8(-1 + i)$ **45.** $\dfrac{1}{2048}(-\sqrt{3} - i)$

47. $2\sqrt{2}\left(\cos \dfrac{\pi}{12} + i \sin \dfrac{\pi}{12}\right), 2\sqrt{2}\left(\cos \dfrac{13\pi}{12} + i \sin \dfrac{13\pi}{12}\right)$

49. $3\left(\cos \dfrac{3\pi}{8} + i \sin \dfrac{3\pi}{8}\right), 3\left(\cos \dfrac{7\pi}{8} + i \sin \dfrac{7\pi}{8}\right), 3\left(\cos \dfrac{11\pi}{8} + i \sin \dfrac{11\pi}{8}\right), 3\left(\cos \dfrac{15\pi}{8} + i \sin \dfrac{15\pi}{8}\right)$

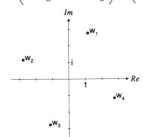

51. $\pm 1, \pm i, \dfrac{\sqrt{2}}{2}(\pm 1 \pm i)$ **53.** $\dfrac{\sqrt{3}}{2} + \dfrac{1}{2}i, -\dfrac{\sqrt{3}}{2} + \dfrac{1}{2}i, -i$ **55.** $\dfrac{\sqrt{2}}{2}(\pm 1 \pm i)$

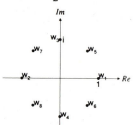   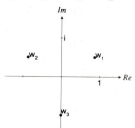   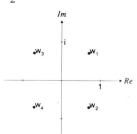

57. $\dfrac{\sqrt{2}}{2}(\pm 1 \pm i)$ **59.** $2\left(\cos \dfrac{\pi}{18} + i\sin \dfrac{\pi}{18}\right), 2\left(\cos \dfrac{13\pi}{18} + i\sin \dfrac{13\pi}{18}\right), 2\left(\cos \dfrac{25\pi}{18} + i\sin \dfrac{25\pi}{18}\right)$ **61.** $\pm 1, \pm \dfrac{1}{2} \pm \dfrac{\sqrt{3}}{2}i$

63. $-3i, -i$

Exercises 6.8

1.    **3.** 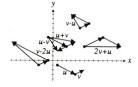   **5.**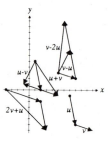

7. $\langle 4,14 \rangle, \langle -9,-3 \rangle, \langle 5,8 \rangle, \langle -6,17 \rangle$ **9.** $\langle 0,-2 \rangle, \langle 6,0 \rangle, \langle -2,-1 \rangle, \langle 8,-3 \rangle$ **11.** $0, -3i + 15j, i - 5j, -4i + 20j$

13. $4i, -9i + 6j, 5i - 2j, -6i + 8j$ **15.** $2i + 2j, -3i + 3j, 2i, -i + 7j$

17. $\sqrt{5}, \sqrt{13}, 2\sqrt{5}, \frac{1}{2}\sqrt{13}, \sqrt{26}, \sqrt{89}, \sqrt{10}, \sqrt{5}, -\sqrt{13}$ **19.** $\sqrt{101}, 2\sqrt{2}, 2\sqrt{101}, \sqrt{2}, \sqrt{73}, 2\sqrt{173}, \sqrt{145}, \sqrt{101} - 2\sqrt{2}$

21. $\langle 5,7 \rangle$ **23.** $\langle -4,-3 \rangle$ **25.** $\langle -6,-4 \rangle$ **27.** $\langle 0,2 \rangle$ **29.** (a) $5, 53.13°$ (b) $\langle 3/5, 4/5 \rangle$

31. (a) $2\sqrt{3}, 300°$ (b) $\langle 1/2, -\sqrt{3}/2 \rangle$ **33.** (a) $13, 157.38°$ (b) $\langle -12/13, 5/13 \rangle$ **35.** (a) $2, 60°$

(b) $\dfrac{1}{2}i + \dfrac{\sqrt{3}}{2}j$ **37.** $20\sqrt{3}i + 20j$ **39.** $-\dfrac{\sqrt{2}}{2}i - \dfrac{\sqrt{2}}{2}j$ **41.** $800\cos 125°i + 800\sin 125°j \approx -458.86i + 655.32j$

43. $4\cos 10° + 4\sin 10° \approx 3.94i + 0.69j$ **45.** $15\sqrt{3}, -15$

Exercises 6.9

1. 427 mi/h, N84.6°E   **3.** 794 mi/h, N26.6°W   **5.** 26 mi/h, N49.1°E   **7.** 7.4 mi/h, 22.8 mi/h   **9.** (a) 93 lb

(b) 32 lb   **11.** (a) $\langle 0,0 \rangle$ (b) none   **13.** (a) $\langle -7,-13 \rangle$ (b) $\langle 7,13 \rangle$ **15.** (a) j (b) $-j$

17. (a) $\langle 2,-4 \rangle$ (b) $\langle -2,4 \rangle$ **19.** (a) $\langle 11,11 \rangle$ (b) $\langle -11,-11 \rangle$ **21.** 2497 lb, 1164 lb   **23.** $23.6°$

25. $-1, 12, -6, \langle -3,2 \rangle, \langle 39,-26 \rangle$ **27.** $-6, -1, -36, -6i + 12j, 5i - 10j$ **29.** $56°$ **31.** $180°$ **33.** $90°$

35. 82 ft-lb

Review Exercises for Chapter 6

31. $0, \pi$ **33.** $\pi/6, 5\pi/6$ **35.** $\pi/3, 5\pi/3$ **37.** $2\pi/3, 4\pi/3$ **39.** $\pi/3, 2\pi/3, 3\pi/4, 4\pi/3, 5\pi/3, 7\pi/4$

41. $\pi/6, \pi/2, 5\pi/6, 7\pi/6, 3\pi/2, 11\pi/6$ **43.** $7\pi/6, 11\pi/6$ **45.** $\pi/6$ **47.** $\frac{1}{2}\sqrt{2 + \sqrt{3}}$ **49.** $-\frac{1}{2}\sqrt{2 + \sqrt{3}}$

51. $\sqrt{2} - 1$ **53.** $\frac{1}{2}\sqrt{2 + \sqrt{2 + \sqrt{3}}}$ **55.** $\sqrt{2}/2$ **57.** $\sqrt{2}/2$ **59.** $(\sqrt{2} + \sqrt{3})/4$ **61.** $2(\sqrt{10} + 1)/9$

63. $(4\sqrt{2} - \sqrt{5})/9$ **65.** $2(2\sqrt{5} + \sqrt{2})/(8 - \sqrt{10})$ **67.** $-1/9$ **69.** $\sqrt{(3 + 2\sqrt{2})/6}$ **71.** $-8\sqrt{5}/81$

73. $\pi/3$ **75.** $1/2$ **77.** $2/\sqrt{21}$ **79.** $7/9$ **81.** $63/65$ **83.** $4\sqrt{2}\left(\cos\dfrac{\pi}{4} + i\sin\dfrac{\pi}{4}\right)$

85. $\sqrt{34}\left[\cos(\tan^{-1}\tfrac{3}{5}) + i\sin(\tan^{-1}\tfrac{3}{5})\right]$ **87.** $4\sqrt{2}\left(\cos\dfrac{3\pi}{4} + i\sin\dfrac{3\pi}{4}\right)$ **89.** $\sqrt{2}\left(\cos\dfrac{3\pi}{4} + i\sin\dfrac{3\pi}{4}\right)$

91. $8(-1 + \sqrt{3}i)$ **93.** $128(-1 + i)$ **95.** $(-1 - \sqrt{3}i)/131{,}072$ **97.** $2\sqrt{2}(-1 + i), 2\sqrt{2}(1 - i)$

99. $2\left(\cos\dfrac{\pi}{9} + i\sin\dfrac{\pi}{9}\right), 2\left(\cos\dfrac{7\pi}{9} + i\sin\dfrac{7\pi}{9}\right), 2\left(\cos\dfrac{13\pi}{9} + i\sin\dfrac{13\pi}{9}\right)$ **101.** $\pm 1, \pm\dfrac{1}{2} \pm \dfrac{\sqrt{3}}{2}i$ **103.** $-1, \dfrac{1}{2} \pm \dfrac{\sqrt{3}}{2}i$

105. $\pm i, \pm\dfrac{\sqrt{3}}{2} \pm \dfrac{1}{2}i$ **107.** **(a)** $\langle 6, 4\rangle, \langle -10, 2\rangle, \langle -4, 6\rangle, \langle -22, 7\rangle$ **(b)**

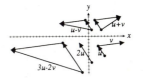

109. **(a)** $3\mathbf{i} - \mathbf{j}, \mathbf{i} + 3\mathbf{j}, 4\mathbf{i} + 2\mathbf{j}, 4\mathbf{i} + 7\mathbf{j}$ **(b)**

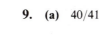

111. $\sqrt{5}$ **113.** $\sqrt{15}$ **115.** $\sqrt{17}$
117. 3 **119.** $10\mathbf{i} + 10\sqrt{3}\mathbf{j}$
121. $3\mathbf{i} - 4\mathbf{j}$ **123.** $3\mathbf{i} + 4\mathbf{j}$
125. $\cos^{-1}(23/5\sqrt{29}) \approx 31.3°$
127. $45°$ **129.** -6

Chapter 6 Test

2. **(a)** $2\pi/3, 4\pi/3$ **(b)** $\pi/6, \pi/2, 5\pi/6, 3\pi/2$ **3.** $\tan\theta$ **4.** $1/2$ **5.** $-33/65$ **6.** **(a)** $\frac{1}{2}(\sin 8x - \sin 2x)$

(b) $-2\cos\tfrac{7}{2}x\sin\tfrac{3}{2}x$ **7.** -2 **8.** $R, [-1, 1]$ **9.** **(a)** $40/41$ **(b)** $\pi/3$ **10.** $2\left(\cos\dfrac{\pi}{3} + i\sin\dfrac{\pi}{3}\right)$

11. $-8, \sqrt{3} + i$ **12.** $-3i, 3\left(\pm\dfrac{\sqrt{3}}{2} + \dfrac{1}{2}i\right)$

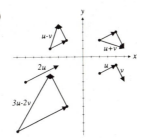

13. **(a)** $\langle -6, 10\rangle$ **(b)** $(1/\sqrt{34})\langle -3, 5\rangle$
14. **(a)** $\langle 19, -3\rangle$ **(b)** $5\sqrt{2}$ **(c)** 0 **(d)** yes
15. 17 mi/h, N53.4°E   **16.** 90

Problems Plus

1. **(a)**

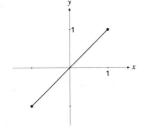

3. $\sqrt{1 + a}$ **5.** **(b)** $p = m^2 + n^2, m, n$ integers

Chapter 7

Exercises 7.1

1. $(4, 0)$ **3.** parallel   **5.** $\left(-\frac{5}{2}, 4\right)$

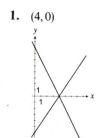

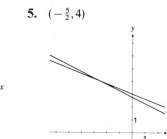

7. $(2, 1)$ **9.** $\left(2, \frac{1}{2}\right)$ **11.** $(10, -9)$
13. $(0, -2)$ **15.** $\left(x, \frac{1}{15}x + \frac{2}{5}\right)$ **17.** $(-1, 4)$
19. $(-3, -7)$ **21.** $(12, -5)$ **23.** $\left(\frac{1}{5}, \frac{2}{5}\right)$
25. no solution   **27.** $\left(\frac{7}{3}, 7\right)$ **29.** $(-1, 2)$
31. $\left(\frac{5}{2}, 4\right)$ **33.** $\left(\frac{2}{3}, -\frac{1}{6}\right)$

35. $\left(\dfrac{1}{a + b}, \dfrac{1}{a + b}\right)$ **37.** $(\cos\theta + \sin\theta, \cos\theta - \sin\theta)$ **39.** $22, 12$ **41.** 5 dimes, 9 quarters

43. plane's speed 120 mi/h, wind speed 30 mi/h   **45.** run 5 mi/h, cycle 20 mi/h   **47.** 200 g of A, 40 g of B
49. $25\%, 10\%$ **51.** 25 mi/h, 5:24 P.M.   **53.** 72   **55.** $y = -5x^2 + 17x$

Exercises 7.2

1. linear   **3.** not linear   **5.** not linear   **7.** $\begin{cases} 2x + 3y = 1 \\ 4x + 2y = 3 \end{cases}$ **9.** $\begin{cases} y = 0 \\ x + z = 0 \\ - 2y + 2z = 7 \end{cases}$ **11.** $(1, 3, 2)$

13. $(1, -1, 3)$ **15.** $(-1, 5, 0)$ **17.** $(10, 3, -2)$ **19.** $(2, 0, -2)$ **21.** $\left(\frac{1}{3}, 1, -\frac{1}{2}\right)$ **23.** $(1, 3, 0, -2)$
25. $(1, -1, 2, -2, 0)$ **27.** $(1, -3, 7)$ **29.** $(1, -1, -3)$ **31.** 2 VitaMix, 1 Vitron, 2 VitaPlus
33. 11 pennies, 4 nickels, 5 dimes, 10 quarters   **35.** standard \$40, deluxe \$55, first class \$90
37. $y = 2x^2 - 5x + 6$ **39.** $(1, -2, 3)$

Exercises 7.3

1. (a) $x = 61, y = 49, z = 5$ **(b)** no solution   **(c)** $x = 12z + 1, y = 10z - 1, z =$ any number   **3.** no solution
5. $x = 7, y = 1$ **7.** no solution   **9.** $x = -11z + 12, y = -12z + 12, z =$ any number
11. $x = -2z + 5, y = z - 2, z =$ any number   **13.** $x = -\frac{1}{2}y + z + 6, y =$ any number, $z =$ any number
15. $x = 2, y = w, z = -2w + 6, w =$ any number
17. $x = -12w + 20, y = -19w + 31, z = 2w - 2, w =$ any number
19. $x = \frac{1}{3}z - \frac{2}{3}w, y = \frac{1}{3}z + \frac{1}{3}w, z =$ any number, $w =$ any number   **21.** no solution
23. $x, y,$ and z are oz of soya, millet, and milk powder respectively: $x = -\frac{1}{3}z + \frac{13}{6}, y = -\frac{2}{3}z + \frac{4}{3}, 0 \le z \le 2$
25. impossible   **27.** 7 pennies, 5 nickels, 4 dimes

Exercises 7.4

1. $\dfrac{1}{x - 2} - \dfrac{1}{x + 2}$ **3.** $\dfrac{3}{x - 4} - \dfrac{2}{x + 2}$ **5.** $\dfrac{-\frac{1}{2}}{2x - 1} + \dfrac{\frac{3}{2}}{4x - 3}$ **7.** $\dfrac{2}{x - 2} + \dfrac{3}{x + 2} - \dfrac{1}{2x - 1}$

9. $\dfrac{2}{x + 1} - \dfrac{1}{x} + \dfrac{1}{x^2}$ **11.** $\dfrac{1}{2x + 3} - \dfrac{3}{(2x + 3)^2}$ **13.** $\dfrac{2}{x} - \dfrac{1}{x^3} - \dfrac{2}{x + 2}$

15. $\dfrac{4}{x + 2} - \dfrac{4}{x - 1} + \dfrac{2}{(x - 1)^2} + \dfrac{1}{(x - 1)^3}$ **17.** $\dfrac{3}{x + 2} - \dfrac{1}{(x + 2)^2} - \dfrac{1}{(x + 3)^2}$ **19.** $\dfrac{x + 1}{x^2 + 3} - \dfrac{1}{x}$

21. $\dfrac{2x - 5}{x^2 + x + 2} + \dfrac{5}{x^2 + 1}$ **23.** $\dfrac{1}{x^2 + 1} - \dfrac{x + 2}{(x^2 + 1)^2} + \dfrac{1}{x}$ **25.** $x^2 + \dfrac{3}{x - 2} - \dfrac{x + 1}{x^2 + 1}$

27. $\dfrac{A}{x} + \dfrac{B}{2x - 5} + \dfrac{C}{(2x - 5)^2} + \dfrac{D}{(2x - 5)^3} + \dfrac{Ex + F}{x^2 + 2x + 5} + \dfrac{Gx + H}{(x^2 + 2x + 5)^2}$ **29.** $A = \dfrac{a + b}{2}, B = \dfrac{a - b}{2}$

Exercises 7.5

1. $\begin{bmatrix} 5 & -2 & 5 \\ 1 & 1 & 0 \end{bmatrix}$　**3.** $\begin{bmatrix} -1 & -3 & -5 \\ -1 & 3 & -6 \end{bmatrix}$　**5.** $\begin{bmatrix} 13 & -\frac{7}{2} & 15 \\ 3 & 1 & 3 \end{bmatrix}$　**7.** $\begin{bmatrix} -14 & -8 & -30 \\ -6 & 10 & -24 \end{bmatrix}$　**9.** impossible

11. $\begin{bmatrix} 3 & \frac{1}{2} & 5 \\ 1 & -1 & 3 \end{bmatrix}$　**13.** $[28 \quad 21 \quad 28]$　**15.** $\begin{bmatrix} 0 & 0 & 0 & 0 & 0 \\ 0 & 0 & 0 & 0 & 0 \\ 0 & 0 & 0 & 0 & 0 \\ 0 & 0 & 0 & 0 & 0 \end{bmatrix}$　**17.** $\begin{bmatrix} 8 & -335 \\ 0 & 343 \end{bmatrix}$　**19.** impossible

21. impossible　**23.** $\begin{bmatrix} 2 & -5 \\ 3 & 2 \end{bmatrix}\begin{bmatrix} x \\ y \end{bmatrix} = \begin{bmatrix} 7 \\ 4 \end{bmatrix}$　**25.** $\begin{bmatrix} 3 & 2 & -1 & 1 \\ 1 & 0 & -1 & 0 \\ 0 & 3 & 1 & -1 \end{bmatrix}\begin{bmatrix} x_1 \\ x_2 \\ x_3 \\ x_4 \end{bmatrix} = \begin{bmatrix} 0 \\ 5 \\ 4 \end{bmatrix}$　**27.** $\begin{bmatrix} 3 & \frac{11}{2} \\ 2 & 5 \end{bmatrix}$

29. impossible　**31.** no　**33.** no　**35.** (a) $A^2 = \begin{bmatrix} 2 & 2 \\ 2 & 2 \end{bmatrix}, A^3 = \begin{bmatrix} 4 & 4 \\ 4 & 4 \end{bmatrix}, A^4 = \begin{bmatrix} 8 & 8 \\ 8 & 8 \end{bmatrix}$

(b) $A^n = \begin{bmatrix} 2^{n-1} & 2^{n-1} \\ 2^{n-1} & 2^{n-1} \end{bmatrix}$　**37.** (a) $\begin{bmatrix} 32{,}000 & 18{,}000 \\ 42{,}000 & 26{,}800 \\ 44{,}000 & 26{,}800 \end{bmatrix}$　(b) $42{,}000　(c) $71{,}600

Exercises 7.6

1. $\begin{bmatrix} 1 & -2 \\ -\frac{3}{2} & \frac{7}{2} \end{bmatrix}$　**3.** $\begin{bmatrix} 5 & -7 \\ -2 & 3 \end{bmatrix}$　**5.** $\begin{bmatrix} 13 & 5 \\ -5 & -2 \end{bmatrix}$　**7.** no inverse　**9.** $\begin{bmatrix} 1 & 2 \\ -\frac{1}{2} & \frac{2}{3} \end{bmatrix}$　**11.** $\begin{bmatrix} -4 & -4 & 5 \\ 1 & 1 & -1 \\ 5 & 4 & -6 \end{bmatrix}$

13. no inverse　**15.** $\begin{bmatrix} -\frac{9}{2} & -1 & 4 \\ 3 & 1 & -3 \\ \frac{7}{2} & 1 & -3 \end{bmatrix}$　**17.** $\begin{bmatrix} 0 & 0 & -2 & 1 \\ -1 & 0 & 1 & 1 \\ 0 & 1 & -1 & 0 \\ 1 & 0 & 0 & -1 \end{bmatrix}$　**19.** $x = 20, y = -8$

21. $x = 126, y = -50$　**23.** $x = -38, y = 9, z = 47$　**25.** $x = -12, y = -21, z = 93$　**27.** $\begin{bmatrix} 7 & 2 & 3 \\ 10 & 3 & 5 \end{bmatrix}$

29. (a) $\begin{bmatrix} 0 & 1 & -1 \\ -2 & \frac{3}{2} & 0 \\ 1 & -\frac{3}{2} & 1 \end{bmatrix}$　(b) 1 oz A, 1 oz B, 2 oz C　(c) 2 oz A, 0 oz B, 1 oz C　(d) no　**31.** $\begin{bmatrix} \sin\theta & \cos\theta \\ -\cos\theta & \sin\theta \end{bmatrix}$

33. $\frac{1}{2}\begin{bmatrix} e^{-x} & e^{-2x} \\ e^{-2x} & e^{-3x} \end{bmatrix}$　**35.** for example, $\begin{bmatrix} 1 & 0 \\ 0 & 1 \end{bmatrix}$ and $\begin{bmatrix} -1 & 0 \\ 0 & -1 \end{bmatrix}$ (There are infinitely many possible answers.)

Exercises 7.7

1. 3　**3.** 6　**5.** no determinant　**7.** 0　**9.** 20, 20　**11.** $-12, 12$　**13.** 0, 0　**15.** -6, has an inverse
17. 5000, has an inverse　**19.** -4, has an inverse　**21.** -18　**23.** 120　**25.** -2　**27.** $(-2, 5)$
29. $(0.6, -0.4)$　**31.** $(4, -1)$　**33.** $(4, 2, -1)$　**35.** $(1, 3, 2)$　**37.** $(0, -1, 1)$　**39.** $(\frac{189}{29}, -\frac{108}{29}, \frac{88}{29})$
41. $(-\frac{49}{80}, \frac{77}{40}, \frac{287}{80})$　**43.** (b) $5x - 6y = -200$　**45.** 0, 1, 2　**47.** 1, -1

Exercises 7.8

1. $(2, -2), (-2, 2)$　**3.** $(4, 0)$　**5.** $(-2, -2)$　**7.** $(6, 2), (-2, -6)$　**9.** $(2, 1), (2, -1), (-2, 1), (-2, -1)$
11. no solution　**13.** $(-\sqrt{6}, 6)$　**15.** $(\sqrt{5}, 2), (\sqrt{5}, -2), (-\sqrt{5}, 2), (-\sqrt{5}, -2)$　**17.** $(3, -\frac{1}{2}), (-3, -\frac{1}{2})$　**19.** $(\frac{1}{5}, \frac{1}{3})$
21. $(9, 4)$　**23.** $(0, 2), (0, -2), (2, 2), (-2, -2), (\sqrt{2}, -\sqrt{2}), (-\sqrt{2}, \sqrt{2})$　**25.** $(2, 0, 0), (\frac{2}{3}, -\frac{4}{3}, \frac{4}{3})$　**27.** $(-1, 1, -1)$
29. $(2, 0, -4), (-2, 0, -4), (0, 1, -1)$　**31.** 8 cm, 15 cm, 17 cm　**33.** 9 in, 12 in, 15 in　**35.** 20, 15　**37.** $y = 3x + 5$
39. $(8, 5), (-5, -8)$　**41.** $(1, 3), (3, 1)$

Exercises 7.9

1.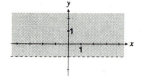

3.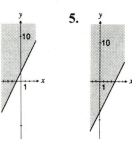

5.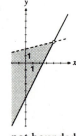

7.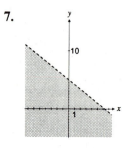

9.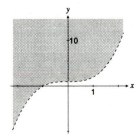

11.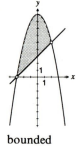

13.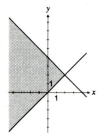
not bounded

15.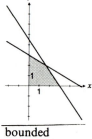
not bounded

17.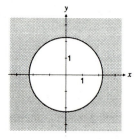
bounded

19.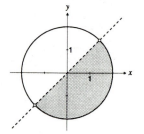
bounded

21.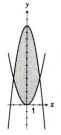
bounded

23.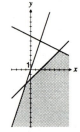
bounded

25.
not bounded

27.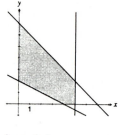
bounded

29.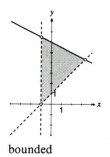
bounded

31.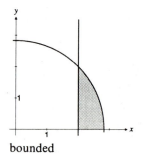
bounded

33.

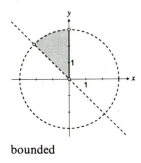

bounded

35. x = number of fiction books
y = number of nonfiction books

$$\begin{cases} x + y \le 100 \\ 20 \le y, \quad x \ge y \\ x \ge 0, \quad y \ge 0 \end{cases}$$

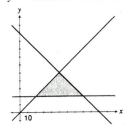

37.

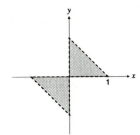

39.

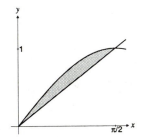

Exercises 7.10

1.

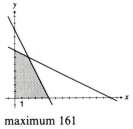

maximum 161
minimum 135

3.

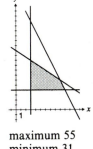

maximum 55
minimum 31

5. 3 tables, 34 chairs   **7.** 30 grapefruit, 30 orange crates
9. 15 Pasadena to Santa Monica,
        3 Pasadena to El Toro, 0 Long Beach to Santa Monica,
        16 Long Beach to El Toro
11. $10\frac{1}{2}$ yd^3 Foamboard, $14\frac{1}{2}$ yd^3 Plastiflex
13. 10 days Vancouver, 4 days Seattle
15. \$7500 in municipal bonds, \$2500 in bank certificates,
        \$2000 in high-risk bonds
17. 4 games, 32 educational, 0 utility

Review Exercises for Chapter 7

1. $(2, 3)$

3. x = any number
        $y = \frac{2}{7}x - 4$

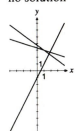

5. no solution

7. $(3, -1, -5)$

9. no solution   **11.** $x = -4z + 1$, $y = -z - 1$, $z =$ any number   **13.** $x = 6 - 5z$, $y = \frac{1}{2}(7 - 3z)$, $z =$ any number

15. \$3000 at 6%, \$6000 at 7%   **17.** $\dfrac{2}{x - 5} + \dfrac{1}{x + 3}$ **19.** $\dfrac{-4}{x} + \dfrac{4}{x - 1} + \dfrac{-2}{(x - 1)^2}$ **21.** impossible

23. $\begin{bmatrix} 4 & 18 \\ 4 & 0 \\ 2 & 2 \end{bmatrix}$ **25.** $\begin{bmatrix} 10 & 0 & -5 \end{bmatrix}$ **27.** $\begin{bmatrix} -\frac{7}{2} & 10 \\ 1 & -\frac{9}{2} \end{bmatrix}$ **29.** $\begin{bmatrix} 30 & 22 & 2 \\ -9 & 1 & -4 \end{bmatrix}$ **31.** $\begin{bmatrix} -\frac{1}{2} & \frac{11}{2} \\ \frac{15}{4} & -\frac{3}{2} \\ -\frac{1}{2} & 1 \end{bmatrix}$

33. $1, \begin{bmatrix} 9 & -4 \\ -2 & 1 \end{bmatrix}$ **35.** 0, no inverse   **37.** $-1, \begin{bmatrix} 3 & 2 & -3 \\ 2 & 1 & -2 \\ -8 & -6 & 9 \end{bmatrix}$ **39.** $(65, 154)$

41. $(\frac{1}{5}, \frac{9}{5})$ **43.** $(-\frac{87}{26}, \frac{21}{26}, \frac{39}{26})$ **45.** $(-3, -3), (\frac{9}{5}, -\frac{3}{5})$ **47.** $(2, 2), (-2, 2)$

49.

bounded

51.

bounded

53. maximum 34, minimum 4
55. (a) 200 acres oats, 200 acres barley
(b) no oats, 360 acres barley
57. $x = \dfrac{b + c}{2}$, $y = \dfrac{a + c}{2}$, $z = \dfrac{a + b}{2}$ **59.** 2, 3

Chapter 7 Test

1. wind speed 60 km/h, airplane 300 km/h   **2.** $(2, -1)$; linear, neither inconsistent nor dependent
3. no solution; linear, inconsistent   **4.** $x = \frac{1}{7}(z + 1)$, $y = \frac{1}{7}(9z + 2)$, $z =$ any number; linear, dependent

5. $(\pm 1, -2), (\pm\frac{5}{3}, -\frac{2}{3})$; nonlinear   **6.** incompatible dimensions   **7.** incompatible dimensions   **8.** $\begin{bmatrix} 11 & 9 \\ 2 & 1 \\ 11 & 2 \end{bmatrix}$

9. $\begin{bmatrix} 46 & 95 \\ 4 & 11 \\ 26 & 55 \end{bmatrix}$ **10.** $\begin{bmatrix} 2 & -\frac{3}{2} \\ -1 & 1 \end{bmatrix}$ **11.** B is not square   **12.** B is not square   **13.** -3 **14.** $(\frac{473}{19}, \frac{90}{19})$

15. $(5, -5, -4)$ **16.** $|A| = 0$, $|B| = 2$ $B^{-1} = \begin{bmatrix} 1 & -2 & 0 \\ 0 & \frac{1}{2} & 0 \\ 3 & -6 & 1 \end{bmatrix}$ **17.**

18. $\dfrac{1}{x - 1} + \dfrac{1}{(x - 1)^2} - \dfrac{1}{x + 2}$ **19.** He should grow $166\frac{2}{3}$ acres of wheat and no barley.

Problems Plus

1. Jason 26, Sacha 14, Vanessa 8   **3.** (a) $x = y = z = 0$
(b) Use the determinant to show that the coefficient matrix is invertible.

Chapter 8

Exercises 8.1

1. focus $F(1,0)$; directrix $x = -1$; focal diameter 4

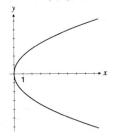

3. $F(0, \frac{9}{4})$; $y = -\frac{9}{4}$; 9

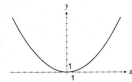

5. $F(0, \frac{1}{20})$; $y = -\frac{1}{20}$; $\frac{1}{5}$

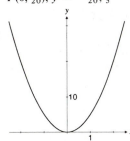

7. $F(-\frac{1}{32}, 0)$; $x = \frac{1}{32}$; $\frac{1}{8}$

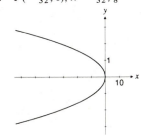

9. $F(0, -\frac{3}{2})$; $y = \frac{3}{2}$; 6

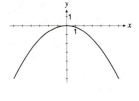

11. $F(-\frac{5}{12}, 0)$; $x = \frac{5}{12}$; $\frac{5}{3}$

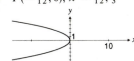

13. $x^2 = -12y$ **15.** $y^2 = -3x$ **17.** $x = y^2$ **19.** $x^2 = -4\sqrt{2}\,y$

21. $y^2 = -12x$ **23.** $y^2 = 4x$ **25.** $x^2 = -24y$ **27.** **(a)** $y^2 = 12x$

(b) $8\sqrt{15} \approx 31$ cm

Exercises 8.2

1. vertices $V(\pm 5, 0)$; foci $F(\pm 4, 0)$; eccentricity 4/5; major axis 10, minor axis 6

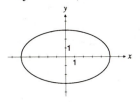

3. $V(0, \pm 3)$; $F(0, \pm\sqrt{5})$; $\sqrt{5}/3$; 6, 4

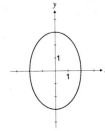

5. $V(\pm 4, 0)$; $F(\pm 2\sqrt{3}, 0)$; $\sqrt{3}/2$; 8, 4

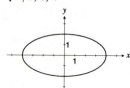

7. $V(0, \pm\sqrt{3})$; $F(0, \pm\sqrt{3/2})$; $1/\sqrt{2}$; $2\sqrt{3}$, $\sqrt{6}$

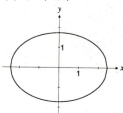

9. $V(\pm 1, 0)$; $F(\pm\sqrt{3}/2, 0)$; $\sqrt{3}/2$; 2, 1

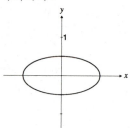

11. $V(0, \pm\sqrt{2})$; $F(0, \pm\sqrt{3/2})$; $\sqrt{3}/2$; $2\sqrt{2}$, $\sqrt{2}$

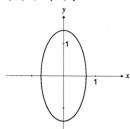

13. $V(0, \pm 1)$; $F(0, \pm 1/\sqrt{2})$; $1/\sqrt{2}$; 2, $\sqrt{2}$

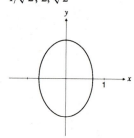

15. $\dfrac{x^2}{25} + \dfrac{y^2}{9} = 1$

17. $x^2 + \dfrac{y^2}{4} = 1$

19. $\dfrac{x^2}{9} + \dfrac{y^2}{13} = 1$

21. $\dfrac{x^2}{100} + \dfrac{y^2}{91} = 1$

23. $\dfrac{x^2}{25} + \dfrac{y^2}{5} = 1$

25. $\dfrac{64x^2}{225} + \dfrac{64y^2}{81} = 1$

27. $(0, \pm 2)$

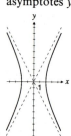

29. $\dfrac{x^2}{2.250 \times 10^{16}} + \dfrac{y^2}{2.249 \times 10^{16}} = 1$

31. $\dfrac{x^2}{1,453,200} + \dfrac{y^2}{1,449,200} = 1$

33. $5\sqrt{39}/2 \approx 15.6$ in.

Exercises 8.3

1. vertices $V(\pm 2, 0)$; foci $F(\pm 2\sqrt{5}, 0)$; asymptotes $y = \pm 2x$

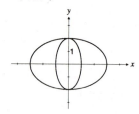

3. $V(0, \pm 1)$; $F(0, \pm\sqrt{26})$; $y = \pm\frac{1}{5}x$

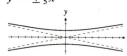

5. $V(\pm 1, 0)$; $F(\pm\sqrt{2}, 0)$; $y = \pm x$

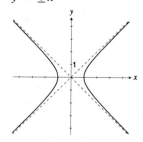

7. $V(0, \pm 3)$; $F(0, \pm\sqrt{34})$; $y = \pm\frac{3}{5}x$

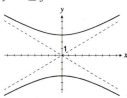

9. $V(\pm 2\sqrt{2}, 0)$; $F(\pm\sqrt{10}, 0)$; $y = \pm\frac{1}{2}x$

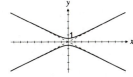

11. $V(0, \pm\frac{1}{2})$; $F(0, \pm\sqrt{5}/2)$; $y = \pm\frac{1}{2}x$

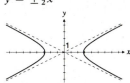

13. $\dfrac{x^2}{9} - \dfrac{y^2}{16} = 1$ **15.** $y^2 - \dfrac{x^2}{3} = 1$ **17.** $x^2 - \dfrac{y^2}{25} = 1$ **19.** $\dfrac{5y^2}{64} - \dfrac{5x^2}{256} = 1$ **21.** $\dfrac{x^2}{16} - \dfrac{y^2}{16} = 1$

23. $\dfrac{x^2}{9} - \dfrac{y^2}{16} = 1$ **25.** **(b)** $x^2 - y^2 = \dfrac{c^2}{2}$ **29.** **(a)** 490 mi   **(b)** $\dfrac{y^2}{60{,}025} - \dfrac{x^2}{2475} = 1$ **(c)** 10.1 mi

Exercises 8.4

1. center $C(2, 1)$; foci $F(2 \pm \sqrt{5}, 1)$; vertices $V_1(-1, 1)$, $V_2(5, 1)$; major axis 6, minor axis 4

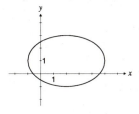

3. $C(0, -5)$; $F_1(0, -1)$, $F_2(0, -9)$; $V_1(0, 0)$, $V_2(0, -10)$; axes 10, 6

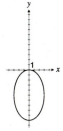

5. vertex $V(3, -1)$; focus $F(3, 1)$; directrix $y = -3$

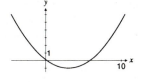

7. $V(-\frac{1}{2}, 0)$; $F(-\frac{1}{2}, -\frac{1}{16})$; $y = \frac{1}{16}$

9. center $C(-1, 3)$; foci $F_1(-6, 3)$, $F_2(4, 3)$; vertices $V_1(-4, 3)$, $V_2(2, 3)$; asymptotes $y = \pm\frac{4}{3}(x + 1) + 3$

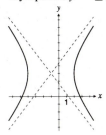

11. $C(-1, 0)$; $F(-1, \pm\sqrt{5})$; $V(-1, \pm 1)$; $y = \pm\frac{1}{2}(x + 1)$

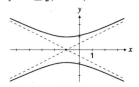

13. ellipse; $C(2, 0)$; $F(2, \pm\sqrt{5})$; $V(2, \pm 3)$;
major axis 6, minor axis 4

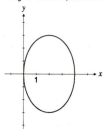

15. hyperbola; $C(1, 2)$; $F_1(-\frac{3}{2}, 2)$, $F_2(\frac{7}{2}, 2)$;
$V(1 \pm \sqrt{5}, 2)$; asymptotes $y = \pm\frac{1}{2}(x - 1) + 2$

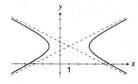

17. ellipse; $C(3, -5)$; $F(3 \pm \sqrt{21}, -5)$;
$V_1(-2, -5)$, $V_2(8, -5)$;
major axis 10, minor axis 4

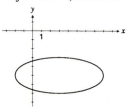

19. hyperbola; $C(3, 0)$; $F(3, \pm 5)$;
$V(3, \pm 4)$; asymptotes $y = \pm\frac{4}{3}(x - 3)$

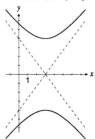

21. degenerate conic (pair of lines), $y = \pm\frac{1}{2}(x - 4)$ **23.** point $(1, 3)$

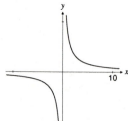

(1, 3)

25. (a) $F < 17$
(b) $F = 17$ (c) $F > 17$

Exercises 8.5

1. $(\sqrt{2}, 0)$ **3.** $(0, -2\sqrt{3})$ **5.** $(1.6383, 1.1472)$ **7.** $X^2 + Y^2 - 2XY - 3\sqrt{2}X + \sqrt{2}Y + 2 = 0$
9. hyperbola, $X^2 - Y^2 = 16$ **11.** parabola, $Y = \sqrt{2}X^2$ **13.** hyperbola, $Y^2 - X^2 = 1$

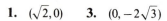

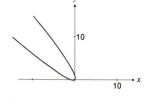

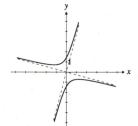

15. hyperbola, $\dfrac{X^2}{4} - Y^2 = 1$ **17.** hyperbola, $3X^2 - Y^2 = 2\sqrt{3}$ **19.** hyperbola, $(X - 1)^2 - 3Y^2 = 1$

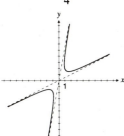 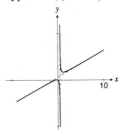

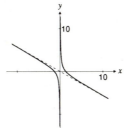

21. ellipse, $X^2 + \dfrac{(Y + 1)^2}{4} = 1$

23. (a) $(X - 5)^2 - Y^2 = 1$
  (b) XY-coordinates: $C(5,0)$; $V_1(6,0)$, $V_2(4,0)$; $F(5 \pm \sqrt{2}, 0)$;
      xy-coordinates: $C(4,3)$; $V_1(\frac{24}{5}, \frac{18}{5})$, $V_2(\frac{16}{5}, \frac{12}{5})$; $F(4 \pm \frac{4}{5}\sqrt{2}, 3 \pm \frac{3}{5}\sqrt{2})$
  (c) $Y = \pm(X - 5)$; $7x + y - 25 = 0$, $x - 7y + 25 = 0$

Exercises 8.6

1.

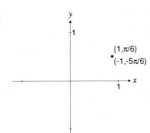

$\left(-3, \dfrac{3\pi}{2}\right), \left(3, \dfrac{5\pi}{2}\right)$

3.

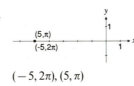

$\left(-1, -\dfrac{5\pi}{6}\right), \left(1, \dfrac{\pi}{6}\right)$

5.

$(-5, 2\pi), (5, \pi)$

7. $(2\sqrt{3}, 2)$ **9.** $(1, -1)$ **11.** $(-5, 0)$ **13.** $\left(\sqrt{2}, \dfrac{3\pi}{4}\right)$ **15.** $\left(4, \dfrac{\pi}{4}\right)$ **17.** $(5, \tan^{-1}(\frac{4}{3}))$ **19.** $\theta = \dfrac{\pi}{4}$

21. $r = \tan\theta \sec\theta$ **23.** $r = 4\sec\theta$ **25.** $x^2 + y^2 = 49$ **27.** $x = 6$ **29.** $x^2 + y^2 = \dfrac{y}{x}$ **31.** $y - x = 1$

33. $x^2 + y^2 = (x^2 + y^2 - x)^2$ **35.**

37.

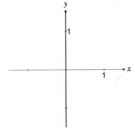

39.

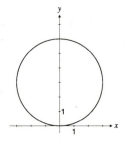

41.

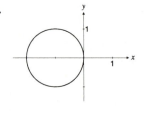

43.

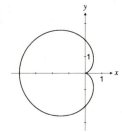

45.

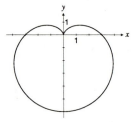

47.

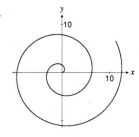

49.

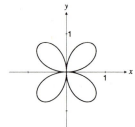

51.

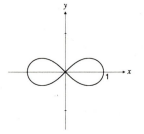

53.

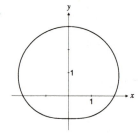

55.

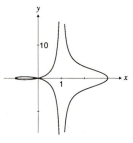

57. (b) $\sqrt{10 + 6\cos\dfrac{5\pi}{12}} \approx 3.40$

Exercises 8.7

1. (a)

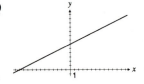

(b) $x - 2y + 12 = 0$ **3. (a)**

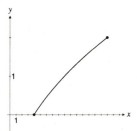

(b) $x = (y + 2)^2$

5. (a)

(b) $x = \sqrt{1 - y}$

7. (a)

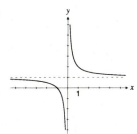

(b) $y = \dfrac{1}{x} + 1$ **9. (a)**

(b) $x^3 = y^2$

11. (a)

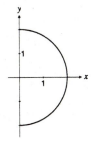

(b) $x^2 + y^2 = 4$ **13. (a)**

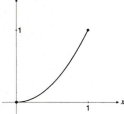

(b) $y = x^2$

15. (a)

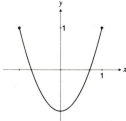

(b) $y = 2x^2 - 1$ **17. (a)**

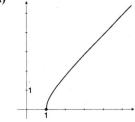

(b) $x^2 - y^2 = 1$

19. (a)

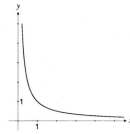

(b) $xy = 1$ **21. (a)**

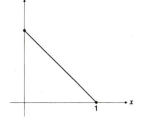

(b) $x + y = 1$

23. $x = 4 + t, y = -1 + \frac{1}{2}t$ **25.** $x = 6 + t, y = 7 + t$ **27.** $x = a\cos t, y = a\sin t$ **31.**

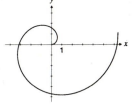

33.
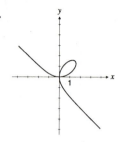

37.
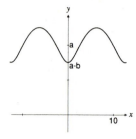

39. (b)
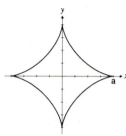

$$x^{2/3} + y^{2/3} = a^{2/3}$$

41. $x = a(\sin\theta\cos\theta + \cot\theta), y = a(1 + \sin^2\theta)$ **43.** $y = a - a\cos\left(\dfrac{x + \sqrt{2ay - y^2}}{a}\right)$

Review Exercises for Chapter 8

1. vertex $V(0,0)$; focus $F(0,-2)$;
directrix $y = 2$

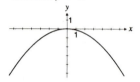

3. $V(-2,2)$; $F(-\frac{7}{4},2)$;
directrix $x = -\frac{9}{4}$

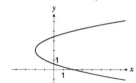

5. center $C(0,0)$; vertices $V(\pm4,0)$; foci $F(\pm2\sqrt{3},0)$;
major axis 8, minor axis 4

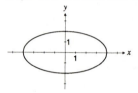

7. $C(0,2)$; $V(\pm3,2)$; $F(\pm\sqrt{5},2)$;
axes 6, 4

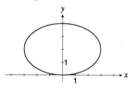

9. $C(0,0)$; $V(\pm4,0)$; $F(\pm2\sqrt{6},0)$;
asymptotes $y = \pm\dfrac{1}{\sqrt{2}}x$

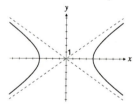

11. $C(-3,-1)$; $V(-3,-1\pm\sqrt{2})$;
$F(-3,-1\pm2\sqrt{5})$;
asymptotes $y = \frac{1}{3}x, y = -\frac{1}{3}x - 2$

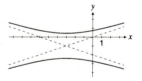

13. parabola; $F(0,-2)$; $V(0,1)$

15. hyperbola; $F(0,\pm12\sqrt{2})$; $V(0,\pm12)$

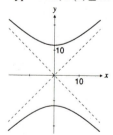

17. ellipse; $F(1, 4\pm\sqrt{15})$; $V(1, 4\pm2\sqrt{5})$

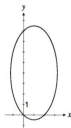

19. parabola; $F(-\frac{255}{4}, 8)$; $V(-64, 8)$

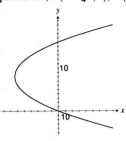

21. ellipse; $F\left(3, -3 \pm \dfrac{1}{\sqrt{2}}\right)$;

$V_1(3, -4)$, $V_2(3, -2)$

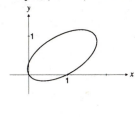

23. has no graph

25. $x^2 = 4y$

27. $\dfrac{y^2}{4} - \dfrac{x^2}{16} = 1$

29. $\dfrac{(x - 1)^2}{3} + \dfrac{(y - 2)^2}{4} = 1$

31. $\dfrac{4(x - 7)^2}{225} + \dfrac{(y - 2)^2}{100} = 1$

33. hyperbola, $3X^2 - Y^2 = 1$

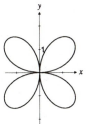

35. ellipse, $(X - 1)^2 + 4Y^2 = 1$

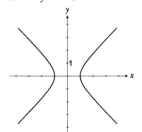

37. $(x^2 + y^2 - 3x)^2 = 9(x^2 + y^2)$

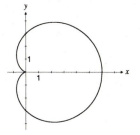

39. $(x^2 + y^2)^3 = 16x^2y^2$

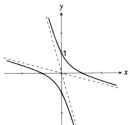

41. $x^2 - y^2 = 1$

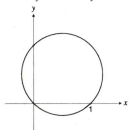

43. $x^2 + y^2 = x + y$

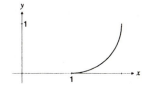

45. $x = 2y - y^2$

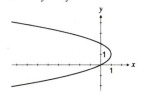

47. $(x - 1)^2 + (y - 1)^2 = 1$

49. (a) $y = x^2$ **(b)**

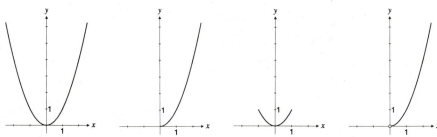

The curves are different parts of the parabola $y = x^2$.

Chapter 8 Test

1. focus $F(0, -\frac{3}{2})$; directrix $y = \frac{3}{2}$

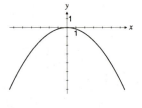

2. vertices $V(0, \pm 2\sqrt{3})$; foci $F(0, \pm 2)$; major axis $4\sqrt{3}$, minor axis $4\sqrt{2}$

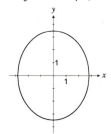

3. vertices $V(0, \pm 7)$; foci $F(0, \pm\sqrt{85})$; asymptotes $y = \pm\frac{7}{6}x$

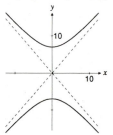

4. $\dfrac{(x-3)^2}{9} + \dfrac{(y+\frac{1}{2})^2}{4} = 1$

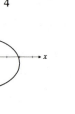

5. $9(x+2)^2 - 8(y-4)^2 = 0$

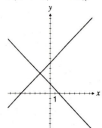

6. $(y+4)^2 = -2(x-4)$

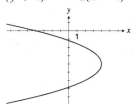

7. $\dfrac{y^2}{9} - \dfrac{x^2}{16} = 1$

8. $x^2 - 4x - 8y + 20 = 0$

9. (a) $\dfrac{X^2}{3} + \dfrac{Y^2}{18} = 1$ **(b)**

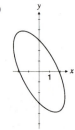

(c) $(-3\sqrt{2/5}, 6\sqrt{2/5}), (3\sqrt{2/5}, -6\sqrt{2/5})$

10.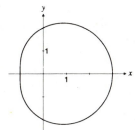

11. $(x - 1)^2 + (y + 2)^2 = 5$, circle

12. (a)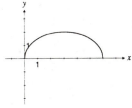

(b) $\dfrac{(x - 3)^2}{9} + \dfrac{y^2}{4} = 1$

Problems Plus

1. 0 points: $k > 3$ or $k < -3$; 1 point: $k = \pm 3$; 2 points: $1 < k \le 3$ or $-3 < k < -1$; 3 points: $k = \pm 1$; 4 points: $-1 < k < 1$; 5 points: never   **3.** The y-intercept of $\overline{PQ}$ is always $4p$.

Chapter 9

Exercises 9.1

1. $1/3, 1/4, 1/5, 1/6; 1/1002$ **3.** $-1, 1/4, -1/9, 1/16; 1/1{,}000{,}000$ **5.** $0, 2, 0, 2; 2$ **7.** $1, 0, -1, 0; 0$
9. $1/2, -2/3, 3/4, -4/5; -1000/1001$ **11.** $-3/\sqrt{2}, 5/\sqrt{3}, -7/2, 9/\sqrt{5}; 2001/\sqrt{1001}$ **13.** $3, 2, 0, -4, -12$
15. $1, 3, 7, 15, 31$ **17.** $0, 1, 1, 0, -1$ **19.** $1, 2, 3, 5, 8$ **21.** 2^n **23.** $3n - 2$ **25.** $(2n - 1)/n^2$ **27.** r^n
29. $1 + (-1)^n$ **31. (c)** $n^2 + (n - 1)(n - 2)(n - 3)(n - 4)(n - 5)(n - 6)$ **33.** $2^{(2^n - 1)/2^n}$
35. $25, 76, 38, 19, 58, 29, 88, 44, 22, 11, 34, 17, 52, 26, 13, 40, 20, 10, 5, 16, 8, 4, 2, 1, 4, 2, 1, \ldots$

Exercises 9.2

1. $5, 24, 4 + 5(n - 1), 499$ **3.** $4, 4, -12 + 4(n - 1), 384$ **5.** $1.5, 31, 25 + 1.5(n - 1), 173.5$
7. $s, 2 + 4s, 2 + (n - 1)s, 2 + 99s$ **9.** $3, 162, 2 \cdot 3^{n-1}$ **11.** $-0.3, 0.00243, (0.3)(-0.3)^{n-1}$
13. $-1/12, 1/144, 144(-1/12)^{n-1}$ **15.** $3^{2/3}, 3^{11/3}, 3^{(2n+1)/3}$ **17.** $s^{2/7}, s^{8/7}, s^{2(n-1)/7}$ **19.** neither
21. geometric, $9\sqrt{3}$ **23.** neither   **25.** arithmetic, $x + 3$ **27.** neither   **29.** arithmetic, 3   **31.** $\frac{1}{2}$
33. $-100, -98, -96$ **35.** $\frac{16}{27}$ **37.** $\frac{25}{4}$ **39.** 30th   **41.** yes, 2985th   **43.** 19 ft   **45.** \$4,714.37
47. $64/25, 1024/625, 5(4/5)^n$ **49.** 3
51. r^2 **53.** $r = 10^d$ **55.** $2, 5, 8$ **57.** 0   **59.** $(\sqrt{5} - 1)/2$
61. $10, 20, 40$ **63.** $15/4$

Exercises 9.3

1. 86   **3.** 12   **5.** 8   **7.** 31   **9.** 549   **11.** 106   **13.** $\sqrt{1} + \sqrt{2} + \sqrt{3} + \sqrt{4} + \sqrt{5}$
15. $\sqrt{4} + \sqrt{5} + \sqrt{6} + \sqrt{7} + \sqrt{8} + \sqrt{9} + \sqrt{10}$ **17.** $1 \cdot x^2 + 2 \cdot x^3 + \cdots + 8 \cdot x^9$

19. $x - x^2 + x^3 - \cdots + (-1)^{n+1}x^n$ **21.** $\sum_{k=1}^{100} k$ **23.** $\sum_{k=2}^{100} \dfrac{(-1)^k}{k \ln k}$ **25.** $\sum_{k=0}^{5} (-1)^k \dfrac{x^k}{3^k}$ **27.** $\sum_{k=1}^{100} (-1)^{k+1} kx^{k-1}$

29. $\sum_{k=1}^{97} k(k + 1)(k + 2)$ **31.** $S_n = \dfrac{1}{2} - \dfrac{1}{n + 2}, \dfrac{250}{501}$ **33.** $S_n = 1 - \dfrac{1}{3^n}, 1 - \dfrac{1}{3^{20}}$ **35.** $S_n = 1 - \sqrt{n + 1}, -9$

37. $S_n = \sqrt{3} - \tan\left(\dfrac{(n + 1)\pi}{3}\right), \sqrt{3}$ **39.** $S_n = -\log(n + 1), -6$

Exercises 9.4

1. 460   **3.** 1090   **5.** 4900   **7.** 315   **9.** 441   **11.** 2.8502   **13.** 20,301   **15.** 832.3   **17.** 46.75
19. 3280   **21.** 1.94117   **23.** 5.997070313   **25.** geometric, 9.2224   **27.** geometric, $(1 + x^{21})/(1 + x)$
29. arithmetic, 250,500   **31.** arithmetic, 4128   **33.** arithmetic, $9(1 + 4\sqrt{2})$ **35.** geometric, $124/(\sqrt[3]{5} - 1)$

37. 50   **39.** 80/3   **41.** 1/2   **43.** 455/27   **45.** 10^{19} **47.** \$10,737,418.23, 37 days
49. (a) 576 ft   **(b)** $16n^2$ ft   **51.** \$5,591.79   **53.** \$245.66   **55.** \$2,601.59   **59.** \$733.76, \$264, 153.60

Exercises 9.5

1. 3/4   **3.** 1/648   **5.** $-27/1300$ **7.** $5^{4/3}/(1 + 5^{1/3})$ **9.** 7/9   **11.** 1/33   **13.** 112/999   **15.** 60
17. $2\pi R^2$ **19.** 1/3   **21.** $-\pi/4$ **23.** 7/24   **25.** \$50,000

Exercises 9.7

3. $2n^2$ **5.** $1 - \dfrac{1}{n+1}$ **7.** $\dfrac{n}{5n+1}$ **9.** $\dfrac{n+1}{2n}$ **11.** $n < 2n$ **13.** $n \le F_n$ for $n \ge 5$

15. $a_n = 5^n - 1$ **17.** $a_n = 5^{(2^n-1)/2^n}$ **19.** $a_n = F_n/F_{n+1}$ **21.** $x^n - 1 = (x-1)(1 + x + x^2 + \cdots + x^{n-1})$
23. $1 + 2 + 3 + \cdots + n + (n-1) + \cdots + 1 = n^2$ **25. (a)** 3, 7, 15, 31   **(b)** 2^{n-1}
(c) $2^{64} - 1 \text{ s} \approx 5.8 \times 10^{11}$ years

Exercises 9.8

1. $x^7 + 7x^6y + 21x^5y^2 + 35x^4y^3 + 35x^3y^4 + 21x^2y^5 + 7xy^6 + y^7$ **3.** $x^4 + 4x^2 + 6 + \dfrac{4}{x^2} + \dfrac{1}{x^4}$

5. $1188 - 684\sqrt{3}$ **7.** $x^{10}y^5 - 5x^8y^4 + 10x^6y^3 - 10x^4y^2 + 5x^2y - 1$ **9.** $8x^3 - 36x^2y + 54xy^2 - 27y^3$

11. $\dfrac{1}{x^5} - \dfrac{5}{x^{7/2}} + \dfrac{10}{x^2} - \dfrac{10}{x^{1/2}} + 5x - x^{5/2}$ **13.** 3,628,800   **15.** 1/462   **17.** 9900   **19.** 10   **21.** $n!$

23. $n(n+1)(n+2)$ **25.** 10   **27.** 4950   **29.** 0   **31.** 32   **33.** $x^4 + 8x^3y + 24x^2y^2 + 32xy^3 + 16y^4$

35. $1 + \dfrac{6}{x} + \dfrac{15}{x^2} + \dfrac{20}{x^3} + \dfrac{15}{x^4} + \dfrac{6}{x^5} + \dfrac{1}{x^6}$ **37.** $x^{20} + 40x^{19}y + 760x^{18}y^2$ **39.** $25a^{26/3} + a^{25/3}$ **41.** $48,620x^{18}$

43. $300a^2b^{23}$ **45.** $100y^{99}$ **47.** $13,440x^4y^6$ **49.** $495a^8b^8$ **51.** 17,920   **53.** $1792c^7$ **55.** 21
57. 11   **59.** $(x+y)^4$ **61.** $(2a+b)^3$ **63.** $a^8 + 4a^7 + 10a^6 + 16a^5 + 19a^4 + 16a^3 + 10a^2 + 4a + 1$
73. (a) 2,598,960   **(b)** 56

Review Exercises for Chapter 9

1. 1/2, 4/3, 9/4, 16/5; 100/11   **3.** 0, 1/4, 0, 1/32; 1/500   **5.** 1, 3, 15, 105; 654,729,075   **7.** 1, 4, 9, 16, 25, 36, 49
9. 1, 3, 5, 11, 21, 43, 85   **11.** 3, 2, 1, 1, 1, 1, 1   **13.** arithmetic, 7   **15.** arithmetic, $5\sqrt{2}$ **17.** arithmetic, $t + 1$
19. geometric, 4/27   **21.** neither   **23.** geometric, ab^4c^{12} **25.** $2i$ **27.** 5   **29.** 81/4   **31.** 64
33. (a) $A_n = (n/2)\sin(2\pi/n)$ **(b)** 1.2990, 2, 3.1395, 3.1415   **37. (a)** 9   **(b)** $\pm 6\sqrt{2}$ **39.** 126   **41.** 384

43. $0^2 + 1^2 + 2^2 \cdots + 9^2$ **45.** $\dfrac{3}{2^2} + \dfrac{3^2}{2^3} + \dfrac{3^3}{2^4} + \cdots + \dfrac{3^{50}}{2^{51}}$ **47.** $\displaystyle\sum_{k=1}^{33} 3k$ **49.** $\displaystyle\sum_{k=1}^{100} k2^{k+2}$ **51.** 4.68559

53. $(1 + 5^{9/2})/(1 + 5^{1/2}) \approx 432.17$ **55.** arithmetic, 1650   **57.** geometric, 9831   **59.** 13   **61.** 29
63. (a) 259,374   **(b)** $100,000(1.1)^n$ **65.** \$30,324.28   **67. (a)** \$428.77   **(b)** \$608.56   **69.** 1/9
71. $a/(1 - b^2)$ **73.** 1/3   **85.** $S_n = (n+1)! - 1$ **87.** $Q_n \le S_n$ **89.** 255   **91.** 12,870
93. $1 - 6x^2 + 15x^4 - 20x^6 + 15x^8 - 6x^{10} + x^{12}$ **95.** $1540a^3b^{19}$ **97.** 5   **99.** a^3

Chapter 9 Test

1. $-10/99$ **2.** -1 **3.** -36 **4.** yes, 6th term   **5. (a)** false   **(b)** true

6. (a) $S_n = \dfrac{n}{2}[2a + (n-1)d]$ or $S_n = n\left(\dfrac{a + a_n}{2}\right)$ **(b)** 60   **(c)** $-8/9, -78$ **7. (a)** $S_n = \dfrac{a(1 - r^n)}{1 - r}$

(b) 6305/6561   **8.** $2 + \sqrt{2}$ **10. (a)** -50 **(b)** 10   **11.** $(a + b)^n = \displaystyle\sum_{k=0}^{n} \binom{n}{k}a^{n-k}b^k$

12. $32x^5 + 80x^4y^2 + 80x^3y^4 + 40x^2y^6 + 10xy^8 + y^{10}$ **13.** $1,293,600\, a^3b^{97}$ **14.** 8064

Problems Plus

3. *Hint:* Try the special cases $n = 1$ and $n = 2$.

Index

To The Student:

Many students experience difficulty with *precalculus* mathematics in calculus courses. Calculus requires that you understand and remember precalculus topics. For this reason, it may be helpful to retain this text as a reference in your calculus course. It has been written with this purpose in mind.

Some References To Calculus in This Text:

Trigonometric Functions of Important Angles

θ	radians	$\sin\theta$	$\cos\theta$	$\tan\theta$
0°	0	0	1	0
30°	$\pi/6$	$\frac{1}{2}$	$\frac{\sqrt{3}}{2}$	$\frac{\sqrt{3}}{3}$
45°	$\pi/4$	$\frac{\sqrt{2}}{2}$	$\frac{\sqrt{2}}{2}$	1
60°	$\pi/3$	$\frac{\sqrt{3}}{2}$	$\frac{1}{2}$	$\sqrt{3}$
90°	$\pi/2$	1	0	—
180°	π	0	-1	0
270°	$3\pi/2$	-1	0	—

Fundamental Identities

$$\csc\theta = \frac{1}{\sin\theta} \qquad \sec\theta = \frac{1}{\cos\theta}$$

$$\tan\theta = \frac{\sin\theta}{\cos\theta} \qquad \cot\theta = \frac{\cos\theta}{\sin\theta}$$

$$\cot\theta = \frac{1}{\tan\theta} \qquad \sin^2\theta + \cos^2\theta = 1$$

$$1 + \tan^2\theta = \sec^2\theta \qquad 1 + \cot^2\theta = \csc^2\theta$$

$$\sin(-\theta) = -\sin\theta \qquad \cos(-\theta) = \cos\theta$$

$$\tan(-\theta) = -\tan\theta \qquad \sin\left(\frac{\pi}{2} - \theta\right) = \cos\theta$$

$$\cos\left(\frac{\pi}{2} - \theta\right) = \sin\theta \qquad \tan\left(\frac{\pi}{2} - \theta\right) = \cot\theta$$

The Law of Sines

$$\frac{\sin A}{a} = \frac{\sin B}{b} = \frac{\sin C}{c}$$

The Law of Cosines

$$a^2 = b^2 + c^2 - 2bc\cos A$$
$$b^2 = a^2 + c^2 - 2ac\cos B$$
$$c^2 = a^2 + b^2 - 2ab\cos C$$

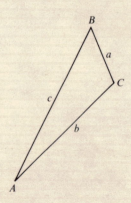

Addition and Subtraction Formulas

$$\sin(x + y) = \sin x\cos y + \cos x\sin y$$

$$\sin(x - y) = \sin x\cos y - \cos x\sin y$$

$$\cos(x + y) = \cos x\cos y - \sin x\sin y$$

$$\cos(x - y) = \cos x\cos y + \sin x\sin y$$

$$\tan(x + y) = \frac{\tan x + \tan y}{1 - \tan x\tan y}$$

$$\tan(x - y) = \frac{\tan x - \tan y}{1 + \tan x\tan y}$$

Double-Angle Formulas

$$\sin 2x = 2\sin x\cos x$$

$$\cos 2x = \cos^2 x - \sin^2 x$$

$$= 2\cos^2 x - 1$$

$$= 1 - 2\sin^2 x$$

$$\tan 2x = \frac{2\tan x}{1 - \tan^2 x}$$

Half-Angle Formulas

$$\sin^2 x = \frac{1 - \cos 2x}{2} \qquad \cos^2 x = \frac{1 + \cos 2x}{2}$$

$$\tan^2 x = \frac{1 - \cos 2x}{1 + \cos 2x}$$

$$\sin\frac{u}{2} = \pm\sqrt{\frac{1 - \cos u}{2}} \qquad \cos\frac{u}{2} = \pm\sqrt{\frac{1 + \cos u}{2}}$$

$$\tan\frac{u}{2} = \frac{1 - \cos u}{\sin u} = \frac{\sin u}{1 + \cos u}$$

Graphs of the Inverse Trigonometric Functions

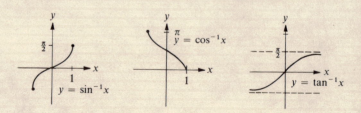